Proceedings of the 2009 International Conference on

Chemical, Biological and Environmental Engineering

CBEE 2009

Proceedings of the 2009 International Conference on

Chemical, Biological and Environmental Engineering

Singapore 9 – 11 October 2009

CBEE 2009

editor

Li Kai

Guizhou University, China

World Scientific

NEW JERSEY • LONDON • SINGAPORE • BEIJING • SHANGHAI • HONG KONG • TAIPEI • CHENNAI

Published by

World Scientific Publishing Co. Pte. Ltd.
5 Toh Tuck Link, Singapore 596224
USA office: 27 Warren Street, Suite 401-402, Hackensack, NJ 07601
UK office: 57 Shelton Street, Covent Garden, London WC2H 9HE

British Library Cataloguing-in-Publication Data
A catalogue record for this book is available from the British Library.

Proceedings of the 2009 International Conference on
CHEMICAL, BIOLOGICAL AND ENVIRONMENTAL ENGINEERING (CBEE 2009)

ISBN-13 978-981-4293-46-4
ISBN-10 981-4293-46-6

Printed in Singapore by Mainland Press Pte Ltd.

IACSIT
WWW.IACSIT.ORG

Conference Chairs

Prof. Li Kai, Macau University, Macau
Prof. Parvinder S. Sandhu, Rayat-Bahra Institute of Engineering and Bio-Technology, India

Program Committee Chairs

Prof. Sushil Kumar Kansal, Panjab University, India
Prof. Iis Sopyan, International Islamic University Malaysia, Malaysia
Prof. Nashat Mansour, Lebanese American University, Lebanon
Dr. Ho Chiung Ching, Multimedia University, Malaysia
Prof. P. Stanley, Allahabad Agricultural Institute-Deemed University, India
Dr. Hossein Ebrahimpour, University of Kashan Kashan, Iran
Dr. Hossein Malekinezhad, Islamic Azad University of Naragh, Iran
Dr. Raúl Rodríguez-Herrera, Universidad Autonoma de Coahuila, Mexico
Dr. R. Sivakumar, RMK Engineering College, India
Dr. Yash Parekh, Padmashree Dr. D.Y. Patil University, India
Dr. Pallavi Awasthi, Delhi University India, India
Dr. Subha Moorthy, Aarupadai Veedu Institute of Technology, India
Dr. R. Jayakumar, Bharathiar University, India
Dr. B. R. Patil, Bharati Vidyapeeth College of Engineering, India
Dr. G. Prasanna Lakshmi, Management and Research Mumbai University, India
Dr. V. S. Sangam, BVBhoomaraddy College of Engineering & Technology, India
Dr. Abhilash, National Metallurgical Laboratory, India
Dr. Ashwani Kumar Thukral, Guru Nanak Dev University, India
Dr. Shishir K. Shandilya, Rukmani Devi Institute of Science and Technology, India

PREFACE

Dear Distinguished Delegates and Guests,

The Organizing Committee warmly welcomes our distinguished delegates and guests to the 2009 International Conference on Chemical, Biological and Environmental Engineering (CBEE 2009), held in Singapore from 9 to 11 October 2009.

The 2009 International Conference on Chemical, Biological and Environmental Engineering (CBEE 2009) is sponsored by IACSIT, Applied Chemical Engineering Society (ACES), Environment Science Society (ESS), and is supported by the University of Florida, East Carolina University and Bioinformatics and Biomedical Engineering Society (BBES). The main objective of CBEE 2009 is to provide a platform for researchers, chemists, biologists, engineers, academicians, as well as industrial professionals from all over the world to present their research results and development activities in chemical, biological and environmental engineering. This conference will provide opportunities for conference delegates to exchange new ideas and application experiences, establish business or research relations and find global partners for future collaboration. The conference's Program Committee is by itself quite diverse and international in membership, with members from the Americas, Europe, Asia, Africa and Oceania.

This proceedings records the fully refereed papers presented at this conference, whose main conference themes and tracks are chemical, biological and environmental engineering. CBEE 2009 aims to bring together researchers, scientists, chemists, biologists, engineers and practitioners to exchange and share their experiences, new ideas and research results on all aspects of the main conference themes and tracks. The practical challenges encountered and the solutions adopted will also be discussed. The main goal of this event is to provide international scientific forums for the exchange of new ideas in a number of interactive fields through discussions with their peers from around the world. Both inward research — core areas of chemical, biological and environmental engineering — and outward research — multi-disciplinary, inter-disciplinary and applications — will be covered during this event.

This conference has solicited and gathered scientific and technical research submissions related to all aspects of major conference themes and tracks. All the submitted papers in this proceedings have been peer reviewed by reviewers drawn from the scientific and technical committee, external reviewers and editorial board members, depending on the subject matter of the paper. The reviewing and initial selection processes were undertaken electronically, after which the submitted papers were selected on the basis of originality, significance and clarity for the purpose of the conference. These selected papers (plus additional late contributions) will be presented as lectures and will make for an exciting technical program. The conference's program is extremely rich, and features high-impact presentations.

The high quality of this program, guaranteed by the presence of an unparalleled number of internationally recognized top experts, can be assessed when reading through the contents of the program. This conference will therefore be a unique event, where attendees will be able to appreciate the latest results in their field of expertise and acquire additional knowledge in other fields. This program has been structured to favor interactions among attendees coming from many diverse horizons — scientifically, geographically, technically, from academia and from industries.

We would like to thank the program chairs, organization staff and the members of the program committees for their work. Thanks are also due to Ms. Yang Dan, for her wonderful editorial service to this proceeding.

We are grateful to all those who have contributed to the success of CBEE 2009. We hope that all participants and other interested readers benefit scientifically from this proceedings, and also find it stimulating in the process. Finally, we would like to wish you success in your scientific and technical presentations and social networking.

We hope you have a unique, rewarding and enjoyable week at CBEE 2009 in Singapore.

With our warmest regards,

CBEE 2009 Organizing Committees
9–11 October 2009 in Singapore

CONTENTS

Session 4

Session 5

Session 6

Session 7

Session 8

Session 9

Session 10

Session 1

THE FUTURE OF BIOPHARMACEUTICS' PRODUCTION

BEHNAZ BAKHSHANDEH
Department of Biotechnology, University of Tehran
Tehran, Iran

ZARVAN SHAHRZAD
Sharif University of Technology, Tehran, Iran

Regarding the augmented pace of protein research and market demand, more efficient and rapid expression of genes in homologous and heterologous expression systems must be developed. All expression systems offer advantages and disadvantages, thus the proper choice of a host organism for the production of a biopharmaceutical has to be made on a case-by-case basis and depends upon the protein under consideration. Here we review some major expression systems used in biopharmaceutical industry.

1. Introduction

During the last years many biopharmaceuticals were approved clinical trials and commercial production resulting in a market of about US$50 billion in 2005. It is also estimated that 400–500 biotech drugs are under clinical development for various disease conditions. Estimated 4,400 biotech companies are globally involved, having brought more than 100 different biopharmaceutical products on the (U.S.) market [1].

Depending on required characteristics, biopharmaceuticals are produced in different host systems. Although transgenic plants and animals are predicted to become commercially competitive for manufacturing therapeutic proteins in the future, every biopharmaceutical on the market today is produced in either a mammalian or microbial cell line [2]. As a result over half (55%) of such products are expressed using microbes, either bacteria (40%) or yeasts (15%), with nearly all those bacteria being a form of E. coli (39%). Another 45% are expressed in mammalian cells primarily CHO cells (35%) [3].

The development of the protein and its laboratory-scale production is just the beginning for a biotechnology company with ambition beyond discovery. Taking a biomanufacturing process from laboratory to clinical production can be a precarious journey, littered with a multitude of possible pitfalls, some of which could prove fatal.

Expression systems are generally patented and available for licensing from their owners or from companies having licensed the patents for bundling (sublicensing) with their own expression systems or components [3]. Here we try to compare some specifications of major expression systems and help you to predict the growth procedure of expression systems in biopharmaceutical industry.

2. Bacteria

Expression of recombinant proteins in E.coli has been investigated extensively over the last 20 years and is one of the most frequently used systems. Clear guidelines by the competent authorities (e.g. EMEA and FDA) for the production of recombinant proteins in E.coli have to be complied [4]. Some of today's blockbusters are expressed in E.coli (table1) because it can be easily scaled-up in high cell-density fermentations yielding protein accumulation levels in the gram per liter scale [5].

An additional benefit is the established rapid cloning and expression procedure. With such technology, expression levels have reached in excess of 50% of total cell protein [6]. Small polypeptide tags, including the hexahistidine (his-tag), the FLAG or the anti-c-myc tag are added either N or C terminally to the protein for detection and purification [7]. Employing leader sequences (e.g. pelB or ompA leader), recombinant proteins are directed into the bacterial periplasm, where the oxidising milieu favours the formation of disulfide bonds by the disulfid-oxidoreductase [8]. Another advantage is the presence of chaperones leading to stabilisation of the recombinant protein. Overexpressed proteins often fail to fold correctly and undergo proteolytic degradation [9].

2.1. *Advantages*

Established regulatory track record; well-understood genetics; cheap and easy to grow; inexpensive media; high expression levels quickly — sometimes within five days; fast-growing (growth time measured in minutes); easy characterization (with few adventitious agents) [10].

Table 1. Biopharmaceutics expressed in bacteria.

Generic	Brand name	Company
Anakinra	Kineret	Amgen
Filgrastim	Neupogen	Amgen
Insulin Glargine	Lantus	Aventis
Insulin Human	Humulin	Eli Lilly
Insulin	Lispro	Eli Lilly
Interfron alfa-2b	Intron-A	Schering Corporation
Interfron Beta-1b	Betaseron	Berlex Corporation
Nesiritide	Natrecor	Scios Inc.
Oprelvekin	Neumega	Wyeth
Pegfilgrastim	Neulasta	Amgen
Peginterfron Alfa-2a	PEGasys	F.Hoffmann-La Roche
Peginterfron alfa-2b	PEG-Intron	Schering-Plough
Retoplase	Retavase	DL BioPharma
Somatropin	Genotropin	Pharmacia
Somatropin	Humatrope	Eli Lilly
Somatropin	Norditropin	Novo Nordisk
Somatropin	Nutropin AQ	Genentech
Teriparatide	Forteo	Eli Lilly

2.2. *Disadvantages*

Proteins are not usually secreted (so cell disruption step complicates harvesting); contain endotoxins; are microheterogeneous; no posttranslational modifications (no glycosylation); possibility of incorrect protein folding; harvesting can damage proteins [10].

3. Yeast

Yeasts are recognized by a growing track record as expression platforms for the production of pharmaceuticals. Yeast expression system includes lower eukaryotes like yeasts Saccharomyces cerevisiae, Schizosaccharomyces pombe, and methylotropic yeasts like Pichia pastoris, and Pichia methanolica have gained importance. They are used in large-scale fermentation processes for beer and other products for hundreds of years (table2). Nowadays the cost of recombinant protein production in yeasts usually runs between $50 and $100 per gram of final product.

Vectors:

YIp Vectors: do not replicate autonomously, but integrate into the genome at low frequencies by homologous recombination.

YEp Vectors (yeast episomal plasmid): replicate autonomously because of the presence of a segment of the yeast 2 μm plasmid that serves as an origin of replication.

YCp Vectors (yeast centromere plasmid): replicate autonomously and contain centromere sequences (CEN), and autonomously replicating sequences (ARS).

Selection Markers:

Dominant selection markers: antibiotic markers such as G418 and cyclohexamide

Complementation markers: marker genes that complement an auxotrophic mutation in the genome like URA3, TRPI, HIS3, and LEU2.

Promoters:

Inducible: like ADH2, SUC2

Constitutive: like GAPDH

Secretion Signals: Prepro alpha factor, HSp150, PHO, SUC2, KILM1 (killer toxin type 1), GGP1

Table 2. Biopharmaceutics expressed in yeast.

Generic	Brand name	Company
Becaplermin	Regranex	Ortho-McNeil
Glucagon	Glucagen	Novo Nordisk
Insulin Aspart	Novolog	Novo Nordisk
Insulin Human	Novolin	Novo Nordisk
Lepirudin	Refludan	Berlex
Sargramostim	Leukine	Berlex

3.1. *Advantage*

"Generally recognized as safe" by regulators; long history of use; genetics well understood; no endotoxins; high expression levels fairly quickly (two to eight weeks); protein is secreted for easy harvesting; fast growth (hours); inexpensive media; proteins usually properly folded; posttranslational modifications [10].

3.2. *Disadvantage*

Overglycosylation can ruin protein bioactivity, safety, activity, potency, or clearance; can contain immunogens or antigens; nonnative proteins are not always properly folded [10].

4. Insect cells

Protein production in insect cells has become more and more relevant for research and development applications (table3). Insect cells are easy to transfect with baculovirus vectors to yield high amounts of protein — with post-translational modifications which are quite close to, but not as complex as, in human systems. Novavax Company work in this filed.

Table 3. Some insect cell lines.

Cell line	Source	Cell density in culture
Sf9	Spodoptera frugiperda ovarian tissue	>12 x 10^6 cells/ml in suspension on day 6-7 with >98% viability
Sf21	Spodoptera frugiperda ovarian tissue	>6 x10^6 cells/ml in suspension on day 6-7 with >98% viability
Tn5B1-4(high five)	Trichoplusia embryonic tissue	>8 x 10^6 cells/ml in suspension on day 6-7 with >98% viability
S2	Drosophila	> 50-100 x 10^6

4.1. *Advantages*

Posttranslational modifications; properly folded protein; product is secreted; fairly high expression levels; expresses within about four weeks; baculoviruses are harmless to humans [10].

4.2. *Disadvantages*

Minimal regulatory track record and short history of use; slow growth, expensive media; baculovirus infection is an extra step in the process; contain immunogenic host cell proteins; some incorrect glycosylation; mammalian viruses can infect the cells in warm culture [103].

5. LEXSY

Leishmania tarentolae, a unicellular, flagellated protozoan organism, belongs to a family of parasites shuttling between sandflies and vertebrates. Since isolation from its host Tarentola mauretanica in 1921, L. tarentolae is kept in axenic culture, and it is not pathogenic to mammalians. The LEXSY expression strains are therefore grown under S1 laboratory safety conditions.

LEXSY strains can be handled like E. coli: they may be grown on agar plates or in liquid media from 96-well plate up to fermenter format. They have doubling times as short as 4 hours under optimum aeration and grow to high density within short time.

6. Mammalian cells

Since 1986, when Genentech announced the market approval of human tPA, the first therapeutic protein from recombinant mammalian cells, this market has increased potentially to several hundred clinical candidates currently in company pipelines (table4). Sixty to seventy percent of all recombinant protein pharmaceuticals are produced in mammalian cells, as mammalian expression reveals the most authentic glycosylation leading to a high percentage of active complex proteins. Some of these cell lines are CHO, NS0, BHK, HEK-29 and PER-C.

Extensive studies have yielded mammalian cell productivity in bioreactors to the gram per litre range (corresponding to a specific productivity of 20 - 90 pg/cell/day), which represents a more than 100-fold yield improvement in the last 25 years. Employing leader sequences (e.g. Ig kappa leader), allows expression of the recombinant protein into the culture supernatant facilitating the purification procedure. Correct folding and functionality of the recombinantly produced proteins is achieved by chaperones and disulfidisomerases located in the ER.

6.1. *Advantages*

Usually fold proteins correctly; usually make correct posttranslational modifications; can secrete protein; good regulatory track record; only choice (except for transgenics) for the largest, most complicated proteins [10].

6.2. *Disadvantages*

Expensive media; slow growth; may contain allergens and contaminants from bovine sources; require extensive characterization; complicated purification; expensive, $500–5,000/gram of final product, very time and labour intensive to reach stable clones [10].

Table 4. Biopharmaceutics expressed in mammalian cell culture.

Generic	Brand name	Company
Abciximab	Reopro	Centocor, Inc &Eli Lilly
Abciximab	Reopro	Centocor, Inc&. Eli Lilly
Alteplase	Activase	Genentech, Inc.
Bevacizumab	Avastin	Genetech
Cetuximab	Erbitux	ImClone Systems Inc.
cyclosporine	RESTASIS	Allergan, Inc
Darbepoetin Alfa	Aranesp	Amgen
Darbepoetin Alfa	Aranesp	Amgen
Dornase Alfa	Pulmozyme	Genentech, Inc.
Drotrecogin Alfa	Xigris	Eli Lilly
Efalizumab	Raptiva	Xoma, Ltd. & Genentech
Epoetin Alfa	Epogen/Procrit	Amgen
Etanercept	Enbrel	Amgen& Wyet
Factor VIIa	NovoSeven	Novo Nordisk
Factor IX	Benefix	Wyeth
Factor VIII	Kogenate FS	Bayer Corp.
Factor VIII	Recombinate	Baxter Healthcare
Factor VIII	Refacto	Wyeth Europe Ltd
Folitropin Alfa	Gonal-FREE	Serono S.A.
Folitropin Beta	Follistim AQ	Organon
Gemtuzumab	Mylotarg	Wyeth
Imiglucerase	Cerezyme	Genzyme Corporation
Infliximab	Remicade	Johnson & Johnson
Interfron Beta-1a	Avonex	Biogen Idec
Interfron Beta-1b	Rebif	Serono
Natalizumab	Tysabri	Biogen Idec& Élan
Omalizumab	Xolair	Genentech / Novartis
Palivizumab	Synagis	MedImmune
Rituximab	Rituxan	IDEC Pharmaceuticals
Somatropin	Saizen	Serono
Somatropin	Serostim	Serono
Tenecteplase	TNKase	Genentech
Thyrotropin Alfa	THyrogen	Genzyme Corp
Urokinase	Abbokinase	ImaRx

7. Transgenic animals

Securing a safe, economical, and reliable supply of a recombinant therapeutic protein for use in clinical trials and commercialization is an important strategic issue that must be addressed early in the development process.

The construction and validation of cGMP compliant facilities requires typically three to five years, which creates challenges in matching capacity to demand. As a result, companies must make manufacturing choices and a substantial upfront investment in capacity while their products are still in the test phases. The uncertainty of regulatory approval and the difficulties in predicting market demand impose a significant business and financial risk. For drug developers, investing large amounts of capital on a drug that is awaiting approval is an expensive and very risky prospect; however, having a drug that is approved without a reliable supply of product can have equally dire consequences.

Transgenic production holds tremendous promise for dealing with the cost, capacity and scale-up limitations faced by traditional systems. Some companies involved in this area are PPL therapeutics, Pharming, Genzyme.

7.1. *Advantages*

Capable of complex protein processing and of very large proteins; very high expression levels; correctly fold proteins; easy scaleup; low-cost production ($20–50/gram of final product) [10].

7.2. *Disadvantages*

Little regulatory experience; unknown potential for viral contamination; variable expression levels; long time scales; unanswered purification questions; continuous production complicates definition of batches and lots; questions regarding observance of cGMPs on the farm; unresolved public image problems [10].

8. Avian transgenics

The eggs, from which the proteins are harvested, are natural protein-production systems. But production of transgenic birds is still several years behind transgenic mammal technology. Intensive animal housing constraints also make them more susceptible to disease. Some companies involved in this area are TranXenoGen, Origen Therapeutics, AviGenics.

8.1. *Advantages*

Short gestation and maturation cycles; prolific reproduction; totally low cost for high yield of production; post-translation modification, no human-virus contamination [10].

8.2. *Disadvantages*

High cost for research stage, susceptible offspring to disease because of sterile environment [10].

9. Transgenic plants

Commercial production of plant secondary products has been used for centuries in human medicine. The progress of molecular plant genetics, and particularly of transgene expression in plants, has spurred the interest of academia and companies to consider them as eukaryotic hosts for biopharmaceutical production (table5). Two arguments favouring plant hosts have frequently been quoted, namely safety – the absence of adventitious mammalian viral or prionic contamination – and the unlimited scalability of production. The costs of producing recombinant proteins in plants might be 90-fold lower than the production in other systems. Additionally, safe biopharmaceuticals can be produced in plants, since plants do not harvest human pathogens compared to mammalian cells. Some companies involved in this area are Croptech, Prodigene, Boyce Thompson.

Table 5. Transgene expression systems in plants.

Expression System	Harvested Biomass
Stable Genome Transformation _ Nuclear transformation Leaves, tubers, seeds _ Chloroplast transformation	Leaves, tubers, seeds Leaves
Transient Expression _ Viral transfection _ Pro-viral transgene amplification	Leaves Leaves
Zygotic Induction of Transgene Amplification	Seeds

9.1. *Advantages*

Shorter development cycles than animals; easy storage of seed banks; easy scale-up; good expression levels (up to 1 kg purified recombinant protein/acre of crop); well-understood genetics; no plant viruses known to infect humans (so viral characterization unnecessary); low-cost production ($10–20/gram of product) [10].

9.2. *Disadvantages*

Potential for new contaminants (soil fungi and bacteria, plant-sourced impurities and metabolites, pesticides, herbicides); posttranslational modifications differ from those made by animal cells; contain possible allergens; unresolved public issues [10].

10. Conclusion

Although traditional expression systems have been advantages is indicative of industry-wide problems. Truly novel expression systems such as fungi, insect cells, plants, and transgenic animals have yet to make much impact on marketed products or those in later-stage development.substantially improved over the years, the failure to adopt new technologies often offering considerable

References

1. FROST&SULLIVAN (A) (2003). Biopharmaceuticals Industry Analysis-Quantification of Supply and Demand of Manufacturing Capacities (B181-39). Mountain View, Frost & Sullivan.
2. THIEL, K. A. (2004). "Biomanufacturing, from bust to boom ... to bubble?" Nat Biotechnol **22**(11): 1365 - 1372.
3. Ronald A. Rader, Expression Systems for Process and Product Improvement, BioProcess International, JUNE 2008, p4-9.
4. Brendan Fish, Richard Williams, "Avoiding pitfalls in scaling up biopharmaceutical production", Pharmaceutical Technology Europe, 2007.
5. ARISTIDOU, A. A., K. Y. SAN and G. N. BENNETT (1999). "Improvement of biomass yield and recombinant gene expression in Escherichia coli by using fructose as the primary carbon source." Biotechnol Prog **15**(1):140-145.
6. MIAO, J., J. WANG, S. PENG, P. TANG, M. ZOU, J. DUAN, C. ZHAO and X. MA (1995). "Expression of human interleukin-11 cDNA in E. coli." Sci China B **38**(10): 1202-1209.
7. NOVAGEN (2002). pET System, Manual.
8. BRADWELL, J. C., K. MCGOVERN and J. BECKWITH (1991). "Identification of a protein required for disulfide bond formation in vivo." Cell **67**: 581 – 589.
9. PLUCKTHUN, A. (1991). "Strategies for the expression of antibody fragments in Escherichia coli." Methods: A Companion to Methods in Ezymology **2**: 88 - 96.
10. The Applications, BioPharm International, Mar 10, 2006.

DENSITY SEGREGATION OF BINARY GRANULAR MIXTURES IN BUMPY VIBRATED BEDS

ELDIN WEE CHUAN LIM

Department of Chemical and Biomolecular Engineering, National University of Singapore
4 Engineering Drive 4, Singapore 117576

Segregation of granular materials by virtue of density or size is a commonly encountered phenomenon in nature. Despite its widespread interest among many researchers in recent years, a complete and unified understanding of granular segregation remains elusive to date. Using molecular dynamics simulations, we report a novel technique of inducing density segregation in a binary mixture of granular materials subjected to vibrations by the use of a bumpy vibrating base. Density segregation in the vertical directions may be induced by oscillating the bumpy base composed of discrete solid particles vertically or horizontally. In both cases, lighter particles tended to rise to the top of the granular bed and form a layer above the heavier particles. We suggest that differences in granular temperature profiles arising from the two different modes of vibrations may play an important role in determining the extent of density segregation occurring in binary granular mixtures.

Keywords: Vibrated bed, granular mixtures, density segregation, discrete element method, numerical simulation.

1. Introduction

A bed of granular materials subjected to vertical vibrations exhibits a variety of interesting phenomena and has been the focus of many research studies in recent years. It is well-established that granular mixtures are capable of undergoing segregation by virtue of a variety of physical property differences. These include segregations induced by particle size,[1–7] density,[2,8–12] and inelasticity or differences in coefficient of restitution.[13] The Brazil Nut Effect and Reversed Brazil Nut Effect whereby a larger intruder in a granular bed of smaller particles either rises or sinks under the effect of external vibrations have also been the subject of considerable interest among research workers in this area.[14–19] In general, these segregation phenomena have been attributed to various physical factors such as friction and shaking properties,[2,20] competition between percolation and condensation,[3] 'void-filling' mechanism,[4] combination of void-filling, sidewall-driven convection rolls and thermal diffusion,[5] competition between buoyancy and sidewall-driven convection,[6] effect of interstitial air,[8,10,12,16] combination of inertia, convection and buoyancy,[9] side-wall convection effects[14] and inertia assisted by Reynolds dilatancy.[15]

Despite the fact that various aspects of the vibrated granular bed system pertaining to granular segregation have been investigated fairly extensively in the research literature, others in contrast have remained conspicuously unexplored so far. One such aspect relates to the nature of the base on which granular materials are oscillated. It has been reported recently that the bulk behaviors of granular materials subjected to vertical vibrations on a flat or bumpy base are qualitatively and quantitatively similar.[21] In the present study, we explore using molecular dynamics simulations the effect of applying a bumpy vibrating base on the segregation behaviors of binary granular mixtures. Both vertical and horizontal vibrations of the bumpy base will be studied and we show that granular temperature profiles developed in the granular beds during vibrations may also be an important parameter determining the extent of segregation of the binary granular mixtures. In the following section, the numerical method applied and physical system of interest to the present study will be described.

2. Model

The molecular dynamics approach to modeling of granular systems, otherwise known as the Discrete Element Method (DEM), has been applied extensively for studies of various aspects of granular behavior. The methodology of DEM and its corresponding governing equations have also been presented numerous times in the research literature[21–25] and only a brief description will be presented here for sake of completeness. For a comprehensive review, the interested reader is referred to a recent review paper by Zhu et al.[26]

The translational and rotational motions of individual solid particles are governed by Newton's laws of motion:

$$m_i \frac{dv_i}{dt} = \sum_{j=1}^{N} \left(f_{c,ij} + f_{d,ij}\right) + m_i g \quad (1)$$

$$I_i \frac{d\omega_i}{dt} = \sum_{j=1}^{N} T_{ij} \quad (2)$$

where m_i and v_i are the mass and velocity of the ith particle respectively, N is the number of particles in contact with the ith particle, $f_{c,ij}$ and $f_{d,ij}$ are the contact and viscous contact damping forces respectively, I_i is the moment of inertia of the ith particle, ω_i is its angular velocity and T_{ij} is the torque arising from contact forces which causes the particle to rotate.

Contact and damping forces have to be calculated using force-displacement models that relate such forces to the relative positions, velocities and angular velocities of the colliding particles. Following Lim[21,22], a linear spring-and-dashpot model is implemented for the calculation of these collision forces. With such a closure, interparticle collisions are modeled as compressions of a perfectly elastic spring while the inelasticities associated with such collisions are modeled by the damping of energy in the dashpot component of the model. Collisions between particles and a wall may be handled in a similar manner but with the latter not incurring any change in its momentum. In other words, a wall at the point of contact with a particle may be treated as another particle but with an infinite amount of inertia. The normal ($f_{cn,ij}$, $f_{dn,ij}$) and tangential ($f_{ct,ij}$, $f_{dt,ij}$) components of the contact and damping forces are calculated according to the following equations:

$$f_{cn,ij} = -\left(\kappa_{n,i}\delta_{n,ij}\right) n_i \quad (3)$$

$$f_{ct,ij} = -\left(\kappa_{t,i}\delta_{t,ij}\right) t_i \quad (4)$$

$$f_{dn,ij} = -\eta_{n,i}\left(v_r x n_i\right) n_i \quad (5)$$

$$f_{dt,ij} = -\eta_{t,i}\left\{\left(v_r x t_i\right) t_i + \left(\omega_i \times R_i - \omega_j \times R_j\right)\right\} \quad (6)$$

where $\kappa_{n,i}$, $\delta_{n,ij}$, n_i, $\eta_{n,i}$ and $\kappa_{t,i}$, $\delta_{t,ij}$, t_i, $\eta_{t,i}$ are the spring constants, displacements between particles, unit vectors and viscous contact damping coefficients in the normal and tangential directions respectively, v_r is the relative velocity between particles and R_i and R_j are the radii of particles i and j respectively. If $\left|f_{ct,ij}\right| > \left|f_{cn,ij}\right| \tan\phi$, then 'slippage' between two contacting surfaces is simulated based on Coulomb-type friction law, i.e. $\left|f_{ct,ij}\right| = \left|f_{cn,ij}\right| \tan\phi$, where tan ϕ is analogous to the coefficient of friction.

The vibrated granular bed systems of interest to the present study are the four types of binary mixtures studied experimentally by Shi et al.[11] recently. We consider a pseudo-three-dimensional equivalent of their system (35 mm width × 140 mm height) of one particle thickness in the spanwise direction. Two vertical walls are imposed on each end of the system in the horizontal direction so as to correspond as closely as possible to the actual physical system used by the previous researchers. Under random packing conditions and with the entire system at rest, this pseudo-three-dimensional volume is estimated to contain a reasonable 1000 particles each of diameter 0.5 mm, amounting to about 14 – 16 layers of granular materials. The base of the system, consisting of discrete 5 mm particles to model a bumpy surface, is simulated to undergo a sinusoidal motion such that its displacement from the equilibrium position is described by Asin(ωt), where A is the amplitude of the oscillatory motion, $\omega = 2\pi f$ is the angular frequency, and t is time. The value of coefficient of restitution for both particle-particle and particle-base collisions is $\varepsilon = 0.9$. The vibrating frequency and amplitude are f = 90 Hz and A = 0.184 mm respectively giving the value of the dimensionless acceleration, $\Gamma = \frac{4\pi^2 f^2 A}{g} = 6.0$

3. Results and Discussion

Figure 1 shows the result of vibrating the four types of binary mixtures of granular materials on a bumpy base. Blue and red particles represent lighter and heavier particles respectively with the former made of aluminum oxide of density 1.31 g cm^{-3}. The heavier particles in the four panels are made of zirconium oxide (2.87 g cm^{-3}), titanium alloy (4.45 g cm^{-3}), cobalt-chromium-molybdenum alloy (8.37 g cm^{-3}) and tungsten alloy (18.0 g cm^{-3}) respectively. There is a tendency for lighter particles to rise through the granular bed to form a layer above the heavier particles. The extent of such density segregation increases with increasing difference in density between light and heavy particles. A more quantitative characterization of the states of segregation may be inferred from Figure 1e which shows the number fraction distributions of light particles along the vertical direction of the bed. It may be observed that the larger the difference in density between light and heavy particles, the more marked the extent of segregation.

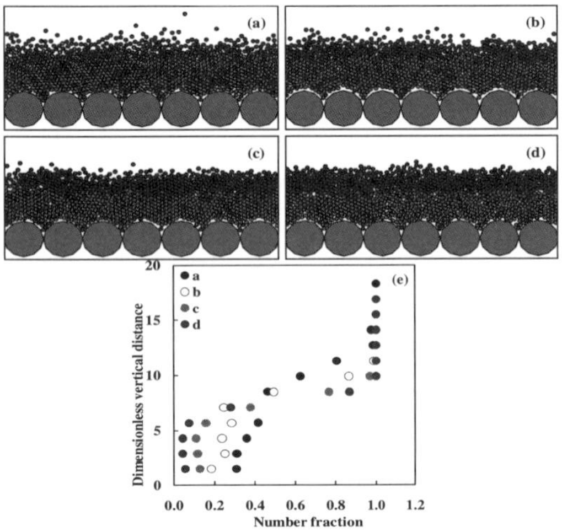

Figure 1. Segregated states of the four types of binary granular subjected to vertical vibrations on a bumpy base at f = 90 Hz, A = 0.184 mm and Γ = 6.0.

More interestingly, we show next the result of changing the mode of vibration of the bumpy base to that of horizontal vibration. The same 5 mm solid particles composing the base are oscillated in the horizontal direction at f = 90 Hz, A = 1.84 mm and Γ = 60 while maintaining all other conditions in the system identical to the previous cases. Due to the vigorous shaking conditions imposed, the granular beds are partially fluidized forming a region of lower solid concentration near the horizontally vibrating base where particles are highly energetic. All four types of binary mixtures undergo significant density segregations (Figure 2).

The two types of oscillatory motions (vertical or horizontal) of the bumpy base as well as the operating conditions imposed will result in significant differences in granular temperature profiles of the vibrated bed.

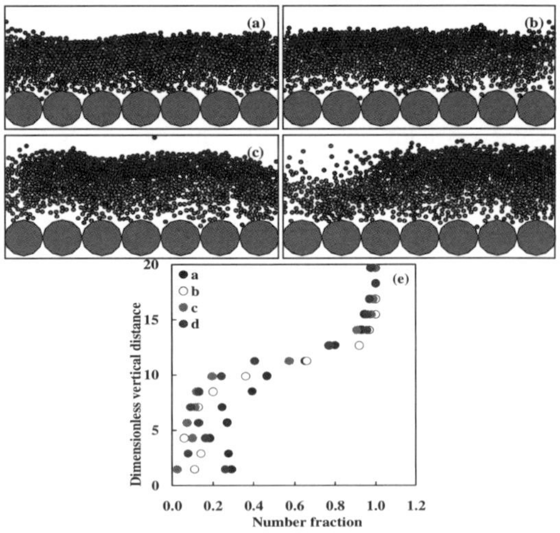

Figure 2. Segregated states of the four types of binary granular subjected to horizontal vibrations on a bumpy base at f = 90 Hz, A = 1.84 mm and Γ = 60.

Figures 3a and 3b show that under vertical oscillations of the bumpy base, the two components of granular temperatures T_y and T_x increase gradually with vertical distance along the bed. Here, $T_y = \frac{1}{2n}\sum_{i=1}^{n}\left(v_{y,i} - \overline{v_y}\right)^2$ and $T_x = \frac{1}{2n}\sum_{i=1}^{n}\left(v_{x,i} - \overline{v_x}\right)^2$ respectively, where $v_{y,i}$ and $v_{x,i}$ are the vertical and horizontal components of velocity of the ith particle, $\overline{v_y}$ and $\overline{v_x}$ are the corresponding mean velocities. Generally, granular temperatures are fairly isotropic for all four types of binary mixtures when subjected to vertical vibrations of the bumpy base, as can be seen from the similar profiles of T_y and T_x. Both components range from about 0 to 1.5 × 10^{-3} m^2 s^{-2} over a vertical distance of about 20 particle diameters. The granular temperature profiles for all four binary mixtures under the more vigorous horizontal vibration condition of Γ = 60 increase to approximately an order of magnitude higher than in the corresponding previous cases as shown in Figures 3c and 3d. Values of both T_y and T_x reach as high as 30 × 10^{-3} m^2 s^{-2} near the vibrating base and decrease smoothly with increasing vertical distance.

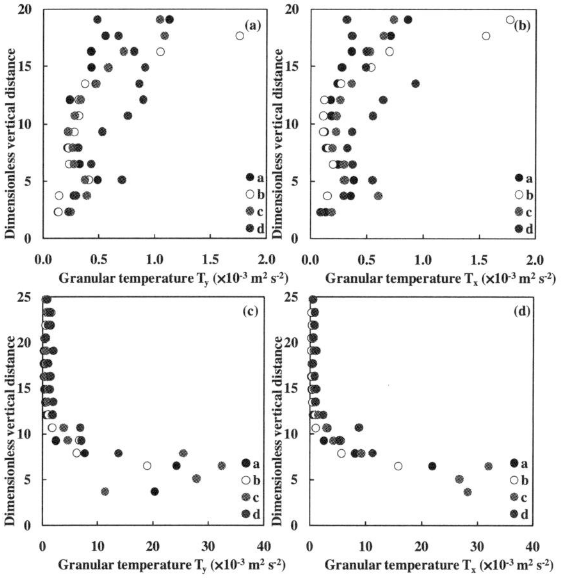

Figure 3. Spatially-averaged granular temperature profiles (a) Ty and (b) Tx for the four types of binary granular mixtures subjected to vertical vibrations on a bumpy base at f = 90 Hz, A = 0.184 mm and Γ = 6.0. With horizontal vibrations at f = 90 Hz, A = 1.84 mm and Γ = 60, both (c) Ty and (d) Tx profiles for the four binary mixtures are approximately an order of magnitude higher.

In comparing the various panels of Figure 3, we observe that granular beds subjected to vertical vibrations exhibit positive granular temperature gradients with particles near the surface having the highest temperatures while beds subjected to horizontal vibrations by a bumpy base are characterized by negative granular temperature gradients with particles near the base having the highest temperatures instead. Secondly, in horizontally vibrated granular beds on bumpy surfaces, higher granular temperatures seem to be more effective at inducing density segregations in binary mixtures. To our knowledge, these relationships between granular temperatures and density segregations that have been observed through the use of a bumpy vibrating base in this study do not seem to have been reported in previous experimental or numerical studies.

4. Conclusions

We have investigated the density segregation behaviors of various binary granular mixtures subjected to vertical and horizontal vibrations on a bumpy vibrating surface using molecular dynamics simulations. The binary mixtures consisted of light and heavy particles with varying density differences. When subjected to vertical vibrations on a bumpy surface, lighter particles tended to rise through the granular bed to form a layer above the heavier particles. The extents of segregation observed increased with increasing density differences between the two types of particles. The same type of segregation behavior could be induced through horizontal vibrations of the bumpy surface. It was observed that the extent of density segregation increased with increasing intensity of vibration characterized by the magnitude of the dimensionless acceleration parameter. This is also associated with a corresponding increase in magnitudes of granular temperatures within the vibrated beds. In comparing between the two modes of vibration, we observed that granular beds subjected to vertical vibrations exhibited positive granular temperature gradients with particles near the surface having the highest temperatures while beds subjected to horizontal vibrations by a bumpy base were characterized by negative granular temperature gradients with particles near the base having the highest temperatures instead. Furthermore, higher granular temperatures seemed to be more effective at inducing density segregations in binary mixtures subjected to horizontal vibrations on bumpy surfaces. We suggest that granular temperature may be a key property of a vibrated granular mixture which governs its segregation behavior. Finally, this study has also demonstrated the possibility of a new method of injecting energy into granular mixtures through the use of horizontally vibrated bumpy surfaces so as to induce density segregations in those mixtures.

Acknowledgements

This study has been supported by the National University of Singapore under Grant No. R-279-000-275-112. The award of a Lee Kuan Yew Postdoctoral Fellowship by the National University of Singapore is gratefully acknowledged.

References

1. Brone D, Muzzio FJ. Size segregation in vibrated granular systems: A reversible process. Phys. Rev. E. 1997;**56**:1059–1063.
2. Ciamarra MP, De Vizia MD, Fierro A, Tarzia M, Coniglio A, Nicodemi M. Granular species segregation under vertical tapping: Effects of size, density, friction, and shaking amplitude. Phys. Rev. Lett. 2006;**96**:058001.

3. Hong DC, Quinn PV, Luding S. Reverse brazil nut problem: Competition between percolation and condensation. Phys. Rev. Lett. 2001;**86**:3423–3426.
4. Rosato AD, Blackmore DL, Zhang N, Lan Y. A perspective on vibration-induced size segregation of granular materials. Chem. Eng. Sci. 2002;**57**:265–275.
5. Schroter M, Ulrich S, Kreft J, Swift JB, Swinney HL. Mechanisms in the size segregation of a binary granular mixture. Phys. Rev. E. 2006;74:011307.
6. Ulrich S, Schroter M, Swinney HL. Influence of friction on granular segregation. Phys. Rev. E. 2007;**76**:042301.
7. Yang X, Zhou K, Qiu K, Zhao Y. Segregation of large granules from close-packed cluster of small granules due to buoyancy. Phys. Rev. E. 2006;**73**:031305.
8. Burtally N, King PJ, Swift MR. Spontaneous air-driven separation in vertically vibrated fine granular mixtures. Science 2002;**295**:1877–1879.
9. Huerta DA, Ruiz-Suarez JC. Vibration-induced granular segregation: A phenomenon driven by three mechanisms. Phys. Rev. Lett. 2004;**92**:114301.
10. Klein M, Tsai LL, Rosen MS, Pavlin T, Candela D, Walsworth RL. Interstitial gas and density segregation of vertically vibrated granular media. Phys. Rev. E. 2006;**74**:010301(R).
11. Shi Q, Sun G, Hou M, Lu K. Density-driven segregation in vertically vibrated binary granular mixtures. Phys. Rev. E. 2007;**75**:061302.
12. Zeilstra C, van der Hoef MA, Kuipers JAM. Simulation of density segregation in vibrated beds. Phys. Rev. E. 2008;**77**:031309.
13. Brito R, Enriquez H, Godoy S, Soto R. Segregation induced by inelasticity in a vibrofluidized granular mixture. Phys. Rev. E. 2008;**77**:061301.
14. Fernando DN, Wassgren CR. Effects of vibration method and wall boundaries on size segregation in granular beds. Phys. Fluids. 2003;**15**:3458–3467.
15. Nahmad-Molinari Y, Canul-Chay G, Ruiz-Suarez JC. Inertia in the brazil nut problem. Phys. Rev. E. 2003;**68**:041301.
16. Naylor MA, Swift MR, King PJ. Air-driven brazil nut effect. Phys. Rev. E. 2003;**68**:012301.
17. Saez A, Vivanco F, Melo F. Size segregation, convection, and arching effect. Phys. Rev. E. 2005;**72**:021307.
18. Shinbrot T, Muzzio FJ. Reverse buoyancy in shaken granular beds. Phys. Rev. Lett. 1998;**81**:4365–4368.
19. Shishodia N, Wassgren CR. Particle segregation in vibrofluidized beds due to buoyant forces. Phys. Rev. Lett. 2001;**87**:084302.
20. Reis PM., Sykes T, Mullin T. Phases of granular segregation in a binary mixture. Phys. Rev. E. 2006;**74**:051306.
21. Lim EWC. Vibrated granular bed on a bumpy surface. Phys. Rev. E. 2009;**79**:041302.
22. Lim EWC. Master curve for the discrete-element method. Ind. Eng. Chem. Res. 2008;**47**:481–485.
23. Lim EWC. Voidage waves in hydraulic conveying through narrow pipes. Chem. Eng. Sci. 2007;**62**:4529–4543.
24. Yang RY, Yu AB McElroy L, Bao J. Numerical simulation of particle dynamics in different flow regimes in a rotating drum. Powder Technol. 2008;**188**:170–177.
25. Yang RY, Yu AB, Choi SK Coates MS, Chan HK. Agglomeration of fine particles subjected to centripetal compaction. Powder Technol. 2008;**184**:122–129.
26. Zhu HP, Zhou ZY, Yang RY, Yu AB. Discrete particle simulation of particulate systems: A review of major applications and findings. Chem. Eng. Sci. 2008;**63**:5728–5770.

STUDY ON PYROLYSIS CHARACTERISTICS OF ELECTRONIC WASTE

JING SUN, WENLONG WANG*, CHUNYUAN MA and YONG DONG

School of Energy and Power Engineering, Shandong University
Jinan, Shandong Province, PR China

The disposal of WEEE (waste electrical and electronic equipment) or "electronic waste"(e-waste) is an important issue in the whole world. Its pyrolysis characteristics under microwave irradiation were studied, and the effects of microwave output power and absorption additive were tested. The experiments prove that over 70% of WEEE would be decomposed through pyrolysis and the gas yield contains about 80% fuel gas such as CO, CH4, H2, etc. Using microwave technology, rapid combustion of WEEE under high temperature could be realized, and dioxin production may also be avoided effectively. After pyrolysis stage, the abundant metals, such as Fe, Cu, etc., would be more convenient to be recovered. Thus, by means of microwave pyrolysis, it is promising to realize fuel gas recovery, metal recovery as well as pollution avoidance for the WEEE treatment.

1. Introduction

Because of the use proliferation and rapid updating of electronic devices, WEEE (waste electrical and electronic equipment) or "electronic waste" (e-waste) is just becoming a big environmental issue puzzling the whole world. For instance, it is estimated that only in China about 17 million PCs, 48 million televisions, 12 million refrigerators and 13 million washing machines will be retired in 2009 and their total weight could amount to 2,692,000 tons [1]. In fact, illegal import from developed countries is another important source of WEEE for the emerging economies like China. Data shows that about 80% of WEEE in developed countries is outputted to Asia and therein 90% enters China [2].

WEEE usually contains a certain amount of metals and plastics, and even some precious and rare metals are included. This is just one reason why the WEEE is inputted to China. For example, in 1 ton of PC waste, there are 272kg plastics, 130kg copper, 0.45kg gold, 41kg iron, 29kg lead, 20kg tin, 18kg nickel, 10kg stibium and 10kg other precious metals such as silver, palladium, and platinum etc. [3,4] Table 1 gives the composition of printed circuit board (PCB) of computer [5].

Table 1. Composition of printed circuit board of computer.

Composition	Contents (%)	Composition	Contents (%)	Composition	Contents (%)
plastics	49.799	Sb	1.825	Be	0.015
Cu	23.728	Zn	0.747	Se	0.014
Fe	7.467	Ag	0.083	Ce	0.008
Br	4.646	Au	0.083	Pt	0.006
Pb	4.480	Cd	0.066	La	0.005
Sn	3.650	Ta	0.032	Hg	0.002
Ni	3.319	Pd	0.021		

Because of the composition complexity and existence of hazardous metals like lead, mercury, chromium, cadmium, etc, WEEE is difficult to dispose of and environmental pollution could easily be caused with improper treatment. This is another reason why WEEE is inputted into China. It was reported that the heavy metal contents were scores or even hundreds of times higher than the WHO guideline in the water or soil of Guiyu, which is one of the biggest WEEE trading and processing site in Guangdong province of China [6].

The majority of WEEE in China is processed in small workshops using basic methods such as manual disassembly and open incineration. The appliances are stripped of their most valuable and easily extracted components and materials, while the remainders, which are actually the real waste, are dumped. The final waste is still mainly made of metals and plastics. To release its heat energy and recycle the left metals is the common idea, and it may be one of the most potential ways through pyrolysis and melting technologies [7].

*Corresponding author. E-mail: wwenlong@sdu.edu.cn

Microwaves are a separate band of electromagnetic radiation with frequencies in the range of 300MHz to 300 GHz. According to their response in microwave field, materials are classified into three groups, conductors, insulators and absorbers. Materials that absorb microwave radiation are called dielectrics. The interaction of dielectric materials with electromagnetic radiation in the microwave range results in energy absorption [8].

Using microwave irradiation to heat materials has many advantages, such as heating quickly, heating selectively, auto–controlling easily, no direct contact between heating sources and heated materials as well as reduction in energy consumption [8,9]. In fact, the electronic waste is a mixture of metals and their alloys, metal oxides and plastics. When it is treated by microwave heating, the above advantages could be exerted sufficiently. Moreover, given the features that metals in WEEE may exist in different forms, and then may result in different performances under microwave irradiation, other advantages could also be made use of. For instance, metals of powder size and their alloys could attenuate electromagnetic wave through magnetic hysteresis loss and eddy current loss[10], while metals in piece forms could reflect microwave and tip discharge may happen at their borders or corners. Normally, the electric spark in microwave oven is disadvantageous, but we can use the heat created by the electric spark to accelerate the pyrolysis here. Thus, microwave-induced pyrolysis of electronic waste might be an effective way to treat WEEE.

In this paper, the study on pyrolysis characteristics of WEEE under microwave irradiation is presented.

2. Experimental materials and methods

To carry out the experiments, some waste keyboards and mice of computers as well as some waste printers and telephones were collected and a Galanz microwave oven model P7021TP-6 was prepared. The obtained electronic waste was crushed and grounded firstly. As a result, the obtained sample was composed of a small part of powder and most of irregular printed circuit boards and plastics, and the sample size was less than 1cm*1cm. The sample placed in an inner quartz cup was put in a relatively sealed outer quartz cup, and then heated in the microwave oven under N_2 atmosphere. The volatile gas produced was washed by a two-stage absorption system firstly, and then dried before it was pumped out. Given that the mass change of the heated sample was a main aspect to study the pyrolysis characteristics, during the experiment, the mass was measured by an electronic balance system and its data was collected every 20 seconds. Here, the mass change is defined as: m_t/m_0 (where m_t stands for mass at t moment, m_0 stands for mass of the original e-waste sample). During the experiments, the flow of N_2 should be regulated to guarantee the anaerobic environment. The experimental system is illustrated in figure 1.

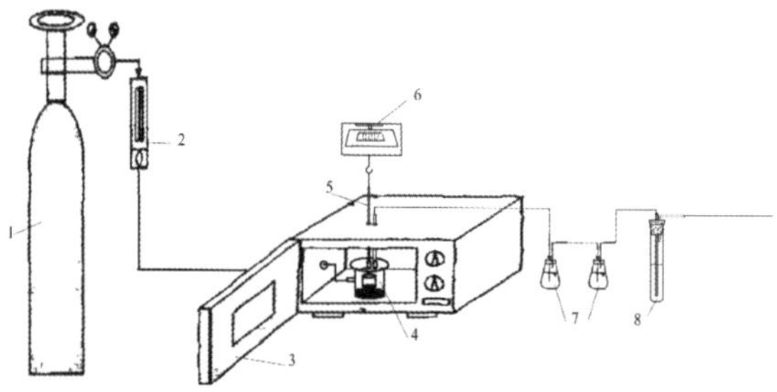

1. N_2 2. rotameter 3. microwave oven 4. quartz reaction device 5. suspension device 6. electronic balance system 7. gas absorption device 8. gas drying device

Fig. 1. The microwave equipment for WEEE pyrolysis.

In order to explore the pyrolysis characteristics of e-waste under microwave irradiation, a series of experiments were carried out. Firstly, to determine the effects of microwave output power, a certain amount of sample (20g) was heated under different Microwave output power. Secondly, e-waste was mixed with different additives and put under irritation of microwave, in this way, the effects of different microwave absorbers were tested. Thirdly, the combined effects of microwave output power and absorption additive were studied with the mix of preferable microwave absorber and under different microwave output power. Finally, a typical sample was illustrated and its products of gas, liquid and solid phases were collected and analyzed.

3. Experimental results and discussions

3.1. *Effects of microwave output power on the pyrolysis characteristics of electronic waste*

In order to investigate the performance of electronic waste at different microwave irradiation intensity, a number of experiments were carried out on the obtained sample by regulating microwave output power. Fig.2 shows that with the enhancement of the microwave output power, the pyrolysis progresses more rapidly, the time needed to start is shortened and the main pyrolysis phase tends to be steeper. This is easy to anticipate. When microwave irradiation imposed on become more intense, the movements of polar molecules and magnetic compounds tend to be more intense and the metal discharge happens more easily. Therefore, the ability of heat absorbing and producing is intensified. When the microwave output power was 700W, only 11 minutes were needed to pyrolyze the 20g e-waste sample. In conclusion, the improvement of microwave output power is beneficial to accelerate the pyrolysis.

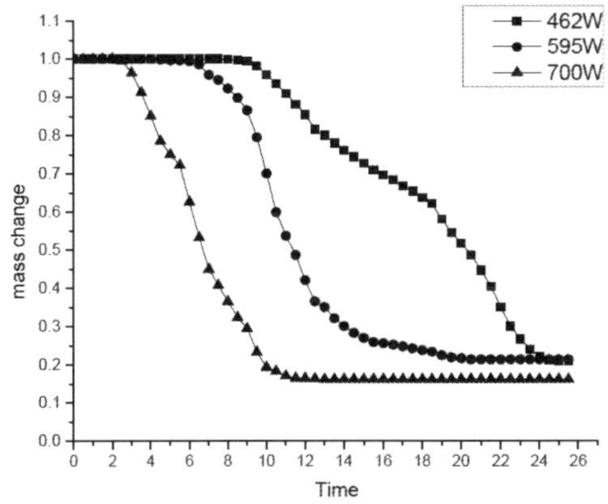

Fig. 2. E-waste sample at different output power.

3.2. *Effects of additives on the pyrolysis characteristics of electronic waste*

In order to investigate the promotion effects of the microwave absorption additive, several experiments were done with different additives. Activated carbon (AC for short) and SiC are two selected additives. At a certain microwave output power of 595W, Fig.4 illustrates that the addition of microwave absorber accelerates the pyrolysis process to a certain extent in comparison with the pyrolysis of pure e-waste sample. This is mainly because the excellent microwave absorption capacity of the additive could be heated more rapidly than e-waste. Then the e-waste nearby could be heated. Thus, the general heating rate could be improved to accelerate the pyrolysis. The effect of AC is more excellent than that of SiC. For example, when the proportion of AC to the 20g e-waste sample is 1/4(5g) or 1/10(2g), the pyrolysis time could be shorten greatly. In contrast, the performance of SiC is far from satisfaction. Given the lager density of SiC, the volume of SiC is less than that of AC when their mass is same. Thus, 8.7g SiC, with the same volume of 5g AC, was mixed with the e-waste sample. However its performance is far behind that of 5g AC. When the addition of SiC is increased to 20g, its performance is better but still inferior to that of AC. One fact necessary to be pointed out is that the microwave-absorbing capacity of SiC is better than that of AC [11]. Therefore, in theory, e-waste mixed with SiC should perform better than that mixed with AC. As to the adverse result, the catalysis effect of AC can be concluded boldly. Further investigations are needed to do in the future. To sum up, AC is a preferable additive to accelerate the microwave pyrolysis of e-waste.

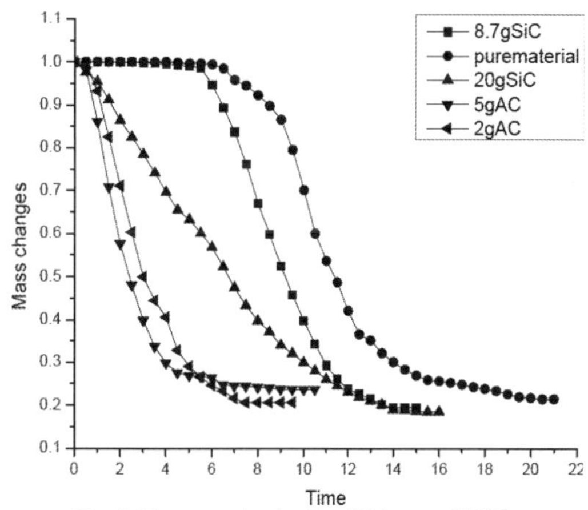

Fig. 3. E-waste mixed with additives at 595W.

3.3. *Combined effects of microwave output power and absorption additive*

After analyzing the effect of microwave output power and additives, a question is proposed. How does e-waste mixed with a preferable kind and amount of additive perform under different microwave power? Fig.4 shows the microwave pyrolysis of the 20g sample mixed with 5g AC at microwave output power of 700W, 595W, 462W and 280W. As a result of the general heating rate improvement caused by microwave absorption additive and the advantages brought by power enhancement, compared to Fig.2, the total pyrolysis time is shortened greatly and the pyrolysis is nearly started at the same time as the microwave irradiation is imposed on. With the enhancement of the microwave output power, the weight loss curve tends steeper in general. Compared the curve of 595W with the curve of 700W, there is a crossing point. This is mainly because the sample is not uniform and the metal existence forms and its quantity are a little different. Of course, the possibility that there is a most favorable microwave output power between

595W and 700W can not be excluded which needs further investigation. In conclusion, compared to the effects of microwave output power, the combined effects of power improvement and AC are more powerful to accelerate the pyrolysis of e-waste.

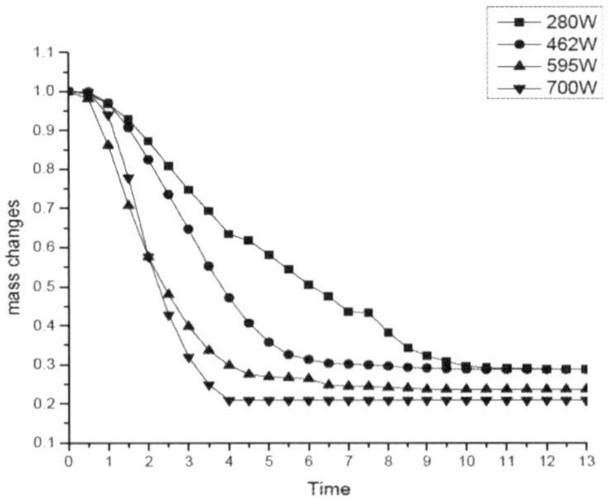

Fig. 4. 20G e-waste mixed with 5G AC at different output power.

3.4. *Yield of three phase output*

In fig.3,4,5, over 70% of WEEE could be decomposed. Take the pyrolysis of e-waste sample under microwave output power of 700W for example, the mass loss of the 20g e-waste sample after reaction is 16.757g which is mainly about 16g gaseous products and 0.757g liquid products. This means that the three phase (Solid, Gas, Liquid) yield rates after pyrolysis are 0.16215, 0.8 and 0.03785 separately. Here, one point necessary to be pointed out is that the gas out of the microwave oven is considered as gas yield, even though condensable gas is included, for the reason that the liquid product stick to the pipe easily and difficult to collect. Take analysis on the gas collected after two-stage absorption and drying by gas chromatography. The result shows that the gas yield contains about 80% fuel gas such as CO, CH_4, H_2 etc.

4. Conclusion

The complexity of the composition determines the feature of WEEE and the difficulty in its treatment. Microwave pyrolysis of WEEE could make good use of the advantages of microwave heating. Moreover, metal performance under microwave irradiation such as tip discharge could also be utilized to accelerate the pyrolysis. Thus, by using microwave technology, rapid combustion of WEEE under high temperature could be realized, and dioxin production could also be avoided effectively. Therefore, it opens a new way to treat WEEE.

Either the microwave power or absorption additive exerts a significant influence on the pyrolysis characteristics. Both the power improvement and suitable additive can accelerate the pyrolysis. In the future, our main task therefore is to explore other influencing factors and the most suitable additive to reduce energy input.

After pyrolysis, the abundant metals, such as Fe, Cu, etc., would be more convenient to be recovered and the gas products can be used as fuel gas after further treatment.

Acknowledgments

The authors thank China Ministry of Science and Technology (2009AA05Z303) for financial assistance for this work.

References

1. Liu Xiaoli, Yang Jianxin, and Wang Rusong, China Population, Resources and Environment, **15**(5), 113-117(2005).
2. Huang Yuanxiang, Journal of Xiangnan University, **29**(5), 55-59, 62(2008).
3. Zhou Cuihong, Lu Maixi, Techniques and Equipment for Environmental Pollution Control, **6**(4), 28-31(2005).
4. Legarth J.B., Resources, Conservation and Recycling, **19**(2), 109-135(1997).
5. Xu Jinqiu, Ma Hong, Wang Jingwei, and Liang Bo, Journal of Shanghai Second Polytechnic University, **24**(4), 263-270(2007).
6. C.Hicks, R. Dietmar, M. Eugster. Environmantal Impact Assessment Review, **25**, 459-471(2005).
7. M.P.Luda, N.Euringer, U.Moratti, M.Zanetti, Waste Management, **25**(2), 203-208(2005).
8. D.A. Jones, T.P. Lelyveld, S.D. Mavrofidis, S.W. Kingman, N.J. Miles, Resources, conservation and Recycling, **34**,75-90(2002)
9. Songhu Yuan, Meng Tian, Xiaohua lu, Journal of Hazardous Materials, B137, 878-885(2006)
10. Zhao Lingzhi, Hu Shejun, Li Weishan, He Qinyu, Chen Junfang, Ru Qiang, Modern Defence Technology, **39**(1), 27-32(2007).
11. Ruidian Tan, Doctorate thesis, Dalian University of Technology, (2007).

MICROWAVE ASSISTED ACTIVATED CARBON PRODUCTION BASED ON BIODIESEL SOLID RESIDUE (RAPHANUS SATIVUS PRESS CAKE)

ADRIANA S. FRANCA
Departamento de Engenharia Mecânica/UFMG, Av. Antônio Carlos 6627
31270-901, Belo Horizonte, MG, Brazil

DIEGO L. NUNES
Programa de Pós-Graduação em Ciência de Alimentos/UFMG, Av. Antônio Carlos 6627
31270-901, Belo Horizonte, MG, Brazil

The objective of this study was to evaluate microwave thermal activation in the production of adsorbents. The produced activated carbon (AC) was tested for methylene blue (MB) adsorption in batch tests. Freundlich model provided the best fit for equilibrium data, indicating heterogeneous adsorption. Adsorption efficiency was similar to the AC obtained by conventional oven processing, showing that microwave treatment is a suitable alternative for AC production, given the significant reduction in processing time.

Keywords: Biodiesel solid residue; microwave treatment; adsorption; dyes.

1. Introduction

Wastewaters generated by industries (textile, leather, paper, etc.) constitute hazardous materials that have severe negative environmental impact when discharged without adequate treatment. The main recalcitrant compounds present in such effluents are synthetic dyes [1]. Aside from aesthetic aspects, such compounds are also of significant health concern even at low concentrations in water.

AC adsorption has been extensively employed for decoloration of aqueous solutions, given its high efficiency. However, it still presents disadvantages associated to high cost of the adsorbent, high initial capital investment, long residence times, and the need for regeneration. The need to overcome such problems has prompted a growing research interest in the production of AC´s from renewable and low-cost materials, mostly industrial and agricultural by-products. Examples include bamboo [2], coconut [3] and coffee husks [4]. Comprehensive reviews on adsorbents from agri-food wastes have been recently published [5-6].

Currently, over 95% of the world´s biodiesel production is derived from edible oils, easily available on large scale from the agricultural industry. However, continuous and large-scale production of biodiesel from edible oil without proper planning may cause problems, including depletion of food supply. A solution would be the use of non-edible oil for biodiesel production. However, this process generates an extensive amount of solid residues (pressed seed cakes), which present environmental problems in terms of adequate disposal, given their limited application as animal feed or silage. A few recent studies have shown that adsorbents prepared from seed press cakes can be employed for adsorption of dyes. Examples include sunflower oil cake [7], defective coffee beans press cake [8] and Raphanus sativus press cake [9]. Recent studies have demonstrated that microwaves can be successfully employed in the production of AC´s [10-11], with both reduction in processing time and increase in adsorption capacity.

In view of the aforementioned, the objective of this study was to evaluate microwave thermal activation as an alternative technique for the production of adsorbents based on Raphanus Sativus press cake. Adsorption studies were conducted with methylene blue (MB) in order to verify the performance of the adsorbent for removal of cationic dyes and to compare with the activated carbon produced by traditional carbonization (1h under N2 flow) in a muffle oven [9].

2. Materials and Methods

2.1. *Materials*

Raphanus sativus (L. Var.) seeds were screw-pressed (ERT60, ScothTech, Brazil) for oil removal. The cationic dye used as target adsorbate was methylene blue (MB - C.I. 52015) 15% hydrated, 95% purity, purchased from Labsynth Lab Products (São Paulo, Brazil).

2.2. *Adsorbent preparation*

The press cake (30g) was submitted to preliminary carbonization tests employing a household microwave (Philco PMS35) for 3, 4 and 5 min at maximum power (900W).

2.3. *Adsorption studies*

Batch experiments of adsorption were performed in 250 mL Erlenmeyer flasks, with the flasks being agitated on an orbital shaker at 100 rpm for pre-determined time intervals. In all sets of experiments, a pre-determined amount of adsorbent was thoroughly mixed with 100 mL MB solution. Effect of particle size (diameter = D) was evaluated in the following ranges: D < 0.43 mm; 0.43 mm < D < 0.85 mm; and D > 0.85 mm. Effects of pH and adsorbent concentration were studied in the respective ranges of 3 to 10 and of 5 to 40 g L^{-1} at a fixed initial dye concentration (200 mg L^{-1}). Effect of contact time was evaluated at time periods ranging from 15 min to 24 hours and initial dye concentrations ranging from 50 to 500 mg L^{-1} at a fixed adsorbent concentration (10 g L^{-1}). All tests were performed in two replicates.

2.4. *Analysis of methylene blue (MB)*

After the specified time periods, 2 mL aliquots were taken from the Erlenmeyer flasks and the concentration of MB was determined by a spectrophotometer (Cole Parmer 1100 RS) at 665 nm. The amount of MB adsorbed was determined by taking the difference between the initial dye concentration and the concentration of the solution at the time of sampling. All determinations were performed in a total of three replicates per experiment and the average values were reported.

3. Results and Discussion

3.1. *Adsorbent preparation*

Preliminary adsorption tests were performed in order to ascertain the effect of the microwave thermal treatment on the adsorbent performance. Results are displayed in Fig. 1. It can be observed that an increase in microwave treatment time resulted in a decrease in adsorption capability. The adsorbent obtained after 3 min carbonization presented a similar adsorption performance in comparison to the one obtained by conventional 1h carbonization in a muffle oven [9] and therefore it was the one chosen for the remaining tests. It is noteworthy to mention that, even though the muffle activated carbon presented slightly better adsorption performance, the mass yield after carbonization was lower than 30%, in comparison to 71% for microwave carbonization.

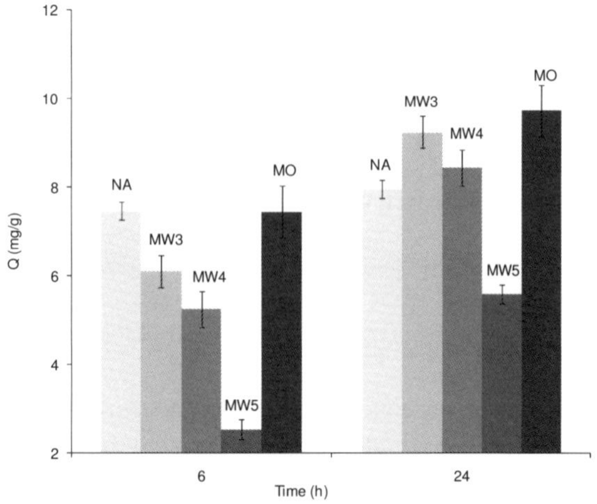

Figure 1. Preliminary adsorption tests: effect of microwave carbonization time and comparison with other activated carbons (NA = no activation – biosorbent; MW3 = 3min microwave activation; MW4 = 4min microwave activation; MW5 = 5min microwave activation; MO = 1h muffle oven activation).

3.2. *Effect of particle size, pH and adsorbent dosage*

Results on the effects of particle size and initial pH are displayed in Fig. 2. MB uptake increased with the decrease in particle size, given the corresponding increase in surface area. The higher adsorption capacity obtained at initial pH 10 (Fig. 2b) is attributed to the adsorbent surface becoming predominantly negatively charged, favoring electrostatic attraction between the surface and MB cations. However, adsorption efficiency was the same after 22h adsorption, with the exception of the lower initial pH value. Based on the results presented in Fig. 2, the remaining tests were conducted employing the activated carbon with D < 0.43 mm and initial solution pH = 7.

The influence of adsorbent dosage on MB removal can be viewed in Fig. 3. Dye removal efficiency increased with the increase in adsorbent dosage, due to the increase in available adsorption sites. The amount of dye adsorbed per unit mass of adsorbent decreased with increasing adsorbent mass, due to the reduction in adsorbate/adsorbent ratio. Based on the results presented in Fig. 3, the remaining experiments were conducted for an adsorbent dosage of 10 g L^{-1}.

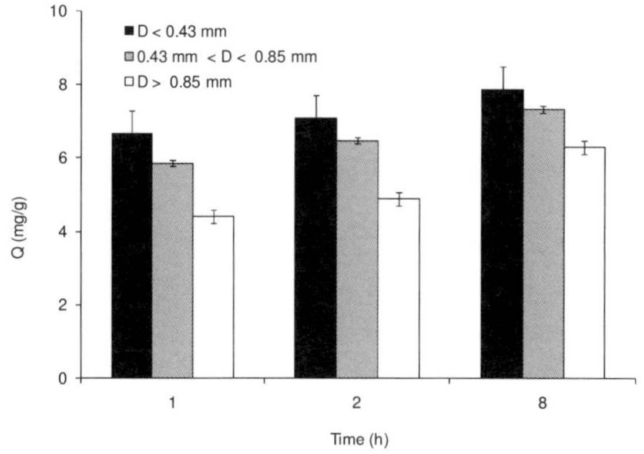

(a)

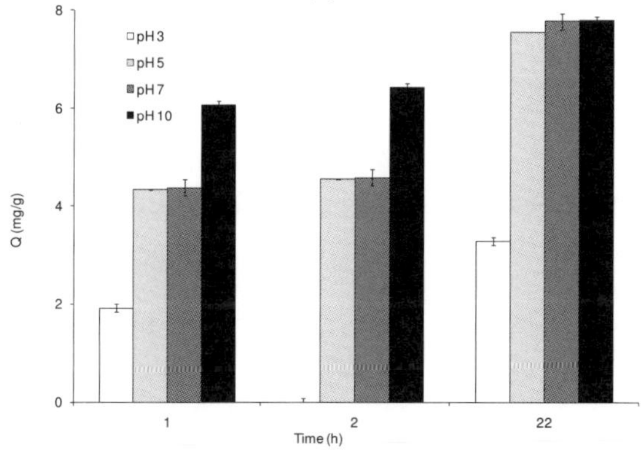

(b)

Figure 2. Effect of (a) particle diameter, D and (b) initial solution pH on MB adsorption by the microwave activated carbon.

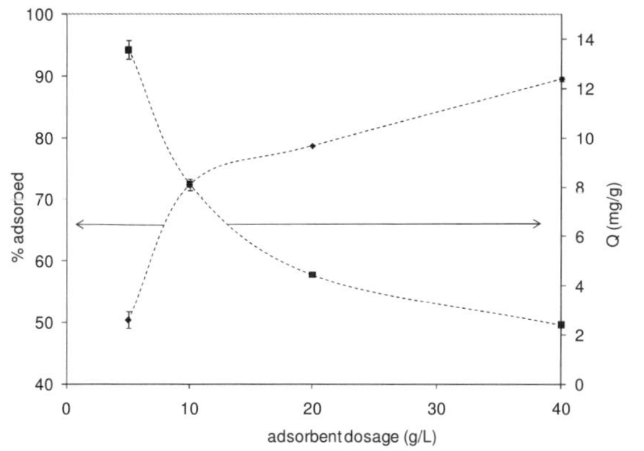

Figure 3. Effect of adsorbent dosage (♦% adsorbed; ■ amount adsorbed per unit mass) on MB adsorption by the microwave activated carbon.

3.3. *Effect of initial MB concentration*

The adsorption data presented in Fig. 4 showed that a contact time of 24 h assured attainment of equilibrium conditions for all initial MB concentrations, except for the highest one (500 mg/L). Results also show that adsorption of MB presents a strong dependency on MB initial concentration. An increase in the initial MB concentration led to an increase in the amount adsorbed, which is attributed to the corresponding increase in driving force (concentration gradient).

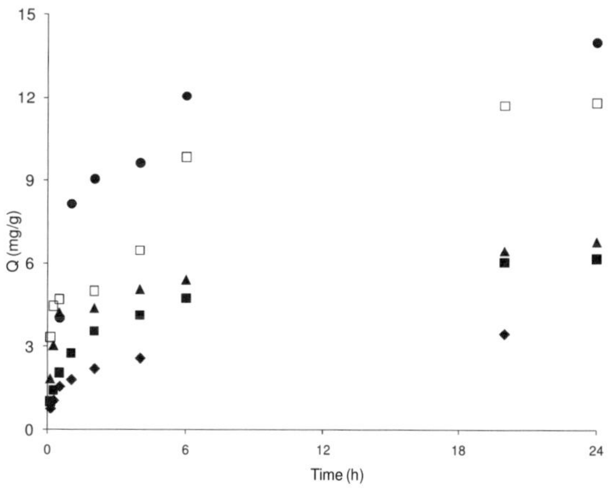

Figure 4. Effect of contact time on MB adsorption by microwave activated carbon. Initial MB concentration: ♦ 50 mg L^{-1}, ■ 100 mg L^{-1}, ▲ 200 mg L^{-1}, □ 300 mg L^{-1}, ● 500 mg L^{-1}.

3.4. *Adsorption equilibrium*

Langmuir and Freundlich models were tested for equilibrium description at ambient temperature [12]. Langmuir model is described by the following equation:

$$q_e = \frac{q_{max} K_L C_e}{1 + K_L C_e} \qquad (1)$$

where q_e (mg g^{-1}) and C_e (mg L^{-1}) represent the amount adsorbed per gram of adsorbent and solute concentration (mg L^{-1}) in the solution, respectively, after equilibrium was reached. q_{max} and K_L are constants related to the maximum adsorption capacity (mg g^{-1}) and the adsorption energy (L mg^{-1}), respectively.

Freundlich's model can be described by the following equation:

$$q_e = K_F C_e^{1/n} \qquad (2)$$

where K_F indicates the relative adsorption capacity ($mg^{1-(1/n)} L^{1/n} g^{-1}$) and n is related to the intensity of adsorption.

The adsorption isotherm at room temperature is displayed in Fig. 5, indicating favorable adsorption. Coefficients of the Langmuir and Freundlich models are displayed in Table 1. MB adsorption was better described by Freundlich, indicating heterogeneous adsorption. Maximum MB uptake capacity was 20.41 mg g-1, which was higher in comparison to adsorbents obtained by thermal and/or chemical treatment of similar types of residues, such as sunflower oil cake (16.4 mg g-1 [7]) and defective coffee beans press cake (14.9 mg g-1[8]). Also, adsorption capacity was similar to the AC obtained by conventional carbonization in a muffle oven of the same residue (19.8 mg g-1 [9]).

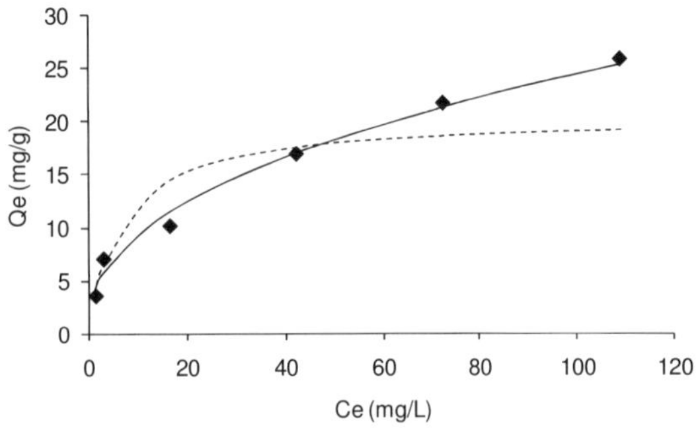

Figure 5. Adsorption isotherm of MB by microwave activated carbon. Solid and dashed lines correspond to Freundlich and Langmuir model fits, respectively.

Table 1. Langmuir and Freundlich isotherm constants for MB adsorption.

Langmuir		Freundlich	
K_L	0.148	K_F	3.48
q_{max}	20.41	1/n	0.42
r	0.01-0.12	R^2	0.970
R^2	0.968		

4. Conclusions

Our results show that this type of residue has potential as an adsorbent for removal of cationic dyes and that microwave activation is a viable alternative for production of activated carbons, given the significant reduction in processing time and energy consumption, without affecting the activated carbon performance.

Acknowledgments

The authors acknowledge financial support from Brazilian Government Agencies: CAPES, CNPq and FAPEMIG.

References

1. S. Asad, M.A. Amoozegar, A.A. Pourbabaee, M.N. Sarbolouki and S.M.M. Dastgheib, *Bioresource Technol.* **98**, 2082 (2007).
2. B.H. Hameed, A.T.M. Din and A.L. Ahmad, *J. Hazard. Mater.*, **141**, 819 (2007).
3. I.A.W. Tan, A.L. Ahmad and B.H. Hameed, *J. Hazard. Mater.*, **154**, 337 (2008).
4. L.S. Oliveira, A.S. Franca, T.M. Alves and S.D.F Rocha, *J. Hazard. Mater.*, **155**, 507 (2008).
5. O. Ioannidou and A. Zabaniotou, *Renew. Sust. Energ. Rev.* **11**, 1966 (2007).
6. L.S. Oliveira and A.S. Franca, "Low-Cost Adsorbents from Agri-Food Wastes", *Food Science and Technology: New Research* (Eds.: L.V. Greco, M.N. Bruno), Nova Publishers, New York, pp. 171-209 (2008).
7. S. Karagoz, T. Tay, S. Ucar and M. Erdem, *Biores. Technol.*, **99**, 6214 (2008).
8. A.A. Nunes, A.S. Franca and L.S. Oliveira, *Biores. Technol.*, **100**, 1786 (2009).
9. D.A. Lázaro, M.B. Mansur, A.S. Franca, L.S. Oliveira and S.D.F. Rocha, *Int. J. Chem. Eng.*, **1**, 147 (2008).
10. M. Hirata, N. Kawasaki, T. Nakamura, K. Matsumoto, M. Kabayama, T. Tamura and S. Tanada, *J. Colloid Int. Sci.*, **254**, 17 (2002).
11. W. Li, L. Zhang, J. Peng, N. Li and X. Zhu, *Ind. Crop Prod.*, **27**, 341 (2008).
12. O. Hamdaoui and E. Naffrechoux, *J. Hazard. Mater.*, **147**, 381 (2007).

ALTERNATIVE USES FOR COFFEE HUSKS — A SOLID WASTE FROM GREEN COFFEE PRODUCTION

ADRIANA S. FRANCA and LEANDRO S. OLIVEIRA

Departamento de Engenharia Mecânica/UFMG, Av. Antônio Carlos 6627, 31270-901, Belo Horizonte, MG, Brazil

Brazil is the largest coffee producer in the world, with an average production of 2.5 million tons per year. Coffee husks (CH) are the major solid residues from the handling and processing of coffee, since for every kg of coffee beans produced, approximately 1 kg of husks are generated. These solid residues pose several problems in terms of adequate disposal, given the high amounts generated and environmental concerns. In view of the aforementioned, the objective of the present study was to present an overview of the works that have been developed in order to find alternative uses for coffee husks. Applications include direct use as fuel in farms, silage, animal feed, fermentation studies, use as adsorbents, and others.

Keywords: Coffee husks; agricultural residues; fermentation; adsorbents.

1. Introduction

Coffee is deemed as a commodity ranking second only to petroleum in terms of currency (usually US dollars) traded worldwide. As such this commodity is quite relevant to the economy of producing countries, including Brazil, Vietnam, Colombia, Ethiopia, Indonesia, Mexico and India [1]. Brazil is the largest coffee producer and exporter in the world, and is the second largest consumer. The production of coffee in Brazil in the last 5 years ranged from 2.0 to 2.7 million tons [2], representing an average of over 2.5 million tons of solid residues being generated every year.

During post-harvest processing, coffee fruits are de-hulled and submitted to drying, with the resulting product being the green coffee beans. There are two major coffee processing methods: dry and wet processing. Dry processing is the simplest technique for processing coffee cherries. After harvesting, the coffee cherries are dried to about 10–11% moisture content. Thereafter, the coffee beans are separated by removing the material covering the beans (outer skin, pulp, parchment and silverskin) in a de-hulling machine. Generated solid residues are denominated coffee husks - CH (outer skin + pulp + parchment) and silverskin. Wet processing, on the other hand, does not require drying of the cherries themselves. In this type of processing, first the outer skin and pulp are mechanically removed, thus generating the solid residue, denominated coffee pulp - CP. CH present moisture contents ranging from 7 to 18%, whereas CP contain ~ 75% moisture and are usually let dry to ~13% moisture [3].

Such residues present high contents of carbohydrates (45-90%), which is expected given their origin, i.e., fruit pulp and outer skin. Some authors also make reference to a specific type of coffee husks known as sticky coffee husks (SCH) [4, 5]. Attributes that differentiate SCH from CH include higher density, protein contents and lower fiber contents. However, the major difference relies on its sugar content, which amounts to ~30% d.b. in comparison to 12-14% d.b. in CH.

Even though CH and CP are rich in organic nature and nutrients, they also contain compounds such as caffeine, tannins, and polyphenols. Due to the presence of such substances, they present toxic nature, which not only adds to the problem of environmental pollution, but also restricts their use as animal feed [6]. A summary description of the specific uses that have been and are currently under evaluation for this type of agricultural solid residue is presented as follows.

2. Alternative Uses

2.1. *Livestock feed*

In the case of coffee husks, its low digestible protein content, in addition to the fact that the starch equivalent is comparable to low quality hay, has prevented its use as animal feed [3]. However, given the high amounts generated, coupled to its availability at low costs throughout the season, recent studies have evaluated its use as a diet supplement for cattle [7, 8], swine [5], fish [9] and sheep [10].

The overall conclusion of the studies developed so far is that CH employment as animal feed is limited due to seasonal availability and perishability due to high water content (CP), but the major limiting factor is the presence of anti-nutritional compounds, thus restricting the amount that can be used. In that regard, some studies have been conducted with the purpose of removing anti-nutritional substances such as caffeine and tannins from CH [6, 11]. However, no in vivo evaluations of such detoxified residues are available.

2.2. *Silage*

Coffee husks and pulp are rich in potassium (~40 g kg^{-1}) and other mineral nutrients, which has induced some studies on its application as organic fertilizers without any treatment or after composting [12]. The use of coffee husks directly as soil coverage is a good option for potassium depleted soils, and can be employed for different types of crops, including coffee. They favor erosion control, decrease temperature fluctuations and water losses by evaporation [12]. CH also present better performance than other agricultural residues [13]. However, there is always the possibility of phytotoxin production.

2.3. *Fuel*

Coffee husks have been deemed as a source of cheap fuel with an approximate calorific value of 16 MJ kg^{-1} [3]. CH combustion, employing single particle combustion techniques and also a pilot-scale fluidized bed facility was evaluated [14]. The low density of coffee husks is an indication that such residue is not appropriate for long distance transportation, given the high costs of transportation and storage, so it was suggested that it should be used in the vicinities of the production area. High values for NO_x emissions (400–500 mg/m^3) were measured, indicating the need for NO_x emission reduction. Another problem is the low melting temperature of ash due to the high content of K_2O, and the consequent problems of agglomeration, fouling, slagging and corrosion. Therefore, the authors suggested further research on the use of additives, proper furnace design and co-firing with coal, in order for the direct use of coffee husks as fuel to become viable [15].

2.4. *Fermentation Studies*

Several applications of coffee pulp and husks in fermentation studies have been reported, including the production of enzymes [16], citric acid [17], gibberellic acid [18] and flavoring substances [19].

Given the high concentration of carbohydrates in coffee husks, it can be also viewed as a potential raw material for bioethanol production. Furthermore, the produced ethanol could be employed for biodiesel production based on coffee oil obtained from defective coffee beans, thus further contributing for the implementation of sustainable development in the coffee and biodiesel production chains [20]. A recent study [4] demonstrated that fermentation of sticky coffee husks lead to a product containing 14% w/v of ethanol. This was only a preliminary feasibility study, employing baker's yeast as a fermentation agent. However, ethanol production was high in comparison to other agricultural residues that are being studied for bioethanol production, including wheat stillage, sweet sorghum, corn stalks and barley straw [4]. Such results indicate that SCH present excellent potential for residue-based ethanol production, given that production levels can be significantly improved by enzymatic hydrolysis, the use of other microorganisms and nutrient supplementation.

2.5. *Substrate for Mushroom Growth*

There are a few recent studies on the use of CH and CP as substrates for mushroom growth [21, 22]. The general conclusion was that, not only the substract was satisfactory in terms of mushroom growth, but it was also concluded that the residues degraded by mushroom cultures might present further post-production value in the elaboration of forages, vermicomposts, organic fertilizers and even animal feed, given the reduction in caffeine and tannin levels.

2.6. *Production of Adsorbents*

The association between coffee husks and the production of charcoal has been thought firstly in terms of fuel, given that the calorific value is practically doubled after carbonization [3]. However, given the significant amount of published information and increasing research interest on the use of agri-food residues in the preparation of activated carbons [23], a few recent studies have dealt with the application of coffee husks as either biosorbents (used as adsorbents without the need for carbonization and activation) or as raw materials for the production of adsorbents [24, 25]. Such residues presented excellent adsorption capabilities in wastewater removal of heavy metals and cationic dyes.

3. Conclusions

Many studies on CH and CP uses have focused on employing them as supplement for animal feed. Several in vivo studies have been carried out with a wide variety of livestock (cattle, fish, pigs, sheep and chicken) and the overall conclusion is that use for animal feed is restricted due to the presence of anti-nutritional factors (caffeine, tannins). Detoxification and fermentation studies have demonstrated that caffeine and tannin levels can be reduced significantly. However, further in vivo studies employing detoxified CH and CP are needed in order to verify if such procedures are sufficient to increase the amount of these solid wastes that can be employed as a supplement or even as a replacement for the currently used animal feed. The use of CH directly as soil coverage seems to be a good option for potassium depleted soils. Composting is also feasible, given that another type of residue be mixed with the husks in order to improve the carbon nitrogen ratio. Even though CH in their original state present a reasonable calorific value and that this value can be further increased by carbonization, its use as a direct solid fuel still presents several problems, including the need to reduce NOx emissions and to avoid pyrolysis in the feeding system, dragging of the coffee husks by the flue gas and agglomeration, fouling and corrosion of the combustion equipment. Among the most recently proposed uses for coffee husks and pulp, fermentation for the production of a diversity of products, use as substrate for growth of mushrooms and use as adsorbents are worth mentioning to be promising alternatives. These alternatives generate more value-added products than the aforementioned ones, requiring, in most cases, less processing steps and simpler technologies, thus, being, in theory, relatively more cost effective alternatives. However, all the studies were carried out at lab scale and were based on attaining a pre-specified target product using the currently available knowledge of the physical and chemical properties of the residues, which is rather superficial. In the majority of these studies, no attempts to work on the scaling-up and to perform economical analyses of the processes were done and further studies are needed in order to proper address these issues.

Acknowledgments

The authors acknowledge financial support from Brazilian Government Agencies: CAPES, CNPq and FAPEMIG.

References

1. A.S. Franca and L.S. Oliveira, "Chemistry of defective coffee beans", *Progress in Food Chemistry*, pp. 105-138 (2008).
2. http://www.conab.gov.br/conabweb/download/safra/2_levantamento_2008.pdf
3. M.R. Adams and J. Dougan. "Waste Products", *Coffee Technology*, pp. 257-291 (1987).
4. B.M. Gouvea, C. Torres, A.S. Franca, L.S. Oliveira, and E.S. Oliveira, *J. Biotech.* **136**, S269 (2008).
5. A.R.P. Parra, I. Moreira, A.C. Furlan, D. Paiano, C. Scherer, C. and P.L.O. Carvalho, *R. Bras. Zootec.* **37**, 433 (2008).
6. A. Pandey, C.R. Soccol, P. Nigam and V.T. Soccol., *Biochem. Eng. J.* **6**, 153 (2000).
7. A.L Souza, R. Garcia, F.S. Bernardino, J.M.S. Campos, S.C.V. Filho, L.S, Cabral and K.F. Gobbi, *R. Bras. Zootec.* **35**, 921 (2006).
8. R.M.A. Teixeira, J.M.S. Campos, S.C.V. Filho, A.S. Oliveira, A.J. Assis, A.J. and D.S. Pina. *R. Bras. Zootec.*, **36**, 968 (2007).
9. J.B. Ulloa Rojas and J.A.J. Verreth, *Aquaculture,* **217**, 275 (2003).
10. A.L. Souza, R. Garcia, F.S. Bernardino, F.C. Rocha, S.C.V. Filho, O.G. Pereiraand A. J. V. Pires, *Rev. Bras. Zootec.,* **33**, 2170 (2004).
11. A.L. Orozco, M.I. Perez, O. Guevara, J. Rodriguez, M. Hernandez, F.J. Gonzalez-Vila, O. Polvillo, O. and M.E. Arias, *J. Anal. Appl. Pyr.,* **81**, 247 (2008).
12. A.T. Matos, "Tratamento de resíduos na pós-colheita do café", *Pós-Colheita do Café,* pp. 161-201 (2008).
13. A.P. Neiva Júnior, J.C. Silva Filho, I.M. Von Tiesenhausen, G.P. Rocha, E.R. Capelle, and C.C.C. Couto Filho, *Cienc. Agrotec.,* **31**, 871 (2007).
14. M. Saenger, E.-U. Hartge, J. Werther, T. Ogadaand Z. Siagi, *Renew. Energy,* **23**, 103, 2001.
15. E.A. Magalhães, J.S. Silva, J.N. Silva, D.O. Filho, S.M.L. Donzeles, S. Martin, S. and L. Dutra, *Rev. Bras. Armazen.* **Esp. Café 10**, 66 (2008).
16. K.N. Niladevi and P. Prema, *Biores. Technol.,* **99**, 4583 (2008).
17. L.P.S. Vandenberghe, C.R. Soccol, A. Pandey and J. -M. Lebeault, *Biores. Technol.,* **74**, 175 (2000).
18. C.M.M. Machado, B.O. Oishi, A. Pandey and C.R. Soccol, *Biotechnol. Prog.,* **20**, 1449 (2004).

19. M. Soares, P. Christen, A. Pandey and C.R. Soccol, *Process Biochem.,* **35**, 857 (2000).
20. L.S. Oliveira, A.S. Franca, R.R.S. Camargos and V.P. Ferraz, *Biores. Technol.,* **99**, 3244 (2008).
21. D. Salmones, G. Mata and K.N. Waliszewski, *Biores. Technol.,* **96**, 537 (2005).
22. L. Fan, A.T. Soccol, A. Pandey, L.P.S. Vandenberghe and C.R. Soccol, *Braz. J. Microb.*, **37**, 420 (2006).
23. L.S. Oliveira and A.S. Franca, "Low-Cost Adsorbents from Agri-Food Wastes", *Food Science and Technology: New Research,* pp. 171-209 (2008).
24. L.S. Oliveira, A.S. Franca, T.M. Alvesand S.D.F. Rocha, *J. Haz. Mater.*, **155**, 507 (2008).
25. W.E. Oliveira, A.S. Franca, L.S. Oliveira and S.D.F. Rocha, *J. Haz. Mater.*, **152**, 1073 (2008).

USE OF *CRAMBE ABYSSINICA* PRESS CAKE AS A BIOSORBENT FOR WASTEWATER TREATMENT

LEANDRO S. OLIVEIRA, ADRIANA S. FRANCA
VICTOR F. OLIVEIRA and AUGUSTO SALLES
Departamento de Engenharia Mecânica/UFMG, Av. Antônio Carlos 6627
31270-901, Belo Horizonte, MG, Brazil

The objective of this study was to verify the feasibility of using *Crambe abyssinica* press cake, a residue from crambe oil and/or biodiesel production, as an adsorbent for removal of basic dyes from aqueous solutions. The pseudo-second order and Freundlich models provided the best description of adsorption kinetics and equilibrium, respectively. Our study showed that this type of waste material is a suitable candidate for use in the production of adsorbents for removal of cationic dyes, thus contributing for the implementation of sustainable development in both the oil and biodiesel production chains.

Keywords: Biodiesel solid residue; adsorption; dyes.

1. Introduction

Crambe abyssinica is a cruciferous oilseed plant, classified under the genus *Crambe* and the family Cruciferae [1]. This crop presents fast growth, tolerance to either low or high temperatures, low nutrient requirements and low cultivation costs. Interest in this specific crop is usually associated to its high oil content (32 – 36%) and also the high erucic acid content of the oil (55–60%). Crambe oil and derivatives present a wide variety of applications including lubricants, hydraulic, dielectric and refrigerating fluids, biodiesel, rubber additives, polymers, base for paints, waxes, and others [1-4].

Crambe meal presents good quality protein, but its use as livestock feed is restricted due to its high contents of glucosinolates and tannins [5]. Some recent studies have verified the feasibility of employing inedible seed press cakes directly as adsorbents, i.e., production of adsorbents with minimal treatment (excess oil removal, drying and grinding) and without submitting the agricultural waste to any thermal and/or chemical activation procedure. Reported applications include adsorption of copper by mustard oil seed cake [6], adsorption of pesticides in cold-pressed rapeseed, moringa and soybean cakes [7] and removal of heavy metals by Jatropha oil cake [8].

Thus, the objective of this study was to investigate the potential of Crambe abyssinica press cake as an adsorbent for the removal of basic dyes from synthetic aqueous media. Methylene blue was selected due to its use for dying cotton, wool and silk [9].

2. Materials and Methods

2.1. *Materials*

Crambe seeds were screw pressed (Ecirtec, Brazil) and submitted to excess oil removal by hexane extraction at 60°C for 1h. The adsorbate was methylene blue (MB - C.I. 52015), 15% hydrated, 95% purity, purchased from Labsynth Lab Products (São Paulo, Brazil).

2.2. *Adsorbent characterization*

Evaluation of the Point of Zero Charge (pH_{PZC}) was based on the mass titration procedure described by Valdés et al. [10]. Surface functional groups determination was based on the Boehm titration method [11].

2.3. *Adsorption studies*

Batch experiments of adsorption were performed in 250 mL Erlenmeyer flasks, with the flasks being agitated on an orbital shaker at 100 rpm for pre-determined time intervals. In all sets of experiments, a pre-determined amount of adsorbent was thoroughly mixed with 100 mL MB solution. Effect of particle size (diameter = D) was evaluated in the following ranges: D < 0.43 mm; 0.43 mm < D < 0.84 mm; 0.84 mm < D < 1.00 mm; and D > 1.00 mm. Effects of pH and adsorbent concentration were studied in the respective ranges of 3 to 10 and of 5 to 40 g L^{-1} at a fixed initial dye concentration (200 mg L^{-1}). Effect of contact time was evaluated at times ranging from 5 min to 8 hours and initial dye concentrations ranging from 50 to 500 mg L^{-1} at a fixed adsorbent concentration (10 g L^{-1}). All tests were performed in two replicates.

2.4. *Analysis of methylene blue (MB)*

At each sampling time, 2 mL aliquots were taken from the Erlenmeyer flasks and MB concentration was determined by a spectrophotometer (Cole Parmer 1100 RS) at 665 nm. The amount of MB adsorbed was determined by taking the difference between the initial dye concentration and the concentration of the solution at the time of sampling. All determinations were performed in a total of three replicates per experiment and the average values were reported.

3. Results and Discussion

3.1. *Adsorbent characterization*

Results for pH_{PZC} determination indicated a pH_{PZC} value of 5.7, and therefore pH values should be maintained above this value to ensure a predominant negatively charged surface and favor electrostatic attraction of the MB cation. The functional groups at the surface of the adsorbent (Boehm) were predominantly basic (0.42 ± 0.01 mmol/gsorbent), followed by carboxylic (0.38 ± 0.03 mmol/gsorbent), phenolic (0.18 ± 0.04 mmol/gsorbent) and lactonic (0.14 ± 0.03 mmol/gsorbent) groups.

3.2. *Effect of particle size, pH and adsorbent dosage*

Results on the effects of particle size and initial pH are shown in Fig. 1. MB uptake increased with the decrease in particle size (Fig. 1a), due to the corresponding increase in surface area, with the exception of the smaller diameter fraction, D < 0.43 mm. Such behavior is attributed to the fact that the finer particles were suspended to the surface of the aqueous MB solution, thus hindering mixing and proper contact between the adsorbent and the solution. These results also indicate that the dye molecules probably adsorb at the adsorbent outer surface. Based on the results presented in Fig. 1a, the remaining experiments were conducted employing crambe press cake ground down to the following particle size range: 0.43 mm < D < 0.84 mm.

The increase in sorption capacity comparing results at pH´s 5 and 7 (Fig. 1b) is attributed to the fact that, above the pH_{PZC} (5.7), the adsorbent surface became negatively charged, enhancing the electrostatic attraction between the surface and MB cations. However, adsorption efficiency decreased at pH 10 (MB dimerization and/or reaction with NaOH [12]). Thus, the remaining tests were conducted at pH = 7, in order to assure a negatively charged surface. It is noteworthy to mention that the pH of the MB solution was measured after all sorption tests, and pH values varied during sorption towards the pH_{PZC} value.

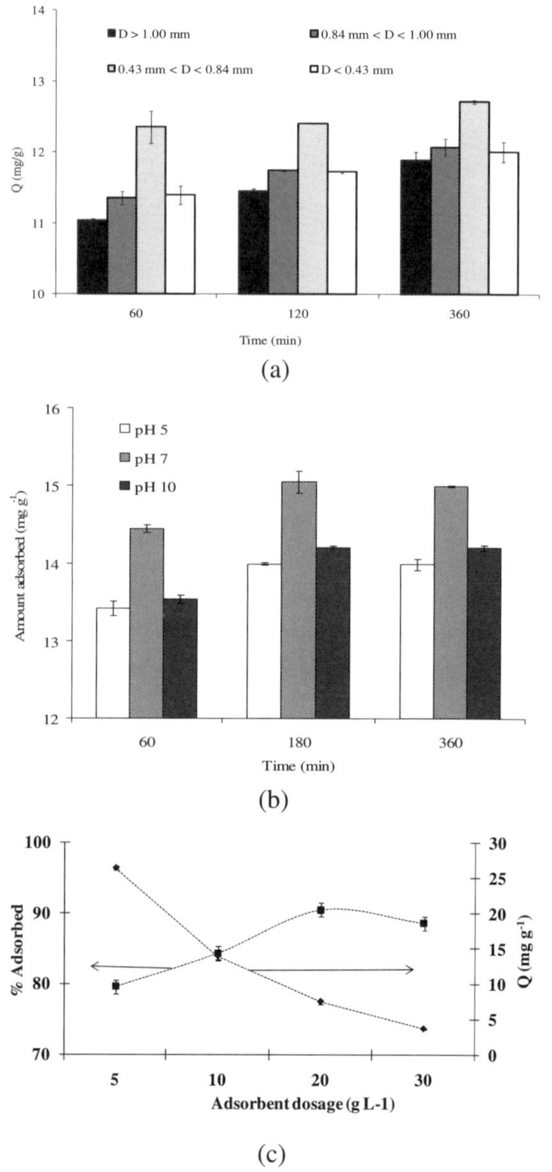

Figure 1. Effects of (a) particle diameter, D; (b) initial solution pH and (c) adsorbent dosage on MB adsorption.

The influence of adsorbent dosage on the efficiency of MB removal can be viewed in Fig. 1c. Dye removal efficiency increased with the increase in adsorbent dosage, given the increase in surface area resulting from the increase in adsorbent mass. The amount of dye adsorbed per unit mass of adsorbent decreased with increasing adsorbent mass, due to the reduction in adsorbate/adsorbent ratio. Thus, further tests were conducted for an adsorbent dosage of 10 g L^{-1}.

3.3. *Effect of initial MB concentration*

The adsorption data presented in Fig. 2 showed that a contact time of 6 hours assured attainment of equilibrium conditions regardless of the initial MB concentrations. Results presented in Fig. 2 also show that adsorption of MB presents a strong dependency on MB initial concentration. An increase in the initial MB concentration led to an increase in the amount adsorbed, which is attributed to the increase in driving force (concentration gradient) with the increase in the initial dye concentration [13]. The amount of MB adsorbed increased from 3.7 to 51.7 mg g^{-1} as the initial concentration was increased from 50 to 700 mg L^{-1}.

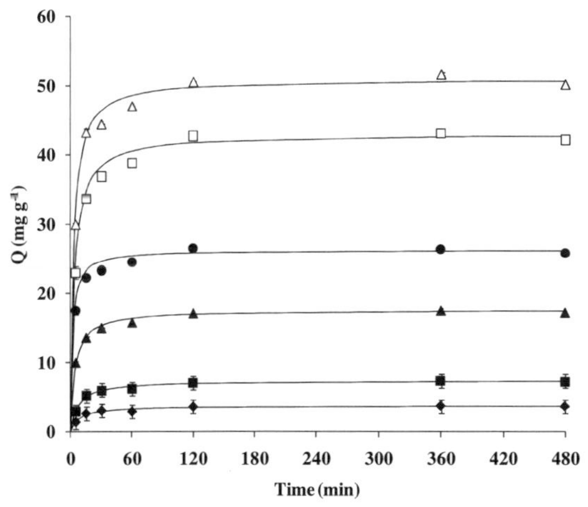

Figure 2. Effect of contact time on MB adsorption by crambe press cake. Initial MB concentration: ♦50 mg L^{-1}, ■100 mg L^{-1}, ▲200 mg L^{-1}, ●300 mg L^{-1}, □500 mg L^{-1}, Δ 700 mg L^{-1}. Solid lines correspond to pseudo-second order kinetics model fits.

The controlling mechanism of MB adsorption was investigated by fitting lumped first and second-order models to the experimental data. The models can be represented by the following equation:

$$\frac{dq_t}{dt} = k_n (q_e - q_t)^n \quad (1)$$

where q_e and q_t correspond to the amount of dye adsorbed per unit mass of biosorbent (mg/g) at equilibrium and at time t, respectively, and k_n is the rate constant for n^{th} order adsorption (k_n units are min^{-1} for n=1 and g $mg^{-1}min^{-1}$ for n=2). The results of the kinetic calculations showed that the first-order model did not provide a good fit, with low R^2 values (0.49-0.98) and underestimated q_e values (70% diff.). An evaluation of both R^2 values (>0.999) and estimated q_e values (1% maximum difference from experiments) indicated that MB adsorption can be satisfactorily described by the pseudo second-order model, indicating chemisorption as the rate-controlling mechanism [13].

3.4. *Adsorption equilibrium*

Langmuir and Freundlich models were tested for equilibrium description at ambient temperature [14]. Langmuir model is described by the following equation:

$$q_e = \frac{q_{max} K_L C_e}{1 + K_L C_e} \quad (2)$$

where q_e (mg g^{-1}) and C_e (mg L^{-1}) represent the amount adsorbed per gram of adsorbent and solute concentration (mg L^{-1}) in the solution, respectively, after equilibrium was reached. q_{max} and K_L are constants related to the maximum adsorption capacity (mg g^{-1}) and the adsorption energy (L mg^{-1}), respectively.

Freundlich's model can be described by the following equation:

$$q_e = K_F C_e^{1/n} \quad (3)$$

where K_F indicates the relative adsorption capacity ($mg^{1-(1/n)}$ $L^{1/n}$ g^{-1}) and n is related to the intensity of adsorption.

The adsorption isotherm is displayed in Fig. 3. The shape of the curve indicates favorable adsorption. Coefficients of the Langmuir and Freundlich models are displayed in Table 1. MB adsorption from aqueous solutions by crambe press cake was better described by Freundlich, indicating heterogeneous adsorption. Maximum MB uptake capacity was 48.9 mg g^{-1}, quite high is comparison to activated carbons obtained by thermal and/or chemical treatment of other press cakes (sunflower, 16.4 mg g^{-1} [15]; defective coffee beans, 14.9 mg g^{-1}[16]; *Rhaphanus sativus*, 19.8 mg g^{-1} [9]).

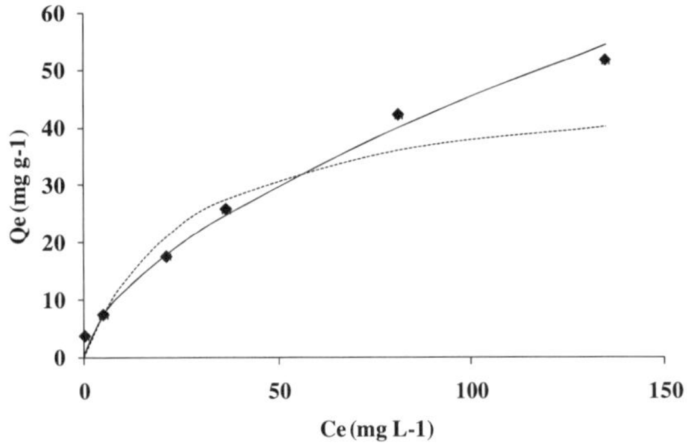

Figure 3. Adsorption isotherm of MB by crambe press cake. Solid and dashed lines correspond to Freundlich and Langmuir model fits, respectively.

Table 1. Langmuir and Freundlich isotherm constants for MB adsorption.

Langmuir		Freundlich	
K_L	0.036	K_F	2.8671
q_{max}	48.31	1/n	0.6
r	0.04-0.36	R^2	0.9968
R^2	0.9837		

4. Conclusions

Our results showed that crambe press cake, without any thermal treatment, can be satisfactorily employed as an adsorbent for removal of methylene blue. This is an indication that this solid residue presents an excellent potential as an adsorbent for removal of cationic dyes. Further studies should be conducted in order to determine the adsorbent capacity with respect to other cationic dyes such as malachite green and neutral red.

Acknowledgments

The authors acknowledge financial support from Brazilian Government Agencies: CAPES, CNPq and FAPEMIG.

References

1. Y.P. Wang, J.S. Tang, C.Q. Chu, J. Tian, *Ind. Crops Prod.* **12**, 47 (2000).
2. L. Lazzeri, O. Leoni, L.S. Conte, S. Palmieri, *Ind. Crops Prod.* **3**, 103 (1994).
3. J.M. Vargas-Lopez, D. Wiesenborn, K. Tostenson, L. Cihacek, *JAOCS* **76**, 801 (1999).
4. G. Steinke, S. Schönwiese, K.D. Mukherjee, *JAOCS* **77**, 367 (2000).
5. L. Cooke, D.A. Konstant, *Agric. Res. USDA Res. Serv.* **39**, 16 (1991).
6. M. Ajmal, R.A.K. Rao, M.A. Khan, *J. Hazard. Mater.* **122**, 177 (2005).
7. J. Boucher, Z. Steiner, I.W. Marison, *Water Res.* **41**, 3209 (2007).
8. U. Garg, M.P. Kaur, G.K. Jawa, D. Sud, V.K. Garg, *J. Hazard. Mater.* **154**, 1149 (2008).
9. D.A. Lázaro, M.B. Mansur, A.S. Franca, L.S. Oliveira and S.D.F. Rocha, *Int. J. Chem. Eng.,* **1**, 147 (2008).
10. H. Valdés, M. Sánchez-Polo, J. Rivera-Utrilla, C.A. Zaror, *Langmuir*, **18**, 2111 (2002).
11. H. P. Boehm, *Carbon,* **32**, 759 (1994).
12. R.A. Shawabkeh, M.F. Tutunji, *Appl. Clay Sci.*, **24**, 111 (2003).
13. L.S. Oliveira, A.S. Franca, T.M. Alves and S.D.F Rocha, *J. Hazard. Mater.,* **155**, 507 (2008).
14. O. Hamdaoui and E. Naffrechoux, *J. Hazard. Mater.*, **147**, 381 (2007).
15. S. Karagoz, T. Tay, S. Ucar and M. Erdem, *Biores. Technol.,* **99**, 6214 (2008).
16. A.A. Nunes, A.S. Franca and L.S. Oliveira, *Biores. Technol.,* **100**, 1786 (2009).

DESIGN AND OPERATION OF A MOBILE BIODIESEL PRODUCTION UNIT

LEANDRO S. OLIVEIRA and ALEX N. BRASIL

Departamento de Engenharia Mecânica/UFMG, Av. Antônio Carlos 6627
31270-901, Belo Horizonte, MG, Brazil

DIEGO L. NUNES

Programa de Pós-Graduação em Ciência de Alimentos/UFMG, Av. Antônio Carlos 6627
31270-901, Belo Horizonte, MG, Brazil

The Brazilian National Program for Production and Use of Biodiesel (PNPB) was launched in 2004 by the Brazilian Government with the objective of guaranteeing an economically viable production of biodiesel, while favoring social inclusion and regional development. Thus, great emphasis was put on small scale production of vegetable oils with attractive fiscal incentives for underdeveloped regions of Brazil. However, no fiscal incentives were planned for the manufacturing of biodiesel production units and their construction and operation are usually costly and out of reach for small-scale producers. Thus, it was the aim of this work to design and construct a mobile biodiesel production unit and verify its feasibility in providing a service to small-scale oil producers in which the oil is processed into biodiesel and the producers costs are only those related to the processing of the oil. In other words, the mobile unit travels from one oil producing farm to another processing the oil into biodiesel which is kept by the oil producers for their own use. The constructed unit is fully operational and has a production capacity of a 100 Liters of biodiesel per hour. The unit was mounted on a truck and it is comprised of a stirred reactor, a decanter and a distillation unit for biodiesel purification.

Keywords: Biodiesel; mobile production.

1. Introduction

Biodiesel production and commercial use in Brazil has greatly expanded in the last couple of years due to the Brazilian Government policy that established a mandatory use of 2% biodiesel in fossil diesel (B2) fuels throughout the country since 2008 [1]. In the last decade, the technologies for producing biodiesel have greatly improved and several different processes are now currently available at different scales [2-4]. However, for small scale production facilities, the preferred technology is still batch transesterification systems [5-7] mostly involving alkali-catalyzed processes. The benefits of installing and operating small-scale biodiesel production units are: (i) low capital investments, (ii) reduction in use of fossil fuels by replacement with local renewable energy sources (e.g., the heat requirements can be supplied by combustion of biomass residues), (iii) electricity needs can be supplied by a generator that runs on biodiesel, and (iv) unit can be operated by inexperienced users not relying on the availability of highly qualified technicians.

The scenario originally considered in the development of the Brazilian Government policy for biofuels was focused on fiscal incentives for small scale oilseed producers, thus, favoring small scale oil extraction plants scattered in remote locations of the less developed regions of Brazil (mostly North and Northeast Regions of the country). However, no fiscal incentives were established for the manufacturing of small-scale biodiesel production units and their construction, and operation costs, although usually lower than the industrial scale ones, are out of reach for small scale oilseed producers and cooperatives. In view of the aforementioned, it was the aim of this work to design/construct a mobile biodiesel production unit and verify its feasibility in providing a service to small-scale oil producers in which the oil is processed into biodiesel and the producers costs are only those related to the processing of the oil.

2. Design Specifications

The mobile biodiesel production unit was built and commissioned at Biominas, a transportation company located in Itauna, Brazil. A drawing of the unit layout is shown in Fig. 1 and a process flowsheet as related to the mobile unit described is depicted in Fig. 2.

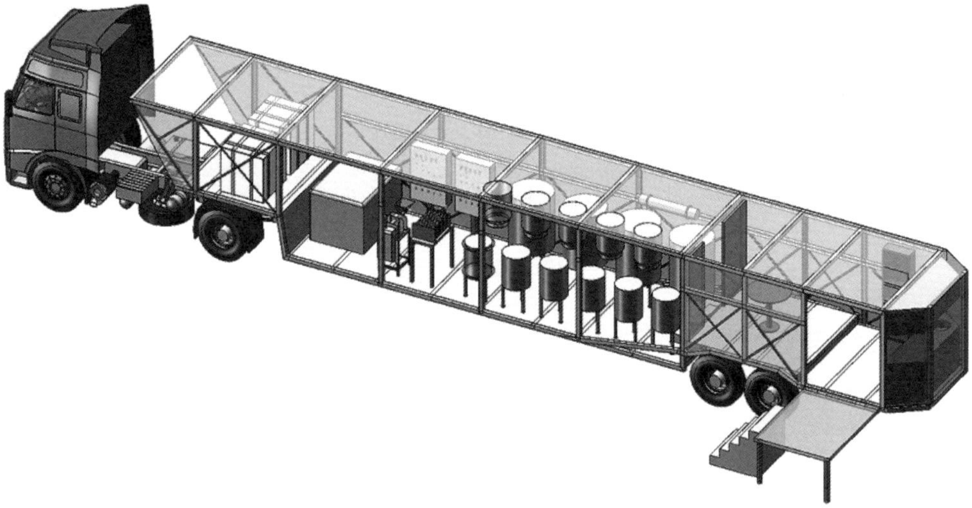

Figure 1. Mobile biodiesel production unit.

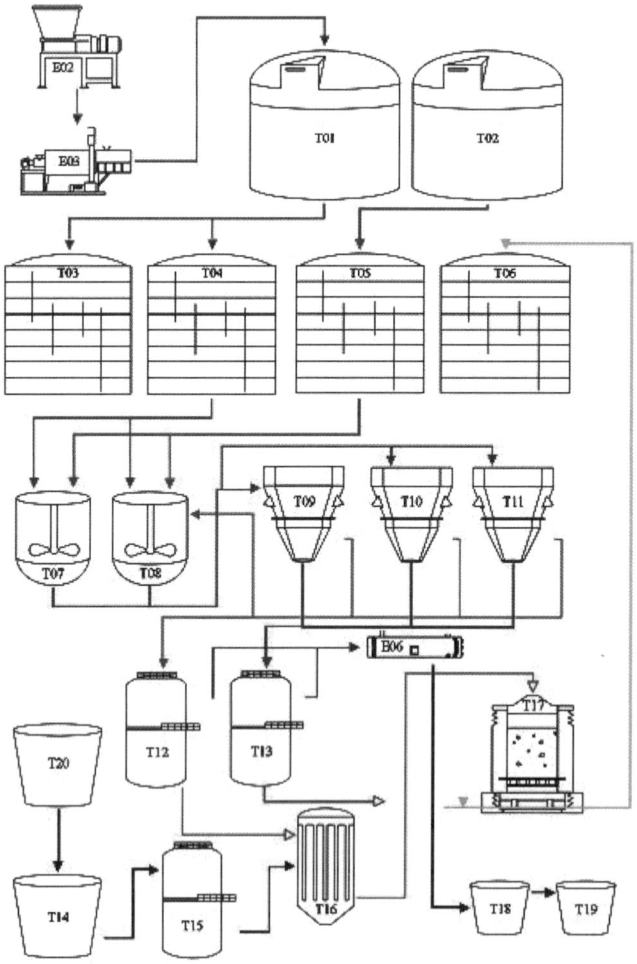

Figure 2. Biodiesel production process flowsheet.

The biodiesel production rig is comprised of two storage tanks (T01 and T02) for the reagents (oil and a mixture of ethanol and sodium methoxide, respectively) during transportation of the unit; one 18 kW power generator (E01), working with a Diesel engine running on biodiesel; a screw-press for oil extraction (E02); a press filter (E03) for the removal of suspended solids in the oil; two control panels (E04 and E05); four tanks, made of stainless steel, for storage and feed of vegetable oil (T03) and of waste frying oil (T04), for storage of ethanol (T05), and for storage of purified biodiesel (T06); two 180 liter-capacity reactors (T07 and T08) for the transesterification reactions, equipped with mechanical stirrers and heating coils (A01 and A02, respectively) that allows for temperatures up to 80°C in the reacting media – reactor T07 is made of acrylic to allow for complete visualization of the reaction for educational purposes; two settler tanks (T09 and T10) for the phase separation (fatty acid ethyl ester and glycerin) – tank T09 is also made of acrylic; a storage tank for glycerin (T12) with a heating coil that allows for temperatures up to 100°C; an evaporation column (T13) equipped with a heating unit (A04) for the removal of the alcohol from the ester phase; a heat exchanger (E06) operated in counter-current mode; a holding tank for process water (T14) and a storage tank for fresh process water (T15) equipped with a heating unit (A05); a water washing column (T16) for the removal of impurities from the alcohol-free biodiesel phase; an evaporation column (T17) equipped with a heating unit (A06), that allows for temperatures up to 100°C, and a vacuum pump (B08) (100 mmHg) for the final purification of the biodiesel phase. A small lab facility (LB) was built in the rear end of the rig, where the quality of the biodiesel produced can be quickly evaluated employing analytical techniques such as near infrared spectroscopy. Since the mobile unit was also commissioned for educational purposes, it has a training room (ES) at the rear end, in between the processing unit and the lab, where a 34" TV was setup for instructional presentations and videos. Safety features include an emergency exit and a fire extinguisher placed at the back end.

3. Process Description

The first step in the process is the extraction of the vegetable oil using the expeller (E02), generating a solid residue (press-cake). After extraction, the oil is filtered (E03) and, if necessary, its pH is adjusted so it can be safely pumped to either the storage tank (T01) or directly to one of the reactors (T07 or T08). The reactors work in parallel, but their configuration allows for them to be run in series in cases when there is a need for the oil to be purified before being subjected to a transesterification reaction. After the oil is pumped (B01) to the reactor and is heated to the desired temperature, the mixture of alcohol and catalyst is pumped (B02) into the reactor where the reaction is carried on under stirring for about 1 h or 1.5 h depending on the conditions established for the reaction to occur. After completion of the reaction, the mixture is transferred to the settler (either T09, T10 or T11) by gravity action so that phase separation can occur by gravity settling in a period that usually range from 0.5 to 1 h. Upon separation, the ester phase is moved to the evaporation column (T13) where the excess alcohol is removed by evaporation (A03); and the glycerin phase is moved to the storage tank (T12) where, upon heating, the alcohol in that phase is also removed by evaporation. The evaporated alcohol (from both T13 and T12) is condensed in the heat exchanger (E06) where it flows counter-currently to the cold fluid, which is comprised of a mixture of water and glycerin recovered in T12 after cooling. The condensed alcohol is returned to the storage tank T05.

After evaporation of the residual alcohol, the ester phase still contains a fraction of glycerin that must be removed to comply with purity specifications established by the Brazilian Biodiesel Standards. Thus, the ester phase is transferred to the water washing column (T16) where water is sprayed over the surface of the ester phase, extracting the glycerin as its tiny drops move down the lighter organic phase. The water with the extracted glycerin settles down at the bottom of the column where it is removed and transferred to tanks T18 and T19. The biodiesel phase, with a bit of water incorporated in it, is transferred to thc distillation column (T17), where the water and residual alcohol are removed as vent gases. The purified biodiesel is then transferred to the fuel storage tank T06.

Research studies are currently being carried on to evaluate the feasibility of applying the water used in the washing of the ester phase (with a small amount of glycerin and other impurities) for irrigation and fertilization purposes in oilseed plantations.

4. Experimental Results

Although the unit has been operated several times for demonstration and educational purposes, only two tests were conducted to verify the quality of the biodiesel produced in the mobile unit. The first test was carried out with refined soybean oil and ethanol in a molar ratio of 1:6, and sodium methoxide (1% w/w, based on the mass of oil) as the alkali catalyst. The biodiesel produced was analyzed by HPLC following the methodology presented by Holcapek et al. [8]. The ester conversion was determined to be 97% and, after purification, the ester content was 99.4%, asserting the functionality of the unit. The second test was carried out with waste frying oil and the reaction conditions were the same as those used for the refined oil. The ester conversion was determined to be ~ 90% and the purified ester was 98.7% pure.

5. Preliminary Economic Feasibility

Capital cost investment (processing equipment and accessories) is one of the most critical steps in economic feasibility studies. Regarding the constructed mobile unit, the capital cost investment can be divided into: (i) cost of the rig where the processing unit was assembled onto; (ii) cost of the processing equipment; (iii) cost of accessories and ancillaries; and (iv) cost of the built-in lab facility.

The cost of the processing equipment, considering a 100 L/h capacity, was estimated to be of 40 to 45 thousand US dollars. The cost of the accessories and ancillary equipment was estimated in the range of 20 to 25 thousand dollars and the cost of the rig, assembled with flooring and walls, was in the range of 30 to 32 thousand dollars. The cost of the lab facility was ten thousand dollars and the total investment cost sums up to US$112,000.00.

6. Conclusions

A mobile biodiesel production unit was designed, constructed and successfully operated. Tests were carried out to evaluate the performance of the unit using both refined and waste frying soybean oil. The ester conversions in both tests were considered satisfactory (higher than 90%) and the purification steps presented a commendable performance, leading to purities higher than 98% for both oils. The technical feasibility of a mobile biodiesel production unit was demonstrated and a preliminary economic analysis showed encouraging prospects for the production, commercialization and operation of such units.

Acknowledgments

The authors acknowledge financial support from Brazilian Government Agencies: CAPES, CNPq and FAPEMIG.

References

1. G.P.A.G. Pousa, A.L.F. Santos and P.A.Z. Suarez, *Energy Policy,* **35**, 5393 (2007).
2. L.C. Meher, D.V. Sagar and S.N. Naik, *Renew. Sust. Energ. Rev.* **10**, 248 (2006).
3. Y.C. Sharma, B. Singh, S.N. Upadhyay, *Fuel,* **87**, 2355 (2008).
4. S.A. Basha, K.R. Gopal and S. Jebaraj, *Renew. Sust. Energ. Rev.* **13**, 1628 (2009).
5. M. Bender, *Bioresource Technol.* **70**, 81 (1999).
6. B. Amigun, F. Müller-Langer and H. von Blottnitz, *Energy for Sustainable Development*, **13**, 5 (2008).
7. I. Sarantopoulos, F. Che, T. Tsoutsos, V. Bakirtzoglou, W. Azangue, D. Bienvenue and F.M. Ndipen, *Phys. Chem. Earth,* **34**, 55 (2009).
8. M. Holcapek, P. Jandera, J. Fischer, J. and B. Prokes, *J. Chromatogr. A*, **858**, 13 (1999).

CHARACTERIZATION AND SEPARATION OF OIL-IN-WATER EMULSION

AJAY MANDAL*, PRADEEP KUMAR and RAJIV CHOUDHARY
Department of Petroleum Engineering

S. K. MAITY
Department of Environment Science and Engineering
Indian School of Mines University, Dhanbad-826004, India

Chemical demulsification is most widely applied in petroleum industries, paining and wastewater treatment technology and involves the use of chemical additives to accelerate the emulsion breaking process. The stability of the emulsion has been characterized and it is observed that the stability depend on oil-water contact time, turbulence and amount of oil in contact with the water. A series of experiments have been carried out with different demulsifiers for separation of oil from oil-in-water emulsion. More than 90% separations are obtained with some demulsifiers under specific operating conditions.

1. Introduction

Oil exploitation is always accompanied by varying quantities of oily wastewater. Other sources of oil-containing wastewaters include petroleum refineries, metal fabrication plants, rolling mills, chemical processing plants, machine shops, and vehicle maintenance shops. Many different types of oils may be present in oily wastewater such as crude oil, diesel fuel, cutting and grinding oils, lubricating oils, water soluble coolants, natural animal or vegetable fats or any other organic immiscible in water. The removal of these oily wastes from wastewater is of importance in preventing pollution and meeting environmental compliance standards. Oily waste removal may also be beneficial for water and oil recovery and reuse. Disposed water should contain less than 40 mg/l of oil [1], and this requirement is becoming more enforced as damaging environmental effects from oily wastewater become more apparent.

Oil in wastewater mostly remain in the form of emulsion. Emulsions are suspensions of droplets, greater than 0.1 μm, consisting of two completely immiscible liquids, one of which is dispersed throughout the other. An emulsifying agent must be present to form stable oil-in-water emulsions [2]. Such agents include clay particles, added chemicals or the crude oil components like asphaltenes, waxes, resins and napthenic acids. The most important characteristics of oil-in-water emulsion are its stability [3]. The four classes of stability are: stable, meso-stable, unstable and entrained water. Out of these four states only stable and meso-stable can be characterized as emulsions.

Chemical demulsification is the most widely applied method of treating oil-in-water emulsions and involves the use of chemical additives to accelerate the emulsion breaking process. The formulation of an emulsion demulsifier for specific oil emulsion is a complicated undertaking. In petroleum system, asphaltenes and resinous substances comprise a major portion of the inetrfacially active components of the oil [2, 4, 5]. Commercial demulsifiers are generally polymeric surfactants such as copolymers of polyxyethylene and polypropylene or alkylphenol-formaldehyde resins or blends of various surface-active substances [6, 7]. Due to more and more severe environmental constraints, there is now a strong need in the oil production to restrict the use of chemicals and to utilize safer formulations, less toxic but at least as efficient as classical demulsifiers.

The present work conducted to study the influence of different chemical demulsifiers on the destabilization of emulsions and hence separation of oil from wastewater. The effects of time and temperature on the demulsification were also studied.

2. Experimental

2.1. Materials

In this study, six different demulsifiers; Polyethylene Glycol-400, Polyethylene Glycol-4000, Polyethylene

*Corresponding author: E-mail: mandal_ajay@hotmail.com

Glycol 6000, Polyethylene oxide, N-hexylamine Amine, N-octylamine were used for demulsification of oil-in-water emulsion. Gear oil (EPX 90) available in the market with sp. Gravity 0.905 and kinematic viscosity 197 cSt at 40°C and 17.3 cSt at 100°C was used for preparation of emulsion in distilled water.

2.2. *Methods*

Oil-in-water emulsion was prepared at room temperature with a standard three blade propeller at speed of various rpm. For oil-in-water emulsion distilled water was used as the continuous phase and oil was used as the dispersed phase. The emulsion is then left to stand in separation flask for 6 hours and the bottom part is separated out. The separation flask can easily lead to the separation of two different density fluids based on their density. For preparation of emulsion, the oil water mixtures were stirred at different rpm and time. It has been found that the stability and hence the characteristics of emulsion is very much dependent on the stirring speed and time.

Emulsions are then analyzed and treated with different demulsifiers for demulsification. The viscosity of the emulsion was measured in Brookfield viscometer and density by conventional method. The Chemical Oxygen Demand (COD) of the emulsion is considered as the oil equivalent in the emulsion. Thus COD of the sample were measured before and after treatment of the emulsion to determine the degree of separation of oil from emulsion. Typical characteristics of four different sets of emulsion have been shown in Table 1.

Table 1. Typical characteristic of oil-water emulsions.

Sample	1	2	3	4
Stirring speed (RPM)	2184	3963	5145	6870
Mixing time (min)	120	120	120	120
pH	7.8	7.96	7.78	7.40
Interfacial tension (Dynes/cm)	34	33	33	32
Viscosity (cp) At 25°C	2.83	4.19	6.45	4.45
Conductivity mho	0.385	0.456	0.422	0.472
COD	24320	27200	33920	30400

3. Results and Discussion

3.1. *Characterization of Emulsion*

The concentration of oil in the oil-in-water emulsion is dependent on the presence of excess oil during formation of emulsion. A series of experiments were performed to characterize the emulsion formation as shown in Fig. 1. From the figure, it may be seen that the emulsification of oil (COD) in water first increases with increase in concentration of oil due to significant interaction between oil and water. However, after a certain oil to water ratio the coalescence of dispersed oil droplets increases in presence of excess oil and hence concentration of oil in the emulsion gradually decreases and then stabilizes at constant value. The stirring speed has a strong effect on the emulsion formation. It has been found (Fig. 2) that the concentration of emulsified oil in water first increases and then decreases due to breaking of emulsion at higher stirring speed.

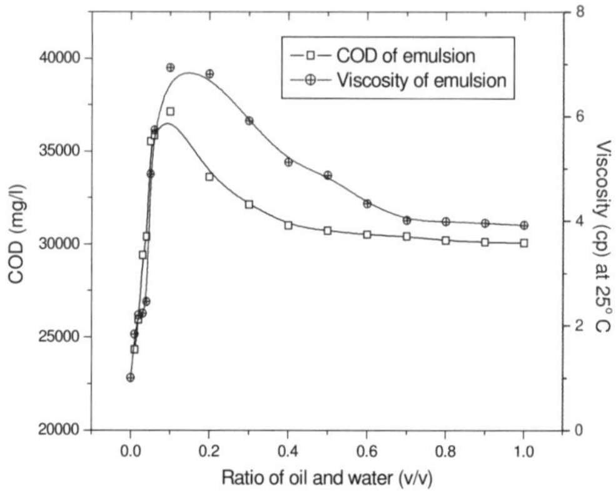

Figure 1. Effect of oil/water ratio on the stability of emulsion.

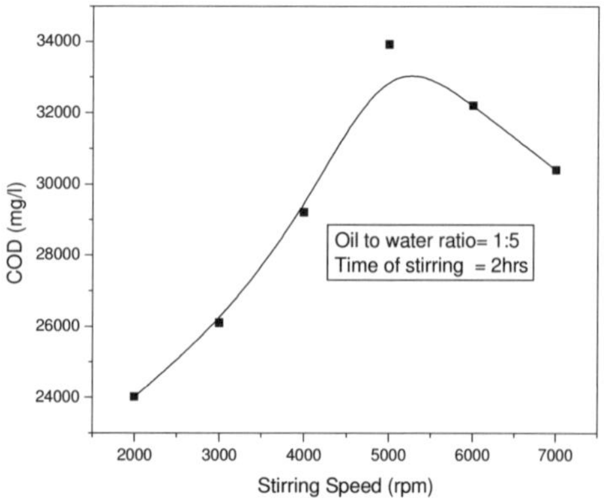

Figure 2. Effect of stirring speed on stability of emulsion.

3.2. Demulsification of Emulsion

In order to investigate the effect of different demulsifier and corresponding dosage on the demulsification of an oil-in-water emulsion, a series of experiments were carried out with different demulsifier. The experiments were performed at a temperature of 30ºC and 1440 minutes retention time was provided for separation. Fig. 3 shows the influence of polyethylene oxide on destabilization of emulsion. The removal of oil increases with increase in amount of polyethylene oxide added due to significant reduction of interfacial tension between oil and water. However after a certain concentration the interfaces are saturated with the demulsifier and stagnation in oil and water separation are occurred. From the Fig. 3 one may see that maximum separation occurs at around 1400 ppm of polyethylene oxide added. Fig 4 shows the effect of different amine and glycol on the separation. The plots show the same trend as that of polyethylene glycol, but percentage separation of oil significantly higher particularly with N-Octylamine, N-Hexyl amine and Polyethylene Glycol-400. The separation capacity of N-Octyl amine (68.39%) was found to be higher than N-Hexyl amine (59.26%) at room temperature due to higher molecular weight which acts as flocculants in adsorption and interaction activities [8]. Polyethylene Glycol-400 was found to have better separation efficiency than Polyethylene Glycol-4000 and 6000 and similar trend was also reported by Abdurahman et al. [8].

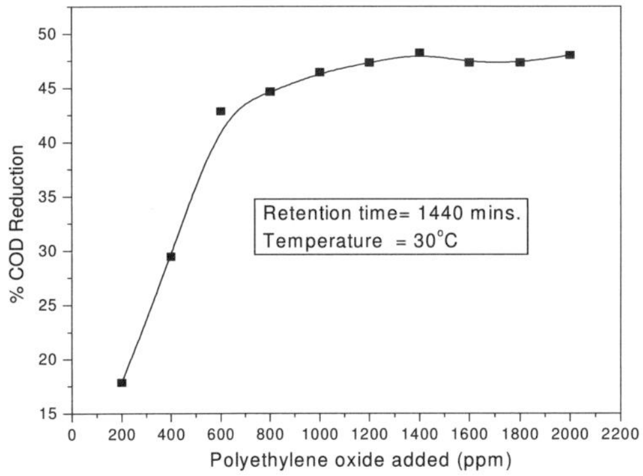

Figure 3. The influence of Polyethylene oxide on separation of oil from emulsion.

The retention time after addition of demulsifier in the emulsion was found to have a strong effect on separation. A minimum retention time should be provided flocculation and adsorption of oil particle. The Fig. 5 shows that a minimum 120 minutes retention time should be provided for significant separation.

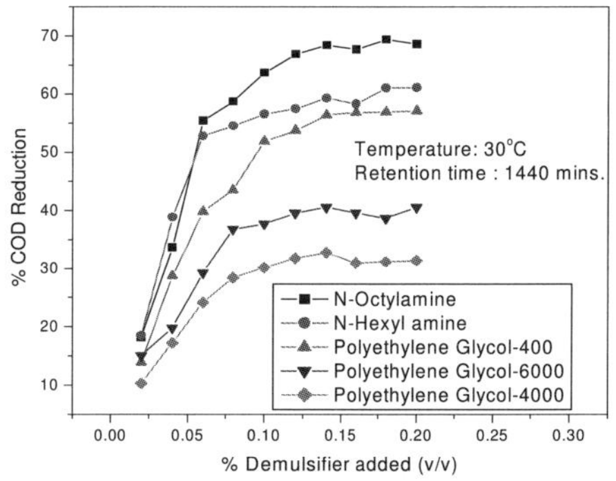

Figure 4. The influence of different demulsifiers on separation of oil from oil-in-water emulsion.

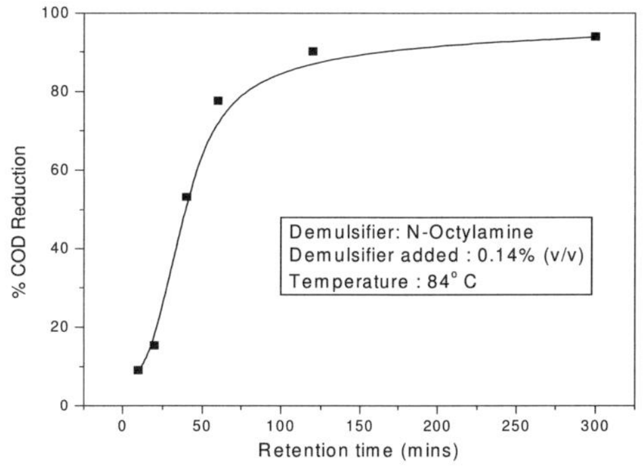

Figure 5. Effect of retention time on separation of oil from oil-in-water emulsion.

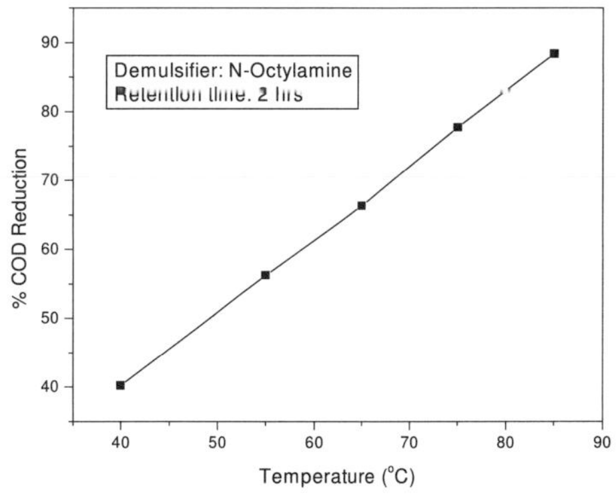

Figure 6. Effect of temperature on separation of oil from oil-in-water emulsion.

Fig. 6 shows the effect of treating temperature on the separation of emulsion. The removal of oil increases with the treating temperature. It is generally believed that the Brownian movement of molecules or finely grounded particles is responsible for this response. Heating up not only intensifies Brownian motion of the particles in oil emulsion to make them collided and coagulated effectively, also falls the internal viscosity of emulsion body vastly to weaken therefore the stability of emulsion.

Microscopic picture of oil-in-water emulsion is shown in Fig. 7 while that of demulsifier treated samples are shown in Fig. 8 & 9 respectively. The figures show the size distribution of droplets of different samples. From the figures it may be seen that the size of the emulsified water droplet in untreated emulsion is higher compared to the demulsifier treated emulsion. A detailed size distributions are shown in Table 2.

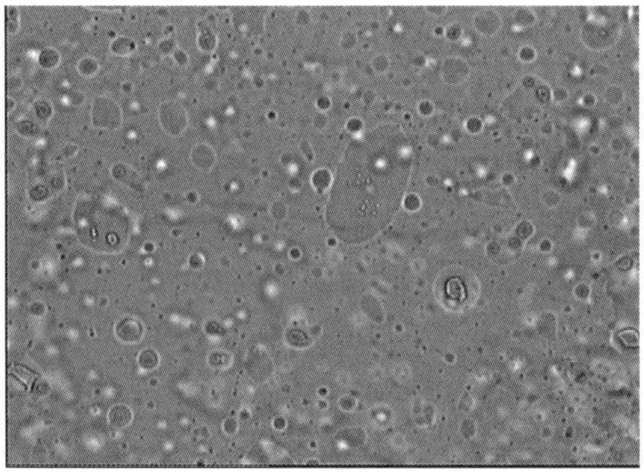

Figure 7. Microscopic picture (Magnification 430x) of untreated oil-in-water emulsion.

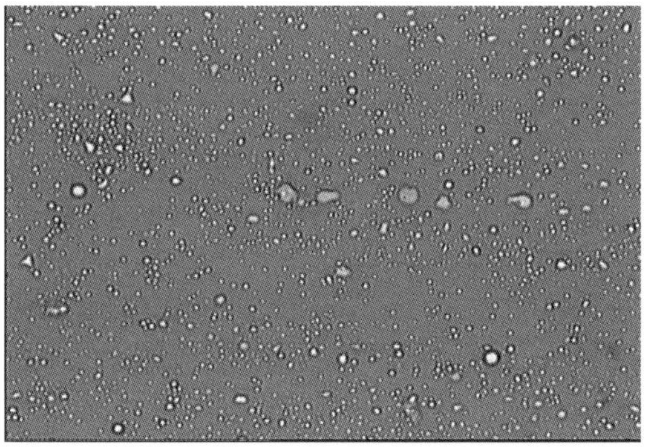

Figure 8. Microscopic picture (Magnification 430x) of oil-in-water emulsion treated with Polyethylene Glycol 400.

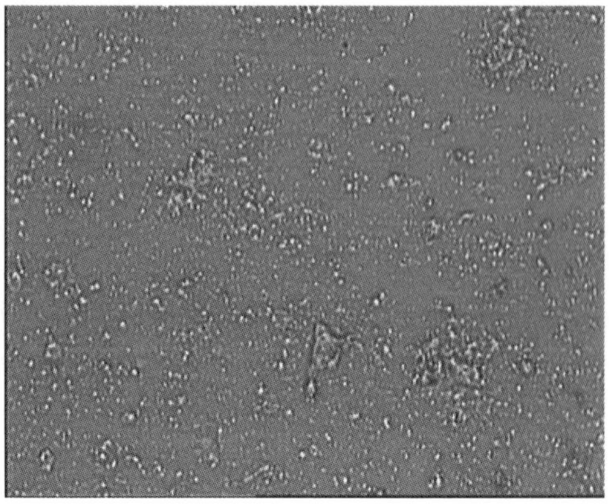

Figure 9. Microscopic picture (Magnification 430x) of oil-in-water emulsion treated with N-Octylamine.

Table 2. Analysis of microscopic pictures of emulsions.

Untreated emulsion (Fig.7)	
Min. Val.	6.955 μm
Max. Val.	121.709 μm
Mean	20.760 μm
Std. Dev	18.478
Sum	830.409
No. of Samples	40
Emulsion treated with Polyethylene Glycol 400 (Fig.8)	
Min. Val.	1.832 μm
Max. Val.	21.871 μm
Mean	6.478 μm
Std. Dev	3.762
Sum	278.562
No. of Samples	43
Emulsion treated with N-Octylamine (Fig.9)	
Min. Val.	0.918 μm
Max. Val.	8.237 μm
Mean	3.116 μm
Std. Dev	2.138
Sum	101.077
No. of Samples	40

4. Conclusion

Based on the experimental study it can be concluded that the stability of emulsion is dependent on the different influencing conditions of the formation, viz. formation time, degree of turbulence and oil to water ratio. Demulsification of the emulsion were studied using different demulsifiers, out of which N-octylamine, N-Hexylamine and Polyethylene glycol-400 were found to have good removal efficiency of oil from water. The results suggest that addition of around 0.15 % (v/v) gives the optimum removal of oil from the emulsion. A minimum retention time of 120 minutes after addition of demulsifiers should be provided for better separation. The separation efficiency was also found to increase with increasing treating temperature.

References

1. R. Moosai L. Richard A. Dawe, "Gas attachment of oil droplets for gas flotation for oily wastewater cleanup, Separation and Purification Technology 33, pp. 303-314, 2003.
2. J. Sjoblom, L. Mingyuan, A. A. Christy and T. Gu, "Water-in-Crude oil emulsions from Norwegian continental shelf 7", Colloids and Surfaces, 66, pp. 55-62, 1992.

3. M. Fingas and B. Fieldhouse, "Formation of water-in-oil emulsions and application to oil spill modelling", Journal of Hazardous Materials, 107, pp. 37-50, 2004.
4. E. J. Johansen, I. M. Skjarov, T. Lund and Sjoblom, "Water-in-Crude oil emulsions from Norwegian continental shelf. Part 1. Formation, Characterization and Stability correlations", Colloids and Surfaces, 34, pp. 353-370, 1989.
5. O. Urdahl, T. Brekke and J. Sjoblom, "13 C n.m.r and multivariate statistical analysis of adsorbed surfaceactive crude oil fractions and the corresponding crude oils", Fuel, 71, pp 739-746, 1992.
6. S.E. Taylor, "Resolving crude oil emulsions", Chemistry and Industry, pp. 770-773, 1992.
7. S. Kokal, "Crude oil emulsions: a state of the art review", In Proceedings SPE ATCE, San Antonio, Tx, SPE paper no 77497, 2002.
8. H. N. Abdurahman, R. M. Yunus and Z. Jemaat, "Chemical demulsification of water-in-crude oil emulsion", J. Appl. Science, 7, pp196-201, 2007.

SYNTHESIS AND RHEOLOGY OF AN ECO-FRIENDLY HYDROFRACTURING FLUID

K. OJHA
Department of Petroleum Engineering Indian School of Mines University
Dhanabd-826 004, India
E-mail: keka_ojha@yahoo.com

P. BAJPAI
Department of Petroleum Engineering Indian School of Mines University
Dhanabd-826 004, India

J. P. SINGH
Department of Petroleum Engineering Indian School of Mines University
Dhanabd-826 004, India

Abstract: Surfactant based fracturing gels are considered as clean gel due to absence of insoluble residues after hydraulic fracturing of tight reservoir. Conventionally used highly viscous polymer gels damage the formation by blocking the fluid path in fractured formation. The present work has been accomplished with the purpose of evaluating an anionic surfactant based gel for hydraulic fracturing. Rheological properties of the synthesized gel were studied as a function of shear rate, temperature and pH which are considered to be the most effective parameters on the gel behavior at the downhole. Proppant carrying capacity was also reported with varying proppant loading.

Keywords: Surfactant, hydro fracturing, rheological properties.

1. Introduction

Hydraulic fracturing is one of the stimulation processes that play a key role in producing unconventional oil and gas resources. Typical unconventional reservoir include tight oil and gas sands, gas shale and coal bend methane, heavy oil , tar sands ,and gas hydrates [1]. In hydraulic fracturing technique, fractures are created in tight sand artificially and widened crevices in the rocks of the bottom-hole zone by injecting a fluid at high pressure and flow rate sufficient to overcome the overburden stress and to initiate a fracture in the formation. The high pressure produces new crevices in the rocks and widens the existing ones. These crevices, connecting with others, become conductors for oil and gas from more remote productive parts of the formation [2]. In order to prevent the collapse of the fractures when pressure declines, coarse-grained sand (also called proppant) is introduced into the fluid that is pumped into the well. High fluid viscosity is required to carry proppants deep into the fracture and prevent proppant settling. The fracture must close on the proppant to prevent formation collapse and to create a long conductive fracture [3]. Fracturing fluid plays a major role in opening the wedge, extend a fracture hydraulically and transporting the proppant along the fracture.

For the last three decades, polymer gels have been used as fracturing fluids either in linear or cross-linked state [4]. A new class of viscoelastic fluids has been discovered to form surfactant based gel which perform fracturing without damage to the formation [5, 6, 1].

The main advantage of these solutions, compared to conventional polymer systems, is the potential for reduced formation damage and proppant pack damage, no additives or breakers are needed to add [7]. These fluids exhibit unexpectedly low high-shear-viscosities resulting in low friction pressures, even in small tubular. In addition, due to the very low viscosity of the broken fluid, faster load recovery of injected fluids is possible.

However, efficient field applications of these new types of fluids require extensive studies about gel. Rheological properties, namely viscosity, are considered as one of the most important properties of fracturing gel. It is influenced by several factors, e.g., temperature, shear rate, pH etc. The gel viscosity as well as proppant carrying capacity largely depends on the concentration of costly surfactant which prevents economic field application of SBG.

In the present work a new surfactant based fracturing gel (SBG) has been synthesized and the rheological properties are evaluated with variation in different parameters like surfactant, cosurfactant concentration, addition of alkali, synthesis time. The synthesized SBG shows very good shear thinning behavior and proppant carrying capacity characteristics under study.

2. Material and Methods

Preparation of surfactant based hydraulic fracturing gel requires one oil phase, aqueous phase, surfactant and co-surfactant as the gel stabilizer. Water was used as aqueous phase, Pine oil as oil phase, Sodium Lauryl Sulphate (SLS) as surfactant and Iso-Amyl Alcohol (IMA) as co-surfactant. In order to identify the gel region in the pseudo ternary phase diagram the oil phase was mixed with the co-surfactant/ surfactant (C/S) phase and the mixture was titrated with water to observe the changes in the gel regions. A number of experiments were carried out with different compositions of oil, water, co-surfactant and surfactant and gel region also identified from that.

The experiments were also carried at different pH by adding KCl or NaOH to the solution. Then the effect of pH on rheological properties was studied. Experiments were also performed at higher synthesis temperatures to study the gel behavior. All concentrations are expressed in wt%.

Rheological Tests

Rheological measurements of all the gels prepared were carried out in Physica US 200 rheometer. Tests were carried out by varying the shear rate from 0.8/sec to 500/sec and at different temperatures.

3. Results and Discussion

3.1. Identification of the gel region

After preparing various samples the following pseudo ternary phase diagram was drawn as shown in the Fig. 1 and the gel region was identified. From the experimental results it was found that no gel was formed with C/S concentration below 18 wt% when salt or alkali was not added to it. Homogeneous gel was formed from 20% concentration and showed good viscosity on addition of alkali.

From the experimental results, it was observed that with increase in the C/S concentration the viscosity is enhanced. However, as addition of surfactant increases, the cost of SBG also increases. So the surfactant concentration not was increased beyond 34%. It was found that gel prepared at this concentration can retain high viscosity even at very high ($500s^{-1}$) shear rate. However, addition of alkali or salt reduced the consumption of costly surfactant and co-surfactant as discussed in later section.

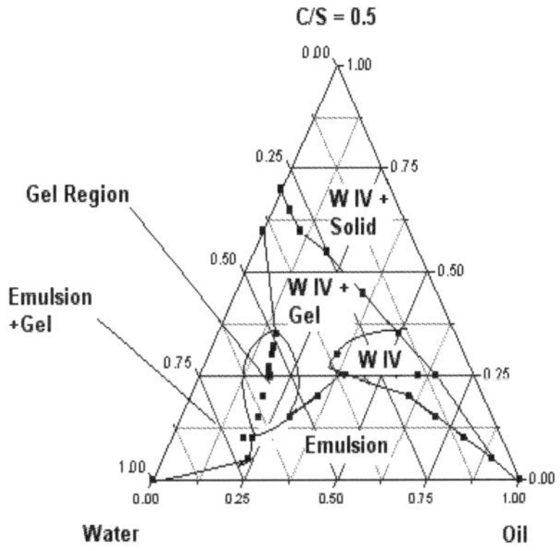

Figure 1. Pseudoternary phase diagram.

3.2. Effect of shear on SBG viscosity

From the experimental results it is observed that the viscosity increases with increasing concentration of surfactant (vide Fig. 2). The viscosity decreases with increasing shear rate. Gels with higher concentration surfactants are more resistant to shear compared to those of lower concentrations.

This rate of reduction in viscosity is more pronounced at higher rate than at lower shear. This non Newtonian flow response and corresponding increase in viscosity with C/S concentration is attributable to the dispersed fraction of micelles in the solution and the structure forming interactions between the aggregates and nature of micelle aggregates. And also at higher concentration of C/S, the spherical micelles will grow and entangle to form long flexible micelles.

3.3. Effect of salt & alkali on SBG viscosity

Addition of Salt: Variation of viscosity as a function of solution pH is shown in Fig. 3. From the graph it can be observed that at higher pH value the SBG is more resistive to higher shear stresses. However, this behavior was observed only up to pH value 10 beyond which the gel rheology was deteriorated. Hence, pH value was optimized at 9.5. Fig. 4 shows the improvement of gel rheology on addition of salt/alkali.

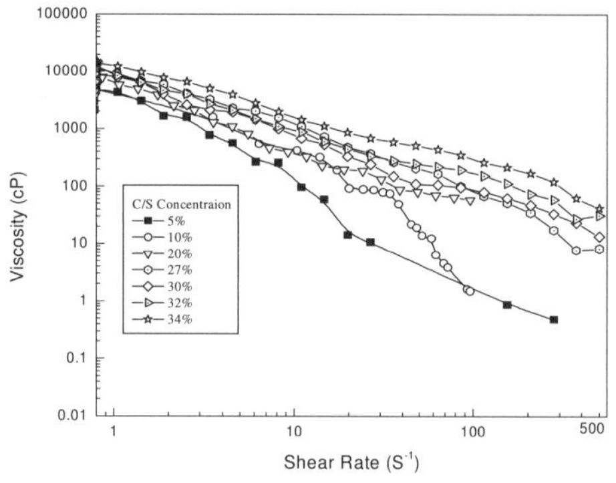

Figure 2. Effect of surfactant, co-surfactant concentration on viscosity.

Addition of Alkali: It was observed from Fig. 4, that addition of NaOH/ alkali to the gel improves the rheological properties. However, beyond a limit of pH or % NaOH concentration, the gel rheology viscosity starts to decline. This phenomena can observed both from Fig. 3 and Fig. 4. This trend is observed at high shear rate and temperature range of the study 30-70°C so the percentage of NaOH is optimized and 0.1 % and 0.5 % gels were chosen.

This can be attributed to the reason that at increased NaOH concentration (hence increased pH value) all the micelles will be aligned in one particular direction forming worm-like or rod shaped micelles leading to resist the shear stress more strongly than the ordinary sample. These micelles resemble polymer chains and can entangle to increase the fluid viscosity. And also higher pH value makes the solution more viscous by eliminating the hydrogen ion in the solution [8].

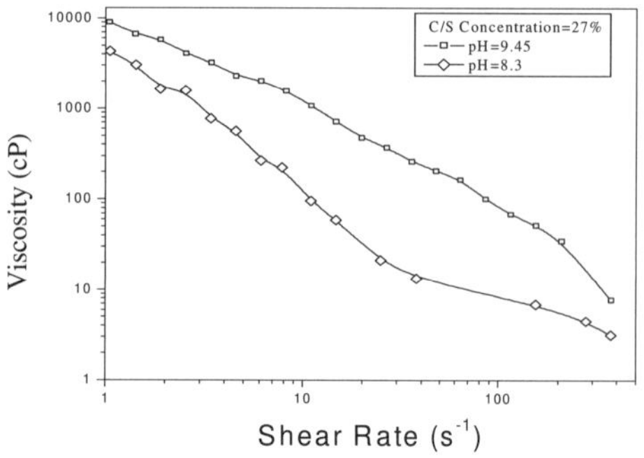

Figure 3. Effect of pH, C/S =27%.

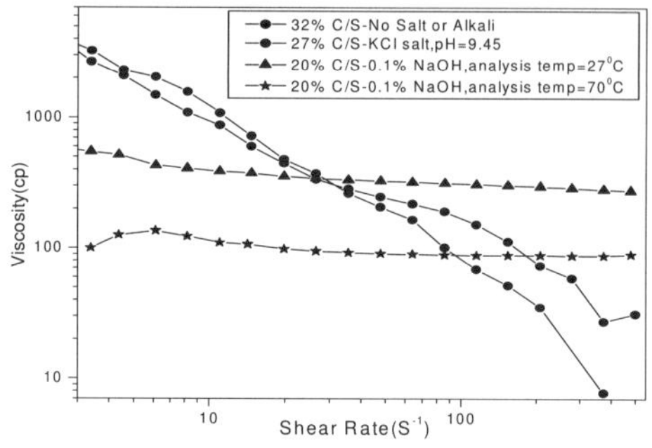

Figure 4. Effect of salt and alkali on viscosity.

3.4. Effect of synthesis temperature

Effect of synthesis temperature on gel properties were tested by varying the synthesis temperature from 30°C to 70°C. Results are shown in Fig. 5.

It can be observed from the figure that with increase in the synthesis temperature the viscosity of SBG increases. This may be because of faster miceller orientation and alignment at higher synthesis temperatures which makes the solution more viscous. However, temperature above 70°C, the micelle formation retarded bacuse of higher kinetic energy and again the viscosity starts to reduce.

3.5. Thermal stability of SBG

The synthesized fracturing gel stability was tested by studying the rheology at various temperatures under shear experiment. The results are shown in Fig. 6. From the figure it can be observed that the synthesized gel is quite resistive even at 70°C, which is similar to the average temperature of most reservoirs.

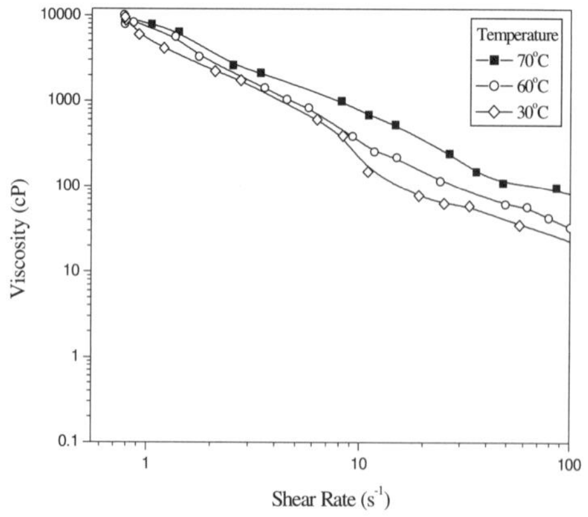

Figure 5. Effect of synthesis temperature, C/S =30%.

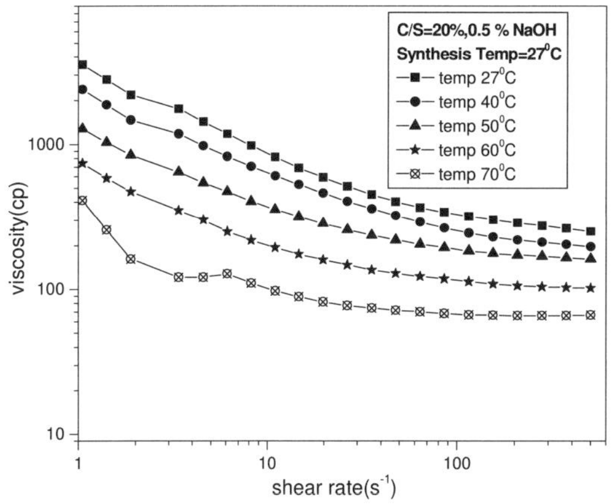

Figure 6. Effect of alkali addition on thermal stability. C/S =20%, NaOH = 0.5%.

3.6. Comparison of SBG with Guar Gum solutions

From the Fig. 7 it is clear that though the viscosity of synthesized gel is lower at low shear rates compared to Guar gum of 1.5% concentration but at higher shear rates (beyond 100 S^{-1}) SBG show better viscosity behavior. So at higher shear rates SBG can be preferred over conventional guar gums as a better hydraulic fracturing fluid with respect to viscoelastic performance, economical point of view and environmental concerns.

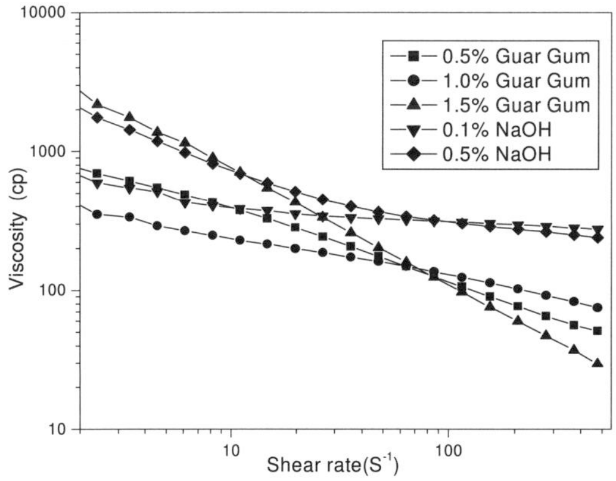

Figure 7. Comparison of viscosity variation with Shear rate for Guar Gums and SBG of different composition at 27°C.

3.7. Proppant carrying capacity of SBG

One of the most important properties of fracturing gels is proppant carrying capacity so that proppant could be carried and placed properly in the formation. Settling velocity was measured for different gels for varying proppant loading in the gels. Experimental results showed that 0.5% 1.0% and 1.5% guar gels and 0.1% NaOH, 20% C/S SBG did not give satisfactory results for proppant carrying capacity as the sand was settled instantly. 0.5% NaOH, 20% C/S Gel showed satisfactory proppant carrying capacity. The measured properties are shown in table 1.

Table 1. Settling Velocities of SBG (for 0.5% NaOH 20% C/S gel, density of gel =0.9852gm/cc).

Proppant concentration in Gel (wt %)	Diameter of the proppant (mm)	Viscosity of the Gel (cp)	Settling Velocity cm/sec
15	0.5925	7800	0.00012
20			0.00026
25			0.00013
30			0.00013

4. Conclusions

In the present work a new surfactant based gel was synthesized and its rheological behaviors were studied. The properties of SBG were observed to change with the concentration of constituents, pH, and preparation temperature. The viscosity of the gel was found to increase with the co-surfactant/surfactant concentration. The synthesized gel also showed superior rheology as a fracturing fluid compared the guar gel.

The experimental results showed that addition of NaOH to solution reduces the requirement of amount of costly surfactant which is the main constraint to filed application of SBG. Also it can be concluded from result that NaOH is more effective additive compared to KCl for improvement of gel properties.

Apart from the above characteristics the gel (C/S = 20%, NaOH = 0.5%) showed very good proppant carrying capacity. The synthesized SBG is soluble in both oil and water. Hence, no breaker or other additives are required. Moreover, it is easy to mix on location, hence it may be considered as an efficient, economic and eco-friendly fracturing fluid on optimization of all the properties.

Acknowledgements

We gratefully acknowledge for the financial assistance provided by Council of Scientific and Industrial Research, India (CSIR, project no: EMR 425/07), and

Department of Petroleum Engineering, Indian School of Mines University, Dhanbad, India. Thanks are also extended to all individual associated with the project.

References

1. Lulla D., Ahmed G.B., Rae P.A., Melani R.A., "Toward zero damage: new fluid points the way" SPE 69453, 1-8 2001.
2. Clark J.B., "A Hydraulic Process for Increasing the Productivity of Oil Wells" Trans., AIME (1949) 186, 1-8.
3. Howard G.C. and Fast C.R., "Hydraulic Fracturing", Monograph Series, SPE, Details (1970).
4. Goel N, Shah S.N. and Grady B.P., "Correlating viscoelastic measurements of fracturing fluid to particles suspension and solids transport" Journal of Petroleum Science and Engineering 35 (2002) 59–81.
5. Welton T.D., Lewis S.J., Funkhor G.P., "Viscoelastic surfactant fluids and associated methods", U. S. Patent Application number 20060180310.
6. Sammual M.M., Card R.J., Nellson E., Brown J.E., Vinod P.S., Temple H.L., Quo Q. and Fu D.K., "SPE Drilling and Completion", 14, 240, 1999
7. Goel Naval, "Correlating Viscoelastic measurements of fracturing fluid to particles suspension and solids transport", Journal of Petroleum Science and Engineering 3, 59-81, 2002.
8. Binks B.P., Espert A., Fletcher P.D.I., L. Soubiran "Phase behaviour of microemulsions stabilised by double chain cationic surfactants and alcohol co-surfactants" Colloids and Surfaces A: Physicochem. Eng. Aspects 212 (2003), 135-/145.

THE ENVIRONMENTAL EFFECTS CAUSED BY CONSTRUCTING MAMLU RESERVOIR DAM IN EAST TEHRAN

SIAMAK BOUDAGHPOUR

Assistant Professor of Environmental Engineering Department, KNT, University of Technology, Tehran, Iran
bodaghpour@kntu.ac.ir

Constructing Mamlu reservoir dam aims at using the potential catchments of Jajrud and Damavand rivers and conveniently utilizing the floods of these two rivers in order to maintain the amount of water required for the fertile and farming lands of Varamin plain and to provide a part of Tehran's drinking water. Latian dam located 27 km fur ther from the site of constructing Mamlu dam was first built in 1968 to supply a part of Tehran's drinking water. The initial plan was to yearly reserve 1/3 drinking water and 2/3 for fields of Varamin plain. The exceeding number of Tehran's population made major changes on the above mentioned relations. Consequently, the water organization determined a certain level for drinking water and more then the above mentioned level to overflow into Varamin plain. Since the water flowing into the plain didn't have enough quality and it was low in volume, farmers had to use the underground water that led into a drastic falling of the underground water level, its mixing with sewage and polluting the layers in some regions. Constructing Mamlu dam can harness, conduct, flow the water and prevent over-using underground water and subsiding earth, more significantly, it can preserve a part of Tehran's environment. Eventually, this article also studies whether the practical environmental purposes around Mamlu dam will be truly proved.

Keywords: Mamlu dam, underground water layer, pollutions, environment.

1. Introduction

Constructing dam can be considered as one of the effective ways providing Tehran's increasing water requirement. In our country dams have been constructed without fundamental studies on the catchments nearby and the environmental conditions. As time dragging, many environmental problems can be easily observed in the reservoir and its surrounding including: water contamination, soil pollution, ground efficiency, inadequate physical existence, biological, economic and social effects [1].

For one point, the yearly piling up sediments have diminished the efficient volume of dam reservoirs providing Tehran's water and for the other point, outfitting the water needed for drinking and fields in the urban and rural areas have all conducted in to constructing Mamlu dam. Therefore, 1/3 of the stored water is attributed to drinking and its 2/3 is flown into the farming lands of Varamin plain. Mamlu dam construction can reduce using underground water of Varamin plain up to 70 million square meters. The cost of using is much more than producing ground water which is already the main resource of maintaining fields of Varamin plain. Constructing Mamlu dam will hold vital influences on agricultural part and consecutive social effects in the area.

The Situation of Dam and Its Technical Specifications. Mamlu dam has been located 2km downward the cross of Jajrud and Damavand rivers with the special feature of 51° 47′ 30″ geographical length and 35° 33′ 7″ geographical width at the gate. The general condition of the dam has been figured out in map (1) according to environmental effects and nearby man-made constructions. The particular situation of Mamlu dam and its location in the protected area of Jajrud and the border of the southeastern of Khajir national park has given it a specific sensitivity. Not only the above mentioned areas are environmentally notable, but also any sort of constructing and capturing in those places have got particular legal conditions and restrictions. Materials used in dam construction are the same as clay's central nucleus. The technical and general specifications of the dam are as the following:

a. Dam's height: 87 meter
b. Dam's length at the head: 772 meter
c. Dam's level of head: 1305 meter
d. Volume of the reservoir at normal level of 1300/5 m: 229 km^2

e. Surface of reservoir at normal level of 1300/5 meter: 7/5 km^2
f. Watershed surface of the dam's reservoir: 1750 km^2
g. Dead volume of reservoir at the level of 1246 meter: 49 million m^2
h. Capacity of the dam's annual regulation: 215 million m^3
i. Yearly average of sediments in the reservoir: 1/23 million m^3
j. Case of constructing a hydroelectric power station

There will be installed 2 turbines of 2×4 (mw). It takes eight years to study and construct the dam up until starting and the dam's useful aging would be fifty years after starting and applying it.

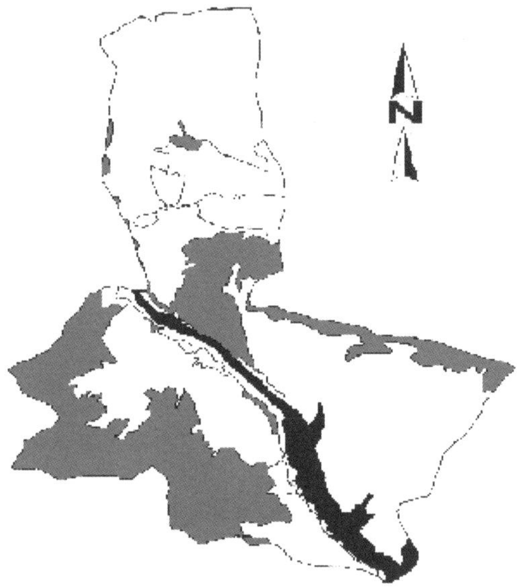

Figure 1. Geographical map of the watershed of Mamlu dam.

1.1. *The goals of constructing Mamlu dam*

The yearly piling up of sediments have diminished the efficient volume of dam reservoirs providing Tehran's water and outfitting the water needed for drinking and farming lands in urban and rural areas of Tehran's provinces all have made it inevitable to explore new water resources. There are several reasons that have made the regional water company of Tehran give a precise consideration to study the significance of maintaining Tehran's water as the capital of Iran and as a developing mega city center and also the leading economic role of agriculture in Varamin plain downward Jajrud river all have compelled the regional water company of Tehran to put constructing a dam over Jajrud river downward Latian dam on top of their agenda. Moreover, several clear-cut and specific researches and studies should be done. Therefore, it is indispensable to carefully study the area to construct a dam.

The water overflowing from Latian dam is usually more than what required for the fields' seasonal irrigation of those locating in lower parts of the dam. As a result of the monthly and yearly unequal extreme dispersion of the flow of Jajrud river and not well-timed controlling by Latian dam for the sake of its limited capacity, a remarkable part of it specially during flooding seasons has no more use and consequently it wouldn't be accessible. Conclusively, constructing another dam over Jajrud river downward Latian dam can highly harness the flood overflowing from Latian dam, control the sediments, distinctly regulate them, retain and store the squandered water out of Latian for drinking and farming applications.

So, the main aim of constructing Mamlu dam is to keep and regulate the overflowing of Latian dam the mid watershed of Jajrud River and Mamlu to provide the needs of Varamin fertile plain. The other purpose of this program is producing electricity by establishing a hydroelectric power station downward the dam and harnessing the flood. Totally, constructing Mamlu dam aims at:

1. Storing the water overflowing from Latian dam and the sediments of Jajrud watershed downward Latian dam and regularly applying it in Varamin plain.
2. Moving a huge amount of floods to the reservoirs of Latian and Mamlu dams and preventing wasting the floods which are not possible to be stored in Lar dam because of water releasing and being impossible to be mended in the near future.
3. Manageably providing water for the fields of Varamin plain from Mamlu dam.
4. Regularly exploiting Latin and Mamlu dams for being exclusively used for Tehran's drinking water.
5. Reducing the probability emptying the reservoir of Latian dam before starting the season of overflowing Jajrud river and making sure of having enough water in this reservoir for a long time as a water resource maintaining the needs of Tehran through making convenient decisions and controllably exploiting Latian dam after constructing Mamlu dam.

2. A Review on the Area's Current Environmental Situation

There are 41 mammal species with unstable and inappropriate dispersal. Sheep and anthelope are the most important migrant mammal species in the area. They migrate through the north axel of the dam.

There has been identified 130 bird species that 81 of them have been protected. Eventually, about 90% of the birds form the watershed is insectivorous and useful and almost 62% of them are protected. There has also been identified 5 valuable species of birds.

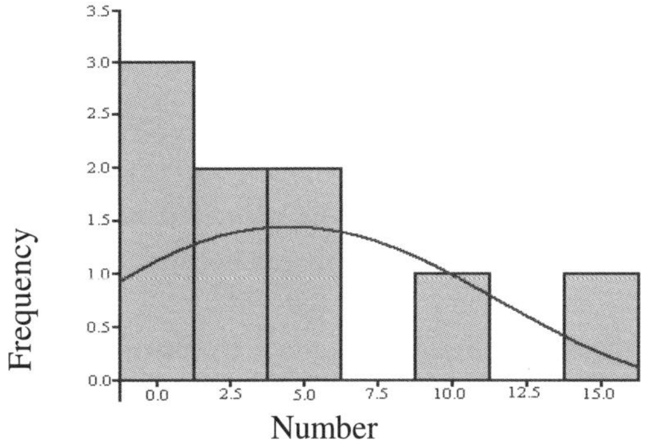

Figure 2. Analysis of the area's identified aquatics in the research center.

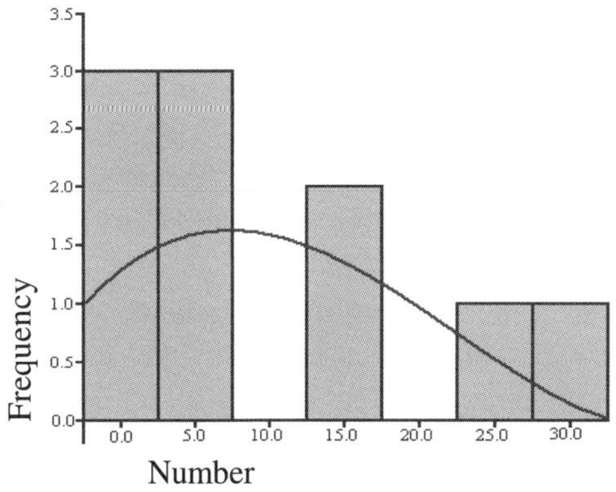

Figure 3. Analysis of the area's identified aquatics in the research center.

This area embraces 2 kinds of reptiles and more than 12 species and subdivisions of lizards and 13 kinds of snakes and sub-species and the viper of Alborz is the very unique type of this sort existing in this area.

The aquatics inhabiting in this area include different types of bentozes, larva of insects, mollusks and fishes. The highly protected and ecologically valuable creatures in this area are: 30 species of different types of bentozes and larva of insects and 11 sorts of fishes such as: rainbow and red spotted trout.

The area has been richly covered with Irano-Turani species holding 785 kinds that 665 of them are bi-cotyledon, 110 sorts are mono cotyledon, one kind gymnospermousand 9 species are vascular cryptogamous. Some of these species are only grown high in mountains and can't be found in low height of valleys, some other plants exist in dry lands and the others are dispersed in aquatic conditions. Rivers of watershed and their banks are the best and most suitable places for aquatic plants. These plants either float on the water or grow in the banks of the rivers. Moreover, they perform a significant role in the healthiness and stability of the aquatic systems of the watershed [6].

3. Sources of Pollutions

3.1. *Waste*

Main resources contaminating Damavand and Jajrud rivers are urban wastes from Damavand, Rudehen, Fasham and Lavasan that dispose a great mass of organic materials floating solid particles in to these rivers and consequently increase the amount of biological need of the river to oxygen. Middle and small workshops are other industrial polluting sources that their wastes are offloaded into the absorbent wells. Parchin sewage is made up of almost the wastes of 159 residential areas and several industrial complexes which contaminate the rivers. Moreover, the wastes of Ab Ali dumping ground and the sewage of Mahisara, Jajrud also contaminate the water. Constructing Mamlu dam itself can contaminate the water as the following way:

- Increasing the amount of floating particles.
- Darkening the water.
- Spreading oily materials produced by different activities.
- Offloading the wastes made by people inhabiting at the bank of the dam.

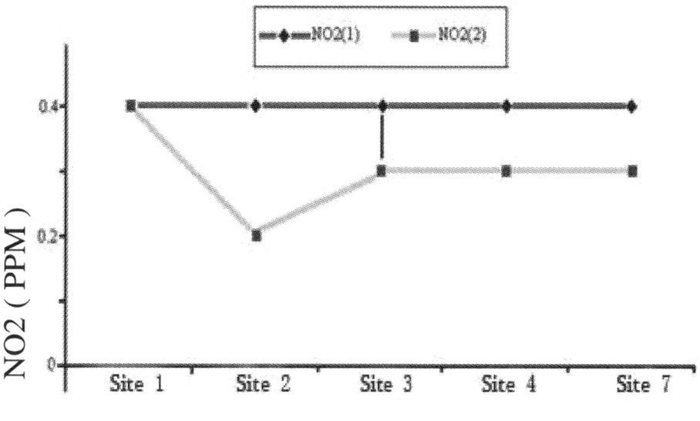

Figure 4. Process of nitrate changes solution in the in the river flow within 2 periods of sample taking.

3.2. *Air*

Constructing operations, mounting materials and trafficking of motor mobilizations are the most probable factors polluting the air. The most important pollutants potentially issuing throughout the area consist of:

- Dusty particles caused by moving soil and materials
- Commuting vehicles
- So_2 rising from igniting heavy fuels and those containing sulfur in machineries
- No_2 caused by machineries and vehicles
- Co_2 made by commuting vehicles and their fuel combustion

3.3. *Sound*

The most permanent sound pollution is made by building machineries, commuting vehicles and digging equipments. Since the site of dam construction is inside the national park and the protected area, it seems impossible to do explosive operations.

3.4. *Soil*

The sewage produced by piling up and land filling in Ab Ali dumping ground and applying fertilizers and poisons to grow farming productions are the chief sources of polluting the soil. Although, the land filling centers of Ab Ali have been stopped, the sewage pollution made by land filled garbage still keeps carrying on.

4. The Negative Consequences of Constructing Dam on Wild Life

The consequences of constructing dam on wildlife are easily predictable that can be divided in two parts of constructing and exploiting as below:
The Effects of constructing process:

a. The sound of vehicles moving and explosions all disturb both the area's peace and the wildlife that would urge animals to migrate for either long or short time.

b. Taking stones and other natural resources from inside the environmental limitations of goats in Mahdi Khani strait and Jujrud protected area and taking clay from inside Khanjir national park all have seriously damaged one of the sensitive passing directions of the wildlife. Significantly, the region is regarded as one of the richest environments which is full of plant coverage and water resources and wildlife. The difficulties and disturbances have obliged animals to disperse and also have endangered the suitable environments of Kafarhumand and Baghshah.

5. The Effects of Exploring Process

Within the process of exploiting and after supplying the dam with water, a main part of the safe and convenient environments of Kafarhumand, Baghshah and Galian neighboring with Jajrud river and the dam will be sunk. During refilling water from the site of the dam construction, almost 6/5 km of Parchin road will be sunk as well. Sinking this part of the road which is belonged to the army urges to build a connecting road. While building the complimentary road, considering the area's environment can absolutely minimize these effects. The consequences made by constructing dam and refilling it with water on the area's wildlife will be as the following:

1. Because of sinking these environments , a part of the wildlife of the national park will be removed from their safe regions and entered in the protected area and as people inhabit there so, animals can not find convenient environments. Conclusively, the wildlife's population will be threatened by diseases made by shared watering –troughs of pets and wild animals, reducing nutritional sources caused by using up shared sources and illegal hunting.

2. One of the major directions of moving and wildlife's migrating is obstructed for the sake of refilling dam with water approximately up until Takhte Yarmir and a sinking part of the environments of Baghshah. The importance of the direction of Kafarhumand and Baghshah is that they both haven't been touched down by human developments. In addition, it is one of the sensitive corridors conducting and supporting the dynamism and stability of the wildlife of Khanjir national park. Wildlife can not access the sunken parts, so they are directed to the upward of Jujrud River or other ways at the time of migrating. Therefore, they are threatened by being exposed to residential areas. This trouble is more perilous for some unique species of Red Sheep that are limitedly dispersed in Iran ranging from the feet of mountains to the area's southern chains of hills of central Alborz. The Red Sheep will absolutely be extinct due to inter-group mating and non-returnable migrations caused by other safe corridors and the complete obstruction of the directions.

3. One of the other effects of constructing dam is increasing the entertaining attraction and if this is not carefully controlled, it will be changed into a disaster and makes the area damaged. Other consequences of the dam construction are changing the applications and

Table 1. The relation between indications and the negative predicted effects.

The Predicted Negative Effects					
The pollution of the river	**Chemical and thermal making layer and changing the water quality of the reservoir and dam**	**Damaging plant growing places and valuable plant species**	**Changing the natural trend of wildlife and damaging the environments**	**Changing the natural migrating trend of aquatics**	
×	×		×	×	Water Contaminating indications of the river and the reservoir
×		×			Quality and quantity indications of the soil
×		×	×		Accumulating indication and plant coverage
×		×	×		Wildlife's statistics specially mammals and birds
×	×			×	The limnologic indication of the water of the river
×		×	×	×	The statistics of area's visitors
×		×	×	×	Changing the blending and quality and the level of the demands
		×	×	×	Verifying the compression of the foundations and linking lines
×			×	×	Hygienic indications

developing the access roads for entertaining the lands around the dam. This can also boost the area's un safety.

4. The security will be threatened by the fact of sinking a part of Parchin road and in case of building a replaced road in the sensational limitations of the park.

6. Conclusions

Studies show that the probable effects of constructing dam on the area's environment have been evaluated after identifying and predicting in the suggested matrix of ICOLD and its consequences have indicated both potential negative and positive ones.

6.1. *The negative effects*

- Changing the natural migrating trend of aquatics
- Damaging plant growing places
- Verifying the natural migrating trend of animals living on the dry lands
- Changing the natural diet situation of the river
- Eroding the slopes directing to the reservoir
- Differing with the management of under controlled and protected areas, making insecure the wildlife's environments
- Potentially destroying both economically and ecologically valuable species

6.2. *Positive effects*

- Improving the management of cultural monuments and historical ruins leading to developing tourism
- Harnessing harmful and perilous floods
- Increasing the income and making advantages
- Developing the cost of grounds nearby the dam
- Developing tourism, fishing and aquatic sports

References

1. Smakhtin, (2004), "Taking into Account Environmental Water Requirement in Global-Scale Water Resources Assessments", IWMI.

2. Sima, S., Tajrishy, M., (2006), "Water allocation for Water land Environmental Water Requirements" World Environment and Water Resources Congress, ASCE.
3. Krogstand F. and Schiess P., (2000). "Cumulative impact estimation for land scale forest planning" 4th International conference on Integration GIS and Environmental Modeling (GIS/EM4): problems, prospects and research needs. Alberta, Canada, September, 2-8.
4. O'sullivan, D. and Unwin, DJ. (2003), "Geographic Information Analysis", John Willey and Sons Press, Hoboken, NJ.
5. Collins M.G, Steiner, F.R. and Rushman, M.J. (2001), Land-use suitability analysis in the united states, historical development and promising technology achievement Environmental Management, **28**(5): 611-621.
6. Lee, C.C., Lin, S.D. (2000), Hand book of Environmental Engineering calculations, McGraw Hill, New York, USA, pp.189-190.
7. Barrow, C.J (1999), Environmental Management: Principle and practice, London and New York: Rout edge Environmental Management series.
8. Reading, H.G. (1996), J (1999), Sedimentary Environments: Process, Facieses and stratigraph, 3rd ed. Black Well.

Session 2

TREATMENT OF DISTILLERY SPENTWASH BY ANAEROBIC BAFFLED REACTORS

A. K. DIKSHIT

Professor, Centre for Environmental Science and Engineering, Indian Institute of Technology Bombay, Powai, Mumbai-400 076, India

M. S. CHAUHAN

Assistant Professor, Department of Civil Engineering, Maulana Azad National Institute of Technology, Bhopal-462007, India

S. S. SINGH

Research Scholar, Centre for Environmental Science and Engineering, Indian Institute of Technology Bombay, Powai, Mumbai-400 076, India

Distillery spentwash is one of the most high strength and recalcitrant wastes having extremely high COD, BOD and color. The present study was undertaken to use anaerobic digestion as complete solution to treat BOD and COD in the same reactor in conjunction with suitable oxidation technique. Two such reactors, namely, Hybrid Anaerobic Baffled Reactor (HABR) and Granular Bed Anaerobic Baffled Reactor (GBABR) were used. The effluents from both the reactors were oxidized by using hydrogen peroxide. The oxidized spent wash was further fed into another set of HABR and GBABR reactors. An overall COD removal efficiency of 92% and 90% was achieved for HABR and GBABR, respectively.

1. Introduction

The biochemical oxygen demand (BOD) and chemical oxygen demand (COD) of raw distillery spentwash (DSW) range between 35,000–50,000 and 100,000–150,000 mg/L, respectively [1]. Distillery wastewater is one of the most difficult waste products to dispose because of low pH, high temperature, dark brown color, high ash content and high percentage of dissolved organic and inorganic matter [2]. The high COD and BOD of molasses spentwash make anaerobic treatment attractive in comparison to direct aerobic treatment. Anaerobic baffled reactor (ABR) was developed by Bachmann et al. [3], who showed its efficiency in treating low to medium strength industrial wastewaters. Several modifications have been made to the original ABR for various reasons like to enhance the solids retention capacity, to treat difficult wastewaters or simply to reduce capital costs [4]. Hybrid anaerobic reactor (HABR) has a solids settling chamber after the final compartment. Packing was also positioned at the liquid surface of each compartment with randomly packed Pall rings in the first two chambers, and a deeper modular corrugated block which had high voidage in the third chamber [5]. Akunna et al. [6] used granular bed anaerobic baffled reactor (GBABR) for the treatment of a whisky distillery wastewater. It combined the advantages of baffled reactor system and upflow anaerobic sludge blanket (UASB) system.

Advanced oxidation processes (AOP) has gained a lot of popularity in wastewater treatment during the past two decades for the treatment of industrial effluent. H_2O_2 is one of the most powerful oxidizers known - stronger than chlorine, chlorine dioxide, and potassium permanganate. Also, through catalysis, H_2O_2 can be converted into hydroxyl radicals (.OH) with reactivity second only to fluorine.

In this study, efficient removal of COD is achieved by using combined anaerobic digestion and advanced oxidation. In the first step, anaerobic baffled reactor (HABR and GBABR) was used with mixed anaerobic consortia to reduce BOD. The recalcitrant COD remaining after the first step was converted to biodegradable organics with the help of hydrogen peroxide treatment. The treated effluent was, then, fed to second stage anaerobic baffled reactor (HABR and GBABR) with mixed anaerobic consortia to achieve maximum BOD and COD reduction.

2. Materials and Methods

2.1. *Wastewater*

The study has been carried out on actual spentwash collected from a Distillery producing rectified spirit using molasses as raw material. Raw distillery spentwash (DSW) was collected after the cooling tower unit. The characteristics of wastewater are given in Table 1 based on the Standard Methods for Examination of Water and Wastewater [7].

2.2. *Sludge*

The seed sludge was also collected from the same distillery plant from where distillery spentwash was collected. The sludge was collected from the existing Upflow anaerobic sludge blanket (UASB) plant in the distillery. This seed sludge collected from the plant was stored in the cold room at 4^0C to prevent any biological activity.

Table 1. Characteristics of raw distillery spentwash.

Parameter	Value
pH	3.67
TS	66,980 mg/l
TDS	14,660 mg/l
TSS	52,320 mg/l
VSS	1,444 mg/l
Kjeldahl Nitrogen	4,704 mg/l
Total Phosphorous	1,534 mg/L
COD	88,000 mg/l
BOD	35,000 mg/l

2.3. *Reactor configuration*

HABR reactor of 10 litres capacity is made of acrylic sheet and it consists of 3 chambers and a final settler. Figure 1 shows the fabricated HABR with its dimensions. The first two chambers had close to the surface level of plastic media while the third chamber was half filled with plastic media. The gas from each chamber was collected to a common rectangular gas collection chamber.

A 10 liter capacity GBABR consists of 10 equal compartments. Figure 2 shows the fabricated lab scale GBABR. Each compartment was further divided into two by slanted edge (45^0) baffles to encourage mixing in each compartment. The influent feed was pumped using variable speed peristaltic pump (Masterflex). The produced gas was determined by water displacement method.

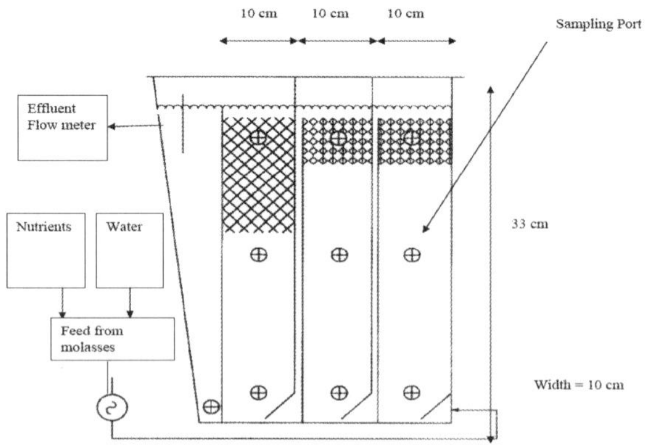

Figure 1. Schematic diagram of Hybrid Anaerobic Baffled Reactor (HABR).

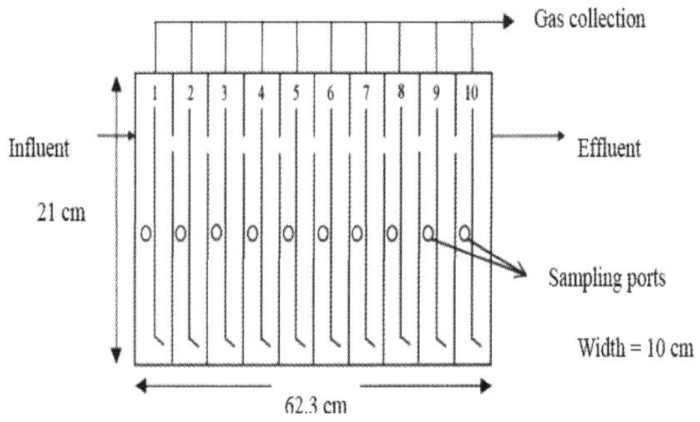

Figure 2. Schematic diagram of Granular Bed Anaerobic Baffled Reactor (GBABR).

3. Results and Discussion

3.1. *First stage anaerobic baffled reactors*

After 60 days of start up operation in the first stage reactors (HABR and GBABR), the organic loading of 5.04 kg COD/m^3/day and 30 day's retention time was maintained. The flow in the reactor was 0.23 ml/min. The biogas generation was 0.5 m^3/kg COD.

3.1.1. *Variation in pH*

In HABR, pH varied from 8 to 9 and stabilized afterwards (Figure 3). pH variation in GBABR was from 7.5 to 8. There was no remarkable drop in pH in both the reactors as drop in pH can cause reactor failure.

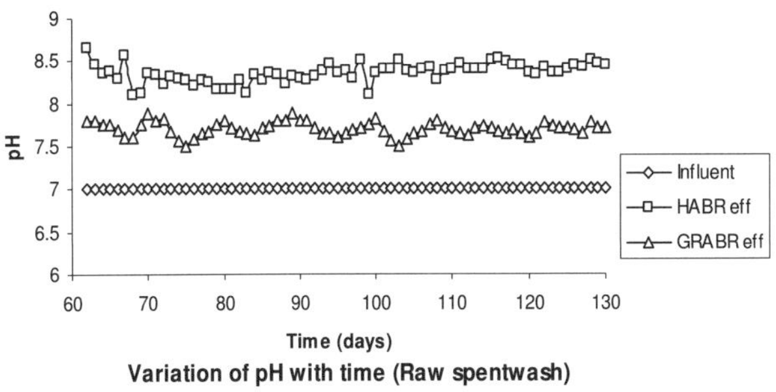

Figure 3. Variation of pH in influent and effluent of first stage reactors with time.

3.1.2. *COD reduction*

The COD reduction in HABR was 79% and that in GBABR was 71% (Figure 4). Further reduction in COD was not possible due to recalcitrant compounds.

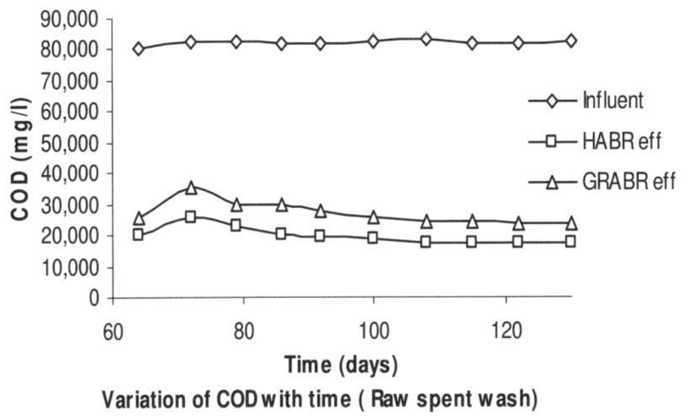

Figure 4. Variation in COD profile in influent and effluent of first stage reactors with time.

3.2. *Hydrogen peroxide treatment*

Hydrogen peroxide treatment was given to improve the biodegradability of digested DSW for feeding it to the second stage HABR and GBABR. Optimum hydrogen peroxide dose was found by varying the dose for maximum color reduction as color is due to the recalcitrant compounds. The optimum dose for maximum colour removal in both the reactors was found to be 8 ml/l. At this dose, color removal for HABR and GBABR was 20% and 17% and COD reduction was 33.5% and 20.8%, respectively.

3.3. *Second stage anaerobic baffled reactors*

After treatment with Hydrogen peroxide, effluent was fed to second stage HABR and GBABR for further reduction in COD.

3.3.1. *Variation in pH*

In HABR, pH varies from 8.5 to 9 and pH variation in GBABR was from 8 to 8.5 (Figure 5). There was no remarkable drop in pH and reactor worked properly.

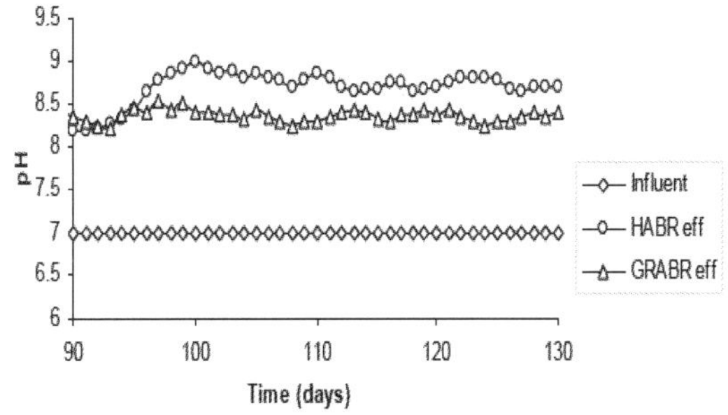

Figure 5. Variation of pH in influent and effluent of second stage reactors with time.

3.3.2. *COD reduction*

Further COD reduction in HABR was 47.8% and 40% in GBABR. The final COD of treated effluent after second stage HABR and GBABR were 6,100 and 8,000 mg/l (Figure 6). The total COD reductions in the effluent second stage HABR and GBABR were 92% and 90%.

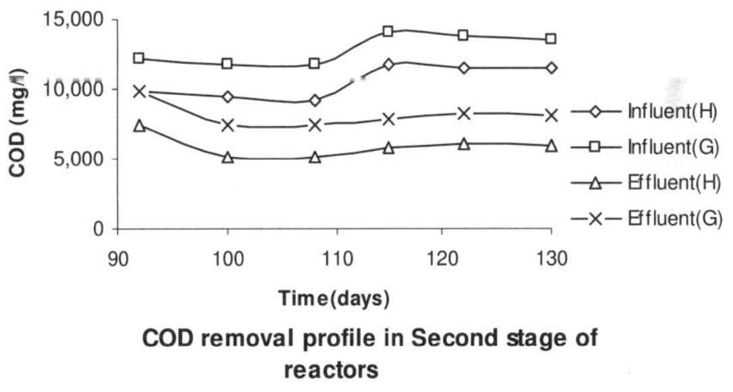

Figure 6. Variation in COD profile in influent and effluent of second stage reactors with time.

4. Conclusion

In this study, two anaerobic baffled reactors HABR and GRABR were used for treating distillery spentwash in conjunction with hydrogen peroxide treatment. In the first stage, COD was reduced to 17,000 mg/l in HABR and 23,000 mg/l in GBABR. The biodegradability of effluent after first stage was increased using hydrogen peroxide oxidation. Effluent after oxidation was fed to another set of ABR's for further reduction of COD. In the second stage, COD of effluent after HABR was further reduced to 6,100 mg/l and that of GBABR was reduced to 8,000 mg/l, thus, giving an overall COD removal efficiency of 92% and 90% for HABR and GBABR, respectively.

References

1. T. Nandy, S. Shastry and S.N. Kaul, *J. Environ.l Manage.t* **65**,25(2002)
2. F. J. Beltran, P. M. Alvarez, E. M. Rodriguez, J. F. Garcia-Araya, and J. Rivas, *Biotechnol. Progr.***17,** 462(2001).
3. A. Bachmann, V.L. Beard, and P.L. McCarty, *Water Resour.* **19(1)**, 99(1985).
4. W. Barber and D. Stuckey, *Water Resour.* **33(7)**, 1559(1999).
5. R. Boopathy and A. Tilche, *Water Resour.* **25(7)**, 785(1991).
6. J.C. Akunna and M. Clark, *Bioresour. Technol.***74 (3)**, 257 (2000).
7. APHA, "Standard methods for the examination of water and wastewater", American Public Health Association, New York, 2001

BACTERIAL AND FUNGAL BIOREACTORS FOR COLOUR REMOVAL FROM DISTILLERY SPENTWASH

S. S. SINGH
Research Scholar, Centre for Environmental Science and Engineering, Indian Institute of Technology Bombay, Powai, Mumbai-400 076, India

A. K. DIKSHIT
Professor, Centre for Environmental Science and Engineering, Indian Institute of Technology Bombay, Powai, Mumbai-400 076, India

Distillery spentwash has very high biological oxygen demand (BOD), chemical oxygen demand (COD), colour and various inorganic chemicals. Colour in spent wash is mainly due to dark brown pigment called melanoidins produced from the non-enzymatic browning reaction between aldose and amino compounds. Melanoidins cannot be degraded in conventional wastewater treatment. New promising technologies are developed in the form of bacterial and fungal reactors using pure microbial cultures for the treatment of Distillery Spentwash. The types of reactor used are batch reactor, sequencing batch reactor, bubble column reactor, rotating biological contactor and two stage reactors. Study done on fungal batch reactor using isolated *Aspergillus niger* gave 70% colour removal.

1. Introduction

Distilleries are one of the most polluting Industries in India. In India, the demand of ethanol is projected to increase because of Central Government notification directing 5% ethanol-blended petrol, as per Bureau of Indian Standards specifications, to be sold in the several States and Union Territories [1]. Most of the distilleries in India use sugarcane molasses for making alcohol. The main source of wastewater generation is the distillation step. Molasses spentwash has very high levels of BOD, COD, COD/BOD ratio as well as high potassium, phosphorus and sulphate content. Colour of spentwash is mainly due to melanoidins which are produced from the non-enzymatic browning reaction between aldose and amino compounds [2]. Conventional wastewater treatment cannot breakdown these coloured pigments and therefore colour remains in the effluent. Many pure cultures of bacteria and fungi were used for the decolourization of distillery effluent. This paper discusses the use of bacterial and fungal reactors in degradation and decolourization of the effluent.

2. Characteristics of spentwash

The characteristics of spentwash generated in Indian distilleries using sugarcane molasses is given in Table 1. The main source of wastewater generation is the distillation step wherein large volumes of dark brown effluent (spentwash) are generated. Molasses spentwash has very high levels of BOD, COD, COD/BOD ratio as well as high potassium, phosphorus and sulphate content. In addition, cane molasses spentwash contains low molecular weight compounds such as lactic acid, glycerol, ethanol and acetic acid [3].

Table 1. Characteristics of spentwash [4].

Characteristics	Value
Colour	Dark brown
COD	65,000-1,30,000 mg/L
BOD	30,000-70,000 mg/L
COD/BOD ratio	2.49
Total solids	30,000-1,00,000 mg/L
Total suspended solids	350 mg/L
Total dissolved solids	80,000 mg/L
Total nitrogen	1,000-2,000 mg/L
Total phosphorus	800-1,200 mg/L
Potassium	8,000-12,000 mg/L
Sulphur as sulphate	2,000-6,000 mg/L
pH	3-5.4

3. Bacterial reactors

3.1. *Air bubble column reactor*

Air bubble column reactor is a type of vertical reactor characterized by good oxygen transfer and a high biological conversion capacity (Figure 1). Petruccioli et al. [5] evaluated COD removal rate and efficiency of winery wastewater in air-bubble column bioreactor using self-adapted microbial populations either free or immobilized on polyurethane particles in a packed-bed bioreactor immobilized on Raschig rings. The highest COD removal rate of 6.6 $kg/m^3/d$ was obtained with free activated sludge in the bubble column bioreactor with HRT of about 0.8 day.

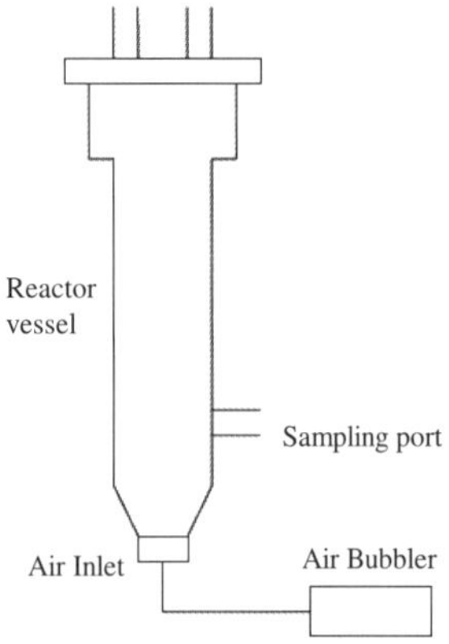

Figure 1. Schematic diagram of an air bubble column bioreactor.

3.2. *Two stage reactor*

Ghosh et al. [6] studied the application of two bacterial strains, *Pseudomonas putida* U and *Aeromonas* strain Ema, in a two-stage column bioreactor to reduce COD and the colour of the anaerobically treated spentwash under aerobic condition. In the first stage, *P. putida* reduced the COD and colour by 44.4% and 60%. The *Aeromonas* strain Ema, in the second stage, reduced the COD by about 44.4%.

3.3. *Sequencing batch reactor*

The Sequencing batch reactor (SBR) is a process designed to operate under non-steady state conditions. An SBR operates in a true batch mode with aeration and sludge settlement both occurring in the same tank.

Thus, there is a degree of flexibility associated with SBR [7]. SBR has normally five steps (Figure 2). In fill step, wastewater is filled in the tank containing biomass that settles in the previous cycle; in react step, aeration is done to facilitate biological growth and degradation; in settle step, mixing and aeration is stopped and solids are allowed to settle in the tank; in decant step, clarified effluent is discharged and in idle step, sludge removal is done, if necessary.

An acetogenic bacteria PB103 was applied to the SBR system for treatment of distillery wastewater. The bacteria showed the highest COD and Melanoidin pigment removal efficiencies of 65.2 ± 2.5%, and 50.2 ± 3.7%, respectively, with 10 times diluted Anaerobic molasses spentwash containing 30 g/L glucose. Good quality bio-sludge (SVI was less than 50 mL/g) was produced from the SBR system at HRT of 7 days [8].

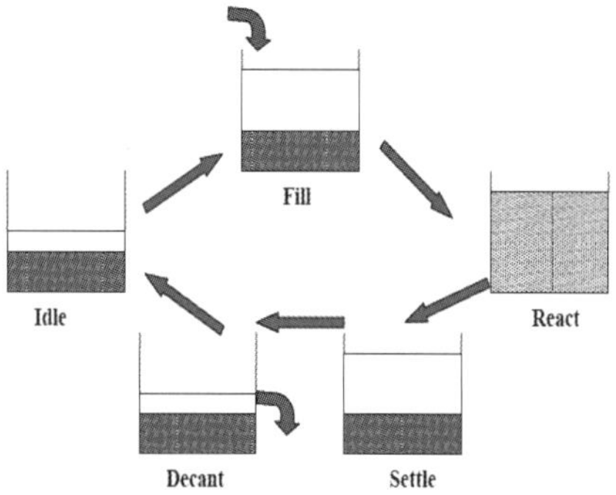

Figure 2. Steps in sequencing batch reactor.

4. Fungal reactors

4.1. *Rotating biological contactor*

This system is based on a microbial biofilm of fungus that develops on the surface of discs mounted onto a horizontal shaft with at least 40% of the discs submerged in the wastewater (Figure 3). Rotation of the shaft results in alternating contact of the discs with wastewater and air that allows for the aerobic growth of the micro-organisms on the surface of the discs. Various operating parameters can be controlled, such as disc rotation speed, recirculation and hydraulic retention time.

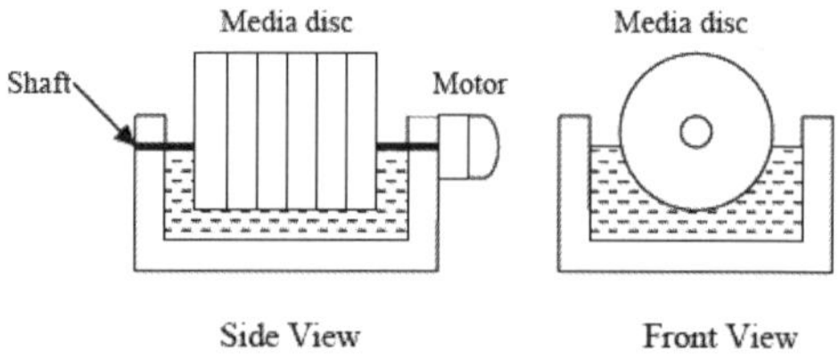

Figure 3. Schematic diagram of a rotating biological contactor.

A rotating biological contactor (RBC) with *P. chrysosporium* immobilized on PUF disks was operated with optimized decolourization medium in continuous mode with a retention time of 3 days. During the experiments, the reductions in colour, total phenols and COD were 55, 63 and 48% respectively [9].

4.2. *Sequencing batch reactor*

Shayegan et al. [10] used an *Aspergillus* species isolated from the soil for decolourization. Both maximum decolourization and COD reduction under optimized conditions were obtained as 84%. A pilot scale study with an activated sludge system for decolourization was conducted using optimum values. The decolourization efficiency of this system with 48 h retention time was about 75% in a limited period of 4 days. Another study was done with SBR. This system with a cycle time of about 4.5 hrs had decolourization efficiency of 70% extended for 12 days, which was more than previous reported works. The colour of undiluted wastewater reduced by only 40% with a cycle time of 17.5 hrs and the system was stable for about 4 days.

5. Study on a fungal batch reactor

Fungal batch reactor is being used by authors to decolourize the anaerobically digested spentwash (Figure 4). The study is done after isolation of microorganism and optimization of nutrients.

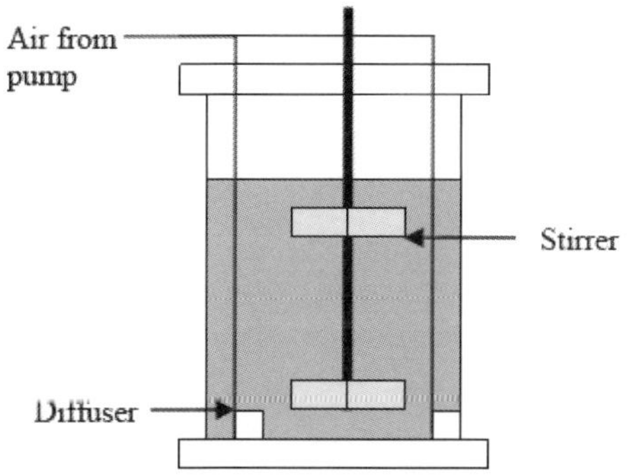

Figure 4. Fungal bioreactor setup.

5.1. *Microorganism*

Aspergillus niger species IITB-V8 was isolated from the sludge of lagoons of the distillery for highest decolourization [11]. This species was maintained on petri dish and slant, prepared in potato dextrose agar (PDA).

5.2. *Optimization of nutrients*

Optimization of nutrients was done using statistical methods. The important parameters selected by screening experiments were glucose, KH_2PO_4 and pH. The optimum values obtained using response surface methodology to achieve maximum decolourization were 5.5 g/L glucose, 1.2 g/L KH_2PO_4 and pH 5 [12].

5.3. *Decolourization*

Maximum decolourization of the effluent in the fungal bioreactor was 70%. This decolourization was attained in 3 days (Figure 5). The maximum decolourization coincides with the maximum growth of fungal biomass.

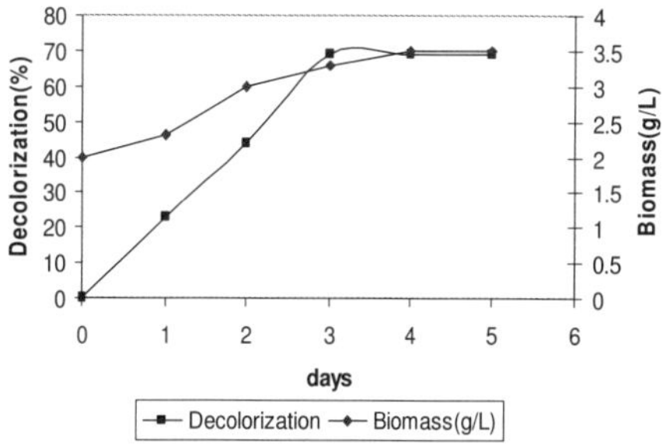

Figure 5. Decolourization by *Aspergillus niger:* changes in biomass in the bioreactor.

6. Conclusion

In the recent years, interest is being directed towards utilizing microbial activity (pure bacteria and fungi) for the decol ourization and mineralization of spent wash. The colour imparting melanoidins are barely affected by conventional biological treatment. Several bacterial and fungal reactors have been used in the decolourization of wastewaters from distilleries. Decolourization of distillery Spent Wash can be effectively done in these reactors. The limitation of these systems is the contamination of reactor, which can lead to lower efficiency of these reactors. Further research is required in this regard.

References

1. http://petroleum.nic.in/gaznew.pdf.
2. S.S. Singh and A.K. Dikshit, *IJEP.* **28(3)**, 200(2008).
3. A.C. Wilkie, K.J. Riedesel and J.M. Owens, *Biomass Bioenerg.***19 (2)**, 63(2000).
4. G.R. Pathade, *Advances in Industrial Wastewater Treatment, Technoscience Publications*, Rajasthan, India, 180(1999).
5. M. Petruccioli, J.C. Duarte and F. Fedrerici, *J. Biosci. Bioeng.* **90**, 381(2000).
6. M. Ghosh, A. Gangul and A.K. Tripathi, *Process Biochem.* **37 (8)**, 857(2002).
7. K.L. Norcross, *Water Sci. Technol.* **26**, 9(1992).
8. S. Sirianuntapiboon and K. Prasertsong, *Bioresour. Technol.***99**, 1806(2008).

9. C. Guimaraes, P. Porto, R. Oliveira and M. Mota, *Process Biochem.* **40**, 535(2005).
10. J. Shayegan, M. Pazouki and A. Afshari, *Process Biochem.* **40**, 1323(2005).
11. M.S. Chauhan, PhD Thesis, CESE, IIT Bombay, Mumbai, India, 2008.
12. S.S. Singh and A.K. Dikshit, *J. Hazard. Mat.* (Under review).

APPLICATION OF NOISE MAPPING ON ENVIRONMENTAL MANAGEMENT

KANG-TING TSAI
Graduate Institute of Rural Planning, National Chung Hsing University, Taichung, Taiwan, Republic of China

MIN-DER LIN
Department of Environmental Engineering, National Chung Hsing University, Taichung, Taiwan, Republic of China

YEN-HUA CHEN
Department of Environmental Engineering, National Chung Hsing University, Taichung, Taiwan, Republic of China

YU-HAO LIN
Department of Environmental Engineering, National Chung Hsing University, Taichung, Taiwan, Republic of China

Spatial noise characteristics of Tainan City in Taiwan were investigated in this study by noise mapping. Noise data were collected from 345 noise monitoring stations. The spatial distributions of the noise levels during different time interval were evaluated and visualized by geographic information systems. Noise standard violations as high as 23dB(A) were frequently observed on summer evenings. Over 90% of the Tainan City population are exposed to unacceptable noise as defined by US Department of Housing and Urban Development. It can be concluded that noise mapping is useful for environment management of urban noise.

Keywords: Noise mapping, noise exposure, geographic information system, Tainan City.

1. Introduction

Noise disturbance is a major environmental complaint in residential areas, especially areas with high population density, and significantly affects the inhabitants in their daily life, sleep, work and study. Proposing noise control policies has become an emergent task in many metropolitan areas.

To propose noise control policies, noise mapping is necessary to survey existing distributions of noise levels, identify primary noise sources and examine noise level regulations. This study utilized geographic information systems (GIS) to analyze noise data collected at 345 monitoring stations in Tainan City, Taiwan. Noise maps were produced to investigate its current noise distribution. Noise exposure levels were also investigated.

2. Developments of noise mapping

The concept of noise mapping has been in development for many years. The most advanced research in noise mapping has been performed in European countries. For example, Germany has conducted relevant research for more than 25 years. Based on the results of previous studies, traffic noise is frequently identified the main noise source. Urban areas near busy road systems are usually selected for initial implementation of mapping systems needed to draft noise control schemes. In 2002, the European Union (EU) passed Directive 2002/49/EC [1] for the Assessment and Management of Environmental Noise, which stipulated that, by July, 2008, all EU members must draw up noise for the main areas of population aggregation and primary road systems for public reference. England then published the London Noise Map in 2004, which is the first noise map produced by a national government, as a reference for London citizens to avoid noise nuisances.

Compared to other environmental issues, noise mapping is a newly emerging issue which is drawing substantial public attention. Steele (2001) argued that researchers conducting Environmental Impact Statements (EIS) should consider not only noise levels, but also other potential factors such as population distribution and surrounding land use [2]. de Kluijver and

Stoter (2003) illustrated and underlined the essence of standardized noise mapping tools, and considerations for the development of these tools, sun as accuracy; cost-benefit criteria for noise measures; and presentation of uncertainties in results, were also described. Furthermore, they concluded that an appropriate use of GIS in mapping noise effects makes it possible to optimize quality and efficiency of noise effect studies [3]. Klæboe *et al.* (2004) investigated cognition of the citizens in two Norwegian cities to different equivalent sound level through questionnaire [4]. The UK Building Research Establishment (BRE) measured environmental noise levels and social surveys of population attitudes toward environmental noise in the UK during 1999-2001. Comparison of these more recent results with those of a similar BRE study in 1990/1991 indicated that, in England and Wales, the proportion exposed above the noise levels of 55dB has decreased between 1990 and 2000. However, the majority of the population of the UK still lives in dwellings exposed to noise levels above those recommended by the WHO Guidelines for Community Noise [5]. Based on the Norwegian socio-environmental survey database and a comprehensive national noise mapping effort, Klæboe *et al.* (2006) developed noise impact maps to provide experts, politicians and the public with high-level impact visualizations of condensed status, "what-if" and scenario information for citywide areas in an easily accessible format. The context sensitive noise impact modeling improves annoyance predictions for individual dwellings and visualizes these impacts in the form of neighborhood soundscape quality areas [6]. Environmental impact assessments measuring daily noise exposure to urban residents have revealed that transportation noise is a major source of noise exposure [7, 8]. Lee *et al.* (2008) proposed the concept of noise impact assessment and developed a noise mapping tool for analyzing how complex acoustic phenomena arise due to the complex topographic configurations of surrounding buildings and terrain. They concluded that noise maps can help determine the main impact of noise sources and are extremely useful for detailed assessment of noise impact influences [9].

In Taiwan, only three administrative districts in Taipei City have been selected for trial establishment of noise maps in 2007. Large-scale noise maps have yet to be established and investigated. Therefore, this study is the first in Taiwan to undertake large-scale measurement of urban noise and to draw up noise maps depicting the time and spatial distribution of noise.

3. Methodology

3.1. *Introduction to the study area – Tainan City*

Tainan City located in southwest Taiwan has a total area of 175 square kilometers and is divided into seven administrative districts (East, Central, West, South, North, Anping and An-Nan districts). It has a population of about 750,000 which meets the EU Noise Policy requirement of "an area with a population over two hundred thousand." Figure 1 shows the geographic location of Tainan City.

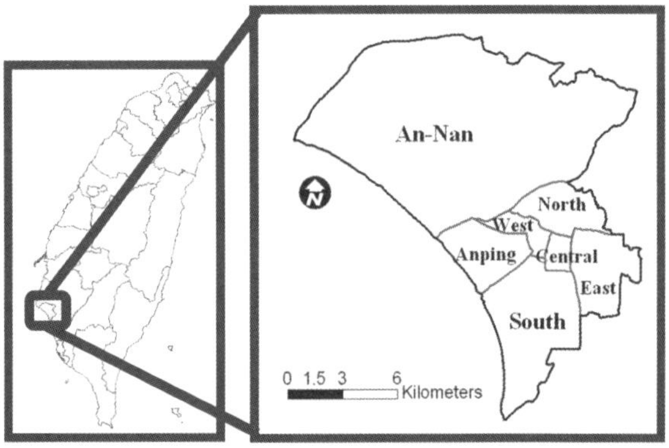

Fig. 1: Geographic location of Tainan City.

This study established 345 temporary noise monitoring stations in Tainan City to investigate noise levels during the morning (08:00 ~ 10:00), afternoon (14:00 ~ 16:00) and evening (20:00 ~ 22:00) in summer and winter. To select the locations of measuring points, the Tainan metropolitan area was divided into standard grids with size of 500m × 500m. One temporary measuring booth was setup within each grid. The instruments employed for noise measurement were the precision integrating sound level meters manufactured by Bruel & Kjaer in Denmark and Rion in Taiwan. The microphone was supported 1.5m above the ground along roads and enclosed in containers to minimize noise generated by wind conditions. To measure actual noise generated in urban environment during times of major activity, the measurements were performed at each monitoring station over two-hour periods. Before starting the measurements, all instruments were calibrated using a B&K 4230 calibrator with a calibration value of 94dB(A) at 1KHz. A total of 2,070 noise data were collected. Data validity was examined, and measuring time errors were also calibrated. Figure 2(a) shows the spatial distribution of the monitoring stations.

Based on the monitoring data, this study analyzed and compared the spatial distribution of noise mapped during different time periods. The GIS tools were used to overlay the noise maps on population distribution maps to evaluate the noise exposure level of the population. The results can be used to justify the rationality of the noise regulation standards as well as to estimate the population exposed to excess noise environments.

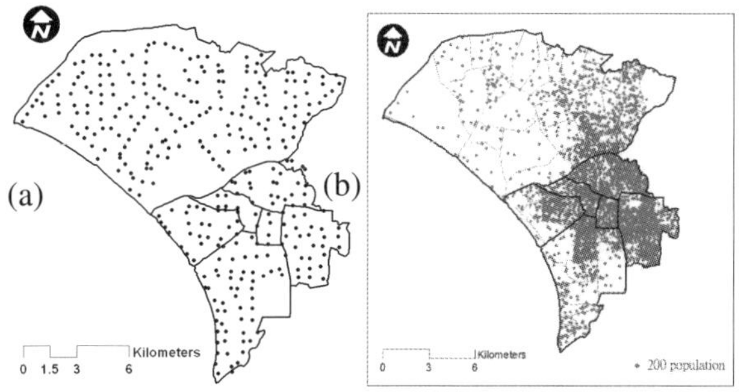

Fig. 2: Spatial distribution of (a) noise monitoring stations and (b) population.

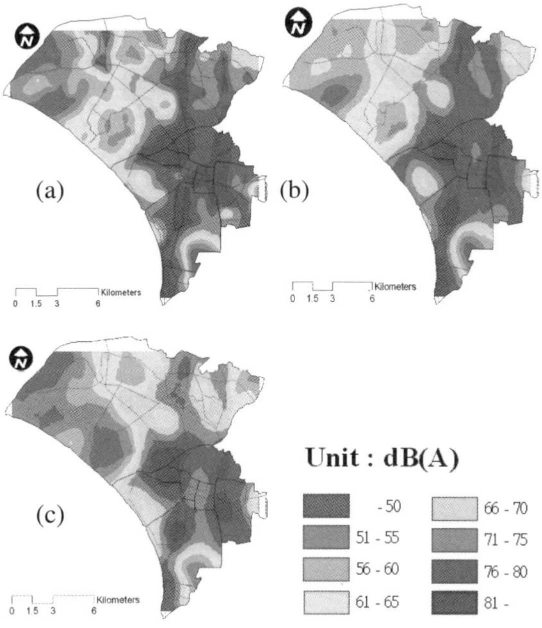

Fig. 3: Summer noise distribution in Tainan City: (a) morning; (b) afternoon; and (c) evening.

3.2. *Noise mapping*

In 1960, the Helsinki Conference published ISO/1996, which suggested the equivalent continuous sound level L_{eq} as an indicator of environmental noise. Later in 1978, ISO 1996-1 proposed the recognition of L_{eq} as an international standard [10]. Therefore, this study also adopted L_{eq} as a standard for measuring environmental noise.

Noise data in Tainan City for the mornings, afternoons and evenings were collected from each noise monitoring station during summer and winter. Noise maps were then constructed by using GIS to integrate the monitoring data with spatial information. Noise level contours were constructed by Kriging method, which was first proposed by D.G. Krige in 1951 and is now the most widely used geo-statistical interpolation method. The ordinary Kriging, one of the Kriging methods, was adopted in this study for spatial interpolation of noise data to measure noise distributions.

4. Methodology

4.1. *Noise maps of Tainan City*

The noise maps got Tainan City present the noise levels for six time intervals: mornings, afternoons and evenings during summer and winter. Figures 3 and 4 demonstrate spatial distributions of noise levels in summer and winter, respectively. The features were analyzed during the following periods:

1. Morning period (08:00 – 10:00):
Areas with high noise levels in the morning were observed in the Central, West and North Districts, which were downtown areas. Noise levels usually remained over 75dB(A) in summer and 70dB(A) in winter. The highest, average and lowest noise levels monitored in Tainan City during summer mornings were 91dB(A), 69.6dB(A) and 51dB(A), respectively. Noise levels during winter were between 80dB(A) to 45dB(A) and averaged 62.4dB(A).

2. Afternoon period (14:00 – 16:00)
High afternoon noise levels were observed in Middle, West and North Districts but were not as high as those in the morning. Areas along Provincial Highway No.1, which is the main road in Tainan City, were exposed to high noise levels. The highest noise level monitored during summer afternoons was 89dB(A) while the average and lowest were 69.3dB(A) and 40dB(A),

respectively. During winter afternoons, the highest, average and lowest noise levels were 79dB(A), 62.3dB(A) and 35dB(A), respectively.

3. Evening period (20:00 – 22:00)
The high noise areas in the evening were mainly distributed along arterial roads. The highest, average and lowest levels of noise monitored in summer evenings were 90dB(A), 66.3dB(A) and 41dB(A), respectively. Additionally, the highest noise level for winter was 80dB(A) while the average and lowest were 59.3dB(A) and 37dB(A), respectively.

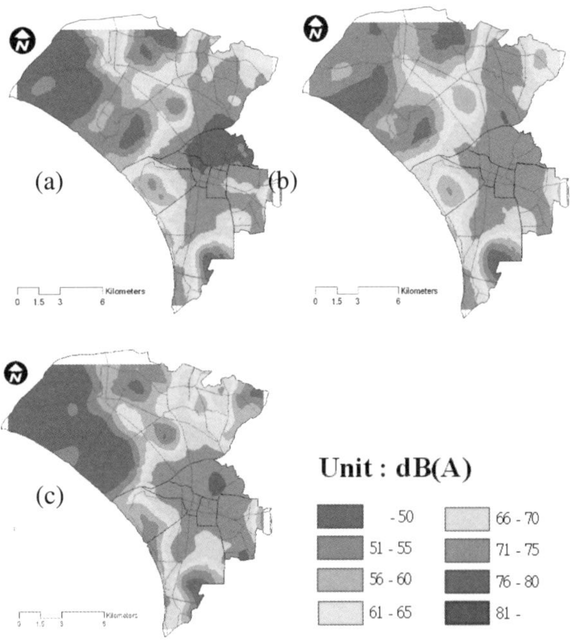

Fig. 4: Winter noise distribution in Tainan City: (a) morning; (b) afternoon; and (c) evening.

The above findings indicate that average noise levels during summer mornings, afternoons and evenings were approximately 7dB(A) higher than those in winter. This difference may have been due to the increased human activity during summer as well as the ecology in Tainan City, which is characterized by sounds generated by birds, insects and frogs.

Based on the aforementioned noise maps, the populations exposed to certain noise levels can be estimated. According to the classification defined by US Department of Housing and Urban Development (HUD), noise levels in residential environments are unacceptable when the L_{eq} is greater than 62dB(A)[11]. By overlapping the population distribution of Tainan City (as shown in Figure 2(b)) and the noise maps, it can be found that over 90% of the Tainan City population are exposed to unacceptable noise environment, and the exposed population in summer (about 97%) is larger than that in winter (about 90%). Such a result indicates the urgent need for improvement of noise pollution in Tainan City.

5. Conclusion

This study constructed noise maps for Tainan City in Taiwan to demonstrate the efficacy of noise mapping to investigate environmental noise. The analytical results revealed that Tainan City suffers serious noise pollution. More than 90% of the Tainan City population are exposed to unacceptable noise environments, indicating that strategies for improving noise control are urgently needed.

Results of this study demonstrate that noise maps are an effective means of understanding noise level distributions in investigated areas. Additionally, when compared with the current regulation standards, noise maps can also be used to identify the areas seriously violated the standards. By using this evaluation method, decision makers can quickly and easily identify the areas most urgently requiring noise improvement as well as reexamine the adequacy of current regulatory standards.

References

1. EU directive COM. Relating to the Assessment and Management of Environmental Noise. Directive (2002).
2. C. Steele, Critical review of some traffic noise prediction models, *Applied Acoustics, 62(3)*, 271-287 (2001).
3. H. de Kluijver, J. Stoter, Noise mapping and GIS: optimizing quality and efficiency of noise effect studies, *Computers, Environment and Urban Systems, 27(1)*, 85-102 (2003).
4. R. Klæboe, A. H. Amundsen, A. Fyhri, S. Solberg, Road traffic noise the relationship between noise exposure and noise annoyance in Norway, *Applied Acoustics, 65(9)*, 893-912 (2004).
5. C. J. Skinner, C. J. Grimwood, The UK noise climate 1990–2001: population exposure and attitudes to environmental noise, *Applied Acoustics, 66(2)*, 231-243 (2005).

6. R. Klæboe, E. Engelien, M. Steinnes, Context sensitive noise impact mapping, *Applied Acoustics, 67(7)*, 620-642 (2006).
7. M. Bonaiuto, F. Fornara, M. Bonnes, Indexes of perceived residential environment quality and neighbourhood attachment in urban environments: a confirmation study on the city of Rome, *Landscape Urban Plan, 65(1-2)*, 43-54 (2003).
8. D. S. Sawicki,, P. Flynn, Neighborhood indicators- a review of the literature and an assessment of conceptual and methodological issues, *J. Am. Plan. Assoc., 62(2)*, 165-83 (1996).
9. S.-W. Lee, S.-I. Chang, Y.-M. Park, Utilizing noise mapping for environmental impact assessment in a downtown redevelopment area of Seoul, Korea. *Applied Acoustics, 69(8)*, 704-714 (2008).
10. E.A.G. Shaw, N. Olson, Theory of steady-state urban noise for an ideal homogeneous city, *J. Acoust. Soc.Am., 51(6)*, 1781-1793 (1971).
11. P. H. T. Zannin, F. B. Diniz, W. A. Barbosa, Environmental noise pollution in the city of Curitiba, Brazil. *Applied Acoustics*, 63(4), 351-358 (2002).

INVESTIGATION ON SEASON EFFECTS ON EFFICIENCY OF INDIVIDUAL SELECTION SYSTEM AND PERFORMANCE RESPONSE IN SILKWORM*

ALIREZA SEIDAVI

Animal Science Department, Islamic Azad University, Rasht Branch, Rasht, Iran
alirezaseidavi@yahoo.com

SEYED ZIAEDDIN MIRHOSSEINI

Animal Science Department, Faculty of Agriculture, Guilan University, Rasht, Iran

MOEINODDIN MAVVAJPOUR

Iran Silkworm Research Center (ISRC), Rasht, Iran

ALIREZA BIZHANNIA

Iran Silkworm Research Center (ISRC), Rasht, Iran

MANI GHANIPOOR

Iran Silkworm Research Center (ISRC), Rasht, Iran

This experiment was conducted for investigation on selection effects silkworm at two spring and autumn seasons. Each line was reared in two selected and random groups. The superior individuals based on cocoon weight were selected as parents of selected group. Meanwhile for each line, individuals who had the cocoon weight averages equal to population average were selected as parents of random groups. Their records analyzed at four successive generations. From obtained results, it was showed lines at spring had higher response than autumn from point of most important traits. Thus breeding programs at pure lines can improve economical gains.

1. Introduction

Silkworm is well-known industrial insects, which produce natural fiber silk. Because of economic importance for silk yarn, an effort to breed new silkworm variety has been made for thousand years. By reviewing the development history of silkworm industry in recent decades and its current situation, it is not difficult to see that, with the sericultural production experiencing twists and turns, ups and downs, there are many problems waiting to be solved (Zhao et al., 2007). Accordingly, in order to realize the great goal of silkworm industry in the 21st century, we must strengthen the position of agriculture as the foundation of international economy, meet new challenges, comply with changes of situation, be initiative, and realize new turning of the industry. Silkworm varieties are the most important items for development of silk industry. They are the key factor determining quality of silkworm eggs. Meanwhile, they directly influence the initiative of silkworm raisers and the quality of cocoon and silk. Through years of efforts made by breeders of our research team, our breeding techniques have been improved a lot (Zhao et al., 2007).

There are approximately 2000 different races and breeds of *Bombyx mori* used in silk production (Reddy, 1986). Many characters of this species are recognized as contributing in silk yield quantitatively or qualitatively (Chatterjee et al., 1990). Because the economic importance of different traits is not similar, hence the main emphasis should be focused on the traits which have more value in economic efficiency of production system. The use of selection for genetic and economic improvement of traits in the shortest time will have the highest improvement (Harris, 1970).

On the other hand, the aim of silkworm breeding is to improve economic traits in order to increase the profit of silk producers and other sections of sericulture industry. In sericulture numerous traits are considered as

*This work is supported by Islamic Azad University, Rasht Branch, Rasht, Iran.

important. The importance of these traits in various parts of this industry is different. The reproductive traits are considered important for egg producers while cocoon producers are interested in improved production potential, cocoon shell percentage and disease resistance (Singh et al., 1998).

Selection is one of the best breeding approaches in silkworm breeding (Seidavi et al., 2008a). Individual selection in parent lines can improve their offspring performance. But, this improvement is affected to environmental factors specially season. Thus this experiment was conducted for estimate and comparing of individual selection effects in six commercial pure lines of silkworm including 31, 32, 103, 104, 107 and 110 at two spring and autumn seasons.

2. Materials and Methods

The present investigation has been performed on six silkworm commercial pure lines which were produced in Iran Silkworm Research Center including 31, 32, 103, 104, 107 and 110. Silkworm rearing technique included single batch rearing system. Feeding and other conditions of larval rearing were conducted following the standard procedure (ESCAP, 1993).

Totally, 27 economical traits were recorded and analyzed. Economical traits included resistance, production and reproduction categories. Resistance traits were alive larvae number, alive pupae number, pupae vitality percentage, pupae vitality percentage in best cocoon, and pupae vitality percentage in middle cocoon. Meanwhile, Production traits were produced cocoon number, best cocoon number, middle cocoon number, low cocoon number, double cocoon number, best cocoon percentage, middle cocoon percentage, low cocoon percentage, double cocoon percentage, best cocoon weight, double cocoon weight, single best cocoon weight, 10000 larvae cocoon weight, larval duration. Also reproduction traits were hatched larvae, un-hatched eggs, unfertilized eggs, hatched eggs percentage, un-hatched eggs percentage, unfertilized eggs percentage, hatchability percentage, and total produced eggs.

For investigation on season effects on efficiency of individual selection system and performance response in six silkworm pure lines, experiment was conducted in two seasons included spring and autumn. All six studied silkworm pure lines were reared during four generations. Each line was reared in two selected and random groups. Individual selection system based on cocoon weight was conducted as selected group in the zero generation. For each pure line, in zero generation, about 500 individuals recorded in each sex from point of cocoon weight trait and the one percentage of individuals in each sex which had the highest cocoon weight selected as parents of selected group at next generation. These individuals mated together and formed selected groups of first generations. Meanwhile for each pure line in zero generation, individuals who had the cocoon weight averages equal to population average were selected as random groups. Individuals and offspring in selected and random groups selected and mated randomly without any phonotypical and genotypic selection in first, second, third and fourth generations. Twelve full sib families from each group and pure line were reared in each generation.

Phenotypic change of resistance, productive and reproductive traits was recorded and evaluated in the studied pure lines at four successive generations collectively. In each pure line and generation, 12 families were reared so that 25 male and 25 female full sib cocoons were individually recorded for cocoon traits per family. Thus, there are 15552 family data included 4 generations × 6 pure lines × 2 treatments (random and selected groups) × 12 families (replications) × 27 family traits. These data recorded in two autumn and spring seasons. All data obtained from the experiment were subjected to ANOVA using the General Linear Model (GLM) procedures of the Statistical Analysis System Institute (SAS, 1998). Treatment means were compared by Duncan New Multiple Range Test (DNMRT).

3. Results and Discussion

From obtained results, it was showed lines at spring had higher response than autumn from point of most important traits. For resistance characters, responses to selection for alive larvae number were 23.08 and 19.72 at spring and autumn respectively. Responses to selection for pupae vitality percentage in best cocoon were -0.304 and -1.486% at spring and autumn respectively. Responses to selection for pupae vitality percentage in middle cocoon were 0.074 and -2.627 at spring and autumn respectively. Thus it is concluded responses to selection for important resistance traits in silkworm pure lines at spring is higher than autumn significantly ($P<0.05$). Thus, commercial pure lines showed higher phenotypic response in spring rearing period compared to autumn one for production traits.

Positive correlations show that as breeders change the mean of one character towards the higher side, the other also goes up with it, while in the negative as the mean value of one character goes up, the value for the other character goes down. In silkworm, most of the correlations between different characters have been worked out by selection experiments. These studies have shown that direct selection for one trait has correlation with the other quantitative characters. The correlation of some characters was found to be negative while in others it was positive (Nagaraju, 2002).

Responses to selection for produced cocoon number were 23.18 and 18.78 at spring and autumn respectively. Responses to selection for best cocoon number were 14.61 and 6.04 at spring and autumn respectively. Responses to selection for middle cocoon number as a negative character was 7.933 and 11.217 at spring and autumn respectively. Responses to selection for low cocoon number as another negative trait was 0.633 and 2.217 at spring and autumn respectively ($P<0.05$). Responses to selection for double cocoon number as a negative trait was 0.0000 and -0.6957 at spring and autumn respectively ($P<0.05$). Responses to selection for low cocoon percentage as a negative trait were -0.0723 and 0.4141% at spring and autumn respectively. Responses to selection for best cocoon weight were 34.21 and 24.13 g at spring and autumn respectively. Responses to selection for single best cocoon weight were 0.0471 and -0.1688 at spring and autumn respectively. Responses to selection for 10000 larvae cocoon weight were 326.1 and 102.5 g at spring and autumn respectively ($P<0.05$). Responses to selection for larval duration were 0.0417 and -2.087 hours at spring and autumn respectively ($P<0.05$).

Individual selection based on cocoon weight had not negative effect on economical and important reproduction traits. Selection response of un-hatched eggs percentage was 0.882 and 1.468% respectively for spring and autumn rearing seasons. Responses to selection for un-hatched eggs percentage were -0.3058 and -0.5891% at spring and autumn respectively. Responses to selection for unfertilized eggs percentage were -0.5770 and -0.8798% at spring and autumn respectively ($P<0.05$).

Results demonstrated that there is a positive genetic correlation between cocoon weight and other economical traits which was in accordance with other studies for our pure lines (Seidavi et al., 2007; Seidavi et al., 2008b; Mirhosseini et al., 2008). Similar results report by other researchers such as He (1998a) and He (1998b). On the other hand, Increasing of selection pressure on cocoon weight will cause total produced eggs to rise. Environmental conditions in autumn season decrease egg lying of moths, significantly. Identification of genetic relationship among productive and reproductive traits is necessary before constructing an effective breeding program.

Meanwhile, heritability have important role in obtained results. Ksham et al. (1995) have reported that the heritability for quantitative traits was comparatively higher and ranged from 48 to 64% with little tendency for the standard error to increase with the estimate, while the range of heritability for fitness traits was much lower (18-25%). The data supported the classical hypothesis that the fitness traits would exhibit lower heritabilities compared to the traits in other categories. Bhargava et al. (1993) have reported that heritability was very high (71.4-86.7%) for larval duration, single-shell weight, filament length, larval weight and single-cocoon weight. Moderate heritability (<70%) was observed for cocoon yield (65.3%) and shell ratio (69.8%), indicating that these 2 characters were influenced by the environment. Malik et al. (1999) and Singh et al. (1998) have reported similar results. Ashoka and Govindan (1990) have showed high heritability as well as genetic gain for cocoon weight and shell weight indicating additive gene action and moderate to high heritability and low genetic gain for shell percentage, implying that this may be predominantly under control of non additive gene action. Rangaiah et al. (1995) indicated high heritability for shell percentage and shell weight while moderate along with high genetic advance for cocoon weight.

Silkworm breeders have uncovered the fact that some characters or aspects of them affect other characters. Each of these characters has its own optimal mean value which is commensurate with the adaptive fitness of the strain (Nagaraju, 2002). The interrelationships between characters are expressed in statistical terms, as correlations, which show how one variable changes as the other changes (Nagaraju, 2002).

As conclusion it is concluded responses to selection for production traits in silkworm pure lines at spring is higher than spring significantly. Also the main reason for improvement of many traits refers to positive correlation between production and other important traits at silkworm opposite some another animals. Thus breeding programs and individual selection for cocoon weight at pure lines can improve total benefit and economical gains.

Acknowledgments

This work was supported by the Islamic Azad University, Rasht Branch, Iran. The authors also acknowledge the kind advice of Mr. Yosef Kheirkhah for valuable comments and Mrs. Khairolnesa Taieb Naeemi, and Mr. Bizhang Jelveh regarding their helpful assistances.

References

1. J. Ashoka, and R. Govindan, "Genetic estimates for quantitative traits in bivoltine silkworm, *Bombyx mori* L," *Mysore Journal of Agricultural Sciences* **24**, 371-374, (1990).
2. SK. Bhargava, V. Thiagarajan, M. Ramesh Babu, and B. Nagaraj, "Heritability of quantitative characters in silkworm (*Bombyx mori* L.)," *Indian Journal of Agricultural Sciences*, **63**, 358-362, (1993).
3. SN. Chaatterjee, CS. Nagaraju, and K. Giridhar, "An approach to silkworm breeding. In workshop on biometrical genetics proceedings" (Ed. By R. K. Datta), *Central Sericultural Research and Training Institute, Mysore*, pp. 11-16, (1990).
4. Escap, "Principles and techniques of silkworm breeding," *New York, United Nations*, (1993).
5. PL. Harris, "Breeding for efficiency in livestock production: Defining the economic objectives". *Journal of Animal Sciences*, **30**, 860-865, (1970).
6. SM. He, "Breeding of summer-autumn using silkworm varieties 415 and 416 and the preparation of their hybrids," *Acta Sericologica Sinica,* **24(4)**, 215-220, (1998a).
7. SM. He, "Breeding of vigorous and hypersilkgeneous silkworm varieties 871 and 872," *Acta Sericologica Sinica,* **24(2)**, 86-90, (1998b).
8. G. Ksham, SN. Kumar, S. Nair, and RK. Datta, "Heritability, genetic and phenotypic correlation studies on fitness and quantitative traits of bivoltine silkworm *Bombyx mori* L," *Indian Journal of Sericulture*, **34**, 22-27, (1995).
9. GN. Malik, MA. Masoodi, AS. Kamili, and M. Aijaz, "Estimation of direct selection parameters in a diallel set of bivoltine silkworm, *Bombyx mori* L," *Entomology*, **24**, 253-257, (1999).
10. SZ. Mirhosseini, AR. Seidavi, M. Ghanipoor, and AR. Bizhannia, "Phenotypical improvement for resistance of Iranian silkworm lines against grasserie disease through one generation of selection," *Journal of Daneshe Keshavarzi*, **18(2)**, 151-159, (2008).
11. L. Nagaraju, "Application of genetic principles for improving silk production. *Current Science*. **83(4)**, 409-414, (2002).
12. S. Rangaiah, R. Govindan, MC. Devaiah, and TK. Narayanaswamy, "Genetic studies for some quantitative traits among multivoltine races of silkworm, *Bombyx mori* L," *Mysore Journal of Agricultural Sciences*, **29**, 248-251, (1995).
13. GS. Reddy, "Genetics and breeding of silkworm, *Bombyx mori* L. In Lectures on Sericulture" (Ed. By Boraiah G.), *Suramaya Publishers, Bangalore, India,* 70-80, (1986).
14. SAS, "SAS/STAT User's Guide," Version 6.12. SAS Inst. Inc., Cary, NC, USA, (1998).
15. AR. Seidavi, SZ. Mirhosseini, AR. Bizhannia, and M. Ghanipoor, "Investigation on selection efficiency for some quantitative cocoon characters at 3P lines and its correlation with reproduction and resistance against diseases parameters of hybrids (F1) silkworm," *Iranian Journal of Biology*, **20(3)**, 262-268, (2007).
16. AR. Seidavi, SZ. Mirhosseini, AR. Bizhannia, and M. Ghanipoor, "Effect of parents selection on the basis of cocoon weight on production and resistance characters at lines of silkworm *Bombyx mori* L," *Journal of Pajouhesh-va-Sazandegi*, **21(1)**, 95-102, (2008b).
17. AR. Seidavi, AR. Bizhannia, M. Ghanipoor, and AA. Qotbi, "Investigation on improvement possibility of resistance, production and reproduction traits in 3P, 2P and P generations in three Japanese pure lines of silkworm *Bombyx mori* L. using individual selection in 3P generation," *Asian Journal of Animal and Veterinary Advances*, **3(6)**, 443-447, (2008a).
18. T. Singh, K. Chandrasekharaiah, and MV. Samson, "Correlation and heritability analysis in the silkworm, *Bombyx mori* L," *Sericologia*, **38**, 1-13, (1998).

19. Singh. ST, K. Chandrasekharaiah, and MV. Samson, "Correlation and heritability analysis in the silkworm, *Bombyx mori* L," *Sericologia*, **38**, 1-13, (1998).

20. Y. Zhao, K. Chen1, and S. He, "Key principles for breeding spring-and-autumn using silkworm varieties: from our experience of breeding 873×874," *Caspian Journal of Environmental Sci*ence, **5(1)**, 57-61, (2007).

STABILIZATION OF HEAVY METALS IN SEWAGE SLUDGE USING SOREL CEMENT

Jianli Ma[1,2]
1. The State Key Laboratory of Pollution Control and Resource Reuse, Tongji University, Siping Road 1239, Shanghai 200092, China

2. Tianjin Academy of Environmental Science, Fukang Road 17, NanKai District, Tianjin 300191, China

Jinmei Wang
School of Environmental Science and Engineering, Hebei University of Science and Technology, Yuhua East Road 70, Shijiazhuang

Youcai Zhao, Tao Liu and Li Wang
School of Environmental Science and Engineering, Tongji University, Siping Road 1239, Shanghai 200092, China

Excessive heavy metals contents in sewage sludge maybe hinder the recycling of the sludge and should be mitigated effectively using engineered processes such stabilization as reported. In this work, Sorel cement is used as additive to stablize heavy metals present in sewage sludge. It was found that the contents of Zn, Cu, As, Cr, and Pb in sewage sludge were stabilized as high as 96%, 87%, 84%, 99% and 92%, respectively in the leachability test. The sequential extraction procedure tests showed that the increase of immobile fractions of heavy metals in mixtures, e.g. organic or residual forms, inhibited the transferability of heavy metals to the solution. In addition, the SEM photos of the stabilized microstructures proved that hydration of Sorel cement was changed with various ratios. Therefore, the increase of pH towards highly alkaline value and the formation of ternary hydration in the mixtures appear to have developed to stable heavy metals in sewage sludge.

Keywords: Sewage sludge; stabilization; sorel cement; heavy metals; leachability.

1. Introduction

Sewage sludge, which is obtained by processing sewage, represents a heterogeneous and diverse material of naturally occurring organic and mineral substances, many kinds of elements, even some pathogenic bacteria [1,2]. Among them, heavy metals are undoubtedly the most crucial and at the same time the most controversial ones as those responsible for pollution [3]. Due to the increasing amount of sewage sludge in China [4], proper disposal of sewage sludge avoiding harm to environment has been one of the most critical environmental issues today [5,6].

According to Little (1991), most of heavy metals can be stabilized in sludge due to pH increase and formation of less soluble metal compounds [7]. In this way the previous alkaline stabilizers, such as Portland cement, lime, fly ash, cement kiln dust or combinations of these materials, are considered as the most common amendment materials for sewage sludge stabilization [8-10]. Significant role in the stabilization process is attributed to the precipitation of metals in pozzolanic reaction, and the binding of metals to minerals by sorption, mixing or substitution [11,12]. For example, the cement-based stabilization retrains the inorganic species mobility through the formation of calcium hydroxide and the hydration of colloidal C-S-H (C: CaO, S: SiO_2, H: H_2O) gel [13]. It is the result of chemical changes in contaminants and their environment (pH>10) that causes the contaminants to be less mobile or less toxic [14].

In this study, Sorel cement would be investigated as an alternative cementitious media more efficiently in the waste immobilization field. Sorel cement is hydraulic cement first produced by Stanislas Sorel in 1867 [15]. Its composition differs greatly from that of common cements. Sorel cement is a combination of magnesia

(MgO) and magnesium chloride ($MgCl_2$). Because of its high alkalinity, MgO has become a feasible option for heavily contaminated sludge [16]. In addition, $MgCl_2$ is known as strong electrolyte and recommended for dissolving selectively the mobile fractions (e.g. exchangeable and carbonate) of heavy metals in the sediments [17]. More importantly, MgO and $MgCl_2$, can react with water from sludge to form hydrations which involve the combined action of additive reducing these metals to their unstable forms [18]. These hydrations of Sorel cement are generated in the ternary gelatinization system of $MgO-MgCl_2-H_2O$ as below [19].

$$5MgO + Mg^{2+} + 2Cl^- + 13H_2O \rightarrow 5Mg(OH)_2 \cdot MgCl_2 \cdot 8H_2O \quad (1)$$

$$3MgO + Mg^{2+} + 2Cl^- + 11H_2O \rightarrow 3Mg(OH)_2 \cdot MgCl_2 \cdot 8H_2O \quad (2)$$

Because of its hydration action and special physical and chemical properties of MgO and $MgCl_2$, Sorel cement is especially impervious to water [20], which, in turn, has led to its consideration for stabilization of hazardous substance in sewage sludge.

Jakubus et al. (2001) have reported that the hazard connected with heavy metals results from not only their total contents, even those exceeding admissible standards, but the quantitative distribution in the fractions easily transferred to the solution [21]. In this sense, the objective of this study is to study the behavior of Sorel cement as well as investigate the fractions of heavy metals which is transported in or bound to sewage sludge during the stabilization of sewage sludge.

2. Materials and methods

2.1. *Materials*

The sewage sludge was obtained from Bailonggang Sewage Treatment Plant (BSTP) which is the largest wastewater treatment plants in Shanghai China. Approximately 2,300,000 m^3 day^{-1} of municipal wastewater from parts of urban and industrial areas of Shanghai are subjected to BSTP, producing about 2040 tonnes day^{-1} of dewatered sewage sludge. The physical and chemical properties of sewage sludge are shown in Table 1. The sludge contains a relatively higher concentration of zinc, copper, chromium, lead and arsenic, but a lower concentration of cadmium and nickel (Table 2). So it is called as chemical sludge (CS).

Table 1. Characteristics of sewage sludge used.

Sludge property	Chemical sludge(CS)
Moisture content (%)	85.4
pH	5.82
Total organic carbon (%)	23
Coefficient of permeability (cm s^{-1})	1.2×10^{-8}
Chemical analysis of chemical sludge	**Values (%)[a]**
Silica (SiO_2)	24.52
Lime (CaO)	16.33
Alumina (Al_2O_3)	35.65
Magnesia (MgO)	2.84
Iron oxide (Fe_2O_3)	16.19
Loss on ignition[b]	9.67

[a]Dry sludge ash
[b]Loss on ignition at 1100°C

Table 2. Mean heavy metal concentration in chemical sludge used (mg kg^{-1} d.s.).

	Heavy metals [a]						MPV [c]	
	1# [b]	2# [b]	3# [b]	4# [b]	5# [b]	6# [b]	pH<6.5	pH≥6.5
Zn	3500	3066	4119	3983	3247	3356	2000	3000
Cu	1984	1903	2033	2306	2149	2083	800	1500
Cr	320	498	533	842	672	439	600	1000
Cd	1.5	2.0	ND	4.4	3.1	ND	5	20
Pb	337	421	790	537	340	312	300	1000
Ni	24.8	21.9	32.6	18.5	ND	ND	100	200
As	107	224	194	286	137	168	75	75

d.s. - Dry sludge; ND - not detected;
[a] Significant different at $p < 0.05$;
[b]Sampling Date; 1#: January 04; 2#: March 11; 3#: May 14; 4#: July 01; 5#: December 05; 6#: November 18;
[c]Maximum permissible values in sewage sludge were limited for agricultural application in China (GB18918-2002).

Magnesia (MgO) and Magnesium chloride ($MgCl_2$) were supplied by Jinshan Chemical Co., Ltd. in Shanghai. Those are current by-products of Mg processing. MgO is main components of Sorel cement, with which a purity of 85%, pH value of 10.8, bulk density of 0.35 g cm^{-3}, Sieve Residue (200 mesh) of 0.05, Ca<0.05%, Fe<0.005%, and Pb<0.004%. $MgCl_2$, with a high hygroscopic adsorption and deliquescencing effect, can be used as excitant in Sorel cement [15].

2.2. *Methods*

The study was conducted on sample 4# (Table 2) and the mixing ratios of materials were shown in Table 3. The experimental procedure as follow: Firstly, $MgCl_2$ was added in CS with centrifugal swinging for 5 minutes.

Then MgO was added and dispersed in CS at higher speed (800 r min^{-1}) in the mixer. The mixtures were solidified in steel molds with a size of 40×40×160 mm for 24 hours, and the resultant products were extruded and cured in airtight condition of 20°Cfor 28 days.

Table 3. Prepared mixtures for stabilizing (% w/w).

Samples	Sorel cement		Sludge
	MgO	*$MgCl_2$*	
CS0	0	0	100
CS1	1	2	97
CS2	4	2	94
CS3	7	2	91
CS4	10	2	88
CS5	13	2	85
CS6	16	2	82

The compression strength of the mixtures was tested using the unconfined compression machine with a maximum load of 5 kN and the loading rate of 1.52 mm/min. SEM imaging of samples were performed with a JSM-5600 LV scanning electron microscope on the condition of 15kV of acceleration voltage. The Toxicity Characteristics Leaching Procedure (TCLP, EPASW-846) was used to evaluate Leachability of mixtures. The heavy metals leaching from the samples were analyzed by ICP-AES (Plasma 400).

Different methods had been proposed for metal speciation but the method developed by Tessier et al. (1979) [17] was the one widely used for the purpose [22] and was used in this work. 1.00 g of dry sludge was treated with 1 mol MgCl2 for 1 h at room temperature and at pH 7.0 extracted the exchangeable phase; 1 mol NaOAc was added and reacted for 5 h at room temperature and adjusted by HOAc to pH 5.0 to extract the carbonate phase. Using 0.175 mol (NH4)2C2O4 and 0.1 mol H2C2O4 for 4 h at 96°C, the fraction bound to reducible phase was extracted. To extract the organic fraction 0.02 mol HNO3 and 30% w/v H2O2 was used for 2 h at 85°C and pH 2.0, followed by the addition of 3.2 mol CH3COONH4 in 20% w/v HNO3 with 30% w/v H2O2 for 3 h at 85°C. The final residue was digested with 10mL of aqua regia. Between each extraction, the samples were centrifuged (5000 rpm for 30 min) and filtered. The filtrates were used for the determination of heavy metal concentration by ICP-AES (Plasma 400).

3. Results

3.1. *Compressive strengths of mixtures*

In Figure 1, comparing the results with the raw sewage sludge, it is evident that the development of compressive strength has a relation with stabilization time and may be affected by the molar ratios of MgO and $MgCl_2$ in Sorel cement. It was found that the compressive strength of mixtures were raised at first 7-days stabilization time and trended to flat after 10 days approximately. The highest value of compressive strength presented when the content of Sorel cement was added as high as 12% in the mixtures and the ratio of MgO and $MgCl_2$ was 5:1. It appears that the appropriate dosage of Sorel cement for efficient sludge stabilization has to be determined based on a case study, taking into account the particular ratios of the corresponding Sorel cement.

3.2. *Toxicity leaching studies*

In order to investigate the influence of Sorel cement on heavy metals of CS, toxic property of samples were evaluated according to leaching method in USA. The TCLP results (Table 4) allow us to estimate the toxicity level taking into account the corresponding limits. Although the leachate concentration of metals in some final products, e.g. CS0, CS1 and CS2, may not reach U.S. Environmental Protection Agency regulations ($Cr < 5$ mg L^{-1}, $Pb < 5$ mg L^{-1}, and $As < 5$ mg L^{-1}), the reduction of the metal concentration in the leachates can be effective when the addition of Sorel cement ratios is higher than 9%. The comparative metal concentration of raw CS and mixtures showed us that metal uptake was highest in CS4 and specifically 96% of zinc, 87% of copper and 84% of arsenic, 99% of chromium and 92% of lead were reduced under the reaction of Sorel cement in CS.

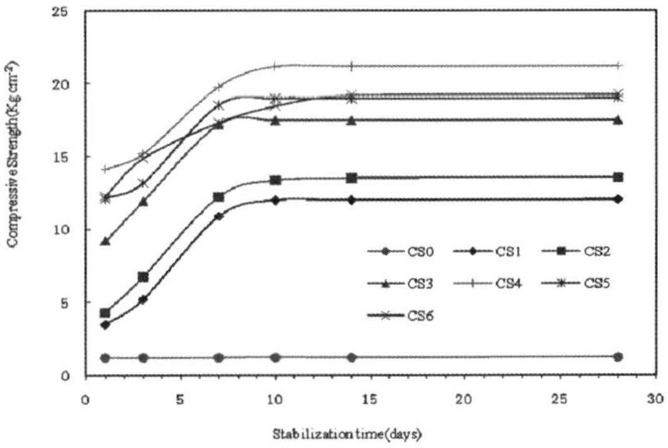

Figure 1. Compressive strengths of different mixtures curing in air for 28 days.

Table 4. Results of toxicity evaluation (TCLP) for the raw CS and mixtures (mg L^{-1}).

	Raw CS	Final products after 28 days						
		CS0	*CS1*	*CS2*	*CS3*	*CS4*	*CS5*	*CS6*
Zn	1174*	1172*	243**	132**	56*	47**	48**	69
Cu	284.3*	282.0**	77*	72*	53**	38**	39**	40**
Cr	47.2	47.2**	5.22**	3.13**	2.47**	1.71**	1.53**	1.64**
Pb	56.0*	55.7**	19.83*	13.42**	5.69*	4.43**	4.43**	4.59*
As	12.3*	11.4**	7.29*	5.74*	3.38**	1.97*	1.99*	2.03

$^*p < 0.05$; $^{**}p < 0.01$

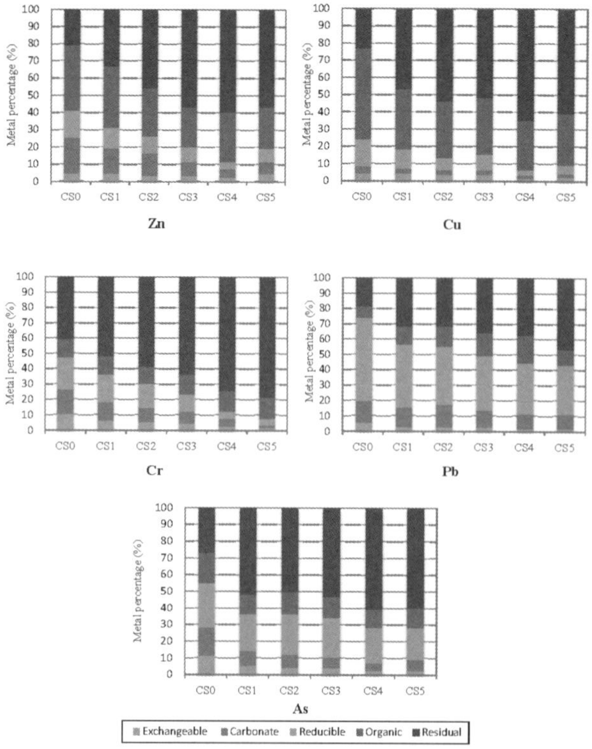

Figure 2. Fraction of Zn, Cu, Cr, Pb and As expressed as percentage of total amount after 28 days.

3.3. *Characterization by sequential extraction procedure*

The metal partitioning of samples, as determined by the sequential extraction, is shown in Figure 2. It may be observed that the increase of immobile fractions of heavy metals in mixtu forms, inhibits the transferability of heavy metals to the solution [23]. The percentages of fractions of heavy metals in CS4 vary apparently with the stabilization time. specifically a significant ($p<0.05$) amount of zinc and copper, 89% and 94% respectively, are bound to the organic and residual fractions, but Almost 79% of chromium is bound in the residual fraction. Lead and arsenic were found to be bound in the reducible and residual fractions, about 79% for lead and 82% for arsenic.

As observed in Table 5, Sorel cement may play an important role in exchanging metal fractions in CS. The metal variations for the different fractions were found to be res, e.g. reducible, organic or residual

Exchangeable: Cr> Zn>Cu>As> Pb;

Carbonate: Zn >Cu >As > Pb> Cr;

Reducible: Zn>Cu> Pb > Cr >As;

Organic: Cu> Zn> Cr >As> Pb;

Residual: Zn>Cu> Cr> Pb >As.

It is also clearly seen that with an increasing amount of residual fraction, the concentration of other metal fractions in the stabilized samples decreases correspondingly (Table 5). A substantial content of zinc is removed from the exchangeable, reducible, carbonate, reducible and organic fractions and 1553 mg kg^{-1} transferred to the residual fraction. Approximately, 968 mg kg^{-1} of the copper and 152 mg kg^{-1} of the chromium was found to be bound in the residual fraction in the final product. The Reducible fraction of lead changed to

Table 5. Values of metal fraction (mg kg^{-1}) compare of CS0 and CS4 studied at 28 days*.

Fractions	Zn	Cu	Cr	Pb	As
Exchangeable	-80	-69	-127	-21	-26
Carbonate	-637	-46	-16	-21	-28
Reducible	-478	-300	-59	-124	-23
Organic	-358	-553	+50	+16	-17
Residual	+1553	+968	+152	+150	+94

$^*p < 0.05$

other fractions, especially residual and organic fractions, the contents of these two fractions increased 150 mg kg^{-1} and 16 mg kg^{-1} respectively. Arsenic seemed to be transformed from the carbonate fraction, the reducible fraction and organic fraction to the residual fraction and 94 mg kg^{-1} of the residual fraction increased.

3.4. *Scanning electron microscopy (SEM) Analysis*

The representative samples were observed using SEM, in order to provide information for the effects of Sorel cement on microstructure. In comparison with discontinuous surface of raw CS (Figure 3a), addition of Sorel cement tended to cause the needle-like hydrations presented in some channels or voids (Figure 3b). It is characteristic products of Sorel cement, as suggested by [24]. While the significant changes in the surface of CS4 were observed (Figure 3c). The needle-like crystals of 2-8 μm in length, identified previously as the hydration of phase 5 ($5Mg(OH)_2{\cdot}MgCl_2{\cdot}8H_2O$) and phase 3 ($3Mg(OH)_2{\cdot}MgCl_2{\cdot}8H_2O$), were distributed at the surface and internal of the sample. In particular, the $Mg(OH)_2$ gel is contained and shown in Figure 3d. It is possible that content of MgO is so excessive in the mixture that the colloidal form of $Mg(OH)_2$ is existed in the hydration. For these changes above, specific surface areas of the mixtures can be expected to approach those for hydrated Sorel cement [25]. Since adsorption of cationic metals on oxy-hydroxide surfaces may be promoted at alkaline condition.

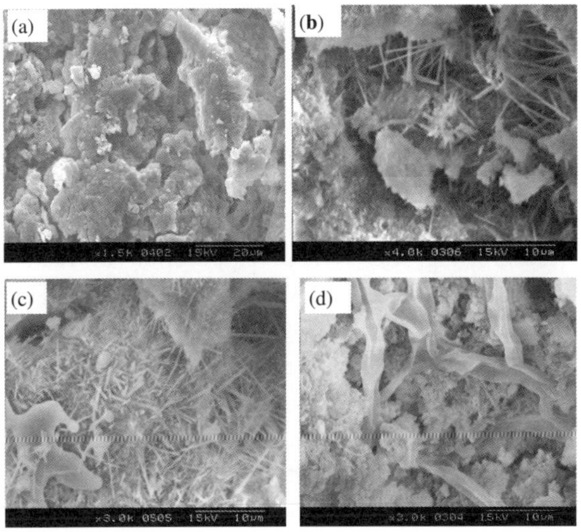

Figure 3. SEM photographs of stabilized CS: (a) CS0, (b) CS2, (c) CS4 and (d) CS6.

3.5. *Sorption isotherm*

According to the results obtained above, the addition of Sorel cement with 12 percent (10% MgO+2% $MgCl_2$) may be an optimum ratio for CS stabilization. Therefore, the single-metal stabilized isotherms in CS4 at room temperature are constructed to represent the stabilized capacity of Sorel cement for heavy metals. The stabilization of Zn, Cu, Cr, Pb and As are described by Freundlich equation [26]:

$$S=KC^{1/n} \tag{3}$$

Where S is concentration of heavy metal stabilized (mg kg^{-1}), C is immobility equilibrium concentration (mg L^{-1}), K is ratio of mobile fractions and immobile fractions of heavy metal and n is a constant.

As shown in Figure 4, the stabilization coefficients (K) of Freundlich stabilization isotherm decrease in the sequence of Zn (3.76) > Cu (3.35) > Pb (2.85) > As (2.70)> Cr (1.13). It indicates that the used Sorel cement has a capacity to stabilize heavy metals in CS on the order of Zn > Cu > Pb > As > Cr. Immobility equilibrium concentration shows that the reduction of heavy metals in leaching tests in which the large amounts of heavy metals are immobilized by Sorel cement and are unavailable to take by leachate.

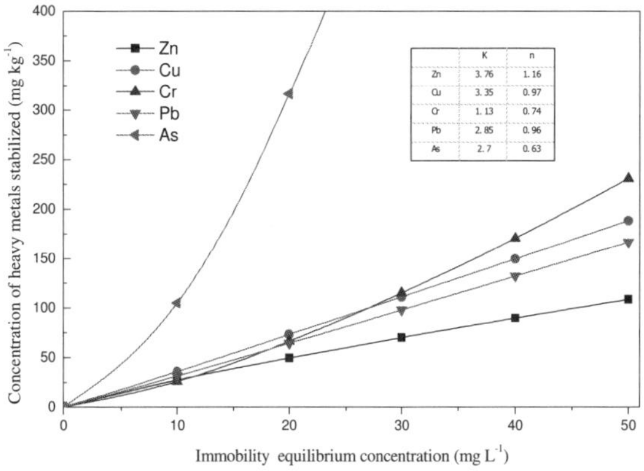

Figure 4. Stabilization and immobility isotherms of heavy metals.

4. Discussion

In the stabilization process, when $MgCl_2$ is dissolved in water from CS, $MgCl_2$ can be dissociated to yield chloride ions and $[Mg(H_2O)_6]^{2+}$. Because of the polarization of magnesian ions, some $[Mg(H_2O)_6]^{2+}$ can be slightly hydrolyzed as equation (4) and equation (5). This hydrolysis yields free hydrogen ions that makes the $MgCl_2$ solution acidic and forms mononuclear aquo hydroxide magnesium complex ions [27].

$$MgCl_2{\cdot}6H_2O \xrightarrow{H_2O} [Mg(H_2O)_6]^{2+} + 2Cl^- \tag{4}$$

$$[Mg(H_2O)_6]^{2+} \underset{H^+}{\overset{OH^-}{\rightleftharpoons}} [Mg(OH)(H_2O)_5]^{+} + H^{+} \tag{5}$$

When MgO is added and mixed with $MgCl_2$ in CS. the possible reactions are shown in equation (6) and equation (7).

$$MgO + 7H_2O \longrightarrow [Mg(OH)(H_2O)_6]^{+} + OH^- \tag{6}$$

$$MgO + 5H_2O \longrightarrow [Mg(OH)(H_2O)_5]^{+} + OH^- \tag{7}$$

These are the neutralizations of MgO and hydrogen ions formed by the hydrolysis in equation (5) in $MgCl_2$

solutions that result in the dissolution of MgO powder and the increase of the pH value in $MgCl_2$ solutions.

Then, the crystal line hydrate phase is converted from the amorphous gel phase over several days or weeks. These crystallizations can be represented as shown in equation (8) and equation (9):

$$6[Mg(OH)(H_2O)_6]^+ + 2Cl^- + 2H^+ + 6OH^- \longrightarrow [Mg_6(OH)_{10}(H_2O)_3]^{2+}2Cl^-5H_2O + 40H_2O \quad (8)$$

$$4[Mg(OH)(H_2O)_5]^+ + 2Cl^- + 2H^+ + 4OH^- \longrightarrow [Mg_4(OH)_6(H_2O)_3]^{2+}2Cl^-5H_2O + 14H_2O \quad (9)$$

Therefore, the formation process of the hydrate phases in Sorel cement pastes can be summarized as the dissolution of MgO powder by the neutralization; the hydrolysing-bridging of magnesian ions to form $[Mg_x(OH)_y(H_2O)_z]^{2x-y}$, that results in the formation of amorphous gel phases and the initial set of Sorel cement pastes; then the conversion of parent compounds in the amorphous form into the crystalline hydrate phases, e.g. phase 5 and phase 3 [19].

5. Conclusions

From the results of the laboratory-scale study on the stabilization of heavy metals (Zn, Cu, Cr, Pb and As) in sewage sludge using Sorel cement, the following conclusions can be drawn:

(1) Sorel cement presents alkaline properties when it is mixed with sewage sludge, and the physical properties of mixtures can be changed in the stabilization procedure.
(2) Toxicity leaching tests show us the metal uptake is effective apparently when the addition of Sorel cement ratios is higher than 9%. Increase of immobile fractions of heavy metals in mixtures, e.g. reducible, organic or residual forms, inhibits the transferability of heavy metals to the solution.
(3) The effect of Sorel cement on adsorption of cationic metals is caused by hydrated procedure and alkaline properties. The capacity for stabilization of heavy metals in sewage sludge follows the order of Zn > Cu > Pb > As > Cr.

The obtained results suggest the possibility to apply Sorel cement for the stabilization of heavy metals (Zn, Cu, Cr, Pb and As) in sewage sludge, obtaining significant stable levels and efficiency percentages. Finally, future work will be focused on scale-up of the stabilization process, aiming to the optimization of parameters for increasing the stabilization efficiency.

Acknowledgment

This work was supported by the Shanghai Expo Foundation (No 05DZ05818) from the Ministry of Science and Technology. Work partially supported by Chinese National Key Projects of Water Pollution Control and Reclamation (2008ZX07314).

References

1. J. Y. Dai, L. Chen, J.F. Zhao et al., Characteristics of sewage sludge and distribution of heavy metal in plants with amendment of sewage sludge, J. Environ. Sci. vol. 6, 2006, pp. 1094-1100.
2. C. Franciszek, M. Tadeusz, Sewage sludge stabilization with calcium hydroxide: Effect on physicochemical properties and molecular composition, Water Res. vol. 40, 2006, pp. 1895-1905.
3. J.L. Wang, J.Z. Wang, Application of radiation technology to sewage sludge processing: A review, J. Hazard. Mater. vol.143, 2007, pp. 2-7.
4. S.R. Smith, A critical review of the bioavailability and impacts of heavy metals in municipal solid waste composts compared to sewage sludge, Environ. Int. vol.35, 2009, pp. 142-156.
5. H. Li, E.X. Wu, Y.X. Liu et al., Heavy metals and PAHs in sewage sludge from twelve wastewater treatment plants in Zhejiang Province, Biomed. Environ. Sci. vol.21, 2008, pp. 345-352.
6. T.J. Logan, B.J. Harrison, Physical characteristics of alkaline stabilized sewage sludge (N-Viro soil) and their effects on soil physical properties, J. Environ. Qual. vol.24, 1995, pp. 153-164.
7. D. Little, R. Reneau, D. martens, Lime-stabilized and chemically fixed sewage sludges as lime amendments. Bioressources Technol. vol. 97, 1991, pp. 93-102.
8. M. Maslehuddin, O.S.B. Al-Amoudi, M. Shameem, Usage of cement kiln dust in cement products-Research review and preliminary investigations, Constr. Build. Mater. vol. 22, 2008, pp. 2369-2375.
9. R. García, R. Vigil de la Villa, I. Vegas et al., The pozzolanic properties of paper sludge waste, Constr. Build Mater. vol. 22, 2008, pp. 1484-1490.

10. M. Sprynskyy, Solid-liquid-solid extraction of heavy metals (Cr, Cu, Cd, Ni and Pb) in aqueous systems of zeolite-sewage sludge, J. Hazard. Mater. vol. 161, 2009, pp. 1377-1383.
11. P. Samaras, C.A. Papadimitriou, I. Haritou et al., Investigation of sewage sludge stabilization potential by the addition of fly ash and lime, J. Hazard. Mater. vol. 154, 2008, pp. 1052-1059.
12. V. Mijno, F. Martin, J.C. Bollinger et al., Stabilization solidification process comparison of synthetic sludge and volcanogenic massive sulphide tailings, Trans IChemE, Part B, Process Saf. Environ. Prot. vol. 85, 2007, pp. 260-264.
13. A.K. Karamalidis, E.A. Voudrias, Cement-based stabilization/solidification of oil refinery sludge: Leaching behavior of alkanes and PAHs, J. Hazard. Mater. vol. 148, 2007, pp. 122-135.
14. B. Batchelor, Overview of waste stabilization with cement, Waste Management. vol. 26, 2006, pp. 689-698.
15. P. Maravelaki-Kalaitzaki, G. Moraitou, Sorel's cement mortars Decay susceptibility and effect on Pentelic marble, Cem. Concr. Res. vol. 29, 1999, pp. 1929-1935.
16. M.A. Garcia , J.M. Chimenos, A.I. Fernandez et al., Low-grade MgO used to stabilize heavy metals in highly contaminated soils. Chemosphere. vol. 56, 2004, pp. 481 491.
17. A. Tessier, P.G.C Campbell, M. Bisson, Sequential extraction procedure for the speciation of particulate trace metals. Analytical chemistry. vol. 51,1979, pp. 844-851.
18. D.A. Hall, R. Stevens, B.E. Jazairi, The effect of retarders on the microstructure and mechanical properties of magnesia-phosphate cement mortar, Cem. Concr. Res. vol. 31, 2001, pp. 455-465.
19. H. Bilinski, B. Matkovic, C. Mazuravic et al., he formation of magnesium oxychloride phases in the system MgO-MgCl2-H2O and NaOH-MgCl2-H2O. J Am Ceram Soc. vol. 67, 1984, pp. 266-272.
20. D. Dehua. The mechanism for soluble phosphates to improve the water resistance of magnesium oxychloride cement. Cem. Concr. Res. vol. 33, 2003, pp. 1311-1317.
21. M. Jakubus, J. Czekala, Heavy metal speciation in sewage sludge. Polish J Environ Stud. vol. 10, 2001, pp. 245-250.
22. A. Tessier, P.G.C. Campbell, Comments on the testing of the accuracy of an extraction procedure for determining the partitioning of trace metals in sediments. Anal Chem. vol. 60, 1988, pp. 1475-1476.
23. S.K. Shrivastava, D.K. Banerjee, Operationally determined speciation of copper and zinc in sewage sludge. Chem Speciation Bioavailab. vol. 10, 1998, pp. 137-143.
24. L. Guozhong, Y. Yanzhen, L. Jianquan et al., Experimental study on urban refuse/magnesium oxychloride cement compound floor tile. Cem Concr Res. vol. 10, 2003, pp. 1663-1668.
25. D.A. Hall, R. Stevens, B.E. Jazairi, The effect of retarders on the microstructure and mechanical properties of magnesia-phosphate cement mortar. Cem Concr Res. vol. 31, 2001, pp. 455-465.
26. C.W. Gray, R.G. Mclaren, A. H. C. Roberts et al., Solubility, sorption and desorption of native and added cadmium in relation to properties of soils in New Zealand. Eur J Soil Sci. vol. 50, 1999, pp. 127-137.
27. D. Dehua, Z. Chuanmei, The formation mechanism of the hydrate phases in magnesium oxychloride cement. Cem Concr Res. vol. 29, pp. 1365-1371.

MODELING OF SUBSTRATE DEGRADATION AND MICROORGANISM GROWTH IN AN UASB REACTOR*

RAUL RODRIGUEZ

Chemical Department, National University of Engineering (UNI), Managua, Nicaragua
Royal Institute of Technology (KTH) Stockholm, Sweden

LUIS MORENO

Royal Institute of Technology (KTH) Stockholm, Sweden

A model for substrate degradation and microorganism growth in an UASB reactor was developed. The model is transient and takes into account advection, dispersion and degradation of the substrate. For the micro-organisms, the growth is described by a Monod type equation including a death constant. The model also assumes that a fraction of the micro-organisms in the reactor may be carried out by the water flow. The UASB reactor is described as formed by many well-stirred reactors in series. Parameters for the model were taken from literature and MATLAB was used to solve the model. The sensibility of the process to the model parameters was also addressed.

1. Introduction

Anaerobic process is widely used to treat wastewater. The major advantages of anaerobic processes over aerobic process are fuel production and less sludge generation.

The anaerobic digestion process has mainly four phases: a) hydrolysis of high molecular weigh compounds into polymers, b) conversion of polymers into alcohols, H_2, CO_2 and organics acid by acidogenic bacteria, c) conversion into acetic acids and H_2 by acetogenic bacteria and d) conversion into biogas mostly composed of CH_4 and CO_2 [1,2] by methanogenic bacteria.

Microorganism growth is usually described by Monod model, which relates the growth rate of microorganisms with the substrate concentration. However, Zwietering et al. [3] compared several models for microorganism growth and concluded that the Gompertz model describes it in a better way than the other tested models (logistic, linear, quadratic, t^{th} power and exponential). They did not consider though the consumption of substrate since they assumed that there was abundance of substrate.

Korsak et al. [4] proposed a model describing the processes in an UASB reactor. In the model, they assumed that the amount of microorganisms had reached a steady state with its concentration decreasing with the height. Since the microorganisms in a UASB form granules, the emphasis of the model was set in the reaction of the substrate within the microorganisms within the granule.

In this paper, the emphasis will be set in the growth of microorganism, i.e., from the start-up of the reactor until a steady state is reached. At present there are no models that account for the main processes for microorganism growth. Therefore, the aim of the paper is the development of a model describing the growth of the microorganism, which follows the Monod kinetics. The model also takes into account the rate at which microorganisms disappear from the reactor by decay (metabolism) and the percentage of microorganisms that could be carried out by the water flow. The model is tested by using data from literature.

Models are important design and operation tools of UASB reactors. The amount of time spent on experiments may be strongly reduced.

2. UASB Reactor

UASB reactor is basically a cylindrical tank through which the wastewater flows upwards. The concentration of sludge or microorganisms is high at the bottom of the reactor, since they form granules which have a tendency to sediment due to their larger density. Part of the substrate is degraded when the liquid passes along the reactor, biogas is then produced and the microorganism population is increases.

*This work is supported by Swedish International Development Cooperation Agency (SIDA/SAREC).

2.1. *Microorganism growth*

Monod equation can be used to describe the change of microorganism concentration in time:

$$\frac{dX}{dt} = X \cdot \mu_{\max} \frac{S}{K_S + S} \tag{1}$$

Where X is the microorganism concentration [mgL^{-1}] and t is the time [d], $\mu_{\max}$ is the maximum specific growth rate [d^{-1}], K_s is the substrate concentration [mgL^{-1}] when the half maximum growth rate takes place, and S is the substrate concentration [mgL^{-1}].

Equation (1) describes the growth at the initial stage, but a steady state should be reached with time. Therefore, Monod equation should include the microorganism decay rate [5-8]:

$$\frac{dX}{dt} = \left(\mu_{\max} \frac{S}{K_s + S} - K_d \right) \cdot X \tag{2}$$

Where K_d is the decay rate of microorganism [d^{-1}]. The decay rate considers death of microorganisms, e.g. due to accumulation of toxic metabolism [9], predation [10] (protozoa are predator of bacteria) and lysis [11]. In addition, Yan et al. [8] evaluating a UASB reactor for wastewater for cheese whey found that, at the initial period, a large amount of sludge left the reactor due to the poor quality of seed sludge. Therefore, the washout of biomass will be also included in the model.

Sponza and Uluköy [12] compared three models to determine the substrate removal kinetics in an UASB reactor treating synthetic water composed of glucose and 2,4 dichlorophenol (2,4 DCP). The models used by Sponza and Uluköy [12] are: Monod, Second-order Grau kinetics and Modified Stover-Kincannon. They concluded the Grau and Stover-Kincannon models are not appropriate for predicting the performance of a UASB reactor for degradation of 2,4 DCP. However Monod model can be used to predict the performance of COD and 2,4 DCP degradation in UASB reactor.

3. Model

The model proposed in this paper considers the substrate degradation along the reactor and the growth of microorganism with time. To describe and solve the problem, the UASB reactor is represented as n small well-stirred reactors in series. Figure 1 shows the reactor i, where q is the flow through each reactor [Ld^{-1}], X is the microorganism concentration [mgL^{-1}], and S is the substrate concentration [mgL^{-1}]. All reactors have the same volume, V/n, where V is the active volume of the UASB reactor and n the number of small reactors used in the model. w is the fraction of microorganisms that is washed out from a reactor into the next one.

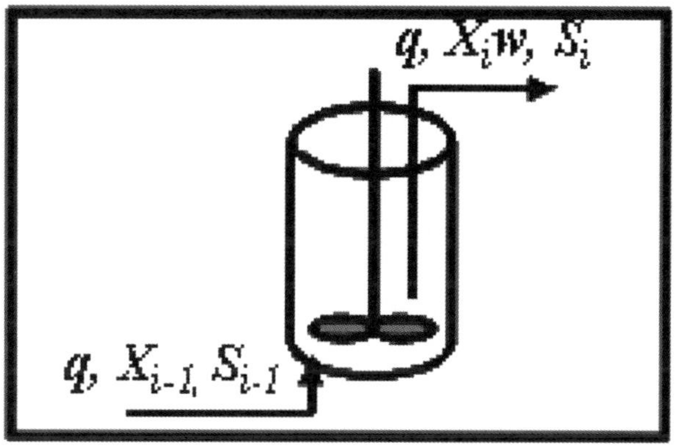

Figure 1. Scheme of i-Reactors.

Initially, the small reactors have a given concentration of substrate (S_{inic}) and microorganisms (X_{inic}). The influent flowrate is q with a substrate concentration of S_o and free of microorganisms.

The model proposed takes into account the variation with time in each reactor (accumulation term), the advective transport from a reactor into the next one, and the reaction term. This may be expressed by a mass balance for the substrate and microorganism for the i-reactor:

$$\frac{dS_i}{dt} = \frac{q}{V_i}(S_{i-1} - S_i) - \left(\frac{\mu_{\max i}}{Y_i} \frac{S_i}{K_s + S_i} \right) \cdot X_i \tag{3}$$

$$\frac{dX_i}{dt} = \frac{q}{V_i}(X_{i-1} w_{i-1} - X_i w_i) + \left(\mu_{\max i} \frac{S_i}{K_s + S_i} - kd \right) \cdot X_i \tag{4}$$

Where Y is the yield coefficient, which represents the mass of biomass produced by the mass of substrate degraded and K_d a constant that takes into account the rate at which the microorganisms disappear [d^{-1}].

In equation (3), the left-hand side term is the accumulation term, the first term on the right-hand side is the advective term and the second one is the reaction term. In equation (4), the first term on the right-hand side is the amount of microorganisms that are washed out from the small reactors. The second term describes the growth and death of microorganisms.

The number of small reactors used in the system accounts for the dispersion in the UASB reactors. A small number of reactors indicate a large dispersion. Calculations were done by MATLAB.

3.1. *Data for the model*

To test the model, data from literature was used. Sponza and Uluköy [12] studied the degradation of glucose (primary carbon source) in an UASB reactor. The reactor worked at six different hydraulic retention time (HRT) between 20h – 2 h, the temperature was 37°C, the height 100 cm, the diameter 6 cm, and the effective volume 2.5 L. Table 1 shows the data reported by Sponza and Uluköy [12] for Monod model for Glucose (as COD).

Table 1. Kinetics parameters (Sponza and Uluköy [12]).

Parameter	COD
V (L)	2.5
S_o (mgL^{-1})	3000
Y	0.000780
k_d (d^{-1})	0.093
μ_{max} (d^{-1})	0.21
Ks (mgL^{-1})	561
X_{inic}(mgL^{-1})	5
S_{inic}(mgL^{-1})	3000

In Table 1 X_{inic} and S_{inic} are the initial values for biomass (microorganism) and substrate concentration. The flow q is determined from the HRT. Results for only two HRT are shown in this paper (20h and 7h). For illustration purpose, the reactor is divided in six small well-stirred reactors. In practice, a greater number of reactors would be used.

4. Results and Discussion

Results show the degradation of the substrate (COD) and the growth of the microorganisms (biomass) in the UASB reactor with time. The study was made for two flow rates (3 and 9 Ld^{-1}) corresponding to HRT of 20h and 7h respectively. The washout was assumed to be 0.002. Figures 2 to 5 show the model response.

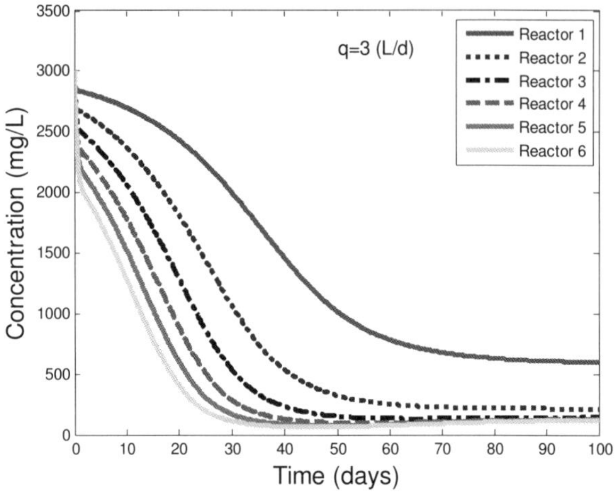

Figure 2. Substrate versus time.

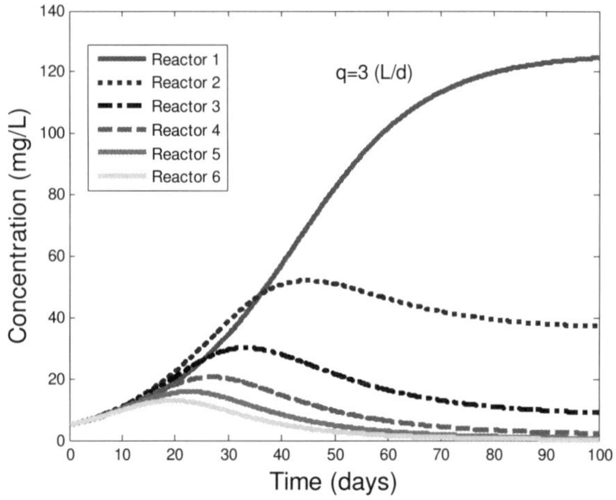

Figure 3. Active biomass versus time.

Figure 2 shows the COD degradation in the six reactors. Reactor one reduces the COD concentration by 82.93%, all reactors present about the same removal and the total COD concentration is reduced by 96% when steady state is reached. Sponza and Uluköy report a reduction of 83%. This difference may be due to the different operation time, their time was 12 days, while our time was very long (100 days) to ensure steady state.

The microorganism behavior is shown in figure 3. In reactor 1 the active biomass grows with a higher rate due to the larger amount of substrate that is degraded in that reactor. The high degradation rate is caused mainly by the substrate concentration as indicated in Eq. 3. The biomass growth rate is lower in the other reactors due to the degradation of the substrate in the reactor near inlet. The steady state is reached at about 100th day.

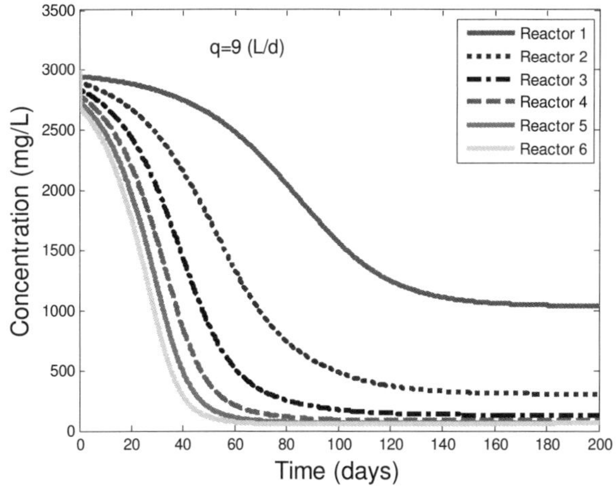

Figure 4. Substrate versus time.

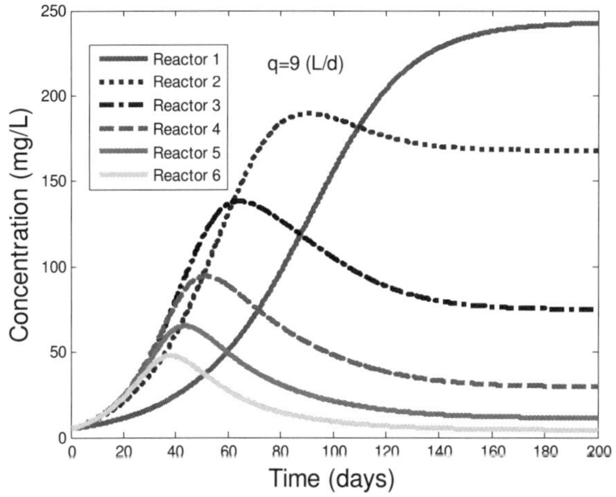

Figure 5. Active biomass versus time.

Figures 4 and 5 show the substrate and active biomass performance for a HRT of 7 days, the flow is then 9 Ld^{-1}. The behavior is similar to the case with a flowrate of 3 Ld^{-1}. The substrate concentration decreases until it reaches steady state and the active biomass increases, but as in this case the flow is higher, it takes more time to reach a steady state. This delay may be due to the water flow that drags a larger quantity of biomass from one reactor to the next one, taking a longer time to reach the steady state. This requires a greater concentration of biomass in the reactor as shown in figure 5 where the concentration of biomass is larger than that in figure 3.

In this model, degradation of the substrate is considered to produce directly CO_2 and CH_4 by the action of the bacteria. As explained above, this is a very complex process with participation of different populations of bacteria and with reaction occurring in several paths. At present there are no data to develop a model taking into account all these reactions.

5. Conclusion

The model has been successfully used to simulate the performance of the UASB reactor studied by Sponza and Uluköy [12].

With this type of model it is possible to study the reactor behavior with time. The model includes substrate degradation, biomass growth and death, and the washout of microorganisms.

References

[1] Tramšek Marko, Andreja Goršek, Peter Glavič, "Methodology for determination of anaerobic digestion kinetics using a bench top digester" Elsevier Resources, Conservation and Recycling pp. 225–236, 2007.

[2] Wheatley A., Fisher M., Grobicki A. "Applications of anaerobic digestion for the treatment of industrial wastewaters in Europe", J Chart Inst Water, pp. 39–46 1997.

[3] Zwietering M., Jongenburger I., Rombouts F. and Riet K., "Modeling of the bacteria growth curve", Applied and Environmental Microbiology, pp. 1875-1881, 1990.

[4] Korsak L., Cabezas A., Moreno L. "A model of kinetics and mass transfer in an UASB reactor", 4th International conference on environmental science and technology 2008.

[5] Seghezzo L., "Anaerobic treatment of domestic wastewater in subtropical regions", PhD. Thesis, Wageningen University The Netherlands, 2004.

[6] Barampouti E. and Vlyssides A. "A mathematical model for the estimation of anaerobic sludge activity", 8th International Conference on Environmental Science and Technology, pp. 72-78, 2003.

[7] Hwang S. and Hansen C., "Biokinetics of an Upflow Anaerobic Sludge Blanket Reactor Treating Whey Permeate", Bioresource Technology pp. 223-230, 1992.

[8] Yan Q., Lo K. and Liao P., "Anaerobic digestion of cheese whey using up-flow anaerobic sludge blanket reactor", Biological wastes pp. 289-305, 1989.

[9] Krylów M., "Kinetics of subsequent phases of the anaerobic processes", Proceedings of a Polish-Swedish seminars, pp. 29-37, 2003.

[10] Bitton G., "Wastewater Microbiology", 3rd Ed. Wiley Editor, 2005.

[11] Kalyuzhnyi S., Fedorovich V., "Dispersed plug flow model for upflow anaerobic sludge bed reactors with focus on granular sludge dynamics", Microbiol Biotechnol pp. 221-237, 2006.

[12] Sponza D. and Uluköy A., "Kinetic of carbonaceous substrate in an upflow anaerobic sludge blanket (UASB) reactor treating 2,4 dichlorophenol (2,4 DCP)", Journal of environmental Management, pp. 121-131 2008.

EFFECTS OF CHEMICAL FERTILIZERS ON THE CR(VI) ADSORPTION ON PADDY SOIL

JIE LIU
The Guangxi Key Laboratory of Environmental Engineering, Protection and Assessment, Guilin University of Technology, Guilin 541004, China

ZHENJIANG JIN
The Guangxi Key Laboratory of Environmental Engineering, Protection and Assessment, Guilin University of Technology, Guilin 541004, China

QIN ZHANG
The Guangxi Key Laboratory of Environmental Engineering, Protection and Assessment, Guilin University of Technology, Guilin 541004, China

SHAOHONG YOU
The Guangxi Key Laboratory of Environmental Engineering, Protection and Assessment, Guilin University of Technology, Guilin 541004, China

Effects of six chemical fertilizers, $Ca(NO_3)_2$, NH_4Cl, NH_4NO_3, $CO(NH_2)_2$, KH_2PO_4, and KCl, on the adsorption isotherm of Cr(VI) on paddy soil were studied using batch equilibrium experiments. The results showed that the adsorption of Cr(VI) on paddy soil was inhibited to some extent by the additions of the six fertilizers. Because of their different effects on Cr(VI) adsorption, fertilizers should be reasonably used during cultivation in order to avoid the Cr mobilization and decrease its bioavailability.

1. Introduction

Chromium is a widespread contaminant in the soil and is released into the environment by a large number of industrial operations such as electroplating, chromate manufacturing, leather tanning and wood preservation [1]. Cr can exists in different valence states ranging from −2 to +6, Cr(III) and Cr(VI) are, however, stable enough to occur in the environment. Cr(VI) is more hazardous than Cr(III) due to its toxicity and carcinogenicity [2]. In general, Cr(VI) exists as anion forms (CrO_4^{2-} and $Cr_2O_7^{2-}$) in the soil environment. Mobility and availability of Cr (VI) in soils strongly depend on the adsorption behavior onto the soil particles. The adsorption reaction of Cr(VI) on soils is influenced by pH, redox potential (Eh), organic matter (OM) and coexisting anion and surface charge [3].

As an important way to increase yield in agricultural production, chemical fertilizers are commonly used worldwide. In China, the amount of fertilizer applied was as high as 4636.8×10^8 kg in 2004, accounting for about 35% of the total consumption in the world [4]. However, excess fertilizer application could alter soil properties such as pH, redox potential, ionic strength, surface charge and so on [5, 6]. These changes of soil properties might greatly affect the adsorption of Cr(VI) on soils. There was a surprising dearth of information concerning the effects of chemical fertilizers on Cr(VI) adsorption by soils. In the present work, a paddy soil was selected to investigate the adsorption of hexavalent chromium in the presence of six chemical fertilizers ($Ca(NO_3)_2$, NH_4Cl, NH_4NO_3, $CO(NH_2)_2$, KH_2PO_4, and KCl). This study provides insights into effects of chemical fertilizers on the Cr(VI) adsorption behavior by soils, which is helpful to safely fertilize and avoid the increasing of mobility and availability of Cr (VI) in soils.

2. Materials and Methods

The soil used in this study was collected from a paddy field located in Guilin, China. Some properties of the soil were measured with routine analytical methods [7], and listed in Table 1. The chromium concentration in the soil was much lower than the environmental quality standard for soils in China [8]. The collected soil was air dried, ground to pass 2 mm sieve and stored in plastic containers.

Table 1. Physical and chemical properties of the soil used in the experiments.

pH (H_2O)	CEC* (cmol /kg)	Eh (mV)	Clay (%)	OM** (%)	Fe_2O_3 (%)	Al_2O_3 (%)	Total Cr (mg/kg)
6.02	7.8	247	22.28	2.7	5.56	13.86	11.97

*Cation exchange capacity; **Soil organic matter.

Seven sets of batch sorption experiments were carried out for the following fertilizers: 200 mg N/kg as $Ca(NO_3)_2$, NH_4Cl, NH_4NO_3 and $CO(NH_2)_2$, 80 mg P/kg as KH_2PO_4, 100 mg K/kg as KCl, and a control with no fertilizer, respectively. For each set, nine 50 ml polypropylene centrifuge tubes were firstly filled with 25ml mixture solutions of fertilizer and $K_2Cr_2O_7$ with different Cr(VI) concentrations (0, 10, 20, 30, 45, 60, 75, 95 and 115 mg/L, respectively). And then, 2.5g soil was added into each tube. The tubes were shaken on an orbital shaker at 25 °C for two hours. The tubes were then centrifuged at 3000 rpm for 10 min. Supernatant solutions were filtered and Cr concentration determined Cr concentration by a flame atomic absorption spectrophotometer (PE-AA700). The amount of Cr adsorbed was calculated from the difference between the initial and final Cr concentration. Each treatment had three replicates.

The Freundlich, Langmuir and De Boer-Zwikker adsorption isotherms [9] were used to describe the results of the batch equilibrium experiments. The Freundlich adsorption isotherm equation is as follows:

$$S = kC_e^{\frac{1}{n}} \tag{1}$$

where S = amount of Cr adsorbed; C_e = final equilibrium concentration of Cr; n = the Freundlich constant; and k = the adsorption coefficient.

The Langmuir adsorption isotherm equation is as follows:

$$S = \frac{bC_e}{k + C_e} \tag{2}$$

where b = the maximum adsorption of Cr; and k = Langmuir constant related to affinity of adsorption.

The De Boer-Zwikker adsorption isotherm equation is as follows:

$$\ln\ln(C_0 / C_e) = \ln(E_0 / RT) - (k / b)S \tag{3}$$

where C_0 = initial concentration of Cr; b = the maximum adsorption of Cr; k = experiential constant; E_0 = adsorption potential; R = the ideal gas constant; and T = absolute temperature.

3. Results

3.1 *Sorption isotherms*

The sorption isotherms of Cr(VI) as influenced by fertilizers were given in Figure 1. Comparison of the sorption isotherms of Cr(VI) as influenced by fertilizers and the control revealed that the presence of the six fertilizers investigated decreased adsorption of Cr(VI) on the paddy soil. At high initial concentrations of Cr(VI) (> 30 mg/L), the differences were more pron- ounced. Moreover, the effects of different fertilizers on the adsorption of Cr(VI) by paddy soil were different (Fig. 1). The effects of NH_4NO_3 and KH_2PO_4 on the adsorption of Cr(VI) were most significant among those six fertilizers. Whereas, the sorption isotherm of Cr(VI) as influenced by NH_4Cl was not obviously different from that of the control, which showed that the effect of NH_4Cl on the adsorption of Cr(VI) by paddy soil was not significant.

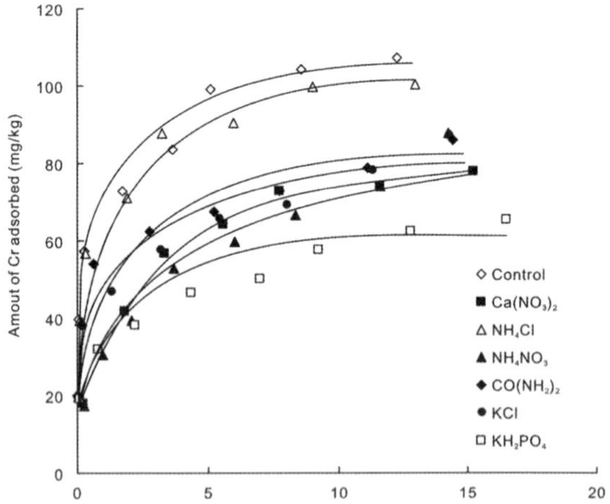

Figure 1. Sorption isotherms of Cr(VI) as influenced by various fertilizers in batch equilibrium experiments.

The experiment results fitted the Freundlich, Langmuir and De Boer-Zwikker well according to the

Table 2. Parameters and correlation coefficients of Cr adsorption equations.

Fertilizers	Freundlich			Langmuir			De Boer-Zwikker	
	k	*n*	r	*k*	*b*	r	E_0	r
Control	67.06	4.76	0.97	0.22	98.6	0.84	23.16	-0.97
$Ca(NO_3)_2$	33.82	2.97	0.99	0.29	85.3	0.91	11.10	-0.99
NH_4Cl	58.03	4.00	0.97	1.96	94.8	0.99	19.11	-0.98
NH_4NO_3	29.9	2.71	0.99	1.15	84.2	0.90	9.84	-0.99
$CO(NH_2)_2$	51.63	4.13	0.97	0.23	88.3	0.87	18.60	-0.98
KH_2PO_4	32.94	4.18	0.99	2.00	70.6	0.93	13.78	-0.96
KCl	45.79	4.43	0.98	2.64	89.4	0.99	17.29	-0.95

$r_{0.05}$= 0.666, $r_{0.01}$=0.798

calculated correlation coefficients (Table 2). The parameter *b* in the Langmuir equation represents the adsorption maxima of Cr(VI). According to the parameter *b*, the inhibition effects of six fertilizers on adsorption of Cr(VI) followed the order: KH_2PO_4 > NH_4NO_3 > $Ca(NO_3)_2$ > $CO(NH_2)_2$ > KCl > NH_4Cl. Furthermore, the adsorption potential E_0 calculated from De Boer-Zwikker equation indicated that the presence of fertilizers also decreased adsorption strength of Cr(VI) on the paddy soil. The order of the adsorption potential E_0 was: no fertilizer (control) > NH_4Cl> $CO(NH_2)_2$ > KCl > KH_2PO_4> $Ca(NO_3)_2$ > NH_4NO_3.

3.2 *Adsorption rate*

The adsorption rate of Cr(VI) by the paddy soil decreased with the increase of initial Cr(VI) concentration (Table 3). The reduction of adsorption rate of Cr(VI) as influenced by fertilizers indicated that the adsorption of Cr(VI) on paddy soil was inhibited to some extent by the additions of the six fertilizers. However, the effects of different fertilizers on the adsorption rate of Cr(VI) by paddy soil were significantly different. For example, when the initial Cr(VI) concentration was 95 mg/L, applying NH_4Cl only decreased the adsorption rate of Cr(VI) by 4.2%, while applications of KH_2PO_4 and NH_4NO_3 decreased the adsorption rate of Cr(VI) by 40.3% and 28.8%.

3.3 *Distribution coefficient*

In the similitude of adsorption rate, the Cr distribution coefficients decreased with increasing initial Cr(VI) concentration, regardless of whether the fertilizers were added or not added. (Table 4). Applications of chemical fertilizers significantly reduced distribution coefficients of Cr(VI). For instance, the distribution coefficient of Cr(VI) in the presence of NH_4NO_3 was 26.7-fold lower than that in control as initial Cr(VI) was 10 mg/L. Whereas, there were significant differences in distribution coefficient of Cr(VI) among the six fertilizers. The inhibitive effectiveness for Cr(VI) distribution coefficient can be classified as: NH_4NO_3, $a(NO_3)_2$ and KH_2PO_4 > $CO(NH_2)_2$ > KCl and NH_4Cl.

Table 3. Effects of fertilizers on Cr(VI) adsorption rate.

Initial concentration (mg/L)	Adsorption rate (%)*						
	Control	$Ca(NO_3)_2$	NH_4Cl	NH_4NO_3	$CO(NH_2)_2$	KH_2PO_4	KCl
10.00	99.59	90.05	98.82	86.15	97.41	92.99	97.99
20.00	99.45	79.96	98.27	76.35	95.76	80.21	95.56
30.00	95.34	69.77	94.74	65.72	89.85	63.65	78.12
45.00	81.09	63.06	78.86	58.75	69.20	51.89	64.11
60.00	69.54	53.51	73.04	49.67	56.25	41.67	54.74
75.00	66.03	48.48	60.21	44.41	48.52	38.36	46.25
95.00	54.85	38.96	52.52	39.03	41.43	32.77	41.10
115.00	46.61	33.94	43.66	38.14	37.32	28.37	37.88

*The percentage of Cr adsorbed by soils to the total Cr.

Table 4. Effects of fertilizers on Cr distribution coefficients in soil and equilibrium solution.

Initial concentration (mg/L)	Distribution coefficients*						
	Control	$Ca(NO_3)_2$	NH_4Cl	NH_4NO_3	$CO(NH_2)_2$	KH_2PO_4	KCl
10.0	1684.95	90.55	586.94	60.78	469.13	189.59	487.40
20.0	1648.44	39.90	226.19	32.29	164.34	40.52	215.04
30.0	204.68	23.08	180.08	19.17	110.23	17.51	35.69
45.0	42.89	17.07	37.30	14.24	31.43	10.79	17.87
60.0	22.83	11.51	27.09	9.87	18.24	7.14	12.09
75.0	19.44	9.41	15.13	7.99	11.06	6.22	8.61
95.0	12.15	6.38	11.06	6.40	8.58	4.87	6.98
115.0	8.73	5.14	7.75	6.17	6.41	4.17	6.10

*The ratio of Cr concentration in solid phase to that in liquid phase.

4. Discussion

In this study, the results showed that the six fertilizers investigated could decrease the adsorption of Cr(VI) on paddy soil to some extent (Fig.1; Table 3). According to the adsorption maxima *b* calculated from Langmuir equation, the depressing extent of six fertilizers was found to follow the sequence: KH_2PO_4 > NH_4NO_3 > $Ca(NO_3)_2$ > $CO(NH_2)_2$ > KCl > NH_4Cl (Table 2). The effect of KH_2PO_4 on the adsorption of Cr(VI) was more significant than those of other five fertilizers. The mainly reason is that application of KH_2PO_4 increase the $H_2PO_4^-$, HPO_4^{2-} and PO_4^{3-} in the soil solution which can compete with $Cr_2O_7^{2-}$ or CrO_4^{2-} for adsorption sites of soil surfaces [3]. The result was agreed with Bartlett and Kimble [10] who reported that KH_2PO_4 could significantly decreased the adsorption of Cr(VI) on soils. The effect of NH_4Cl on adsorption of Cr(VI) was not significant because of the low competitive capacity of Cl^- which also could explain that the application of NH_4Cl depressed Cr(VI) sorption less than that of NH_4NO_3. Another possible reason was that NH_4Cl could reduce the soil pH [5]. It was reported that Cr(VI) existed as anion forms in soil and its adsorption by soil decreased with increasing pH [11]. Decrease of Cr(VI) adsorption rate by $CO(NH_2)_2$ was also ascribed to the changes of soil pH. In acid soil, hydrolyzation of urea would consume hydrion in the soil solution, and thus increased soil pH [12].

The six fertilizers investigated not only decreased the adsorption of Cr(VI), but also decreased the adsorption potential of Cr(VI) on the paddy soil. The order of the adsorption potential E_0 from De Boer-Zwikker equation is as follow: control > NH_4Cl> $CO(NH_2)_2$ > KCl > KH_2PO_4> $Ca(NO_3)_2$ > NH_4NO_3 (Table 2). Generally, anion adsorption by soils is divided into two types [13]. One is non-specific adsorption. Non-specifically adsorbed anions are held on the soil surface through electrostatic attraction and their adsorption potential is relatively low. Another is specific adsorption, i.e. anion adsorbed by soil through surface complexation. Due to electron transfer or shared electron occurs in specific adsorption process, the adsorption potential of specific adsorption is higher than non-specifically adsorption. The present results which KH_2PO_4, $Ca(NO_3)_2$ and NH_4NO_3 significantly depressed the adsorption potential of Cr(VI) could attributed to that NO_3^- and $H_2PO_4^-$ decreased the specific adsorption of Cr(VI). Our suggestion was supported by Chen et al. [3] who considered phosphate and nitrate could effectively occupy the specific adsorption sites of Cr(VI) and prevented Cr(VI) adsorption. On the contrary, effects of NH_4Cl and KCl on the adsorption potential of Cr(VI) were little because Cl^- was considered to be a non-specifically adsorbed ions and could not affect the specific adsorption of Cr(VI) [14].

It was well documented that applying fertilizers could alter the mobility and availability of heavy metals in soils [15, 16]. In the present work, it was found that the presence of KH_2PO_4, NH_4NO_3, $Ca(NO_3)_2$, $CO(NH_2)_2$, KCl and NH_4Cl could decreased the adsorption rate and adsorption potential of Cr(VI) on the paddy soil, indicating that applications of the six fertilizers investigated may increase the mobility and availability of Cr(VI). These results were consistent with Cr(VI) distribution coefficients in soil and equilibrium solution (Table 4). The reduction of Cr(VI) distribution coefficient in the presence of fertilizers implied that fertilizers increased the Cr(VI) in liquid phase and might enhance the mobility and availability of Cr(VI) in soils. In order to avoid the Cr mobilization and decrease its bioavailability, fertilizers should be reasonably used during cultivation in the soil contaminated by chromium because of their different effects on the Cr(VI) adsorption. For example, according to the result of this study, the application of N fertilizer as NH_4Cl is superior to NH_4NO_3 using in the soil contaminated by Cr(VI) due to the less effect of NH_4Cl on C(VI) adsorption on soils.

Acknowledgements

The authors thank the financial supports from the Scientific Research and Technological Development Project of Guangxi (GuiKeGong 0816003-1-1), the Provincial Natural Science Foundation of Guangxi (GuiKeNeng 0701K005, GuiKeZi 0728222) and the Program to Sponsor Teams for Innovation in the Construction of Talent Highlands in Guangxi (GuiKeRen 2007-71, 2004217).

References

1. K. R. Krishna and L Philip, *J. Hazard. Mater.* **B121**, 109 (2005).
2. J. Kotas and Z. Stasicka, *Environ. Pollut.* **107**, 263 (2000).
3. Y. Chen, Z. Zhu and Z. He, *Pedosphere* **1(4)**, 325 (1991).
4. L. Huang, *Fortune World* **4**, 101 (2006). (in Chinese)
5. J. Liu, Q. Duan, Y. Zhu, X. Zhang and C. Wang, *Environ. Geol.*, **52**, 1601 (2007).
6. S. S. Malhi, M. Nyborg and J. T. Harapiak, *Soil Till. Res.* **48**, 91 (1998).
7. Agrochemistry Commission (AC), "Routine methods for soil and agrochemical analyses, soil science society of China". Beijing: Science Press. 1983. (in Chinese)
8. State Environmental Protection Administration of China (SEPAC), "Environmental quality standard for soils. GB 15618-1995" Beijing. 1995. (in Chinese)
9. D. Zhou, X. Li, Y. Dong and F. Xu, *J. Huazhong Agr. Univ.* **14(1)**, 58 (1995). (in Chinese)
10. R. J. Bartlett and J. M. Kimble, *J. Environ. Qual.*, **5**, 382 (1976).
11. B. Fonseca, A. Teixeira, H. Figueiredo and T. Tavares, *J. Hazard. Mater.* doi: 10.1016/j.jhazmat. 2009.01.049. (2009).
12. C. Tu, C. Zheng and H. Chen, *Chemosphere* **41**, 133 (2000).
13. D. Dong, X. Zhao, X. Hua, J. Liu and M. Gao, *J. Hazard. Mater.* **162**, 1261 (2009).
14. Y. Zhang, *Environ. Chem.* **9(4)**, 43 (1990). (in Chinese)
15. J. T. Sims, *J. Environ. Qual.* **19**, 669 (1990). C. D. Tsadilas, A. Karaivazogloun, N. C. Tsotsolis, S. Stamatiadis and V. Samaras, *Environ. Pollut.* **134**, 239 (2005).

A STUDY ON THE EFFECT OF FRACTIONAL COMPOSITION OF NANO SIZE PARTICLES IN COATINGS OF CACO$_3$ ON RELEASE RATES OF ACTIVE CONSTITUENTS FROM FERTILIZER PELLETS

W. J. N. FERNANDO, A. L. AHMAD and N. M. SOON

School of Chemical Engineering, University Sains Malaysia, Nibong Tebal, 14300 Pulau Pinang, Malaysia

Different sizes of cylindrical pellets of two types of fertilizers, namely; calcium dihydrogen phosphate ($Ca\ (H_2PO_4)_2$) and ammonium sulfate ($(NH_4)_2SO_4$) of typical compositions were coated with layers of $CaCO_3$ consisting of different proportions of nano-$CaCO_3$ of 0%, 33.33%, 66.67% and 100% in analytical grade micro sized $CaCO_3$ powder. The pellets are exposed to de-ionized distilled water in an experimental rig designed for this purpose, so that transfer of nutrients occurs through the coated layer to the water. The extents of overall transfer of nutrients through the coats are measured at different intervals of time. The rates of mass transfer are evaluated from the data. Parameters that determine mass transfer rates across surface layers are evaluated for different compositions of nano-$CaCO_3$. Results show that inclusion of nano $CaCO_3$ in surface coated layers of micro sized $CaCO_3$ could be used to regulate the mass transfer rates showing possibilities to engineer coated fertilizes using mixes of nano and micro sized $CaCO_3$.

Keywords: Coating; diffusion; fertilizer pellets; transfer rates.

Nomenclature:

C_o	Original concentration of component in the pellet (kg/m^3)	kc	Apparent liquid gas film mass transfer coefficient for external surface of the coat based on superficial surface area.
D	diffusion coefficient of component within membrane(m^2/min)	r	$K\sqrt{\dfrac{D}{D_f}}$
D_f	diffusion coefficients of component within the bulk(m^2/s)		
dC/dt	rate of concentration of fertilizer released over one period (mg/L.min)	t	time (min)
l	thickness of the $CaCO_3$ (m)	Z	thickness of fertilizer pellet (m)
K	permeability coefficient		

1. Introduction

Use of coated fertilizers (slow release fertilizers or controlled release fertilizers) is a new trend to reduce fertilizer wastage and to minimize environmental pollution [1, 6]. Coated fertilizers are made to release their nutrient contents gradually preferably in synchronization with the nutrient requirement of plants. These fertilizers can be physically prepared by coating granules of conventional fertilizers with various materials that reduce their dissolution rate [7]. The release and dissolution rates of water-soluble fertilizers depend on the coating materials. Suitable fertilizers for coating are those containing any type of water soluble fertilizer compounds [9]. These may also comprise a core fertilizer material comprising at least one soluble salt of nitrogen (N),phosphorus (P) or potassium (K) [5]. The combination of compositions may also suitable such as N-P, N-K, P-K, and N-P-K fertilizers [10]. Numerous coating materials for slow and controlled release fertilizers are those having good water barrier properties as proposed by Russell et al. [8], coating powders such as silica compositions and silicates, diatomaceous earth, clays and several inorganic materials. .In this study fertilizer pellets containing coats consisting of different proportions of analytical grade calcium carbonate ($CaCO_3$) and nano-$CaCO_3$ of are examined for the release rates of fertilizer constituentsproposed by Backlund et al. [2], sulphur, polymer and latex coating materials as proposed by Tijsma et al. [9], a medium density polyethylene wax and a tackifying resin as proposed by Gossett et al. (1993), as well as the carboxyl-carrying ethylene copolymer with some or all of carboxyl groups may also be in the form of fertilizer salts such as zinc, sodium, ammonium salts, etc. as proposed by Von Locquenghien et al. [10].

2. Materials and methods

2.1 Materials

Two fertilizers namely, calcium dihydrogen phosphate Ca $(H_2PO_4)_2$ and ammonium sulfate $((NH_4)_2SO_4)$ were employed in this study. These chemicals were of Aldrich analytical grade. Sand of uniform average size of 0.075mm sorted out by a sieve separator was used as the inert material for the preparation of the pellet. The materials used to coat the fertilizer were nano-$CaCO_3$ of average 160nm particle size with 98% purity. Analytical grade CaO supplied by Sigma Aldrich was used to produce micro sized $CaCO_3$ by means of reaction of CaO with 99.9% CO_2 in a stainless steel vessel until complete conversion of CaO to $CaCO_3$ is achieved according to the stoichiometric ratio. An average particle size of 10microns was observed for the $CaCO_3$ so produced.

2.1.1 Equipment

Table 1 lists the equipment employed in this study.

2.2 Methods

2.2. 1 Preparation of fertilizer and micro sized $CaCO_3$

The average percentage of phosphorus in commercial phosphate fertilizers is 16.63 %wt. However, the analytical grad of calcium dihydrogen phosphate ($Ca(H_2PO_4)_2$) is 26.50 %wt. Therefore, every 16.06g of $Ca(H_2PO_4)_2$ is mixed with 10.00g of sand in order to achieve the commercial phosphate composition in test sample, The average nitrogen content in ammonium sulphate fertilizers is 21 %wt which was found to be identical with the analytical grade and therefore no inert material was required.

2.2.2 Experimental procedure

A desired amount of fertilizer was first poured into a 2cm cylindrical mould made of stainless steel followed by $CaCO_3$ powder. The fertilizer and the $CaCO_3$ were pressed into a single pellet by a hydraulic press. The pressures employed for the preparation of fertilizer pellets n this study was 180psi.

A desired thickness of pellet was obtained by adjusting the amounts of fertilizer and $CaCO_3$ determined by trial test conducted previously. Each of the coated fertilizer pellets together with the respective mould were attached to the bottom of a cylindrical mass transfer vessel by means of a screw joint, By this way the coated surface is positioned upwards towards the cylinder. Distilled water is carefully filled on to each cylinder avoiding any turbulence at the top of the pellet, ten molds and cylinders were used for a given experiment. Sample solutions were withdrawn from each cylinder at different times. Each sample solution was analyzed (i.e. to measure its concentration) by a spectrophotometer manufactured by HACH Co., model DR2800. The reagent applied in the measurement of concentration of Ca $(H_2PO_4)_2$ and $(NH_4)_2SO_4$ were PhosVer 3 Phosphate and SulfaVer 4 reagent respectively. Meanwhile, the method used to measure the sample solutions of $Ca(H_2PO_4)_2$ and $(NH_4)_2SO_4$ fertilizer were Method 8007 (Per sulfate UV Oxidation Method) and Method 8051 (SulfaVer 4 Method) respectively. A dilution method is necessary when the concentrations of sample solutions are over the maximum concentration that could be measured by this spectrophotometer.

Table 1. List of equipment components used and their function.

Component	Model name	Purpose of use
Hydraulic press	-	To produce pellet of $CaCO_3$ and fertilizer powders.
Portable spectrophotometer	HACH DR2800	To analyze phosphate, sulphate and nitrate content in water.
Digital balance	Mettler Toledo	To weight the $CaCO_3$, Ca $(H_2PO_4)_2$, and $(NH_4)_2SO_4$ powder.
Cylindrical cells for mass transfer(ID= 2cm)	Custom made	To contain the coated fertilizer with water for the mass transfer experiment.
Oven	Memmert	For drying purpose of material(s).
Sieve shaker	Recsch AS200	To obtain a uniform size of sand used as the inert material in the experiment.

3. Analysis of results and discussion

Fig. 1 shows the plot of concentration of phosphate in milligrams per litre versus time in minutes for the fertilizer pellets coated with a thin layer of 10 microns $CaCO_3$ (i.e. l = 0.0006m) without mixing of nano-$CaCO_3$. The concentrations of sample solutions are measured by a spectrophotometer with suitable method and reagent. In Fig. 1, it shows the trend-line of different pellet thicknesses (Z in m) is about similar to each other, where the concentration of phosphate in distilled water is increasing with time. The similar trend-line for the plot of concentration of phosphate (mg/L) versus time (min) for the fertilizer pellets coated with a thin layer of $CaCO_3$ with mixing of nano-$CaCO_3$ (0%, 33.33%, 66.67% and 100%) shown in Fig. 2. The nano-$CaCO_3$ used has particle size of less than 160 nm. Different percentages of nano-$CaCO_3$ were used to study how the nano-$CaCO_3$ will be performed differently with the larger size of $CaCO_3$ (i.e. 10 micron). In Fig. 2, it is obvious that the more percentages of nano-$CaCO_3$ added, the slower the transfer rates or the concentration of phosphate releases into the environment. Fig. 3 shows the plot of concentration of sulphate in milligrams per litre versus time in minutes for the fertilizer pellets coated with a thin layer of 10 microns $CaCO_3$ without mixing with nano-$CaCO_3$. In Fig. 3, it shows the trend-line of different thickness (Z in m) is about similar to each other, where the concentration of sulphate in distilled water is increasing with time. Similarly, in Fig. 4, it is obvious that the more percentages of nano-$CaCO_3$ added, the slower the transfer rates or the concentration of sulfate releases into the environment.

The experimental data were examined with the diffusion model presented by Fernando et al. [4] for mass transfer across surface layers. Here the model equation for the mass transfer rate across a layer [3] taking into consideration the concentration boundary layers at the surface coat can be presented as

$$Z\frac{dC}{dt} = k_p H.C_o.\frac{\sqrt{Dt}}{l}\left[\frac{2r}{(r+1)} \times \frac{1}{\pi} - \frac{4r^2}{(r+1)^2}.ierfc\left(\frac{l}{2\sqrt{Dt}}\right)\right]$$

$$\frac{-Zd(C_{Av})}{dt} = k_c.C_0\frac{\sqrt{Dt}}{l}\left[\frac{2r}{(r+1)} \times \frac{1}{\sqrt{\pi}} - \frac{4r^2}{(1+r)^2} \times ierfc\left(\frac{l}{2\sqrt{Dt}}\right)\right] \quad (1)$$

Equation (1) is derived on the assumption that the primary mass transfer resistance occurs across the surface layer/coating and surface boundary layers and that subsequent mass transfer takes place through a liquid film at the outer surface of the coat represented by $k_c A \Delta C$.

Tables 2 and 3 summarize the values of parameters $K_c.C_o$, r, l/D obtained by using Matlab™ program. With a knowledge of the value of l/D, the parameter D was predicted.

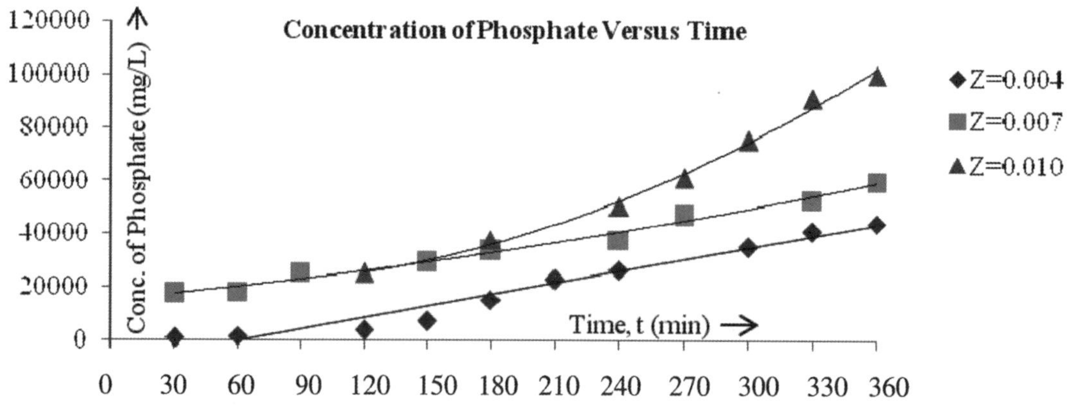

Fig. 1. The concentration of phosphate (mg/L) versus time (min) for the fertilizer pellets coated with 0.0006m $CaCO_3$.

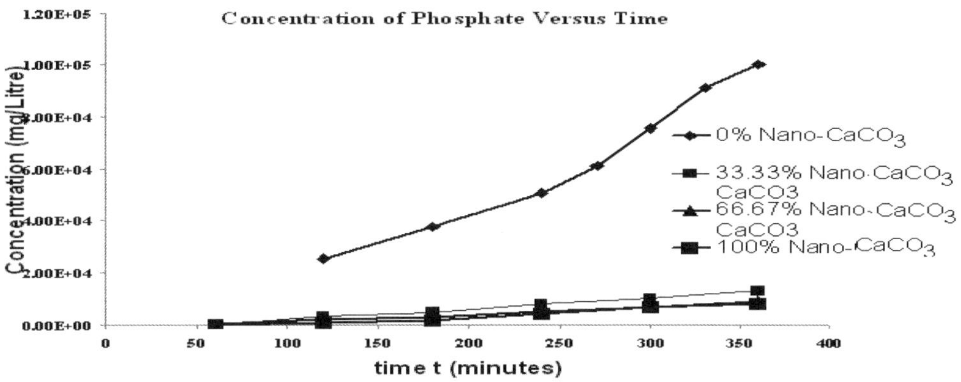

Fig. 2. The concentration of phosphate (mg/L) versus time (min) for the fertilizer pellets coated with different percentages of nano-$CaCO_3$.

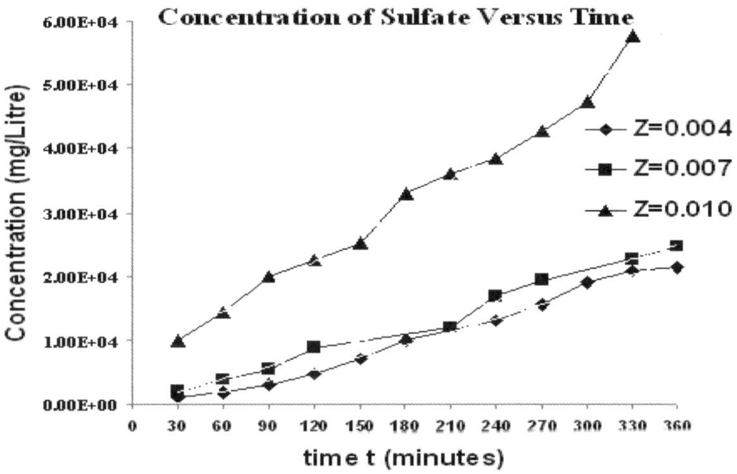

Fig. 3. The concentration of sulphate (mg/L) versus time (min) for the fertilizer pellets coated with 0.0006m microns $CaCO_3$ without nano-$CaCO_3$.

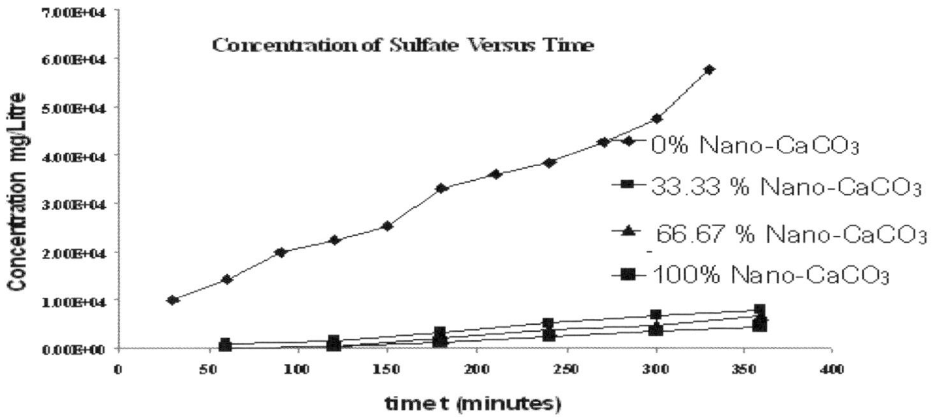

Fig. 4. The concentration of sulphate (mg/L) versus for the fertilizer pellets coated with different percentages of nano-$CaCO_3$.

Table 2. Values of parameters Kc.Co, r, l/D and D for Ca $(H_2PO_4)_2$ fertilizer pellets (l=0.0006m, Z=0.01m) obtained from $Matlab_{TM}$.

$Ca(H_2PO_4)_2$							
0% nano-$CaCO_3$				33.33% nano-$CaCO_3$			
$K_c.C_o$ *(kg.min/m⁴)*	*r*	*l/D (m/min)*$^{0.5}$	*D* (m^2/min).	$K_c.C_o$ *(kg.min/m⁴)*	*r*	*l/D (m/min)*$^{0.5}$	*D* (m^2/min).
0.9821	0.9807	2.7367	0.0045	0.8295	0.9985	1.5896	0.0015
66.67% nano-$CaCO_3$				100.00% nano-$CaCO_3$			
0.2840	0.8179	0.8352	0.0004	0.6800	0.0200	1.3369	0.0011

Table 3 Values of parameters Kc.Co, r, l/D and D for $(NH_4)_2SO_4$ fertilizer pellets (l=0.0006m, Z=0.01m) obtained from Matlab^{TM}.

$(NH_4)_2SO_4$							
0% nano-$CaCO_3$				33.33% nano-$CaCO_3$			
$K_c.C_o$ $(kg.min/m^4)$	*r*	*l/D* $(m/min)^{0.5}$	*D* (m^2/min).	$K_c.C_o$ $(kg.min/m^4)$	*r*	*l/D* $(m/min)^{0.5}$	*D* (m^2/min).
1.7969	0.9667	2.7427	0.0045	0.3931	0.8727	0.4516	0.0001
66.67% nano-$CaCO_3$				100.00% nano-$CaCO_3$			
0.0865	0.1629	1.6433	0.0016	2.1002	0.0017	2.7610	0.0046

4. Conclusion

An analysis of the above results show that the diffusion coefficient of the phosphate as well as ammonium sulphate decreased when the nano-content of the coat increased up to about 66.7% of nano $CaCO_3$ and thereafter increased. This is expected because mico as well as nano packings could be expected to have a similar void fraction whereas the mixtures of nano with micro could be expected to produce packings of lesser void fractions due to inclusions of nano particles in micro sized pores leading to lesser migration of constituents. A similar trend was observed for the parameter r which behavious could be explained on similar lines as above. The parameter kcCo showed a lower value at mid ranges of nano content (i.e. at 33.33% and 66.77% $CaCO_3$) indicating a dominant resistance of the liquid film to mass transfer in the mid ranges compared with respective resistance of the relatively pure components in the coat . In the overall, it can be observed that nano and micro $CaCO_3$ could be nixed in order to generate mixtures with specific parameters of Kc.Co, r and D. Selection of a coat thickness (l) determines the parameter l/D which eventually determines the dissolution rate presented by equation (1) thereby enabling possibilities to engineer a coat of a nano-micro mix of $CaCO_3$ in coats of fertilizers.

Acknowledgements

The authors wish to acknowledge the financial support and facilities received from the e-Science grant of Ministry of Science, Technology and Innovation of Malaysia and the FRGS grant.

References

1. A. Akelah, Novel utilizations of conventional agrochemicals by controlled release formulations, Materials Science and Engineering, C4, 83–98 (1996).
2. A. Backlund, Method for producing Granular Coated Fertilizer, U.S. Patent 4565464 (1986).
3. A. Dworecki, B. Slezak, S. Ornal-Wasik, Wasik, Evolution of concentration field in a membrane system, J. Biochem. Biophys. Methods 62 (2), 153 (2005).
4. W.J.N. Fernando, A.L. Ahmad, S.R. Abd Shukor and Lok, Y.H., A model for constant temperature drying rates of case hardened slices of papaya and garlic, J. Food. Engineering, 229-238 (2008).
5. Jr. Gossett, S. Kirk, Coating Compositions for Slow Release Fertilizer Formulations U.S. Patent 5264019 (1993).
6. A. Jarosiewicz, & M. Tomaszewska, M., Controlled-release NPK fertilizer encapsulated by polymeric membranes. Journal of Agricultural and Food Chemistry, 51, 413–417 (2003).
7. M.V. Hopewell, Vacuum coated particulate fertilizers, U.S. Patent 6080221 (2000).
8. Russell et al., Apparatus and Process for Coating Particles (to ICI Canada Inc, Ontario, Canada), U.S. Patent 5399186 (1995).
9. Tijsma, Antonius, Aalto, and Adrianus van K., Controlled Release Fertilizer Compositions and Process for the Preparation Thereof, U.S. Patent 5993505 (1999).
10. Von Locquenghien, Engelhardt, Kleinbach, Muller Coated Fertilizer Granules, U.S. Patent 6187074 (2001).

A CASE STUDY OF REGIONAL MUNICIPAL SOLID WASTE MANAGEMENT PROGRAM IN TAIWAN

YAO-JEN CHANG

Department of Environmental Engineering, National Chung Hsing University, Taichung, Taiwan, Republic of China

MIN-DER LIN

Department of Environmental Engineering, National Chung Hsing University, Taichung, Taiwan, Republic of China

SHUANG-FU YEH

Department of Environmental Engineering, National Chung Hsing University, Taichung, Taiwan, Republic of China

Municipal solid waste management (MSWM) is an important, practical and challenging environmental subject. Regionalization of MSWM program is more complicated due to the involvement of multi-municipality and multi-facility issues. Therefore, an effective technique capable of solving regional MSWM problems is necessary for decision makers. This paper employed linear programming techniques to establish a optimization model which is able to achieve regional MSWM strategies with the minimum costs. The results of investigating a real world case study in central Taiwan indicate that regionalization of MSWM program is more economical and efficient.

Keywords: Municipal solid waste, regionalization, optimization, linear programming.

1. Introduction

Municipal solid waste management (MSWM) is a major concern of highly urbanized societies. Over the past 10 years, several studies have utilized different techniques, such as linear programming (LP), nonlinear programming, mixed integer linear programming, and multi-objective programming, to investigate MSWM problems.[1, 2, 3, 4, 5, 6, 7] Due to the involvement of multi-municipality and multi-facility issues, regionalization of MSWM program becomes more complicated, and the development of an effective analytical tool will be very helpful for decision makers.

This paper employed LP techniques to establish a MSWM optimization model that is capable of solving large scale regional MSWM problems. Furthermore, a case study of central Taiwan area will also be investigated. The optimal strategies obtained by realistic operation parameters will be presented to demonstrate the effects of the regionalization of MSWM program.

2. Case study: central Taiwan area

The central Taiwan area, as shown in Fig. 1, is located in middle Taiwan and has a total area of 10,506 km^2. The area is composed of six municipalities including Miaoli County, Taichung city, Taichung county, Nantou county, Changhua county, and Yunlin county, which are divided into 18, 8, 21, 13, 26, and 20 administrative districts, respectively. The daily MSW generation of Miaoli County, Taichung city, Taichung county, Nantou county, Changhua county, and Yunlin county is approximately 325, 482, 806, 270, 775, and 370 tons, respectively. Currently, there are six incinerators, namely Jhunan, Nantun, Houli, Wurih, Sijhou, and Linnei, whose maximum treatment capacity is 500, 900, 900, 900, 900, and 600 ton/day, respectively[8]. The incineration ashes are disposed by private treatment plants authorized by government.

3. Model development

The optimization model developed in this study aims at analyzing the most economical strategies for the waste stream allocation of large scale regional MSWM problems. The MATLAB^R software package by Mathworks was used in this work for the development of the optimization models.

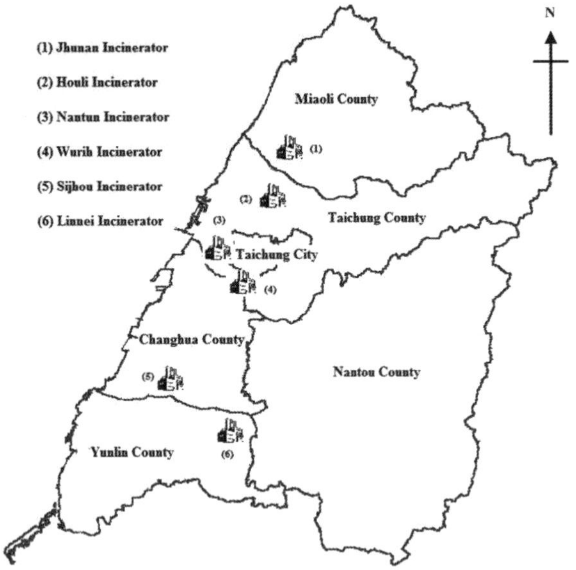

Fig. 1. Locations of the incinerators and municipalities in the central Taiwan.

3.1. *Optimization model*

This section describes the mathematical formulation of the optimization model used in the study. The items considered in the optimization model include: (1) transportation costs of the MSW from collection station to incinerators; (2) treatment costs for incinerating MSW; (3) waste-to-electricity efficiencies of incinerators; and (4) disposal costs of incineration ashes. The mathematical model is established using the Optimization toolbox of Matlab^R software package.

The objective function of the optimization model is to minimize the total cost of the MSWM strategy, it can be expressed as the following:

$$Minimize\ CT + CO - EI \tag{1}$$

$$CT = \sum_{i=1}^{n} \sum_{j=1}^{m} T_t \times D\_di_{ij} \times MSW_{ij} \tag{2}$$

$$CO = \sum_{j=1}^{m} C_{Inc} \times MSW_j + C_{Ash\_auth} \times ASH\_auth \tag{3}$$

$$EI = \sum_{j=1}^{m} P \times Coef\_elect_j \times MSW_j \tag{4}$$

where CT is the transportation cost (NT\$/day) of shipping MSW and incineration ash from their sources to incinerators; CO is the total treatment costs (NT\$/day) including the treatment cost of incinerating MSW and disposal cost of incineration ash; EI is the revenues (NT\$/day) from incinerators by selling the generated electricity; T_t is the unit transportation cost (NT\$/ton-km); D_di_{ij} is the distance (km) from administrative district i to incinerator j; MSW_{ij} is the amount of MSW (ton/day) transported from district i to incinerator j; C_{Inc} is the unit treatment cost (NT\$/ton) of incinerator j; MSW_j is the amount of MSW (ton/day) treated in incinerator j; C_{Ash_auth} is the unit treatment cost (NT\$/ton) of incineration ash; ASH_auth is the quantity of incineration ash (ton/day); P is the unit price (NT\$/kwh) of electricity sales; $Coef\_elect_j$ is the waste-to-electricity transformation coefficient (kwh/ton MSW) of incinerator j. The objective function above is conditioned by the following constraints.

For each district, the MSW it generates should be shipped to the available incinerators. For each incinerator, the mass conservation law should be satisfied, and their designed treatment capacity should not be exceeded:

$$Gene_i = \sum_{j=1}^{m} MSW_{ij} \quad \forall i \tag{5}$$

$$MSW_j = \sum_{i=1}^{n} MSW_{ij} \quad \forall j \tag{6}$$

$$MSW_j \leq CAP\_inc_j \quad \forall j \tag{7}$$

where $Gene_i$ is the MSW generated by district i (ton/day); MSW_{ij} is the quantity of MSW transported from district i to incinerator j (ton/day); CAP_inc_j is the designed treatment capacity (ton/day) in incinerator j.

For each incinerator, the incineration ash should be shipped to available ash disposal facilities:

$$\sum_{j=1}^{m} Coef_j \times MSW_j = ASH\_auth \quad \forall j \tag{8}$$

where, $Coef_j$ is the incineration ash generated rate per unit mass of MSW incinerated at incinerator j (ton/ton); ASH_auth is the total quantity of ash (ton/day).

3.2. *The input/output interface*

The input/output interface was developed using a graphical user interface design environment (GUIDE) of Matlab^{R}. This module has graphic and interactive facilities that offer a friendly environment for the user to define all needed data, and to run the optimization model.

As the input/output interface shown in Fig. 2, users can key in the parameters required for the optimization model through input interface, and the output interface will present the computation results. The input parameter includes the unit transportation cost, administrative district selection, MSW generation, treatment cost, incinerator design capacity, incineration ash generation rate, waste-to-electricity transformation coefficient, sell price of electricity, the selection of incineration ash disposed pattern, and incineration ash disposal cost.

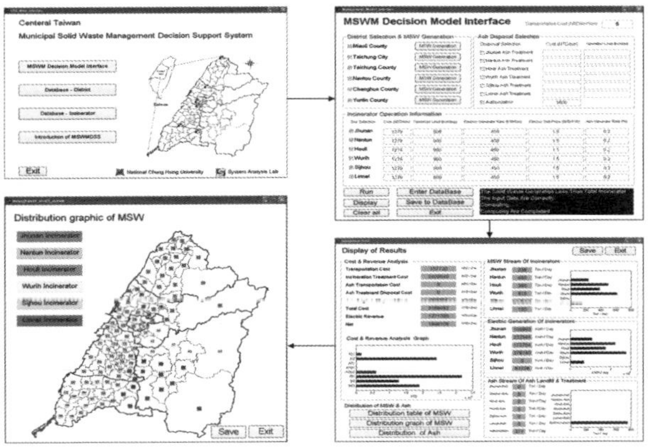

Fig. 2. The interface of the optimization model, including the main, input, and output interfaces, and spatial analysis graphs of MSW stream allocation.

The output interface includes the following items: (1) the cost/revenue analysis; (2) the spatial distribution of MSW; (3) the electricity generation of incinerators; and (4) the amount of incineration ash treated. In addition, the interface also provides the spatial analysis capability to illustrate stream/allocation of MSW from administrative districts to incinerators. For large scale regionalization MSWM problems, the optimization model not only raises the efficiency of decision-making process, but also provides more information for decision makers to deeply understand the management strategies. Furthermore, the decision makers can easily notify the changes when management strategies are modified.

4. Alternative strategy

According to the current circumstances in the central Taiwan area, this study proposed two strategies from practical points of view to assess their economic impacts to the six municipalities. The scenario of each strategy is described below:

Case 1: each municipality treats its own MSW, and the MSW of Nantou county is treated in Wurih incinerator.

Case 2: each municipality can pay other municipalities treatment cost to treat its MSW.

The influence of the multi-municipality MSWM program can be demonstrated by comparing the results of cases 1 and 2.

5. Results and discussion

The treatment cost used in this study is 1,279 NT$/ton and the unit transportation cost is 5NT$/ton-km. The waste-to-electricity transformation coefficients of Jhunan, Nantun, Houli, Wurih, Sijhou, Linnei are 522.73, 414.87, 571.40, 592.07, 512.58, and 522.73 KWh/ton, respectively, and the ash generation rates are 20.86%, 18.37%, 22.50%, 20.41%, 22.16%, and 20.86%, respectively.

The cost/revenue analysis data of the multi-municipality MSWM operational program are summarized in Table 1. It can be found that the net daily costs of cases 1 and 2 are NT$3,044,322 and NT$2,849,413, respectively. These results indicate that the regional multi-municipality MSWM program (case 2) is more economical and can achieve a daily cost saving of 6.5%. It is due to the regional MSWM program ignores the constraints of municipality boundaries, so not only a more efficient allocation of MSW stream can be achieved (reduce 9% transportation cost), but also increase the electricity sale revenue (increase 7.5% revenue).

6. Conclusion

A prototype optimization model that is designed to help decision makers drawing up feasible MSWM strategies has been developed. It includes numerous unit processes related to transportation, treatment, waste-to-electricity, and disposal of MSW. The primary components of the optimization model that provide this decision support environment have been presented. These components include an LP optimization model, input interfaces, and

Table 1. The summary of cost/revenue and MSW stream allocation for different strategies.

Alternative		Case 1	Case 2
Cost / Revenue (NT$/day)	Transportation cost	333,113	303,619
	MSW treatment cost	3,872,938	3,872,938
	Ash disposal cost	1,037,013	1,037,013
	Total cost	5,243,064	5,213,570
	Electric revenue	2,198,742	2,364,157
	Net	3,044,322	2,849,413
MSW treated in incinerators (Ton/day)	Jhunan	325	238
	Nantun	482	0
	Houli	402	851
	Wurih	674	900
	Sijhou	775	505
	Linnei	370	534
Incineration ash generated in incinerators (Ton/day)	Jhunan	68	50
	Nantun	88	0
	Houli	90	191
	Wurih	138	184
	Sijhou	172	112
	Linnei	77	111

a graphical output interface. This study, based on real-world data, demonstrates that a regionalization policy is a better strategy for metropolitan MSW management. Therefore, this tool and results is very beneficial for regional MSW management policymaking.

References

1. Chang, N.B., Wang, S.F. Solid waste management system analysis by multi-objective mixed integer programming model. *Journal of Environmental Management* 48, 17-43(1996).
2. Chang, N. B., & Lin, Y. T. Economic evaluation of a regionalization program for solid waste management in a metropolitan region. *Journal of Environmental Management, 3*(51), 241-274(1997).
3. Lin, M. D., Wang, C., & Lin, C. Evaluation of solid waste management strategies in the Taipei metropolitan area of Taiwan. *Journal of the Air and Waste Management Association, 56*, 650-656(2006).
4. Wang, C., Lin, M. D., & Lin, C. Factors Influencing Regional Municipal Solid Waste Management Strategies. *Journal of the Air and Waste Management Association, 58,* 957-964(2008).
5. Chang, N.B., Lin, Y.T. Optimal Siting of transfer station locations in a metropolitan solid waste management system. Journal of Environmental Science and Health A 32 (8), 2379-2401(1997).
6. Harrison, K. W., Dumas, R. D., Solano, E., Barlaz, M. A., Brill, D. J., & Ranjithan, S. R. Decision support tool for life-cycle-based solid waste management. *Journal of Computing in Civil Engineering, 15*(1), 44-58(2001).
7. Chang, N. B., Davila, E., Dyson, B., & Brown, R. Optimal design for sustainable development of a material recovery facility in a fast-growing urban setting. *Waste Management, 25*(8), 833-846(2005).
8. Yearbook of Environmental Protection Statistics, Taiwan Area, the Republic of China. R.O.C.; Environmental Protection Administration: Taiwan, Republic of China, 2007. pp 2-82-2-106.

AN EXPERIMENTAL STUDY ON THE EFFECT OF ENGINE SPEED VARIATION ON PREMIXED RATIO, EQUIVALENCE RATIO AND EMISSIONS OF DUAL FUEL HCCI ENGINE

M. R. KALATEH and M. GHAZIKHANI

Department of Mechanical Engineering, Ferdowsi University of Mashhad, P.O. Box No. 917751111, Mashhad, Iran

In this study, effect of engine speed variation on premixed ratio, equivalence ratio and emissions of a dual fuel HCCI engine are investigated. The experiments were conducted on a variable compression ratio (VCR) single-cylinder research engine with compression ratio of 17.5:1. Premixed gasoline was introduced to inlet manifold through a carburetor which is equipped with a needle screw to adjust the gasoline mass flow rate. Results show that the CO, HC and soot emissions increases by increasing the engine speed due to less homogeneity in the mixture and decreasing the overall reactivity, also the premixed fuel and therefore the equivalence ratio increases as a result of engine speed increase in HCCI combustion.

Keywords: Dual fuel HCCI engine, engine speed, equivalence and premixed ratios, emissions.

1. Introduction

Reducing exhaust emissions and increasing the fuel efficiency of internal combustion engines are of global importance. In this regard, diesel engines are attractive because of their relatively high efficiency. Unfortunately, they produce greater NO_X emissions and particulate matter (PM) than spark-ignition engines. However, because diesel combustion involves turbulent diffusion flames that produce locally rich low-temperature and locally lean high-temperature regions, it is difficult to reduce both NO_X and PM simultaneously and still maintain high efficiency. One way to achieve both of these objectives would be to use homogeneous charge compression ignition (HCCI) combustion [1]. HCCI was identified as a distinct combustion phenomenon about 30 years ago. The initial papers of what may be considered as the modern era of HCCI research was issued by Onishi et al. and subsequently by Noguchi et al. These researches first successfully applied HCCI concept to a two-stroke gasoline engine with high levels of recycled or trapped residual gas [2].The main concepts of HCCI are breathing premixed air/fuel mixture, as in conventional spark ignition (SI) engines, and ignition without a spark plug, as in conventional compression ignition (CI) engines [3]. Ideal HCCI combustion is characterized by the lean and low-temperature reactions that are initiated at multiple sites simultaneously without any flame propagation [4]. HCCI combustion is the process in which a homogeneous mixture is auto-ignited by the compression from the piston motion, so the fuel chemical kinetics plays a dominate role during the whole combustion process. Also, auto-ignition occurs at various locations in the combustion chamber (multi-point ignition), which may cause extremely high rates of heat release, and consequently, high rates of pressurization [5, 6]. Depending on the fuel, the auto-ignition can proceed in one-stage, two-stage or multi-stage ignitions. This confirms that the fuel chosen is crucial for the combustion process, since the chemical kinetic pathways can be considerably different [7]. Since the HCCI combustion is lean burn and occurs without flame propagation, it provides much lower combustion temperature than that of conventional SI and CI engines. As a result, HCCI combustion produces very low levels of NOx and particulate matter (PM) while maintaining high thermal efficiency. However, greater amounts of hydrocarbon (HC) and carbon monoxide (CO) emissions are released in HCCI engine relative to conventional SI and CI engines [3, 8-11].

HCCI combustion has been considered as a potential substitute for traditional engines. However, the HCCI combustion system has several challenges, including the controlling auto-ignition timing over a wide range of speeds and loads, the improvement of cold-start capability (especially for the HCCI engine with a low compression ratio or with high octane rating fuel), the expansion of operating range and meeting emission standards [12-14]. The lack of a well-defined ignition timing control has led researchers to explore a

Nomenclature	
HCCI	Homogeneous Charge Compression Ignition
EGR	Exhaust Gas Recirculation
BTDC	Before Top Dead Center
DI	Direct injection
CI	Compression Ignition
SI	Spark Ignition
r_p	Premixed ratio
h_{up}	Heating value of premixed fuel (kJ/kg)
$\dot{m}_p$	Mass flow rate of premixed gasoline (kg/s))
$\dot{m}_d$	Mass flow rate of injected fuel (kg/s)
h_{ud}	Heating value of diesel fuel (kJ/kg)
$\dot{m}_{aEGR}$	Mass flow rate of intake air with EGR (kg/s)
$\dot{m}_{EGR}$	Mass flow rate of EGR gases (kg/s)
ϕ	Equivalence ratio
MON	Measured octane number
$(A/F)_s$	Stoichiometric air-fuel ratio
UHC	Unburned hydrocarbon
CO	Carbon monoxide
CO_2	Carbon dioxide
NO_X	Oxides of nitrogen
PM	Particulate matter

range of control strategies, such as variable valve timing (VVT), variable compression ratio (VCR), EGR and multi-pulse injection [12,15]. Now, it is generally accepted that the EGR technique is an efficient way to control HCCI combustion. HCCI combustion can occur in internal combustion engines by varying the inlet air temperature and exhaust gas recirculation (EGR) fraction over a range of equivalence ratios [15,16].

EGR is widely used as the main method to depress the NO_X emission from diesel engines. Currently, EGR is also used as the basic method to control the ignition timing and burn rate of HCCI combustion [16]. EGR consists of many gaseous chemical species, which includes the main components of burned gases, CO_2, H_2O, N_2 and O_2, partial burned gases such as CO, particular matters, HCs and high temperature combustion products NO_X. Different species has different heat capacity and chemical reactivity, therefore has different effect towards ignition timing and heat release rate of HCCI combustion [17]. The application of EGR on HCCI combustion engine has a number of effects on the combustion process and emissions. The effect of EGR on HCCI combustion can be divided into three parts: a dilution effect (inert gasses present in the EGR), a thermal effect (heat exchange, thermal loss to the wall, EGR ratio mixture quality, EGR temperature) and a chemical effect. The diluting effect influences the auto-ignition process, by influencing directly the concentrations along with the kinetics. With the thermal effect one can influence the overall kinetics directly by the reaction rates. The chemical effect, however, is more complex. It not only influences the overall kinetics, but it also can change a specific reaction path [6, 16-18].

The objective of this study is to investigate the effect of engine speed variations on premixed ratio, equivalence ratio and emissions of a dual fuel HCCI engine using premixed gasoline.

2. Experimental setup and procedure

2.1. *Experimental setup*

In this study all experiments were conducted on a four stroke VCR single-cylinder naturally-aspired research engine with a displacement volume of 582 cm^3, which the test rig is TD43 model equipped by Techquipment Co. The engine has a bowel type piston with a bowel diameter of 41.5 mm. The base engine can operate as SI engine with the compression ratio in the range of 7 to 11, and also can convert to diesel type (spark plug is replaced by an injector) in the range of 14 to 18 for the compression ratio. Premixed gasoline is introduced to the inlet manifold by means of a carburetor which is mounted on 180 mm from intake valve upstream and equipped with a fuel adjustment needle screw. Diesel fuel is injected into the cylinder through an injector. The specifications of the test engine are listed in the table1.

The experimental apparatus is composed of electrical dynamometer, AVL-415 smoke meter, RE 205 Plint exhaust analyzer, which measures HC (as C_6), K-type thermocouples, 220 V single-phase 2kW intake air heater which placed in the air flow stream, external EGR system, adjustable coolant system and air mass flow meter (surge tank and orifice system). Fig. 1 shows the schematic diagram of the experimental apparatus.

The electrical dynamometer which is connected directly to the engine plays the role of rotating initiator in starting mode, and when running the engine, consumes the power output for generating electricity. The stator of dynamometer can rotate freely around its shaft and as a result of engine torque during power

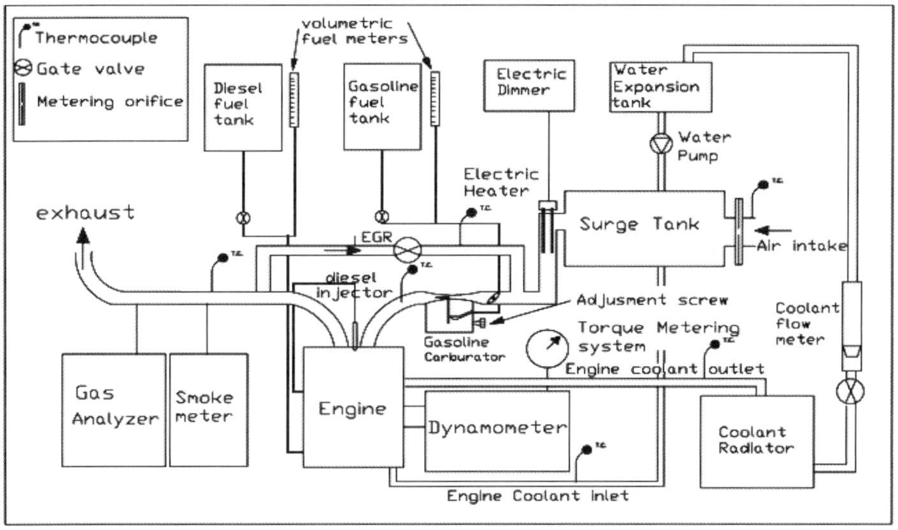

Figure 1: Schematic diagram of experimental apparatus.

Table 1: Engine specifications

Engine	Four stroke
Number of cylinders	1
Compression ratio	17.5:1
Displacement volume	582 cm^3
Bore	82 mm
Stroke	95 mm
Injection mode	DI
Number of injection holes	4
Injection pressure (bar)	250
Gasoline fueling	Carburetor

generation; it is forced out from the horizontal equilibrium position. Using a Newton-meter on a known-length beam, for bringing the dynamometer back to the horizontal, the engine torque is measured.

2.2. *Experimental procedure*

In this work, the diesel engine was firstly started at idle position in the compression ratio of 17.5:1. Then intake charge temperature was adjusted on 115 ^{0}C. After a few minutes to achieve steady condition at constant cooling temperature of 50 ^{0}C, premixed gasoline was introduced to engine and torque increased gradually. Then by using an external EGR system different rates of EGR were introduced to get HCCI auto-ignition close to TDC. Table 2 shows the specifications of the fuels used in dual fuel HCCI engine. It should be noted that using early injection (35BTDC) in HCCI engine in compare with conventional CI engine was the necessity of initiating the auto-ignition of dual fuel HCCI engine.

To show the effect of fuels in dual fuel HCCI engine, the premixed ratio (r_p) is defined as a ratio of premixed fuel energy (Q_p) to total energy (Q_t). It can be obtained from the following equation [1]:

$$r_p = \frac{\dot{Q}_p}{\dot{Q}_t} = \frac{\dot{m}_p h_{up}}{\dot{m}_p h_{up} + \dot{m}_d h_{ud}} \quad (1)$$

In the equation (1), $\dot{m}_p$ represents mass flow rate of premixed gasoline, $\dot{m}_d$ is mass flow rate of injected fuel, hup is heating value of premixed fuel and hud is heating value of diesel fuel. Therefore, rp=1 corresponds to single fuel HCCI combustion and rp =0 corresponds to typical CIDI combustion.

EGR rate also is calculated as follow [25]:

$$EGR(\%) = \frac{\dot{m}_{EGR}}{\dot{m}_{EGR} + \dot{m}_{aEGR}} \times 100 \quad (2)$$

Where, $\dot{m}_{aEGR}$ is mass flow rate of intake air with EGR, and $\dot{m}_{EGR}$ is mass flow rate of EGR.

Different EGR rates were applied in each test by using a simple computer code. The code can estimate the orifice pressure drop at specific EGR rate in term of engine speed, ambient conditions and intake air properties at orifice.

In this study the intake charge temperature was increased up to 115 ^{0}C, which helps the fuel to overcome its activation energy and improves the pre-ignition chemical reactions. Altering the coolant temperature at the 40 to 70 ^{0}C, HCCI-DI combustion showed better results at 50 ^{0}C. Therefore, coolant temperature was maintained 50 ^{0}C throughout all the tests.

Engine tests were carried out at the operating conditions shown in Table 3, trying to modify every one of the main parameters affecting the oxidation kinetics of any fuel, such as the engine speed, injection timing and the charge composition (quantified by the EGR percentage). All these parameter combinations provide different levels of pressure and temperature in the cylinder. As it can be observed in Table 3, HCCI conditions were achieved by an early start of injection (SOI) diesel fuel, which promote the auto-ignition of gasoline premixed fuel due to heat releases by diesel fuel before HCCI combustion of premixed fuel. Having reached HCCI combustion, emissions and other data were recorded.

Table 2: Fuel specifications [19].

Fuel type	Gasoline	Diesel fuel
MON	87	-
Cetane number	-	54
Higher Heating Value (kJ/kg)	47300	46100
Lower Heating Value (kJ/kg)	44000	43200
Heat of vaporization (kJ/kg), at 1 atm, 25 ^{0}C)	305	270
Density (kg/m^3)	720	780
(A/F)s	14.6	14.5

Table 3: Test conditions.

Speed (rpm)	1200-1700
Intake charge temperature (^{0}C)	110-115
Coolant temperature (^{0}C)	50
EGR rate (based on mass flow rate of intake air)	0-15%
Injection timing	35 BTDC
Premixed ratio (r_p)	0-1

3. Results

3.1. *Variation of equivalence ratio with engine speed variation*

The variation of equivalence ratio due to engine speed variation in different EGR rates of a dual fuel HCCI engine is shown in Fig. 2. As it can be seen, the equivalence ratio has increased with increasing the engine speed. Increasing the engine speed reduces the time needed to prepare a homogeneous mixture, also the probability of mixing the EGR species with the air-fuel mixture reduces. Therefore the reactivity of low temperature reactions is reduced and retarded, which causes lower advanced premixed gasoline HCCI combustion. In this situation, for the main HCCI combustion to take place, the premixed fuel or the equivalence ratio has been increased, which increases the total reactivity and energy produced by the low temperature reactions, though the main HCCI combustion is advanced.

The important point in Fig. 2 is the comparison between the range of engine speed and equivalence ratio of a dual-fuel HCCI engine, in presence of EGR and the case of no EGR used. Increasing the EGR and air dilution causes the maximum temperature of the cylinder to decrease. In comparison with the case of no EGR used, the knock happens in higher speeds and equivalence ratio, so the EGR expands the speed range of dual fuel HCCI engine. Also EGR expands the range of equivalence ratio in dual fuel HCCI engine.

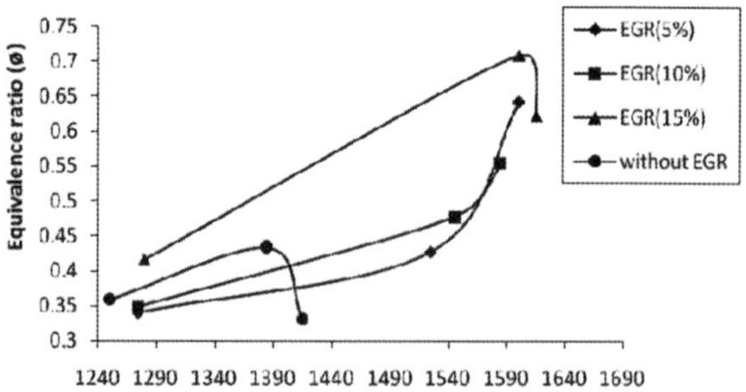

Figure 2: Variation of equivalence ratio with variation of engine speed in different EGR rates.

3.2. *Variation of premixed ratio with engine speed variation*

In Fig. 3 the variation of premixed ratio in a dual fuel HCCI engine is shown on the basis of engine speed variations for different EGR rates. It can be seen that by increasing the engine speed, the premixed ratio and therefore the premixed fuel is increased to catch up the target speed. Higher premixed ratio provides more homogeneous charge before the premixed auto-ignition happens. The upper limit of premixed ratio is restricted by knock phenomenon. As it is shown in Fig. 3 by introducing EGR higher premixed ratios is obtained, due to lower maximum temperature of in cylinder charge. In other words, by increasing the premixed ratio higher homogeneity and better HCCI combustion is obtained.

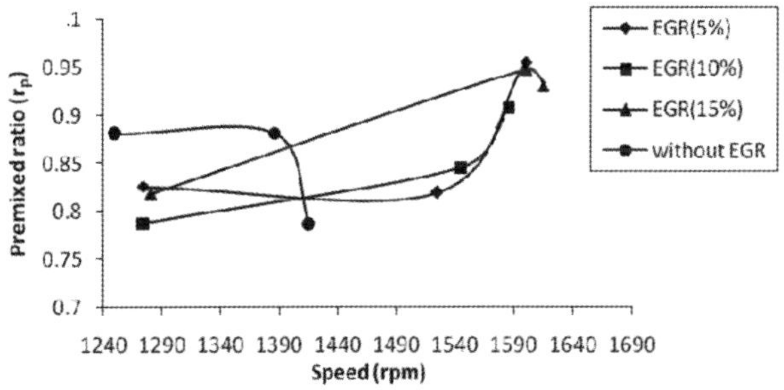

Figure 3: Variation of premixed ratio with variation of engine speed in different EGR rates.

3.3. *CO emission*

Fig. 4 shows the variations of equivalence ratio and CO emission due to speed variations, in a dual fuel HCCI engine. Fig. 4 shows that increasing the engine speed provides less time for the homogeneous mixture to form, so at first the CO emission increases a bit due to incomplete combustion. But increasing the equivalence ratio causes lower CO emission levels due to more radicals produced (including OH radicals) and also more reactivity of the system.

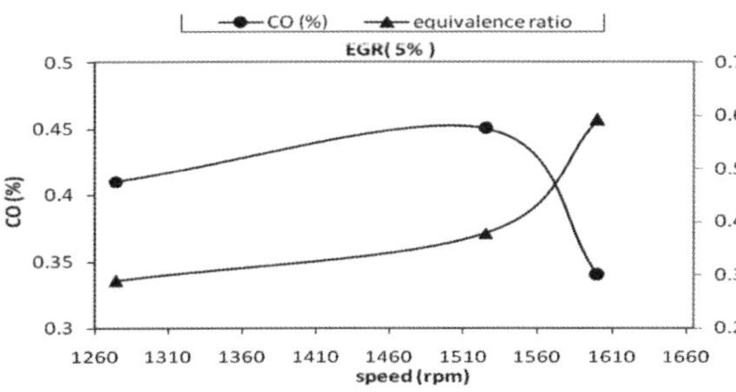

Figure 4: Variations of equivalence ratio and CO emission with engine speed variation.

3.4. *HC emission*

The variations of equivalence ratio and HC emission is shown in Fig. 5 on the basis of engine speed variations. Is can be seen that, increasing the engine speed, due to less homogeneity in the mixture and incomplete combustion, the HC emission increases. Also increasing the equivalence ratio provides more HC emission, because of more premixed fuel trapped in boundary layers and crevice which is hardly oxidized in low temperature HCCI combustion.

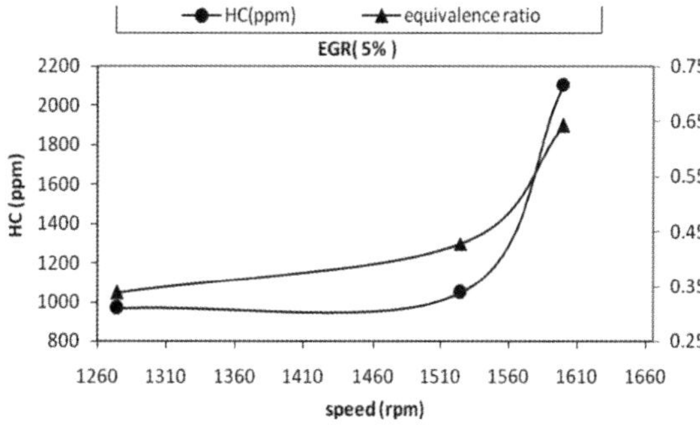

Figure 5: Variations of equivalence ratio and HC emissions with engine speed variation.

3.5. *Soot emission*

Fig. 6 shows the variations of equivalence ratio and soot emission on the basis of engine speed. It is shown that increasing the engine speed; the soot emission is increased due to less homogeneity and incomplete combustion.

In dual fuel HCCI engine the soot emission is produced in the first stage of the combustion, which is the diesel fuel auto-ignition. Increasing the engine speed provides less time for this product of diesel combustion to re-burn or participate in the premixed gasoline auto-ignition. So the soot emission of the engine increases.

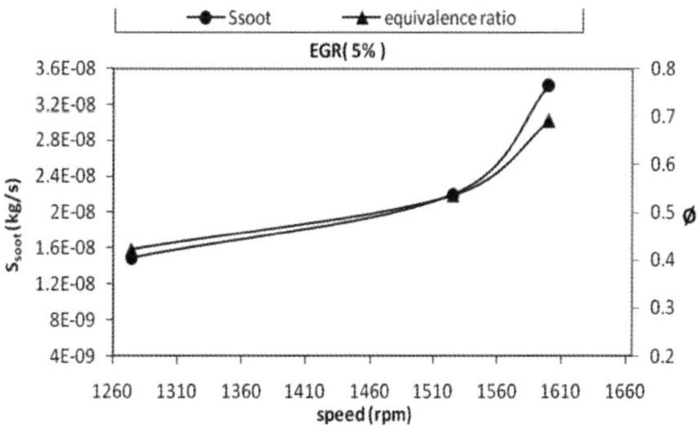

Figure 6: Variations of equivalence ratio and soot emission with engine speed variation.

4. Conclusions

In this study, the effect of engine speed variation on premixed ratio, equivalence ratio and emissions of dual fuel HCCI engine were investigated. Results can be summarized as:

1) Increasing EGR rate expands the range of speed and equivalence ratio in dual fuel HCCI engine due to dilution effect and lower maximum cylinder temperature.
2) The premixed fuel and therefore the equivalence ratio increases as a result of engine speed increase in HCCI combustion.
3) Increasing EGR rate causes higher premixed ratio and more homogeneity of the mixture, and improvement in HCCI combustion is obtained.
4) Increasing engine speed causes higher level of HC, CO and Soot emissions, due to less homogeneity in the mixture.
5) Increasing the equivalence ratio in parallel with the engine speed, causes lower level of CO emission as a result of increasing reactivity of the in-cylinder mixture, but HC emissions increased due to more premixed fuel trapped in crevice and boundary layers.

References

1. Sh. Tanaka, F. Ayala, J. C. Keck, J. B. Heywood, Two-stage ignition in HCCI combustion and HCCI control by fuels and additives, Combustion and Flame 132 (2003) 219–239.
2. X. C. Lu¨, W. Chen, Z. Huang, A fundamental study on the control of the HCCI combustion and emissions by fuel design concept combined with controllable EGR. Part 1. The basic characteristics of HCCI combustion, International Journal of Fuel, 84, 1074-1083, 2005.
3. K. Yeom, J. Jang, Ch. Bae, Homogeneous charge compression ignition of LPG and gasoline using variable valve timing in an engine, Fuel 86 (2007) 494–503.
4. D. S. Kim, Ch. S. Lee, improved emission characteristics of HCCI engine by various premixed fuels and cooled EGR, International Journal of Fuel, 85, 695–704, 2006.
5. M. Jia, M. Xie, A chemical kinetics model of iso-octane oxidation for HCCI engines, Fuel 85 (2006) 2593–2604.
6. A. C. Alkidas, Combustion advancements in gasoline engines, Energy Conversion and Management 48 (2007) 2751–2761.
7. H. Machrafi, S. Cavadias, Ph. Gilbert, An experimental and numerical analysis of the HCCI auto-ignition process of primary reference fuels, toluene reference fuels and diesel fuel in an engine, varying the engine parameters, Fuel processing technology 89 (2008) 1007-1016.
8. D. Ganesh, G. Nagarajan, M. M. Ibrahim, Study of performance, combustion and emission characteristics of diesel homogeneous charge compression ignition (HCCI) combustion with external mixture formation, Fuel 87 (2008) 3497–3503.
9. J. Andrae, D. Johansson, P. Bjrnbom, P. Risberg ,G. Kalghatgi, Co-oxidation in the auto-ignition of primary reference fuels and n-heptane/toluene blends, Combustion and Flame 140 (2005) 267–286.
10. H. Machrafi, K. Lombaert, S. Cavadias, P. Guibert, J. Amouroux, Reduced chemical reaction mechanisms: experimental and HCCI modelling investigations of autoignition processes of iso-octane in internal combustion engines, Fuel 84 (2005) 2330–2340.
11. J. Ma, X. Lü, L. Ji, Z. Huang, an experimental study of HCCI-DI combustion and emissions in a diesel engine with dual fuel, International Journal of Thermal Sciences, 47, 1235-1242, 2008.
12. M. Yao, Z. Chen, Z. Zheng, B. Zhang, Y. Xing, Study on the controlling strategies of homogeneous charge compression ignition combustion with fuel of dimethyl ether and methanol, Fuel 85 (2006) 2046–2056.
13. S. Soylu, Examination of combustion characteristics and phasing strategies of a natural gas HCCI engine, Energy Conversion and Management 46 (2005) 101–119.
14. W. Zeng, M. Xie, M. Jia, Numerical investigation on the application of catalytic combustion to HCCI engines, Chemical Engineering Journal 127 (2007) 81–93.
15. L. Shi, Y. Cui, K. Deng, H. Peng, Y. Chen, study of low emission homogeneous charge compression ignition (HCCI) engine using combined internal and external exhaust gas recirculation (EGR), Energy 31 (2006) 2665–2676.
16. X. C. Lü, W. Chen, Z. Huang, a fundamental study on the control of the HCCI combustion and emissions by fuel design concept combined with controllable EGR. Part 2. Effect of operating conditions and EGR on HCCI combustion, International Journal of Fuel, 84, 1084-1092, 2005.
17. R. Chen, N. Milovanovic, A computational study into the effect of exhaust gas recycling on homogeneous charge

compression ignition combustion in internal combustion engines fuelled with methane, International Journal of Thermal Sciences 41 (2002) 805–813.

18. H. Machrafi, "Experimental validation of a kinetic multi-component mechanism in a wide HCCI engine operating range for mixtures of n-heptane, iso-octane and toluene: Influence of EGR parameters,'' Energy Conversion and Management 49, 2008, 2956–2965.

19. J.B. Heywood, Fundamentals of Internal Combustion engine,'' McGraw Hill, New York, 1988.

ANALYSIS OF A SINGLE BIOMOLECULE TRANSITING WITH NANOPORE*

GANG CHENG[1,2] and JIA WANG[3]

1, Key Laboratory of Marine Chemistry Theory and Technology, Ministry of Education, College of Chemistry and Chemical Engineering, Ocean University of China, Qingdao, China 266100

2, Marine School, Harbin Institute of Technology at Weihai, Weihai, China 264209

3, State Key Laboratory for Corrosion and Protection of Metals, Shenyang, China 110016

With the development of science and technology, much more aspects of nano-technology have been used for studying macrobiomolecules, but seldom for single-molecular detection (SMD).We reviews recent results in the research of DNA molecules with nanopores, introduce basic concept and technology which can be used to study the disassembling dynamics of nucleosomes. We present here that the two transport parameters of blockade current and duration yield the sequence and the structural information of the collagen-like peptides and α-helix peptides. As the nanopore is the only channel the conducts ions, when the large biomolecules is being translocated through the pore under the applied electric field, a passing ion current blockade will be observed, which is based on the relationship of the protein's conformation feature with the change of ion current parameters. Nanopore analysis clearly showed the blockade current and duration is related to the tertiary structures of peptides. The new method could be possible to develop a highly sensitive analysis for the conformation change of protein related to human disease.

Keywords: Single biomolecule, single-molecular detection, nanopore.

1. Introduction

In nature, the transport of biopolymers (DNA/RNA, peptides and proteins) is a ubiquitous process done at the nanoscale level usually involving nanopores embedded in lipid membranes[1]. The long-term goals of this research are the complete structural characterization of peptides and proteins by analyzing the signature current of transport. This technology will permit the single-molecule detection of mutants in proteins in their "native" environments, which is imperative for examining disease related proteins.

Bacterial α-hemolysin (α-HL), a heptameric protein that spontaneously assembles in a lipid membrane, has proven useful for the analysis of DNA translocation. The crystal structure of α-HL was solved in 1996 and consists of a vestibule with a diameter of about 30 Å which leads into a constricted pore with an internal diameter of about 16 Å, therefore allowing relatively small linear polymers to transport[2]. When inserted into a planar lipid membrane, α-HL has well-behaved single-channel conductance. In general, a single pore in a buffer containing 1 M KCl, at an applied potential of 100 mV across the membrane, gives a current of 100 pA, which is conveniently measured by a commercial "patch-clamp" amplifier. The amplifier has the ability to measure currents as low as 0.01 pA at a maximum acquisition rate of 500 kHz. Under open pore conditions, the current is due to the enhanced transport of K+ ions compared to the much larger Cl- ions. When iopolymers transit the channel by the driving electric field, the ionic current is interrupted, indicating the presence of a biopolymer inside the channel. Typical transport events are on the time scales of micro- to milli- seconds. As well, it is common for one pore to remain open for 3 hours, allowing for thousands of measurements to be made on a single pore assembly. In general, two parameters can be measured when a molecule traverses the nanopore; the magnitude of the channel current blockade (i_{block}) and its duration (t_{block}). Based on those parameters, the biopolymer properties, such as the composition, length and secondary structure can be elucidated.

*This work is supported by the National Nature Science Foundation of China (No. 50671097).

The narrowest portion of the α-HL is 16 Å in diameter and thus limits the biopolymers that thread through the pore. For example, experiments have shown that single strand DNA (ss-DNA) can enter the pore and produce the blockade events, but double strand DNA (ds-DNA) with a diameter of 26 Å does not transit through the pore. In the pioneering work of Deamer (UC at Santa Cruz), Branton (Harvard) and colleagues, ss-DNA or RNA was studied in detail, and the effects of length, sequence, temperature and applied potential on the transport kinetics have all been reported[3]. More recently, studies involved ds-DNA containing is matched or ss-DNA hairpin loops transporting through a nanopore and it was shown that DNA hairpins and mismatched DNA had to unzip in order to transit the pore[4-5] .The findings of the ds-DNA transport were that the less stable double helical structures of DNA resulted in a faster rate of transit due to the lower activation energy of unzipping. Therefore, the nanopore technique also offers the possibility of gathering structural information of biopolymers.

The particular focus of my research has been studying the conformation of peptides and proteins by the nanopore technique. Preliminary experiments have been performed with oligopeptides (Figure 1) in Saskatoon. Peptides were synthesized containing one, two or three repeats of the collagen-like sequence [-CSA-(Gly-Pro-Pro)n]2 where n = 1, 2 or 3. Extended repeats of this sequence form a stable collagen triple-helix with a crystallographically-determined diameter of 12 Å, which is expected to give a large blockade current[6]. It was shown that shorter repeats n = 1 or 2 also form dimers and monomers, which exhibit smaller blockade currents and shorter transit times. In general, the two transport parameters, blockade current (i_{block}) and duration (t_{block}), yield sequence and structural information, which is difficult to obtain by bulk spectroscopic techniques, such as CD or NMR.

We also investigated another simple peptide, CT7n, having the sequence Arg(Tyr-Ser-Pro-Thr –Ser–Pro-Ser)5-Gly. The peptide is found as a tandem repeat at the C-terminus of RNA polymerse II from eukaryotes and provides sites for protein-protein interactions that regulate mRNA synthesis. The structure of CT7n is thought to be extended, very flexible and plastic in the sense that the conformation it adopts is determined by the protein that binds to it. One striking feature of CT7n nanopore transit was that the blockade duration was very short (< 2 ms) and comparable to [-CSA-(Gly-Pro-Pro)$_2$]$_2$. On the other hand, the i_{block} values tended to be as large as for [-CSA-(Gly-Pro-Pro)$_3$]$_2$. Yet [-CSA-Gly-Pro-Pro]$_2$ is only 8 amino acids in length whereas CT7n is 37 amino acids. The simplest explanation is that the peptide is multiply-folded as it passes through the pore.

Other preliminary work involving protein and substrate interactions have been performed. We studied ss-DNA with a 5'-biotin tag that produced 15 transport events/sec. Upon addition of avidin this was reduced to about 3 events/sec, even when avidin was in two fold excess. Since 95% of the ss-DNA contained the biotin tag, the residual events are attributed to disassociation of the avidin/biotin complex even though the binding constant is 10^{14} M^{-1}. Similar results were obtained when an antibody specific for ss-DNA, ***Hed10***, was added in excess, although, in this case the number of residual events was about 8 events/sec. The binding constant of the antibody is about 10^6 M^{-1}. Therefore, the nanopore technique can provide kinetic and thermodynamic

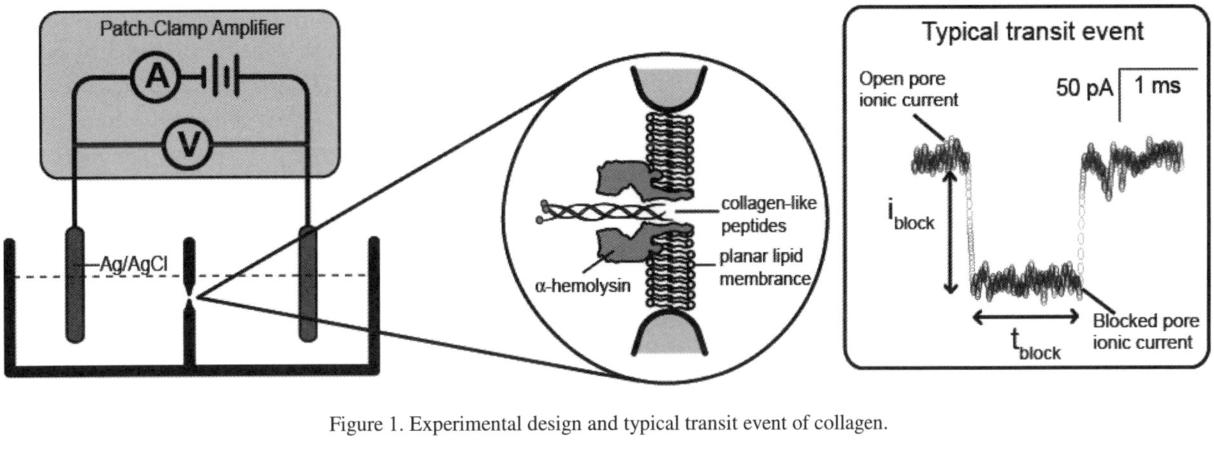

Figure 1. Experimental design and typical transit event of collagen.

information of the interaction between biomolecules at the single-molecule level.

It is clear from the above discussion that single-molecule events can be analysed by the nanopore technology. In the case of the collagen-like peptides, the pore analysis showed the presence of intermediate conformations, which were not apparent from the CD spectra. Therefore, we hypothesize that structural features of biopolymers such as the size, length, charge, α-helix, β-sheet and β-turn content of peptides and proteins can be achieved by nanopores.

2. Results and Discussion

2.1. *Systematic investigation of the transit of defined peptides through the pore*

2.1.1. Charge

The basic building block will be the peptide Asp3Ala10Lys3 which is easy to synthesize and purify, soluble, does not aggregate and gives a distinct α-helical CD spectrum. Since the diameter of an α-helix is about 7 Å, it should give a blockade current of about 50 pA. The related peptides Lys3Ala10Lys3 and Asp3Ala10Asp3, Lys6Ala10 and Asp6Ala10 will be prepared and checked for α-helical content by CD. The number of events/sec is largely dominated by concentration and diffusion but the duration of the blockade duration will be in the order Lys6Ala10> Lys3Ala10Lys3> Asp3Ala10Lys3> Asp3Ala10Asp3> Asp6Ala10 since the positively charged peptides will be transiting up the voltage gradient. Furthermore, the information inherent in the transit event may provide sequencing information of a polypeptide as indicated in Figure 2.

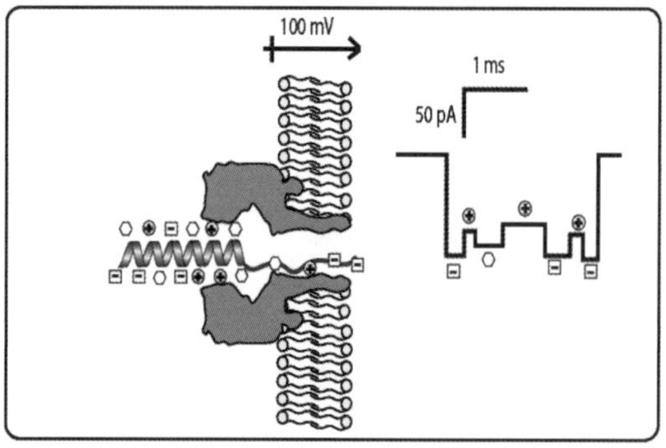

Figure 2. Idealized stepped blockade current for charged peptides for use in sequencing applications.

The standard voltage is 100 mV (unblocked pore current of 100 pA) but the membrane remains stable even at 200 mV (current of 200 pA) and at 50 mV the signal to noise is still acceptable (current of 50 pA). Therefore, the relationship between voltage and peptide charge can also be studied. For example, the t_{block} for the neutral peptide Asp3Ala10Lys3 is expected to show a shallow relationship with increased voltage. On the other hand, the t_{block} for Asp3Ala10Asp3 should decrease and for Lys3Ala10Lys3 should increase as the voltage is increased. In long term, it should be possible to use the value of $d\, t_{block} /dV$ to determine the charge on a peptide or the segment of a protein as it traverses the pore. This information is invaluable in understanding the repulsion forces and dipole properties in polypepetides.

2.1.2. Length

The peptides of section [a] are all of the same length. The effect of length will be studied with Asp3Ala10Lys3, Asp3Ala14Lys3, Asp3Ala18Lys3 and Asp3Ala22Lys3, which will contain 3, 4, 5 and 6 turns of α-helix, respectively. It is anticipated that there will be a linear relationship between length and transit time with a constant current blockade. However, if the rate limiting step is orientation of the peptide to align itself with the pore before transit then the relationship may be more complex. Again the effect of changing the voltage may be very informative because the peptides have an intrinsic dipole moment. Thus, the longer the peptide the larger the dipole moment and the faster it will orient as the voltage is increased. Therefore, a linear relationship between length and t_{block} may only be observed at low voltages.

2.1.3. Helix breaker

Proline residues are known to disrupt α-helices and cause a kink between two α-helices. Thus, the peptide Asp3Ala3ProAla3ProAla3Lys3 is expected to exist mostly as a random coil although this will be confirmed by CD. Compared to Asp3Ala10Lys3, Asp3Ala3ProAla3ProAla3Lys3 is anticipated to give a longer t_{block} with a smaller i_{block} because it will be longer and thinner than the parent peptide. As described in the introduction, the repeating heptapeptide CT7*n* gave a shorter t_{block} because of folding but this is unlikely to be the case with the much shorter $Asp_3Ala_3ProAla_3ProAla_3Lys_3$. In this way, the standard values for the transit of α-helix and random coil can be established.

Perhaps of more interest will be Asp3Ala10ProAla10Lys3 and related peptides in which the proline will force a kink between two stable α-helices. It is very difficult to predict what effect the kink will have on the transit parameters. By analogy, a straight stick can be pushed through a pipe whereas a bent one cannot unless the pipe or the stick can be also bent. Bent duplex DNA moving in an agarose gel is retarded relative to a linear molecule. Therefore, it seems likely that a bent peptide will have increased value for t_{block} but will still be able to transit because of flexibility in both the peptide and the α-HL pore. These results will be important for understanding the transit of larger proteins described below.

2.1.4. β-turn

If a Glycine and a third residue such as Asparagine are added to the helix breaker then a tight turn is usually the result. Thus, the peptide Asp3Ala10GlyProAsnAla10Lys3 will have the opportunity to fold back on itself to form a primitive antiparallel coiled-coil. Because the transit experiments are performed in 1 M KCl these ionic bonds will be quite weak. Therefore, Asp3Ala10GlyPro-AsnAla10Lys3 may exist as two conformations; in one the α-helices are joined by a flexible linker and will transit with similar parameters to Asp3Ala22Lys3 or Asp3Ala10ProAla10Lys3. In the second, the folded conformation will give a larger i_{block} and a shorter t_{block}. If this is the case, two distinct populations will be observed in the contour plots as was found for the collagen-like peptides. The ratio of the two conformations can be used to estimate the equilibrium constant for folding which in turn can be used to calculate the free energy of folding (ΔGf). More importantly, the ratio of folded to unfolded states can be adjusted in a predictable way. For example, Asp3Ala10GlyProAsnAla10Asp3 is predicted to be more unfolded than AspAla10GlyProAsnAla10Lys because of increased electrostatic repulsion. Alternatively, introduction of hydrogen bond forming residues or hydrophobic interactions, which bridge the two α-helices, will stabilize the folded state. Thus, Asp3Ala10GlyProAsnAla10Lys3 is expected to be more folded than the parent peptide. Again this highlights the advantages of being able to analyze the stricture of individual molecules. In the long term, it will be possible to evaluate the contribution of individual amino acids to the free energy of folding.

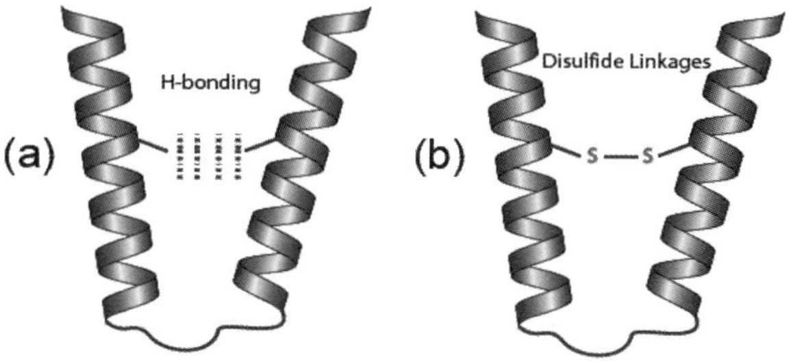

Figure 3. Protein contains (a) H-bonding, (b) disulfide bonds between parts of the same chain.

2.1.5. Disulfide bonds

Disulfide bonds are very important cross-linkers between different chains or between parts of the same chain in polypeptides. Intracellular proteins usually lack disulfide bonds, whereas extracellular proteins often contain several. By the comparison of two peptides, shown in Figure 3, we can understand the effects of H-bonding and disulfide in secondary structure.

2.2. *Investigation into the use of the pore as a device to detect protein mutations*

Many neurodegenerative diseases are caused by the presence of extended polyglutamines (PolyGln) tracts in the expressed protein. To date there are nine such diseases attributed to PolyGln with Huntington's Disease (HD) receiving the most attention[7]. In HD, the protein responsible, Huntingtin, contains greater than 35 repeats of Gln which is coded for by (CAG)n in exon 1 of the gene. In the normal protein, there are fewer than 34 Gln and in HD the number of repeats is strongly and inversely correlated with the age of onset. There is little sequence similarity between the HD gene and other disease genes of this group except for the presence of the (CAG)n repeats. Therefore, the focus for understanding the pathogenesis has been on the structure of the polyGln repeats. It is now clear that the hallmark of these diseases is the pathological aggregation of the polyGln-containing protein into β-sheet aggregates. Presumably, more repeats promote greater aggregation and thus an earlier age of onset. Several recent studies have addresses the properties of polyGln. Of particular relevance is the use of peptides with periodic substitution of Pro-Gly which prevents the formation of extended β-chains and encourages incorporation of a turn. For example, Q45 aggregates rapidly as does the same sequence with Pro-Gly insert every 9 or 10 Q residues. On the other hand, peptides with inserts every 7 or 8 residues aggregates are much less readily. Therefore, it would appear that the basic repeat of the polyGln is an extended β-chain with at least 9 residues. As a good example, the structural features of polyGln will be identified by nanopore experiments. It is should be noted that the pore experiments require nanomolar concentrations of peptides so that aggregation will be very slow, if at all. Therefore, we will be able to investigate the folding of individual peptides before they aggregate. The detection method of polyGln at different aggregation states can be developed.

3. Conclusion

A striking characteristic of proteins is that they have well-defined three dimensional structures. A stretched out or randomly arranged polypeptide chain is without biological activity. Function arises from conformation in polypeptides. The most important conformations of polypeptides and proteins in biological aqueous medium can be analysed at single-molecule level by the nanopore technique. We can characterize the H-bonding, hydrophobic interactions, disulphide bridges and electrostatic interactions. Second, not only will the basic physical properties of polypeptides and proteins such as size, charge, length can be studied, but also the chemical secondary structures such, as α-helix, β-sheet, β-turn and disulfide bonds. Nanopore technology may become a tool to understand the relationship between structures and function in polypeptides and proteins. The ultimate goal, with the achieved single-molecule current signature information, is to detect protein-based diseases at extremely low concentrations.

Acknowledgement

The authors are grateful for the support from the National Nature Science Foundation of China (No. 50671097).

References

1. Lodish, H., Balimtore, D., Berk, A., Zipursky, S. L., Matsudaira, P. and Darnell, J., Molecular Cell Biology 1996 (New York: Freeman).
2. Song, L. Z., Hobaugh, M. R., Shustak, C., Cheley, S., Bayley, H. and Gouaux, J. E., *Science* 1996, 274, 1859-1866.
3. Deamer, D. and Branton, D., *Acc. Chem. Res.* 2002, 35, 817-825.
4. Vercoutere, W., Wiinters-Hilt, S., Olsen, H., Deamer, D., Haussier, D., and Akeson. M., *Nat. Biotechnol.* 2001, 19, 248-252.
5. Meller, A., Nivon, L., Branton, D., *Proc. Natl. Acad. Sci.* 2000, 97, 1079-1084.
6. Sutherland, T. C., Long, Y. T., Stefureac R. I., Bediaiko-Amona I., Kraatz, H. B., and Lee, J. S., *Nano. Letts.* 2004, 4, 1273-1277.
7. Thakur, A. K., and Wetzel, R., *Proc. Natl. Acad. Sci.* USA, 2002, 99, 17014-17019.

CHARACTERISTICS AND TRANSPORT PROPERTIES OF TWO MODIFIED ZERO VALENT IRON

YU-HAO LIN
Department of Environmental Engineering, National Chung-Hsing University, 250, Kuo Kuang Road Taichung 402, Taiwan

MIN-DER LIN
Department of Environmental Engineering, National Chung-Hsing University, 250, Kuo Kuang Road Taichung 402, Taiwan

CHENJU LIANG
Department of Environmental Engineering, National Chung-Hsing University, 250, Kuo Kuang Road Taichung 402, Taiwan

This study synthesizes two modified nanoscale zero valent iron (NZVI) with poly acrylic acid (PAA) and Carboxymethyl Cellulose (CMC) to reduce particles aggregation, obtain smaller particle sizes and enhance the migration ability of NZVI in porous media. The average particle size of NZVI modified by PAA and CMC are 10 nm and 40–100 nm, respectively. The results from Fourier transform infrared spectroscopy test suggest that the PAA and CMC adsorb onto NZVI via chemisorption, and the carboxylate group forms the binding in bidentate bridging mode. The laboratory column test illustrated that the modified NZVI could transport as a reactive material in water-saturated porous media and the transport can be enhanced by increasing flow rate. The steric repulsions afforded by the modifier allow the modified NZVI to be transportable even as the electrical double layer (EDL) repulsions are screened due to increase the background Na+ concentration.

1. Introduction

1.1. *Producing hard copy using MS-Word*

Injection of nano-scale zero valent iron (NZVI) was evidenced as a promising treatment for the remediation of contaminated aquifers. However, transport properties of NZVI are poor owing to the Brownian motion, density of iron, and magnetic attractive forces which increase the aggregation phenomena of particles [1], and result in clogging of pores in soils. To overcome this limitation, various approaches had been devoted to improve the transport capability through modifying the NZVI surface properties via stabilizer such as triblock polymer, Carboxymethyl Cellulose (CMC), and poly acrylic acid (PAA) [2–4]. Cao et al. revealed that the bimetals did not enhance the reduction of perchlorate [5], so reducing the iron size to enhance remediation is possibly necessary for certain contaminants. Therefore, the objective of this study is to (1) utilize the modifier PAA and CMC to synthesize the stabilized and finer-sized NZVI (titled PAA-NZVI and CMC-NZVI); (2) characterize the PAA-NZVI and CMC-NZVI morphology and their stabilization mechanism; and (3) investigates their transport ability in porous media. The transmission electron microscopy (TEM) and Fourier transform infrared spectroscopy (FTIR) were used in this work to analyze the characteristics of the modified NZVI. Besides, several column tests were also conducted by varying the background Na+ ionic strength (IC) and pore velocity of the fluid to identify and observe the transport ability of PAA-NZVI and CMC-NZVI in porous media.

2. Materials and methodology

2.1. *Materials*

All chemicals used in this study were reagent grade. NaBH4 and PAA (M.W. = 2,500,000) were purchased from Sigma-Aldrich Chemical Co, and CMC (mean M.W. = 90,000) is from Acros Organics. Ferric and ferrous iron was obtained from Alfa Aesar and Riedel-de Haën. The concentration of PAA solution was diluted by deionized water from original 35 wt% to 3.5 wt%.

2.2. *Synthesis of modified NZVI slurry*

The synthesis process of PAA-NZVI and CMC-NZVI employed in this work were modified from previous studies [6, 7]. The 8% PAA (v/v) and CMC (with a mole ratio to Fe equals 0.0124) were respectively added to iron precursors to form iron complex, followed by mixing with NaBH4 under an N2 (99.9%) atmosphere associated with ice bath and vigorously stirred until hydrogen evolution ceased to form stabilized PAA-NZVI and CMC-NZVI slurry, which we added palladium ions (mole ratio of Pd to Fe equals 1:1000) as seeds for iron particle nucleation and NH4OH to adjust the pH of solution in PAA-NZVI synthesis. All resultant slurry was then adjusted to 2.0 Fe-g/L and directly employed in the transport experiment.

2.3. *Transport experiment of NZVI*

A glass column with an inner diameter of 2.66 cm and 10.8 cm length, packed with quartz sand (30–40 mesh, dp = 500 μm), was used to perform the transport experiments. The porosity of porous media was 0.374. A HPLC pump (LABAL-LIANCE Series I) was then used to feed background Na+ solution for at least 10 pore volumes (PV) to condition the collector. Two PV of the modified NZVI slurry was introduced upward into the column using a peristaltic pump (Longer pump BT 100-2J). Effluent samples were collected using a fraction collector (ISCO retriever 500) and an aliquot of the effluent was dissolved in 0.1 N HCl (1 to 1 vol/vol ratio) and mixed for 24 hours and digest. Then the total iron was analyzed using Hach methods 8008. A nonreactive tracer test was also conducted by introducing Bromophenol blue solution. Different linear pore water velocity and background Na+ concentration were applied in the column tests to evaluate their impacts on the transport of NZVI slurry.

3. Result and discussion

3.1. *PAA-NZVI and CMC-NZVI nanoparticle Characterizations*

The morphology of the PAA-NZVI and CMC-NZVI identified from TEM (JEOL JEM-1200CX II systems) images are shown in Fig. 1 and 2. The individual particles of PAA-NZVI and CMC-NZVI appear spherical, sized about 7–12 nm (average 10 nm) and 40–100 nm diameter, respectively. All of them are aggregated in chain like structures due to magnetic properties. The particles of PAA-NZVI are more uniform and smaller than the unmodified NZVI as a result of adapted PAA dosage, higher pH and the palladium acted as the seeds for larger amount of nuclei of iron [6]. The fine-grained surface of the CMC-NZVI is observed in TEM film, and we suggest those were the iron primary particles assemble-self to spherical secondary particle, probably due to the excess CMC existence and Ostwald ripening [8].

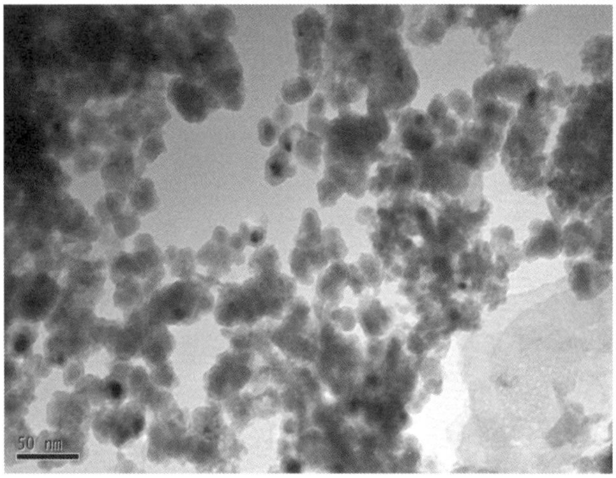

Figure 1. TEM images of PAA-NZVI.

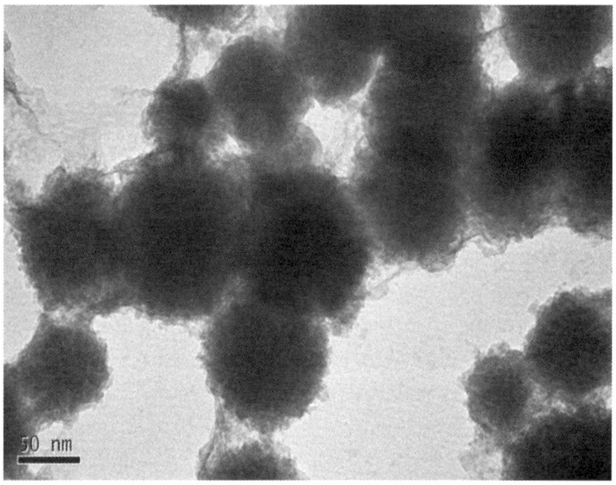

Figure 2. TEM images of CMC-NZVI.

3.2. *The stabilization mechanism and configuration of modified NZVI*

To elucidate the stabilization mechanisms, FTIR (Jasco 410) measurements were carried out on PAA, CMC and the modified NZVI. It was observed that there was no apparent assignment peak in the NZVI. However, several characteristic peaks appear in modified NZVI surface such as 2348 cm^{-1} in PAA-NZVI and 2361 cm^{-1} in CMC-NZVI, both indicate that PAA and CMC are absorbed onto NZVI surface. In addition, the separation (Δv) of the symmetric and asymmetric carboxylate stretches for the modifier with/without NZVI adsorbed, can be used to identify the adsorbed configuration between modifier and particle. The rules to determine the adsorbed configuration are summarized as three fashions and shown in Fig. 3: (a) If there is C=O character in the spectrum and Δvadsorbed is greater than Δvsalt, then the adsorbed structure is monodentate chelating; (b) If there is no C=O character in the spectrum and Δvadsorbed is smaller than Δvsalt, then the adsorbed structure is bidentate chelating; and (c) If there is no C=O character in the spectrum and Δvadsorbed is similar to Δvsalt, then the adsorbed structure is bidentate bridging.

It is evident from the FTIR result (Fig. 4) that there were no characteristic C=O band in the PAA and CMC spectrums, and the Δvadsorbed (1558-1397=161 cm^{-1}) is similar to Δvsalt (1559-1408=151 cm^{-1}) for the PAA-NZVI. Similarly, the Δvadsorbed (1624-1418=206 cm^{-1}) in CMC-NZVI is also similar to Δvsalt (1603-1419= 184 cm^{-1}). It can be suggested that the PAA and CMC adsorbed onto NZVI via chemisorption and both the PAA and CMC utilize bidentate bridging fashion to complex with NZVI. However, it's also observed that the O-H stretching frequency (at 3358 and 3445 cm^{-1}) and bandwidth were changed in PAA-NZVI and CMC-NZVI, implying that the hydrogen bonding occurs [9] on NZVI surface due to the fact that partially hydroxylated NZVI may contain negatively charged Fe-O-.

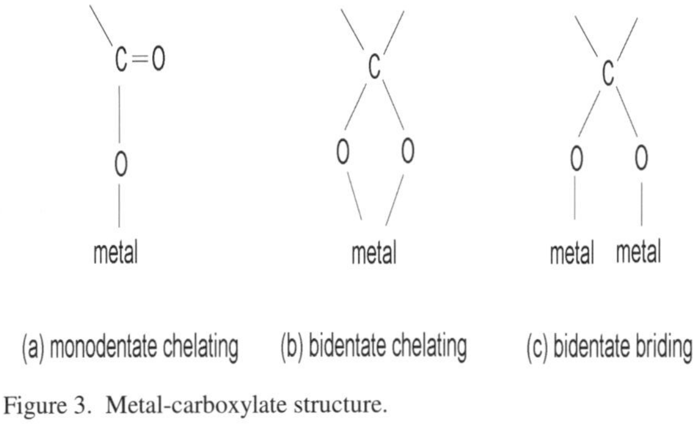

Figure 3. Metal-carboxylate structure.

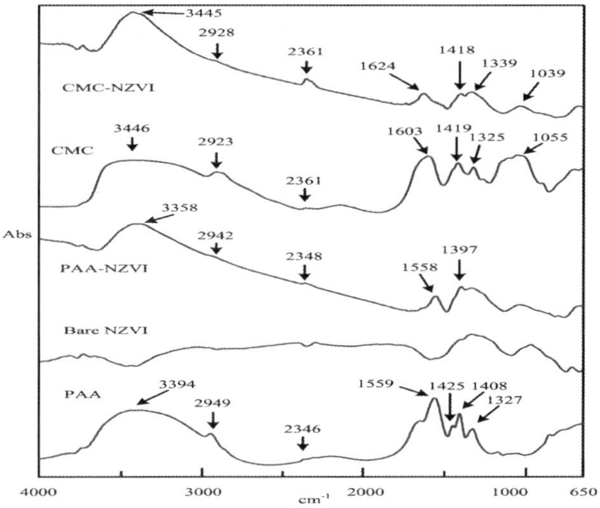

Figure 4. FTIR spectra of modifier, NZVI, PAA-NZVI and CMC-NZVI.

3.3. *Transport of modified NZVI in porous media*

The column tests were conducted under two Na+ ionic strengths, 40 mM Na+ and 100 mM Na+, and two different velocities, 7.11 m/day and 32.7 m/day, respectively. The results of column tests were illustrated in Fig. 5 and 6. The symmetrical shape and absence of tailing in the nonreactive tracer breakthrough curves (BTC) indicate that the physical nonequilibrium processes and preferential flow paths were not present in the column. Fig. 6 also reveals that the breakthrough of PAA-NZVI with low and high pore velocity in same IC appeared after injecting about 1.4 PV and 1.3 PV slurries, respectively. The plateaus of concentration (C/C0) both approach 1. Slight retardation effect is observed after 3 PV and seems to be reduced when pore velocity increased. In additional, increasing the IC would reduce electrical double layer (EDL) repulsion and theoretically result in lower colloidal stability. However, with the Na+ IC condition increasing from 40 mM to 100mM, it just slightly increased the retardation phenomenon after 3 PV, but the plateau still reach C/C0=1. It was possibly due to the steric repulsions afforded by the PAA polymer allows it to be transportable even as the EDL repulsions are screened.

Similarly, in low flow velocity, the breakthrough of CMC-NZVI appeared after injecting about 1.7 PV and the plateau of concentration (C/C0) approaches 0.9. It was possibly due to the higher viscosity of CMC-NZVI.

Further, when the flow velocity increases to 32.75 m/day, the breakthrough of CMC-NZVI appeared earlier to 1.3 PV and the retardation effect was slightly reduced. The retardation of CMC-NZVI aforementioned could be due to the detachment of the polymer adsorbed NZVI [10, 11]. Additionally, the recovery of all modified NZVI in column tests within five PV is up to 95% and that could also demonstrate the detachment effect of the polymer adsorbed NZVI.

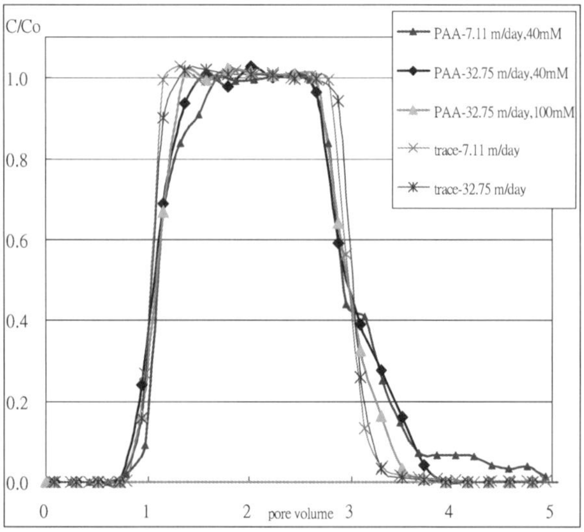

Figure 5. PAA-NZVI and tracer breakthrough curves for different iconic strengths and flow velocities.

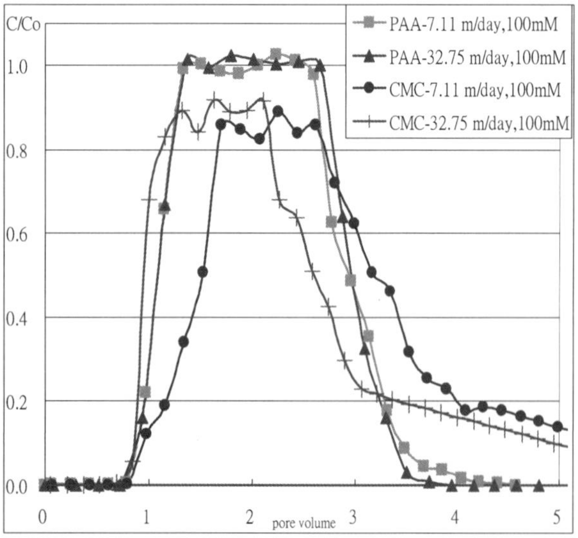

Figure 6. PAA-NZVI and CMC-NZVI breakthrough curves for different iconic strengths and flow velocities.

4. Conclusion

This study utilizes the modifiers PAA and CMC to synthesize the stabilized and fine size NZVI. The result shows that the adapted PAA dosage, higher pH and the palladium as the seeds for iron nuclei successfully produced uniform and smaller NZVI particles. Besides, both PAA-NZVI and CMC-NZVI could be smoothly delivered in porous media, and higher flow rates accelerated the transport of modified NZVI and reduced the retardation effect. In additional, as the background Na+ IC increases, the retardation effect increases slightly, but the modified NZVI can still be transported through porous media. It was possible due to the steric repulsions afforded by the polymer and detachment of the polymer adsorbed NZVI.

References

1. N. Saleh, H.-J. Kim, T. Phenrat, K. Matyjaszewski, R.D. Tilton, G.V. Lowry, "Ionic Strength and Composition Affect the Mobility of Surface-Modified Fe0 Nanoparticles in Water-Saturated Sand Columns," Environ. Sci. Technol., pp. 2008.
2. S.R. Kanel, R.R. Goswami, T.P. Clement, M.O. Barnett, D. Zhao, "Two dimensional transport characteristics of surface stabilized zero-valent iron nanoparticles in porous media," Environ. Sci. Technol., 42, pp. 896-900, 2008.
3. F. He, D. Zhao, J. Liu, C.B. Roberts, "Stabilization of Fe - Pd nanoparticles with sodium carboxymethyl cellulose for enhanced transport and dechlorination of trichloroethylene in soil and groundwater," Ind. Eng. Chem. Res., 46, pp. 29-34, 2007.
4. N. Saleh, T. Phenrat, K. Sirk, B. Dufour, J. Ok, T. Sarbu, K. Matyjaszewski, R.D. Tilton, G.V. Lowry, "Adsorbed triblock copolymers deliver reactive iron nanoparticles to the oil/water interface," Nano Letters, 5, pp. 2489-2494, 2005.
5. J. Cao, D. Elliott, W.-X. Zhang, "Perchlorate reduction by nanoscale iron particles," J. Nanopart. Res., 7, pp. 499-506, 2005.
6. K.C. Huang, S.H. Ehrman, "Synthesis of Iron Nanoparticles via Chemical Reduction with Palladium Ion Seeds," Langmuir, 23, pp. 1419-1426, 2007.
7. F. He, Preparation, characterization and applications of polysaccharide-stabilized metal

nanoparticles for remediation of chlorinated solvents in soils and groundwater, Environmental Engineering, Auburn University, Auburn, 2007.

8. J.S.H.H.P.L.A.M.C.W.G.S.L.J.W. L.-S. Zhong, Self-Assembled 3D Flowerlike Iron Oxide Nanostructures and Their Application in Water Treatment, 2006, pp. 2426-2431.
9. F. Jones, J.B. Farrow, W. van Bronswijk, "An Infrared Study of a Polyacrylate Flocculant Adsorbed on Hematite," Langmuir, 14, pp. 6512-6517, 1998.
10. H.-J. Kim, T. Phenrat, R.D. Tilton, G.V. Lowry, "Fe0 Nanoparticles Remain Mobile in Porous Media after Aging Due to Slow Desorption of Polymeric Surface Modifiers," Environ. Sci. Technol., 43, pp. 3824-3830, 2009.
11. K.M. Sirk, N.B. Saleh, T. Phenrat, H.-J. Kim, B. Dufour, J. Ok, P.L. Golas, K. Matyjaszewski, G.V. Lowry, R.D. Tilton, "Effect of Adsorbed Polyelectrolytes on Nanoscale Zero Valent Iron Particle Attachment to Soil Surface Models," Environ. Sci. Technol., 43, pp. 3803-3808, 2009.

Session 3

SYNTHESIS OF VISIBLE LIGHT ACTIVE N-DOPED TITANIA PHOTOCATALYST

CAHYORINI KUSUMAWARDANI
Chemistry Department, Yogyakarta State University, Karangmalang Yogyakarta, 55283, Indonesia

INDRIANA KARTINI, NARSITO
Chemistry Department, Gadjah Mada University, Bulaksumur Yogyakarta, Indonesia

Nitrogen-doped TiO_2 (N-doped TiO_2) photocatalyst with nanosize has been synthesized through the hydrolysis of N-substituted titanium isopropoxide precursors. The light absorption onset shifts to the visible region at 550 nm with N-doped TiO_2. A clear decrease in the band gap and the nitrogen 2p states on the top of the valence band on N-doped TiO_2 (compared to Degussa P25) is deduced from the optical absorption spectroscopy results. Photocatalytic decomposition of methylene blue has been carried out in the visible region, which N-doped TiO_2 shows higher activity than the Degussa P25.

1. Introduction

Photocatalysis has been thought to be a fascinating approach for resolving the embarrassments of energy, environment and resource for several decades [1,2]. The wide-bandgap TiO2 has become the dominant photocatalyst in the field of air and water detoxification because of its high stability, low cost, high oxidation potential and chemically favorable properties [3]. However, the intrinsic wide band-gap nature (3.0 – 3.2 eV) impairs TiO2 from playing such an important role, because it only allows TiO2 to absorb UV light, which accounts for merely 5% of total solar energy [4,5]. In order to overcome this limitation, extending the absorption of TiO2 into the wide visible light region seems to be a possible solution, which enables it to utilize as much as 50% of total sunlight that reaches the surface of the earth.

Early attempts on the shifting of TiO_2 absorption into visible light region mainly focus on the doping with transition metals [6,7] but shortcomings of metal doped TiO_2 such as thermally instability, its tendency to form charge carrier recombination centers, as well as the expensive ion implantation facilities make metal-doped TiO_2 impractical. It is not until 2001, that Asahi et al. [8] successfully intrigued the interest in the anion doping (such as N, C, S, P and F) of TiO_2. Among these anion dopants, nitrogen seems to be the most effective dopant due to its similar size to oxygen and small ionization energy [9,10]. Therefore, N-doped TiO_2 became the most studied system in the field and a lot of efforts have been put on the synthesis, characterization and application of N-doped TiO_2.

While Asahi *et al* [8] reported the synthesis of N doped titania in 2001, by the sputtering of TiO_2 targets with nitrogen and argon gas mixtures, there have been several reports thereafter which have discussed several ways to synthesize the same. N-doped titania has been synthesized via sputtering techniques [9], CVD [10] as well as mechanochemical mixing of titania with ammonium carbonate [11] via a solvothermal process using $TiCl_3$-hexamethylamine-alcohol solution. It has also been synthesized via a hydrazine precursor route [12]. However, all these processes either require a special apparatus for synthesis or require tight control of experimental conditions. While all these methods brought about the incorporation of N in the lattice, there is no clear indication about the mechanism of the same.

In this paper, we report a simple sol gel method to synthesize nanometer-sized visible-light N-doped TiO_2 photocatalysts. The prepared photocatalysts show an enhancement in the photodegradation efficiency of methylene blue under visible light (wavelength ≥390 nm) irradiation compared to commercially available TiO_2 catalyst.

2. Experimental Section

2.1. *Synthesis of N-doped TiO_2*

For the typical synthesis of the N-doped TiO2 powders, a mixture of 3 mL of Ti(OPr)4 (Aldrich 97%), 10 mL dodecylamine, and 80 mL of etanol absolute solution was refluxed until the solution became clear. This precursor solution was cooled to room temperature and 5 mL of CH3COOH was added to neutralize the excess of dodecylamine. Hydrolysis process was then achieved by adding 40 mL of distilled water dropwise into the solution under vigorous stirring for 24 hours. The resulting yellowish precipitate was centrifuged and washed subsequently with distilled water and ethanol. Finally, the N-doped TiO2 NPs were vacuum-dried for 12 h and calcined for 30 min in air atmosphere at 200, 300, 400, 600 and 800 °C.

2.2. *Characterization*

The N-doped TiO2 powder was analyzed by thermogravimetric analysis (TGA) with a Q500TGA Instrument using a 10oC/min ramp up to 800°C. The structure of the products was examined with X-Ray powder diffractometer (XRD, Shimadzu, XRD-6000) with Cu Kα radiation (λ = 0.15406 nm) and N2 adsorption-desorption measurements at 77 K (Belsorp28, Bel.). UV-visible diffuse reflectance spectra were obtained for the dry-pressed disk samples using a UV-visible spectrophotometer (UV-2550, Shimadzu).

2.3. *Photodegradation of Methylene Blue*

The photocatalytic activity of the N-doped TiO2 was determined by measuring the decomposition of methylene blue (MB) under the irradiation with visible light (>400 nm). A 150 W high-pressure Xenon arc lamp was used as light source. The concentration of methylene blue (MB) was determined by measuring the absorbance of the MB solution in a Shimadzu UV-Vis spectometer. For comparison, a photocatalytic reaction was carried out using commercial titania (Degussa P25) nanoparticles.

3. Result and Discussion

N-doped TiO2 nanoparticles can be prepared in different ways. We found that complexation of organic amines on the Ti metal center creates highly efficient precursors for N-doped TiO2 nanoparticles. Advantages of this route are high doping levels, controllable doping concentrations, and aqueous chemistry that results in a hydrophilic surface at low cost. The N-doped TiO2 nanoparticles were synthesized by hydrolysis of N-substituted titanium isopropoxide precursors in alcohol solution. In the synthesis the nitrogen has been doped into the lattice and/or attached to the nanoparticles surface. However, at the same time, there is water and organic residue adsorbed on the surface and enclosed into the amorphous and porous N-doped TiO2 powder.

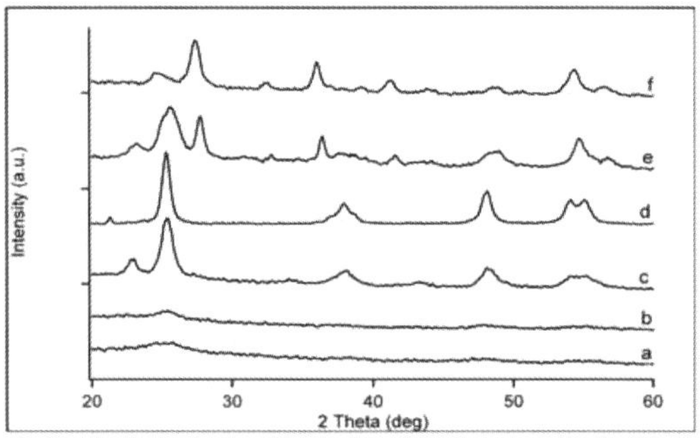

Figure 1. XRD Spectra of N-doped TiO2 after calcination at (a) 200; (b) 300; (c) 400; (d) 450; (e) 600 and (f) 800°C.

Powder X-ray Diffractometry

The crystal structure of a photocatalyst is an important property for catalytic activity of the nanoparticles. The crystal structure of the synthezised N-doped TiO_2 were studied by X-ray powder diffraction (XRD), which showed in Figure 1. It is found that N-doped TiO_2 nanoparticles are amorphous when sintered at temperatures below 400°C. The XRD patterns of N-doped TiO_2 calcined at 400°C correspond to the anatase TiO_2. The anatase phase has been transformed to rutile after increasing the calcination temperature to 600°C. It can also be seen from the XRD patterns that the N-doped TiO_2 samples show broad peak which indicates the formation of nanoparticles. Following the Debye-Scherrer formula, it could be calculated that the particle size were around 5–8 nm.

Thermogravimetric Analysis (TGA)

The above results show that the properties of the N-doped TiO_2 change with calcining temperatures. To understand the loss of organic residues on the surface of the N-doped TiO_2 under calcining, we used the thermogravimetric analysis to study the weight loss of the N-doped TiO_2. The TGA results of the N-doped TiO_2 in Figure 2 show that there is less weight loss, about 20% from room temperature up to 150°C, which is mostly due to the loss of physically adsorbed or embedded water and solvent on the surface of the

N-doped TiO_2. From 150 to 200°C, there is about 40% weight loss of The N-doped TiO_2 sample. From 250 to 400°C there is much slower (about 5%) weight loss of the N-doped TiO_2 sample in association with the crystallization of these NPs, as confirmed by XRD. The latter weight loss is about the same as the nitrogen content in the N-doped TiO_2 calcined at 400°C. Above 400°C, there is very little, about 3–5% weight loss of the N-doped TiO_2 sample in association with the crystallization of these nanoparticles, as confirmed by XRD.

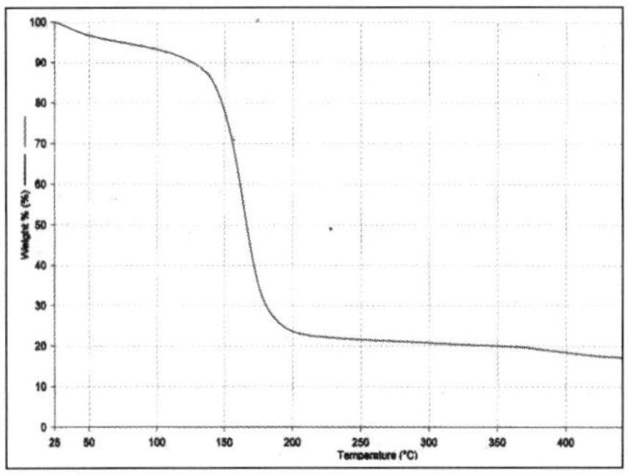

Figure 2. TGA graph of N-doped TiO_2.

N_2 Adsorption-Desorption

Figure 3 shows the typical nitrogen isotherm of N-doped TiO2 spheres. It exhibits a type-IV isotherm curve with an H2 hysteresis loop according to IUPAC classification [13,14], which means the spheres have mesoporous structure. From BJH distribution, the N-doped TiO2 show high BET surface areas of 147 m2/g. It was because of the mesoporous structure and the large amount of nanometer crystallites.

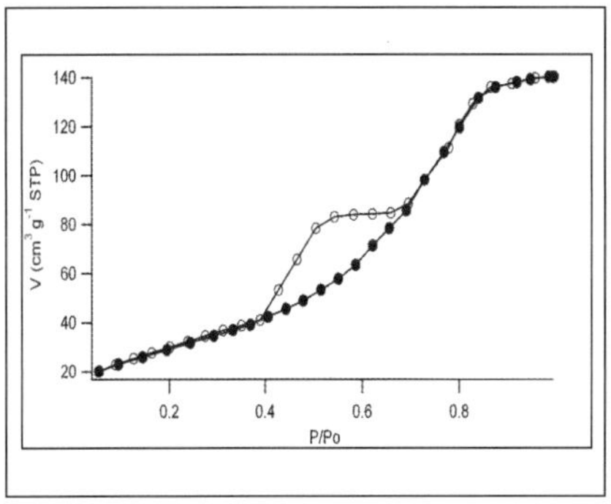

Figure 3. Isotherm curve of N-doped TiO_2.

Diffuse Reflectance Spectroscopy

The optical absorbance and reflectance was used to study the capability to photosensitize the TiO2 nanoparticles. The absorbance shift of the N-doped TiO2 NPs can be observed from the reflectance spectra of undoped (Degussa P25) and N-doped TiO2 NPs in Figure 4. The yellowish N-TiO2 sphere powders show good absorbance of visible light. It can be seen from Figure 3 that the visible light absorption is high and extended up to 550 nm in the case of N-doped TiO2 calcined at 400°C compared to that of pure TiO2 which could only absorb light in the UV range. It may be due to that nitrogen species occupy some of the oxygen positions in the lattice. This also rules out the occupancy of N in any other positions such as interstitial sites, which should give rise to a mid gap band/level between valence and conduction bands.

The band gaps of N-doped TiO_2 and Degussa P25 were calculated by the equation [12]

$$E_g = 1239.8/\lambda \tag{1}$$

where E_g is the band gap (eV) and λ (nm) is the wavelength of the absorption edges in the spectrum. The band gap for N-doped TiO_2 and Degussa P25 as calculated for the absorption edge in the visible region was found to be 2.25 eV and 3.09 eV respectively.

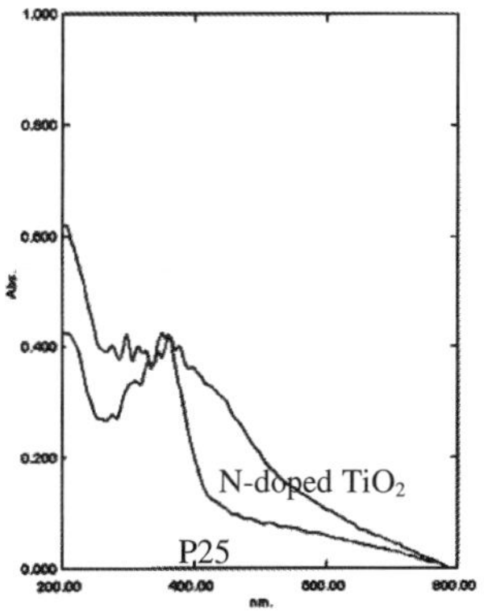

Figure 4. DRUV Spectra of N-doped TiO2 compared to Degussa P25.

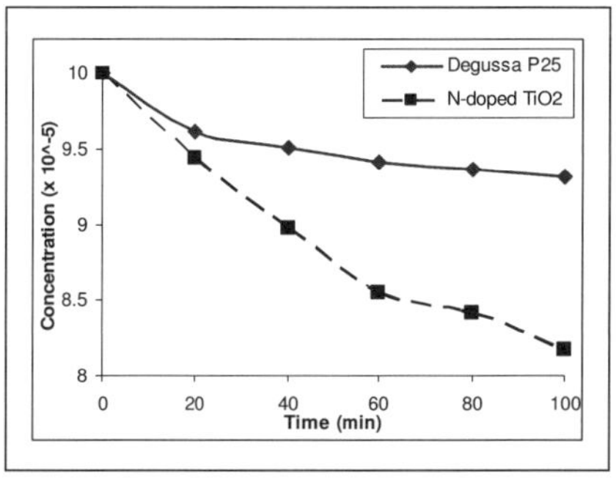

Figure 5. Photodegradation of Methylene Blue.

Photocatalytic Activity

The photocatalytic activity of N-doped TiO2 and Degussa P25 samples have been studied by measuring the percentage decomposition of methylene blue (MB) aqueous solution. It can be seen from Figure 5 that the photocatalytic activity of N-TiO2 (calcined at 450°C) sample shows higher activity to that of Degussa P25 catalyst in the visible region. The enhanced photocatalytic activities of N-TiO2 in visible region may be due to the increase of the surface deficiency after the introduction of nitrogen into the TiO2 structure [15], which could be the evidence to confirm the role of nitrogen in the lattice for improvement of the visible-light response of N-TiO2 (Figure 4).

4. Conclusion

N-doped TiO2 photocatalyst with a nanosize (14 nm) has been prepared by a simple chemical route and calcined at 400°C. The light absorption onset of N-doped sample in the visible region (550 nm) is clearly observed. The advantages of this method over other methods of preparation are (i) use of inexpensive chemical precursors for the synthesis of N-doped TiO2 and (ii) formation of TiO2 as uniformly sized nanoparticles. Optical absorption studies clearly identified the substitutional N-doping and localized N-states in the TiO2 lattice. A higher photocatalytic activity for the decomposition of methylene blue in the visible region has been obtained for the N-doped TiO2 sample compared to Degussa P25.

5. Acknowledgement

Financial support from Indonesian Government by Directorate General of Higher Education through project contract no. 01/J.35/KTR.HP/2009 is gratefully acknowledged.

References

1. Linsebigler, A. L.; Lu, G. Q.; Yates J. T. Chem. Rev. 95, 735 (1995).
2. Fujishima, A.; Hashimoto, K.; Watanabe, T. TiO2 Photocatalysis, Fundamentals and Applications; BKC, Inc.: Tokyo (1999).
3. Lindgren, T.; Mwabora, J. M.; Jonsson, J.; Hoel, A.; Granqvist, C. G.; Lindquist, S. E. J. Phys. Chem. B 107, 5709 (2003).
4. Qiu, X. and Burda, C. Chem. Phys. 339, 1 (2007).
5. Burda, C.; Lou, Y. B.; Chen, X. B.; Samia, A. C. S.; Stout, J.; Gole, J. L. Nano Lett. 3, 1049 (2003).
6. Irie, H.; Watanabe, Y.; Hashimoto, K. J. Phys. Chem. B 107, 5483 (2003).
7. Ihara, T.; Miyoshi, M.; Iriyama, Y.; Matsumoto, O.; Sugihara, S. Appl. Catal., B 42, 403 (2003).
8. Asahi, R.; Morikawa, T.; Ohwaki, T.; Aoki, K.; Taga, Y. Science 293, 269 (2001).
9. Suda, Y.; Kawasaki, H.; Ueda, T.; Ohshima, T.; Thin Solid Films.453, 162 (2004).
10. Yin, S.; Aita, Y.; Komatsu, M.; Sato, T. J. Eur. Ceram. Soc. 26, 2735 (2006).
11. Yamashita, H.; Ichihashi, Y.; Takeuchi, M.; Kishiguchi, S.; Anpo, M. J. Synchrotron Radiat. 6, 451 (1999).
12. Chen, H.; Nambu, A.; Wen, W.; Graciani, J.; Zhong, Z.; Hanson, J. C.; Fujita, E.; Rodriguez, J. A. J. Phys. Chem. C 111, 1366 (2007).
13. Rouquerol, F.; Rouquerol, J.; Sing, K. Adsorption by Powders and Porous Solids: Principles, Methodology, and Applications; Academic Press: San Diego (1999).
14. Sing, K. S. W.; Everett, D. H.; Haul, R. A. W.; Moscou, L.; Pierotti, R. A.; Rouquerol, J.; Siemieniewska, T. Pure Appl. Chem. 57, 603 (1985).
15. Nowotny, J.; Bak, T.; Nowotny, M. K.; Sheppard, L. R. J. Phys. Chem. B 110, 18492 (2006).

SYNTHESIS AND ACTIVITY TEST OF TWO ASYMMETRIC DIBENZALACETONES AS POTENTIAL SUNSCREEN MATERIAL

SRI HANDAYANI

Chemical Education Department, Yogyakarta State University, Karangmalang Yogyakarta, 55281, Indonesia

Synthesis and activity test as potential sunscreen material of two asymmetric dibenzalacetones have been conducted. The first compound, anisalbenzalacetone, was synthesized from benzaldehyde, acetone and p-anisaldehyde. The second, benzalveratralacetone was synthesized from benzaldehyde, acetone and veratraldehyde. Those compounds were synthesized by crossed aldol condensation in base condition with water-ethanol solvent. The synthesis of both compounds using ice bath throughout the stirring. The precipitate was purified by coloumn chromatography. Each product was characterized by FTIR, ^{1}H-NMR, ^{13}C-NMR, HMQC and HMBC. Activity test as UV protection was done by Walter method. The result showed that anisalbenzalacetone was active in UV-A area, while benzalveratralacetone protection is in UV-B area.

1. Introduction

Global warming issue have been attained most attention during last decade. The “holes” in ozone layer have lead more UV radiation emitted to earth’s surface. There are three types of UV light namely UV-A (320-400nm), UV-B (280-320nm) and UV-C (100-280nm). UV-A radiation could cause collagen breaken that lead to premature aging such as wrinkle skin. UV-B could make skin damaged which lead to sun-burned and cancer skin. The most dangerous radiation UV-C light, fortunately could be absorp by ozone layer, so it could not reach earth’s surface [1]. Therefore, there is a need a compound that used for protection of danger sun light radiation.

Some of reported compound that used for sunscreen are benzalacetone and its derivatives. Handayani and Arty have synthesized 1(E),4(E)-1.5-diphenyl-1,4-pentadiene-3-one and its derivatives which known as symmetrical dibenzalacetone that have been made by crossed aldol condensation between acetone:benzaldehyde (1:2). It have been tested as a radical hydroxyl scavenger [2], and as potential sun screen material [3]. Another synthesized of dibenzalacetone was report by Guofeng et.al give an excellent yields [4]. Tutik D has synthesized symmetrical dibenzalacetone which have similar structure with the cinnamic acid derivative [5]. From its structure, it could be estimated that benzalacetone and dibenzalacetone, which have similar structure with cinnamic acid or its derivative, have ultraviolet absorption in the same range. Thus, asymmetric dibenzalacetone was predicted as a sunscreen. Asymmetric dibenzalacetone is dibenzalacetone that its two benzene rings have different number, kind and/or position of substituent. In this research two of asymmetric dibenzalacetones, compounds **5** and **6** namely anisalbenzalacetone (1(E),4(E)-1-(4’-methoxy-phenyl)-5-phenyl-1,4-pentadiene-3-one) and benzal-veratralacetone (1(E),4(E)-1-(3’,4’-dimethoxy-phenyl)-5-phenyl-1,4-pentadiene-3-one) will be synthesized and tested as sunscreen material.

2. Experimental Section

2.1. *Materials*

All materials used from Merck, among other acetone, 4-methoxybenzaldehyde (p-anisaldehyde), 3,4-dimethoxybenzaldehyde (veratraldehyde), chloroform ethanol, benzaldehyde, hexane, and ethyl acetate. TLC was carried out using 0.25-mm plate Silica gel Merck 60 F254, column chromatography were performed by Silica gel 60 (230-400 mesh).

2.2. *Instrumentation*

The ^{1}H, ^{13}C-NMR, HMQC and HMBC Spectra were recorded on 500 MHz Jeol spectrophotometer. IR spectra were conducted using a Shimadzu 8300 FTIR spectrometer.

2.3. *Synthesis of Compound 5*

Into a solution of NaOH (0.025 mol, 1g) in aqueous ethanol (1:1) that was prepared at ambient temperature,

1 benzaldehyde (0.01 mol, 1,06 g), **2** acetone (0,01 mol, 0,58 g) and **3** 4-methoxybenzaldehyde (0,01 mol, 1,66 g) were added drop wise alternately. After additional stirring for 60 minutes, water (20 ml) was added to the reaction mixture which then filtered. The extract was washed with water (20 ml x 3) and separated by column chromatography (d 2.5 cm, h 50 cm), with silica gel 60 (230-400 mesh) as the stationary phase and ethylacetate-hexane 1 : 9 as the eluent. The target compound (**5**) was identified using thin layer chromatography with ethylacetate-hexane 5:1 as the mobile phase.

2.4. *Synthesis of Compound 6*

The similar procedure was repeated for 3,4-dimethoxybenzaldehyde to replace 4-methoxybenzaldehyde in order to synthesize compound **6** (Figure 1). Four fractions obtained from the column chromatography, and the target compound was identified using thin layer chromatography with chloroform-hexane 4: 6.

2.5. *In Vitro Test as Potential Sunscreen Material*

Activity test as sunscreen using in vitro method have been done by: Sample was dissolved in ethanol by various concentration 0f 1-50 ppm. The concentration of 1 ppm was used to measure optimum wavelength. The absorbances all variation concentration solution have been measured at optimum wavelength between 240-400 nm. The SPF (Sun Protection Factor) can be calculated following Walter formula as

$$SPF = 10^{A} \qquad (1)$$

Where A = absorbance of each solution. SPF level and UV light type could be determined, also C value which related to concentration that gives ultra protection at more than SPF 15.

3. Result and Discussion

Separation of crossed aldol condensation product between benzaldehyde, acetone and 4-methoxy-benzaldehyde was done by coloumn chromatography. The product from coloumn chromatography separation yielded 3 fractions, which were identified by TLC with hexane-ethylacetate 5:1 as the eluent. Retardation factor datas from TLC scanner showed that fraction II was the target of compound **5** (10,6%) determined as yellow residue. Fraction I and fraction II supposed to be benzalacetone and dianisalacetone as the side products of crossed aldol condensation reaction.

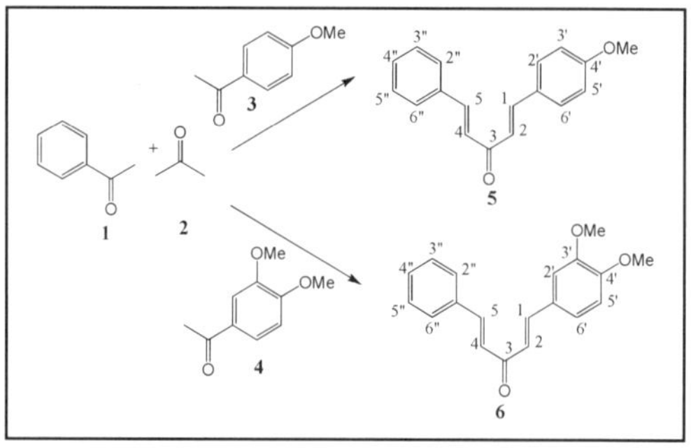

Figure 1. Scheme of cross aldol condensation to synthesized of compound **5** and **6**.

Characterization of compound **5** by FTIR (KBr) resulted peaks on 3035; 2922; 2842; 1668; 1423; 1446; and 1175 cm^{-1}. A series of one and two dimensional NMR spectroscopic experiment using HMQC and Heteronuclear Multitiple Bond Coherence (HMBC) patterns were performed to assign the proton and carbon resonance correlation of the com compounds. The signal pattern of the aromatic ring showed the influence of methoxy (OMe) in δ 3,8 (3H, s) ppm (Figure 1; Table 1). Therefore compound **5** is (1(E),4(E)-1-(4'-methoxyphenyl)-5-phenyl-1,4-pentadiene-3-one) or anisalbenzalacetone.

The preparation of compound **6** was initiated by the mixing of **1, 2** and **4** to give **6** (Figure 1). The yield of crossed aldol condensation between acetone, 3,4-dimethoxybenzaldehyde and benzaldehyde was a mixture consist of 4 compounds. It was separated by Column Chromatography (AcOEt-hexane, 1:9) to provide the asymmetric dibenzalacetone **6** (15,53%) determined as pale yellow oil.

The multiple bond correlation of HMBC supported the structure (Figure 1; Table 2). In the ^{1}H-NMR spectrum (500 MHz, $CDCl_3$), two patterns singlet, nine doublet and three double dublet were observed. The double dublet at δ = 7,2; 7,61; and 7,41 was assignable to H2", H3" and H6" respectively. Two equivalence methoxy signals at δ 3,91 and 3,9 were assigned to C3' and C4'. Support spectra data provided by the IR (KBr), which indicates the existence of C=O (1645cm^{-1}), aromatic C=C (1514-1417 cm^{-1}) and CO ether (1255-1139 cm^{-1}). Therefore, the structure of **6** was 1(E),4(E)-1-(3',4'-dimethoxyphenyl)-5-phenyl-1,4-pentadiene-3 one or benzalveratralacetone.

Table 1. ^{1}H, ^{13}C-NMR and HMBC datas of compound **5** ($CDCl_3$).

C no.	Δ H (∑H; m) ppm	δC ppm	HMBC (500 MHz)
1	7,70 (1H; d)	143	C6'', C2,C3
2	6,97 (1H; d)	123,4	C3, C2'
3	-	189	-
4	7,07 (1H; d)	125,7	C3, C1"
5	7,74 (1H; d)	143,3	C3
2',6'	7,58 (2H; d)	129	C1, C3', C4'
3',5'	6,93 (2H; d)	114,6	C4'
4'OMe	3,8; (3H, s)	56	C4'
4'	-	162	-
2"	7,61 (1H; d)	130,3	C5, C6"
3",5"	7,4 (2H; d)	127,6	C4", C2"
4''	7,41 (1H; d)	128,5	C1", C2"
6''	7,62 (2H; d)	130,5	C5, C2"

Table 2. ^{1}H, ^{13}C-NMR and HMBC datas of compound **6** ($CDCl_3$).

C no.	δH (∑H; m) ppm	δC ppm	HMBC (500 MHz)
1	7,69 (1H; d)	143,6	C2,C3
2	6,94 (1H; d)	124	C3
3	-	188	-
4	7,10 (1H; d)	125,6	C3, C1"
5	7,74 (1H; d)	143	C3
2'	7,06 (1H; d)	110	C3'
3'	-	151	-
3'OMe	3,9 (3H; s)	56	C4'
4'	-	150	-
4'OMe	3,91 (3H; s)	56	C3'
5'	6,83 (1H; d)	111	C4', C6', C1'
6'	6,89 (1H; d)	120	C1', C5', C4'
2"	7,61 (1H; dd)	130	C5
3"	7,2 (1H; dd)	123,4	C1", C4"
4"	7,14 (1H; d)	110	C3"
5"	7,33 (1H; d)	129	C1"
6''	7,41 (1H; dd)	128	C1"

Anisalbenzalacetone (**5**) showed absorbance at UV-A area (λmax 350 nm). Minimal concentration of this compound that can give ultra protection is 15 ppm at SPF 26 (Figure 2). Benzalveratralacetone (**6**) absorb radiation at UV-B area (λmax 303 nm). Minimal concentration of this compound for ultra protection is 30 ppm at SPF 16,143 (Figure 3).

Between those compounds, **5** more potential use as sunscreen, because it can give more SPF value at lower concentration than **6**.

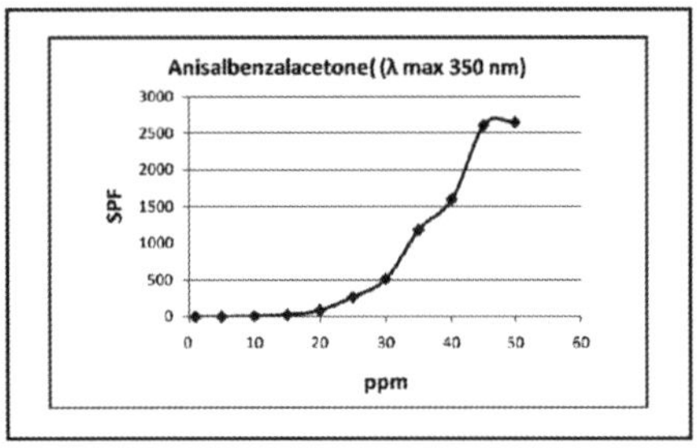

Figure 2. SPF vs C of compound **5**.

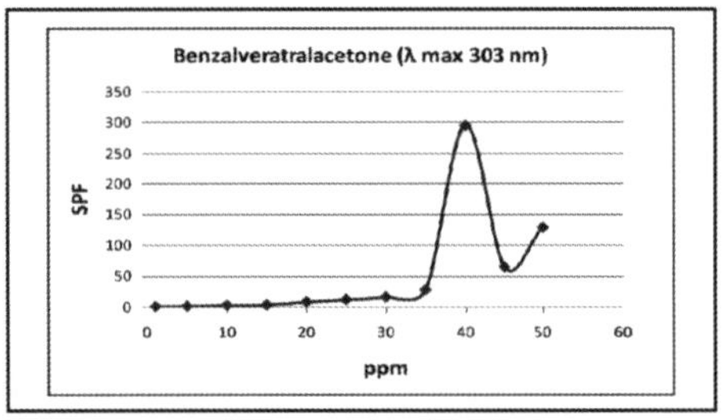

Figure 3. SPF vs C of compound **6**.

Figure 2 and 3 showed that some concentrations have absorbance above 3, giving SPF value higher than 1000. Sunscreen with higher SPF rating block slighty more UV-B rays, but none offers 100% protection [6] The data from high concentration could not be conciderate because device condition was not valid to measure compound with high absorbance. Also, for commercial application, it prefer minimal concentration that giving ultra protection.

4. Conclusion

Synthesis of asymmetric dibenzalacetone through crossed aldol condensation reaction has been done successfully. Low yields of the product was because there some other side products of the reaction. In this research, compound **5** more potent as sunscreen than **6** because it can give ultra protection at lower concentration.

Reference

1. Moloney F.J., Collins S., Murphy G.M., 2002, Am. J. Clin. Dermatol., 3(3), 185-191
2. Handayani, S. and Arty, I.S., 2008, Journal of Physical Science, 19(2), 61-68
3. Handayani, S., and Arty I.S, "Synthesis and Activity Test of Some Compounds 1,5-diphenyl-1,4-pentadiene-3-one as Potensial Sun Screen Material", Proc. of the 1st International Seminar on science and Technology (ISSTEC), Yogyakarta, Indonesia, 2009
4. http://www.chemistrymag.org/cji/2004/061007ne.htm
5. Tutik, D., Oksidasi Anetol dan Kajian Pengaruh Gugus Metoksi Turunan benzaldehide Terhadap Reaksi Kondensasi Benzoin dan Aldol Silang, Master Thesis, Gadjah Mada University, 1996
6. http://www.webmd.com/skin-problems-snd-trearments

CALCIUM CARBIDE RESIDUE-GROUND FLY ASH MIXTURE AS A NEW MATERIAL TO ENCAPSULATE ZINC

KIATSUDA SOMNA

Department of Environmental Engineering, Faculty of Engineering, King Mongkut's University of Technology Thonburi, Bangkok, 10140, THAILAND

CHAI JATURAPITAKKUL

Department of Civil Engineering, Faculty of Engineering, King Mongkut's University of Technology Thonburi, Bangkok, 10140, THAILAND

PUANGRAT KAJITVICHYANUKUL

National Center of Excellence for Environmental and Hazardous Waste Management, Department of Environmental Engineering, Faculty of Engineering, King Mongkut's University of Technology Thonburi, Bangkok, 10140, THAILAND

This paper proposed using calcium carbide residue-ground fly ash (CCR-GFA) mixture as a new material to encapsulate zinc instead of using Portland cement. A ratio of calcium carbide residue (CCR) to ground fly ash (GFA) of 30 to 70 by weight was used as a binder without Portland cement and had a water to binder ratio of 0.3. Zinc was added into CCR-GFA paste at 1, 5 and 10% by weight of binder. Crystalline phase of CCR-GFA paste and leaching behavior of zinc from CCR-GFA paste were determined by XRD and TCLP test, respectively. In addition, the compressive strengths of CCR-GFA pastes were also determined. The results showed that the compressive strengths of pastes containing zinc were dramatically decreased with the increased of zinc. However, the compressive strength were high enough to be used in landfill. Calcium zincate (CaZn2(OH)6.2H2O) can be observed in XRD technique. In addition, the leaching concentrations of zinc from CCR-GFA pastes containing zinc are much lower than the allowable limit (500 mg/l). This result suggested that CCR-GFA paste can be used effectively to encapsulate zinc.

1. Introduction

Encapsulation is not a method to remove heavy metal from the pollution waste but has a purpose to physically as well as chemically fix it in the solid matrix in order to reduce its mobility and to ensure compliance with existing regulatory standard [1]. It is noted that Portland cement is the most famous used for encapsulation of heavy metals and it has been shown to be the excellent material for fixing the heavy metal. However, it is a good idea to use waste material to fix the heavy metal since this concept not only encapsulates the heavy metal but also reduces the use of cement thus conserves the energy as well as be good for the environment.

Calcium carbide residue (CCR) is a by-product from acetylene gas production process and is rich in $Ca(OH)_2$ (see Eq.1) [2]. It is obvious that $Ca(OH)_2$ obtained from acetylene gas production process is similar to that obtained from hydration of cement. In general, $Ca(OH)_2$ from hydration of cement can react with a pozzolanic material such as fly ash or rice husk ash to form calcium silicate hydrate (CSH) which can develop strength [2].

$$CaC_2 + 2H_2O \rightarrow C_2H_2 + Ca(OH)_2 \quad ---(1)$$

In 1995 to 1998, 74,000 tons of calcium carbide were used in Thailand to produce acetylene gas and provided 86,000 or about 21,500 tons/year of calcium carbide residue (CCR). CCR was hardly utilized in any work due to its high base and all of it went to a disposal landfill area [3].

Fly ash is a by-product from the electric thermal power plant and is a good source of pozzolanic material. Since CCR is rich in $Ca(OH)_2$ thus it can react with SiO_2 and Al_2O_3 in fly ash to form CSH. It is anticipated that the mixture of fly ash and CCR can develop strength and strong enough to encapsulate heavy metal. In Thailand, more than 4 million tons of fly ash has been produced annually and more than half of its has to be disposed off.

Zinc (Zn) has been identified as one of the hazardous elements to human health and is a widespread contaminant in many hazardous waste sites. Zinc is an essential element in our diet. Too little zinc can cause problems, but too much zinc is also harmful [4].

The aim of this work is to encapsulate zinc by using paste obtained from CCR and GFA mixture. Both CCR and GFA are considered as wastes and need to be eliminated. This mixture does not contain any Portland cement, thus calcium carbide residue-fly ash paste is environmentally friendly material.

2. Materials and Methods

2.1. *Materials*

Fly ash used in this study was obtained from Mae Moh power plant, Thailand. Calcium carbide residue (CCR) was obtained from an acetylene gas production system factory at Samutsakorn province, Thailand. A lab grade zinc oxide (ZnO) was used as the source of Zn2+ contamination and was added to CCR-GFA paste at a concentration level of 1, 5 and 10% by weight of binder (CCR plus ground fly ash). Distilled water was used throughout of the test.

2.2. *Experimental designs and methods*

Fly ash and calcium carbide residue were ground separately by ball mill until the particles retained on a 45-μm sieve were about 2%. This process is to accelerate the reactivity of fly ash and calcium carbide residue [5]. The ratio of CCR to GFA of 70:30 by weight was used as a binder with a water to binder (W/B) ratio of 0.3 (CF-CT in shorthand). Zinc was added into CCR-GFA mixture at 1, 5 and 10% by weight of binder (1ZnCF, 5ZnCF and 10ZnCF in shorthand, respectively). After finishing mixing, the samples were cast in plastic molds with a diameter of 30 mm and a height of 60 mm and then were kept at an ambient temperature. CCR-GFA pastes were removed from the molds after casting for 24 h, then they were sealed by plastic sheet to prevent the water evaporation until the testing ages.

2.3. *Analyzing methods*

Particle size distribution of GFA and CCR were performed by using Mastersizers Malvern Instruments. Major components of GFA and CCR were analyzed by X-ray fluorescence spectroscopy (XRF). Compressive strengths of the pastes were determined following by ASTM D1633[6]. Crystalline phase of the specimen was identified by X-ray diffractrometer (XRD). Leaching test was determined by toxicity characteristic leaching procedure (TCLP US EPA) testing and 0.1 M acetic acid, L/S ratios of 20 and rotate 360° for 18 hours was performed in the test.

3. Results and Discussion

3.1. *Physical and chemical properties of materials*

Physical properties of CCR and GFA are illustrated in Table 1. The retained particles on a 45-μm sieve of CCR and GFA were 2.3 and 2.0%, respectively. After grinding, CCR and GFA had mean particle size, d50, of 8.8 and 10.4 micron, respectively. Table 2 shows the chemical composition of CCR and GFA. CCR is rich in CaO of 54.5% with LOI of 41.2%. The high LOI of CCR is due to the material mainly consisting of Ca(OH)2 which decomposes into CaO and H2O at approximately temperature of 550°C [3]. According to the chemical composition, GFA has SiO2+Al2O3+Fe2O3 of 66.6%, SO3 of 4.1% and LOI of 1.8% and can be treated as Class C fly ash as prescribed by ASTM C 618 [7].

Table 1. Physical properties of CCR and GFA.

Materials	Retained on a 45-μm sieve	Mean particle size, d_{50} (micron)
CCR	2.3	8.8
GFA	2.0	10.4

3.2. *XRD analysis*

Fig. 1 shows diffraction pattern of CF-CT and 10ZnCF. This pattern was detected between 20 – 80° at 2θ. For CF-CT paste, when CCR was mixed with fly ash calcium silicate hydrate (CSH) was formed which was represented by Ca5(SiO4)2(OH)2. Moreover, it could be observed that CaCO3 was slightly changed throughout the test. Diffraction pattern changed noticibly after zinc was added in CCR-fly ash pastes. Zincate (CaZn2(OH)6.2H2O) and zinc oxide (ZnO) were observed. Ca(OH)2 in CCR caused the extensive dissolution of Ca+ and OH-. This led to the rapid resolubilization of the amorphous zinc oxide and later transformed to calcium zincate (CaZn2(OH)6.2H2O).

Table 2. Chemical compositions of CCR and GFA.

Composition (%)	CCR	GFA
SiO_2	1.9	31.2
Al_2O_3	0.8	18.9
Fe_2O_3	0.4	16.5
CaO	54.5	20.8
MgO	0.6	1.86
Na_2O	-	1.53
K_2O	-	2.8
SO_3	0.6	4.1
LOI	41.2	1.8

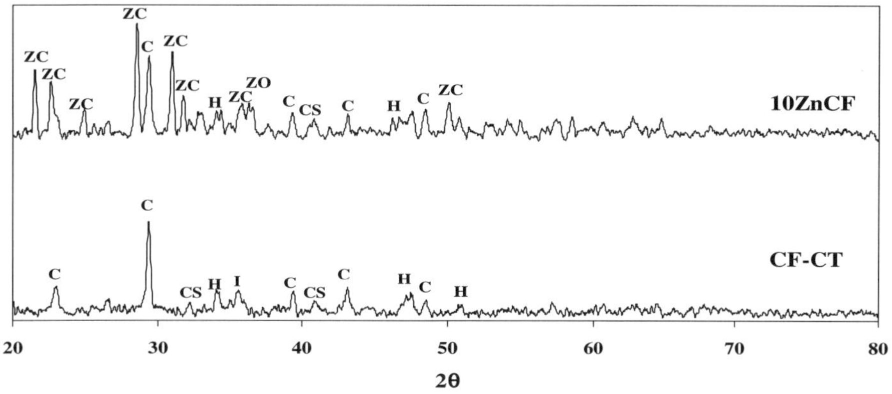

$C = CaCO_3$ $H = Ca(OH)_2$ $CS = Ca_5(SiO_4)_2(OH)_2$ $I = SiC$

$ZO = ZnO$ $ZC = CaZn_2(OH)_6{\cdot}2H_2O$

Fig. 1. X-Ray diffraction pattern of CCR-GFA pastes with and without adding zinc.

4. Compressive Strength

Table 3 presents the compressive strengths of CF-CT and 1ZnCF, 5ZnCF and 10ZnCF. Fig. 2 shows the relationship between the compressive strength of CCR-GFA pastes containing zinc and ages. The compressive strengths of the specimens were investigated in order to ensure that their strengths exceeded the standard value for landfill (3.5 ksc) [8]. The strength developments of all samples increased with age up to 60 days. It is seen clearly that the compressive strengths of CCR-GFA pastes containing zinc of 1, 5 and 10% by weight of binder were dramatically decreased with the increased of zinc. Their had compressive strengths were much lower than that of CF-CT. This suggested that zinc oxide can rapidly react with Ca(OH)2 from CCR to form calcium zincate (CaZn2(OH)6.2H2O) which is shown in XRD results. Moreover, the addition of zinc in Portland cement based solidification caused lower the strength development of the paste. Asavapisit et al. [9] reported that heavy metal such as zinc and lead retarded hydration of Portland cement at high concentration.

Table 3. Compressive strength of CCR-GFA pastes with and without zinc.

Paste	Compressive Strength (ksc)				
	7 Days	14 Days	28 Days	42 Days	60 Days
CF-CT	72.3	77.7	84.7	89.9	91.5
1ZnCF	17.8	20.0	23.0	23.6	24.0
5ZnCF	13.9	17.0	19.5	20.0	21.0
10ZnCF	11.0	12.7	17.4	18.0	19.0

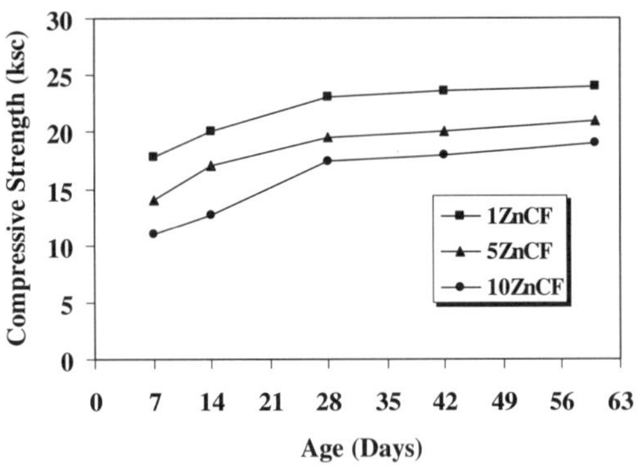

Fig. 2. Relationship between compressive strength of CCR-GFA pastes containing zinc and age.

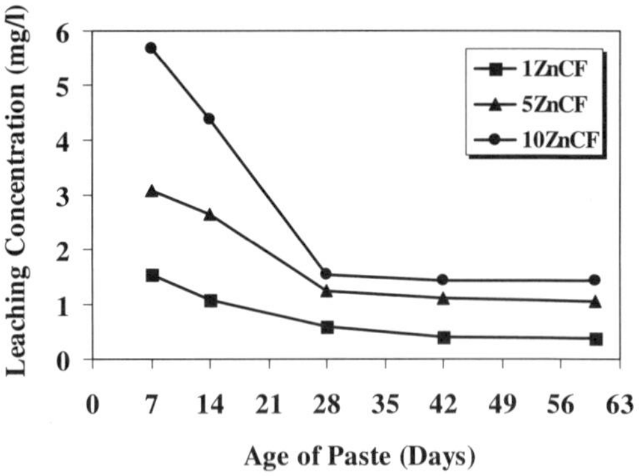

Fig. 3. Leaching concentration of zinc from CCR-GFA pastes all at specified ages.

3.4 TCLP US EPA Leaching Test

Table 4 shows leachate pHs in TCLP leachates. As the age increased, the leachates pH of all samples slightly increased. This suggested that CaZn2(OH)6.2H2O had base form and represented by high pHs.

The results of the TCLP US EPA tests on CCR-GFA pastes mixed with zinc of 1, 5, and 10% by weight of binder are illustrated in Fig. 3. Leaching concentration of zinc was substantially decreased from 7 to 28 days and slightly decreased from 28 to 60 days. It can be interpreted that ZnO nearly become insoluble in water but become soluble in base [10]. ZnO reacted with $Ca(OH)_2$ from CCR as represented by $CaZn_2(OH)_6.2H_2O$ (as shown in Fig. 1) and thus zinc was encapsulated in CCR-GFA paste.

Based on TCLP US EPA guideline, the regulatory limit for zinc is not available. However, the limit of this value was chosen by multiplying the maximum permissible concentration for potable water (5 mg/l) by one hundred. From the selected regulatory, leaching of zinc should not be higher than 500 mg/l of US EPA MCLs [11]. In summary, CCR-GFA pastes can be used effectively to encapsulate zinc since all leaching concentrations of zinc are not more than 6 mg/l (see Fig. 3).

Table 4. pH of leachate by TCLP test.

Age of Paste (days)	pH		
	Zinc Added 1%	Zinc Added 5%	Zinc Added 10%
7	11.6	10.9	10.7
14	11.7	11.1	10.8
28	12.6	11.2	10.9
42	12.6	11.3	10.9
60	12.6	11.4	11.0

5. Conclusion

This research used calcium carbide residue and ground fly ash which were wastes to encapsulate zinc. Both CCR and GFA were mixed together to form a cementitious material which is environmentally friendly material. The results showed that zinc rapidly reacts with Ca(OH)2 from CCR to form a new crystalline phase of CaZn2(OH)6.2H2O. This can be seen by the dramatically decrease in the compressive strength of CCR-GFA paste containing zinc since it reduced the pozzolanic reaction between CCR and ground fly ash. It is interesting to see that after zinc reacted with Ca(OH)2, it was fixed in the CCR-GFA matrix. For TCLP test, the leaching concentration of zinc from CCR-GFA paste was much lower than that the limit of multiplying the maximum permissible concentrations for potable water or 500 mg/l.

Acknowledgement

The authors gratefully acknowledge the financial supports from the Thailand Research Fund (TRF) under TRF Senior Research Scholar, Contact No. RTA5080020 and the Commission on Higher Education, Ministry of Education, Thailand.

References

1. S. Srivastava, R. Chaudhary and D. Khale, J. Hazard. Mater. 153, p.1103 – 1109 (2008).
2. J. M. John, "Inorganic chemicals handbook," Vol.1, Marcle Dekker, New York, 302 – 309 (1993).
3. C. Jaturapitakkul and B. Roongreung, J. Mater. Civil. Eng. ASCE 2003. 15, 470 – 475 (2003).

4. http://www.atsdr.cdc.gov/toxfaq.html [2009 May 1].
5. P. Chindaprasirt, C. Jaturapitakkul and T. Sinsiri, Const. Build. Mater. 21, 1534 –1541, 2007.
6. ASTM. D1633. Annual Book of ASTM Standard 1996; 04.08: 152-154.
7. ASTM. C618. Annual Book of ASTM Standard 2001; 04.02: 310 – 313.
8. Ministry of Industry. 2003b. Notification of Ministry of Industry No. 6 B.E. 2540 (1997) [Online]. Available from: http://www.diw.go.th/w_laws/law13.html, [2009 May 31].
9. S. Asavapisita, S. Naksrichumb, and N. Harnwajanawongb, Cem. Concr. Res. 35, 1042–1049 (2005).
10. http://en.wikipedia.org/wiki/Zinc_ oxide [2009 May 31].
11. V. Ribe, E. Nehrenhein, M. Odlare and S. Waara, J. Hazard. Mater. 163, 1096 – 1100 (2009).

EFFECT OF CR_2O_3 IN CEMENT CLINKER ON MICROSTRUCTURE AND LEACHABILITY OF CEMENT MORTAR

SUTHATIP SINYOUNG

Department of Environmental Engineering, Faculty of Engineering, King Mongkut's University of Technology, Bangkok 10140, Thailand

SUWIMOL ASAVAPISIT

Environmental Technology, School of Energy and Materials, King Mongkut's University of Technology Thonburi, Bangkok 10140, Thailand

PUANGRAT KAJITVICHYANUKUL

National Center of Excellence for Environmental and Hazardous Waste Management, Department of Environmental Engineering, Faculty of Engineering, King Mongkut's University of Technology Thonburi, Bangkok 10140, Thailand

PRAYOON SONGSRIRIRITTIGUL

National Synchrotron Research Center, P.O. Box 93 Nakhon Ratchasima, 30000, Thailand

Effect of Cr_2O_3 in cement clinker on microstructure and leachability of cement mortar was studied. Cement clinker was synthesized in the presence of 0.1, 0.5, 1, 2, 5 and 10 % by weight of Cr. The mixture of raw meal and Cr_2O_3 was burned in electric furnace at 1,450°C for 90 min. The synthesized clinker was mixed with 4.5 % gypsum and hydrated for 28 days. Microstructure of hydrated cement was investigated using XRD, SEM and TG/DTA techniques and leaching behavior of chromium was evaluated using ANS 16.1. XRD analysis showed that Cr_2O_3 was present in the hydrated cement in the form of $Al_2(OH)_4CrO_4$, $Ca_5(CrO_4)._3OH$, $CaCrO_4.2H_2O$, and $CaCrSi_4O_{10}$. In addition, chromium (VI) was present in the leachate at high proportion.

1. Introduction

Disposal of hazardous waste by co-incineration in cement kiln during the production of cement clinker is a common management practice. Under the high operating temperature between 1,400 and 1,600 °C of cement kiln, the principle organic hazardous constituents are highly destroyed. Several types of hazardous waste include used oil, spent solvent, scrap tyres, waste plastic, and sludge from wastewater treatment plant are normally disposed of in the cement kiln. These wastes consist of various heavy metals such as cobalt, cadmium, lead, zinc, nickel, and chromium [1,2].

Chromium is one of heavy metals which are present in many industrial wastes. Previous work reported that chromium could either be incorporated into clinker phases or formed new crystalline phases during reactions at high temperature [3]. Little is known about the newly formed chromium compounds during clinkering process on hydration reactions and leachability of chromium from the hardened cement mortar.

In this research work, the effect of Cr_2O_3 in cement clinker on microstructure and leachability of hardened cement mortar was studied. XRD and SEM techniques were used to examine the crystalline phases and morphology of hardened cement mortar hydrated for 28 days. The amount of $Ca(OH)_2$ present in the sample was analyzed using TG/DTA technique and release of total chromium and chromium (VI) from hardened cement paste was also determined by ANS 16.1 leach test method.

2. Research Methodology

2.1. *Raw material*

Raw meal was brought from Siam City Cement, Co., Ltd. (Thailand). Raw meal consists of limestone, limestone high grade, clay mix, and iron ore. Oxide composition of raw meal was analyzed using X-ray fluorescence and is shown in Table 1.

Table 1. Composition of raw meal.

Compound	Quantity (wt.%)
SiO_2	14.33
Al_2O_3	3.26
Fe_2O_3	2.17
CaO	43.13
MgO	1.14
K_2O	0.47
Na_2O	0.03
SO_3	0.17
*LOI	35.31

** LOI = loss on ignition*

2.2. *Clinker preparation*

Cr_2O_3 (BDH) was mixed with raw meal at concentration of 0.1, 0.5, 1, 2, 5 and 10 % by weight of Cr. The mixture was burned in an electric furnace at 1450 °C for 90 min [4]. The temperature was increased at the rate of 10 °C/min. The synthesized clinker was ground to particle size less than <125 μm [3].

2.3. *Sample preparation*

The synthesized clinker was mixed with gypsum at 4.5 % by weight and mixed with Ottawa sand at the ratio between binder and sand of 1:2.75. Water to cement (binder and gypsum) ratio of 0.45 was used in this work. Cement mortar slurry was then transferred to cylindrical plastic mold. After demolded, the samples were wrapped with cling film and continue curing under ambient condition.

Samples containing 2, 5 and 10 % by weight of Cr and hydrated for 28 days were used for XRD, SEM and TG/DTA analysis. Hydration was stopped by soaking the sample in acetone. The crystalline phases present in the samples were examined by XRD technique using a Philips PW 3710 diffractometer with Cu K-alpha. The surface morphology of the hydrated sample was analyzed by SEM technique using a JEOL JSM-5800LV Link ISIS series 300. The development of $Ca(OH)_2$ was investigated using TG/DTA (Netzsch-409). 15 mg sub-samples, ground to<150 μm were heated to 600°C in the alumina crucibles under a N_2 atmosphere at rate of 5 °C/min. Under these conditions $Ca(OH)_2$ dehydrates in the temperature range between approximately 370 and 430 °C. DTA data was calibrated by analyzing varying amounts of GPR grade $Ca(OH)_2$ mixed with Al_2O_3[5].

Leachability of chromium was assessed using ANS 16.1 leaching test method on samples containing 0.1, 0.5 and 1.0 % by weight of Cr. Synthetic acid rain containing sulfuric and nitric acid solutions at the weight ratio of 80:20 was used as leachant. Initial pH of the leachant was maintained at 3 and the weight ratio between leachant and monolith sample was 10:1. Leachate was renewed after contacted with the sample for 0.08, 0.29, 1, 2, 3, 4, 5, 7, 14, 19, 28, 48, 68, and 90 days [6]. Total chromium and chromium (VI) were determined using ICP-AES and UV-visible spectrometer at a wavelength of 540 nm.

3. Results and Discussion

3.1. *XRD patterns*

XRD patterns of hydrated cement with and without chromium after 28 days of hydration are shown in figures 1a-d.

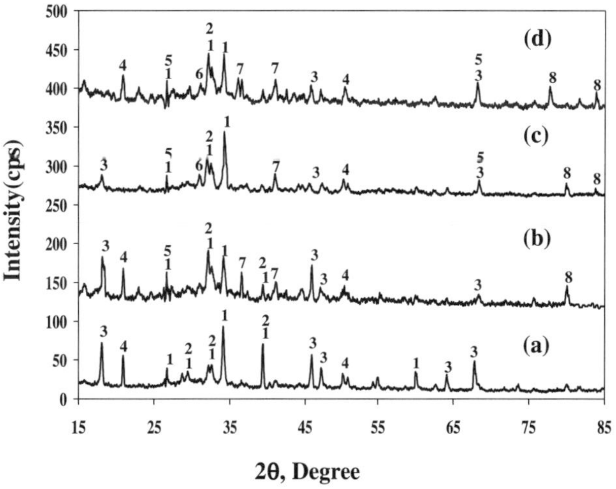

1 = C_3S, **2** = C_2S, **3** = $Ca(OH)_2$, **4** = Ettringite,
5 = $Al_2(OH)_4CrO_4$, **6** = $Ca_5(CrO_4)_3OH$,
7 = $CaCrO_4.2H_2O$, **8** = $CaCrSi_4O_{10}$

Figure 1. XRD patterns of hydrated cements at the age of 28 days (a), control (b), 2 % Cr (c), 5 % Cr (d), 10 % Cr.

$Ca(OH)_2$ which is one of the hydration products was found at 2θ of 18.1°, 47.3°, 64.1°, and 68.9° but for ettringite was identified at 2θ of 20.8°, and 50.7°. The remaining of the peaks appeared in the control sample were identified as the unhydrated C_3S and C_2S. In the presence of Cr_2O_3, it is observed that the intensity of the peak present at the position for $Ca(OH)_2$ decreased with increasing concentration of Cr_2O_3. In addition, $Al_2(OH)_4CrO_4$, $Ca_5(CrO_4)._3OH$, $CaCrO_4.2H_2O$, and $CaCrSi_4O_{10}$ were found in all samples containing Cr_2O_3. This implies that the added Cr_2O_3 can react with CaO, SiO_2, and Al_2O_3 present in the raw meal during synthesis of clinker and results in the formation of new compound containing chromium. Similar chromium containing compounds were reported during the preparation of C_3S in the presence of $Cr(NO_3)_3.9H_2O$ [7].

3.2. *DTA results*

Table.2 shows the amount of $Ca(OH)_2$ present in samples containing 2.0, 5.0, and 10.0 % by weight of Cr during the synthesis of clinker. Results showed that the amount of $Ca(OH)_2$ present in the control sample (without chromium) hydrated for 28 days was 18.86 %. In the present of 2.0, 5.0, and 10.0 % by weight of Cr, the amount of $Ca(OH)_2$ detected during 28 days of hydration decreased to 14.1, 10.06, and 6.56 %, respectively.

Table 2. The amont amount of Ca(OH)2 present in cement mortar containing chromium at 28 days of hydration.

Initial conc. of chromium	Area of Peak	% $Ca(OH)_2$
0	41.15	18.86
2	29.14	14.10
5	21.96	10.06
10	14.81	6.56

3.3. *SEM micrograph*

Surface morphology of hydrated cements with and without chromium is shown in figures 2 a-d.

A dense microstructure consists of amorphous CSH gel which is the main hydration product was observed from mortar hydrate for 28 days (fig. 2 a). In the presence of an increasing amount of Cr_2O_3, it is observed that less hydration products especially amorphous CSH gels were present. This is in agreement with XRD and DTA results which showed that the present of $Ca(OH)_2$, one of the hydration products, decreased with increasing the amount of Cr_2O_3 (fig. 2 b - d).

Figure 2. SEM micrographs of hydrated cements at the age of 28 days (a), control (b), 2 % Cr (c), 5 % Cr (d), 10 % Cr.

3.4. *Leaching behavior*

The cumulative release of total chromium from cement mortar as a function of exposure time is presented in figure 3. It was found that the rate of chromium leaching was high during the first 7 days of exposure to synthetic acid rain solution. This could be caused by the wash out of chromium compounds present at the surface of the solid phases or in the pore solution. However, the rate of leaching decreased thereafter as the exposure time increased. In addition, the cumulative release of total chromium in the leachate after 90 days of leaching was 2.6, 13.3 and 26 mg/L after contacting with samples containing 0.1, 0.5 and 1.0 % by weight of Cr, respectively (fig. 3).

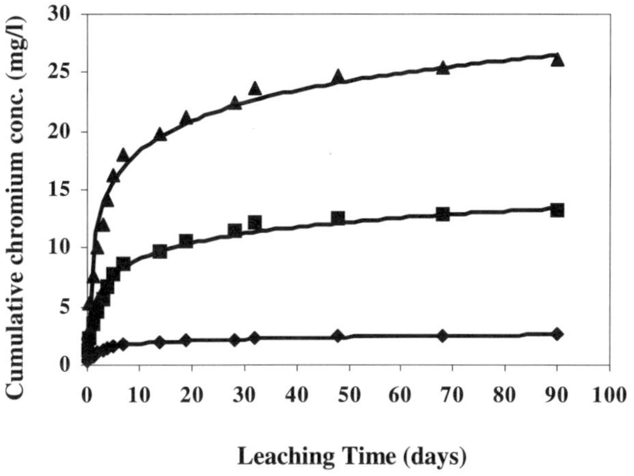

Figure 3. Cumulative concentration of total chromium during 90 days.

◆ 0.1 % Cr ■ 0.5 % Cr ▲ 1.0 % Cr

The calculated leachable fraction of chromium from cement mortar (c_t/c_0) in the synthetic acid rain solution was observed in the range of 0.30-0.31. Experimental results showed that about 87-92 % of chromium present in the leachate was in the form of chromium (VI). This could be because of the present of chromium (VI) in the newly formed chromium compounds such as $CaCrO_4.2H_2O$, and $Al_2(OH)_4CrO_4$ during the synthesis of clinker under high temperature. This chromium (VI) is generally known to have high solubility.

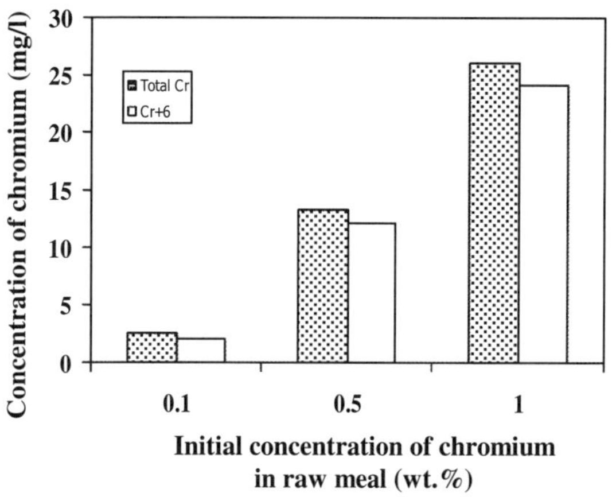

Figure 4. Comparison between total chromium and heaxavalent chromium present in leachate.

4. Conclusions

The investigation of the effect of chromium on microstructure and leachability of cement mortar using various techniqures can be concluded as follow.

- Crystalline phases of $Al_2(OH)_4CrO_4$, $Ca_5(CrO_4)._3OH$, $CaCrO_4.2H_2O$, and $CaCrSi_4O_{10}$ were found in all samples during synthesis of clinker in presence of Cr_2O_3 at high temperature.
- Amorphous CSH gel and $Ca(OH)_2$ which are the main hydration products of hydraulic cement were found to decreased as the amount of chromium in the clinker increased.
- The leachable fraction of total chromium from the monolith sample during 90 days of exposure to synthetic acid rain was in the range of 0.30-0.31. About 87-92 % of chromium present in the leachate was in the form of chromium (VI).

Acknowledgement

This work was supported by the National Synchrotron Research Center (Thailand) GS-49-D06 and GRANT 1-2549/PS03. The authors would like to appreciate the Siam City Cement Public Company Limited for their assistants in the experimental works.

References

1. F. Willitsch and G. Sturm, 2003, Use and Preparalation of Alternative Fuels for the Cement Industry, *Cement Plant Environmental Handbook,* pp. 151-157.
2. I. Serclerat, P. Moszkowicz, and B. Pollet, 2000, Retention mechanism in mortars of the trace metals contained in Portland cement clinkers, *Waste Management*, Vol. 20, pp. 259-264.
3. D. Stephan, H. Maleki, D. Knofel, B. Eber, and R. Hardtl, 1999(b), Influence of Cr, Ni and Zn on the properties of pure clinker phases: Part II. C_3A and C_4AF", Cement and Concrete Research, Vol.29, pp.651– 657.
4. M.A. Trezza, and A.N. Scian, 2000, Burning wastes as an industrial resource their effect on Portland cement clinker, Cement and Concrete Research, Vol.30, pp. 137-144.
5. S. Asavapisit, G. Fowler, and C.R. Cheeseman, 1997, Solution chemistry during cement hydration in the presence of metal hydroxide wastes, Cement and Concrete Research, Vol.28, No.8, pp. 1249-1260.
6. ANS, 1986, American national standard for the measurement of the leachability of solidified low-level radioactive wastes by a short-term test procedure. ANS/ANS-1601, American National Standard Institute, New York.
7. C.K. Lin, J.N. Chen, and CC. Lin, 1997, An NMR, XRD and EDS study of solidification/stabilization of chromium with Portland cement and C_3S, Journal of Hazardous Materials, Vol. 56, pp. 21-34.

PREPARATION OF ZIRCONIUM DISPERSED IN ALUMINIUM PILLARED MONTMORILLONITE AND ITS ACTIVITY IN PHENOL HYDROXYLATION

IS FATIMAH

Chemistry Dept., Islamic University of Indonesia, Kampus Terpadu UII, Jl. Kaliurang Km 14, Sleman, Yogyakarta Indonesia 55581. email: isfatimah@fmipa.uii.ac.id

NARSITO, KARNA WIJAYA

Chemistry Dept., Gadjah Mada University, Sekip Utara, Bulaksumur Bls-21 Yogyakarta, Indonesia

The preparation of zirconia dispersed in aluminium pillared montmorillonite and its application as catalyst in phenol hydroxylation is presented in this paper. Characterization of material by using XRD, DRUV-Vis and Gas Sorption Analyzer (GSA). Its activity in the phenol hydroxylation reaction was investigated. It was observed that prepared material increase significantly the reaction rate. Effect of reaction parameter to the conversion and product selectivity is also presented in this paper.

1. Introduction

Zirconia (ZrO_2) is an important material in heterogeneous catalysis. Due to its nature as acidic solid it has been considered recently as catalyst for many heterogeneous reactions. ZrO_2 becomes bifunctional catalyst in several reactions such as reduction reaction according to its thermal stability and oxygen storage capacity. However, compared with other metal oxide, commercial ZrO_2 has a rather low specific surface area. Many attempts have been made to prepare nanomaterial ZrO_2 with a large surface are by the sol-gel method. The improvement of catalytic activity as function of particle size, surface area and homogeneous distribution can be achieved by the dispersion onto inorganic matrix. Inorganic material with high thermal stability, specific surface area and available pore distribution is appropriate to improve the catalytic character [1].

Pillared clays are semi-synthesized materials characterized as zeolite-like materials. Due to its mechanism in formation of pillar, pillared clays are porous materials having high specific surface area and exchangeable cations in the pore. Recent advance in clays modification are the use of pillared clays as solid support for metal and metal oxide catalyst [2]. These reports describe a new challenge for pillared clays act as support for metal oxide, such as ZrO_2. Higher dispersion and specific surface area tend to increase molecular interaction in surface mechanism.

In this paper, the ZrO_2 dispersed into pillared clay was prepared. Among other metal oxide, the large number of papers are concerning with pillared clays (PILCs) with pillars of Al. Despite some reports of its application as catalyst in several reactions, the study on the use of aluminium pillared clays as solid support is limited. These are the important reason for the utilization of aluminium as pillaring metal in this research. The aim of the work described here is to synthesize and characterize the ZrO_2 dispersed in aluminium pillared clays. Indonesian natural montmorillonite was used as smectite clay. Role of physicochemical characters to the improvement of catalytic activity and kinetics is investigated. Phenol hydroxylation reaction was employed as reaction model for evaluating the surface evolution during the preparation.

2. Materials and Method

2.1. *Materials*

Zirconium oxide chloride hydrate ($ZrOCl_2.8H_2O$) and aluminium chloride hexahydrate ($AlCl_3.6H_2O$) were purchased from E. Merk Company. Natural montmorillonite from PT. Tunas Inti Makmur, Semarang, Indonesia was used as natural clays used as received.

2.2. *Equipment and Analytical Instrument*

Equipment in material preparation consists of calcination reactor, adsorption and reduction reactor, oven and controlled stirrer. For characterization purpose, X-ray diffraction (XRD) Shimadzu X 6000, Surface area analyzer NOVA 1200e, and Diffuse Reflectance UV-Visible spectrophotometer JASCO V 670 were used. Cation exchange capacity of materials were measured by copper (Cu^{2+}) ion exchanged method. Exchanged copper was analyzed by Atomic Absorption Spectrophotometer (AAS) Perkin Elmer.

2.3. *Pillarization and ZrO_2 Dispersion*

Procedure of aluminium pillarization for montmorillonite has been reported in previous research [4]. ZrO_2 dispersion into APM was engaged by impregnation method with the theoretic zirconium concentration was equal 4 times to CEC. Zirconium oxide chloride solution added to the ammonium hydroxide solution and kept at stirring for 3 hours. The solution was then dispersed into the suspension made from five grams of aluminium pillared clays powder (particle size is 200 mesh) in distilled water followed by stirring at room temperature for 1 h. Solvent was evaporated by heating at temperature of 50°C. Dry powder resulted in this step then calcined at 400°C for 3h with heating rate of 1-3°C/min. The synthesized material was called as zirconia dispersed in aluminium pillared montmorillonite abbreviated as ZrO_2/APM.

2.4. Catalyst Activity

Hydroxylation of phenol with prepared catalyst was carried out in a two-neck glass reactor (100 ml) fitted with a condenser and a septum. Hydrogen peroxide (30%, w/v) was added slowly through the septum at once to the magnetically stirred solution of phenol containing catalyst kept at the reflux temperature. The kinetics of the reaction was monitored by taking periodically small samples (0.05 cm^3) which were analyzed by high performance liquid chromatography (HPLC) fitted with SE-80 packed column and UV detector. The conversion of phenol was measured as the percentage of reacted phenol and the product selectivity was determined as percentage of selected product among all products.

3. Result and Discussion

3.1. *Characterization*

Figure 1 shows XRD patterns of ZrO_2/APM compared to APM and the raw montmorillonite. Reflections of natural montmorillonite show the significant characteristic peaks at $2\theta = 5.92^o$ (d = 14.47 Å) and $2\theta = 19.91^o$ (d = 4.45 Å) with the valuable intensity and other reflections at 2θ=20.18°, 23.57° and 26.57°. The reflection at $2\theta = 5.92^o$ is correlated to the d_{001} basal spacing of smectitic clay structure. Other reflections corresponding to the impurities minerals of quartz (encoded by q) and crystoballite (encoded by cr) are also identified (a). As seen in the figure, the pillarization process shows the shift of d_{001} reflection in the region of 4.5-5.0 Å to the lower angle correspond to the increase of basal spacing d_{001}. On the aluminium pillarization, there is an improvement of d_{001} basal spacing from 14.47Å to 19.57Å which is also the evidences for aluminium oxide pillar formation in the interlayer space of montmorillonite (b). By the dispersion of zirconia, main structure of clay is still maintained with insignificant differences on the reflection patterns and intensities, however, the reflection correspond to basal spacing d_{001} is positioned in the same as the reflection of raw montmorillonite. This may caused by structure delamination during the dispersion by ZrO_2 aggregation in the matrix pore. This assumption is also strengthened by new peak at around $2\theta = 28.5^o$ (encoded by *) in ZrO_2/APM which is probably the evidence for zirconia formation in the matrix in monoclinic phase [1,3]. This assumption is also in agreement with the spectra obtained by the DRUV-Visible measurement depicted in Figure 2.

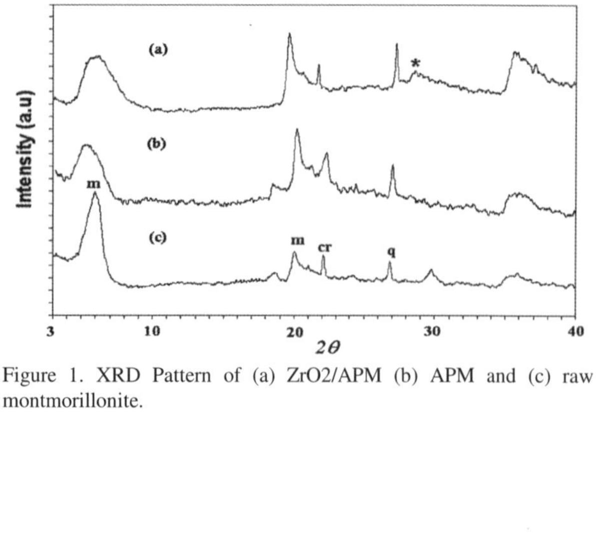

Figure 1. XRD Pattern of (a) ZrO2/APM (b) APM and (c) raw montmorillonite.

The appreciable zirconia formation in the matrix is revealed by the shift of absorbance spectra. Edge wavelength of ZrO_2/APM is estimated to be around 380 nm suggesting the zirconia formation in nanosize in the APM matrix.

Due to the curve, there is significant alteration of N_2 adsorption isotherm of montmorillonite to APM and ZrO_2/APM. At overall range of pressure, APM and ZrO_2/APM have better sorption ability than natural montmorillonite.

Zirconia dispersion into APM seems to slightly decrease the adsorption-desorption profile. These data are agreement with the specific surface area and pore volume data as presented in Table 1.

Table 1. Specific Surface Area, Pore Volume and Zr content data of Materials.

Parameter	Montmorillonite	APM	ZrO_2/APM
Specific Surface area (m^2/g)	45,68	169.06	120.09
Pore Volume (10^{-3}cc/g)	0.037	116.27	97.90
Zr content (%w/w)	-	-	3.19

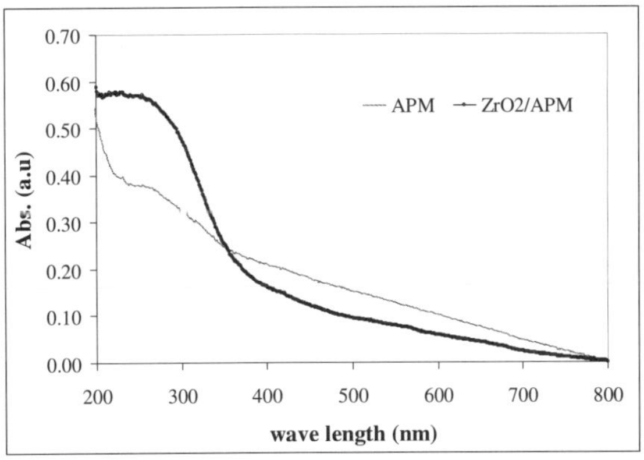

Figure 2. Diffuse Reflectance Spectra of ZrO2/APM and APM.

Nitrogenadsorption–desorption isotherms of materials are presented in Figure 3.

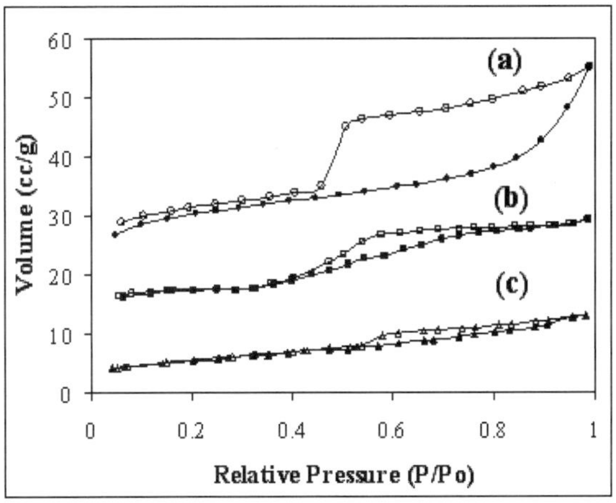

Figure 3. Nitrogen adsorption–desorption isotherms of (a) ZrO2/APM (b) APM, and (c) raw montmorillonite.

3.2. *Catalytic Activity Test*

Figure 4 presents the kinetic curve regarding to the phenol concentration as the function of reaction time. Due to the concentration reduction profile, it is demonstrated that ZrO_2/APM exhibits higher activity as shown by the higher phenol consumption in the reaction correspond to the reaction rate.

Physicochemical characters of materials are the main reason contributing for this improvement data

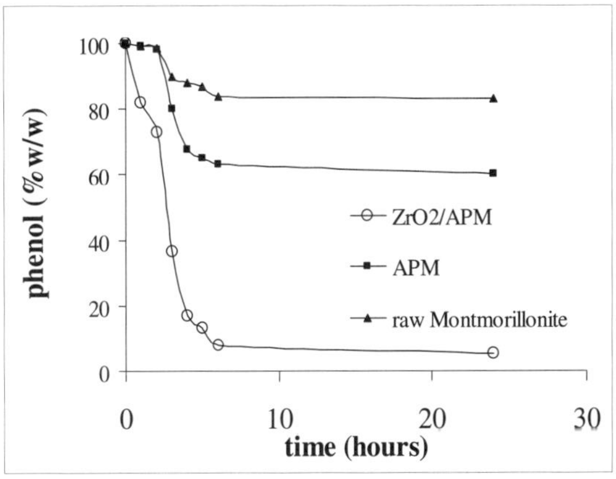

Figure 4. Kinetics of Phenol Hydroxylation by Different Catalyst.

Raw clays; montmorillonite itself has the lowest activity compared to the modified materials. It suggest that the reaction rate is not controlled by only the surface area, but also the active surface [5].

The simulation to the kinetics data of phenol hydroxylation with the prepared materials reveals that reactions obey second order reaction. Kinetics constant k derived by the second order linear plot shows the effect of zirconium dispersion to increase k value in the factor of 29.44 to APM and 98.43 to raw montmorillonite.

Furthermore, effect of several reaction parameters in of phenol hydroxlation reaction with ZrO_2/APM catalyst was studied by the variation of H_2O_2 to phenol mol ratio and catalyst dosage.

Mechanism of phenol hydroxylation presented in Figure 5.

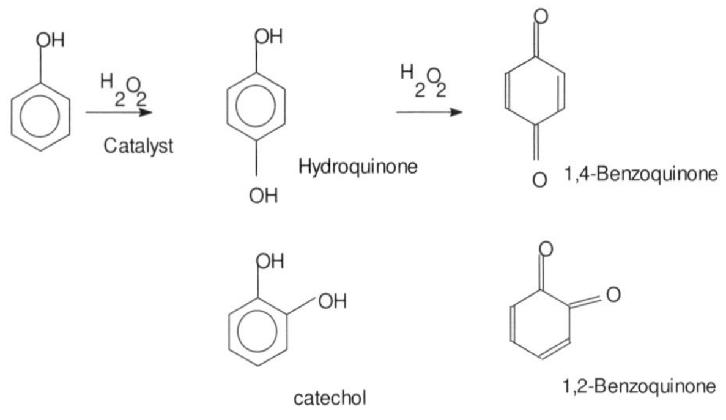

Figure 5. Mechanism of Phenol Hydroxylation.

Due to mechanism in Figure 5, there are four products possible in reaction: hydroquinone (HQ), catechol (CAT), 1,2-benzoquinone and 1,4-benzoquinon. Due to the HPLC analysis of product, HQ, CAT and 1,4-benzoquinone (further called as BQ) are the derived.

Two factors contributing to the selectivity and conversion; H_2O_2 to phenol mol ratio and also catalyst dosage is investigated. Table 2 shows the effect of H_2O_2 to phenol mol ratio to the conversion and product selectivity compared to the use of APM and raw montmorillonite. In the same objective, the effect of catalyst dosage is presented in Figure 6.

Table 2. Effect of H_2O_2 to phenol mol ratio and Catalyst to Conversion and Selectivity.

Catalyst	H_2O_2:phenol mol ratio	Selectivity (%)			Conversion (%)
		HQ	CAT	BQ	
ZrO_2/APM	1:1	88.57	11.43	0	16.60
	5:1	83.68	15.87	0.44	86.40
	10:1	0	44.50	55.49	100
APM	5:1	98.00	2.00	0	37.00
Montmorillonite	5:1	99.05	0.05	0	16.00

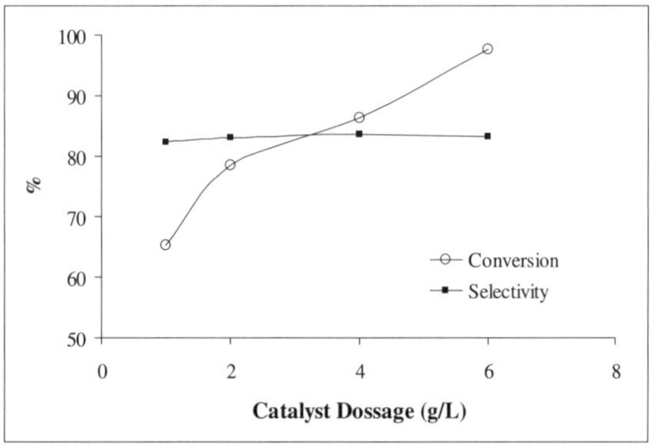

Figure 6. Effect of Catalyst Dosage to Conversion on Selectivity to HQ.

Data in Table 2 implying the different pattern of selectivity by utilizing different catalyst also by the different H_2O_2 to phenol mol ratio. The conversion increase as H_2O_2 to phenol ratio increase. At the ratio of 1:1, the product is dominated by hydroquinone in almost 100% selectivity in low conversion and elevating the ratio to 5:1 result in improved conversion, however the selectivity to HQ decreased because of the other products obtained as a result of the reaction. On the ratio of 10:1, the product is dominated by the BQ. This is because H_2O_2 play role in advanced oxidation mechanism to BQ. At this situation, selectivity to HQ decreased. In this mechanism, it can be said that the HQ is the intermediate product.

From Figure 6 it is found that catalyst dosage tends to be positive affecting the reaction conversion, but not give any influence to the product selectivity.

4. Conclussion

The zirconia dispersed in aluminium pillared montmorillonite has been successfully synthesized and characterized. The catalytic activity of the synthesized material was examined in phenol hydroxylation. The study suggests that prepared material helps in significantly enhancing the rate of reaction. It was found that H_2O_2 to phenol mol ratio is important factor to the reaction rate and product selectivity.

Acknowledgements

Authors wish to thank DP2M-DIKTI for financial support through Hibah Mahasiswa Doktor 2009, Gadjah Mada University.

References

1. Xiong, Y., Yu, K.N., and Xiong, C., *Phys. Rev. B* ***49***, 5607 (1994).
2. Gil, A.,Korili, S.A., and Vicente, M.A., *Catal Rev.*, **50**:2, 153(2008).
3. Rao and Sahu, *Proc. Indian Acad. Sci.* (*Chem. Sci.*), 113(5), 651 (2001).
4. Fatimah, I., Narsito, and Wijaya, K, *Indonesian Journal of Chemistry*, **9**(1):6(2009).
5. Rives, V., Prieto, O., Dubey, A., and Kannan, S., *J.Catal.***220**, 161(2003).

COPPER(II) EXCHANGED TIO_2-PILLARED CLAY PREPARATION FROM INDONESIAN MONTMORILLONITE: EFFECT OF TITANIA PRECURSOR

IS FATIMAH

Chemistry Department, Islamic University of Indonesia, Kampus Terpadu UII, Jl. Kaliurang Km 14., Besi, Sleman, Yogyakarta 55581, Indonesia. email: isfatimah@fmipa.uii.ac.id

YULI ROHYAMI

Diplome Program of Chemistry, Islamic University of Indonesia, Kampus Terpadu UII, Jl. Kaliurang Km 14., Besi, Sleman, Yogyakarta 55581, Indonesia

BAYU WIYANTOKO and TUTY ALAWIYAH

Chemistry Department, Islamic University of Indonesia, Kampus Terpadu UII, Jl. Kaliurang Km 14., Besi, Sleman, Yogyakarta 55581, Indonesia

This article presents preparation and characterization of copper (II) exchanged TiO_2-pillared clay from Indonesian montmorillonite. Two kinds of titania precursors: $TiOCl_2$ and titanium tetraisopropoxide were used for investigating the effect of titania precursor to the physicochemical character. The materials were analyzed by several techniques such as XRD, Gas Sorption Analysis, Diffuse-Reflectance UV-Vis and FTIR. Salient feature of this study relates to the result that the physicochemical character of material is dependant on the titania precursor used as titania pillaring precursor.

1. Introduction

Clays are known as aluminum/magnesium silicates having 1:1 or 2:1 layer structures. Among 2:1 clay minerals, smectite type clays are used for wide variety applications in the various fields. Pillared interlayered clays (PILCs), or pillared clays, are twodimensional zeolite- like materials. They are prepared by exchanging the native cations of the parent clay with partially hydrated polymeric or oligomeric metal cation complexes which, upon heating, undergo dehydration/dehydroxylation resulting in thermally stable pillars composed of metal oxide clusters [1]. PILCs show interesting properties towards adsorption and catalysis. The fields of application are very broad due to the possibility to intercalate different clay types with different types of pillars, making them suitable for specific adsorption and catalytic processes. Titanium dioxide (titania) pillared clays (Ti-PILCs) has been reported exhibiting high ordered porosity and high surface area provided by increase of inter-pillar gallery spacing in pillarization process [2,3]. Due to these properties, attention has has been focused recently on the Ti-PILCs performance in selective catalytic reduction for reducing NO_x pollutant. Recent of Ti-PILCs in SCR reaction is the improvement of catalytic activity via copper (II) immobilization in Ti-PILCs pore [3,4]. As other metal oxide pillarization of clays, the titania pillared clays having cationic sites which are possible to be replaced by other cations such as copper (II). Since natural montmorillonite is abundant in Indonesia, investigation on the synthesis of Copper(II) exchanged titanium pillared montmorillonite is an interesting one. This paper describes the synthesis of titanium pillared montmorillonite with different titania precursors.

2. Material and Method

2.1 *Material*

The starting material used was Na-montmorillonite (Kunipia F), kindly supplied by PT. Tunas Inti Makmur Semarang (Indonesia). The montmorillonite was dispersed in water and aged for two days to make the clay suspension. Two kinds of titania precursor: titanium oxide chloride ($TiOCl_2$) was supplied by E. Merck and titanium tetraisopropoxide was supplied by Sigma Aldrich.

2.2 *Catalyst Preparation*

Detail of titanium pillared montmorillonite synthesis by different precursor were conducted as follows:

a.titanium oxide chloride: Titanium pillaring precursor was prepared by dissolving titanium oxide chloride in HCl solution under strong stirring for 4 hours. The clear solution from this step was then dispersed slowly into montmorillonite suspension with the appropriate amount to obtain Ti content of 5 mmol/g of clay. The mixture was stirred at room temperature for a night. After the obtained slurry was filtered, the solid was washed with distilled water several times, and then dried in a oven at 60°C. Dry solid was calcined at 400°C for 4 h.

b.titanium tetraisopropoxide: titanium pillaring precursor with titanium tetraisopropoxide precursor was prepared by dilluting into HCl solution until the HCl/Ti mol ratio of 4 and white suspension was obtained. The dispersion of pillaring precursor into clay suspension was similar to the use of titanium oxide chloride.

Materials obtained by these synthesis step were designated as TiM-OCl_2 and TiM-TIP respectively. Ion exchange process of Cu (II) to the titanium pillared montmorillonite was enganged in the reflux system. Cu (II) acetate solution was mixed with titanium pillared montmorillonite with theoritic Cu^{2+} content is fourth times to CEC. This exchange process was done twice and for each process followed by washing and neutralization by using deionized water. Solid obtained by these step was then named as copper (II) exchanged titanium pillared montmorillonite. By the variation of titanium precursor, each of material encoded as Cu/ TiM-OCl_2 and Cu/TiM-TIP.

2.3 *Characterization*

The X-ray diffraction (XRD) patterns were obtained using a Shimadzu X6000 diffractometer with Ni-filtered Cu Ka radiation using a powder sample. DRUV–vis spectra of the samples have been recorded with a Jasco V-670 spectrophotometer equipped with a diffuse reflectance accessory; the % R values have been transformed via the Kubelka Munk formula. Gas Sorption Analyzer NOVA 1200e was used to determine specific surface area and adsorption-desorption profile by using N_2 as the sorbate at 77 K. Chemical composition of Na-montmorillonite was determined by gravimetric method and Atomic Absorption Spectrophotometry Perkin Elmer. The solid acidity of each sample was measured by n-butylammine and ammonia adsorption. FTIR spectra was recorded using Nicolet Avatar series.

2.4 *Activity Test*

In order to evaluate the contribution of physicochemical character materials to the catalytic properties, experiment on NO reduction at the temperature of 200°C was conducted. The investigation was enganged by flowing NO and H2 gas into tubular stainless steel reactor (dimension: 1 dm in inner diameter and 25 cm in length) supported by tubular furnace. The reduction reaction of occurred in the system. The catalyst activity is determined by analyzing NOx concentration in gas before and after passing the reactor with following equation:

$$activity = \frac{[NO]_i - [NO]_f}{[NO]_f} x100\%$$

with

$[NO]_i = initial\ concentration\ \ of\ NO$

$[NO]_f = final\ concentration\ of\ NO$

3. Result and Discussion

Table 1 lists the components of natural montmorillonite used in the preparation. Silica and alumina are major components in the mineral due to the montmorillonite structure. According to the data, it is observed that natural montmorillonite is in the form of natrium montmorillonite in that sodium available in the higher percentage than calcium.

Table 1. Chemical composition of raw montmorillonite.

PARAMETER	CONTENT (% WT.)
SiO_2	26.14
Al_2O_3	10.74
Mg	5.38
Ca	0.58
Na	5.25

To study the effect of titania pillarization and Cu dispersion in titania pillared montmorillonite, XRD pattern of materials were recorded. The difractograms are presented in Figure 1.

Reflections of raw montmorillonite (a) show the significant characteristic peaks at 2θ = 5.93° (d = 14.89 Å) and 2θ = 19.91° (d = 4.45 Å) with the valuable intensity and other reflections at 2θ = 20.18°, 23.57° and 26.57°. The reflection at 2θ = 6.30° is correlated to the d_{001} basal spacing of smectitic clay structure. Other reflections corresponding to the impurities minerals of quartz (Q) and crystoballite (Cr) are also identified. Specific alteration to the montmorillonite reflection was observed, which shows a shift of d_{001} to the lower angle after the titania pillarization.

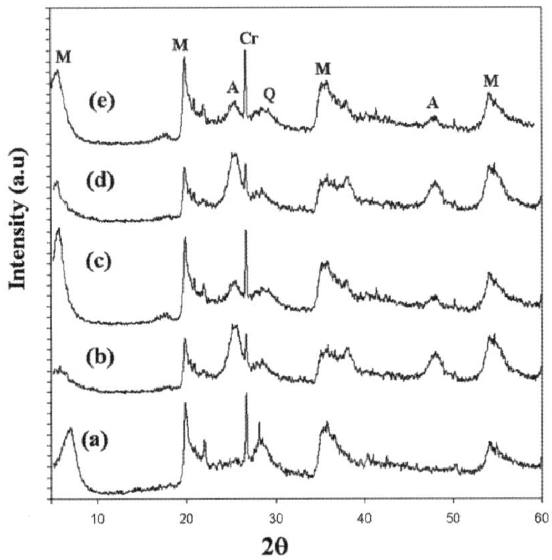

Figure 1. XRD pattern of (a) raw montmorillonite (b) TiM-OCl2 (c) TiM-TIP (d) Cu/TiM-OCl2 (e) Cu/TiM-TIP. M stands for montmorillonite, Cr for crystoballite, Q for Quartz and A for anatase.

An interesting finding of this work is the increase of d_{001} reflection upon the pillarization. The d_{001} reflection of TiM-TIP is observed at 5.20° (16.98) implying a larger d_{001} value compared to the d_{001} of raw montmorillonite as well as the reflection of TiM-OCl_2 that is observed at 2θ = 5.60 (d = 15.77Å). The characteristic reflections for anatase are observed in samples b, c, d and e. A similar result has also been reported by Sun *et al.* (2002), Liu *et al.* (2007) and Miao *et al.* (2006) [5-7].

The anatase reflections in the pillared montmorillonite assumed that this was due to the formation of nanosize titania particles in the galleries of the montmorillonite. Moreover, it is found that copper exchanged titanium pillared montmorillonite were not significantly show the different pattern with the each matrix. It means that neither the effect of acidity nor the copper in oxide formation during the exchange process occurred. Important data regarding to the change of material character are summarized in Table 2.

DRUV-Visible spectra of materials compared to the spectra of raw montmorillonite is presented in Figure 2.

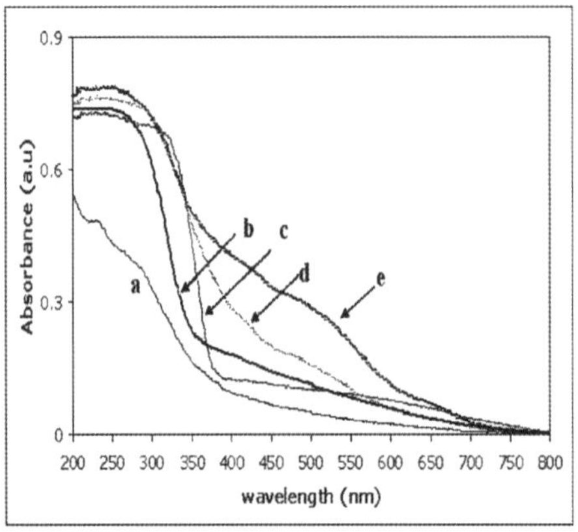

Figure 2. DRUV-Visible Spectra of (a) raw montmorillonite (b) TiM-OCl2 (c) TiM-TIP (d) Cu/TiM-OCl2 (e) Cu/TiM-TIP.

As demonstrated in Figure 2, the raw MMT has low absorbance indicate that material was transparent in the wavelength range longer than 300 nm. The spectrum of TiM-OCl_2 shows the smaller edge wavelenght compared to the TiM-TIP samples suggest that the particle size anatase in TiM-OCl_2 is smaller than in TiM-TIP composite. The reason for this data is mainly due to the titania precursor aggregation which is probably formed with the use of titanium tetraisopropoxide precursor. The similar effect of red shift upon the copper (II) exchange is depicted by both titania pillared montmorillonite. This shift may correlate to the interaction between Cu^{2+} and H_2O which lies at visible region.

Figure 3 show nitrogen adsorption–desorption isotherms of materials. The corresponding surface area and pore volume determined using BET isotherm is presented in Table 2. The specific surface area of modified materials are higher than raw montmorillonite.

These values were in agreement with the d_{001} data for each material in that the improvement of d_{001} affect to the increasing of specific surface area. The higher dimension of interlayer space and interlamellar space formation by pillarization process make up the adsorption sites surface.

The isotherms on the pillared montmorillonite with both precursors are similar and seems to obey the type IV indicating both micro and mesopores (BDDT classification). These data show the pore evolution from raw montmorillonite which is dominated by micropores. Both materials from different precursors suggest the decrease in specific surface area and also pore volume, but these changes can be said as not significant.

Copper exchanged form of both titania pillared montmorillonites have lower specific surface area compared to its hosts indicate that there is an acid or chemical effect during the ion exchange process.

Determination of total acidity and Lewis acidity experiments were performed to characterize the acidic properties of raw, pillared samples and copper exchanged pillared samples. Gravimetric analysis of ammonia adsorption was engaged to measure the Lewis acidity and n-butylammine titration was performed to measure the Bronsted acidity. From Table 2, it is observed that pillared montmorillonite and its copper exchanged form having higher Lewis acidity compared to raw montmorillonite sample. It is suggested that titanium oxide pillar formed in the interlayer space contribute to the adsorption interaction with NH_3. However, the Bronsted acidity data seems to be not affected by copper exchanging process. Hydroxylation process always included in the pillarization during the calcinations step.

Table 2. Result of Material Characterization.

Sample	Specific Surface area (m^2/g)	Pore Volume (10^{-3} cm^3/g)	CEC (meq/100g)	Ti (%wt.)	d_{001} (Å)	Lewis Acidity (meq/g)	Bronsted Acidity (meq/g)
Raw Montmorillonite	45.68	0.037	67-80	-	14.97	30.09	2.59
TiM-OCl_2	207.69	198.30	45-65	9.80	15.77	33.38	3.57
Cu/TiM-OCl_2	127.06	156.0	-	9.80	15.75	37.15	3.52
TiM-TIP	152.97	195.40	40-56	10.09	16.98	38.9	2.97
Cu/TiM-TIP	98.87	113.6	-	9.98	16.92	41.08	2.98

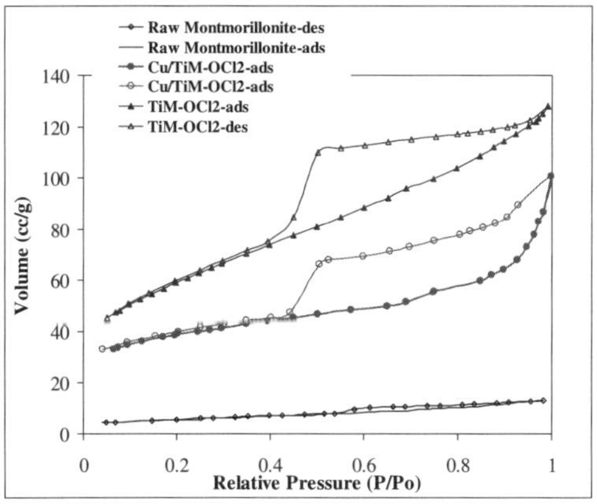

(a)

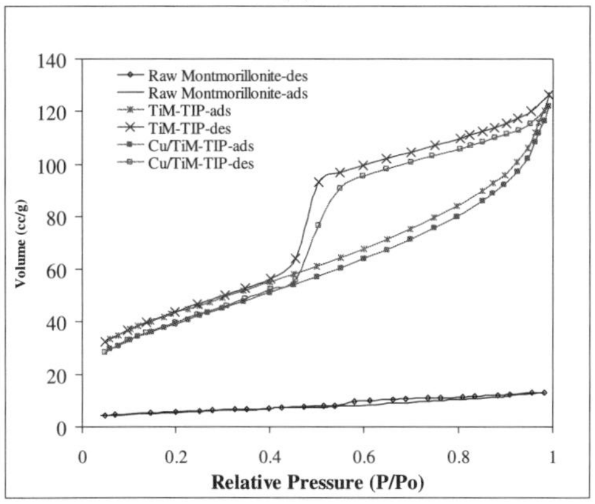

(b)

Figure 3. Adsorption-Desorption Profile of (a) TiM-OCl2 and Cu/TiM-OCl2 (b) TiM-TIP and Cu/TiM-TIP compared to raw montmorillonite.

Basically, upon thermal treatments, the intercalated titanium polyoxocations was converted to form titanium dioxide and H^+ was generated. The relative constant value of Bronsted acidity by the copper exchanging process correlate with the cation exchange mechanism itself in that there is no H^+ generation involved in the reaction.

FTIR analysis on pyridine-adsorbed materials was also used to study the presence of both acid sites in surface. Pyridine species formed via surface interaction are shown in Figure 4.

Species A (Lpy) is formed via coordinately bound pyridine through its lone pair of electrons on nitrogen atom to Lewis acid center of Al or Ti cations present on clay mineral surface. The recorded spectra of materials is shown in Figure 5. The similar bands found at all samples are the band at 3444-3445 cm^{-1} correspond to the hydroxylic water adsorbed by materials, the band at 916-920 cm-1 and at around 1020-1040 cm^{-1} correlate to the vibration of Si-O and also Al-O. The band at 2360 cm^{-1} is found for pillared montmorillonite and its copper exchanged form attributed to Ti-O-Ti vibration as evidence for titania formation in montmorillonite matrix. The presence of copper ion in the Cu/TiM-OCl_2 and Cu/TiM-TIP is proven by the shift of bands at around at 3444-3445 cm^{-1} and at 2360 cm^{-1} to the higher wave number. These shift assigned to the higher vibration energy associate to the copper interaction with H_2O and with Ti-O bond.

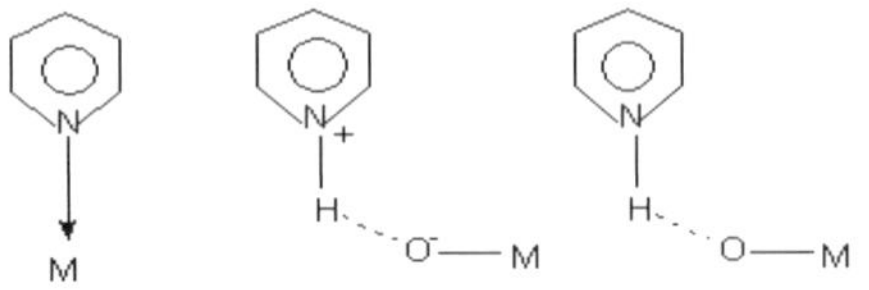

(I) (II) (III)

Figure 4. Pyridine Specieses in surface.

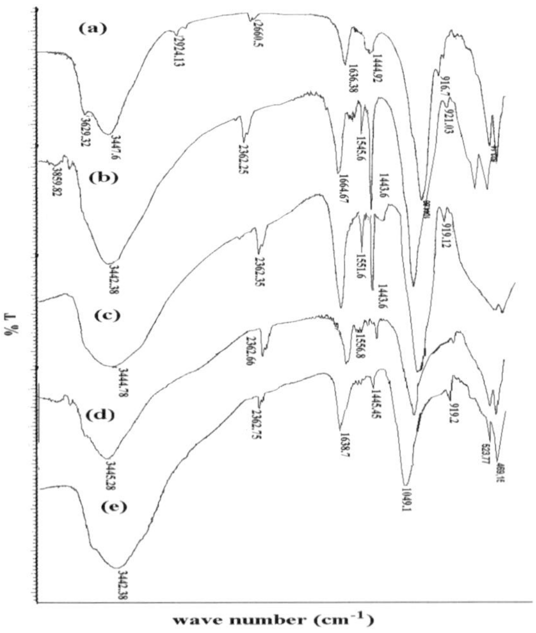

Figure 5. FTIR Spectra of (a) raw montmorillonite (b) TiM-OCl2 (c) TiM-TIP (d) Cu/TiM-OCl2 (e) Cu/TiM-TIP.

Pyridine bands appear in the IR region of 1400-1700 cm-^{1}. The band at around 1443-1445 cm^{-1} found in all materials indicate the Lewis acid site interaction with pyridine. The IR bands at the range 1635-1645 cm^{-1} and 1545-1556 cm^{-1} correspond to Bpy acid sites. Due to the intensity of band, it is suggested that Lewis acidity improvement are more pronounced in TiM-OCl_2 and Cu/TiM-OCl_2 .The activity as function of catalyst is presented in Table 3. What can be concluded as the important analysis from the acidity and FTIR spectra is that the titania precursor significantly affect to the material character as well as found in other comparative study of characterization and NO reduction activity data. The most active material, Cu/TiM-TIP suggest that the total acidity strong affect to the activity. Intensive interaction between NO and Lewis site in surface is probably the main reason for these indication. However, the further analysis needed in more intensive study.

Table 3. Comparative data on NO reduction activity of materials.

Catalyst	Activity in NO Reduction (%)
Raw montmorillonite	56.63
TiM-OCl_2	66.08
Cu/TiM- OCl_2	75.44
TiM-TIP	67.99
Cu/TiM-TIP	77.67

4. Conclusions

The present investigation illustrates the following:

a. Cu(II) exchanged titania pillared montmorillonite has been successfully prepared from Indonesian montmorillonite.
b. XRD, gas sorption analysis, FTIR, acidity and CEC measurements suggest that the synthesis parameter of titania precursor significantly affect to the physichochemical character of material.

Acknowledgement

Authors wish to thank DP2M-DIKTI for financial support through Hibah Bersaing 2009.

References

1. Gil, A., Korili, S.A., and Vicente, M.A., *Catal Rev.*, **50**:2, 153 (2008).
2. Vicente, M. A. Banares, M. A., Munos, Toranzo, R., Gandia, L. M. And Gil, A., *Clay Miner.* **36**, 125(2001).
3. Kim, B.S., Lee, S., Park, Y., Ham, S., Chae, H., Nam, I.S., *Korean J. Chem. Eng.*, **18**(5): 704(2001).

4. Chae, H., Nam, I.S., Ham, S., Hong, S., *App. Cat. B: Environmental* , **53**, 117(2004).
5. Sun, S.M., Jiang, Y.S., Yu, L., Li, F.F., Yang, Z.W., Hou, T.Y., Hu, D.Q., Xia, M.S., *Mater.* Chem. Phys. **98**, 377 (2006).
6. Liu, J.J., Li, X.P., Zuo, S.L., Yu, Y.C., Appl. Clay Sci., **37**, 355(2007).
7. Miao, S., Liu, S., Han, B., Zhang, J., Yu, X., Du, J and Sun, A., J. Mater. Chem., **16**: 579 (2006).

PHOSPHATE AND NITRATE REMOVAL IN A LABORATORY-SCALE BATCH REACTOR BY SELECTED WASTEWATER PROTOZOA

AKPOR OB and MOMBA MNB

Department of Environmental, Water and Earth Sciences, Tshwane University of Technology, Pretoria, South Africa

The present study was aimed at investigating the phosphate and nitrate removal efficiency of three wastewater protozoa in a laboratory-scale water treatment system operated over four batches. For each experimental batch, aliquot samples were taken from each reactor at time zero and every 24 h, for the analysis of phosphate and nitrate, using standard methods. The results revealed phosphate concentrations in the reactor decreasing from initial ranges of 53.03 to 55.57 mg/L to final ranges of 2.88 to 52.74 mg/L in the reactors. In the case of nitrate, concentrations in the reactors were observed to decrease from initial ranges of 19.31 to 20.97 mg/L to final ranges of 0.10 to 6.59 mg/L in the reactors, respectively. The present study was able to reveal the phosphate and nitrate removal efficiency of the test protozoan isolates in laboratory-scale system, under aerated and non-aerated conditions.

1. Introduction

The world is faced with problems related to the management of wastewater due to extensive industrialization, increasing population density and highly urbanized society (EPA, 1993; McCasland et al., 2008). A large percentage of effluents generated from domestic and industrial activities constitute great sources of natural water pollution load. This is a great burden in terms of wastewater management, consequently can lead to a point-source pollution problem. Point-source pollution not only increases treatment cost considerably but also introduces a wide range of chemical pollutants and microbial contaminants to water sources (EPA, 1993; Eikelboom and Draaijer, 1999).

Nutrient-induced production of aquatic plants in receiving water bodies has the following detrimental consequences: (i) algal clumps, odours and decolouration of water sources, thus interfering with recreational and aesthetic use, (ii) extensive growth of rooted aquatic life which can interfere with navigation, aeration and channel capacity, (iii) dead macrophytes and phytoplankton settling to the bottom of water bodies, thus stimulating microbial breakdown processes that require oxygen, thus causing oxygen depletion, (iv) extreme oxygen depletion, which can lead to the death of desirable aquatic life, (v) siliceous diatoms and filamentous algae may clog water treatment plant filters, thus resulting in reduced backwashing, (vi) algal blooms may shade and submerged aquatic vegetation, thus reducing or eliminating photosynthesis and productivity (Atlas and Batha, 1987; Ratsak *et al.*, 1996; Kurosu, 2001; Alm, 2003; Mbewele, 2006; McCasland *et al.*, 2008).

Biological phosphorus and nitrogen removal systems have been extensively investigated for municipal wastewater treatment over the past two decades. However, most investigations have centred mainly on the importance and role of bacteria in the engineering (design and technology/ implementation/operation) models of the activated sludge system (Momba and Cloete, 1996; Caron e*t al.*, 1998; Mino, 2000). The few studies on protozoa have been carried out under shaking flasks conditions (Akpor *et al.*, 2007).

This study is therefore aimed at investigating the efficiency of three selected protozoan isolates *Aspidisca, Trachelophyllum* and *Peranema*) in the removal of phosphate and nitrate in a laboratory-scale batch reactor fitted with an aerated and a non-aerated reactor.

2. Materials and Methods

Three wastewater protozoa (*Aspidisca,Trachelophyllum* and *Peranema*) were used for this investigation. The isolates have previously been ascertained to have phosphate and nitrate removal ability (Akpor *et al.*, 2007). The wastewater used for this investigation was activated sludge mixed liquor collected from the anaerobic zone of Daspoort Wastewater Treatment in Pretoria, South Africa. The mixed liquor was filtered using Whatman No1 filter paper and then supplemented with the following salts: sodium acetate (5 g/L) magnesium sulphate (0.5 g/L) and potassium nitrate (0.17 g/L).

The laboratory-scale reactor used for the study was made up of two 5-L capacity glass reactors (aerated and non-aerated reactors). Each of the reactors was fitted with a stirrer that operated at 100 rpm. The schematic diagram of the batch reactor used for this study is shown in Fig. 1.

The experimental setup was as follows. In both reactors, 3 L of sterile mixed liquor was added and inoculated with approximately 10^3 cells/mL of the test protozoan isolate. Each experimental setup was then operated for four batches, with each batch lasting 120 h. During each sampling, aliquot samples of mixed were collected from each reactor at time zero and every 24 for the determination of phosphate and nitrate, using standard methods (APHA, 2001).

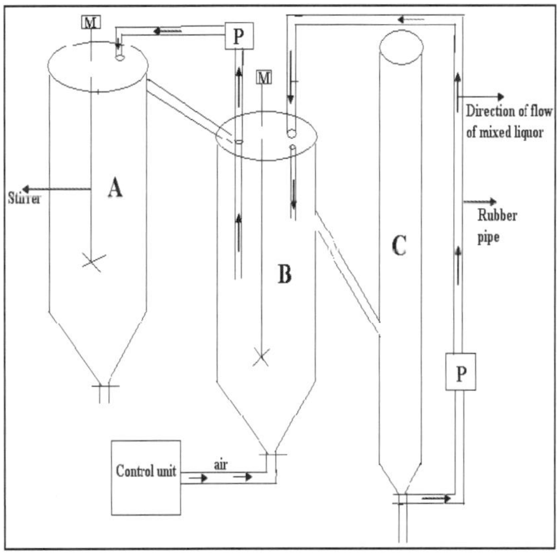

Fig. 1: Schematic diagram of the laboratory-scale batch reactor. A B, C, M and P represent the aerated and non-aerated reactors, auxiliary tank, electric motorand pump respectively.

3. Results

As shown in Fig. 2 and 3, the trend in phosphate removal in the reactors revealed phosphate concentrations in the aerated reactor decreasing from initial ranges of 54.22 to 55.57 mg/L, from 54.73 to 56.83 mg/L and from 54.29 to 56.63 mg/L to final ranges of 2.88 to 34.22 mg/L, from 6.89 to 24.87 mg/L and from 19.77 to 50.12 mg/L, in the presence of *Aspidisca, Trachelophyllum* and *Peranema*, respectively.

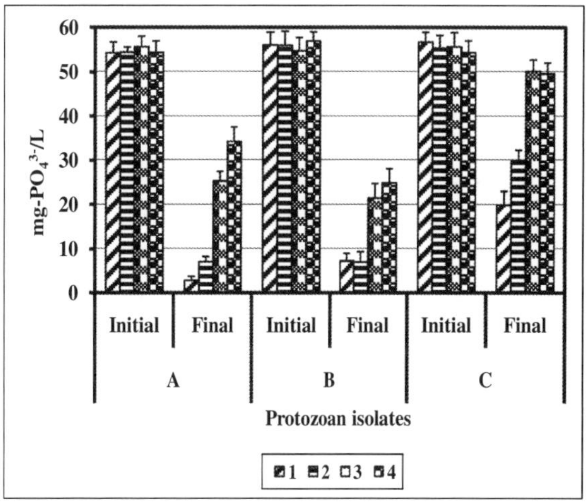

Fig. 2: Phosphate removal in the aerated reactor. Initial and final represent phosphate concentrations at time zero and after 120h. A, B and C refer to Aspidisca, Trachelophyllum and Peranema, respectively while 1, 2, 3 and 4 represent the different experimental batches.

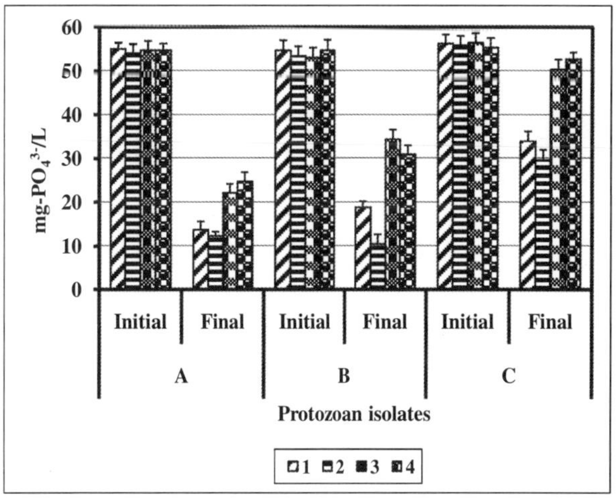

Fig. 3: Phosphate removal in the non-aerated reactor. Initial and final represent phosphate concentrations at time zero and after 120h. A, B and C refer to Aspidisca, Trachelophyllum and Peranema, respectively while 1, 2, 3 and 4 represent the different experimental batches.

In the non-aerated reactor, phosphate concentrations were observed to decrease from initial ranges of 54.73 to 54.97 mg/L, from 53.03 to 54.71 mg/L and from 55.26 to 55.35 mg/L to final ranges of 12.16 to 24.63 mg/L, 10.37 to 34.34 mg/L and 29.567 to 52.74 mg/L. This trend translates to percentage removals in the aerated reactor ranging from 36.97 to 94.69%, from 56.24 to 87.67% and from 8.79 to 65.09%, respectively. In the non-aerated reactor, percentage removals were observed to range from 54.95 to 84.88%, from 35.19 to 80.60% and 2.80 to 39.62% respectively.

Similarly, nitrate concentrations in the aerated reactors were observed to decrease from initial ranges of 19.93 to 20.66 mg/L, from 20.14 to 20.93 mg/L and from 19.79 to 20.97 mg/L to final ranges of 0.36 to 4.54 mg/L, from 0.29 to 4.46 mg/L and from 0.44 to 6.59 mg/L, in the presence of *Aspidisca*, *Trachelophyllum* and *Peranema*, respectively (Fig. 4).

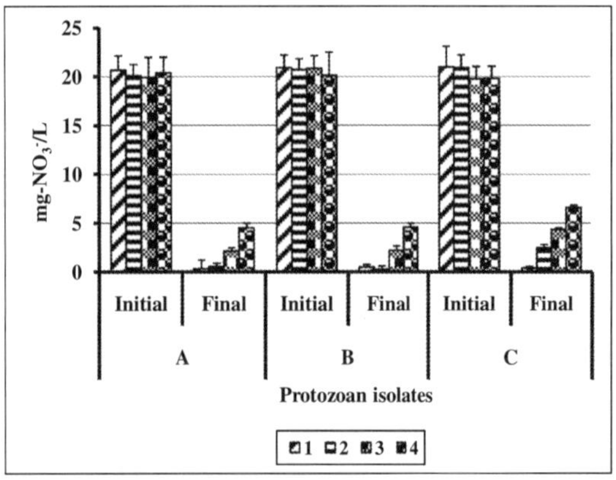

Fig. 4: Nitrate removal in the aerated reactor. Initial and final represent phosphate concentrations at time zero and after 120h. A, B and C refer to Aspidisca, Trachelophyllum and Peranema, respectively, while 1, 2, 3 and 4 represent the different experimental batches.

For the non-aerated reactor, nitrate concentration in the presence of the protozoan isolates decreased from initial ranges of 19.38 to 20.57 mg/L, from 19.31 to 20.06 mg/L and from 19.48 to 19.89 mg/L to final ranges of 0.29 to 0.72 mg/L, from 0.25 to 0.86 mg/L and from 0.10 to 0.38 mg/L, respectively. This translates to removal percentages that ranged from 77.76 to 98.26%, from 77.86 to 98.60% and from 66.73 to 97.90%, in the aerated reactor. In the on-aerated reactor, removal percentages ranged from 96.28 to 98.55%, from 91.39 to 98.71 and from 98.05 to 99.50%, respectively (Fig. 5).

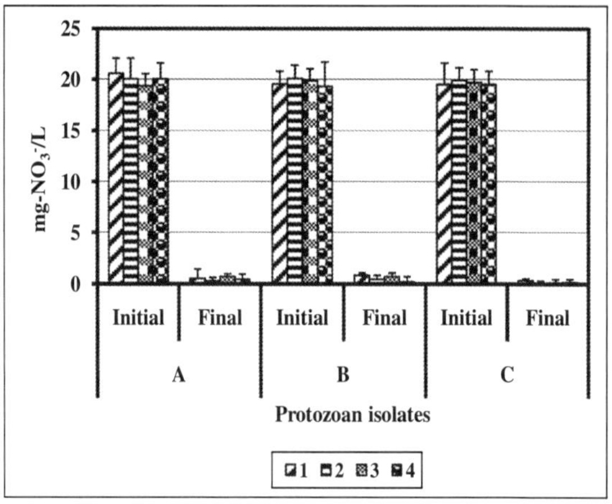

Fig. 5: Nitrate removal in the non-aerated reactor. Initial and final represent phosphate concentrations at time zero and after 120h. A, B and C refer to Aspidisca, Trachelophyllum and Peranema, respectively, while 1, 2, 3 and 4 represent the different experimental batches.

4. Discussion and Conclusion

In the present study, mixed liquor used was supplemented with sodium acetate to serve as carbon source. The choice of acetate was deliberate, since it has been reported to be a preferred carbon source for phosphate removal from wastewaters by luxury uptake organisms (Kargi and Uygur, 2003). Another reason for the choice of acetate was because nutrient removal process in activated sludge is known to depend on the availability of easily biodegradable carbon sources, such as acetate (Ekama *et al.,* 1986).

Our findings showed simultaneous increases in phosphate and nitrate removal with time in the presence of the test isolates. However, the pattern of nitrate removal in mixed liquor was slightly different rom that of phosphate. A comparison of the two reactors revealed a higher phosphate removal in the aerated reactor while a higher nitrate removal was observed in the non-aerated reactor.

Also, higher phosphate and nitrate removals were observed at the earlier batches in both reactors, except for the non-aerated reactor were nitrate removal was remarkable at all experimental batches. The low phosphate removal observed in the later batches could be attributed to a probable low cell synthesis in such batches. Moosavi and co-workers (2005) have ttributed phosphorus removal to cell synthesis, hence suggested that increased phosphorus removal could be obtained in reactors by possibly adding a biological suspended growth system. Factors such as composition of wastewater, oxygen supply, and active sludge/active age have been recognized to have a positive and negative impact on biological nutrient removal (Rybicki, 1997).

The pattern of nutrient removal in this study was observed to vary slightly among the isolates in the presence of the respective carbon supplements. This was irrespective of initial biomass concentration. Rates of nutrient uptake, empirically described as a simple hyperbolic function of ambient nutrient concentrations, are not necessarily fixed for respective organisms, but variable depending on nutrient conditions where cells are growing. This could suggest that nutrient uptake/release is maintained by regulatory mechanism (Ikeya *et al.*, 1997)

The present study revealed a higher nitrate removal under non-aerated conditions. Brdjanovic and co-workers (1996) have revealed that dissolved concentrations greater than 4 mg/L do not appear to further stimulate biological nutrient removal, hence excessive aeration tends to negatively affect biological nutrient removal processes. Although the findings of this study cannot be considered to be exhaustive, it has been able to reveal the phosphate and nitrate removal efficiency of the test protozoan isolates in laboratory-scale system, under aerated and non-aerated conditions. Overall the study has revealed the need to create an environment for the proliferation of the test protozoan isolates in activated sludge systems. This will help to enhance effective biological nutrient removal. Knowledge of this will enhance unpolluted wastewater discharges.

References

1. Akpor, O.B.; Momba, M.N.B. and Okonkwo, J. 2007. Phosphorus and nitrate removal by selected wastewater protozoa isolates. Pakistan Journal of Biological Sciences, 10(22): 4008-4014.
2. Alm, E. 2003. Implication of microbial heavy metal tolerance in the environment. Reviews in Undergraduate Research, 2: 1-5.
3. Atlas, R.M. and Batha, R. 1987. Microbial Ecology: Fundamentals and Applications. Benjamin/Cummings Publishing Co. Menl Park, USA.
4. APHA. 2001. Standard Methods for the Examination of Water and Wastewater, 20th Edition. APHA, Washington D.C.
5. Brdjanovic, D.; Hooijmans, C.M.; Van Loosdrecht, M.C.M; Alaerts, G.J. and Heeijen, J.J. 1996. The dynamic effect of potassium limitation on biological phosphorus removal. Water Research, 30(1), 303-313.
6. Caron, D.A.; Joel, C.G. and Dennette, M.R. 1998. Experimental demonstration of the roles of bacteria and bacterivorous protozoa in plankton nutrient cycles. Hydrobiologia, 159(1), 27-40.
7. Eikelboom, D.H. and Draaijer, A. 1999. Activated sludge information system (ASIS). Available from http://www.asissludge.com. Accessed 09/11/2007.
8. Ekama, G.A.; Dold, P.L. and Marias, G.R. 1986. Procedures for determining influent COD fraction and the maximum specific growth rate of heterotrophs in activated sludge system. Water Science and Technology, 18(6), 91-114.
9. EPA. 1993. Constructed wetlands for wastewater treatment and wildlife habitat. Available form http://www.epa.gov/owow/wetlands/construct. Accessed 14/01/2008.
10. Ikeya, T.; Ohki; K.; Takahashi, M. and Fujita, Y. 1997. Study on phosphate upake of the marine cyanophyte Synechococcus sp NIBB 1071 in relation to oligotrophic environments in the open ocean. Marine Biology, 129, 195-202.
11. Kargi, F. and Uygur, A. 2003) Effect of carbon source on biological nutrient removal in a sequencing batch reactor. Bioscience Technology, 89(1), 89-93.
12. Mbewele, L. 2006. Microbial phosphorus removal in wastewater stabilization pond. A Licentiate Thesis from the School of Biotechnology: A Royal Institute of Technology, Albanova, Stockholm, Sweden, ISBN91-7179-280-X.
13. McCasland, M.; Trautmann, N.; Porter, K. and Wagenet, R. 2008. Nitrate: Health effects in drinking water. Available from http://pmep.cee.com.ell.edu/facts.slides–self/facts/nitheef-grw85.html. Accessed 05/04/2008.
14. Mino, T. 2000. Review: Microbial selection of polyphosphateaccumulating bacteria in activated sludge wastewater treatment processes for enhanced biological removal. Biochemistry, 63(3): 341-348.
15. Momba, M.N.B. and Cloete, T.E. 1996. The relationship of biomass to phosphate uptake by Acinetobacter junii activated sludge mixed liquor.

Water Research, 30(2): 364-370.

16. Moosavi, G.H.; Naddafi, K.; Mesdaphinia, A.R. and Nabizadeh, R. 2005. Simultaneous organics and nutrients removal from municipal wastewater in an up-flow anaerobic/aerobic fixed bed reactor. Journal of Applied Sciences, 5(3): 503-507.
17. Ratsak, C.H.; Maarsen, K.A. and Kooijman, S.A.L. 1996. Effects of protozoa on carbon mineralization in activated sludge. Water Research,30(1):1-12.
18. Ratsak, C.H.; Maarsen, K.A. and Kooijman, S.A.L. 1996. Effects of protozoa on carbon mineralization in activated sludge. Water Research 30(1): 1-12.

DISSOLVED AIR FLOTATION (DAF) PROCESS FOR COLOUR AND CHEMICAL OXYGEN DEMAND (COD) REMOVAL IN LANDFILL LEACHTE TREATMENT*

PUGANESHWARY PALANIANDY†

PhD Research Student, School of Civil Engineering, Engineering Campus, Universiti Sains Malaysia, 14300 Nibong Tebal, Pulau Pinang, Malaysia

MOHD NORDIN ADLAN

Associate Professor, School of Civil Engineering, Engineering Campus, Universiti Sains Malaysia, 14300 Nibong Tebal, Pulau Pinang, Malaysia

HAMIDI ADBUL AZIZ

Professor, School of Civil Engineering, Engineering Campus, Universiti Sains Malaysia, 14300 Nibong Tebal, Pulau Pinang, Malaysia

DAF was utilized in leachate treatment as a batch process and focused in colour and COD removal. Using DAF alone, 36 and 33 percent removal for COD and colour was obtained. Nevertheless, the removal increase up to 77 and 93 percent for COD and colour using coagulation process prior to DAF. This was obtained at pressure (600kPa), flow rate (6LPM) using 625mg/L $FeCl_3$ as a coagulant with 120 seconds injection time (IT) and 20 minutes retention time (RT). However, study on RT indicates 10 minutes was enough to float the particle to the surface which proves rapid start-up in DAF.

1. Introduction

Landfill leachate is highly polluted and hazardous wastewater. The complex characteristic of the leachate makes it's very difficult to treat and fulfil the discharge requirement to water body. Various type of wastewater treatment method had been drawn to reduce the concentration of the pollutant in the leachate, [1]. Corresponding to the age of the landfill and the biochemical reaction, the multifaceted characteristics of the leachate varies to produce high chemical oxygen demand (COD), and colour intensity, [2]. The high concentration of COD in leachate signify that the leachate contain a great deal of biodegradable organic compounds, non-biodegradable compounds and inorganic oxidizable compounds [3]. The feature of the landfill leachate is very dark brown coloured liquid. The main cause of this characteristic is due to the present of humic acid in the leachate. This is an organic macromolecules substance and presenting yellow to black appearance [4]. Due to aforementioned problems cause by the present of the organic and non-organic substance, it will be a great positive environmental impact in removing them from leachate. The common method of accomplish this is by coagulation/flocculation and sedimentation process [1], [5]. However this process will produce high amounts of sludge, time-consuming because of gravitational settling, and required more space for the conventional settling clarifiers [6], [7].

Therefore in the present study, the research was focused on the application of dissolved air flotation (DAF) on the removal colour and COD from the landfill leachate.

DAF is a physical treatment method that is successful in removing low density particle and highly coloured waters where light flocs are produce. DAF offers a low energy, low maintenance alternative to sedimentation and filtration [8]. These systems also are modular components which are ease for installation [7]. To this end the objective of this study is to evaluate the efficiency of DAF process in leachate treatment. The leachate was collected from Pulau Burung Landfill Site (PBLS), Nibong Tebal, Pulau Pinang. The landfill is about more than 20 years old. The characteristic of this leachate are shown in Table 1.

*This work is supported by School of Civil Engineering, Universiti Sains Malaysia.
†Work partially supported by grant 6013309 of the e-Science Fund, MOSTI Malaysia

Table 1. Characteristic of raw leachate from Pulau Burung Landfill Site.

Parameters	Range	Mean*
COD (mg/L)	2270-3390	2718
Colour (PtCo Unit)	3611-4932	4005
pH	7.76-8.24	8.05
Turbidity (FAU)	178-386	246
Alkalinity (mg/L as $CaCO_3$)	8117-11000	10130
Dissolved Oxygen (mg/L)	0.09-4.10	0.56
Ferum (mg/L Fe)	1.93-2.80	2.35

*Average of samples taken from Sept 2007 to March 2009

2. Materials and Methods

A bench scale of DAF unit was set in the lab. This unit consists of compressor, water tank, pump, saturator and flotation cell. The cell was equipped with a nozzle through which air saturated water was injected. Sampling point is located at 7.5cm from the floor of the flotation cell. A scour valve was installed at the bottom of the cell for cleaning purpose. In the first study, the leachate was treated using DAF process alone, this followed by the second study, coupling of DAF with coagulation process using ferric (III) chloride (FeCl3) as the coagulant agent. Here, the two steps were conducted in the same DAF units, coagulation and flotation. The experimental steps for the leachate treatment using DAF alone (first study) were executed as follows; (i) leachate was introduced into the flotation cell, (ii) in the saturator the tap water with air from the compressor was saturated under desired pressure, (iii) the supersaturated water from the saturator was injected through nozzle into the flotation tank. Retention time will be given before sample taken from the sampling point.

In the second study, the experimental work was carried out as follows: (i) 4 liters of pH adjustment leachate was introduced into the flotation cell, (ii) coagulant was added to the flotation cell, (iii) the leachate was rapidly mixed (470rpm for 5 minutes). Flotation process was carried out immediately after the coagulation process. The operating and fixed parameters that were used in this research work are shown in Table 2.

Table 2. Operational parameters.

Operational Parameter	Range	
	First Study	Second Study
Water flow rate	4 and 6 LPM	4 LPM
Saturator pressure	400 and 600 kPa	600 kPa
Injection time (IT)	120 seconds	120 seconds
Volume of leachate	8 liters	4 liters
Retention Time (RT)	10 minutes	10-60 minutes
Dosage of Coagulant	-	125-1500mg/L
pH	-	5

3. Results and Discussion

3.1. *DAF in Leachate treatment*

Saturator pressure (400kPa and 600kPa) with flow rate (4LPM and 6 LPM), [9], were chose and four different combination of leachate treatment were carried out. The first treatment was using high pressure, 600kPa, with low flow rate, 4 LPM; second treatment, 600kPa with 6 LPM; third treatment, 400kPa, with 4 LPM; fourth treatment, 400kPa, with 6 LPM. Using analysis of variance (ANOVA) it indicates that there are significant differences between these four treatments with p value less than 0.05, (Table 3). However in term of percentage removal, the results were fairly low for both of the studied parameter, (Figure 1). Generally the results indicate that the removal for the both parameters is lower than 40 percent.

On the other hand, comparison between these four treatments, second treatment gives the highest removal, 36 and 33 percent for COD and colour respectively. This is because of the bubble volume concentration (BVC). In a preliminary study for determining BVC, (results not discussed), higher pressure (600kPa) and flow rate (6 LPM) gave higher bubble volume concentration (41522 ppm) compare to other combination of pressure and flow rate. This is due to the increasing in flow rate and pressure up to 600 kPa and 6 LPM would increase the air available for flotation, resulted in an improvement in the treated leachate, [10]. Since the removals are very low, coagulation process using $FeCl_3$ was carried out first followed by DAF in the next study. According to [11], DAF successfully reduce the concentration of COD, BOD, and turbidity when the alum or ferric chloride was used.

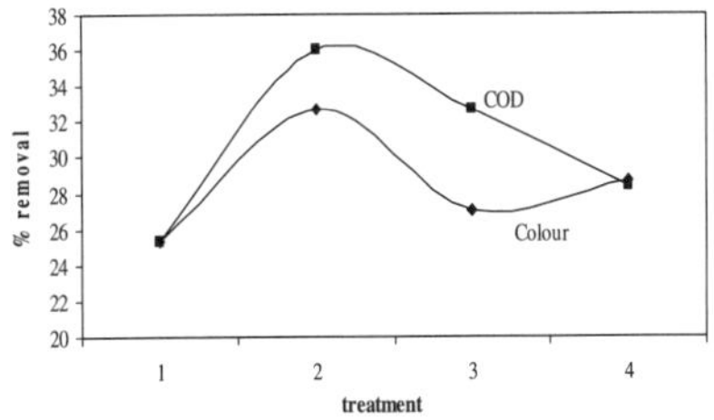

Fig. 1. Colour and COD removal at different treatment.

3.2. *Coupling of DAF with coagulation*

Ferric chloride was chose as the coagulant agent due to its superior in leachate treatment, [12]. The leachate pH was adjusted to pH 5 before coagulation process was proceed, [1]. Pressure and flow rate were fixed at 600kPa and 6 LPM since it gives better removal in previous study. Dosage concentration was varies from 125mg/ to 1500mg/L, and after 20 minute the sample were analyzed. The results show the removals of the studied parameters were increased, (Figure 2).

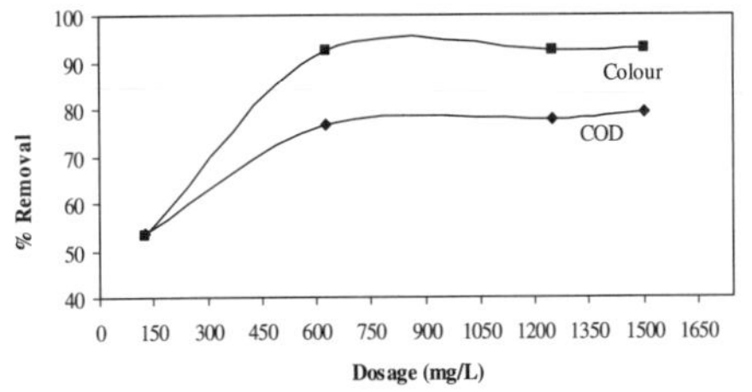

Fig. 2. Percentage removal of colour and COD at 20 minute retention times (RT), 2 minute injection time (IT) with different dosage.

The highest removal of COD and colour was 77 and 93 percent respectively at 625mg/L using this coupling process. This result accordance with [13], [14], [15], which explain that coagulation is an important process to achieve high efficiency in the DAF. The reason is because to obtain pin-point size particles that will increase the bubble-particle attachment, during flotation process. Comparisons between DAF with coagulation/flocculation-sedimentation process indicate that 1500 mg/L $FeCl_3$ was needed to achieved 75 % removal of COD, [5]. Aziz et al., 2007 [12], reported that 1200mg/L of $FeCl_3$ was required for 51 % of COD removal. Therefore, coagulation/flocculation-sedimentation process will produce more sludge compare to DAF.

In order to study on the flotation time in flotation column, the retention time had been varies from 10 minutes to 60 minutes. The results indicate that 10 minutes retention time was enough float the particle to the surface, and increasing the retention time does not give any significant differences, (Figure 3).

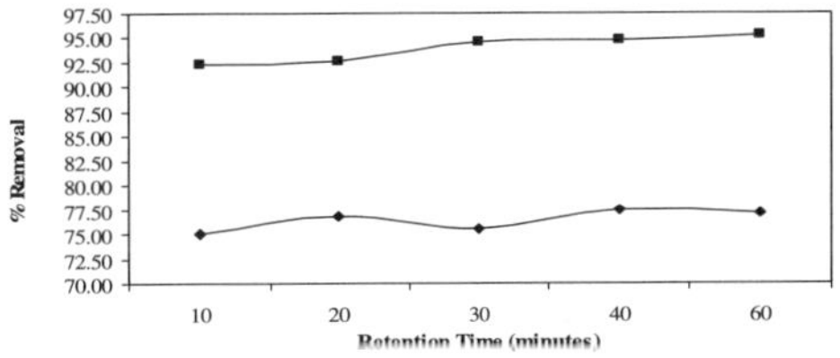

Fig. 3. Percentage removal of colour and COD at 625mg/L FeCl3, 2 minute injection time (IT) with different retention time.

Study by Al-Shamrani et al., 2002 [9], required 7 minutes for the flotation could proceed before samples were taken for analysis. Normal batch sedimentation process will take 30 to 60 minutes before the samples are taken for analysis, [5], [16].

Table 3. One-way ANOVA analysis between four different treatments: Colour and COD removal at different treatments.

Source	Degrees of freedom	Sum of squares	Mean square	*F*-Value	P
Colour					
Treatments	3	890.42	296.81	34.39	0.00
Error	116	1001.17	8.63		
Total	119	1891.59			
COD					
Treatments	3	1574.2	524.7	20.88	0.00
Error	116	2915.0	25.1		
Total	119	4489.2			

4. Conclusion

DAF is a physical treatment method for water and wastewater. Application of this method for leachate treatment was a new approach, cause up to this stage no study has been reported in literature review. The results shows that using DAF as a single treatment gives a fairly low removal in COD and colour (36 and 33 percent respectively). Nevertheless, coupling DAF with coagulation enhance the removal up to 77 and 93 percent for COD and colour. This was achieved at highest pressure and flow rate (600kPa; 6LPM) using 625mg/L FeCl3 with 120 seconds injection time and 20 minutes retention time. However study on the retention time shows that within 10 minutes 75 and 92 percent of COD and colour removal was achieved. This proves that DAF process offer rapid start –up.

Acknowledgments

The authors acknowledge the Ministry of Science, Technology and Innovation (MOSTI) Malaysia for the Science Fund grant. Thanks due to the contractor Idaman Bersih Sdn. Bhd., Penang for their cooperation during the study.

References

1. A. Amokrane, C. Comel, and J. Veron, Landfill leachates pre-treatment by coagulation flocculation, Water Res. **31** (1997) 2775-2782.
2. J. Crawford and P.G. Smith, Landfill Technology, Anchor-Bendon Ltd, Triptee, Essex, 1985.
3. H.A. Moreno Casillas, D.L. Cocke, J.A.G. Gomes, P. Morkovsky, J.R. Parga and E. Peterson, Electrocoagulation mechanism for COD removal, Sep. Purif. Technol. 56 (2007) 204-211.
4. A. I. Zoubolis, X. L. Chai, and I. A. Katsoyiannis, The application of bioflocculant for the removal of humic acids from stabilized landfill leachates, J. Environ. Manage. **70** (2004) 35-41.
5. A.A. Tatsi, A.I. Zouboulis, K.A. Matis, P. Samaras, Coagulation-flocculation pretreatment of sanitary landfill leachates, Chemosphere **53** (2003) 737-744.
6. S.E. Burns, S. Yiacoumi, and C. Tsouris, Microbubble generation for environmental and industrial separation, Sep. and Purif. Technol. **11** (1997) 221-232.
7. L.A. Feris, J. Rubio, Dissolved air flotation (DAF) performance at low saturation pressure, Filtration and separation conference (1999), 1999: 61-65.
8. A.L. Macfarlane, R. Prestidge, M.M. Farid, and J.J.J. Chen, Dissolved air flotation: A novel approach to recovery of organosolv lignin, Chem. Eng. J. **148** (2009) 15-19.
9. A.A. Al-Shamrani, A. James, H. Xiao, Separation of oil from water by dissolved air flotation, Colloids and Surf., A **209** (2002) 15-26.
10. R.A. Hyde, Design and experimental studies on a dissolved-air flotation pilot plant treating 8.2 m^3/h of Thames River water, In: Water Clarification by Flotation, Water Research Centre Technical Report (TR 13), November, 1975.
11. S.I. Abo-Elela and S.S. Nawar, Treatment of wastewater from an oil and soap factory via dissolved air flotation. Environ. Int. **4** (1980) 47-52.
12. H.A. Aziz, S. Alias, M.N. Adlan, F.A.H. Asaari, and M.S. Zahari, Colour removal from landfill leachate by coagulation and flocculation processes, Bioresour. Technol. **98** (2007) 218–220.
13. A.I. Zouboulis and A. Avranas, Treatment of oil-in-water emulsions by coagulation and dissolved-air-flotation, Colloids and Surf., A, **172** (2000) 152-161.
14. A.A. Al-Shamrani, A. James, H. Xiao, Destabilisation of oil-water emulsions and separation by dissolved air flotation, Water Res. **36** (2002) 1503-1512.
15. K. Bensadok, M. Belkacem, G. Nezzal, Treatment of cutting oil/water emulsion by coupling coagulation and dissolved air flotation, Desalination **206** (2007) 440-448.
16. S. Ghafari, H. A. Aziz, M. H. Isa, and A. A. Zinatizadeh, Application of response surface methodology (RSM) to optimize coagulation–flocculation treatment of leachate using poly-aluminium chloride (PAC) and alum, J. Hazard. Mater. **163** (2009) 650-656.

MOLECULAR DOCKING OF 3-O-MICAROCYLMICAMINOCYL-5-O-FOROSAMINYLERYTHRONOLIDE-B (MMFE) TO RRNA 23S DEINOCOCCUS RADIODURANS AND THE PREDICTION OF ITS ANTIBIOTIC POTENCY

WINARTO HARYADI[1]
[1]*Department of Chemistry, Gadjah Mada University, Sekip Utara Yogyakarta 55281, Indonesia*

UMAR ANGGARA JENIE[2], RETNO S SUDIBYO[2], HARNO DWI PRANOWO[1] and FAJAR RAKHMAN WIBOWO[3]
[2]*Faculty of Pharmacy, Gadjah Mada University, Sekip Utara Yogyakarta 55281, Indonesia*
[3]*Department of Chemistry, Sebelas Maret University, Surakarta, Indonesia*

MMFE is a new derivative of erythromycin which is synthesized through hybrid biosynthetic technique. The molecular docking to rRNA 23S *Deinicoccus radiodurans* is accomplished to determine the model and strength of binding to the target macromolecule. The molecular docking of erythromycin-A to the same macromolecule is used as a control. The docking result of MMFE shows that it occupies the same cavity as of the experimental erythromycin-A in the same macromolecule. The binding position of MMFE is not exactly the same as erythromycin-A due to the presence of glycon dimer of MMFE molecules. According to free energy of binding, MMFE is predictably a higher potency than erythromycin-A.

1. Introduction

Erythromycin is a broad spectrum macrolide antibiotic. This compound can be used to substitute penicillin because of the resistence of penicillin to microbes and its sensitivity to some patients. The disadvantage of the usage of erythromycin is the instability in stomach acid [1]. The instability is caused by internal nucleophilic attack of hydroxyl group(C-6) to protonated carbonyl (C-9) in macrolide ring [1]. This reaction causes decomposition of erythromycin and lack of its activity. The decomposition can be prevented by structural modifications, because these modifications can prevent the internal nucleophilic attack [2]. Some erythromycin derivative antibiotics produced by structural modification are clarithromycin, roxythromycin and azythromycin [3].

3-*O*-micarocylmicaminocyl-5-O-foro-saminyleritr-onolid-B (MMFE) is new derivative of erythromycin which is synthesized using hybrid biosynthetic [4]. This compound was predicted to be more stable in acidic condition because of the steric hindrance of internal nucleophilic attack but at the present time their antibiotic potency has not been investigated. The investigation will be conducted using molecular docking method.

Macrolide antibiotics able to inhibit the elongation of oligopeptide synthesis in the cavity of rRNA 23S named peptide exit tunnel [5]. The complex structure of macrolide binding to rRNA 23S D. *radiodurans* has been found through X-ray crystallography [6]. Based on the complex structure, molecular docking of MMFE is accomplished to determine the mode and strength of binding to rRNA 23S. As references, molecular docking erythromycin-A has also been conducted to the active site of rRNA 23S *D. radiodurans* experimentally. The binding position of these molecules resulting from the docking are compared with the site of attachment of erythromycin-A produced by X-ray crystallography based on RMSD (root mean square deviation) calculations. The strength of binding of these molecules from the docking to rRNA 23S macromolecule could be computated based on the free energy of binding [6].

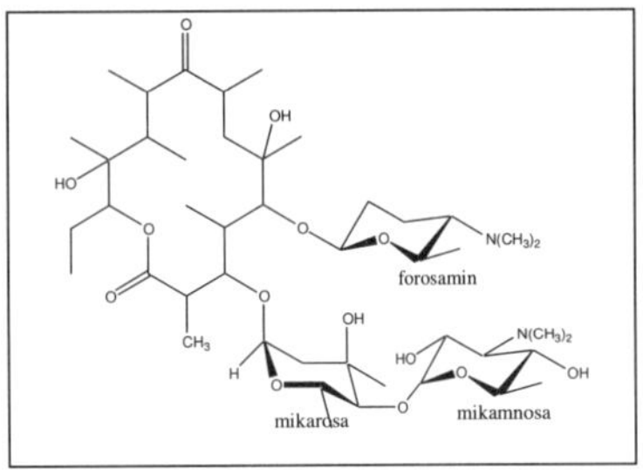

Fig. 1. The structure of MMFE.

2. Experimental Section

2.1. *Material*

Software which is used in this research are: CentOS, Molden, Gaussian98, Chimera, Rasmol and Autodock Tools version 4.01, Protein data bank site (PDB) www.RCSB.org and other sites which are relevant.

2.2. *Method*

The complex structure of erytromycin-A with rRNA 23S Deinococcus radiodurans macromolecule is obtained through Protein Data Bank searching via www.RCSB.org. The erythromycin-A structure is seperated from antibiotic macromolecule with the use of CHIMERA programme. The structure of MMFE was drawn using MOLDEN programme and optimalized with ab initio method using Gaussian98 software. The docking of erythromycin-A and MMFE to rRNA 23S macromolecule using Autodock version 4.0. The docking parameters used in this research are Lamarckian Genetic Algoritm, grid box 60 and total energy evaluation amounted to 2,5 x 108. The Genetic Algoritm parameter amongst others are 150 initial population, 27.000 maximum generation, cross over rate of 0.8, and mutation rate of 0.02. Local search Solis parameter and Wet including 300 iterations of local search, consecutive successes before changing and 4 consecutive failures before changing.

3. Result and Discussion

The docking model of MMFE and erythromycin-A molecules are executed ten times at the active site of 23S macromolecule by X-ray crystallography [6]. The RMSD of erythromycin-A and MMFE to erythromycin-A position experimentally and their free energy of binding are shown in table 1.

Table 1.RMSD and free energy of binding through molecular docking of erythromycin-A and MMFE.

	Erythromycin-A		MMFE	
Ligand	RMSD (Å)	Free energy of binding (kkal/mol)	RMSD (Å)	Free energy of binding (kkal/mol)
1	8.30	-8.78	6.54	-18.40
2	8.85	-9.41	6.66	-18.44
3	8.27	-9.20	2.67	-18.96
4	8.25	-9.19	**2.55**	**-18.88**
5	8.85	-9.40	2.64	-18.96
6	7.68	-8.66	5.28	-17.88
7	6.92	-8.25	2.66	-18.94
8	**5.83**	**-8.14**	4.38	-17.50
9	8.84	-9.41	2.67	-18.96
10	8.24	-9.17	6.25	-18.11

Based on RMSD calculations, the 8th ligand of erythromycin-A (5.83 Å) and 4th ligand of MMFE (2.22 Å) are molecules which occupies the closest position to experimental erythromycin-A. The relative position of these docking molecular models against X-ray crystallography erythromycin-A in the active site of RNA 23S macromolecule is shown in figure 2.

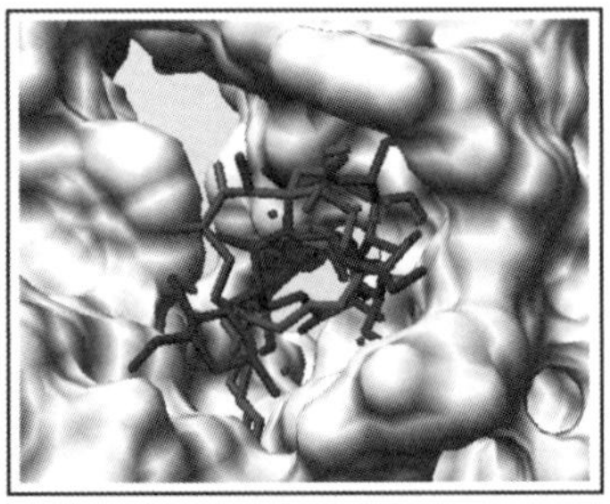

(a)

= erythromycin-A (X Ray)
= erythromycin-A (*docking*)

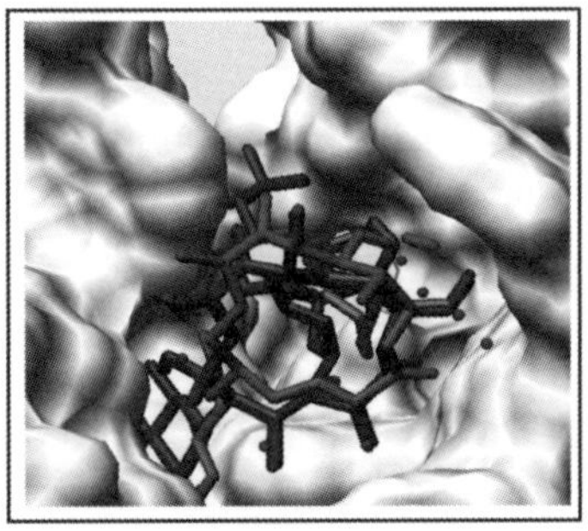

(b)

= erythromycin-A (X Ray)
= MMFE (*docking*)

Fig. 2. The relative position of erythromycin-A mode (a) and MMFE (b) against erythromycin-A position experimentally.

According to table 1 and fig2, the binding position of MMFE to rRNA 23S is not exactly match to experimental position of erythromycin-A, however their binding position are closer to binding position of experimental erythromycin-A than docking position of erythromycin-A. It means that MMFE binding position in rRNA 23S are similar to binding position of real erythromycin-A. Therefore, MMFE are predicted able to inhibit the elongation of oligopeptide in peptide exit tunnel [5].

Based on the free binding energy data, the binding free energy of MMFE is lower than binding free energy of erythromycin-A. The binding free energy difference is more than 2 kkal/mol [7], therefore the binding free energy of MMFE is significantly lower than erythromycin-A and it means that the binding strength of MMFE to rRNA 23S is higher than erythromycin-A. The higher strength of MMFE could be predicted that the dimer of MMFE gycon at C-5' causes the additional of MMFE binding to active site of rRNA 23S. From this fact, it could be predicted that antibiotics potency of MMFE is higher than erythromycin-A.

4. Conclusion

The binding position of MMFE to rRNA 23S are similar to binding position of real erythromycin-A. Therefore, MMFE are predicted able to inhibit the elongation of oligopeptide in peptide exit tunnel. Based on the free binding energy data, the free energy of MMFE is lower than erythromycin-A. It could be predicted that antibiotics potency of MMFE is higher than erythromycin-A.

References

1. Sakakibara H. and S. Omura, 1984, Chemical Modification and Structure-Activity Relationship of Macrolides in Satoshi Omura (Ed) : Macrolide Antibiotics : *Chemistry, Biology and Practise, Academic Press*, Orlando
2. Omura S. and Y. Tanaka, 1984, Biochemistry, Regulation and Genetics of Macrolide Production in Mura S (ed) : Macrolide Antibiotics : *Chemistry, Biology and Practise*, Academic Press, Orlando
3. Fass, R.J., 1993, Erythromycin, Clarithromycin and Azithromycin: Use of Frequency Distribution Curves, Scattergrams and Regression Analyses to Compare in vitro Acvtivities and Describe Cross-Resistence, *Antimicrob*. Agents Chemother., 37, 10, 2080-2086
4. Jenie, U.A., R.S. Sudibyo, and A. Yanuar, 1998, *Development of New Erythromycin Derivatives using Hybrid Biosynthetic Technique*, Majalah Farmasi Indonesia, Vol. 9, No.2, 50-67
5. Lovmar, M, K. Nilson, V. Vimberg, T. Tenson, M. Nervall and M. Ehrenberg, 2005, The Molecular Mechanism of Peptide –mediated Erythromycin Resistence, *The Journal of Biological Chemistry* Vol 281. pp6742-6750
6. Schlünzen F, R Zarivach, J Harms, A Bashan, A Tocilj[1], R Albrecht, A Yonath and F Franceschi, 2001, Structural Basis For The Interaction Of Antibiotics With The Peptidyl Transferase Centre In Eubacteria, *Nature* , **413**, 814-821
7. Morris, G.M., D.S. Goodsell, R.S. Halliday, R. Huey, W.E. Hart, R.K. Belew and A.J. Olson, 1998, Automated Docking Using a Lamarckian Genetic Algorithm and an Empirical Binding Free Energy Function, *Journal of Computational Chemistry*, Vol. 19, No 14, 1639-1662

TREATMENT OF DIMETHYL-SULFOXIDE-CONTAINING OPTOELECTRONICS WASTEWATER USING AIRLIFT BIOREACTOR WITH PVA-IMMOBILIZED CELL BEADS

SIN-YI HE

Department of Civil Engineering, Chung Hua University, 707, Sec. 2, Wufu Road, Hsinchu City 300, Taiwan, R.O.C.

SZ-CHWUN JOHN HWANG

Department of Civil Engineering, Chung Hua University, 707, Sec. 2, Wufu Road, Hsinchu City 300, Taiwan, R.O.C.

YUN-HUIN LIN

Energy & Environment Laboratories, Industrial Technology Research Institute, Bldg. 64, 195, Sec. 4, Chung Hsing Rd., Chutung, Hsinchu 310, Taiwan, R.O.C.

KUAN-YUN HOU

Department of Civil Engineering, Chung Hua University, 707, Sec. 2, Wufu Road, Hsinchu City 300, Taiwan, R.O.C.

DMSO (dimethyl sulfoxide) is a useful and inexpensive environment benign solvent easy to recycle. Its industry-wide adoption has revealed an odorous problem to the industrial parks and their adjacent residential areas due to its decomposition product, DMS (dimethyl sulfide). Our research goal is to develop a feasible biological treatment technology to effectively treat the DMSO into oxidative pathway instead of going to the DMS pathway. We have adopted the specific activated sludge as bacterial inoculums to decompose the DMSO into $DMSO_2$ pathway from the wastewater treatment plant of a DMSO-producing chemical plant. In this year, we focused on the research results from the immobilization technology, the repeated batch process and the effect of sucrose addition. Finally, we have found that the best pH range in biodegradation of DMSO using a PVA-immobilized cell beads is 5.0-8.5. The best dose of sucrose is 0-50 mg L^{-1} that helps bacteria to tolerate the toxicity of DMSO. From the repeated batch process, we found the sucrose helps improve the treatment efficiency only for the raw activated sludge rather than the acclimated sludge. In the performance of airlift bioreactor, the PVA-immobilized cell beads can degrade the 1,200-mg L^{-1} DMSO within 45 hr, in comparison to the 10 hr for the free cell system in decomposition of 850-mg/L DMSO.

1. Introduction

Dimethyl sulfoxide (DMSO; chemical formula, $(CH_3)_2SO$) has been widely used for many industrial especially opto-electronic and IC packaging industry. It purposes because of its high hygroscopic property, low flammability, good thermal stability, and versatile solubility in a range of organic solvents or their aqueous solutions, e.g. ethanol, benzene and chloroform, etc. DMSO is wasted in water with other compounds such as mono-ethanol amine (MEA), butyl diglycol (BDG), tetramethylammonium (TMAH), isopropyl alcohol (IPA), film developer, photo-resistor, and detergent, etc [ITRI, 2004][1]. In addition, the DMSO will further be biodegraded through the reduction pathway to produce odorous dimethyl sulfide (DMS) [Lomans *et al.*, 2002][3]. DMS has been reported to be an odorous compound because of its very low odor threshold value, and minute amounts of DMS pervading the atmosphere could cause a major complaint by the adjacent residents or inside-park factory workers [Leonardos et al., 1969][4]. Glindemann *et al.* [2006][5] found the link between the intermittent occurrence of DMSO residues in influent of the wastewater treatment plant and the malodorous DMS in the activated sludge aeration tank. Therefore, DMSO must be removed or retrieved from wastewater, and the relevant treatment technology must be urgently developed in the near future.

The method used in treatment of DMSO is really dependent upon the level of the DMSO concentration in the liquor. The wastewater containing DMSO exceeding 1,000 mg L^{-1} is most concentrated and recycled back for reuse in the original processes. The wastewater, derived from the washing or rinsing processes in the manufacture of semiconductors or LCD/TFT, contains 500-800-mg L^{-1} DMSO whose annual total waste amount is comparable to that of the aforementioned DMSO-recycled liquor. Thus, the chemical decomposition and oxidation of the low concentration of DMSO by UV or Fenton processes are claimed as the preferred solutions to solve the odor problem of wastewater treatment plant [Muratani, 1999; Scaduto, 1995; Koito *et al.*, 1998; Park *et al.*, 2001][6,7,8,9].

However, the huge amount of chemicals consumed by the physical chemical process is not cost effective in treating the relatively low DMSO concentration in wastewater discharge. Koito et al. [1998][8] proposed a chemical oxidative pathway for DMSO oxidation to $DMSO_2$ followed by aerobically biological treatment. Conversion of DMSO to $DMSO_2$ by chemical oxidation in the beginning could avoid the malodorous problem and reduce the cost of investment by the following biological treatment process. Their large capacity of biological treatment technology is very comparable to physio-chemical treatment technologies (as seen in Table 1). Now, if we could develop a feasible biological treatment technology for DMSO removal instead of screening for a specific microorganism, we would have solved both the odorous DMS problem and the wastewater pollution problem simultaneously for both the opto-electronic industry and IC packaging industry, too.

Table 1. The common treatment technologies for LCD/TFT wastewater.

Waste water treatment technologies	Flow rate, CMD
Physio-chemical treatment method	300~900
Biological treatment method	900~1700
Anaerobic-aerobic membrane biological treatment	700~1500
Anaerobic bioreactor	600~1200

Airlift has been applied to many applications such as organic synthesis, wastewater treatment, and fermentation [Hinks *et al.*, 1996; Adinarayana *et al.*, 2004][12,13], but less applied to environmental gaseous pollutant treatment processes [Zuber *et al.*, 1997; Lo and Hwang, 2004; Wen *et al.*, 2005][14,15,16]. Using airlift to treat acetone-containing airstream was to solve all the problems in biofiltration, including clogging in bio-trickling filter and low mass transfer efficiency in bio-scrubber. Using our previous developed polyvinyl alcohol- (PVA-) entrapped cell beads will protect the enclosed microorganisms from toxicity [Chen et al., 2002][17]. The main goal of this study was to achieve an optimal biodegradation efficiency of dimethyl sulfoxide using a high efficient bioreactor with the activated sludge (Sludge 2) selected from a specific chemical plant. The purpose of using an airlift bioreactor is to furnish the supply of dissolved oxygen for DMSO biodegradation. Many influencing factors in this new application were also considered, e.g. pH, temperature, oxygen supply, and acclimation.

2. Materials and Methods

2.1. *Microorganisms and cultivation*

According to our previous report, the aerobic activated sludge (Sludge 2) which was obtained from a chemical plant (Chang Chun Group, Miaoli County, Taiwan) was chosen as an inoculum, because it did not release odorous compounds during biodegradation [Hwang et al., 2007][18]. It was incubated with a medium as described by Yang et al. [2003][19].

2.2. *PVA immobilized cell beads*

All the microorganisms inside cell beads were co-immobilized in phosphorylated PVA gel beads according to the method presented in our previous work [Chen and Lin, 1994][20]. A mixture containing concentrated mixed cultures (0-150 g dry cell L^{-1}) was thoroughly mixed with an equal volume of PVA (9.5%, w/w). The mechanical strength of the immobilized beads was measured according to our previous work [Chen and Lin, 1994][20].

2.3. *Airlift bioreactor*

The airlift bioreactor was built using an annular acrylic column with height of 47 cm and internal diameters (I.D.) of 11 cm and 5 cm for an external and internal column, respectively, in which there was a 33.5 cm long as a drag tube [Hwang et al., 2007][18].

2.4. *Experimental design*

Many environmental influencing factors such as the initial pH, the content of carbon sources, the amount of inoculums, and the aged sludge source were considered in the biodegradation of 100-mg L^{-1} DMSO using PVA-entrapped cell beads. For example, the pH was set at 3, 5, 7, 8.5 and 10; the amounts of inoculums seeded were 0, 15, 30, 45 and 60 g. The make-up of the buffering agents for various pHs was made as described by Dawson et al [1968][21]. The content of sucrose was 0, 50, 100, 300, 500, 1000, and 1500 mg with a constant amount of inoculums (30 g; 0 g for control). According to our previous report, DMSO degradation needs an appropriate acclimation time to allow the mono-oxygenized enzymes being synthesized [Hwang *et al.*, 2007][18]. Therefore, a fresh and aged DMSO-degrading culture of 300 mL was used as inoculums for biodegradation of 200 mg DMSO L^{-1} under the condition of whether or not adding sucrose as a carbon source (0 or 50 g of sucrose). In comparison of DMSO or $DMSO_2$ degradation results, the aged activated sludge culture was chosen (although no $DMSO_2$ was detected).

2.5. *Analytical methods*

Two milliliters of water sample were filled into a 2-ml aliquot and centrifuged at 10,000 rpm for 10 min. After the supernatant was filtered out through a 0.22 μm filter as the report described by Murakami et al. [2002]2, a needle was employed to suck out 0.1 μL of the filtered liquid for measuring DMSO concentration using High Performance Liquid Chromatography (HPLC) with UV or RI detectors (Chromatograph Acme 9000, Younglin Instrument, Korea) equipped with a Hypersil ODS column (5 μL, 250×4.6 mm, England). The pH and optical density (O.D.) were measured using a pH meter and spectrophotometer at 600 nm, respectively.

3. Results and Discussion

3.1. *The effect of acclimation on the degradation of DMSO*

The Sludge 2 inoculum originally used in our previous study was compared with the fresh sludge inoculum obtained from the same source for removal efficiency in the degradation of DMSO. This study used the 300-mL shake-flask results for biodegradation of 200 mg L^{-1} of DMSO from both acclimated and fresh Sludge 2 with the alternative of adding carbon source (sucrose). Obviously, the acclimated sludge had the better performance than the fresh sludge, and addition of sucrose would delay the time for complete degradation of DMSO while using the fresh sludge as the inoculums. The result indicates that sucrose was a more degradable compound than DMSO for the fresh sludge inoculums, but there was no difference while using the acclimated sludge inoculums. The acclimated sludge was finally chosen to make the PVA-entrapped cell beads for the following experiments.

3.2. *The effect of initial pH on the degradation of DMSO using PVA-entrapped cell beads*

The effect of initial pH on the degradation of DMSO in the shake-flask culture using the PVA-entrapped cell beads was shown in Figure 3. The removal efficiencies were 92.8%, 100% and 71.7% for the initial values of pH 5, 7 and 8.5, respectively. After the first repeated batch experiment, the time to reach the removal efficiency of 100% at pH 7 was decreased from 72 hr to 48 hr. In general, the 90% removal efficiency was achieved at pH 5-8.5 within 56 hr. In the earlier experiments, white turbid floc was formed after 24 hr at pH 3.0, 5.0 and 7.0 rather than pH 8.5 and 10.0. It indicates that one of buffering agent, citric acid, could provide the nutrient for the bacteria which were on the surface of cell beads to grow. However, the boric acid could be used as disinfectant, so there was no white sediment formed. Thus, the cell beads were then rinsed with boric acid for 10 seconds to control the surface bacteria between each repeated batch.

3.3. *The effect of carbohydrate on the degradation of DMSO using PVA-entrapped cell beads*

The PVA-entrapped cell bead system was to be carried in treatment of a multiple substrates. According to the Yang's report [Yang and Myint, 2003][19], adding 50 mg L^{-1} of sucrose to the DMSO-containing wastewater will enhance the biodegradation of DMSO. The TOC removal efficiency would increase from 80-84% to 92-95%. The additional sucrose compound was the main cause to enhance the microbial capability in detoxification of the toxic DMSO. The similar reaction mechanism was also found in the co-metabolism [Rittman and McCarty, 2001][24].

The 100 mg L^{-1} of DMSO was biodegraded with various sucrose contents under repeated batches. In the earlier experiments, the addition of extra carbon source seems to have a positive effect on the initial rates of biodegradation of DMSO. Consequently, the sucrose content was no longer a major influencing factor when the cell beads were acclimated. However, the reactions with 200 and 300 mg L^{-1} of sucrose were far slower than those with 0-100 mg L^{-1} of it. In order to quantify the threshold of sucrose on the biodegradation of DMSO, the sucrose content was increased up to 1,500 mg L^{-1}. Obviously, the system was still maintained a high removal efficiency of 100% within 12 hour even with 500 mg L^{-1} of sucrose added.

3.4. *The effect of inoculant biomass on the degradation of DMSO using PVA-entrapped cell beads*

The amount of inoculums or the microbial activity was one of the reasons to increase the biodegradation rate in a kinetic study. There was an optimum biomass loading,

30 g or 45 g, occurring in the series of operations, and it was a conflict with the theory of microbial metabolism. In light of mass transfer efficiency, the orbital rotation speed of 120 rpm was too slow for 60 g inolums to drive the oxygen entering the media of 300 mL simulated wastewater.

3.5. *Airlift performance in repeated batch mode*

Before the airlift was performed, the background check experiment was implemented for examining the quantity of DMSO loss in aeration. According to Figure 8, the DMSO was not vaporized during air bubbling through the 700-mg L^{-1} DMSO solution with air flow rate of 10 L min^{-1} at 25°C. In the beginning of experiment (< 24 hr), the loss of DMSO concentration was lower than 10 mg L^{-1} due to the experimental error or minor DMSO evaporation. The DMSO solution was in concentrated after 40 hr due to the water evaporation by air bubble. Thus, the loss of DMSO concentration by air bubbling was ignored.In the suspended sludge system, the airlift performance was illustrated in Figure 1a. Eight hundred fifty mg L^{-1} of DMSO was removed in 10 hr with the removal efficiency of near 99%. Yang and Myint [2003][19] have also reported that 600-mg L^{-1} DMSO needed 6.66 hr for 95% TOC removal.Using the PVA-entrapped cell beads, the airlift performance with many repeated batches was illustrated in Figure 1b.

Due to the mass transfer limitation, treating 1,200 mg L^{-1} of DMSO required more than 45 hr for complete biodegradation, compared to treating 800 mg L^{-1} of DMSO only requiring 21 hr for complete degradation. Thus, the microorganisms in the PVA-entrapped cell beads could stand at double loading of DMSO. Znad et al. (2006) adopted airlift with suspended activated sludge or immobilized cell beads at air flow rate of 3.5 L min^{-1} under repeated batches to treat 2,184 ppm of S -ethyl dipropylthiocarbamate (EPTC). Using immobilized cell beads could avoid the toxicity of EPTC and its removal efficiency was higher than using the suspended activated sludge system.

4. Conclusions

This study has revealed many environmental influencing factors which are very important in treatment of DMSO using PVA-immobilized cell beads for Opto-electronics industry and Semiconductor industry, e.g. the carbon source addition and the acclimation effect. The optimal pH range was in 5.0-7.0, but the extended pHs could be 3.0 to 8.5. Minor extra carbon source of 50 mg L^{-1} would help fresh sludge to degrade DMSO, but the decomposition products from the high concentration of sucrose would inhibit the biodegradation of DMSO. In addition, the repeated batch airlift performance has shown the capability of PVA-immobilized cell beads to treat high concentration of DMSO (100-1,200 mg L^{-1}) with about 100% removal efficiency. The complete degradation time was less and less after each repeated batch operation. Thus, the stability of running PVA-immobilized cell bead system has revealed the promising feature in biofiltration of odorous compounds such as DMSO.

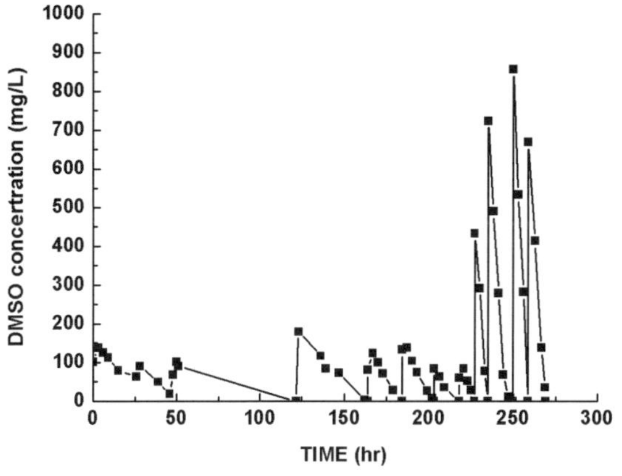

(a)

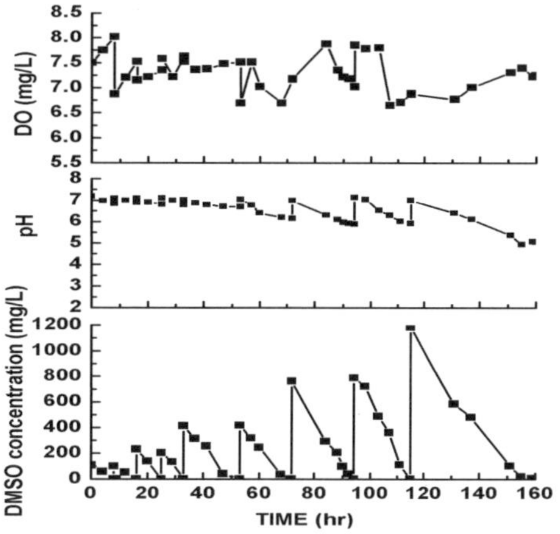

(b)

Figure 1. DMSO biodegradation in airlift bioreactor under repeated-batch mode; (a) suspended sludge system, (b) PVA-immobilized cell bead system. Operational conditions: initial pH, 7.0 ± 0.2; Temp., 30°C; air flow rate, 10 L min^{-1}.

Acknowledgments

This work was fully supported by the National Science Council, Taiwan (NSC 96-2221-E-216-004).

References

1. Industrial Technology Research Institute of Taiwan, R.O.C. (ITRI) (2004) Flat Panel Display Industry Infrastructure Project, MOEA-Authorized Projects in FY2004.
2. Murakami-Nitta T, Kurimura H, Kirimura K, Kino K, Usami S. (2002) Continuous degradation of dimethyl sulfoxide to sulfate ion by *Hyphomicrobium denitrificans* WU-K217. Journal of Bioscience and Bioengineering. 94(1): 52-56.
3. Lomans BP, van der Drift C, Pol A, Op den Camp HJM. (2002) Microbial cycling of volatile organic sulfur compounds. Cellular and Molecular Life Sciences. 59: 575-588.
4. Leonardos G, Kendall D, Barnard N. (1969) Odor threshold determinations of 53 odorant chemicals. Journal of Air Pollution and Control Association. 19(2): 91-95.
5. Glindemann D, Novak J, Witherspoon J. (2006) Dimethyl sulfoxide (DMSO) waste residues and municipal waste water odor by dimethyl sulfide (DMS): the North-East WPCP plant of Philadelphia. Environmental Science & Technology. 40: 202-207.
6. Muratani T. (1999) Biological treatment of wastewater containing DMSO. Sharp Giho. 73: 20-25 (in Japanese).
7. Scaduto Jr. RC. (1995) Oxidation of DMSO and methanesulfinic acid by the hydroxyl radical. Free Radical Biology & Medicine. 18(2): 271-277.
8. Koito T, Tekawa M, Toyoda A. (1998) A novel treatment technique for DMSO wastewater. IEEE Transactions on Semiconductor Manufacturing. 11(1):3-8.
9. Park SJ, Yoon TI, Bae JH, Seo HJ, Park HJ. (2001) Biological treatment of wastewater containing dimethyl sulphoxide from the semi-conductor industry. Process Biochemistry. 36: 579-589.
10. Murakami-Nitta T, Kirimura K, Kino K. (2003) Oxidative degradation of dimethyl sulfoxide by *Cryptococcus humicolus* WU-2, a newly isolated yeast. Journal of Bioscience and Bioengineering. 95(1): 109-111.
11. Kino K, Murakami-Nitta T, Oishi M, Ishiguro S, Kirimura K. (2004) Isolation of dimethyl sulfone-degrading microorganisms and application to odorless degradation of dimethyl sulfoxide. Journal of Bioscience and Bioengineering. 97(1): 82-84.
12. Hinks JW, Cawte H, Sanders DA, Hudson A, Dockree CN. (1996) Prediction of flow rates and stability in large scale airlift reactors. Water Science and Technology. 34(5-6): 51-57.
13. Adinarayana K, Srinivasulu B, Bapi Raju KVVSN, Ellaiah P. (2004) Continuous neomycin production by immobilized cells of *Streptomyces marinensis* NUV-5 in an airlift bioreactor. Process Biochemistry. 39: 1407-1414.
14. Zuber L, Dunn IJ, Deshusses MA. (1997) Comparative scale-up and cost estimation of a biological trickling filter and a three-phase airlift bioreactor for removal of methylene chloride from polluted air. Journal of Air and Waste Management Association. 47: 969-975.
15. Lo CS, Hwang SJ. (2004) Degradation of waste gas containing toluene in an airlift bioreactor. Environmental Science and Technology. 38(7): 2271-2280.
16. Wen J, Chen Y, Jia X, Chen D. (2005) Simultaneous removal of ethyl acetate and ethanol in air streams using a gas-liquid-solid three-phase flow airlift loop bioreactor. Chemical Engineering Journal. 106(2): 171-175.
17. Chen KC, Lin YH, Chen WH, Liu YC. (2002) Degradation of phenol by PAA-immobilized *Candida tropicalis*. Enzyme and Microbial Technology. 31: 490-497.
18. Hwang SCJ, Wu JY, Lin YH, Wen IC, Hou KY, He SY. (2007) Optimal dimethyl sulfoxide biodegradation using activated sludge from a chemical plant. Process Biochemistry. 42: 1398-1405.
19. Yang PY, Myint TT. (2003) Integrating entrapped mixed microbial cell (EMMC) technology for treatment of wastewater containing dimethyl sulfoxide (DMSO) for reuse in semiconductor industries. Clean Technol Environ Policy 6(1): 43-50.
20. Chen KC, Lin YF. (1994) Immobilization of microorganism with phosphorylated polyvinyl alcohol (PVA) gel. Enzyme and Microbial Technology. 16: 79-83.
21. Dawson RMC, Elliot DC, Elliot WH, Jones KM. *pH, buffers and physiological media.* In: *Data for biochemical research, 2nd ed.* Oxford: Clarendon Press, 1968, pp. 427-438.
22. Pierce chemical company. BCA Protein Assay Kit. US Patent No: 4839295, 1997.
23. Eckenfelder WW. (1967) Theory of biological treatment of trace wastes. Journal of Water Pollution Control Federation. 30: 240-250.
24. Rittmann BE, McCarty PL. *Environmental biotechnology: principles and application.* New York: McGraw-Hill Inc., 2001.

Session 4

ON SITE INTEGRATED LANDFILL LEACHATE TREATMENT: RECIRCULATION AND UPFLOW ANAEROBIC SLUDGE BLANKET REACTOR (UASBR)

RAJESH TANEJA

Municipal Corporation of Delhi, New Delhi, India

MANISH KUMAR

Technology Information, Forecasting and Assessment Council (TIFAC), Vishwakarma Bhavan Shaheed Jeet Singh Marg, New Delhi 110016, India

A. K. RAGHAV

Department of Applied Mechanics, IIT Delhi, New Delhi 110016, India

ATUL K. MITTAL

Department of Civil Engineering, IIT Delhi, New Delhi 110016, India

The paper deals with an integrated leachate treatment through recirculation column and Upflow Anaerobic sludge Blanket Reactor. Leachate produced at the Okhla landfill in Delhi, India, was treated on site through a simulated landfill recirculation column. Reactor in line. One pass leachate from the recirculation column was combined with raw municipal sewage. One pass leachate was analyzed and found to have high in pH, COD and heavy metals as compared to raw leachate. COD removal was 74 – 85%. The integrated treatment offers feasibility to treat the leachate in combination with municipal sewage.

1. Introduction

Delhi, with its urban and semi-urban population of 13.6 million is feeling growing urbanization, improvements in living standards, exploding population pressure which are just some of the contributing factors in accelerated waste generation in the city. The daily total solid waste collection in Delhi, ranges from 6600 to 7500 tonnes. Due to lack of engineered landfill within city, the conventional way (open dumping) of solid waste disposal is still practiced making the scenario worse. These dumps generate the leachate, a highly dangerous and toxic liquid substance. Landfill leachate is a complex wastewater with considerable variation in both quality and quantity. In general, leachate is highly contaminated with organic contaminants measured as chemical oxygen demand (COD) and biochemical oxygen demand (BOD), and also with high ammonium nitrogen concentration. Biological processes have been found ineffective for leachate from relatively old landfill.

The monthly production of leachate in Delhi has been estimated to be 81.5m3. Actual production may be more than this which is not properly estimated. There is no treatment facility in place for leachate in Delhi. Recirculation is an attractive alternative for landfill operation as it is an inexpensive and offers the advantage of faster stabilization of waste in the landfill This paper presents a leachate treatment process using leachate recirculation and Upflow Anaerobic Sludge Blanket Reactor (UASBR) for the leachate from Okhla, New Delhi.

2. Materials and Methods

2.1. *Experimental set-up*

The experimental set up was installed at the Okhla landfill site, which is spread over an area of 56 acre, out of which approximately 41 acre is filled up and remaining 15acre is vacant land. The landfill is in operation since 1994. The waste fill height varies from 4 to 20 m. The set-up consists of a recirculation column and UASB Reactor. The recirculation column was made up of UV stabilized Linear Low Density Polyethylene (LLDEP). The total height of the recirculation column was 2.6m. This column was packed with MSW in three layers having different densities. The bottom part of the recirculation column was conical in shape to facilitate the collection of the recirculated leachate. The volume of the top portion of the column above the upper sieve was 300L, which could be used to store the leachate. Experimental set-up is presented in figure 1.

3. Methodology

The UASB Reactor volume was 177.1 L with an internal diameter 23.75 cm, and a height of 100cm. Three evenly distributed sampling ports were installed at 15, 55, and 65cm from the bottom sieve of the reactor. An internal diameter of top portion (gas liquid separator) of the reactor was 1.08m with a height of 80cmLeachate was passed through a recirculation column. After single pass, leachate was diluted using raw sewage from the Okhla Sewage Treatment Plant. COD of the raw sewage ranged from 325- 470 mg/L. The influent to the UASB reactor was pumped through a Peristaltic pump.

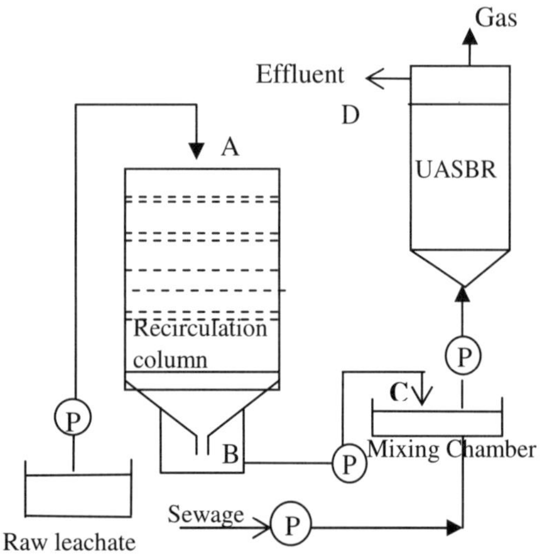

Figure 1: Schematic diagram of experimental set-up. (Sampling point for A. Raw Leachate, B. Single pass Leachate, C. Mixed influent (raw sewage + one pass leachate), D. Treated effluent, P. Pump).

The reactor was operated at varying organic loading rate (OLR). Characteristics of leachate after passing through the recirculation column, and after UASBR treatment were determined with the help parameters like COD, alkalinity, pH, TKN, Ammonia Nitrogen and heavy metals as in Figure 1.

3.1. *Excavation and characteristics of MSW from landfill*

Around 750kg of MSW was excavated from Okhla landfill and was packed in the recirculation column. Wastes were segregated to remove plastics, polythene, glass pieces and demolition waste. For the better simulation, waste was taken from different depths having different densities. Re-circulation column was compacted with waste in three layers. Density of waste in column was kept similar to the one in actual landfill.

Characterization of the MSW was carried out by first drying the excavated MSW naturally. The larger non-biodegradable materials such as stones, glass bottles, plastic films (bags), rubber, etc were removed and weighed. For analysis, the finer fractions were broken into small pieces by hammer, ball mill, and grinded if necessary, until the particles were smaller than 2 mm. Fines of MSW were used for heavy metal analysis and to determine the characteristics of leachate.

4. Results and Discussion

4.1. *Waste characterization*

The content of solid waste were analyzed by weight and were found to be high in moisture content (30 to 35 %), 12% plastic films, bags, bottles and other products, 44% organic matter such as wood, and cooking wastes and about 34% inorganic matter such as stone, sand, coal ashes, and glass bottles.

4.2. *Leachate characteristics*

Raw leachate was analyzed for COD, BOD, nitrogen content and heavy metals. It is evident from Fig. 2 that all the heavy metals are quite high, which makes Okhala

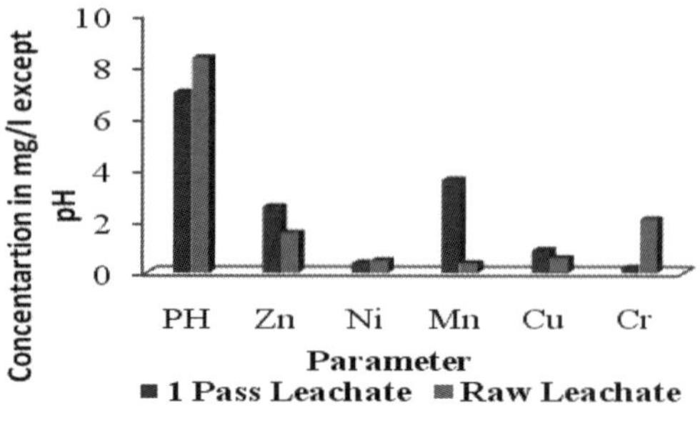

Figure 2: Heavy metals at the Okhla landfill.

landfill leachate much polluted as compared to the leachate from other landfills reported by Heyer et al, 1998.

4.3. *Performance of the UASBR*

In the initial 7 days, HRT was kept at 35 h. Latter on, it was changed to 48 h. During entire study, the lowest OLR was 0.533 Kg COD/m3.day and maximum was 2.039 Kg COD/m3.day. The variation in OLR is shown in Fig. 3. The variation of pH in combined influent and treated effluent is shown in the Fig. 4. The pH of the treated effluent throughout the study varied most of the time between 6.92 -7.75. Sometimes it went up to 7.95 and 7.85 due to higher pH of the domestic sewage. Whereas, the combine influent pH ranged between 6.72 -7.75. The COD removal efficiency of the reactor is illustrated in Fig 5. Initially, reactor was acclimatized for 5 weeks before feeding the leachate mixed with the sewage. Reactor was fed with combine waste on 35th day. Efficiency is presented after the UASBR started receiving leachate mixed with domestic sewage. Initially, the efficiency is more than 85%.

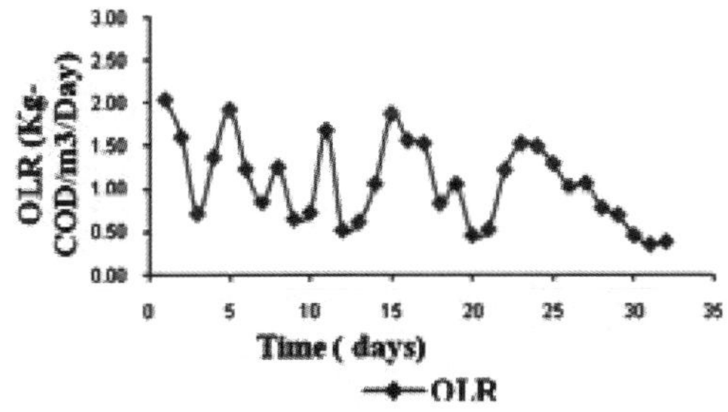

Figure 3: OLR used for the operation of the UASBR at the Okhla landfill.

For most of the time COD removal efficiency was between 70 to 87%. From the study, it is clear that efficiency increased with increase of OLR and vice versa. Variation in alkalinity in influent and effluent is shown in Fig 6. In the beginning the influent alkalinity is 1300mg/l and effluent alkalinity is 600mg/l, with time, alkalinity started increasing and sometimes it was more than the influent. It shows that methenogens are very active in methane forming. And the production of methane gas along with the CO_2 is responsible for increase in alkalinity. CO_2 in system forms bicarbonate alkalinity to increase the total alkalinity in effluent.

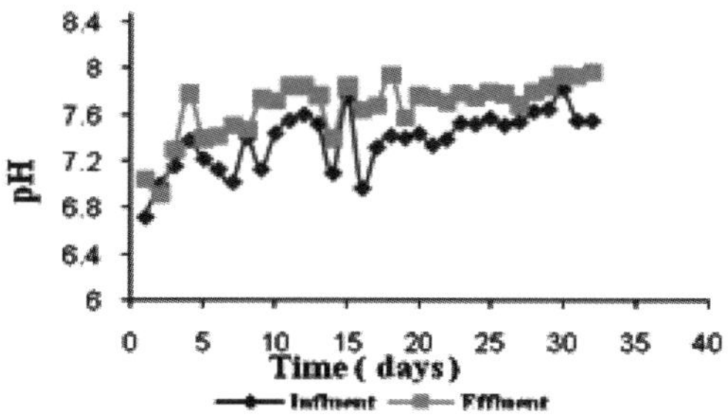

Figure 4: Variation in pHs at the Okhla landfill.

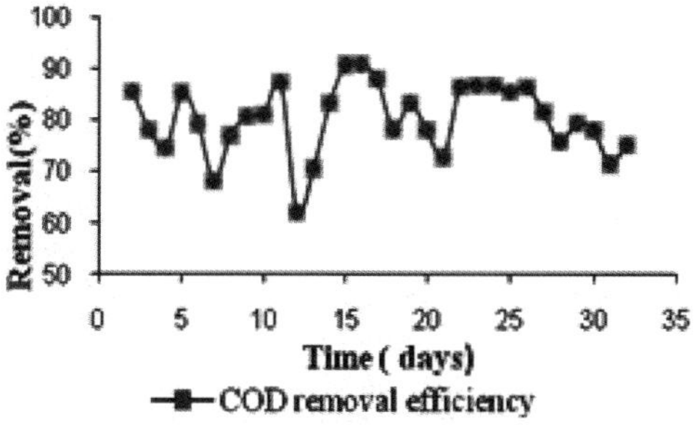

Figure 5: Variation in COD removal efficiency at the Okhla landfill.

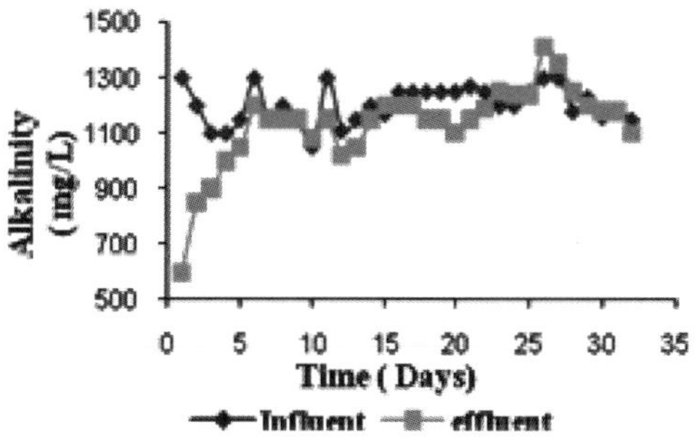

Figure 6: Variation in alkalinity at the Okhla landfill.

5. Conclusions

In the present study, a pilot plant was set-up at the Okhla landfill site. A number of heavy metals like Pb, Cu, Cr, Zn, and Ni are present in solid waste at the Okhla site. Though, Cd and Co were below the detection limit. The leachate at the Okhla Landfill is quite hazardous in nature. It is high in COD, with many heavy metals present in it. Notable, heavy metals are Pb, Cu, Cr and Ni. There is a potential of recirculating the leachate through the landfill. UASBR is quite effective (85%) to remove the COD. It is not possible to effectively remove the Ammonia-N by this unit. It needs to be further investigated. The concentration of heavy metals, both in raw leachate and after passing it through the re-circulation column remains quite high. However, in the present study, the leachate was passed only once. The present study shows a positive approach for the management of MSW and treatment of leachate. The treatment of leachate in combine form with raw domestic sewage is an attractive option to manage it at a single platform. It is economical as well as technically feasible also. But further study is needed to find the toxic effect of treated effluent by UASBR for final disposal.

References

1. Census 2001, *Census of Delhi 2001 Report*, *Delhi Government*, (2001).
2. D. Kimar, and B.J. Alappat, *9th International Waste Management and Landfill Symposium*, Cagliari, Italy, (2003).
3. Debra R. Reinhart, *Waste Management and Research*, **14**, 347 (1996).
4. Li Ye, Li Ling, Zheng Shizhong. *J. of Wuhan University of Technology*, **26**(12) (2004).
5. Zhang Xiangdan, and Wang Jiaming, *J. Water Supply and Drainage*, 26(1), 12 (2000).
6. K.-U. Heyer, R. Stegmann, and H.-J. Ehrig, *International Seminar, Present and Future of MSW Landfilling*. 24.-26. Cittadella, Italia, Proceedings Edizioni Euro Waste (1998).

THE APPLICATION OF VARIED IMPREGNATION METHODS ON THE CHARACTERS OF NON-SULFIDED NIMO/USY CATALYST

KHOIRINA DWI NUGRAHANINGTYAS†

Department of Chemistry, Faculty of Maths and Natural Sciences, Gadjah Mada University, Sekip Utara Yogyakarta, 55281, Indonesia

Department of Chemistry, Faculty of Maths and Natural Sciences, Sebelas Maret University, Jl Ir Sutami No 36 A Kentingan Surakarta, 57126, Indonesia

WEGA TRI SUNARYANTI, TRIYONO, NURYONO

Department of Chemistry, Faculty of Maths and Natural Sciences, Gadjah Mada University, Sekip Utara Yogyakarta, 55281, Indonesia

DIAN MARUTO WIDJONARKO, ARI YUSNANI

Department of Chemistry, Faculty of Maths and Natural Sciences, Sebelas Maret University, Jl Ir Sutami No 36 A Kentingan Surakarta, 57126, Indonesia

Hydro-desulphurization catalysts are widely used in petroleum industry. However, those catalysts still work as *specialist* and, work ineffectively when removing sulphure. Yet, this research is aimed to search for the alternative. A variation of concentration of Ni metal precursor is made. Co-impregnation and sequential impregnation method is performed. All catalysts obtained were verified according to the metal content, surface area, acidity, total pore volume and pore radius. According to this research the best catalyst preparation is sequential impregnation method by adding MoO_3, followed by NiO 2 (% w). This catalyst is expected to be multifunction and capable to remove the sulphure.

Keywords: NiMo/USY bimetal catalyst, co-impregnation method, sequential impregnation method.

1. Introduction

Hydro-desulphurization catalysts are important in petroleum industry[1]. Those help hydro-treatment reaction to undergo impurities removal. The sulfided-catalyst NiMo/Al_2O_3 and Co Mo/Al_2O_3 are of hydro-desulphurization catalyst that is very common in hydro-treatment reaction[1,4,6,7]. Indeed, those catalysts have performed a series of good catalytic performance to support hydro-treatment reaction. However, those are working as specialist and, ineffective when removing sulphure out of the produced petroleum[5].

Further researches have focused to replace Al_2O_3 as its inert state is of possible factors on causing such ineffective work. Replacing Al_2O_3 by USY has resulted a better catalyst for hydro-treatment[5]. However, unsolved problem about its failure to remove the sulphure does still exist.

Yet, this research is aimed to search the more effective-alternative catalyst for hydro-desulphurization and, provide a multifunction catalyst. In this research, Al_2O_3 is replaced by USY as according to[5].

2. Materials and Methods

2.1. *Materials*

The major material used in this research is NH4-USY that is provided by Tosoh Inc (Japan). Other materials are ammonium heptamolybdate and nickel nitrate hexahydrate provided by Merck (Germany). Both materials are the solution of salt-metal precursor to be supported toward the USY.

2.2. *Methods*

NH_4-USY was calcined at 550°C for 1 hour to obtain USY. Bimetal catalyst is synthesized by co-

†Work supported by *Intensif Riset Dasar* grant of the MENRISTEK.

impregnation method[5] and sequential impregnation method. Co-impregnation method was performed by providing different concentration of NiO (2, 4, 6, and 8% w). Sequential impregnation method was conducted trough two reversed-mechanisms. The first was by supporting the precursor of Ni, followed by ammonium heptamolybdate and the second is conducted trough the other opposite sequence.

All catalysts obtained then were verified according to several criteria of characters. The degree of crystalinity was analyzed using the XRD from Philips (MMCR mode). The metal content was measured by using atomic adsorption spectroscopy (AAS) from Perkin-Elmer. Total acidity was determined using ammonia adsorption method. Specific surface area was measured trough BET method using surface area analyzer from NOVA Data Analysis Package Ver.200.

3. Result and Discussion

Catalyst activity is determined by the chemical structure and its properties. In the hydrodesulphurization catalysts, the degree of acidity is one of characters that determining the catalytic performance. The higher degree of acidity often gives results to the higher catalytic activity respectively[3]. Despite degree of acidity, the catalytic activity is determined by many characters e.g. the metal content and, the surface area. The metal content and surface area is a simple and relatively more accurate indicator for the degree of active site dispersion. That is why, this research choosing those criteria to verify the catalysts obtained. In addition, pattern of XRD from obtained catalyst, totally pore volume and average pore radius are presented in order to complement the verification criteria of a good catalyst[3].

3.1. *The USY Characters*

The spectra of FTIR (see fig.1) from both NH_4-USY and USY are presented in comparison. This is in order to show the successfulness on removing NH_3 without changing the main structure of USY. Cation exchange then becomes facilitated, allowing more metal cation to be impregnated.

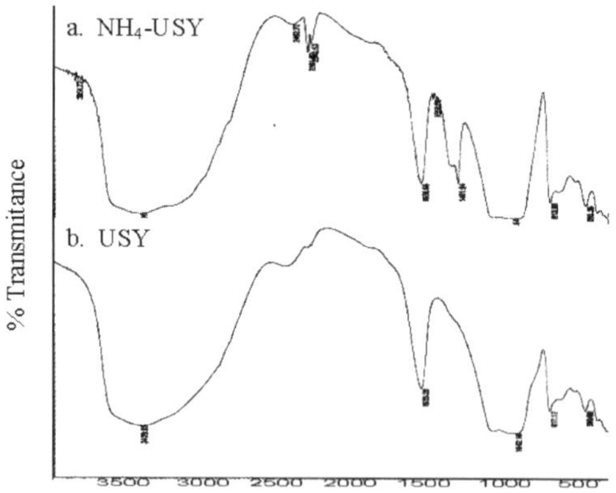

Fig. 1. Spectra of FTIR: a. NH4USY, b.USY.

Above FTIR, spectra tell that calcinations process at 550°C is successfully worked, causing the NH_3 to remove out of zeolite structure. This phenomenon causes the transformation of NH_4USY into H-USY/USY. That removal processes is also shown by several data below:

The successfulness on removing NH_3 molecules as shown by the spectra of FITR is supported by the data of XRD (fig.2) and, several characteristics of the supporting agent (table 1).

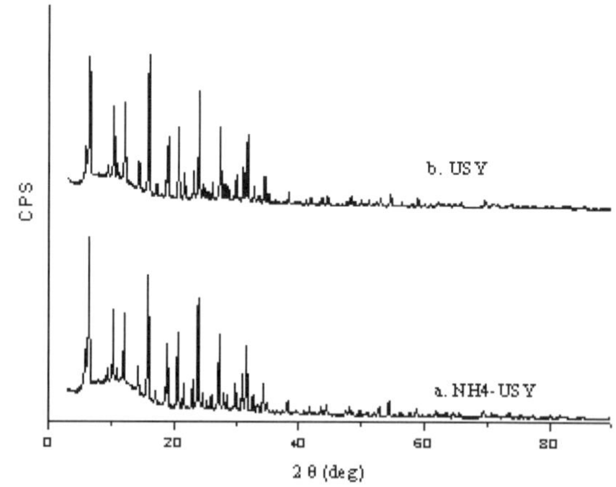

Fig. 2. Pattern of the XRD: a. NH_4-USY, b. USY.

Indeed, the XRD analysis towards USY (fig.2) is in order to identify the mineral constituent of the USY (this is performed trough assessing the d value as compare with the standard of JCPDS Powder Diffraction File) as well as the amount of it in percent. The results of XRD also tell similar phenomenon of NH_3 removal.

Table 1. Characteristics of the supporting agent.

Parameters	NH_4-USY	USY
Faujacite content (% w)	93	94
Acidity (mmol/g)	2.9	5.66
Specific surface area (m^2/g)	238	409
Totally pore volume ($10^{-3}m^3/g$)	111	208
Average pore radius (Å)	9	10
Oxide compounds (% w):		
• SiO_2	60	60
• Al_2O_3	15	15
• NiO	0.16	0.16
• MoO_3	0.42	0.42

Table 1 show that transformation from NH_4USY into USY has been followed by increasing several characters of such catalyst i.e. the specific surface area, totally pore volume and the average pore radius. These phenomenons's are because of NH_3 removal always results to the pore of USY to open.

Yet, specific surface area, totally pore volume and the average pore radius of the USY are tend to increased respectively.

3.2. *The Co-impregnation and Impregnation Methods in Comparison*

3.1.1. *The metal content*

Indeed, the metal content is one of criteria of a good catalyst. The metal content for each method of impregnation is presented in fig.3.

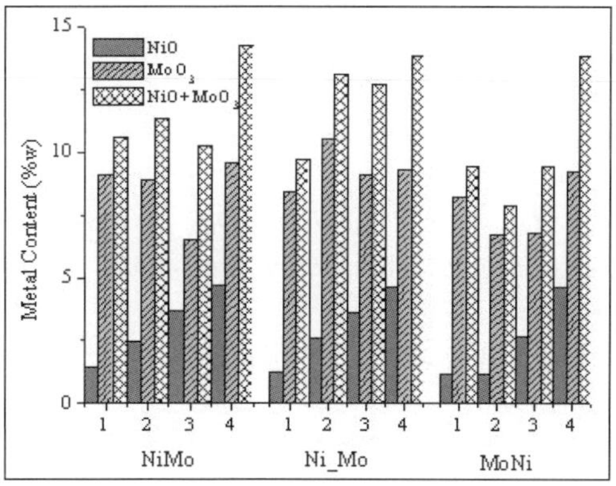

Fig. 3. Totally impregnated metal ($NiO+MoO_3$) for each impregnation method.

Generally, above figure tells that applying more concentration of nickel precursor tends to be followed by the increase of Ni content that is impregnated as well as of the total metal content ($NiO+MoO_3$). Yet, a given treatment of varied nickel concentration is considered as unsaturated. Hence all nickel concentration applied is effectively impregnated upon the USY. According to figure 3, sequential impregnation method by adding Mo precursor (8% w) followed by Ni (8% w) give result to the best character in term of its metal content. Indeed, this method is not allowing the competition mechanism between Mo and Ni to exist as occur in the co-impregnation method. Such sequential impregnation method is performing a greater metal content character. This is because of Ni is smaller than Mo; hence, Ni does easily impregnated upon the USY after Mo does.

3.1.2. Total acidity character

The acidity for each applied method and each given concentration of Ni are presented in fig.4.

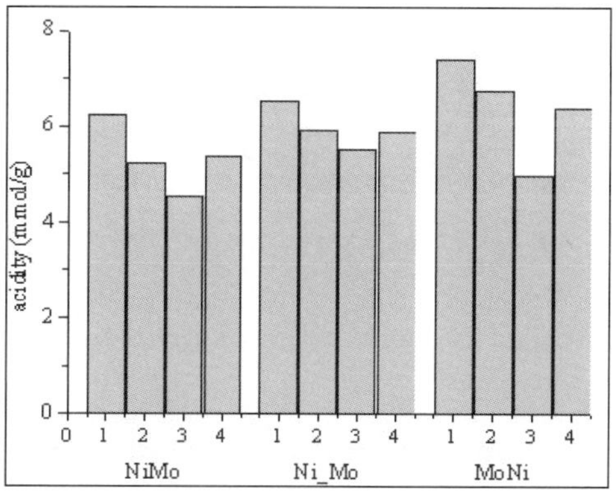

Fig. 4. The total acidity for each given treatment.

Figure 4 shows that the total acidity value is not always in line with total metal content. There is a phenomenon of unjust distribution of the impregnated metal upon the surface of USY. Sintering phenomenon does also occur. Such phenomenon causes the Ni less capable to donate its acid sites. According to figure 4, sequential impregnation method by adding Mo precursor (8% w) followed by Ni precursor (2% w) give result to the best character in term of its total acidity.

3.1.3. *Specific surface area*

The impregnation of metal precursor toward particular supporting agent is mainly, in order to increase its surface area[2,3]. However, this expectation is sometimes goes beyond the hope. The metal that is impregnated upon the surface area may cause negative externalities. The more metal is accumulated on the surface of supporting agent, the more pores of it will be covered. Hence, decreasing of the surface area is a main consequence. This is the sintering phenomenon[3]. This sintering phenomenon does occur in this research. Hence, all catalyst produced has smaller surface area as compare to USY. The specific surface area of the catalyst obtained in this research is diagrammatically presented in Fig 5.

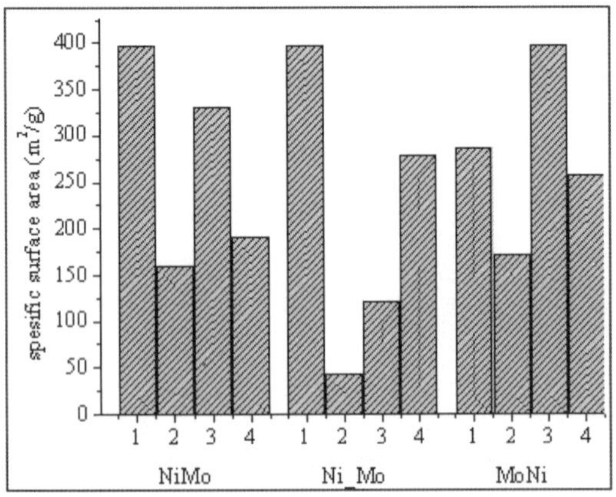

Fig. 5. The specific surface area of the produced.

3.1.4. *Total pore volume and average pore radius*

Specific surface area is closely related with total pore volume and average pore radius. Hence, the total pore volume and the average pore radius can also show a sintering phenomenon. The result of both, total pore volume and average pore radius is presented in table 2.

Table 2. Total pore volume and average pore radius.

No	Sample	V (10^{-3} cc/g)	R (Å)
1	USY	208	10
2	NiMo 1	193	10
3	NiMo 2	14	2
4	NiMo 3	111	7
5	NiMo 4	59	6
6	Ni1Mo	113	6
7	Ni2Mo	33	16
8	Ni3Mo	48	8
9	Ni4Mo	134	10
10	MoNi 1	138	10
11	MoNi 2	83	10
12	MoNi 3	189	10
13	MoNi 4	120	10

Based on table 2, three important points are identified namely:

1. Different metal concentration applied lead to the differences of total pore volume and average pore radius.
2. The total pore volume and average pore radius of produced catalyst are smaller than in the USY. Both metal (Ni and Mo) seems like to cover the pore of USY leading to the sintering phenomenon.
3. At the sequential impregnation MoNi/USY, the average pore radius is relatively similar with that of USY. However, the total pore volume is decrease.

4. Conclusion

Based on all results, the best catalyst preparation is sequential impregnation method by adding in the first MoO_3 8% w followed by Ni 2% w. The use of NiMo and USY is expected to be a multifunction catalyst as well as its capability in sulphure removal.

Acknowledgments

This reserach is funded by Insentif Riset Dasar 2007/2008 grant. Therefore, I would like to thanks to the MENRISTEK for funding this reserach.

References

1. Absi-Halabi, M., Stanislaus. A, Al-Dolama, K., *Fuel*, **77**, 787 (1998).
2. Augustine, R.L., *Heterogeneous Catalysis for the Synthetic Chemist*, Marcel Dekker Inc., New York (1996).
3. Hegedus, L.L., *Catalysts Design Progress and Perspectives*, John Wiley and Sons, New York (1987).
4. Isoda, T., Nagao, S., Ma, X., Korai, Y., Mochida, I., *Energy & Fuels*, **10**, 487 (1996).
5. Li, D., Xu H., Guthrie, G. D. Jr., *J. Cat*, **189**, 281 (2000).
6. Martinez, M. T., Benito, A. B., Callejas, M. A., Trasobares, S., *Energy & Fuels*, **12**, 365 (1998).
7. Sugioka, M., Kanda, Y., Kobayashi, T., Uemichi, Y., 2004 *J. Mem. Muroran Inst. Tech.*, **54**, 41 (2004).

PREPARATION AND CHARACTERIZATION OF NIMO/ACTIVE NATURAL ZEOLITE CATALYSTS*

KHOIRINA DWI NUGRAHANINGTYAS

Department of Chemistry, Faculty of Maths and Natural Sciences, Sebelas Maret University, Jl Ir Sutami No 36 A Kentingan Surakarta, 57126, Indonesia

DIAN MARUTO WIDJONARKO

Department of Chemistry, Faculty of Maths and Natural Sciences Sebelas Maret University, Jl Ir Sutami No 36 A Kentingan Surakarta, 57126, Indonesia

WEGA TRI SUNARYANTI, TRIYONO

Department of Chemistry, Faculty of Maths and Natural Sciences Gadjah Mada University, Sekip Utara Yogyakarta, 55281, Indonesia

Most previous research on catalyst's preparation of NiMo/ANZ (active natural zeolite) catalyst has long been done by focusing on one simple method of impregnation. In this research co-impregnation method and sequential impregnation method are used. The aim of this research therefore, is to find the best performance catalyst which is reflected by its characters. For each method, four categories of NiO concentration and single concentration of MoO_3 are applied, followed by standard method of preparation i.e. calcinations, oxidation and reduction. The performance of catalyst produced then verified by several criteria of characteristics i.e. mineral composition, metal content, and acidity. Based on the previous criteria, catalyst Ni2Mo/ANZ performs many ideal criteria.

Keywords: NiMo/ANZ bimetal catalyst, catalyst's preparation, co-impregnation method, sequential impregnation method.

1. Introduction

Many researches on catalyst HDS has published by many authors[1,2,4]. Those researches suggest that compatible metal combination as well as the supporting agent is needed to achieve the best catalytic performance of the HDS. Indeed, most HDS catalyst uses a combination of $NiMo/Al_2O_3$ and $CoMo/Al_2O_3$[1,4,6]. $NiMo/Al_2O_3$ has a better catalytic performance when it used on petroleum industry 6. Despite its good catalytic performance however, using Al_2O_3 is considered as *specific*. This is because of the inert state of the Al_2O_3, which make this supporting agent less contributes to donate the active site during catalytic reaction. Yet, this research provides natural zeolite as an alternative altering the Al_2O_3.

The use of USY, NaY zeolite, mordenit and ZSM-5 as supporting agent toward bimetal NiMo are already studied[3]. The results showed that NiMo/USY has well demonstrated as having good catalytic activities during HDS reaction. Both, Ni and Mo work in a synergetic manner. In this research, natural zeolite is used altering the USY zeolite. Indeed, this research uses natural zeolite as it has better acidity performance than the USY[7].

2. Materials and Methods

2.1. *Materials*

Local natural zeolite is the main materials used in this research. HCl, Ammonium Heptamolybdat and nickel nitrate Hexahydrate (MERCK; Germany) are used as reagent.

2.2. *Methods*

Local natural zeolite was activated using HCl (6M). The zeolite was heated at 130°C in oven to produce its active state (ANZ).

Active catalyst NiMo/ANZ was prepared by supporting ANZ toward bimetal (NiMo). This, is performed trough two mechanisms i.e. the Sequential

*This work is supported by PEKERTI grant 035/SP2H/PP/DP2M/ III/2007 of the Department of Human resources Project, DIRJEN DIKTI DIKNAS.

impregnation method and the simple co-impregnation method. Indeed, this research tries to compare those two mechanisms to find the best catalytic performance. The co-impregnation method used in this research is performed by making a variation on concentration of NiO from 2 to 8% w[5]. As pertain to this research, sequential impregnation is performed by both, supporting the precursor of salt-metal solution of nickel nitrate Hexahydrate (MERCK; Germany) first, followed by adding the precursor of salt-metal solution of Ammonium Heptamolybdat and, supporting the precursor of salt-metal solution of Ammonium Heptamolibdat first, then followed by adding the precursor of salt-metal solution of nickel nitrate Hexahydrate. Those two mechanisms are also compared.

All catalyst synthesized is characterized by several criteria i.e. the degree of crystalline using XRD analysis (MMCR Philip's x-ray), the metal content analysis using atomic-absorbent Spectroscopy (Perkins Elmer), the total acidity value and, the surface area (BET method) using surface area analyzer NOVA data Analysis package Ver.200.

3. Result and Discussion

3.1. *Catalyst for Hydrodesulphurization*

Most hydrodesulphurization catalysts are characterized by several standard of measurement. The value of acidity of the catalyst used has been well known to affects the catalytic activity. Significantly, the increase of the value of acidity tends to be followed by the increase of its catalytic activity respectively. Despite that degree of acidity, the high degree of dispersion has also contributed to affects the catalytic activity trough their contribution as donor for the active sites. That high degree of dispersion corresponds with the larger amount of metal content and, corresponds with the larger total surface area of the catalysts. Yet, this research focuses on those standards of measurement to assess the bimetal catalyst NiMo/ANZ produced. Despite those assessments, however, the additional data (the patterns of XRD) are also presented.

3.2. *The Active Natural Zeolite*

The use of HCl is in order to remove any impurity from the natural zeolite frame which then may lead to its active state, the active natural zeolite (ANZ). The XRD analysis toward natural zeolite and the ANZ are presented in fig.1.

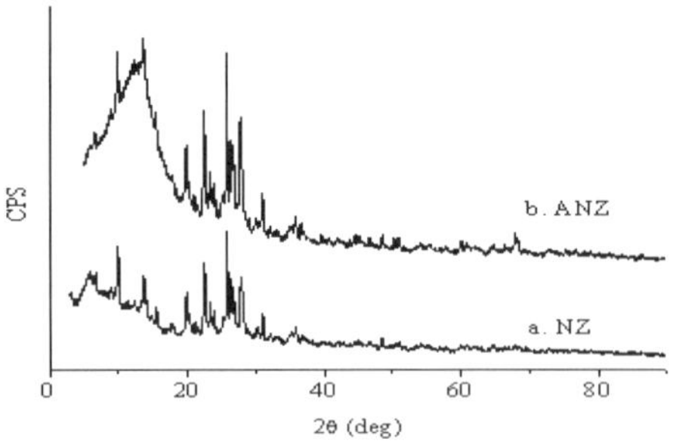

Fig. 1. The result of XRD sample (a) NZ, (b) ANZ.

HCl successfully removes impurities from the natural zeolite frame. That removal allow the catalytic site of ANZ to open, making this catalyst reach its active state. Removal process also increases degree of crystal as well. The increase of crystal degree is shown by the emerging mordenit's peak that tends to rise as according to Fig.1. An active state of the catalyst then reached.

Table 1. Characteristics of ANZ as compared to NZ.

Parameters	NZ	ANZ
Acidity (mmol/g)	1.42	1.72
Oxide compounds (% w):		
• SiO_2	60.27	64.80
• Al_2O_3	12.79	11.99
• CaO	3.12	1.84
• Na_2O	1.68	0.88
• K_2O	1.02	1.00

Indeed, this active catalyst is not only verified by degree of crystalline criteria. That catalyst must also meet with common standard of characters as already presented. Table 1 show several characters of ANZ as compared to natural zeolite (NZ)

Removing impurities has caused the increase of acidity value on the ANZ after activation (see table 1). The hydrogen replacement against metal cation also contributes to this increase. Despite its impact on the value of acidity, activation process is also increasing the surface area. Large amount of impurities oxide compounds (CaO, Na_2O and K_2O) are also removed. Yet, XRD analysis towards ANZ shows that activation process is successfully remove most of impurities.

3.3. *The Characteristic of Catalyst from Co-impregnation Method*

A variation of NiO concentration is made during performing the co-impregnation method. All catalysts produced from each variation of NiO concentration is analyzed by XRD, see figure 2.

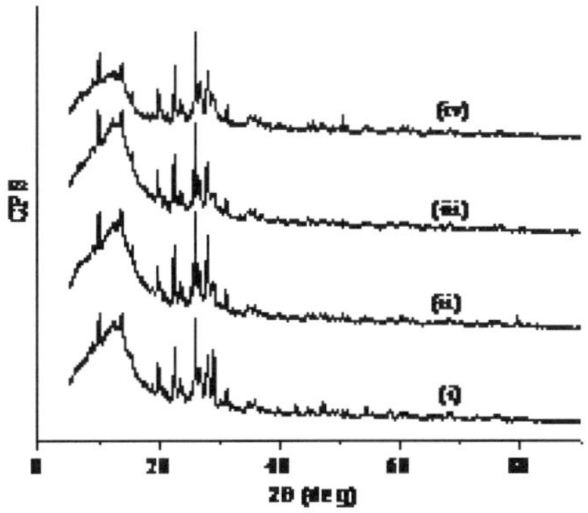

Fig. 2. The XRD analysis of (i) NiMo/ANZ 1, (ii) NiMo/ANZ 2, (iii) NiMo/ANZ 3, (iv) NiMo/ANZ 4.

Fig 2 looks similar with fig.1. According to both figures, no different pattern and intensity is resulted. Only slightly different on d 10.510 Å – 14.018 Å and d: 3.324 Å, and 3.082 Å, and in some planes that fit with the alloy of Ni and Mo. Yet, it guesses that most of the Ni and Mo supported into the ANZ are in an alloy state.

3.4. *Characteristic of Catalysts Resulted from Sequential Impregnation Method*

This research is also aimed to compare which the sequential impregnation method is resulting the best characters of catalyst. The result of sequential co-impregnation method where the precursor of salt-metal solution nickel nitrate Hexahydrate is firstly supported presented in figure 3.

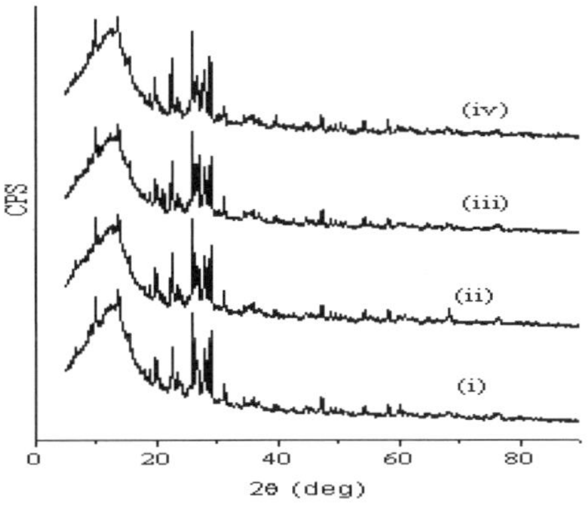

Fig. 3. The XRD Pattern of (i) Ni1Mo/ANZ; (ii) Ni2 Mo/ANZ; (iii) Ni3Mo/ANZ dan (iv) Ni4Mo/ANZ.

Fig. 3 shows similar pattern as shown by fig.2. Ni and Mo is found in alloy state. There is only slightly different in term of its intensity.

The result of sequential impregnation method where the precursor of salt-metal solution of Ammonium Heptamolybdate is firstly supported first is presented in fig 4.

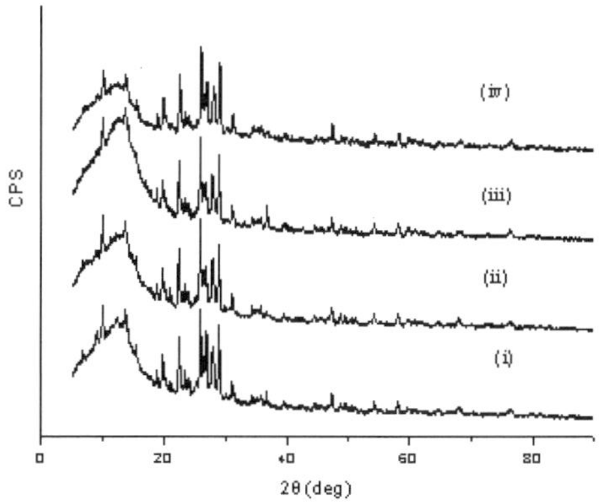

Fig. 4. The XRD Pattern of (i) MoNi/ANZ 1; (ii) MoNi/ANZ 2; (iii) MoNi/ANZ 3 and (iv) MoNi/ANZ 4.

Both, fig 3 and fig 4 shows that Ni and Mo is in alloy state. According to fig 4 only small amount of Ni and Mo is found in a monometal state.

The content of metal-oxide content (NiO and MoO_3)) in those three methods are presented diagrammatically (see figure 5).

Fig.5 indicates that total metal content (NiO + MoO_3) is considered to increase as the precursor NiO is added. Ni The capacity for cation exchange is greater in Mo than Ni. Therefore, Ni will tend to decrease if Mo is added into ANZ.

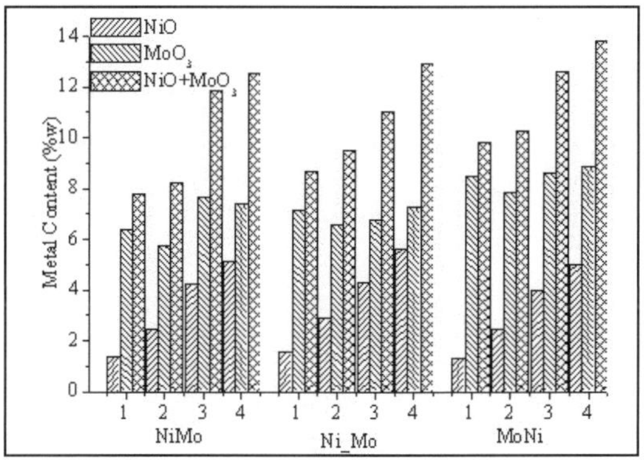

Fig. 5. Metal content (all catalysts are presented).

In this research, catalyst's acidity is assumed to be similar for every Ni added. This is shown by fig.6.

Fig. 6 shows that catalyst produced by using co-impregnation method performing a particular phenomenon. The increase on NiO compound is followed by the increase of acidity respectively. This phenomenon is absent in the sequential impregnation method.

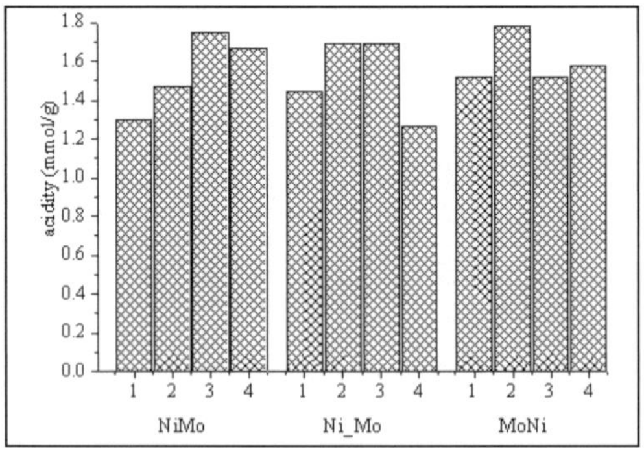

Fig. 6. Total acidity (all catalysts are presented).

There are three reasons to answer why the acidity is increased in line with the amount of NIO. The explanations are: Unequal metal dispersed on the surface of ANZ, a case of sintering and, the presence of alloy state of the metal. The surface area then analyzed in order to get the exact explanation. The result of the analysis is presented in fig 7.

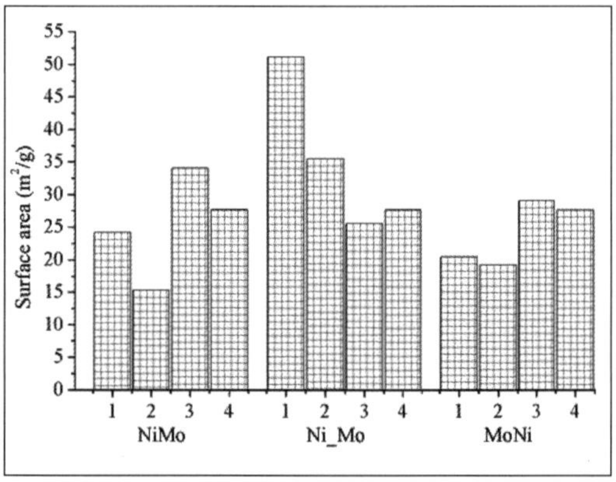

Fig. 7. The surface area (all catalysts are presented).

Fig. 7 shows that sintering case and the presence of alloy is affecting the catalyst's acidity.

Indeed; crystalline, metal content, acidity, and surface area are characters of a good catalyst. However, as pertain to this research, not all criteria are fulfilled. Yet, the best catalyst is determined by looking that the at the average value for each characters. Yet, Ni2Mo/ANZ regards as the best catalyst.

4. Conclusion

Best catalyst performance is produced on the sequential impregnation method by adding Ni, followed by Mo.

Acknowledgments

This reserach is funded by PEKERTI grant 035/SP2H/PP/DP2M/ III/2007. Therefore, I would like to thanks to the Department of Human resources Project, DIRJEN DIKTI DIKNAS for funding this reserach.

References

1. Absi-Halabi, M., Stanislaus. A, Al-Dolama, K., *Fuel*, **77**, 787 (1998).
2. Isoda, T., Nagao, S., Ma, X., Korai, Y., Mochida, I., *Energy & Fuels*, **10**, 487 (1996).
3. Li, D., Xu H., Guthrie, G. D. Jr., *J. Cat*, **189**, 281 (2000).
4. Martinez, M. T., Benito, A. B., Callejas, M. A., Trasobares, S., *Energy & Fuels*, **12**, 365 (1998).
5. Nugrahaningtyas, K.,D., Widjonarko, D.,M.,Muhammad F.J.P, Wijaya, D., *Proseding Konggres dan Simposium ke II MKICS,* **A3**, 76 (2007).
6. Steiner, P., 2002, Kinetic and deactivation studies of hydrodesulfurization catalyst, *Dissertation*, The Norwegian University of Science and Technology, Norwegia.
7. Trisunaryanti, W., *Skiyu Gakkaishi (J.P.I)*, **39**, 20 (1996).

MULTI-OBJECTIVE OPTIMIZATION FOR THE DESIGN OF A RIVER MONITORING NETWORK

HUEI-TAU OUYANG
*Civil Engineering, National Ilan University, 1, Sec. 1, Sheng_Lung Rd.
Yilan City, 26047, Taiwan*

JIHN-SUNG LAI
*Hydrotech Research Institute, National Taiwan University, 1, Sec. 4, Roosevelt Rd.
Taipei, 106, Taiwan*

HSIN YU
*Civil Engineering, National Ilan University, 1, Sec. 1, Sheng_Lung Rd.
Yilan City, 26047, Taiwan*

CHENG-PIN LIN
*Civil Engineering, National Ilan University, 1, Sec. 1, Sheng_Lung Rd.
Yilan City, 26047, Taiwan*

The efficiency of a river pollutant detection network is strongly related to the numbers and locations of the sampling stations in the network. A methodology by means of a multi-objective optimization procedure for the determination on both the optimal numbers and spatial distribution of a river sampling network is described. The methodology is based on Pareto Analysis and Genetic Algorithms (GAs). The performance of the developed methodology is evaluated by a case study on a hypothesized river system consisting of 20 reaches. A set of optimal designs is obtained which provides useful information for the decision making in river monitoring network designs.

1. Introduction

One major objective of river monitoring program in a river system is to regularly screen the water quality and to locate the pollutant source as soon as possible if a pollution is detected in the flow. For that, it is best to have as many sampling stations in the monitoring network as possible. However, due to budgetary constraints, the numbers of the sampling stations are usually limited. With costs in consideration, a question arises on how to decide the number of sampling stations in a network that will suffice the surveillance function yet without much sacrifice on the pollutant detection efficiency. To answer the question, the selection of the spatial distribution for the sampling stations becomes a decisive factor, since the efficiency of the monitoring network is strongly related to the location of the sampling stations. This study aims to develop a methodology based on a multi-objective optimization analysis to determine the optimal numbers of the monitoring stations and to locate the optimal spatial distribution of these stations in a river monitoring network such that the network has high efficiency on detecting a point source of pollution in a river system.

2. Problem Formulation

Following Sanders et al.'s suggestion (1983), as the first stage of design for the spatial distribution of a monitoring network, this study will focus on the determination of the macrolocations only, i.e., to decide which river reach, or link, of a river should place a sampling station. For simplicity, two assumptions have been made: (1) the flow is fully mixed at the downstream end of each river reach, and (2) there is no backwash upstream into the river and the pollutant materials flow downstream only. Based on these assumptions, the sampling stations will be placed at the downstream end of a reach, immediately before the next confluence. Also, one sampling station is always located at the outlet of the river system ensuring that all the river reaches are within the surveillance range of the monitoring network.

2.1. *Geometrical data of the river reaches*

Basic geometrical data for the river system is first extracted for later use in the optimization process, including (1) the connection relations of the river reaches in the river system; and (2) the lengths of each river reaching the system.

Following Ouyang et al.'s procedure (2008), each river reach is labeled with an integer serial code number. The length of each reach and the associated serial code number for the downstream link are then recorded and exported as an ASCII file to serve as a data base for the optimization process. Table 1 gives an example for the data structure extracted from a simple hypothetical river system shown in Figure 1.

2.2. *Objective functions*

Two costs are involved in the optimization problem: namely (1) the construction cost and (2) the operation cost. The construction cost is the required cost for setting up the monitoring stations of the network at the beginning of the surveillance program. This cost is related to the numbers of the monitoring stations in the surveillance network. Also, in practical situations, it is not possible to have every river reach monitored and usually one sampling station has to monitor a number of river reaches. Thus, if pollution is detected at a sampling station, there will be an operation cost for the subsequent investigation in the area monitored by the station to locate the pollutant source. With more sampling stations in the network, the construction cost will increase, but the subsequent operation cost will generally decrease. The situation goes vice versa. Therefore, these two costs are basically contradicted to each other.

The expected construction cost is defined in Eq. (1). For simplicity, it is assumed that the construction cost for each monitoring station is the same.

$$E(construction\ cost) = n_s W \tag{1}$$

where n_s is the number of the monitoring stations; W is the construction cost for each station.

The operation cost is related to the works or samplings required to locate the pollutant source if pollution is detected by a monitoring station. Following Dixon et al.'s study (1999), the operation cost is defined as follows.

$$E(operation\ cost) = \sum_i (L_i / L_0) \log_2 n_i \tag{2}$$

In which n_i and L_i are the numbers and the accumulative length of the river reaches in the surveillance range of the ith sampling station, respectively; L_0 is the total length of all reaches in the river system.

On economical point of view, both the construction cost and the operation cost should be as low as possible. To decide the weighting between these two design goals, specific information about the construction cost and operation cost is required. This information is usually varied in different cases. In this study, instead, a Pareto analysis is utilized to search for a set of optimal solutions for the problem (the Pareto Front). Although the Pareto analysis does not give a single answer, it might be of help for the river authority to make a rational decision on the number of monitoring sites.

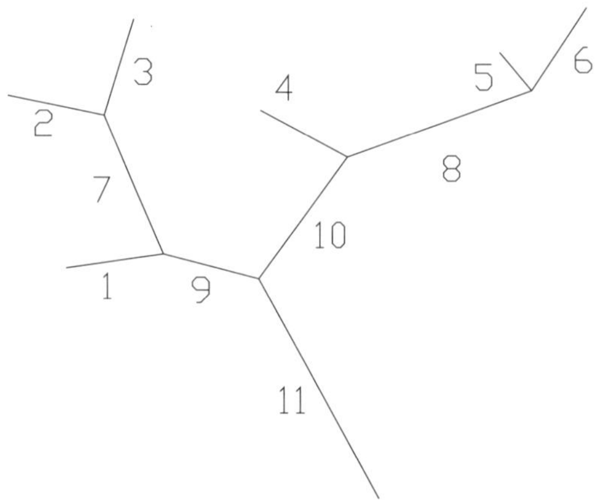

Figure 1. A hypothetical river system comprising 11 reaches.

Table 1. Example geometrical data extracted from the river system in Figure 1.

Reach No.	Downstream link	Reach Length
1	9	2
2	7	2
3	7	2
4	10	2
5	8	1
6	8	2
7	9	3
8	10	4
9	11	2
10	11	3
11	0	5

3. Pareto Analysis

The number of sampling stations is enumerated from 1 to n_0. For each number of sampling stations, the optimal spatial distribution is search for by a Genetic Algorithm (Goldberg, 1989). Following Ouyang et al. (2008), the integer serial code numbers of the selected river reaches which will be a monitoring site are codified into a binary chromosome. The performance of each candidate monitoring network is evaluated by Eq. (2). The next generation of monitoring networks is then generated according to the performance of the monitoring networks in present generation. The population of monitoring networks evolves and gradually improves their performance. After several generations of evolution, the optimal monitoring network might eventually emerge out of the others. The iteration procedure of the search is as follows:

Step 1. The chromosomes for the first generation of monitoring networks are generated randomly.

Step 2. The chromosome segments for each candidate monitoring network are then decoded back to the N_s integer serial code numbers of the river reaches selected as monitoring sites.

Step 3. For each candidate monitoring network, the river system is divided into several subcatchments based on the spatial distribution of the monitoring sites. The accumulative length of the river reaches in each subcatchment is estimated based on the geometrical data of the river system extracted by GIS.

Step 4. The performance of each candidate monitoring network is then evaluated using Eq. (2).

Step 5. The chromosomes of the next generation of monitoring network are then generated by applying the GA operators based on the performance of the monitoring networks in the current generation.

Repeat Step 2 to Step 5 until the stopping criteria are reached.

With the number of sampling stations enumerated, the optimal spatial distribution for each network is located by GA search. The iteration procedure for the Pareto analysis is illustrated in Figure 2.

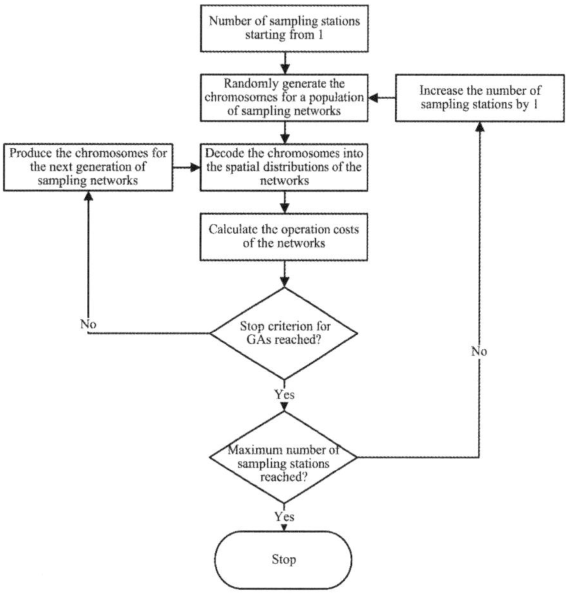

Figure 2. Flow chart for the Pareto analysis.

4. Case Study

A case study on a hypothesized river system consisting of 20 reaches was undertaken to demonstrate the use of the methodology in river monitoring network design. The river system is shown in Figure 3. The probability of crossover and mutation for the GA runs are set to be 0.6 and 0.01, respectively. The iteration stops if there are no changes for 100 consecutive generations.

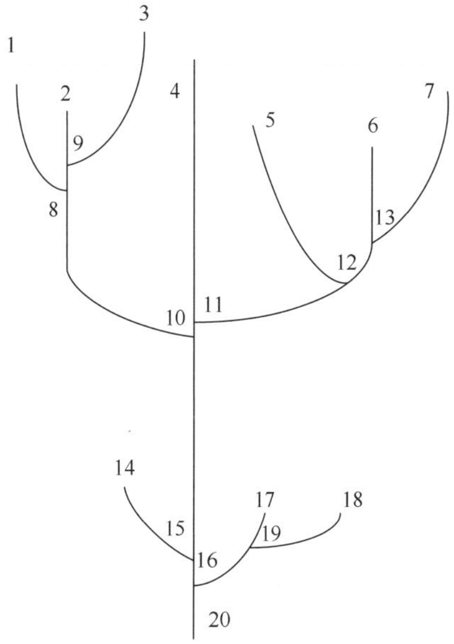

Figure 3. A hypothesized river system consisting of 20 reaches.

The set of optimal solutions obtained by the Pareto analysis is shown in Figure 4. It is seen that the expected operation cost is inverse-proportional to the expected construction cost. More sampling stations require higher E(construction cost) at the beginning of the monitoring program but has lower E(operation cost) in the subsequent investigation. The decision on the exact number of sampling stations involved in the monitoring program will rely on a further detailed economic analysis. However, Figure 4 has revealed certain information which is useful for the decision making in the first stage of design. As seen, when the expected construction cost (number of sampling stations) is larger than 9, the expected operation cost has dropped down to less than 1. The expected operation cost represents the water quality samplings required to locate the pollutant source. In reality, the number of water quality samplings is always an integer. Therefore, for a monitoring network with more than 9 sampling stations, the improvement on pollutant detecting efficiency by adding more sampling stations will be relatively small.

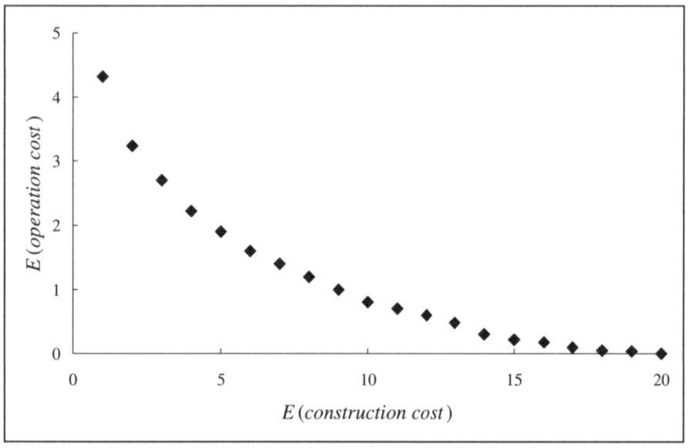

Figure 4. Pareto front of the optimization problem.

On the other hand, when the number of sampling stations is less than 5, the expected operation cost is higher than 2 and is increasing drastically. Therefore, the pollutant detecting efficiency for a monitoring network with less than 5 sampling stations will be relatively low, and the efficiency deteriorates fast when the number of sampling stations is decreased. Based on the above analysis, a rational suggestion on the number of sampling stations in this case study will be between 5 to 9. As an example, the optimal spatial distribution of a monitoring network with 6 sampling stations is presented in figure 5. The expect operation cost for this network to locate the pollutant source is 1.6.

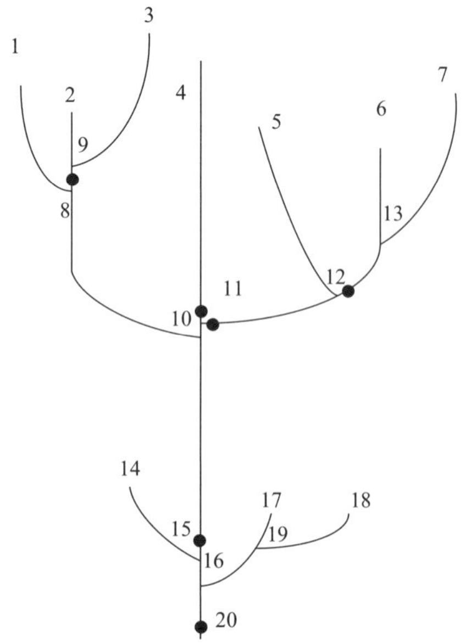

Figure 5. The optimal spatial distribution for a network with 6 sampling stations.

5. Conclusions

A methodology based on the Pareto analysis and a Genetic Algorithm to search for the optimal number and locations of the sampling stations in a river monitoring network for pollutant source detection in a river system is described. A case study is presented to demonstrate the use of the methodology for the design of a monitoring network. The results provide first stage information for the decision making on the number of the sampling stations. The optimal spatial distributions for different number of sampling stations are also obtained by utilizing Genetic Algorithm. The plausible results suggest the possibility of utilizing the methodology at the early design stage of a river monitoring network.

Acknowledgments

This research is based upon work supported by the National Science Council, Taiwan, Grant NSC 98-2625-M-197-001.

References

1. T.G. Sanders, R.C. Ward, J.C. Loftis, T.D. Steele, D.D. Adrian, and V. Yevjevich, *Water Resources Publications, Littleton, CO.* (1983).
2. H.T. Ouyang, H. Yu, C. Lu and Y. Luo, *Journal of Water Resources Planning and Management, ASCE*, **134(1)**, 83 (2008).
3. W. Dixon, G.K. Smith, and B. Chiswell, *Water Research*, **33(4)**, 971 (1999).
4. D.E. Goldberg, *Genetic algorithms in search, optimization, and machine learning*, Addison-Wesley Reading, MA (1989).

SOLUBILITY MEASUREMENTS OF IBUPROFEN IN SUPERCRITICAL CO_2 BY A DYNAMIC METHOD*

MEHRDAD MIRZAJANZADEH

Department of Chemical Engineering, Islamic Azad University, Tehran South Branch, Tehran, Tehran state, Iran

MEHDI ARDJMAND

Department of Chemical Engineering, Islamic Azad University, Tehran South Branch, Tehran, Tehran state, Iran

FATEMEH ZABIHI

Department of Chemical Engineering, Islamic Azad University, Ayatollah Amoli Branch, Amol, Mazandaran state, Iran

We used a dynamic apparatus to determine the solubility of ibuprofen in supercritical carbon dioxide at two temperatures, 35 and 40°C and in the pressure range from 80 to 130 bar and the solubility curves was generated. We used Mendez-Santiago-Teja equation to check the consistency of experimental data. Peng-Robinson thermodynamic model was used to predict the obtained experimental data on solubility of ibuprofen in supercritical CO_2 system, which showed a good accordance with our experimental results (%AARD=16.58 at 35°C and %AARD=11.62 at 40°C). In addition to thermodynamic model, we used a group contribution method as well to estimate the interaction parameter in Van der Waals mixing rule (k_{12}). Also, the Lydersen group contribution method, was used in predicting of the physicochemical and critical properties of the ibuprofen.

1. Introduction

A supercritical fluid has both liquid-like and gas-like characters. In the supercritical region, molecules are tiny enough to penetrate into any other media and occupy minimum spaces. So, in spite of their liquid-like density, supercritical fluids are intensively compressible. That is with a little change in pressure and temperature, a significant change in density and solubility power can be accrued.

Actually, operational conditions as temperature and pressure, can be controlled for a supercritical fluid easily [1,2,3]. Taking in to these abilities, supercritical fluids' technology presents a new and interesting approach for many of the applications such as nano-particles formation, extraction, separation, purification and chemical reactions [4].

2. Experimental method and chemical used

2.1. *Materials*

Ibuprofen (Sina Daru, 99.99% purity) was used as original compound and CO_2 (Roham Gaz, 99.95%) was also used as solvent. Ethanol (99.8%, Sigma Aldrich) used in analytical grade form.

2.2. *Solubility measurements*

We used a dynamic apparatus to determine the solubility (Fig. 1). Condensed CO_2 was fed to an HPLC pump through a 0.5μm filter, to reach the desired pressure. The compressed fluid entered into a preheating coil contained in a temperature-controlled electronic oven to achieve thermal equilibrium and passed into a solubility cell that was packed with pure Ibuprofen powder. The solute laden purged to the atmosphere in order to ensure the operational condition consistency. The sampling part was a 1/8 inch stainless steel coil installed with two unilateral needle valves. This pass was kept closed during the purging step. After three minutes the purge valve was closed and the needle valves were opened orderly. A sample of supercritical solution was trapped into the sampling pass by closing the needle valves.

The sampling equipment (coil and valves) then separated from the system and marinated in the pure ethanol. The valves were opened then and the CO_2 was

*This work is supported by Islamic Azad University, Tehran South Branch.

vented out and all the solute was washed out by ethanol. The ethanol-Ibuprofen solution was prepared for GC test to determine the solute amount. The used CO_2 amount was calculated by measuring the volume of the sampling equipments and the CO_2 density in the operational pressure and temperature.

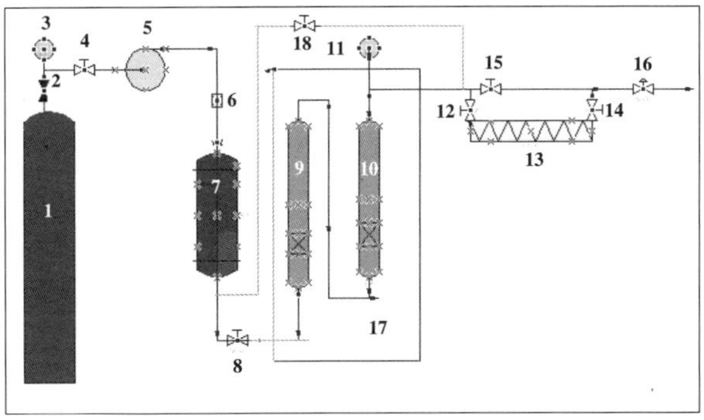

Figure 1. Schematic diagram of the experimental apparatus 1) gas tank, 2) ball valve, 3,11) pressure gauge, 4,8,12,14,15,18) needle valve, 9,10) equilibrium sells, 5) pump, 6) filter, 7) preheater, 16) throttling valve, 17) constant temp. bath.

3. Results and discussion

3.1. *Experimental results*

The solubility of ibuprofen was measured at pressures in the range of 80-130 bar, and temperatures of 35 and 40°C.

With this experimental region, the mole fraction of ibuprofen in fluid phase was in the range from 0.0428 × 10-3 to 3.261 × 10-3. The effect of pressure on the solubility is illustrated in Fig. 2, and it can be seen that the solubility increased as the pressure was increased. On the other hand, at constant temperature the solubility increases with the pressure and this effect is greater for 40°C than 35°C.

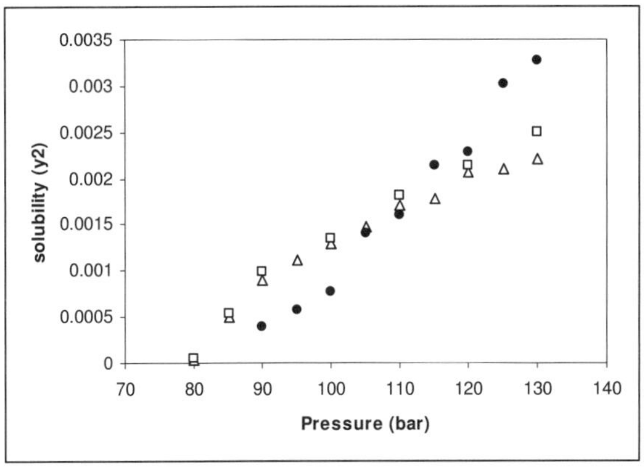

Figure 2. Solubility isotherms of ibuprofen in supercritical CO_2.

Fig. 2, also, includes solubility data found in the literature for ibuprofen by Foster et al. at 35°C.

As pressure is increased, carbon dioxide density increases and the intermolecular mean distance of carbon dioxide molecules decreases. Therefore, the specific interaction between the solute and solvent molecules is increased. The crossover region for the ibuprofen-CO_2 system is in the range of pressure between 105 and 115 bar.

3.2. *Modeling with a density-based correlation*

In order to check the consistency of experimental data of solubilities of ibuprofen in SC carbon dioxide as a function of absolute temperature and solvent density, we used semi-empirical Mendez-Santiago-Teja equation (Fig. 3).

The Mendez-Teja equation can be expressed as [5]:

$$T \ln E = T \ln \frac{y_2 P}{P_2^{sub}} = A + B\rho_1 \qquad (1)$$

where T is absolute temperature, E is the enhancement factor, y_2 is the solute mole fraction, P is the total pressure, P_2^{sub} is the sublimation pressure of the solute, ρ_1 is the density of the supercritical fluid, and A and B are regressed constants.

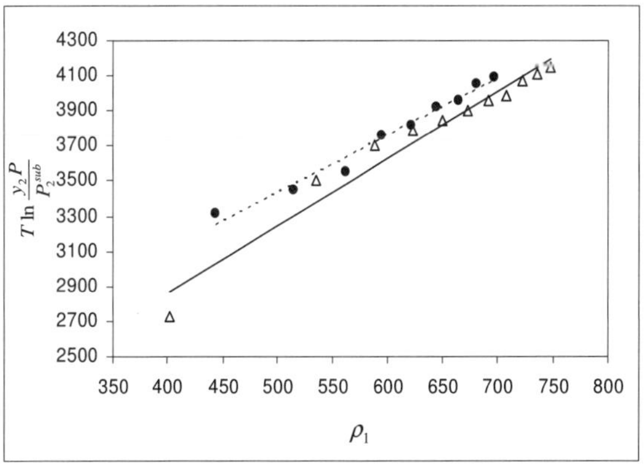

Figure 3. Modeling of solubility of ibuprofen in CO_2 using a density based correlation for 35 (Δ –) and 40°C(• · · ·).

Mendez-Santiago-Teja equation parameters along with correlation facto r(R^2) have been shown in Table 1.

Table 1. Mendez-Santiago-Teja equation constants along correlation coefficient at 35 and 40°C.

constant	Temperature	
	35°C	40°C
A	1332.2	1803.9
B	3.8278	3.2592
R^2	0.9612	0.9784

3.3. *Modeling with a cubic equation of state*

If the solid phase does not dissolve the SC-CO_2, the solubility of the solute in the CO_2-rich fluid phase at equilibrium can be calculated using the isofugacity criteria [6]:

$$y_2 = \frac{P_2^{sub}}{P} E \qquad E \equiv \frac{\exp\left(\frac{v_2^s (P - P_2^{sub})}{RT}\right)}{\phi_2^{SCF}} \qquad (2)$$

In these equations, P is the equilibrium pressure, T the equilibrium temperature, P_2^{sub} the sublimation pressure of the solute, ϕ_2^{SCF} the fugacity coefficient of the solute in the fluid phase, E the enhancement factor and v_2^s is the saturated molar volume of solid solute at the equilibrium temperature. The Peng-Robinson EoS has been selected to calculate ϕ_2^{SCF}, because it has proved to combine the simplicity and accuracy which are required to model the solubility of some solids in SCFs.

The Van der Waals mixing rules may be used to calculate the mixture parameter 'a' and 'b' in PR-EoS (Table 2).

In Table 2, we have summarized the cubic EoS and the mixing and combining rules and fugacity coefficient (ϕ_2^{SCF}) used in this work.

The Correlation is performed by minimizing the objective function average absolute relative deviation (AARD) defined by usually as [7]:

$$\text{AARD}(\%) = \frac{100}{\text{N}} \sum_{i=1}^{N} \frac{\left| y_{i,calc} - y_{i,ex} \right|}{y_{i,ex}} \qquad (3)$$

where N is the number of experimental data.

Also, the Lydersen group contribution method [8], was used in predicting of the physicochemical and critical properties of the ibuprofen. Physicochemical and critical properties of the ibuprofen and CO_2 are shown in Table 3.

Literature values for the interaction parameters were obtained from the work by Haselow [9] who correlated experimental measurements of solubilities of a variety of compounds with the Peng-Robinson equation of state and standard mixing rules.

Table 2. Summary of the cubic EoS, mixing and combining rules and fugacity coefficient.

Peng-Robinson EoS
$P = \frac{RT}{v-b} - \frac{a(T)}{v(v+b)+b(v-b)}$
$a(T) = \frac{0.45724 R^2 T_c^2}{P_c} \times \alpha(T_r, \omega)$
$\alpha(T_r, \omega) = \left[1 + K\left(1 - \sqrt{T_r}\right)\right]^2$
$K = 0.37464 + 1.54226\omega - 0.26992\omega^2$
$b = \frac{0.07780 RT_c}{P_c}$
VdW1 mixing and combining rules
$a = a_{11} y_1^2 + a_{22} y_2^2 + 2a_{12} y_1 y_2$ $b = b_{11} y_1^2 + b_{22} y_2^2 + 2b_{12} y_1 y_2$
$a_{12} = (1 - k_{12})\sqrt{a_{11} a_{22}}$ $b_{12} = \frac{(b_{11} + b_{22})}{2}$
fugacity coefficient
$\ln \phi_2^{SCF} = -\ln\left(\frac{P\left(ZRT/P\right) - B}{RT}\right) + \frac{Z-1}{B}(B_{22}) - \frac{1}{2\sqrt{2}RT} \ln \frac{v + B(1+\sqrt{2})}{v + B(1-\sqrt{2})} \times \left[\frac{2A_{22} y_2 + 2\sqrt{A_{11} A_{22}}(1 - k_{12})(1 - y_2)}{B} - \frac{A}{B^2}(B_{22})\right]$

Table 3. Physicochemical and critical properties of the ibuprofen and CO_2.

Substance	$T_c(K)$	$P_c(MPa)$	ω
Ibuprofen[a]	749.7	2.33	0.819
CO_2	304.2	7.38	0.239
v^s	$P^{sub}(Pa) \times 10^{-2}$		
$\times 10^6 m^3 mol^{-1}$	35°C	40°C	45°C
182.1	4.95	8.97	1.60
[a] Estimated by Lydersen method			

These values were used to determine the group contributions in the simple model as:

$$k_{12} = c + \sum_{i=1}^{k} \Delta_i \qquad (4)$$

The estimated group contributions, Δ_i, are tabulated in Table 4. These values are for the PR EoS where CO_2 is the solvent.

Due to the limited data set and no variation of temperature incorporated in the model, the contributions for several chemical groups are grouped together as one and the calculated interaction parameters are assumed independent of temperature. The parameter c, the intercept, was found to be 0.03155.

Table 4. Group contributions to interaction parameter (by Haselow).

Increments	Δ_k
$-C_6H_6$ *Phenyl*	0.02539
$-H$ (replaced from phenyl groups)	0.01139
$>C=,-CH<,-CH=,>CH_2$	0.00888
$-CH_3$	0.00131
$-COOH$ (caboxyl)	-0.01885
$>C=O$	-0.01024
$>CH-OH$	0.04135

Under these conditions, the optimized interaction parameter (k_{12}), that gave the best fit between the experimental data and the modeling with PR-EoS and Van der Waals mixing rules and group contribution method is presented in Table 5.

Table 5. Optimized interaction parameter for the modeling of ibuprofen solubility in CO_2 with the PR-EoS and the van der Waals mixing rules.

Model	Temp. (°C)	k_{12}	%AARD
EoS	35	8.5	16.58
	40	7.3	11.62
Group contribtion	35	9.14	25.52
	40	9.14	37.08

A comparison between the solubility data for ibuprofen at 35°C and 40°C is for two model is illustrated in Fig. 4 and Fig. 5.

4. Conclusion

Peng-Robinson EoS with Van der Waals mixing rule are used to calculate the solubility of ibuprofen+CO_2 system and inconsistency in experimental data checked. The correlation may be improved by using different models which are not dependent on the estimated ibuprofen critical properties.

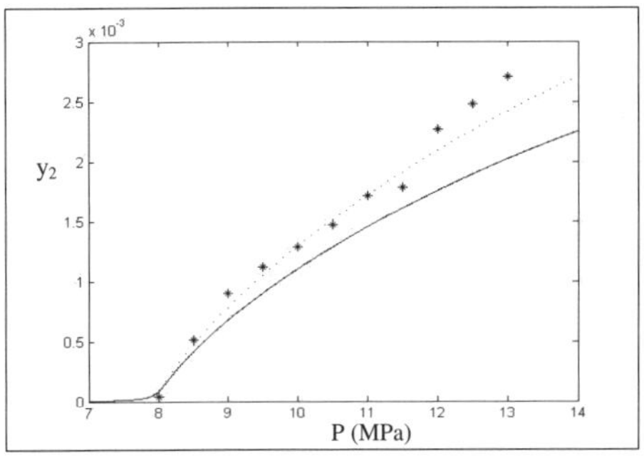

Figure 4. Experimental solubility of ibuprofen in SC-CO2, at 35°C, and correlation results for two model. EoS model(….); Group contribution model(—).

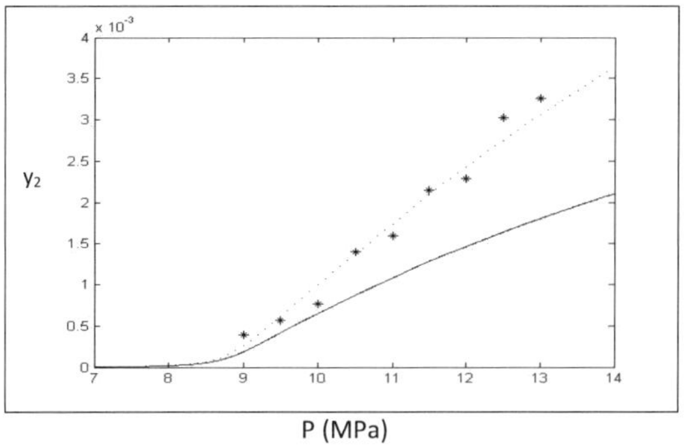

Figure 5. Experimental solubility of ibuprofen in SC-CO_2, at 40°C, and correlation results for two model. EoS model(….); Group contribution model(—).

References

1. S.S. Chen, A. Kreglewski, "Application of the augmented van der Waals theory of fluids. I. Pure Fluids", *Ber. Bunsenges*, **81**, 1048 (1977).
2. P. Subra, P. Justin, "Powder elaboration in supercritical media: Comparision with conventional routes", *J. of Powder Technology,* **103**, 2 (1999).
3. B.E. Poling, J.M. Prausnitz, J.P. O'Connell, "The Properties of Gases and Liquids", fifth ed., McGraw-Hill, NY, USA, **255** (2001).
4. C.J. Chang, A.D. Randolph, "Precipitation of microsize organic particles from supercritical fluids", *J. AIChE*, **35**, 1876 (1989).
5. D. L. Sparks, R. Hernandez, A. Estevez, "Evaluation of density-based models for the solubility of solids in supercritical carbon dioxide and formulation of a new model", *J. Chem. Eng. Science*, **63**, 4249 (2008).
6. P. Coimbra, C.M.M. Duarte, H.C. de Susa, "Cubic equation-of-state correlation of the solubility of some anti-inflammatory drugs in supercritical carbon dioxide", *J. of Fluid Phase Equilibria*, **239**, 188 (2006) .
7. M.R. Housaindokht, M.R. Bozorgmehr, "Calculation of solubility of methimazol, phenazopyridine and propanol in supercritical carbon dioxide", *J. Sup. Fluids*, **43**, 390 (2008).
8. M. Charoenchaitrakool, F. Dehghani, N.R. Foster, "Micronization by Rapid Expansion of Supercritical Solutions to Enhance the Dissolution Rates of Poorly Water-Soluble Pharmaceuticals", *J. of Ind. Eng. Chem. Res.*, **39**, 4794 (2000).
9. J.S Haselow, R.A. Greenkorn, K.C. Chao, "In Equations of State-Theories and Applications", *ACS Symposium Series 300*, American Chemical Society: **156** (1986).

CORRELATION OF THE SOLUBILITY OF IBUPROFEN IN SUPERCRITICAL CARBON DIOXIDE*

MEHDI ARDJMAND

Department of Chemical Engineering, Islamic Azad University, Tehran South Branch, Tehran, Tehran state, Iran

MEHRDAD MIRZAJANZADEH

Department of Chemical Engineering, Islamic Azad University, Tehran South Branch, Tehran, Tehran state, Iran

The solubility of a solute in the supercritical fluid is the most important thermophysical property that needs to be determined and modeled as a first step to develop any supercritical fluids application. The solubility data was correlated using seven equations of state along with two mixing rules. The equations of state are Van der Waals, Redlich-Kwong, Soave-Redlich-Kwong, Peng-Robinson, Stryjek-Vera, Patel-Teja-Valdrama and Pazuki and some others. Mixing rules are VdW1 and VdW2. Also, three estimated physical properties of ibuprofen using various group contributions methods are predicted. Correlations were compared and discussed on the basis of the employed equation-of-state, mixing/combining rules and three sets of physical properties.

1. Introduction

The main part of the investigations around the supercritical fluids technology contains evaluating the fluid phase ability to solve the component which is supposed to be processed. Carbon dioxide is a promising solvent since it is inexpensive, nontoxic, inflammable and environmentally acceptable and has a low critical temperature and a moderate critical pressure. The design of the extraction processes with SCF's depend on the ability to model and predict the solubility of solid in supercritical fluids.

To predict the solubility of a solute in supercritical fluids, EoS models are widely used. Cubic EoS's are the simplest equations capable of predicting and representing fluid phase equilibrium.

This research was undertaken to develop a mathematical model to compute the solubility of ibuprofen in supercritical carbon dioxide [1,2].

2. Methodology

If the solid phase does not dissolve the SC-CO_2, the solubility of the solute in the CO_2-rich fluid phase at equilibrium can be calculated using the isofugacity criteria [3]:

$$y_2 = \frac{P_2^{sub}}{P} E$$

$$E \equiv \frac{\exp\left(\frac{v_2^s (P - P_2^{sub})}{RT}\right)}{\phi_2^{SCF}} \qquad (1)$$

In these equations, P is the equilibrium pressure, T the equilibrium temperature, P_2^{sub} the sublimation pressure of the solute, E the enhancement factor and v_2^s is the saturated molar volume of solid solute at the equilibrium temperature, ϕ_2^{SCF} the fugacity coefficient of the solute in the fluid phase, which can be derived, from any EoS, by the new useful expression for deriving component fugacity coefficient from the mixture fugacity coefficient [4]:

$$RT \ln \hat{\phi}_i = \int_0^P \left(\bar{v}_i - \bar{v}_i^{ig} \right) dp = \int_0^P \left(\frac{\bar{v}_i}{RT} - \frac{1}{P} \right) dp$$
$$= -RT \ln Z + \int_V^\infty \left[\left(\frac{\partial P}{\partial n_i} \right)_{T,V,n_j \neq n_i} - \frac{RT}{V} \right] dV \qquad (2)$$

In Tables 1 and 2 we have summarized the cubic EoS's and the mixing and combining rules used in this work. T_c, P_c and ω, calculated with three methods of group contribution by Joback, Lydersen and Ambrose.

*This work is supported by Islamic Azad University, Tehran South Branch.

A consistent approach of sublimation pressure estimations is proposed by Lyman et al. The estimated physicochemical and critical properties of ibuprofen and using group contribution methods, are shown in Table3 [5]. Modeling programs were developed in MATLAB (version 7.4, R-2007-A).

3. Results and discussion

In this work, the solubility of ibuprofen at 35, 40 and 45°C in supercritical carbon dioxide, is correlated as a function of pressure, using the VW, RK, SRK, PR, SV, PTV and Pazuki EoS's with VdW1 and VdW2 mixing and combining rules.

Table 1. Summary of the cubic EoS's used in this work.

Authors	Analytical expressions
Van der Waals (VW)	$P=\dfrac{RT}{v-b}-\dfrac{a}{v^2}$
Redlich-Kwong (RK)	$P=\dfrac{RT}{v-b}-\dfrac{a}{T^{0.5}v(v+b)}$
Soave-Redlich-Kwong(SRK)	$P=\dfrac{RT}{v-b}-\dfrac{a(T)}{v(v+b)}$
Peng-Robinson (PR)	$P=\dfrac{RT}{v-b}-\dfrac{a(T)}{v(v+b)+b(v-b)}$
Stryjek-Vera (SV)	$P=\dfrac{RT}{v-b}-\dfrac{a(T)}{v(v+b)+b(v-b)}$
Patel-Teja-Valdrama(PTV)	$P=\dfrac{RT}{v-b}-\dfrac{a(T)}{v(v+b)+c(v-b)}$
Pazuki-Ghotbi-Taghikhani-Dashtizadeh	$P=\dfrac{RT}{v-b}\cdot\dfrac{v+b}{v}-\dfrac{a}{v(v+b)}$ $=\dfrac{RT}{v-b}-\dfrac{a(v-b)-bRT(v+b)}{v(v-b)(v+b)}$

Table 2. Summary of the mixing and combining rules used in this work.

Authors	Mixing rules
Van der Waals-1 paramter (VdW1)	$a=\sum_i^n\sum_j^n y_i y_j a_{ij}$ $b=\sum_i^n\sum_j^n y_i y_j b_{ij}$ $c=\sum_i^n\sum_j^n y_i y_j c_{ij}$ $a_{ij}=(1-k_{ij})(a_{ii}a_{jj})^{0.5}$ $b_{ij}=\dfrac{(b_{ii}+b_{jj})}{2}$ $c_{ij}=\dfrac{c_i+c_j}{2}$
Van der Waals-2 paramter (VdW2)	$a=\sum_i^n\sum_j^n y_i y_j a_{ij}$ $b=\sum_i^n\sum_j^n y_i y_j b_{ij}$ $c=\sum_i^n\sum_j^n y_i y_j c_{ij}$ $a_{ij}=(1-k_{ij})(a_{ii}a_{jj})^{0.5}$ $b_{ij}=\dfrac{(1-l_{ij})(b_{ii}+b_{jj})}{2}$ $c_{ij}=\dfrac{c_i+c_j}{2}$

Original experimental solubility data for the ibuprofen were already published by Foster et al.

The fitness function at each temperature is the percentage average absolute relative deviation (%AARD) between the calculated and experimental solubility [6]:

The results of this work (just, most relevant cases), will be discussed with the graphical analysis of the correlation results.

Table 3. Estimated properties of ibuprofen.

	$T_c(K)$	$P_c(MPa)$	ω
Set 1[a]	891.2	2.25	0.788
Set 2[b]	749.7	2.33	0.819
Set 3[c]	753.6	2.18	0.749
v^s [d]	$P^{sub}(Pa)\times10^{-2}$ [e]		
$\times10^6 m^3 mol^{-1}$	35°C	40°C	45°C
182.1	4.95	8.97	1.60

[a] Estimated by Joback method
[b] Estimated by Lydersen method
[c] Estimated by Ambrose method
[d] Estimated by Immirzi and Perini method
[e] Estimated by Lyman et al. method

To have an overall view of the results, %AARD values versus EoS's along with mixing rules plot for ibuprofen+CO_2 system were shown in Figs. 1, 2 and 3.

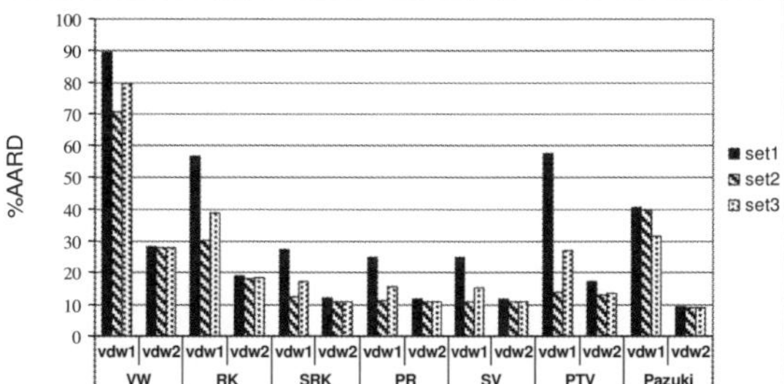

	VW vdw1	VW vdw2	RK vdw1	RK vdw2	SRK vdw1	SRK vdw2	PR vdw1	PR vdw2	SV vdw1	SV vdw2	PTV vdw1	PTV vdw2	Pazuki vdw1	Pazuki vdw2
set1	89.39	28.17	56.61	18.85	27.19	11.95	25.07	11.71	24.9	11.67	57.75	17.21	40.43	9.35
set2	70.48	27.8	30.42	18.12	12.27	10.76	11.06	10.69	10.86	10.71	13.88	12.81	39.9	8.92
set3	79.51	27.85	38.62	18.21	17.24	10.83	15.53	10.79	15.44	10.89	26.89	13.41	31.66	9.01

Figure 1. %AARD values versus EoS's along with mixing rules plot for ibuprofen+CO_2 system at 35°C.

The following results can be concluded:

- From the complete analysis of the results, we can conclude that VdW2 mixing rule, always produces better correlation results for all EoS's than VdW1. This fact is probably due to the two available adjustable parameters, which can offer a higher flexibility to the EoS model fit of the experimental solubility data (Figs. 4a, 4b, 4c).

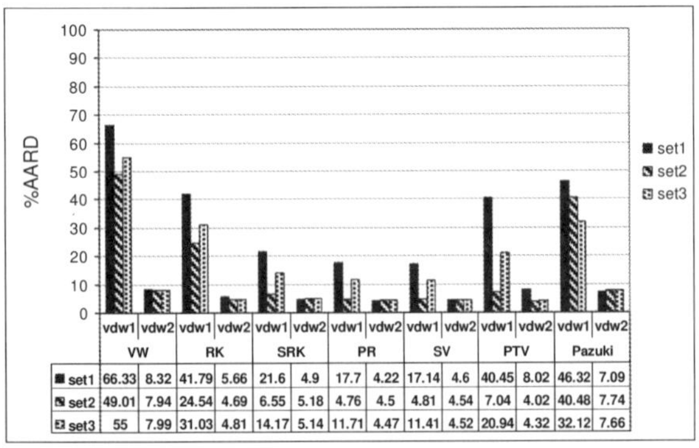

	vdw1	vdw2	vdw1	vdw2	vdw1	vdw2	vdw1	vdw2	vdw1	vdw2	vdw1	vdw2	vdw1	vdw2
	VW		RK		SRK		PR		SV		PTV		Pazuki	
set1	66.33	8.32	41.79	5.66	21.6	4.9	17.7	4.22	17.14	4.6	40.45	8.02	46.32	7.09
set2	49.01	7.94	24.54	4.69	6.55	5.18	4.76	4.5	4.81	4.54	7.04	4.02	40.48	7.74
set3	55	7.99	31.03	4.81	14.17	5.14	11.71	4.47	11.41	4.52	20.94	4.32	32.12	7.66

Figure 2. %AARD values versus EoS's along with mixing rules plot for ibuprofen+CO_2 system at 40°C.

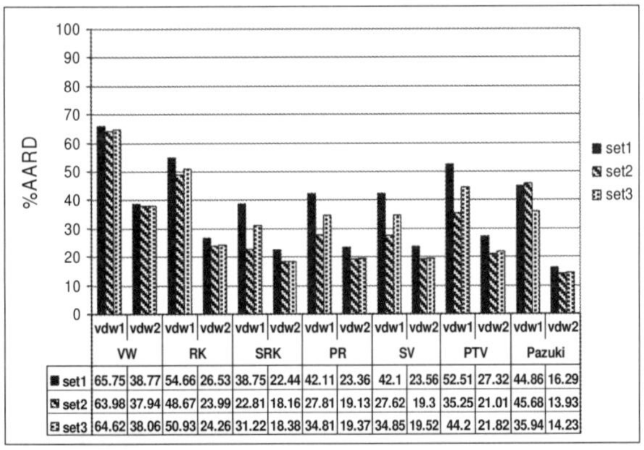

	vdw1	vdw2	vdw1	vdw2	vdw1	vdw2	vdw1	vdw2	vdw1	vdw2	vdw1	vdw2	vdw1	vdw2
	VW		RK		SRK		PR		SV		PTV		Pazuki	
set1	65.75	38.77	54.66	26.53	38.75	22.44	42.11	23.36	42.1	23.56	52.51	27.32	44.86	16.29
set2	63.98	37.94	48.67	23.99	22.81	18.16	27.81	19.13	27.62	19.3	35.25	21.01	45.68	13.93
set3	64.62	38.06	50.93	24.26	31.22	18.38	34.81	19.37	34.85	19.52	44.2	21.82	35.94	14.23

Figure 3. %AARD values versus EoS's along with mixing rules plot for ibuprofen+CO_2 system at 45°C.

- While using the estimation methods, one should be very cautious. Also, sublimation pressure is one of the significant factors in the exactness of the calculations and it is very crucial to be careful in choosing a suitable amount for it [7]. It could be said that the Lydersen method at 35 and 45°C almost works better than the other two for all EoS's while using VdW1 and VdW2. For 40°C with VdW1, the best revenue belongs to the Lydersen method but at this temperature for VdW2, this is not so far all the EoS's (Fig.5).

- We can conclude that almost in all isotherms, when we use VdW1, VW and RK EoS's (to a great deal) and PTV and Pazuki EoS's (to a lesser amount), we don't have an acceptable operation in estimating the solubility and this causes large errors. In contrast, SRK, PR and SV EoS's with a similar behavior, will cause smaller errors than other EoS's. If we use VdW2, this is the sole VW EoS which is not capable enough to predict the solubility while other EoS's report a more logical %AARD.
- In ibuprofen-SC CO_2 system, at 35 and 45°C, using VdW2 mixing rule, for all the different applied sets of properties, Pazuki EoS makes the least %AARD and can fit the experimental data with the present models very well. For 40°C in all EoS's, little errors are observed and using set2 and set3, PTV EoS along with VdW2, leads to a successful correlation and using set1, PR EoS is more accurate than other equations of state.

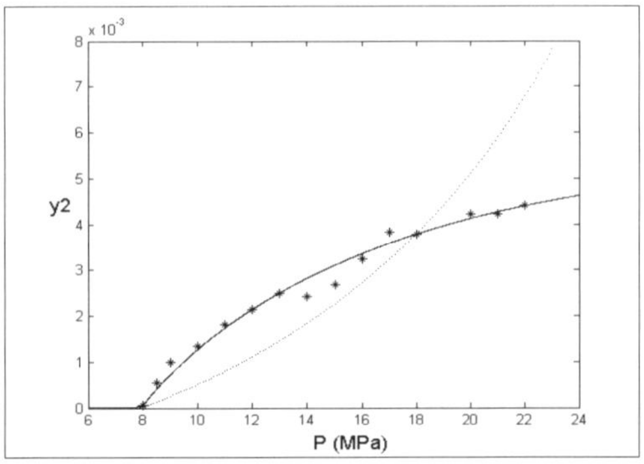

Figure 4a. Experimental solubility of ibuprofen in SC-CO_2 at 35°C and correlation results obtained with Pazuki EoS using the properties of set2; VdW1(···);VdW2(—).

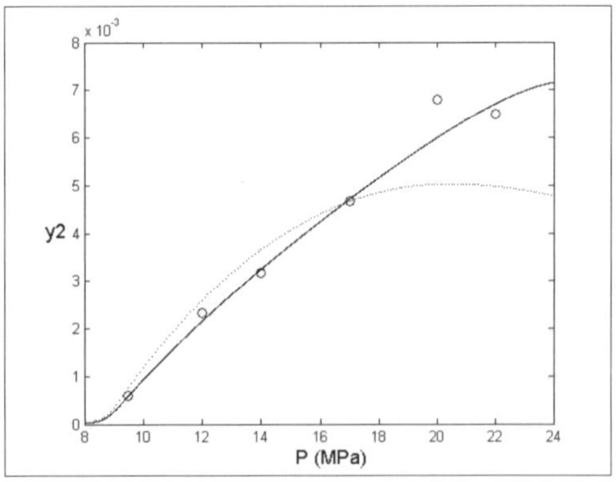

Figure 4b. Experimental solubility of ibuprofen in SC-CO_2 at 40°C and correlation results obtained with with PR EoS using the properties of set1; VdW1(···); VdW2(—).

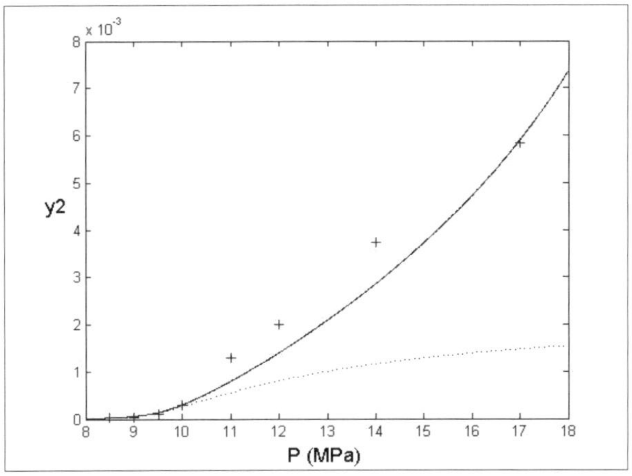

Figure 4c. Experimental solubility of ibuprofen in SC-CO_2 at 45°C and correlation results obtained with PTV EoS using the properties of set3; VdW1(···); VdW2(—).

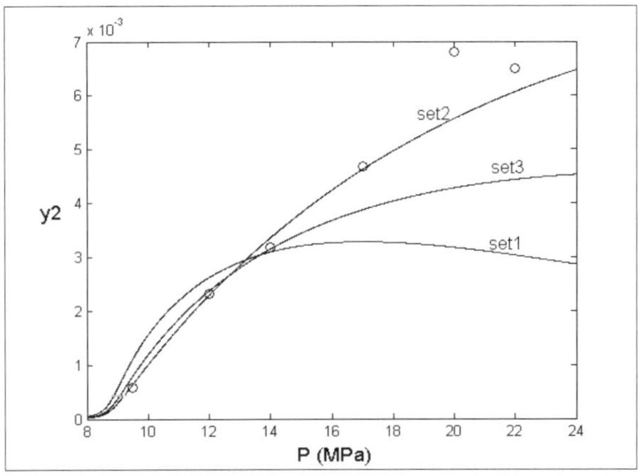

Figure 5. Experimental solubility of ibuprofen in SC-CO_2 at 40°C and correlation results obtained with PTV EoS combined VdW1 mixing rule for the three different sets of estimated properties of the solid.

4. Conclusion

VdW2, always produces better correlation results for all EoS's than Vdw1. In most cases, the VdW2 mixing rule seems to be sufficient to generate very good experimental fits.

The importance of the accurate knowledge of the values for the critical and physicochemical solid's properties required for the applied EoS models was investigated using several different sets of solid properties, estimated by different methods. Generally, it is suggested that we apply the Lydersen method while the Joback method has caused more errors in most cases.

At 35 and 45°C, Pazuki equation along with VdW2 mixing rule is more accurate than other equations and %AARD Values calculated by Pazuki EoS less than %AARD values calculated by other EoS's. In Pazuki EoS, a and b parameters depend on temperature whereas in other EoS's only a parameter depends on temperature Parameter b expresses the volume of the molecules so it plays an important role in calculation liquid densities. In asymmetric size systems like the one studied here, b parameter is more important, and we can relate better prediction of %AARD values to this characteristic of Pazuki equation of state. Also, at 40°C, using set 2 and set3, PTV EoS along with VdW2, leads to a successful correlation.

References

1. T.Y. Kwak, "Van der Waals mixing rules for cubic equations of state, Applications for supercritical fluid extraction modeling", *J. of Chemical Engineering Science*, Vol.41, No5, pp. 1303-1309, (1986).
2. A. Anderko, "EoS Methods for the Modeling of Phase Equilibria", *J. of fluid Phase Equilib.*, **61** (1990) 145-225.
3. P. Coimbra, C.M.M. Duarte, H.C.de Susa, "Cubic equation-of-state correlation of the solubility of some anti-inflammatory drugs in supercritical carbon dioxide", *J. of Fluid Phase Equilibria*, **239** (2006) 188-199.
4. J. Hu, R. Wang, S. Mao, "Some useful expressions for deriving component fugacity coefficients from mixtue fugacity coefficient", *J. of Fluid Phase Equilibria*, **268** (2008) 7-13.
5. M. Charoenchaitrakool, F. Dehghani, N.R. Foster, "Micronization by Rapid Expansion of Supercritical Solutions to Enhance the Dissolution Rates of Poorly Water-Soluble Pharmaceuticals", J. *of Ind. Eng. Chem. Res.,* **39** (2000) 4794-4802.
6. M.R. Housaindokht, M.R. Bozorgmehr, "Calculation of solubility of methimazol, phenazopyridine and propanol in supercritical carbon dioxide", *J. of Supercritical Fluids*, **43** (2008) 390-397.
7. T.E. Carleson, S. Chandra, C. Wai, L. Wai, S.Huang, "Group Contribution Method for Estimation the Solubility of Selected Hydrocarbon Solutes in Supercritical Carbon Dioxide", *J. Ind. Eng. Chem.* **49** (1993) 125.

IDENTIFICATION AND MONITORING OF PESTICIDES IN A WELL FIELD IN DELHI, INDIA

PRAVIN K. MUTIYAR
Department of Civil Engineering, Indian Institute of Technology Delhi, New Delhi-110016, India

ATUL K. MITTAL
Department of Civil Engineering, Indian Institute of Technology Delhi, New Delhi-110016, India

ASAF PEKDEGER
Institute of Geological Sciences, Freie Universität of Berlin, Malteserstr. 74-100, 10965 Berlin, Germany

Pesticide residues in drinking water have appeared as major challenge over the last few years. In this work efforts are made to assess the level of pesticides in public water supply resources in national capital territory i.e. New Delhi, India. A total of 16 different organochlorine pesticides (OCPs) were identified from potable water samples. The majority of pesticides were isomers of HCH, DDT, endosulphan, endrin, aldrin, dialdrin and heptachlore. The concentration of pesticides varied from below detectable limit (BDL) to higher than maximum permissible limits. Results thus suggested that continuous consumption of contaminated can pose severe health threats to local residents of this area.

Keywords: Organochlorine pesticides, groundwater, contamination, GCMS.

1. Introduction

The contamination due to agro-chemicals in surface as well as ground water resources is a big issue for developing world. It is well known fact that intensive use of agrochemicals (synthetic fertilizers, pesticides, nematicides, fungicides, weedicides etc.) over the years has created b environmental pollution issues. It is interesting that pesticides are applied to protect crops from pests and diseases; only around 15% of the preparation hits the target. The rest is distributed in the soil and air and consequently creates environmental contaminations. Agrochemicals in soil can move from the surface when they are dissolved in runoff water, or when they percolate down through the soil. Those that have infiltrated the soil will eventually reach the groundwater [1]. As the environment and human health are closely interrelated, no one can have a better health in polluted environment. Organochlorine pesticides (OCPs) are an important potential chemical pollutants used extensively over the globe especially in developing and underdeveloped countries as they are comparatively cheap and effective. The continuous uses or misuses has resulted the contamination of every sector of environment especially hydrosphere. These persistent organic compounds such as HCH isomers, Endosulphan, DDT and its metabolites are the predominant chemical contaminants found in water resources. Due to their persistent and lipophilc nature they accumulate in food web with high risk to human beings. Some of these compounds are considered to act as hazardous environmental hormones, which disrupt reproductive cycles of humans and wildlife and acts as an endocrine disrupter also. Although specific upper limits to concentrations in drinking water have been proposed for different pesticides, a general norm is 0.1 ug/L for a single pesticide and 0.5 ug/L for multiple pesticides [2], but still it seems on the higher side as the water is not the only source from which we are consuming the contaminates. So the availability of safe drinking water in metro cities like Delhi is a big challenge to authorities managing the water supply and the industrial sector engaged in water business. Aim of this study was to assess the level of pesticide contamination in a well field in Delhi NCT, India.

2. Materials and Methods

2.1. *Field site description*

The sampling site is located between latitude 28°46'56.69"N North, longitude 77°11'58.62"E East and latitude 28°50'34.94"N North, longitude77° 12'34.91"E

East. National Capital Territory (NCT) of Delhi spreads over a total area of almost 1500 km^2, of which more than 60% is urbanized. Sampling points are located in the northern part of Delhi, on upstream flood basin of the river Yamuna, Delhi. It is surrounded by a field of abstraction wells operated by Delhi Jal Board (DJB) for drinking water production and privately owned borewells.

2.2. *Sample collection*

A total of 21 samples were collected during the entire study period. The sampling campaign was carried out twice to catch the dry and wet period. Samples were collected in 1 L amber coloured, food grade, virgin quality plastic bottles. The bottles were washed with tap water and properly rinsed with distilled water. Before sampling, the bore-wells were run for at least 10 minutes to clear the casing of standing water and to bring the fresh water from the aquifer, to represent the actual quality of aquifer. The pumping rate was around 20 lps. A standard protocol described by APHA [3] was followed to collect water samples for pesticide residues analysis. The collected sample bottles were kept in airtight large plastic ice-cold containers and were transported to laboratory within 6 h of their collection and stored in deep refrigerator at 40 C until further analysis.

2.3. *Reagents and standards*

Analytical grade (AR) chemicals (Merck, Germany) were used throughout the study without any further purification. To prepare all the reagents and calibration standards, double glass distilled water was used. The glassware were washed with dilute nitric acid (1.15) followed by several portions to distilled water. Standards of organochlorine pesticide i.e. EPA 8081 Pesticide Standard Mix (46845-U) was procured from Sigma-Aldrich USA. The working standards of pesticides were prepared by diluting EPA mixture standard in n-Hexane and were stored at 4°C.

2.4. *Extraction of pesticide residue*

One liter water sample was taken into a 2L capacity separating flask, mixed with 30 g of NaCl and 50 ml of n-hexane. The sample was shacked properly and hexane layer was separated. This process was repeated thrice with same water sample and hexane portion was transferred to a Buchi Condenser flask through a column of anhydrous sodium sulphate. The extracted was condensed to around 5 ml on Rotavac and further reduced to 1 ml under gentle N_2 stream.

2.5. *Qualitative and quantitative analysis*

Sample analysis was conducted using a Thermo® Trace GCMS equipped with a Thermo® A2000 auto sampler and DB-5 fused silica capillary column (30 m × 0.25 mm ID, film thickness 0.25um). Pure helium was used as carrier gas with a flow rate of 1.0 ml/min. One micro liter of sample was injected into the GCMS in split-less mode with an injection time of 1 min. Injection and MS transfer line temperatures were set to 250°C and 280°C respectively. The oven temperature for analyses of OCPs was programmed: 90 to150°C at a rate of 15°C min^{-1}, 150 to 220°C at a rate of 3°C min^{-1} and 220 to 270°C at a rate of 5°C min^{-1}. The analysis was done in full scan mode and specific ions were targeted for identification. To ensure the quality of extraction and detection procedure, 5 different concentration of each OCP was mixed with distilled water, extracted by the same method and recovery percentage was checked. Table 1 illustrates the retention time for different pesticides.

Table 1. Retention time and selected ion for the different pesticides.

Pesticide	Retention time (min)	Selected Ion
α BHC	12.29	181, 183
β BHC	13.55	
γ BHC	13.77	
δ BHC	14.99	
Heptachlor	16.34	100,272
Aldrin	17.23	263
Heptachlor Epoxide	18.86	353
Endosulphan-I	19.00	195,241
Dieldrin	20.60	79
DDE	21.21	246,279
Endrin	23.15	81, 263
Endosulphan -II	24.62	195,241
DDD	24.81	235,237
Endrin aldehyde	25.89	67
DDT	26.37	235
Methoxychlor	30.53	227

A chromatogram (Fig. 1) was obtained of water sample showing organochlorine pesticides.

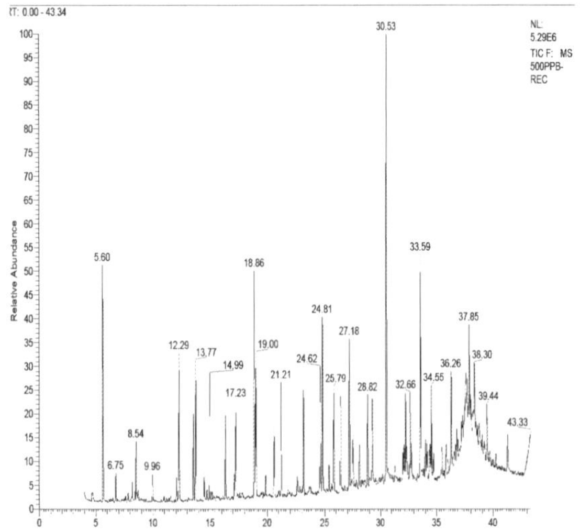

Fig. 1. Chromatogram of spiked water sample containing a mixture of organochlorine pesticides.

3. Results and Discussion

The level of pesticide residues are presented in Table 2. A total of 21 water samples were collected form borewell and about 38 % samples showed the presence of Σ HCH, 24 % sample showed the presence of Σ endo I and II and 20 % samples showed the presence of Σ DDT. A total of 16 different organochlorine pesticides (OCPs) were identified from potable water samples. The analysis showed that the water sample contains isomers residues of HCH (α, β, γ, and δ), DDT, DDD, DDE, Endosulphan (I and II), endrin, aldrin, dialdrin and heptachlor. The residues concentration (ng l-1) ranged: BDL to 1.79 for α HCH, BDL to 7.45 for β HCH, BDL to 4.53 for γ HCH, BDL to 8.45 for δ HCH, BDL to 14.08 for heptachlor, BDL to 12.39 for aldrin, BDL to 9.72 for heptachlor epoxide, BDL to 11.86 for endosulphan-I, BDL to 10.32 for dieldrin, BDL to 11.48 for DDE, BDL to 27.06 for endrin, BDL to 29.82 for endo- II, BDL to 15.37 for DDD, BDL to 41.27 for endrin aldehyde, BDL to 36.86 for DDT, and BDL to 35.16 for methoxychlor (Table 2). The high concentrations of organochlorine compounds indicate agriculture as well as household articles. Among the HCH, δ HCH was found frequently in drinking water resources. HCH is widely used in Indian agriculture the heavy use of pesticides in local and adjoining area of Delhi. The maximum concentration was of endrin aldehyde which popularly used for termite control in systems. It is a broad spectrum pesticide mainly used for standing crop pest controlling. The maximum occurrences was of aldrin (67 % of total sampling site), while the individual concentration was noted for endrin aldehyde. These are popular termite control and their highest concentration in water indicated the consumption of this formulation in this region. Thus, the use of endrin aldehyde and aldrin in agriculture soils are completely restricted in the country, but their higher content in groundwater indicates the continuous use of them. Jayshree and Vasudevan [4] reported relatively higher contents of isomers of HCH and DDT in groundwater of Thiruvallur district of Tamil Nadu. They attributed the high level of such formulations in groundwater to intensive cultivation. Since isomers of HCH, in general have greater solubility and leaching properties [5]. Organochlorine pesticides are extremely used for control of agricultural and vector born diseases and their prolonged presence as residues is not only to cause serious disease in humans but are detrimental to soil microflora [6]. Results thus indicate that persistence of toxic pesticide formulations in groundwater, which is major source of local residents, may pose serious health hazardous if water is continuously consumed. Attention is required immediately to halve the risks of these chemicals on human and other biota of the biosphere. The major health effects due to pesticide may be neuropathy headache, memory loss, bradycardia, jaundice, respiratory disorders, birth defects etc.

4. Conclusions

This work illustrates the level of organochlorine pesticides in a well field which is used for the public water supply in NCT Delhi, India. A total of 16 different formulations of pesticides were obtained from groundwater samples. The maximum occurrences of isomers of HCH and DDT indicate the heavy use of these formulations in this area, although uses of some pesticides are completely restricted by regulating authorities in the country. The leaching of pesticides is mainly depends upon the solubility and leaching process, soil profile, land use patterns and even contamination history. Further studies are required to trace the actual source of pesticides in groundwater.

Table 2. Concentration of pesticides (ng/L) detected in groundwater samples of Delhi.

Site	Pesticides Detected (Out of 16 targeted)
S1	ALD(10.28), ENDO-1(11.86)
S2	METH CHL(34.06)
S3	γ HCH(3.67), ALD(8.93)
S4	Not Detected
S5	β HCH(7.45), γ HCH(3.89), δ HCH(8.44), ALD(8.88), HEPT EPOX(9.72), DIE(10.32), DDE(11.47), NDRIN(14.81), ENDO-II(15.84), DDD(15.37), DDT(36.9), METH CHL(35.18)
S6	γ HCH(3.62),
S7	ALD(9.67)
S8	ENDRIN(27.06), DDD(11.77)
S9	α HCH(1.36), ALD(5.58)
S10	ALD(14.21), ENDO-II(15.70)
S11	ALD(8.26), ENDR ALD(19.42)
S12	ALD(11.79)
S13	ALD(7.06)
S14	ALD(9.14), DIE(4.47), DDE(8.91), DDD(11.22) METH CHL(33.19)
S15	β HCH(6.50), γ HCH(4.53), δ HCH(7.22), HEPT CH(13.87) ALD(12.39), HEPT EPOX(8.46), DIE(7.80), DDE(10.31), ENDRIN(14.48), DDD(12.26), ENDR ALD(41.27), DDT(34.26), METH CHL(34.26)
S16	β HCH(4.21),
S17	HEPT CH(14.08), ALD(8.10), HEPT EPOX(7.84), DDE(8.99), ENDO-II(29.82), METH CHL(33.21)
S18	δ HCH(6.30), ALD8.76), HEPT EPOX(7.83), DIE(4.84), DDE(9.35), DDD(12.21), DDT(33.96), METH CHL(33.77)
S19	DIE(5.01), ENDRIN(21.02)
S20	Not Detected
S21	α HCH(1.79), β HCH(4.56), ALD(11.88)

HEPT CH – Heptachlor, ALD – Aldrin, HEPT EPOX – Heptachloro epooxide, DIE – Dieldrin, ENDR ALD – Endrin aldehyde, METH CHL – Methoxychlor

Parenthesis shows the concentration of individual OCP in ng/L

Acknowledgments

One of the authors PKM is thankful to University Grant Commission (UGC) Delhi for providing him the junior research fellowship (JRF) as financial assistance.

References

1. FAO, "Assessing Soil Contamination: A Reference Manual. FAO Pesticide Disposal Series", Food and Agriculture Organization of the United Nations, Rome, 2000.
2. M. Leistra and J.J.T.I. Boesten, "Pesticide contamination of groundwater in western Europe", Agril. Ecosys. Environ., pp. 369-389, 1989.
3. APHA, Standard methods for the examination of water and wastewater (18th ed.), American Public Health Association, Washington, DC, 1998.
4. R. Jayashree and N. Vasudevan, "Organochlorine pesticide residues in groundwater of Thiruvallur district, India", Environ. Monito. Assess., pp. 209-215, 2007.
5. M.K. Pillai, "Pesticide pollution of soil, water in Delhi area, India", Sci. Tot. Environ., pp. 321-327, 1986.
6. S.P. Mohapatra, M. Kumar, V.T. Gajbhiye, N.P. Agnihotri, M. Kumar, "Groundwater contamination by organochlorine insecticide residues in rural areas in the Indo gangetic plain", Environ. Monito. Assess., pp. 155-164, 1995.

WATER TREATMENT SCHEME FOR SPECIFIC REQUIREMENT OF COOLING: CASE STUDY OF GROUND WATER TREATMENT AT DELHI

SANTOSH S. KOLTE

Department of Civil Engineering, Indian Institute of Technology Delhi, New Delhi-110016, India

ATUL K. MITTAL

Department of Civil Engineering, Indian Institute of Technology Delhi, New Delhi-110016, India

The present study aimed at identifying and developing the most appropriate water treatment scheme for ground water to be used for air conditioning/cooling, in large quantity for commercial utility in Delhi (India). High values of TDS, hardness and alkalinity up to 600, 400 and 400 mg/L, respectively, are reported in the source water. Dealkalization using weak acid cation exchanger and degasification coupled with multi-media and activated carbon filtration emerged as the optimal treatment scheme, efficiently reducing hardness and alkalinity. The maximum hardness and alkalinity reduction was 99.42 and 100%, respectively. The scheme has an added credential of eliminating handling of large quantity of brine solution for regeneration.

Keywords: Hardness; alkalinity; softening; degasification; weak acid cation.

1. Introduction

Appropriate water treatment system and the subsequent desired water quality are lifelines of the commercial and industrial air conditioning/cooling systems. Quality and quantity of the available water sources are the constraints and play a key role aspect in formulating the appropriate water treatment scheme for the specific end use applications. The main considerations for water treatment scheme are operating economics, quality and reliability of its performance, and minimizing environmental costs. A constant and reliable supply of water with the desired quality is must to ensure maximum life, control excessive fuel consumption due to scaling, obtain and maintain optimal thermal performance and reduce maintenance expanses of the cooling systems [1, 2]. Dissolved salt concentrations imparting hardness and alkalinity leading to corrosion and salt depositions (scaling) on heat transfer surfaces, equipments and piping are issues of concern for air conditioning / cooling systems, as these ultimately result in large damages to the whole system [3]. Ion exchange (softening, demineralization), reverse osmosis and their sub-stage combinations are the proven treatment technologies for reduction of total dissolved solids (TDS) and hardness in raw waters [4]. Capacitive deionization is a potentially attractive method for desalination of low salinity sources such as brackish and surface waters [5]. The use of ion exchange processes affords numerous efficient and effective means of conditioning feed water. The proper selection of the specific ion exchange process depends on water quality needs, operating convenience, and economic considerations. For effective results, the system must be carefully selected, designed, operated and maintained. The reverse osmosis process, though being good on quality side is not suitable for large quantities on economy and production quantity issues. In reverse osmosis, 50% (up to 75% for small capacity plants) of the water is reject that is to be drained off and may pose problems for the sewerage system [6].

The present study considered various treatment options of ion exchange coupled with multi-media pressure filtration and activated carbon filtration; to identify and develop the most appropriate water treatment scheme for ground water in the metropolitan city of Delhi (India). The ground water with high TDS, hardness and alkalinity is required to be treated to meet the air conditioning/cooling requirements demanding hardness and dissolved salts reduction, to be used in large quantity of the order of 5 mld; for commercial application in the city.

Experimental studies were performed for softening, and dealkalization. Complete demineralization with cationic and anionic exchange resins would prove uneconomical and also not desired for the intended use and thus not recommended. Reverse osmosis is not feasible for very high costs and smaller yields. Laboratory experiments were performed for softening with strong acid cation (SAC) exchange resins (H^+ and Na^+) and dealkalization with weak acid cation exchange resin (WAC); and subsequent gasification.

2. Materials and Methods

2.1. *Experimental system and operating conditions*

Fig. 1 shows the schematic diagram of the experimental apparatus consisting of the reaction vessel, peristaltic pump for feedwater and regeneration solution, a backwash water source and a backwash pump. The service flow, regeneration and backwash were facilitated with flow lines and valves as shown.

The reaction vessel is in the form of a glass column of 0.027 m inner diameter facilitated with bottom drain of cotton strainer supported on glass beads to retain the resin.

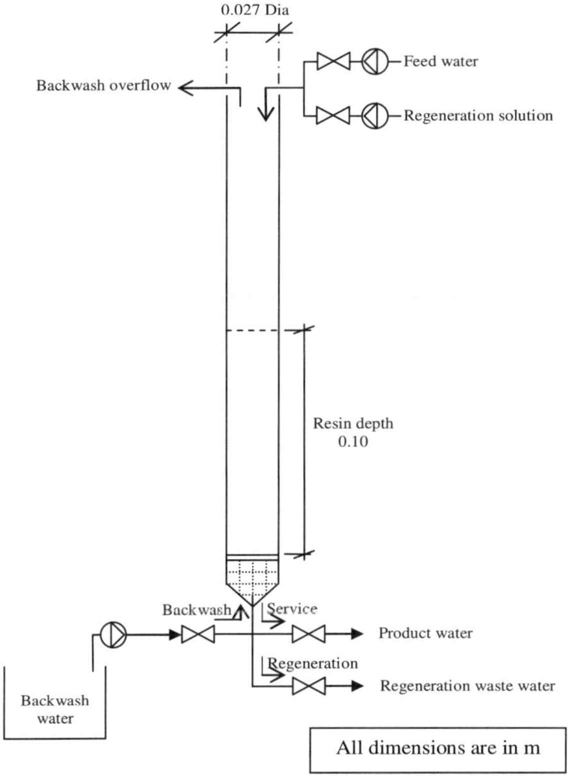

Fig. 1. Schematic diagram of the experimental system.

The experimental operating conditions for service flow and regeneration rate and regeneration solutions for the SAC and WAC are as given in table 1 below.

Table 1. Experimental operating conditions for the ion exchange treatment.

Item	**Value**
Vessel diameter	0.027 m
Resin Depth	0.1 m
Resin volume	0.00005725 m^3
Service flow rate:	
SAC	0.00036 m^3/hr
WAC	0.0006m^3/hr
Regeneration rate:	
SAC	0.00012 m^3/hr
WAC	0.00023 m^3/hr
Regeneration Sol.:	
SAC (H^+)	Hydrochloric acid, 10%
SAC(Na^+)	Sodium chloride, 10%
WAC	Hydrochloric acid, 2%
Regeneration flow	Cocurrent

2.2. *The feed water*

The feed water was the ground water from Yamuna vihar location, from the bank of the river Yamuna in Delhi (India). It is the source of the water to be used for the large scale commercial application.

The maximum reported hardness and alkalinity for the collected samples were 342 and 356 mg/L as $CaCO_3$, respectively.

2.3. *Experimental procedure and methodology*

The reactors were sequentially operated for SAC and WAC, respectively. The service and regeneration flows were co-current in the downward direction.

Four major classes of resins exists for industrial water treatment viz, strong acid cation, weak acid cation, strong base anion and weak base anion [7]. These resins are phenol aldehyde condensation products, which on sulphonation produces resinous mass having baseexchange properties [8]. Each of these major resin classes has several physical or chemical variations within the class. The variations impart different operating properties to the resin. SAC resins are the most widely employed resins for water treatment for hardness reduction.

2.3.1. *SAC resins*

SAC resins in the hydrogen form and sodium form were used with the objective of removing the calcium and magnesium cations by exchanging with the respective hydrogen and sodium. The resins had a finite capacity for removal/exchange of calcium and magnesium ions. The regeneration capacity was assumed to be exhausted when hardness of the product water reversed after attaining the minimum value. The resins were regenerated with 10% hydrochloric acid and sodium chloride solutions, respectively to restore their exchange capacities; as per the following reactions with 'R' being the resin material matrix.

2.3.1(a). *SAC in the hydrogen (H^+) form*

i) During service cycle:

$\underline{R\text{-}H} + Ca^{++} = \underline{R\text{-}Ca} + H^{+}$
$\underline{R\text{-}H} + Mg^{++} = \underline{R\text{-}Mg} + H^{+}$

ii) During regeneration cycle:

$\underline{R\text{-}Ca} + HCl = \underline{R\text{-}H} + CaCl_2$
$\underline{R\text{-}Mg} + HCl = \underline{R\text{-}H} + MgCl_2$

2.3.1(b). *SAC in the hydrogen (Na^+) form*

i) During service cycle:

$\underline{R\text{-}Na} + Ca^{++} = \underline{R\text{-}Ca} + Na^{+}$
$\underline{R\text{-}Na} + Mg^{++} = \underline{R\text{-}Mg} + Na^{+}$

ii) During regeneration cycle:

$\underline{R\text{-}Ca} + NaCl = \underline{R\text{-}Na} + CaCl_2$
$\underline{R\text{-}Mg} + NaCl = \underline{R\text{-}Na} + MgCl_2$

The reactors were operated at the mentioned service flow rates and the product water hardness was monitored for periodical volumes.

2.3.2. *WAC in the hydrogen (H^+) form*

The weak acid cation exchange resins are recommended when hardness to alkalinity ratio is greater than 60% and maximum removal is obtained when this ratio is 100%. Normally, the unit service cycle is terminated for recharge at a hardness breakthrough of 5 ppm $CaCO_3$ [7, 9]. The resin is regenerated with dilute acid using near stoichiometric quantity [9]. A WAC resin based on a cross linked acrylic-divinyl benzene (DVB) matrix containing carboxylic acid functional groups was used.

The regeneration capacity was assumed to be exhausted when hardness of the product water reversed after attaining the minimum value. The resins were regenerated with 2% hydrochloric acid solution. WAC exchange resin can efficiently remove all the cations associated with alkalinity, thus having a dual benefit of taking care of hardness as well as alkalinity [10]. The alkalinity gets converted into carbonic acid that is removed by the following unit of degasser. The degasser is employed downstream of the WAC, wherein the carbonic acid is split in water and carbon dioxide; and the carbon dioxide so formed is released in the atmosphere along with the induced air.

The typical reactions of WAC exchanger are as follows.

i) During service cycle:

$\underline{R\text{-}H} + Ca(HCO_3) = \underline{R\text{-}Ca} + H_2CO_3$
$\underline{R\text{-}H} + Mg(HCO_3) = \underline{R\text{-}Mg} + H_2CO_3$

$H_2CO_3 \xrightarrow{\text{Degasser}} H_2O + CO_2$

ii) During regeneration cycle:

$\underline{R\text{-}Ca} + HCl = \underline{R\text{-}H} + CaCl_2$
$\underline{R\text{-}Mg} + HCl = \underline{R\text{-}H} + MgCl_2$

3. Results and Discussions

The regenerated resins yielded better performance than the fresh resins. The hardness reduction was very fast, but yield was poor for the SAC (H^+) with a maximum 95.92% reduction at 200 mL output and acidic pH, after which the hardness was observed to be higher implying the exhaustion of the resin. The same reduction was observed for the SAC (Na^+) but for 2800 mL output of the product water at neutral pH.

The regenerated WAC resin performance was highly admirable. For WAC; pH, hardness, alkalinity and electrical conductivity (EC) were monitored for periodical volumes. Intermediate outlet samples were monitored for pH before and after degassing to determine the effect of gasification on the pH of the sample. The very small increase in pH so observed in the intermediately aerated samples necessitated in addition of weak NaOH solution to raise the pH of the final product to the neutral value.

The maximum hardness reduction observed was 99.42% and alkalinity was nil; at 1400 mL output volume and pH of 3.38, after which the hardness constantly increased indicating exhaustion of the resin. The operation was then terminated for backwashing and regeneration.

The above observations for WAC and their comparison in a nutshell are given in table 2 and figure 2, respectively.

Table 2. WAC resin treatment.

Outlet Volume (Cummulative)	WAC resin			
	pH	Hardness (mg/L)	Alkalinity (mg/L)	EC (mS)
0.00	7.44	342	356	1.160
200	2.84	10	Nil	0.984
400	2.90	10	Nil	0.817
600	3.18	06	Nil	0.729
Before aeration	3.12	-	Nil	-
After aeration	3.18	-	Nil	-
800	3.2	03	Nil	0.713
1000	3.29	04	Nil	0.653
1200	3.32	04	Nil	0.670
Before aeration	3.27	-	Nil	-
After aeration	3.32	-	Nil	-
1400	3.38	02	Nil	0.664
1600	3.45	03	Nil	0.639
3000	3.70	14	14	0.643
3250	3.72	14	08	0.625
3500	3.75	14	10	0.618
Before aeration	3.75	-	10	-
After aeration	3.90	-	Nil	-
3900	3.79	12	10	0.606
4300	3.85	12	10	0.619
4800	3.95	12	10	0.616
6200 *	4.16	10	10	0.623

[* These last samples were analyzed for the NaOH consumption required to raise the pH to neutral.]

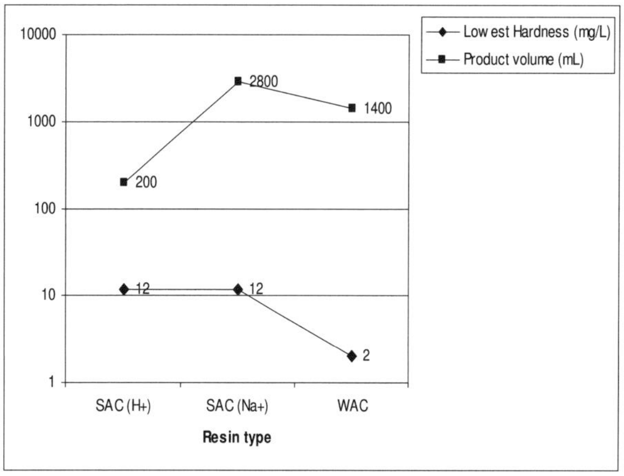

Fig. 2. Comparison of results of SAC (H+), SAC (Na+) and WAC.

The product water of the regenerated WAC resin was further treated with 0.02N NaOH to make it neutral. Nonaerated and aerated samples in duplicates were treated with 0.02N NaOH for attaining the final product water pH around 7 and alkalinity of this neutral water was noted.

The aerated/degasified samples proved better as compared to the non aerated samples. The aerated/degasified samples required 0.40 mL of 0.02N NaOH to raise the pH to an average value of 8.06 and the respective alkalinity was 13 mg/L; against the non-aerated samples requiring 2.2 mL of 0.02N NaOH raising the pH to an average value of 7.07 and respective alkalinity of 44 mg/L. It indicates that final aeration after neutralization would act as a polishing process resulting in the stabilized water.

4. Conclusions

The following conclusions in terms of performance and development of the appropriate water treatment scheme are supported by the experimental results.

- A promising yield of 1400 mL of product water was obtained with 99.42% reduction of the hardness and no alkalinity using WAC, though the pH was required to be raised to neutral by gasification in the final stage.

- The use of SAC requires large brine solutions for regeneration. WAC regeneration by dilute acid eliminates large brine solution handling and its associated possible hazards thus making the whole system environmentally compatible.
- The water treatment scheme of multi-media pressure filtration and activated carbon filtration, followed by ion exchange treatment using weak acid cation and final gasification would be the most appropriate solution for the intended use, considering the product water quality, quantity, operating economics and environmental costs.

References

1. Vogt Power International Inc. Technical Bulletin. 4000 Dupont Circle, Suite 400 Louisville, Kentucky 40207, pp. 1-5, January 2001.
2. http://www.watertreatmentboks.com/
3. water-treatment-boiler.asp
4. E. Rakanta, E. Daflou, and G. Batis, "Evaluation of corrosion problems in a closed air-conditioning system: a case study," Desalination 213, pp. 9-17, 2007.
5. Nalco Chemical Company, "Ion Exchange Process," One Nalco Center Naperville, Illinois 60563-1198, pp. 1-8, 1998.
6. Yoram Oren, "Capacitive deionization (CDI) for desalination and water treatment — past, present and future (a review)," Desalination 228, pp. 10-29, 2008.
7. Hartford, CT 06134. Publication No. 21: Reverse Osmosis Treatment of Private Drinking Water Systems. Department of Public Health, Environmental Health Section, Private Well Program, 450 Capitol Avenue, MS#51REC, PO Box 340308, pp. 1-3, March 2009.
8. http://www.wwdmag.com/Ion-Exchange-Resins-and-Processesfor-Industrial-Water-reatmentarticle4224
9. Santosh Kumar Garg. Water Supply Engineering (Environmental Engineering, Volume-I). Khanna Publishers, Nai Sarak, Delhi 110006, 1999.
10. Ion Exchange, "Role of Weak Acid Cation Resin in Water Treatment," Water Tech, Ion Exchange (India) Limited, Dr. E. Moses Road, Mahalaxmi, Mumbai 400 011, India, pp. 1-4, March 2004. http://www.rsc.org

CFD-PBM MODELING OF VERTICAL BUBBLY FLOWS

MAHMOOD REZA RAHIMI
Chemical Engineering Department, Yasouj University, Yasouj, 75914, Iran, mrrahimi@mail.yu.ac.ir

HAJIR KARIMI
Chemical Engineering Department, Yasouj University, Yasouj, 75914, Iran

A 3-D Eulerian-Eulerian multiphase Computational Fluid Dynamics (CFD) model combined with Population Balance Modeling (PBM) were presented for two phase bubbly flow in vertical pipes. The discrete bubble sizes prescribed in the dispersed phase were tracked by solving an additional set of transport equations, which these equations were progressively coupled with the flow equations during the simulations. Important flow quantities such as local void fraction, liquid velocity and normal turbulent stresses were calculated and compared against experimental data from literature. Good agreement with the experimental data was obtained. It was found that void fraction profile exhibited a sharp peak near the wall. The liquid velocity profile was flattened by the presence of the vapor phase (i.e. the bubbles), and all three normal fluctuations were affected by the presence of the vapor phase. These fluctuations do not increase monotonically as the void fraction increases. Thus this CFD-PBM modeling can be used for prediction of wall peaking and coring phenomena, and radial void distribution.

Keywords: CFD, bubbly flow, vertical pipe, population balance modeling, gas void fraction, liquid velocity, normal turbulent stresses.

1. Introduction

Gas–liquid two-phase flow exists in a wide variety of forms, depending on the flow rates of phases, the physical properties of the phases, the geometry and orientation of the pipe. The different interfacial structures are called flow patterns or flow regimes. Vertical two-phase flows are usually classified into four basic flow regimes consist of bubbly flow, slug flow, churn turbulent flow and annular flow [1]. In the bubbly flow, liquid phase is continuous and small dispersed bubbles flow within the liquid. This flow regime is widely take place in many devices in chemical, petroleum, mining, food and pharmaceutical industries. The uncertainty in bubble flow occur from a lack of essential understanding of the local hydrodynamics and rate processes, which govern bubble size and thus the interfacial area between phases. In dispersed gas-liquid flows, the bubble size distribution plays an important role in the phase structure and interphase forces, which, in turn determine the multiphase hydrodynamic behaviours, including the spatial profiles of the gas fraction, gas and liquid velocities, and mixing and mass-transfer behaviours. These influences must be taken into account for obtaining good predictions in wide operating conditions when using computational fluid dynamics (CFD) simulation. The population balance model (PBM) is an effective technique to simulate the bubble size distribution. The PBM was first formulated for chemical engineering purposes by Hulburt and Katz [2]. The population balance model has been used extensively for modeling of different processes. The discrete bubble sizes prescribed in the dispersed phase were tracked by solving an additional set of transport equations, which these equations were progressively coupled with the flow equations during the simulations. Few attempts have been reported on the modeling of turbulent two phase bubbly flows [3-9]. The main aim of this work is to examine the ability of proposed CFD-PBM modeling, especially for showing packing and coring phenomena. In this study turbulent bubbly air/water two-phase flows in a circular pipe were investigated. The internal phase distribution of air-water bubbly flow in a 57.15 mm i.d. 3.06 m length vertical pipe has been modeled using the 3-D Eulerian-Eulerian multiphase flow approach combined with Population Balance Modeling (PBM). Important flow quantities such as local void fraction, liquid velocity and the Reynolds stresses were calculated and compared against experimental data of Wang et al. [10].

2. Modeling

The numerical simulations presented here are based on the two-fluid, Eulerian–Eulerian model where is based on ensemble-averaged mass and momentum transport equations for each phase. Regarding the liquid phase as the continuous and the gas phase (bubbles) as the dispersed phase, these equations, without interface heat and mass transfer, can be written as follows: Continuity equation of the liquid phase and gas phase with a source term that takes into account the death and birth of bubbles caused by coalescence and break-up processes: the momentum conservation where in which the total interfacial force acting between two phases is the sum of several independent physical effects: drag force, lift force, wall lubrication force and turbulence dispersion forces are considered into account. The break-up of bubbles in turbulent dispersions employs the model developed by Luo and sevendsen [11] and the coalescence rate considering turbulent collision by Prince and Blanch [12]. The drag coefficient has been modeled using drag model of Ishii-Zuber [13]. The lift force is modeled as [14]. The turbulent dispersion force, derived by Lopez de Bertodano [15]. Wall lubrication force is expressed by Antal et al. [16]. Turbulence is taken into consideration for the continuous phase using k- ε model and the influence of the dispersed phase on the turbulence of the continuous phase is taken into account with the Sato's bubble-induced turbulent viscosity model [17].

3. Solution method

An air/water loop was used by Wang et al. [10] had a 57.15 mm i.d. and 3.06 m length test section for measurements of both up and down flows. A single sensor cylindrical hot-film probe was used to measure the mean and fluctuations in the axial liquid velocity and the local void fraction. Reynolds stress components in the liquid phase, measured using a special 3-D conical probe. The grid structure is shown in Fig. 1. CFD analysis was carried out using CFX 10 of Ansys, Inc. The simulations were carried out as 3-D upflow in a vertical pipe based on the Eulerian–Eulerian description combined with Population Balance Modeling (PBM). Conservation equations are discretised using finite volume Technique, upwind scheme was used for all equations, SIMPLEC algorithm was used for pressure-velocity coupling. Water was considered as the continuous phase, and air was considered as the dispersed phase. In the present study, bubbles are equally divided into 10 classes. The PBM model has been used to account for the non-uniform bubble size distribution in a gas–liquid mixture. Several simulations were carried out using progressively larger number of grid points starting from about 40000, there was seen practically no change in the gas volume fraction and liquid velocity profiles when the grid size increased beyond 115000. At the pipe inlet, uniform gas and liquid velocities and average volume fractions have been specified; a relative average static pressure of zero was specified at the pipe outlet. Average volume fraction and uniform liquid velocity profile are specified for initiating the numerical solution. No slip boundary conditions were used at wall.

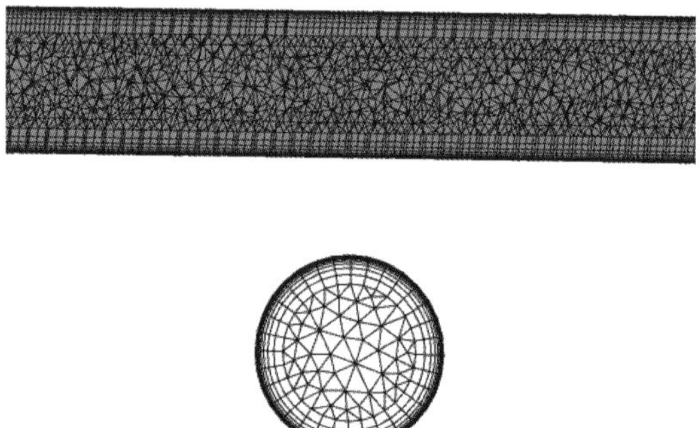

Figure 1. Grid structure.

4. Results and discussion

The proposed model was tested by using this model for an air/water flow system used by Wang et al. [10] had a 57.15 mm i.d. The predicted values of gas void fraction were presented in Figure 2. The results are follows the correct trend, as can be seen in this figure. It was found that the void fraction profile showed a distinct peak near the wall that means bubbles tended to migrate toward the wall. It was also shown that the observed wall peaking and coring phenomena, and thus the radial void distribution, can be predicted. Fig. 2 shows the void fraction distributions obtained from the model comparing with the measured data at the dimensionless axial position z/D = 35. From the phenomenological view point, the phase distribution patterns along the radial direction of the bubble column exhibits four basic types of distributions: "wall peak", "intermediate peak", "core peak" and "transition", as categorized by Serizawa and Kataoka [12]. In the bubbly flow regime, maximum void fraction located close to the wall demonstrated the flow phase distributions typically known as the "wall peak" behavior, which was mainly due to the positive lift force pushing the small bubbles toward the pipe wall. As depicted in Fig. 2, a well-developed wall peaking behavior was recorded in the experiment and had been successfully captured by the model. For higher liquid flows it was found that "wall-peaking" became more pronounced. The predictions are in good agreement with measurements within the error bands, but the discrepancy occurred close to the wall. Overally the peaking and coring of void fraction are well-predicted using this combined CFD- PBM model of this study. Fig. 3 shows the local radial liquid velocity distributions at z/D = 35 for two cases involving liquid single phase flow and two phase flow with specified gas and liquid superficial velocities. Although the liquid velocities at the core were still over predicted for the case of two-phase flow with Jg=0.1m/s. Normal turbulent fluctuations predicted by this modeling is shown in Fig. 4 in comparison with measured data. All Reynolds stress components were measured using a special 3-D conical probe by Wang et al. [10]. In two-phase flows, the normal Reynolds stress components (i.e. u'^2, v'^2 and w'^2) showed nearly flat profiles in the core region *(r/R* < 0.8) and, except near the wall, the turbulence structure was more anisotropic compared to single-phase flows[10]. Normally, the presence of the bubbles increased the level of turbulence in the flow. However, because the bubbles in turbulent two-phase flow enhance dissipation as well as promoting the production of turbulence kinetic energy, it was found that for higher flow rates the presence of bubbles suppressed the level of turbulence. In two-phase flows, the presence of voids tends to flatten the liquid velocity profile for both up and down flows. Moreover, for high flows in the upward direction the location of the maximum liquid velocity occurred off the pipe's centerline. But the maximum liquid velocity and the void peak did not occur at the same location, probably because of the counteracting effect of high shear stress near the wall as shown by Wang et al. [10]. The void fraction profile exhibited a sharp peak near the wall. The liquid velocity profile was flattened by the presence of the vapor phase (i.e. the bubbles). However, a "chimney effect", in which the maximum liquid velocity occurs away from the pipe's center, was observed. Moreover, all three normal fluctuations were affected by the presence of the vapor phase. These fluctuations do not increase monotonically as the void fraction increases.

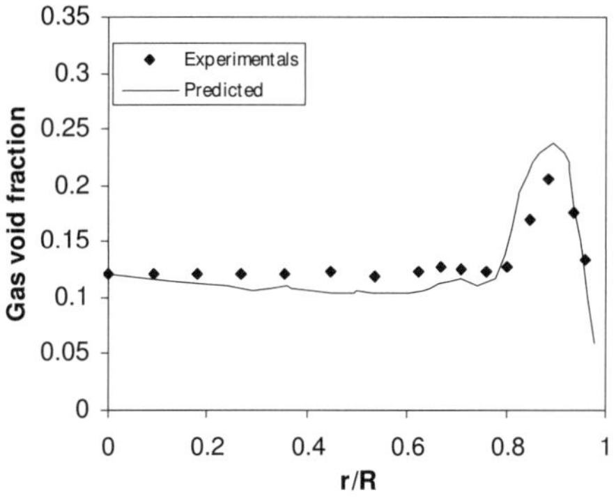

(a) Jg=0.1 m/s and Jl=0.43 m/s

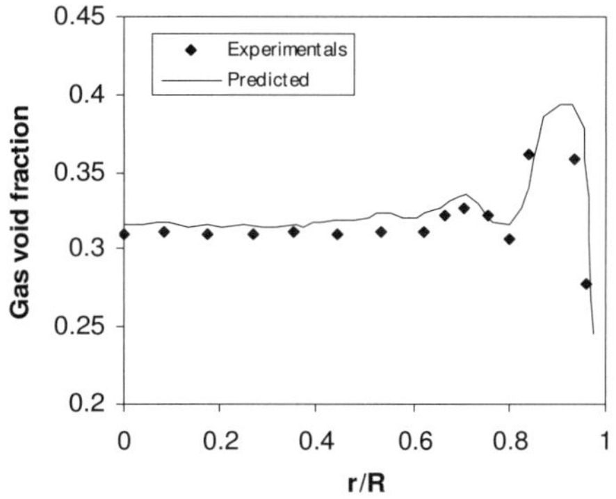

(b) Jg=0.1 m/s na Jl=0.94 m/s

Figure 2. Gas void fraction in comparison with experimental data of Wang et al. [10].

All Reynolds stress components were measured using a special 3-D conical probe by Wang et al. [10]. In two-phase flows, the normal Reynolds stress components (i.e. u'^2, v'^2 and w'^2) showed nearly flat profiles in the core region *(r/R* < 0.8) and, except near the wall, the turbulence structure was more anisotropic compared to single-phase flows [10]. Normally, the presence of the bubbles increased the level of turbulence in the flow. However, because the bubbles in turbulent two-phase flow enhance dissipation as well as promoting the production of turbulence kinetic energy, it was found that for higher flow rates the presence of bubbles suppressed the level of turbulence. In two-phase flows, the presence of voids tends to flatten the liquid velocity profile for both up and down flows. Moreover, for high flows in the upward direction the location of the maximum liquid velocity occurred off the pipe's centerline. But the maximum liquid velocity and the void peak did not occur at the same location, probably because of the counteracting effect of high shear stress near the wall as shown by Wang et al. [10]. The void fraction profile exhibited a sharp peak near the wall. The liquid velocity profile was flattened by the presence of the vapor phase (i.e. the bubbles). However, a "chimney effect", in which the maximum liquid velocity occurs away from the pipe's center, was observed. Moreover, all three normal fluctuations were affected by the presence of the vapor phase. These fluctuations do not increase monotonically as the void fraction increases.

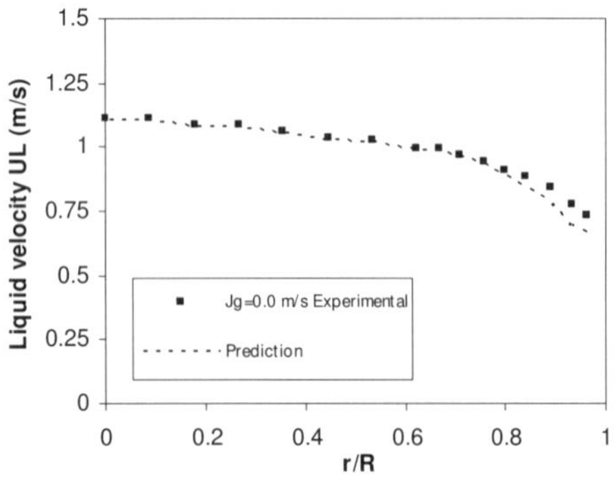

(a) single phase liquid flow with Jl=0.94 m/s

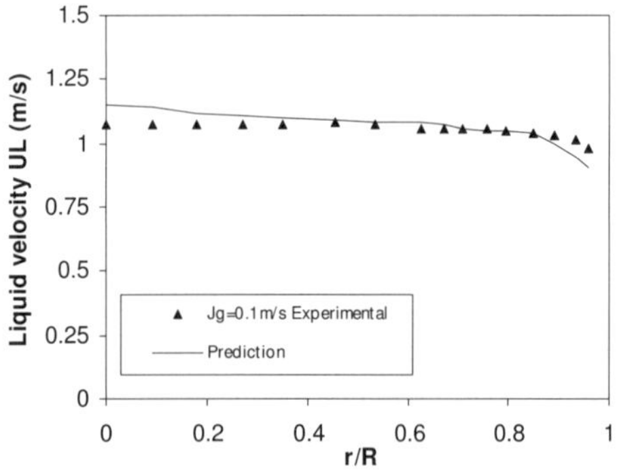

(b) two-phase flow with Jl=0.94 m/s and Jg=0.1 m/s

Figure 3. Local radial liquid velocity distributions at z/D = 35.

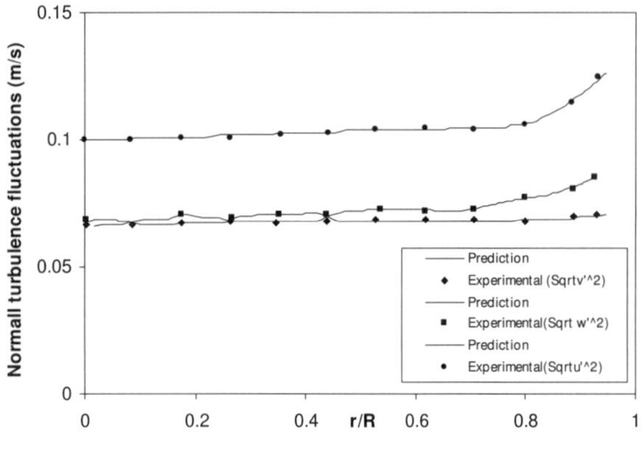

Figure 4. Predicted normal turbulent fluctuations in comparison with measured data of Wang et al [10], z/d=35, Jl=0.43 m/s Jg=0.0 m/s.

5. Conclusion

Turbulent bubbly air/water two-phase flows in a circular pipe were investigated. The internal phase distribution air-water bubbly flow in a 57.15 mm i.d. 3.06 m length vertical pipe has been modeled using the 3-D Eulerian-Eulerian multiphase flow approach combined with Population Balance Modeling (PBM). Important flow quantities such as local void fraction, liquid velocity and the Reynolds stresses were calculated and compared against experimental data of Wang et al. [10]. The void fraction profile exhibited a sharp peak near the wall. The liquid velocity profile was flattened by the presence of the vapor phase (i.e. the bubbles). However, a "chimney effect", in which the maximum liquid velocity occurs away from the pipe's center, was observed. Moreover, all three normal fluctuations were affected by the presence of the vapor phase. These fluctuations do not increase monotonically as the void fraction increases. In the core region, the normal turbulent fluctuations and the void fraction frequently showed flat profiles.

References

1. H. Karimi, M. R. Rahimi, 11th Int. Conf. on Multiphase Flow in Industrial Plants, 443-451 (2008).
2. H. Hulburt, S. Katz, Chem. Eng. Sci. 19, 555 (1964).
3. G. Kocamustafaogullari, M. Ishii, Int. J. Heat Mass Trans. 38 (3), 481–493 (1995).
4. T. Wang, J. Wang, and Y. Jin, Ind. Eng. Chem. Res., 44 19, (2005).
5. T. Hibiki, M. Ishii, Z. Xiao, Int. J. Heat Mass Trans. 44, 1869–1888 (2001).
6. T. Hibiki, M. Ishii, Int. J. Heat Mass Trans. 45, 2351–2372 (2002).
7. K. Ekambara, R.S. Sanders, K. Nandakumar, J.H. Masliyah, Chem. Eng. J. 144 277–288 (2008).
8. C.P. Cheung, G.H. Yeoh, J.Y. Tu, Chem. Eng. Process. 46 (8), 742–756 (2007).
9. S. Sarı, S. Ergün, M. Barık, C. Kocar, C. N. Sökmen, Annals of Nuclear Energy doi:10.1016/j.anucene.2008.11.016 (2008).
10. S. K. Wang, S. J. Lee, O. C. Jones and R. T. Lahey, Int. J. Multiphase flow, 13, 3, 327-343 (1987).
11. H. Luo, H. Svendsen, AIChE J. 42 1225–1233 (1996).
12. M.J. Prince, H.W. Blanch, AIChE J. 36 1485–1499 (1990).

13. M. Ishii, N. Zuber, AIChE J. 25 843–855(1979).
14. I. Zun, Int. J. Multiphase Flow 6, 583–588 (1980).
15. M.A. Lopez de Bertodano, Ph.D. dissertation, Rensselaer Polytechnic Institute, 1992.
16. S.P. Antal, R.T. Lahey, J.E. Flaherty, Int. J. Multiphase Flow 7, 635(1991).
17. Y. Sato, M. Sadatomi, K. Sekoguchi, Int. J. Multiphase Flow 7,167–178 (1981).

MODELING OF SEAWATER INTRUSION FOR A SMALL TROPICAL ISLAND AQUIFER IN EAST MALAYSIA*

PRAVEENA, S. M.†
School of Science and Technology, Universiti Malaysia Sabah Locked Bag 2073
Kota Kinabalu, 88999, Sabah, Malaysia

ARIS, A. Z.
Department of Environmental Sciences, Faculty of Environmental Studies, Universiti Putra Malaysia,
UPM Serdang, 43400, Selangor, Malaysia

ABDULLAH, M. H.
School of Science and Technology, Universiti Malaysia Sabah Locked Bag 2073
Kota Kinabalu, 88999, Sabah, Malaysia

This paper provides a numerical modeling output in Manukan Island aquifer using numerical model, SEAWAT-2000. The modeling output showed that the low lying area of Manukan Island has been affected by seawater intrusion. The upcoming process is taking place beneath the pumping well and is attributed to seawater intrusion occurring in the low lying area of Manukan Island. There was about 14.5% of fresh-water and seawater mixing ratio in the low lying area of Manukan Island. Simulation output provided by this work using numerical model (SEAWAT-2000) can be used by local water management authorities in this case (Sabah Parks authorities), to understand the seawater intrusion in Manukan Island. The modeling output of this study is a foundation to formulate the optimal strategies for mitigation of possible future saline groundwater contamination through management and planning.

1. Introduction

Seawater intrusion can be defined as the seawater inflow into freshwater aquifer. In small islands, seawater intrusion is of-ten the consequence of freshwater aquifers overexploitation [1,2]. Number of tourist arrivals to this island increased about 400% from 1997 to 2004 which attracted by famous dive sites in Manukan Island [3]. This resulted in remarkable increased of groundwater extraction to meet the groundwater supply and domestic needs. Eight dug wells in low lying area which have been used for groundwater extraction have been shut down due to incursion of seawater into the island's aquifer. Only one dug well is currently operating as a pumping well in groundwater extraction to meet the domestic supply. With this current situation, large pumping of groundwater will lead to the depletion of groundwater and deterioration in its quality.

A complete understanding of processes occurring in Manukan Island's aquifer is crucial to development of a sustainable scheme for groundwater management. This study extends previous the findings of Aris et al. [4] in Manukan Island by attempting to study the possible process attributed mixing mechanisms of saline water and fresh groundwater in Manukan Island and investigates the current status of seawater intrusion in a coastal aquifer of Manukan Island using SEAWAT-2000.

2. Materials and Methods

2.1. *Study area*

Manukan Island (5° 57'-5° 58'N and 115° 59'-116° 01'E) in Figure 1 covers an area of 206,000 m^2, about one and half kilometers long and three kilometer wide in the middle. The island has a warm and humid climate and receives annual rain-fall between 2,000 and 2,500 mm. The island consists of unconfined sandy aquifer underlain sedimentary rock (sand-stone and shale). The sedimentary rock dips towards the low lying area (east-northeast) with dipping angles of 15° – 45°. The sedimentary rock forms a slight symmetrical syncline in the low area and can be observed in several locations

*This work is supported by Ministry of Science, Technology & Innovation, Malaysia (ScienceFund: 04-01-10-SF0065).
†The primary author sincerely acknowledges her National Science Fellowship scholarship by Ministry of Science, Technology & Innovation, Malaysia (ScienceFund: 04-01-10-SF0065).

around the island [5,6]. Abdullah et al. [7] conducted a study on the morphological of the island found that the thickness of the aquifer from the ground surface to bedrock are approximately, 5.7 m (northern part), 11 m (southern part) and 12 m (at the middle). Generally, the profiles at low lying area are flatter and thinner than hilly area. Almost 80% of the area is covered by dense vegetation on high relief area (eastern part), while the rest of the area is located on the low lying area of the island (western coastline). On the low lying area of Manukan Island, the small area and low elevations lead to very limited water storage [7]. The flat low lying area of the island has been developed for tourism activities. There are also accommodation available with excellent facilities (i.e., 20 unit chalets, a clubhouse, restaurant, souvenirs, diving centers and recreational facilities such as a swimming pool, foot-ball field, squash and tennis courts), and good infrastructure support (i.e. water, electricity, desalination plant, sewerage system, and even a solar powered public telephone) [3, 4].

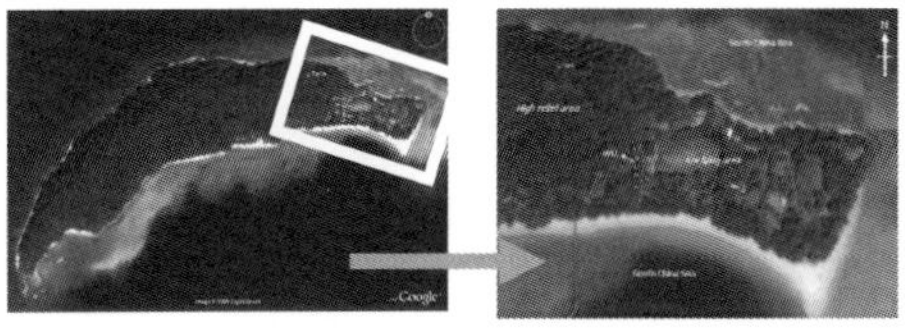

Figure 1. Location of study area (a) Manukan Island (b) high and low lying area.

2.2. *Model design*

Low lying area was selected for this study (Figure 1) as it has been developed for tourism activities. Moreover, all the wells for groundwater extraction are located on the low lying area of the island (8 wells have been shut down and only 1 well is currently operating). Ten boreholes (in-stalled by hand auger manually) were constructed across the study area (Figure 2) to provide a good horizontal and vertical spatial distribution of hydrologic data. Hydraulic heads and chloride concentrations data were collected from October 2008-March 2009 from the wells and boreholes. Water samples were pumped through silicon tubing to polyethylene sampling bottles via a peristaltic pump. Analysis for chloride was done using argentometric method [8].

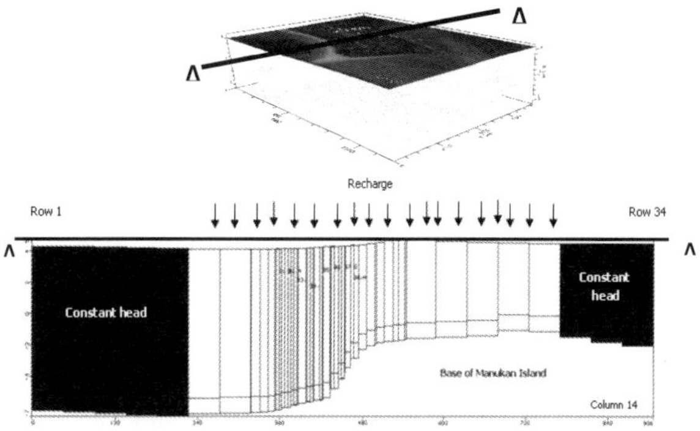

Figure 2. Model grid and boundary conditions for Manukan Island.

SEAWAT-2000 is the latest modeling software available in ground-water modeling that couples flow and transport together [9]. SEAWAT-2000 couples the flow and transport equations of two widely accepted codes MOD-FLOW [10,11] and MT3DS [12]. The SEAWAT-2000 package is very flexible and user friendly. This package has been very useful to simulate variable density flow through complex geological. One advantage of SEAWAT is that it uses MT3DMS to represent solute-transport and an implicit finite-difference method. The model grid (Figure 2) consists of 34 columns and 79 rows with grid spacing of 1093 (x-direction) and 906 m (y-direction). The unconfined layers divided into 2 layers based on the hydrogeological information. Representative properties of the Manukan Island aquifer (Table 1) were assigned to active model cells. Constant head cells were assigned along the bottom of the ocean. A chloride concentration of 19,999 mg/L was assigned for constant concentration while 0 mg/L for recharge concentration. Based on groundwater data availability, one day was chosen as the time step in this study and to be increased by a multiplier factor by 1.2 within which all the hydrological stresses can be assumed constant. Time step is required for transient state. Generally, the smaller the time step, the more accurate the predicted results although it will require excessive computation time. In this study, model calibration is achieved through trial and error approach by adjusting the zonation, values of hydraulic conductivities, dispersivity and effective porosity until the calculated values match the observed values to a satisfactory degree at the end of simulation. It has been done in various groundwater modeling studies [13].

3. Results and Discussion

The overall correlation coefficients (r) are 0.97 and 0.95 for the hydraulic heads and chloride concentrations, respectively. The model calibrations indicate a reasonably good match between the observed and calculated hydraulic heads and chloride concentrations. This indicates the model can be used for further simulation

Aris et al. [4] used the PHREEQC code to understand the mixing mechanisms of saline water and fresh groundwater attributed by seawater intrusion in Manukan Island. The study revealed that the groundwater of Manukan Island experienced mineralization, an indication of seawater intrusion into the island's aquifer which is marked by the presence of Na–Cl water type. Seawater intrusion process was found to be one of the factors that control the saturation states of the carbonate minerals. The mixing of freshwater–seawater creates a diversity of geochemical processes that has altered the theoretical composition of the fresh-water and seawater mixture in the Manukan island's aquifer. The processes involved were complex and varies, spatially and temporally. In this study it ex-tends the findings of Aris et al. [4] to show the possible process attributed to the mixing mechanisms of saline water and fresh groundwater and current status of seawater intrusion in a coastal aquifer of Manukan Island, Sabah.

According to a study done by Mao et al. [14] in Ardeer (Scotland), seawater intrusion occurs in two ways: one is the advancement of the seawater at the bottom of the aquifer result from the density difference between seawater and freshwater. The other was the infiltration of seawater through the beach. Intensive and unregulated exploitation of groundwater from unconfined layer of Manukan Island by pumping wells contributed to the upconing process leads the advancement of the seawater at the bottom of the aquifer result from the density difference between seawater and freshwater. The upconing process showed in Figure 3 by the numerical model output is the possible process attributed to the mixing mechanisms of saline water and fresh groundwater in low lying area of Manukan Island. The upconing process is taking place at the beneath of the pumping well. Saltwater intrusion into a freshwater aquifer occurs when pumping of wells lower the groundwater levels enough to cause seawater to move towards the well. In addition, the model output showed current status of seawater intrusion in Manukan Island is about 14.5% of freshwater and seawater mixing ratio in low lying area of Manukan Island. According to Simpson [15], the size and shape of the upconing will grow and shrink as the rate and duration of well pumping change. This numerical model output is in agreement with study done by Aris et al. [3] in Manukan Island. The groundwater quality of Manukan Island is affected by seawater intrusion attributed to overextraction of freshwater from the aquifer. When a well is pumped in an unconfined aquifer, it can induce a vertical flow component from the underlying saltwater regions, causing the saltwater to migrate upwards and become intercepted by the pumping.

Table 1. Input parameters for the model and simulation strategies for this study.

Parameter	Value
Hydraulic conductivity (K_x, K_y & K_z)	
Layer 1	5.42E-4, 1E-5, 1E-5
Layer 2 (Sedimentary rock)	3.4E-7, 1E-5, 1E-5
Total porosity	0.30
Specific storage	0.0014
Specific yield	0.35
Longitudinal dispersivity	1 m
Horizontal transverse dispersivity	0.1m
Vertical transverse dispersivity	0.01 m
Constant head	0 m
Recharge concentration	0 mg/l
Constant head concentration	19999 mg/l
Initial Concentration	Based on October 2008
Initial head	Based on October 2008

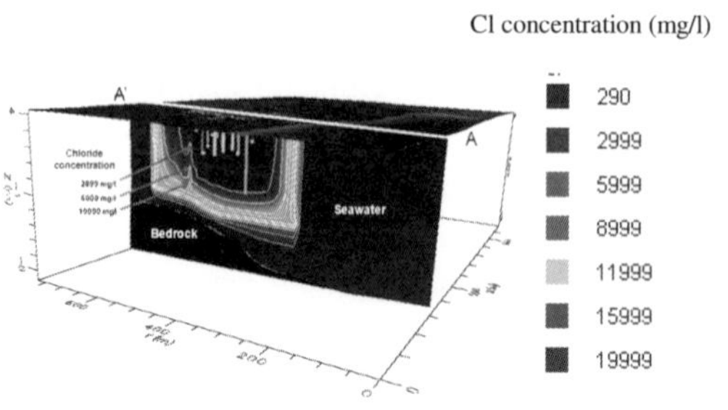

Figure 3. Upcoming process in Manukan Island attributed to seawater intrusion.

4. Conclusions

Seawater intrusion has been affected in the low lying area of Manukan Island. The upconing process is taking place at the beneath of the pumping well and attributed to the seawater intrusion occur-ring in low lying area of Manukan Island. The model simulation indicated that there is about 14.5% of freshwater and seawater mixing in low lying area of Manukan Island. Groundwater quality of Manukan Island is affected by seawater intrusion attributed to overextraction of freshwater from the aquifer. The results of this study are foundation in order to work out comprehensive strategies for mitigation of possible future saline groundwater contamination through management and planning.

References

1. E.D.P. Perera, K. Jinno, A. Tsutsumi, and Y. Hiroshiro, Memoirs of the Faculty of Engineering, Kyushu University 68, 93-106 (2008).
2. M.G. Khublaryan, A.P. Frolov, and I.O. Yushmanov, Water. Resour. 35, 274-286 (2008).
3. A.Z. Aris, M.H. Abdullah, A. Ahmed, and K.K. Woong, Int. J. Environ. Sci. Technol. 4, 441-450 (2007).
4. A.Z. Aris, M.H. Abdullah, K.K. Kim, and S.M. Praveena, Environ. Geo. 56, 1721-1732 (2009).
5. J. Basir, T. Sanudin, and F.F. Tating, Warta Geologi 14, 1-15 (1991).
6. M.H. Abdullah, M.B. Mokhtar, S. Tahir, and A. Awaluddin, Environ. Geo. 29,112–117 (1997).
7. M.H. Abdullah, M.A. Kassim, and M.N. Hanapi, Borneo Island 12, 1-22 (2002).
8. APHA, 19th Ed. American Water Works Association, Washington (1995).
9. W. Guo, and C.D. Langevin, Technique of Water Resources Investigations, Book 6 (2002).
10. M.G. McDonald, and A.W. Harbaugh, Geological Survey Technique of Water Resources (1988).
11. A.W. Harbaugh, E.R. Banta, M.C. Hill, and M.G. Macdonald, US Geological Survey (2000).
12. C. Zheng, U.S. Environemntal Protection Agency, Oklahoma (1990).
13. R. Rejani, M.K. Jha, S.N. Panda, and R. Mull, Water. Resour. Manage. 22, 23-50 (2008).
14. H. Simpson, http://www.omafra.gov.on.ca (2006).
15. X. Mao, P. Enot, D.A. Barry, L. Li, A. Binley, and D.S. Jeng, J. Hydrol. 327, 110-127. 2005.

A NOVEL STEADY-STATE TEST CYCLE FOR EMISSIONS CHARACTERISATION OF A LIGHT-DUTY DIESEL ENGINE FUELLED WITH BIODIESEL

NG JO-HAN, NG HOON KIAT, JOURABCHI SEYED AMIRMOSTAFA, YOU CING MIN

Department of Mechanical, Materials and Manufacturing Engineering, The University of Nottingham Malaysia Campus, Jalan Broga, 43500 Semenyih, Selangor, Malaysia

GAN SUYIN

Department of Chemical and Environmental Engineering, The University of Nottingham Malaysia Campus, Jalan Broga, 43500 Semenyih, Selangor, Malaysia

The environmental impacts of diesel emissions coupled with the imminent depletion of petroleum, has led to extensive research efforts on alternative fuels. Biodiesel has emerged as the favoured alternative fuel for diesel engine. With this in mind, comprehensive emissions mapping would be required to ascertain the effectiveness of biodiesel in emissions reduction. In this experimental study, a modified steady-state heavy-duty diesel engine test cycle was used. Test point reduction was proposed to reduce the time required within an acceptable reduction in accuracy of the test data. Using the reduced test cycle, the effects of biodiesel from palm, soybean and coconut on tailpipe emissions were compared against that of conventional diesel.

1. Introduction

The mounting worldwide crises of crude oil reserves exhaustion and global warming has necessitated the search for an alternative fuel that is sustainable, economically feasible and environmental friendly for mass adoption. In the context of on-road diesel engines, biodiesel appears to have an edge over other alternative fuels due to its compatibility with present day conventional diesel engines and infrastructure [1].

The emissions characteristics from the combustion of biodiesel in compression ignition engines have been widely researched, mainly on heavy-duty diesel engines. The reported potential benefits include sustainable energy supply, cleaner combustion with reduced exhaust species such as carbon dioxide (CO_2), carbon monoxide (CO), sulphur dioxide (SO_2), particulate matters (PMs) and unburnt hydrocarbons (UHCs) [2,3].

There is an emerging trend favouring light-duty diesel engine to power passenger car, where market share has exceed 50% in European countries especially Austria, Spain, France and Italy [4]. Thus, concentrated research efforts are required on light-duty diesel engines, preferably following a reference test cycle, where it can be either transient or steady-state depending upon the experimental set up and test conditions needed. Test for light-duty diesel engines usually require the engine to be tested on a prescribed 'drive cycle' on a chassis dynamometer which is of transient type [5]. In the EU, the New European Driving Cycle (NEDC) urban driving cycle is used to simulate driving conditions of a typical light-duty vehicle. The US has its own light-duty exhaust emissions test procedure in the form of the Federal Test Procedure-75 (FTP-75). However, there is a lack of certified mainstream steady-state light-duty diesel engine emissions cycles. Furthermore, not all light-duty diesel engines could be tested under transient conditions due to the configuration of the test stand, dynamometer and engine combination. Hence, a steady-state light-duty diesel engine emissions cycle could be used as a tool for engine research.

With that in mind, the main objectives of the study is to assess the capability of biodiesels from feedstocks such as palm, soybean and coconut in the reduction of tailpipe emissions under a modified steady-state heavy-duty diesel engine test cycle adapted for use in light-duty diesel engines. As experimental engine testings are time consuming and costly, a method for test point reduction is also proposed. Using the proposed method, time required for the experiment is halved with acceptable trade-off in the accuracy of the results.

2. Experimental Setup

The test engine used in this study is a single cylinder four-stroke diesel engine. It is air-cooled and has a swept volume of 347cm^3 with a compression ratio of 22:1. It has a rated power of 4.6 kW at 3500 rev/min.

3. Test Cycle

The 13-mode European Stationary Cycle (ESC) for heavy-duty diesel engines is modified for use with a light-duty diesel engine in the experimental test programme. The conversion of a heavy-duty test cycle to that of a light-duty equivalent is not expected to pose any problems as the 'drive cycle' for both the light-duty and heavy-duty diesel engines in real road conditions are similar. The properties of the idle and 12 non-idle test points for the test engine are tabulated in Table 1.

Table 1. Test speeds, loads and weighting factors according to the ESC 13-mode test cycle.

Mode	Test speed (rev/min)	Test load (kW)	Weighting factor (%)
1(Idle)	1500	Minimum load	15
2	2069	2.65	8
3	2344	1.53	10
4	2344	2.30	10
5	2069	1.33	5
6	2069	1.99	5
7	2069	0.66	5
8	2344	3.06	9
9	2344	0.77	10
10	2618	3.35	8
11	2618	0.84	5
12	2618	2.51	5
13	2618	1.68	5

3.1. *Test cycle reduction*

For the purpose of time and cost reduction, it is proposed that the test cycle be reduced from 13-mode to a non-idle 6-mode, while redistributing the weight factor of each points to mimic that of the original 13-mode case. The idle mode is removed as the fuel consumption measuring device used could not accurately measure the minute amount of fuel used in the idling mode. The idle mode would cause alteration in the simulation of typical driving conditions, but would still remain acceptable as emissions could still be tested extensively over the entire speed-load map of the engine.

Six points from the ESC test cycle, namely modes 2, 3, 4, 7, 10 and 11 are selected for the test engine as shown in Figure 1. The extreme corner points of 2, 7, 10 and 11 are chosen as they encompass the largest possible non-idle area of the engine speed-load map. The points are complemented by the selection of points 3 and 4 where significant weights are given to the mid-load and mid-speed points.

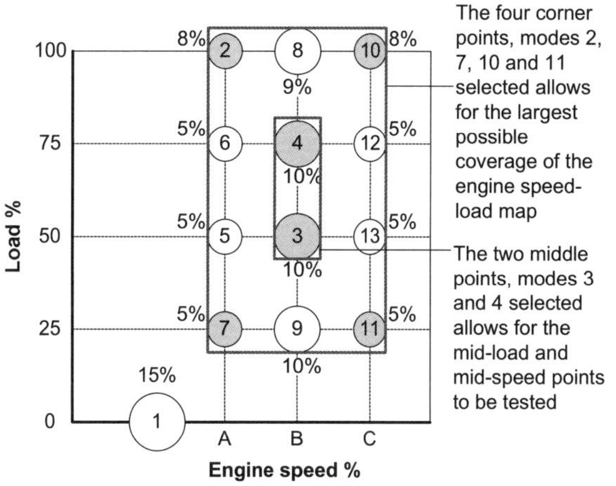

Figure 1. Selected points for the reduced ESC non-idle 6-mode test cycle.

3.2. *Determination of weighting factor on test points*

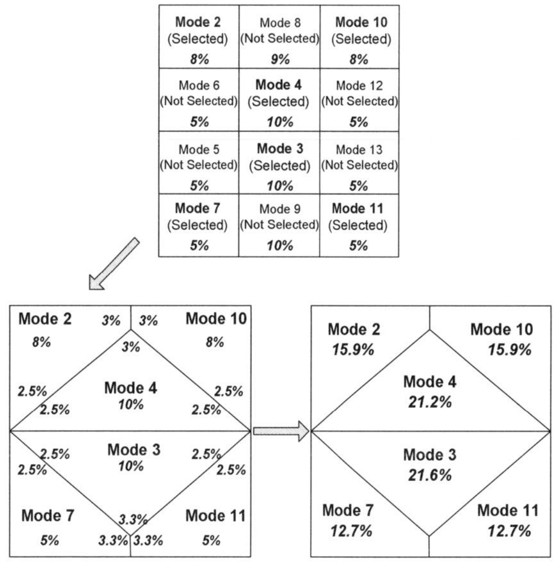

Figure 2. Converting weighting factor of ESC points into a reduced ESC non-idle 6-mode test cycle.

The removal of points requires the redistribution of weighting factors to match that of the original ESC test cycle. Disregarding the weighting factor of idle mode, the non-idle weighting factor is adjusted by redistributing the value of the weighting factor from the eliminated points equally among adjacent points of the retained modes. Figure 2 illustrates a two-step method of converting the weighting factors of the points in ESC into a reduced ESC 6-mode test cycle, where weighting factors of non-selected points are divided equally among adjacent selected points before they are normalised in the final step.

This will also ensure that the delineation of the test area within the speed-load map from the 12 non-idle points to the reduced 6-point test cycle are done evenly without bias. The test points using the reduced ESC 6-mode test cycle are summarised in Table 2.

Table 2. Test speeds, loads and weighting factors according to the reduced ESC 6-mode test cycle.

Mode	Test speed (rev/min)	Test load (kW)	Weighting factor (%)
1	2069	2.65	15.9
2	2344	1.53	21.6
3	2344	2.30	21.2
4	2069	0.66	12.7
5	2618	3.35	15.9
6	2618	0.84	12.7

3.3. *Validation of the reduced ESC non-idle 6-mode test cycle*

The reduction of test points from 13-mode to the non-idle 6-mode test cycle would only be valid if the time saved outweighs the trade-off in accuracy. As the reduced form only include non-idle modes, a fair comparison would be to compare the weighted values of the 12 non-idle modes from the original 13-mode cycle against that of an adjusted 6-mode reduced test cycle. As the idle mode is assigned a weighting factor of 15%, the original non-idle 12-mode remained would only have a total weighting factor of 85%. Thus, the reduced non-idle 6-mode weight value would need to be adjusted by dividing the value with 0.85 to compensate for the omission of the idle mode.

Table 3. Weighted values and absolute errors from the non-idle form of the original ESC and reduced ESC test cycle for all the pure blends of the fuels tested.

Emission type	Original ESC non-idle 12-point weighted values	Reduced ESC non-idle 6-point weighted values	Absolute error (%)
CO (% vol)	0.0242	0.0259	6.9724
CO_2 (% vol)	3.9809	4.0064	0.6397
HC (ppm vol)	9.3300	9.8456	5.5257
O_2 (% vol)	12.237	12.197	0.3315
NO (ppm vol)	329.35	327.47	0.5721

A comparison of the weighted values obtained from the non-idle form of the original ESC and reduced ESC test cycle for baseline diesel is tabulated in Table 3. Clearly, experimental test time required can be reduced by half with a trade-off of not more than 10% error on the average for all engine-out emissions. Thus, the reduced ESC non-idle 6-mode test cycle is validated. It is expected that as the resolution of the measurement unit improves, the average absolute error will be decreased.

4. Results

This set of results reflect the effects of biodiesel from various feedstocks on the emissions characteristics on a light-duty diesel engine tested on a reduced steady-state, ESC, non-idle, 6-mode test cycle. Figure 3 shows the normalised emissions value of the various biodiesels against diesel, where diesel is set as the baseline.

It is observed that the UHC for biodiesels of any feedstock are noticeably lower as compared to fossil diesel, in spite of the higher fuel consumption of biodiesels. The lower engine-out UHC concentration is an indication that the combustion of biodiesels is more efficient than that of fossil diesel. The improved biodiesel combustion can be highly attributed to the fuel-bound oxygen content, where locally fuel rich in-cylinder regions are reduced significantly.

Since O_2 is needed for complete combustion, a lower O_2 concentration at engine tailpipe meant that more O_2 is consumed and involved in the combustion process. Hence, the superior combustion of biodiesel can be further reinforced by comparing the tailpipe O_2

concentration. Lower concentration is observed for all biodiesel except for palm B100 which is higher by 0.01% as compared to diesel. Despite the higher tailpipe O_2 for palm B100, the overall consumption of O_2 would have been higher if the O_2 content within the biodiesel fuel is taken into account.

The production of thermal NO which forms the majority of NO released from diesel engines are dependent on local peak in-cylinder temperature, where high temperatures are the main factor for the production of thermal NO. Thus, the improved combustion efficiency of biodiesel leads to a higher in-cylinder temperature, where NO formation rate is potentiallly increased. This is shown where the NO concentration for palm B100 and soy B100 as compared to baseline diesel are 6.44% and 19.85% higher, respectively. However, the NO concentration for coconut B100 is reduced by 7.47% as compared to conventional diesel. A closer investigation on the order of gross energy content of the pure biodiesels tested yield a trend where energy content coincides with the amount of NO released. The energy content for coconut B100, palm B100 and soy B100 are 38.21 MJ/kg, 39.68 MJ/kg and 39.89 MJ/kg, respectively and the corresponding engine-out NO concentration are 359 ppm, 413 ppm and 465 ppm by volume. Hence, the lower heating value of coconut as compared to the other two biodiesels meant that reduced heat is released from the fuel, hence lowering the in-cylinder temperature which in turn retarded the formation of NO.

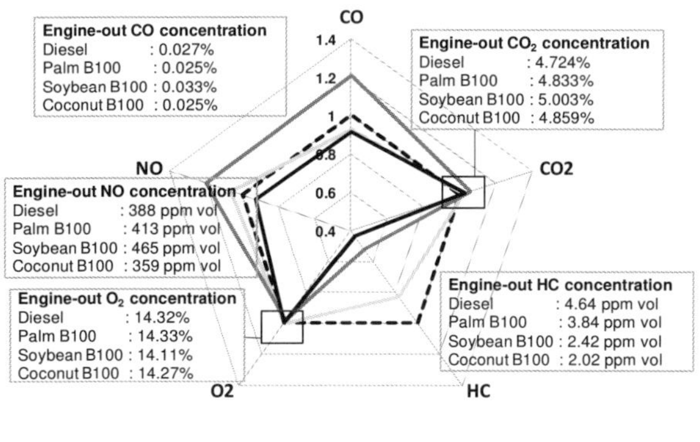

Figure 3. Comparison of the reduced ESC non-idle 6-mode test cycle between diesel and B100 biodiesel of various feedstocks.

CO_2 released from conventional diesel is the lowest among all fuel tested. This is due to the weaker combustion of diesel to convert the incomplete combustion product, CO into complete combustion product which is CO_2. This is further indicated by the lower CO emitted from palm B100 and coconut B100 which further proves that the combustion process of biodiesel fuels is complete.

5. Conclusions

The ESC steady-state heavy-duty test cycle is adapted for use in a light-duty diesel test engine. A reduction in test points which reduces experimental time by half was proposed and successfully validated with less than 10% degradation in the accuracy of the results obtained. The trade-off between time saved and accuracy is justified.

Results obtained showed that the oxygen content in biodiesels plays a crucial role in dictating the combustion process which in turn affects the emissions characteristics. The oxygen content in biodiesel is especially useful in limiting locally fuel rich regions, resulting in improved combustion. The improved biodiesel combustion over fossil diesel produced a reduction in engine-out UHCs. The intermediate combustion product, CO is also more readily converted into complete combustion product, CO_2, hence CO for biodiesel reduces as compared to fossil diesel. The reverse occurs for CO_2. However, as a consequence of the improved combustion, in-cylinder temperature is raised, increasing engine-out NO concentration. However, NO concentration is reduced as exhibited by coconut B100 if energy content is too low, as heat released is lower. As demonstrated in this experimental study, the use of biodiesels can potentially reduce engine-out emissions.

Acknowledgments

The Department of Mechanical, Materials and Manufacturing Engineering from the University of Nottingham Malaysia Campus and the World Federation of Scientists are acknowledged for contributions towards this research project.

References

1. S. Kalligeros, F. Zannikos, S. Stournas, E. Lois, G. Anastopoulos, Ch. Teas and F. Sakellaropoulos, *Biomass & Bioenergy* **24**(2), 141-149 (2003).
2. S.C.A. de Almeida, C. R. Belchior, M. V. G. Nascimento, L. D. S. R. Vieira and G. Fleury, *Fuel* **81**(16), 2097-2102 (2002).
3. EPA, "A comprehensive analysis of biodiesel impacts on exhaust emissions." EPA420-P-02-001. *United States Environmental Protection Agency* (2002).
4. H. Eichselder and A. Wimmer, *Atmospheric Environment* **37**(37), 5227-5236 (2003).
5. A. J. Martyr and M. A. Plint, "Engine Testing: Theory and Practice," 3rd ed. *Butterworth-Heinemann* (2007).

Session 5

INFLUENCE OF TEMPERATURE ON THE REMEDIATION OF PAH-CONTAMINATED SOIL USING VEGETABLE OIL*

E. V. LAU†

Faculty of Engineering, The University of Nottingham Malaysia Campus
Jalan Broga, 43500 Semenyih, Selangor Darul Ehsan, Malaysia

S. GAN and H. K. NG

Faculty of Engineering, The University of Nottingham Malaysia Campus
Jalan Broga, 43500 Semenyih, Selangor Darul Ehsan, Malaysia

This paper aims to treat PAH-contaminated soil with vegetable oil, a more environmentally-friendly alternative to other organic solvents. The difference in PAH removal rate using palm kernel oil (PKO) and soybean oil (SO) is compared. An empirical model is employed to describe the kinetics of PAH dissolution and predict equilibrium concentrations of extracted PAHs in oil. A comparison of different temperatures (30, 50 and 70°C) indicated that increasing temperature enhances the extraction efficiency of PAHs with optimum extraction efficiencies of >88% occurring at 50°C and 70°C for PKO and SO respectively. Thus, the potential of vegetable oil as a promising extraction agent for PAHs in soil is highlighted.

1. Introduction

Polycyclic aromatic hydrocarbons (PAHs) in soils pose potential risks to human and ecological health due to their hydrophobic nature and high resistance to environmental degradation. Various countries around the world including the Netherlands, Denmark and Canada have stipulated guidelines for PAH concentrations in soils. Former industrialised sites which are heavily contaminated with PAHs have to undergo remediation prior to conversion to housing areas or parklands. Remediation techniques for PAH-contaminated soils cover a wide range of separation and/or destruction processes including extraction, oxidation, incineration and bioremediation.

Soil washing/flushing is a solvent extraction method that is widely used in the treatment of contaminated soils. Water and various solvent mixtures including cyclohexane and ethanol mixture (3:1), dichloromethane, ethanol, methanol, methyl ethyl ketone, N,N-dimethylacetamide, N,N-dimethylformamide, N-butyl acetate, N-propyl acetate and toluene have been tested at laboratory-scale to remove PAHs from highly contaminated soil [1,2]. Similarly, surfactants have also been tested as additives to enhance the solubility of PAHs in water by partitioning them into the hydrophobic cores of surfactant micelles. Several non-ionic surfactants such as Tween 40, Tween 80, Brij 30, Brij 35, T-Maz 80, T-Maz 20, CA 620 and TerraSurf 80 as well as anionic surfactants including Dowfax 8390, sodium dodecyl benzene sulfonate, sodium dodecyl sulphate and Steol 330 have been shown to be effective for PAH removal from soil [3,4].

However, due to the toxicity of these organic solvents and surfactants, other non-toxic and biodegradable alternatives such as cyclodextrin [5] and vegetable oil [6-9] have been proposed. Although the addition of cyclodextrin in soil washing solutions enhances the extraction efficiency of PAHs from soil, the availability of this compound and its cost effectiveness are factors yet to be considered in wide-scale adoption of this method.

As such, recent studies advocate the use of vegetable oil as an environmental-friendly yet cost-effective solvent. Vegetable oil is a strong sorption medium for hydrophobic compounds such as PAHs. Furthermore, its free fatty acids with molar solubility ratios for PAHs are similar to those of synthetic chemical surfactants [6]. The advantages of using peanut oil and sunflower oil to remove PAHs from spiked soil and manufactured gas plant (MGP) soil have been reported by Pannu [7] and Gong [8,9] respectively with

*This work is supported by Faculty of Engineering, The University of Nottingham Malaysia Campus.
†Work partially supported by the New Researchers Fund, The University of Nottingham UK.

extraction efficiencies above 80%. It was also suggested that the remaining oil in the soil has the potential to enhance biodegradation of residual contaminants by aiding bioavailability and by being a substrate for growth of microorganisms which can biodegrade contaminants [7].

Thus, this study aims to further extend these works by comparing the performances of saturated palm kernel oil (PKO) and unsaturated soybean oil (SO) as extraction agent for PAHs in soils. Additionally, the influence of temperature on the kinetics of PAH dissolution into these oils are also investigated.

2. Material and methods

2.1. *Materials*

Phenanthrene (technical grade purity >90%) and fluoranthene (purity >98%) were purchased from Sigma-Aldrich (Steinheim, Germany). Reagent grade solvents (>98%) including n-pentane, dichloromethane (DCM), methanol and n-hexane were purchased from Sigma Aldrich while acetonitrile (HPLC, Ultra Gradient Grade) was purchased from Baker. ACS reagent grade (>85%) potassium hydroxide (KOH) in pellet form was purchased from Sigma Aldrich. Fine sand was collected from a beach site and sieved (<500μm) to uniform consistency. Palm kernel oil (PKO) was donated by Lee palm oil mill, Kapar while soybean oil (SO) was commercially purchased.

2.2. *Classification of oil type*

PKO is a major saturated oil with high levels of lauric acid (C12:0) while SO is classified as highly unsaturated due to its high oleic acids (18:1 9*c*) and linolenic acids (18:2 9*c*12*c*) composition. These two oils were chosen based on their world-wide abundance and large annual production [10].

2.3. *PAH removal with vegetable oils*

Sand samples were spiked with phenanthrene (PHEN) and fluoranthene (FLUO) in DCM to a final concentration of 200 ppm in the sand. Extractions were performed in 250 ml Erlenmeyer flask at a soil/oil ratio of 50 g:50 ml in a hooded horizontal shaker with water bath set at temperatures of 30, 50 and 70°C. The sand slurry samples were shaken for a total duration of 24 hours.

2.4. *Oil sample analyses*

A 1 ml aliquot of oil was sampled at time intervals of 0 min, 15 min, 30 min, 45 min, 1 hr, 2 hr, 3 hr, 4 hr, 6 hr and 24 hr and analysed following a saponification and liquid-liquid phase extraction method [11]. The PAH-contaminated oil sample was shaken with 8 ml of methanol and 0.4 g of KOH prior to an overnight incubation at 55°C. PAHs were recovered by liquid-liquid extraction of oil sample, performed by addition of 6 ml of hexane and 6 ml of distilled water, to fraction the extract into two phases.

The upper hexane phase (insaponifiable) was then transferred through a manually prepared silica gel 40 (70-230 mesh, Merck) glass column chromatography (6cm x 10mm diameter) and eluted with 15 ml of pentane, followed by 15 ml DCM:pentane (2:3, v/v). The final eluate was evaporated to dryness using a rotary evaporator and dissolved in 0.1ml of acetonitrile for GC analysis.

2.5. *Gas chromatography (GC) analysis*

PAH samples were analysed in a splitless mode using a GC-FID (Clarus 500, Perkin-Elmer, USA) using a fused silica column of 30 m length, 0.25 m diameter and 0.25μm film thickness (DB5-MS, Agilent Technologies, USA). The injector and detector temperatures were set at 290°C and 300°C respectively. PAH analysis temperature program was set with an initial temperature of 180°C, held for 6 mins and ramped to 280°C at 10°C/min. The final temperature was held for 2 min. All experiments were carried out in triplicates.

2.6. *Dissolution kinetics of PAHs*

An empirical first-order model was employed to describe the kinetics of PAH dissolution from soil (solid phase) into oil (liquid phase). The data was fitted according to the Freundlich-type Eq. (1), which is widely used to study oil processing adsorption [12]

$$C_o = C_e\left[1-\exp(-kt)\right] \tag{1}$$

where C_o is the PAH concentration in the oil phase at time t, k is the mass transfer coefficient, C_e defines the equilibrium oil-phase concentration and t is the PAH contact time with oil [13]. Data fitting curves were plotted using the least squares approach.

3. Results

3.1. *Effect of temperature on PAH extraction*

The kinetics of PAH (PHEN and FLUO) dissolution from soil/oil slurry versus extraction time at 30, 50 and 70oC were plotted against time for both SO and PKO. Curve-fitting was then carried out to determine the total PAH extraction efficiencies from the oil-phase equilibrium concentrations. As can be seen in Table 1, the total PAH extraction efficiencies for both types of vegetable oils in all cases were greater than 77%. The extraction efficiency using SO increased with increasing temperature from an average of 83.1% at 30°C to 88.9% at 70°C. In contrast, the maximum extraction efficiency of 91.4% using PKO occurred at 50°C. This was an increase of 13.8% from 77.5% at 30°C. Further increasing the temperature of extraction from 50°C to 70°C resulted in a 4.7% decrease in efficiency.

Table 1. Extraction efficiencies of SO and PKO at different temperatures.

Oil type	SO			PKO		
PAH	PAH extraction efficiencies at temperature (%)					
	30°C	50°C	70°C	30°C	50°C	70°C
PHEN	84.7	85.4	91.0	79.4	90.6	80.8
FLUO	81.5	81.0	86.9	75.6	92.1	92.4
Average	83.1	83.2	88.9	77.5	91.4	86.6

3.2. *Rate of PAH dissolution*

The mass transfer coefficient, k derived from the fitted curves using the first-order empirical model suggested an increasing trend of k against temperature as shown in Figure 1. It is also noted that the coefficients for both PHEN (3-ring) and FLUO (4-ring) do not differ significantly.

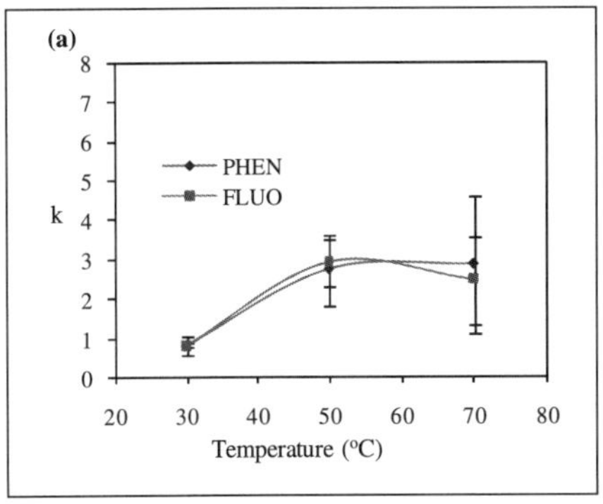

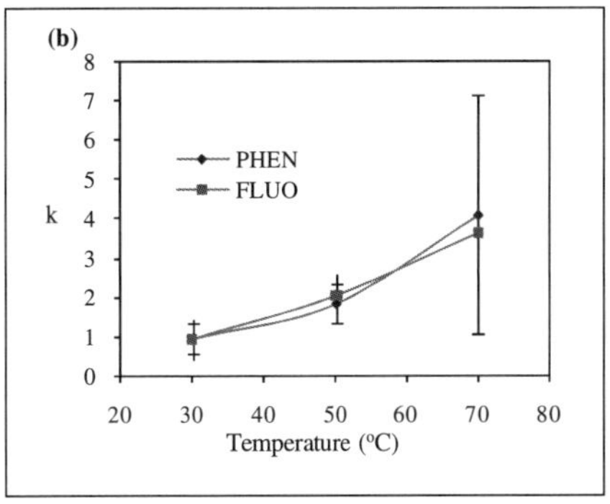

Figure 1. Mass transfer coefficient, k against temperature using (a) soybean oil (SO); (b) palm kernel oil (PKO).

4. Discussions

The results obtained demonstrated the capability of SO and PKO to remove PAHs in a rapid dissolution manner from soil within 24 hours. The increase in temperature enhances PAH solubility by lowering both surface and interfacial tension which allows more dissolution of PAHs into oil. The increase in extraction efficiency with increasing temperature using SO displays similar trends with peanut oil used by Pannu [7], whereby the total PAH extraction efficiencies increased with temperature from 51.5% at 20°C to 81.4% at 60°C. Luers [14] noted that sorption of PAHs into dissolved organic carbon (DOC) decreases with increasing temperature which may explain the decrease in extraction efficiency above 50°C using PKO.

The increase in mass transfer coefficient, k with increasing temperature indicates that similar PAHs extraction efficiencies can be achieved at higher rates. It was also observed that a higher k value was achieved using PKO at 70oC compared to SO. This may be due to the larger surface tension of SO compared to PKO due to the longer chain lengths of SO [15]. Nonetheless, the effect of temperature on the rate of PAH dissolution also depended on the physico-chemical properties of the PAHs and the soil type [16].

5. Conclusions

This present study reveals that both PKO and SO are suitable candidates to replace organic solvents in PAH-contaminated soil washing techniques. The optimum extraction temperatures using SO and PKO occurred at 70°C and 50°C respectively. Extraction efficiencies of greater than 77% can be achieved with mechanical shaking in 24 hours. The increasing mass transfer coefficient, k with increasing temperature suggests that the optimum PAH removal rate can be achieved in less than 24 hours.

Although the results showed that both SO and PKO are promising alternatives, the feasibility of this study should be further extended into cleaning PAH-contaminated native soil with a combination of silt and clay fractions which are smaller in particle size than sand.

References

1. A. P. Khodadoust, R. Bagchi, M. T. Suidan, R. C. Brenner and N. G. Sellers, *J. of Haz. Mat.* **B80**, 159 (2000).
2. A. Rababah and S. Matsuzawa, *Chemosphere.* **46**, 39 (2002).
3. C. K. Ahn, Y. M. Kim, S. H. Woo and J. M. Park, *J. of Haz. Mat.* **154**, 153 (2008).
4. S. Deshpande, B. J. Shiau, D. Wade, D. A. Sabatini and J. H. Harwell, *Wat. Res.* **33**, 351 (1999).
5. C. Viglianti, K. Hanna, C. Brauer and P. Germain, *Env. Poll.***140**, 427 (2006).
6. B. W. Bogan, V. Trbovic and J. R. Paterek, *Chemosphere.* **50**, 15 (2003).
7. J. K. Pannu, A. Singh and O. P. Ward, *Proc. Biochem.* **39**, 1211 (2004).
8. Z. Gong, K. Alef, B.-M. Wilke and P. Li, *Chemosphere.* **58**, 291 (2005).
9. Z. Gong, B.-M. Wilke, K. Alef, P. Li and Q. Zhou, *Chemosphere.* **62**, 780 (2006).
10. C. Scrimgeour, *John Wiley & Sons*, Bailey's Industrial Oil and Fat Products 6th Edition (2005).
11. A. Meudec, J. Dussauze, M. Jourdin, E. Deslandes and N. Poupart, *J. of Chrom. A.* **1108**, 240 (2006).
12. A. Proctor and D. D. Brooks, *John Wiley & Sons*, Bailey's Industrial Oil and Fat Products 6th Edition (2005).
13. P. J. Woolgar and K. C. Jones, *Env. Sci. Tech.* **33**, 2118 (1999).
14. F. Luers and Th. E. M. ten Hulscher, *Chemosphere.* **33**, 643 (1996).
15. R. Ohlson, *American Oil Chemist's Society*, Dietary Fats and Health (AOCS Monograph), (1983).
16. B. Maliszewskz-Kordybach, *Env. Poll.* **79**, 15 (1993).

THE STRENGTH AND DIRECTION OF THE LINKAGE BETWEEN RURAL ASSETS AND ENVIRONMENTAL DEGRADATION IN UPLAND AGRICULTURAL PRACTICES

PUGUH KARYANTO

Dept. of Biological Education, Faculty of Teacher's Training and Education, Universitas Sebelas Maret, Surakarta 57126, Indonesia

Dept. of International Studies, College of Law Goverment and International Studies, Universiti Utara Malaysia, Sintok, Kedah Darulaman 06010, Malaysia

MOHAMMAD HANAPI MOHAMMAD and FARIZA HASHIM

Dept. of International Studies, College of Law Goverment and International Studies, Universiti Utara Malaysia, Sintok, Kedah Darulaman 06010, Malaysia

Upland bear many diffferent social properties as compare to lowland. It also plays an important ecological role in conserving and buffering its surrounding area. This area has been threatened by agricultural practices that threat its sustainability[6]. Indeed, environmentally unfriendly agriculture has resulted an environmental degradation. Such degradation is related with five types of capital of the rural/livelihood assets[3]. Framework of capacity-environmental degradation was used to asses this linkage[8]. Data is collected by questionnaire. Multiple-linear regression was the tool of analyses used. The multiple-linear regression model does a good job in explaining the linkage between rural assets and agricultural-environmental degradation. Human capital and financial capital are the dominant factors that affect farmer's decision and, agricultural-environmental degradation respectively.

Keywords: Livelihood asset, environmental degradation, upland agriculture.

1. Introduction

Many studies performing ecological approach to assess environmental degradation often give too late results. Societal approaches provide a more holistic approach and give an earlier warning signal1. Hence, focusing on societal aspects is of alternative approaches to prevent environmental degradation.

Five types of capital have been introduced as social factors affecting farmer's decision to act in an environmentally friendly manner[3]. These capitals belong to a newly introduced construct, the capacity[9]. Capacity is defined as a construct of assessing the rural poverty from different side of poverty definition[2]. Indeed, poverty still regard as the main factor causing the environmental degradation[5]. Hence, it is possible to make a linkage between capacity/five type of rural capital and environmental degradation. However, the strength and direction of each capital within this linkage needs to be verified[2,9]. Yet the aim of this research is to examine the relationship between capacity and the extent of agricultural-environmental degradation.

2. Research Methodology

This research is conducted at two sub districts surrounding mount Lawu as a representative upland area. Sample was taken randomly toward 408 upland farmer trough a five scaled-Likert's questionnaire. Such questionnaire is measuring the extent of the five types of rural capital/the capacity i.e. social capital, human capital, physical capital, financial capital and natural capital. These five types of capital are the independent variables as according to this research. As the extent of agricultural-environmental degradation is resulted from farmer's decision and action, hence, questionnaire is also built to assess the farmer's decision as a dependent variable.

The result of the questionnaire was tabulated and analyzed using correlation and linear regression analysis provided by SPSS release 16. Basic assumption to enter the linear regression analysis is also fixed before running the regression analysis. The significant t value (less than 0.05) would be and indication for the strength of the correlation. The direction of the correlation is showed by the standardized beta coefficient[4]. A fit regression model of the linkage between capacity and agricultural-environmental degradation is also presented.

3. Result and Discussion

The *p* value of Kolmogorov-Smirnov normality test is greater than 0.05. Therefore the null hypothesis is accepted, that the error of the dependent variable (the farmer's decision in performing environmentally friendly farming practices) is considered as normal. Other basic assumptions for linear regression analysis (linearity, no multicollinearity, no autocorrelation, and homoscedasticity) are fulfilled (the detail of the result are not presented).

Using SPSS 16, the multiple linear regression analysis is performed, and giving the following results:

Table 1. The R square value resulted from regression analysis.

R	R square	Adj. R Square	Std. error of the estimate
0.806	0.649	0.643	2.285

The value of R square is 0.649, indicating that the regression does a good job. More than 60% of the dependent variable is explained linearly by the independent variables, with only about 35% of the variation of the dependent variable is explained by other factor that not covered by this research. Yet, the chosen independent variables are considered as a good predictor to the change of the dependent variable.

The ANOVA table reports a significant F statistic as follows:

Table 2. The R square value resulted from regression analysis.

ANOVA[b]				
Model	Sum of Squares	Df	Mean Square	F
Regression	3864.219	7	552.031	105.714
Residual	2088.779	400	5.222	
Total	5952.998	407		

The F statistic shows that F = 105,714, with p (0.000) < 0.05. Yet, the model of regression does a good job in explaining the relationship between variables, both, the dependent (farmer's decision) and the independent one (capacity).

Based on the value of R square and the ANOVA test, the regression model is built as according to the table of coefficient as follows:

Table 3. The coefficient table.

Model	Standardized Coefficients (Beta)	t	Sig.
1 (Constant)		8.007	.000
Soccapital	.093	2.369	.018
Humancapital	.296	6.191	.000
Financialcapital	.110	2.771	.006
Physicalcapital	.226	6.927	.000
Naturalcapital	-.181	-4.153	.000

Beta standardized coefficients tells the unique effects size for each independent variable to the dependent variable. The value of *p* must be less than 0.05 for each beta value to make the independent variable is significantly to affect the dependent variable.

Previous table shows that from the chosen five predictors, all predictors are significantly affecting the farmer's decision in positive direction i.e. social capital (0.93; p = 0.018), human capital (0,296; p = 0.000), financial capital (0.110; p = 0.006), physical capital (0.226; p = 0.000). Natural capital (0,181; p = 0.000) is significantly affecting the farmer's decision in negative direction. The regression model then presented in this following equation:

$$Y=12.488+0.93X_1+0.296X_2+0.110X_3+0.226X_4 0.181X_5+e$$

Social capital refers to the value of connectedness and trust between people that characterized by three type of connectedness i.e. bonding, bridging and linking[7]. Social capitals are featured by four aspects i.e. relation/trust; reciprocity and exchanges; common rules, norm and sanction; and connectedness, network and group. Indeed, social capital is strongly affecting farmer's decision to reduce agricultural-environmental degradation. This factor gives a confidence toward the farmer to act in collective activities, knowing that others will also do so. Hence, the affect of this capital is depending on the value and norm embedded within the rural society.

Despite above capital, human capital also plays significant role to affect farmer's decision. Human capital acts as a basic resource for any societal activities. Its degree of education and health contribute significantly to create ignorance[5]. Moreover, ignorance leads to injudicious practices when managing the agricultural environment. Hence, human capital also acts as a factor that determines the degradation.

Financial capital is very critical to support any activities of managing the environment. It provides a financial capability and incentive to invest in the environmentally friendly manner. Indeed, agricultural management always emphasizes on mechanism to reduce the risks, including the financial risk.

Physical capital provides the manufactured capital to farmer to conduct in the environmentally friendly manner. If it is not available and inaccessible to farmer, they will not able to work properly to perform an environmentally friendly practice.

Natural capital is a capital given by nature. Here, in this sense, natural capital is as a set of natural services provided by nature e.g. land fertility and the climatic condition. When farmer know about the comparative value of the nature, they will perform particular managements that fit with its resource limitation.

As pertain to this research, natural capital is significantly affecting the farmer's decision in negative direction. This means that farmer tends to perform over-exploitation agricultural activities towards their given nature and land when they feel that their natural capital is able to do so.

4. Conclussion

According to previous result and discussion, the multiple-linear regression model does a good job in explaining the linkage between rural assets and agricultural-environmental degradation. Human capital and financial capital are the dominant factors that are affecting farmer's decision and, agricultural-environmental degradation respectively.

References

1. Azar C., John Holmberg, Kristian Lindren, *Ecological Economics,* **18**, 89 (1996).
2. Bahamondes M. *Journal of World Development,* **31** (11), 1947 (2003).
3. Bebbington, A., Journal of World Development, **27** (12), 2021 (1999)
4. Kutner M.H., C.J. Nachtseim and J. Neter, *Applied Linear Regression Models*, 4th edition, Mc Graw-Hill Company. Inc., New York (2004).
5. Levine R., 2006. A Whole-System View of Agriculture, People and the Rest of Nature, Russel Press, Nottingham, UK.
6. Partap T., *Sustainable Farming Systems in Upland Areas. Report of the APO Study Meeting on Sustainable Farming System in Upland Areas,* New Delhi 15-19 January 2001.
7. Pretty J., *Social Capital and connectedness: Issues and Implication for agriculture, rural development and natural resource management in ACP countries.* The ACP-EU publisher (2003).
8. Reardon T., Stephen A. Vosti. *Journal of World Development,* **23** (9), 1495 (1995).
9. Swinton S.M., German Escobar, and Thomas Reardon, *Journal of World Development*, **31**(11), 1865 (2003).

PERFORMANCE OF THE CONTINUOUS FLOW OF UNVEGETATED CONSTRUCTED WETLAND WITH DIFFERENT FILTER MEDIA FOR REMOVAL OF CHEMICAL OXYGEN DEMAND AND IRON IN LANDFILL LEACHATE TREATMENT

AHMAD M. D. NOOR and LOU XIAO YING
School of Chemical Sciences, Universiti Sains Malaysia, Minden, Penang, 11800 Malaysia

H. P. S. ABDUL KHALIL
School of Industrial Technology, Universiti Sains Malaysia, Minden, Penang, 11800 Malaysia

S. SURYANI
School of Distant Education, Universiti Sains Malaysia, Minden, Penang, 11800 Malaysia

A study was conducted to investigate the efficiency of unvegetated subsurface flow constructed wetland in leachate treatment. Three reactors (I, II and III) filled with different media of granite, gravel and sand, respectively, were placed in descending order height in order to allow gravity flow of leachate to each reactor. Raw leachate was let to flow continuously from Reactor I to III at a predetermined rate. The leachate was analyzed for chemical oxygen demand (COD) and iron before and after the treatment. The results show that COD removal efficiency was 15.4, 22.6 and 29.2% for Reactor I, II and III, respectively. The removal efficiency for iron in Reactors I, II and III were 20.3, 35.2 and 38.5%, respectively. The overall average COD and iron removal at the end of process were 53.6 and 68.1%, respectively.

Keywords: Chemical oxygen demand, constructed wetland, iron, leachate.

1. Introduction

Landfill is constructed for the ultimate disposal of solid wastes. It is a low cost method of solid waste disposal in which rubbish is buried between layers of soil to fill in lay-lowing land. The municipal solid wastes in Malaysia that have gone for landfilling is approximately 95% and only 5% are recycled [1].

Leachate is produced when water is allowed to come into contact with waste and generation of gases from biodegradation of the waste at landfill sites may cause environmental pollution [2]. Leachate is defined as a liquid that has percolated through soil or solid waste and the dissolution of certain constituent in the water. The production of leachate is affected by numerous factors, including the composition of the waste, precipitation characteristics of the site, site hydrogeology, and climate at the site [3]. Leachate initially is a high-strength water waste, containing high concentration of organic and inorganic matters, nutrients and heavy metal [4].

Therefore, the disposal of landfill generated leachate is widely recognized as matters of concern because of the potential for negatives impacts to the environment [5]. Hence, leachate treatment is needed before discharged into surface water bodies. Constructed wetland has the potential to provide a low cost, natural system for treating wastewater discharged from municipal, agricultural and industrial sources [6]. A constructed wetland is a designed and man-made complex of saturated substrates, emergent and submerging vegetation, animal life, and water that simulate natural wetland for human use and benefits [7]. Constructed wetland may provide a treatment system for leachate that decreases contaminant concentration and total leachate volume and also requires little maintenance, management and energy [8].

The basic types of constructed wetland treatments system are free water surface (FWS) and subsurface flow (SSF) types [1]. FWS consists of basins or channels and barrier such as constructed clay layer or geotechnical material to prevent seepage. In FWS, water surface is exposed to the atmosphere and the intended flow path through the system is horizontal. Conversely, water level is maintained below the surface of a media supporting emergent vegetation in SSF. Both FWS and SSF constructed wetlands have been identified as promising technologies for the treatment of landfill leachate [9].

One of the main concerns in treating leachate is the high concentration of organic matter and heavy metal presence in the water stream [10]. Nonvolatile organic matters are often lumped as Chemical Oxygen Demand (COD) because they can create oxygen deficiencies in receiving water [6]. COD is the amount of oxygen required for the complete oxidation of organic substances in a water sample into CO_2 and H_2O. Decomposable organic matter results in the consumption of dissolved oxygen and thus decreases the amount of dissolved oxygen in the receiving streams.

Iron is also a common constituent of municipal landfill leachate. When left uncontrolled, these constituents can leach into and seriously degrade groundwater aquifers and aquatic habitats [10]. Relatively little work had been done on metals removal in constructed wetland system [6]. The processes of metal removal in constructed wetland include sedimentation, filtration, adsorption, complexation, precipitation, cation exchange, plant uptake, and microbially-mediated reactions.

The main objective of this study was to investigate the capability of SSF unvegetated constructed wetland system consisting of different media (granite, gravel, and sand) that placed in sequence order in treating landfill leachate. The removal of COD and iron will be analyzed before and after the treatment. The overall performance of SSF will be analyzed at different sampling location and at the end of process.

2. Experimental

Methods of parameter analysis were based on the Standard Methods (APHA, 1992) [11].

2.1. *Filter media*

Granite, gravel and sand were obtained from construction materials franchised store and used as constructed wetland filter media in this study. The diameter size of granite was 10 mm $< \phi <$ 25 mm while for gravel and sand was 4 mm $< \phi <$ 10 mm. The media were washed thoroughly with tap water 4 times repeatedly before being used to remove all the debris. Then, the media were sieved with sieve in order to obtain the appropriate size.

2.2. *Reactors*

The tanks with dimensions 50 cm wide, 150 cm long, 60 cm depth were used as reactors. Three reactors, namely Reactors I, II and III were filled with different media namely, granite, gravel and sand, respectively until the depth reached 45 cm for Reactor I and II and 30 cm for Reactor III. Each reactor was fixed with 5 perforated tubes labeled as Inlet 1, Point 2, Point 3, Point 4, Point 5 and overflow (Figure 1).

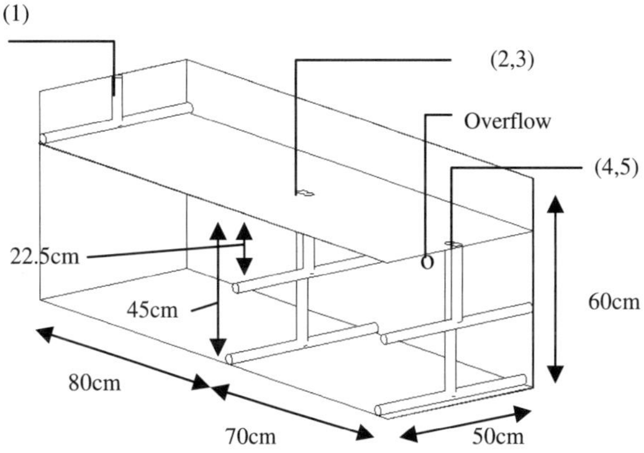

Figure 1. Schematic diagram of a reactor.

Reactors I, II and III were placed in descending order height in order to allow gravity flow of leachate to each reactor. The leachate was let to flow continuously from Reactors I to III. The arrangement of reactors is shown in Figure 2.

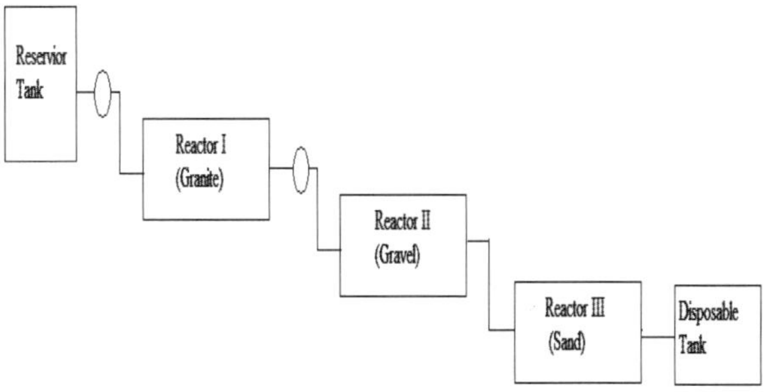

Figure 2. Arrangement of reactors.

2.3. *Sampling*

Leachate sample was taken from the Pulau Burong Sanitary Landfill, Penang and stored in the plastic tank. The sample for analysis will be taken from Inlet 1, Point 2, Point 3, Point 4, Point 5 and overflow.

The observation of the CW performance was carried out in four series of treatment which has taken time about four weeks and for each sample collection was done in triplet.

2.4. *Hydraulic retention time (HRT)*

HRT plays an important role in the performance of constructed wetland for wastewater treatment. It was determined by using lab-scaled reactor with different filter media by analysing COD and iron at different HRT. Based on the performance, HRT for the experiment process was set to be 7 and 5 days for Reactors I, II and III, respectively. The volume of leachate to be fed into the reactor was calculated by ratio of volume of lab-scaled reactor with experiment reactor.

2.5. *Flow control*

The flow rate was set to be 20 mL/min for every reactor based on the void volume of media in reactors. The flow rates must be controlled properly to prevent over flow of leachate which will cause a decline in system efficiency.

2.6. *COD determination*

Determination of COD followed the closed reflux, titrimetric method. The culture tube with the dimension of 16 mm x 100 mm was used. An amount of 2.5 mL of diluted sample was placed into each culture tube followed by 1.5 mL digestion solution and 3.5 mL of sulphuric reagent. The culture tube was tightly capped and inverted several times to mix the solution. Two blanks were prepared by using the distilled water as sample.

The culture tube was placed at the preheat COD reactor at 150°C for 2 hours to perform digestion process. Then, the sample was let to cool and transferred to 250 mL volumetric flask for titration. One or two drops of ferroin indicator solution was added to the sample and titrated with ferrous ammonium sulfate (FAS) solution. Sharp colour changes from yellow → green → blue → reddish brown. The end point was indicated by reddish brown colour. The volume of standard FAS used was recorded. Calculation of COD is based on the following formula:

$$COD\ (mg\ O_2/L) = \frac{(A - B) \times M \times 8000}{Sample\ (mL)}$$

where A - FAS used for blank (mL)
B - FAS used for sample (mL)
M - Normality of FAS

2.7. *Iron determination*

100 mL of sample was digested by wet digestion process using 69% (v/v) HNO_3 on a hotplate. 5 mL of HNO_3 and few boiling chips were then added to 100 mL sample. The mixture was boiled gently and evaporated on a hotplate to the lowest volume possible (about 10 to 20 mL) before precipitation occurs. HNO_3 was added as necessary until complete digestion as shown by light-colour, clear solution. Fe was analyzed with the flame AAS using standard addition method.

3. Results and Discussion

3.1. *COD removal*

The result of COD removal for series of treatment is shown in Figure 3. From the result, all reactors show capabilities in removing COD by the increase trend. This is due to the leachate become decrease in amount after passing through each reactor.

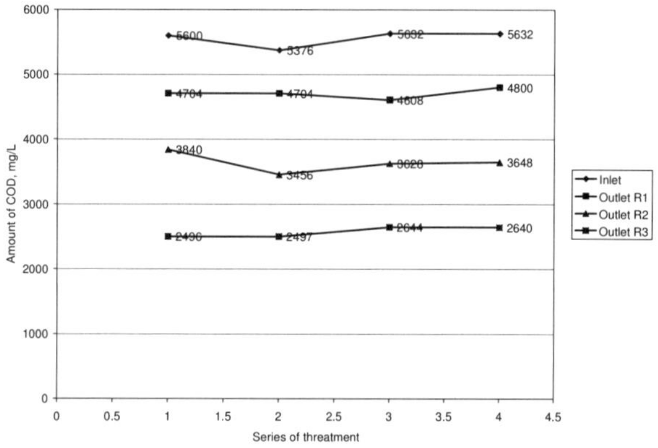

Figure 3. Amount of COD before and after treatment through CW system.

The average COD value of effluent for all four series of treatment are 4704, 3643 and 2580 mg/L resulting in 15.4, 22.6 and 29.2% removal for Reactor I, II and III respectively.

Figure 4 shows the plot of the average COD removal observed at different sampling points for Reactors I, II and III. The result shows decreasing amount of COD from inlet to effluent. COD value for Reactor I has decreases from 5560 ± 124 to 4704 ± 78 mg/L whereas the COD value for Reactor II has decreases from 4704 ± 78 to 3643 ± 157 mg/L. For Reactor III, COD value has decreases from 3643 ± 157 to 2580 ± 99 mg/L. The overall average COD removal efficiency was 53.6%. Reactor I which contains granite shows the lowest efficiency with 15.4%. Reactors II and III with 22.6% and 29.2%, respectively, in removal efficiency show better performance than Reactor I.

Microbial degradation plays an important role in the removal of soluble or biodegradable organic matter (BOD/COD) in waste water [6]. Microbial degradation is the breaking down of organic matter into nutrients that can be used by other organism. From Figure 4, the COD concentrations at Point 3, Point 5 and effluent were lower than the Inlet, Point 2 and Point 4. The arrangements of sampling points are shown in Figure 1. The sampling point of Point 3, and Point 5

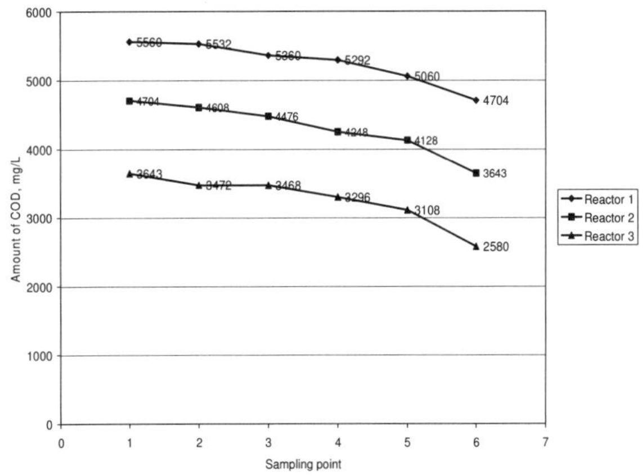

Figure 4. Amount of COD determined at different sampling point.

which were half depth of media (22.5 cm for Reactor I, II and 15 cm for Reactor III) was higher than other sampling point (Point 2 and Pont 4) which were placed at same depth of media (45 cm for Reactor I, II and 30 cm for Reactor III). Hence, organic compounds were degraded aerobically by bacteria attached to the sediment surfaces at Point 3, 5 and Overflow whereas anaerobic degradation taken place at Point 2 and Point 4. Therefore, the results show that organic contaminant comprising COD was removed more efficiently in aerobic zone than anaerobic. This result is also in accordance to Guan [12].

Reactor III with shallower media depth (30 cm) has shown greater COD removal rate (29.2%) compared to Reactor I and II (15.4 and 22.6%) where the media depth was 45 cm. This shows that great proportion of readily degradable organic matter will be removed in the presence of greater oxygen diffusion from the atmosphere through the media surface [8]. The media placed inside the reactor allowed the accumulation of immense amount of bacteria that can help in catalyzing chemical reactions [13]. The higher porosity of gravel and sand which was 30–40% than granite which was only 0.003–0.009% porosity can provide larger attach surface area for bacteria [14]. Thus, gravel (Reactor II) and sand (Reactor III) with 35.2 and 38.5% show higher removal rates than granite (Reactor I) which was 20.3%.

From Figure 3, the COD removal for individual reactors was considered low, ranging only from 12.5 to 35%. This is due to the absence of aeration system [15]. The main treatment mechanism for aerobic removal of organic material is oxygen-limited. In this study oxygen is supplied to media in reactor through oxygen diffusion from atmosphere. The treatment efficiency of COD can be increase with sufficient oxygen supply [15]. Furthermore, this may also due to biodegradation and leaching process [13]. When the bacteria grow, die and decay, organic compound comprising COD can be leach into water flowing through wetlands.

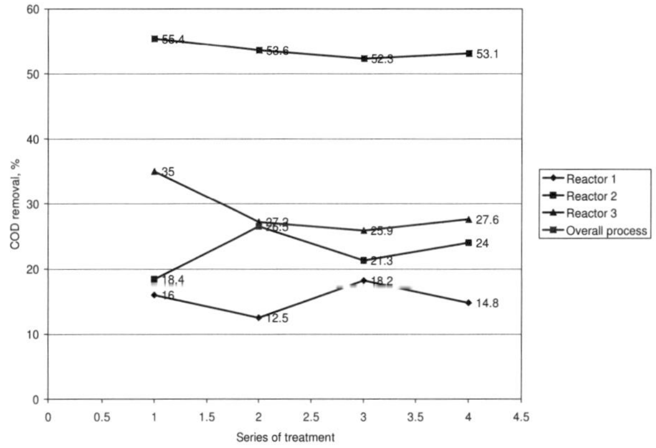

Figure 5. Percentage removal of COD through the CW system.

The overall average removal of COD after passed through all reactors was 53.6%. Figure 5, shows COD removal percentage for each reactor and overall process. It shows that the COD removal percentage after passed through all reactors was 55.4, 53.6, 52.3 and 53.1% for series of treatment 1 to 4, respectively. This shows that the continuous flow of sample can treat the sample more effectively. This may due to the leachate undergoes repeated removal mechanism such as reaeration for each reactor and also by the dilution process of leachate when passed through from one reactor to another.

From the statistical analysis by ANOVA to compare the COD removal of Reactor I, II and III, it found that the COD removal efficiency of three reactors was significantly different.

3.2. *Iron removal*

Figure 6 shows the overall performance of Fe removal of the system through the series of the treatment and found that Fe was removed in all reactors. It can be seen the average initial concentration of Fe which was 4.52 mg/L has decreased to 3.61, 2.35 and finally to 1.8 mg/L for the effluent of Reactors I, II and III respectively. This amounts are represented to 20.3, 35.2 and 38.5% of removal efficiency for Reactors I, II and III, respectively.

The iron rapidly precipitated and was removed in contact with the atmosphere [16]. The presence of metal oxidizing bacteria in the aerobic zones and sulphate reducing bacteria in the anaerobic zones cause the precipitation of metal oxides and sulphate respectively [17]. Higher removal efficiency of Fe was achieved by Reactor III with 38.5% compared to Reactor I, 18.9% and Reactor II 31.5%. This is due to more reaeration process in Reactor III as the media depth, 30 cm was lower than Reactor I and II which was 45 cm.

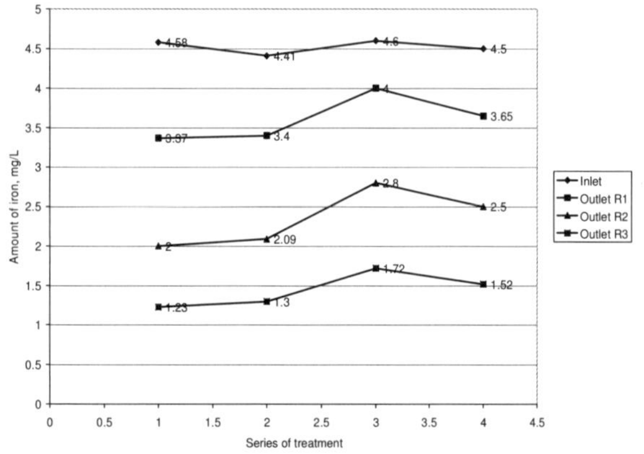

Figure 6. Amount of iron before and after treatment by CW system.

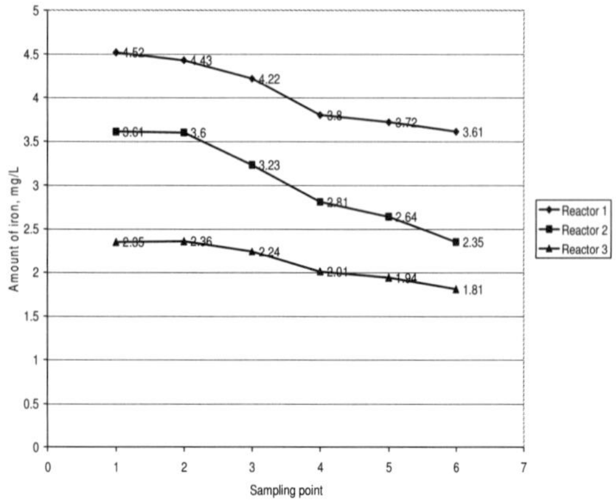

Figure 7. Amount of iron at different sampling point of CW system.

Reaeration is the solubilization of atmospheric oxygen at the water surface from the atmosphere [18]. The lower media depth of Reactor III causes higher efficiency of dissolvation of oxygen from atmosphere. From the Figure 7, it is shown that iron concentration of Point 3 and 5 are higher than Point 2 and 4. The factor mentioned above can also explain the higher removal rate of iron in Point 3 and 5 in all reactors where more reaeration occurs at upper surface of media.

Metals also removed in wetlands through adsorption onto solid-phase surfaces and chelation or complexation by organic material and biofilm [13]. The iron entering the system was oxidized and bound to the sediment [13]. Studies on treatment of effluent bearing heavy metal have revealed adsorption to be a highly effective technique for the removal of heavy metals from waste stream [19]. Gravel and sand with porosity about 30–40% has larger surface area than granite which has 0.003–0.009% porosity [14]. Therefore, Reactor II and III which consist of gravel and sand have larger surface area available for adsorption, resulting in 35.2 and 38.5% of iron removal.

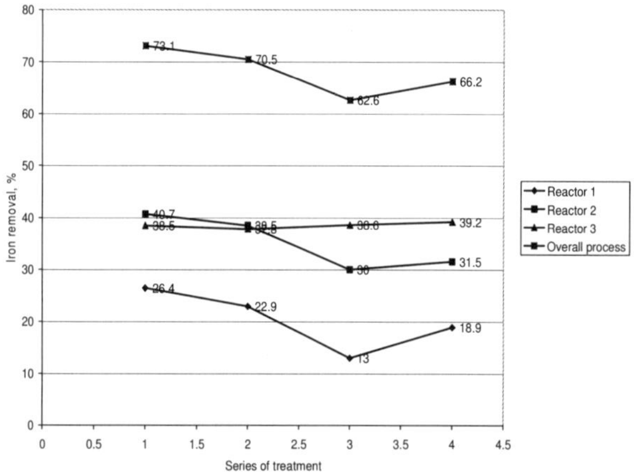

Figure 8. Iron removal through the CW system.

Figure 8 shows the percentage removal of Fe for each reactor and also for overall process. Fe removal rate for Reactor I and II has decreased over a period of time in the experiment. The average removal of iron has decreased from 26.4 to 19.0% for Reactor I and 40.7 to 31.5% for Reactor II after ran through four trial. This is because the sorption and cation exchange capacity of the wetland substrate for metal removal will decrease after some period of time [10]. This phenomenon was also due to accumulation of metal in media [8]. However, the average iron removal rate of Reactor III has increased from 38.9 to 39.2%. This may due to the dilution of sample by rainfall. Although significant metals removal mechanism has been observed for most of the wetland studied, the removal mechanisms are poorly understood [6]. The removal efficiency of iron after passed through Reactor I, II and III was 73.1, 70.5, 62.6 and 66.2%. This suggests that higher efficiency of iron removal when the sample passed through to all reactors continuously.

Again from the statistical analysis by ANOVA for the performance of Reactor I, II and III show that the iron removal efficiency of three reactors is significantly difference.

4. Conclusion

All reactors show the capability in the treatment of landfill leachate. The results show that the gravel and sand which was used as the media in Reactor II and III has better performance compared to granite due the higher porosity and adsorption capacity. Furthermore, the depth of media in the reactor and the placement of the sampling points play a major role in the removal rate of COD and iron. The sequential treatment by continuous flow of leachate sample through Reactor I, II and III show a better result than by the individual reactors.

Acknowledgment

The authors would like to express their gratitude to the USM for providing the research grant under Research University Grants NO. 1001/PKIMIA/811007 and 1001/PAWAM/814021 and also to Idaman Bersih Sdn. Bhd. for supplying the leachate sample.

References

1. Nazaititulshila Rasit, M. Sc Thesis, UTM (2006).
2. D. C.Wilson, Waste Management; Planning, Evaluation and Technologies. New York: Oxford University (1981).
3. A. D. Ward, W. S. Trimble, 2nd edition, Environmental Hydrology, Lewis Publisher, Boca Rato, 369-370 (2004).
4. S. R. Quasim, & W. Chiang, Sanitary Landfill Leachate: Generation, Control and Treatment. Teachnomic Publication: Texas (1994).
5. G. Mulamootill, A. E. McBean, F. Rovers, Constructed Wetlands for the Treatment of Landfill Leachates, Lewis Publishers, Boca Rato and London N. Y Washington, D.C. (1999)
6. M. Mansor, P. E. Lim, R. B. E. Shutes, Constructed Wetland Design, Management and Education, Publisher Universiti Sains Malaysia, Pulau Pinang (2002).
7. D. A. Hammer, (ed.), Constructed Wetlands for Waste Water Treatment: Municipal, Industrial and Agricultural. Michigan: Lewis Publisher, Inc. 831 (1990).
8. Moshiri, A. M., Constructed Wetlands for Water Quality Improvement, Lewis Publisher: Boca Raton, Ann Arbon, London and Tokyo (1993).
9. Kadlec, R. H. and Knight, R. L., Treatment Wetlands. CRC Press, Lewis Publisher, Boca Raton, Florida. 893 (1996).
10. Kadlec RH, Constructed Wetlands for Treating Landfill Leachates. In: Mulamoottil G, McBean E, Rovers FA, editors. Constructed Wetland for the Treatment of Landfill Leachates. Lewis Publisher, Boca Raton, Florida (1999).
11. American Public Health Association, APHA Standard Methods for the Examination of Water and Wastewater (18th ed.). APHA, American Water Works Association and Water Environment Federation. Washington D.C (1992).
12. Guan, T.M., M. Sc. Thesis, USM (2002).
13. Artsanti, P., M. Sc. Thesis, USM (2005).
14. Joseph A., Salvato, Neson L., Nemerow, and Franklin J. A., 5th Edition, Enviromental Engineering, John Wiley and Sons, 267 (2003).
15. Nivala, J., M. B. Hoss, C. Cross, S. Wllace and G. Parkin, Journal of Science of the Total Environment, 380, 1-3, 19-27 (2007).
16. Robinson, H., Harris, G., Carville, M., Barr. M., and Last, S., The Use of an Engineered Reed Bed System to Threat Leachate at Monument Hill Landfill Site – Southern England. In: Mulamootill, G., McBean, E.A., Rovers, F. (Eds.) Constructed Wetlands for the Treatment of Landfill Leachates. Lewis Publishes. Boca Raton, 71-97 (1999).
17. Thian, S. H., M. Sc. Thesis, UTM (2005).
18. Des, W. C., Gregory, J. M., Chemistry and Ecotoxicology of Pollution, Wiley-Interscience, 107 (1984).
19. Jusoh, A., Lam S. S., Nora'aini Ali, M. J. M. M. Noor, Journal of Science of the Total Environment, 206, 1-3, 9-16 (2007).

A PRIMARY MATHEMATICAL MODEL TO DIFFERENTIATE BACTERIAL GROWTH FROM RESUSCITATION OF DORMANT BACTERIA

RAMAIAH SACHIDANANDHAM

School of Civil and Environmental Engineering, Nanyang Technological University
Block N1, 50 Nanyang Avenue, Singapore 639798,
Tel. +65 67905948, Fax: (65) 6792 7319, E-mail: ram@ntu.edu.sg

Regeneration of dormant bacterial cells has been considered significantly important both in the environmental and biotechnology processes. Major efforts have gone into unveiling these aspects involving rigorous experimentation with the use of microbiological and molecular methods. Here we show that the use of appropriate mathematical model(s) can rationally and accurately unveil these processes. These mathematical models have been developed using the law of conservation of mass, bioreactor material balance and growth kinetics. Hence, it can be used in a wide variety of systems involving both environmental (e.g. microcosm based experimentations) and bio-logical processes (e.g. bioreactors, fermentation processes).

1. Introduction

Integrating mathematical models into the findings of biotechnology have accelerated our intrinsic discovery capabilities significantly. In an earlier study, we had shown that a universal mathematical model has been used to scale-up the production of proteolytic enzymes by *Bacillus amyloliquefaciens* [9]. In a subsequent study, we have demonstrated using an unstructured mathematical model that maintenance coefficient (m), an index which has been widely used to describe the maintenance requirement in bacterial bioprocesses, can be computed using total cell retention cultures in *B. sphaericus.* A mathematical model was also proposed and validated for the computation of maintenance coefficient in complex substrate cultivations involving *B. thuringiensis* and *B. mojavesis.* Some of these models used in bioprocess engineering have been summarized in Table 1. Wide majority of mathematical models that describe bioprocesses have been designed using law of conservation of mass. Hence, this law has been extensively used in the past for the rationalized understanding of chemical, environmental and biotechnology processes. The objective of the present study was to revisit regeneration process of non-culturable or dormant cells and demonstrate the importance of mathematical models in the characterization of environmental and bioprocesses.

2. Mathematical Model of Resuscitation of Dormant or VBNC cells

In a conventional batch culture intended for bacterial proliferation, pre-grown bacterial cells (also called as inoculum) is transferred to a sterile culture medium consisting of substrate which mainly supply the carbon and nitrogen, and other nutrients. Bacterial cells are allowed to multiply spontaneously. During the progression of the culture, bacterial cells grow by cell multiplication and/or die depending upon the prevailing conditions. Note that (i) dead cell is a product of the process 'cell death' and (ii) the process of cell death can occur only when there are live cells in the culture. We use 'law of conservation of mass' as the basis to describe this process which states:

$$Input - Output = Accumulation \; --Eq.(1)$$

Using the above boundary conditions together with Eq. (1), cell death in a batch culture can be described as:

$$\frac{dN_v}{dt} = -k * N_v \; --Eq.(2)$$

k = specific death rate (a constant); t = time; N_v = viable cell count.

When the process of cell multiplication and cell death occur in a batch culture, in spite of the presence of cell multiplication, a population will grow only when the growth rate is greater than death rate. If growth occurs in stressed bacterial cultures which also experiences cell death simultaneously, the rate of growth of viable cells can be described taking into account of the cell death as,

Table 1. Examples of successful mathematical modeling applications in bioprocess.

Mathematical Model	Application	
$k_L a^* = 9.8*10^{-5}\left[\frac{P_g}{V}\right]^{0.4} / B^{-0.6} + 0.81*10^{-0.65/B}$	*Used to characterize protease production from B. amyliliquefaciens*	*See ref (9) for the expansion of abbreviations and model details*
$\left[\frac{ds}{dt}\right] = D*S_{in} - D*S_{out} - \frac{\mu*x}{Y_{x/s}} - m*x$	*Predict maintenance metabolism in bacteria*	*See ref (8) for the expansion of abbreviations and model details.*
$m_c = \left[\frac{\mu_c * x_c}{Yx_c/s_c} - \frac{\mu_t * x_t}{Yx_t/s_t}\right]$	*Predict intensity of maintenance metabolism in bacterial fermentations*	*See ref (7) the expansion of abbreviations and model details.*

$$\frac{dN_v}{dt} = (\mu - k)*N_v \text{ --}Eq.(3)$$

μ = specific growth rate. Therefore, integrating Eq. (3) yields,

$$\ln N_v = \ln N_{vo} + (\mu - k)*t \text{ --}Eq.(4)$$

If specific growth rate is greater than the death rate, cells can grow exponentially. In an environmentally stressed batch culture, dead cells are created by a transformation process wherein live cells irreversibly lose their capacity to regenerate growing cells. Therefore, this process cannot contribute to the increase in population size, if any. However, dead cells represent a fraction of the overall population. Therefore, growth of the population can also be written as:

$$\frac{dN_T}{dt} = \mu * N_v \text{ ---} Eq.(5)$$

Substituting for N_v from Eqn (4)

$$\int_{N_{To}}^{N_T} dN_T = \mu * N_{vo} \int_{o}^{t} e^{(\mu-k)t} dt \text{ --}Eq.(6)$$

$$N_T = N_{T0} + \frac{\mu}{\mu - k} * N_{v0} * (e^{(\mu-k)t} - 1) - Eq.(7)$$

N_T = total viable and dead cell count (cell number) at a given time point in a batch culture

N_{T0} = viable and dead cell count inoculated into the batch culture at the beginning of the batch; μ = specific growth rate; k = specific death rate; N_{v0} or N_v = viable cells that can multiply or die at a specific growth rate μ or death rate k respectively at the beginning of the batch or at a given time. In contrast to the conventional batch cultures used for the proliferation of bacterial cells, usually no substrate addition is made at the beginning or during the progress of environmentally stressed cultures. Therefore, no cell growth can occur during the progression of the culture. Cells were shown to be transformed from culturable state to nonculturable state (e.g. [5], [10], [11]). Model Eq. (7) can be used to describe and compute the levels of viable and dead cells at any time point in the microcosm. Usually inoculum used in these studies on environmental stress is developed in nutritionally rich medium. Hence, the level of dead cells will be negligible when compared to the live cells (i.e $N_{T0} \approx N_{v0}$). Then Eq. (7) can be written as,

$$N_T = N_{v0} + \frac{\mu}{\mu - k} * N_{v0} * (e^{(\mu-k)t} - 1) \text{ --}Eq..(8)$$

For example, when bacterial cells are subjected to environmental stress, if loss of culturability is the only cellular event that occurs in the culture, the cell count, when the culture completely loses its culturability and the cell count upon completion of resuscitation should be the same. Therefore, terms 'μ' and 'k' become zero in Eq. (8) and it becomes

$$N_T = N_{v0} \text{ -----} Eq.(9)$$

Eq. (7), (8) & (9) are generic mathematical equations that can be used describe, compute and simulate environmental microcosms exhibiting cell growth, survival and/or death.

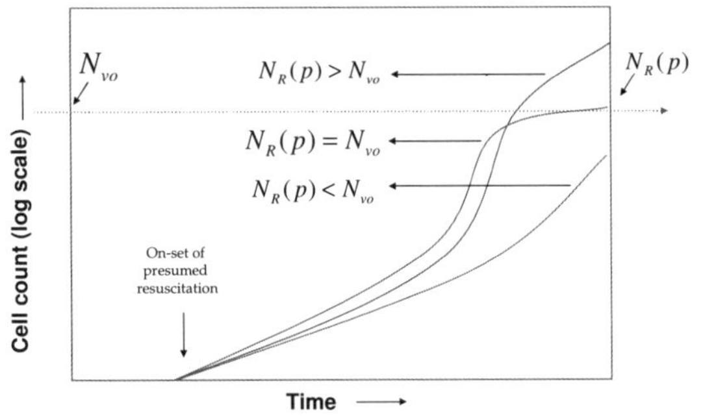

Figure. 1. Diagrammatic representation of profiles of input and output cell counts in the growth or resuscitation processes of microcosm batch experiments.

Study described in the text below implies that $N_R(p) > N_{vo}$. The original Figures and Table in the published work can be viewed using the following hyperlink. Fig 1 of Nilsson and Colleagues (1991). http://jb.asm.org/cgi/reprint/173/16/5054. Fig. 1 and Fig. 2 of Whitesides and Oliver (1997). http://aem.asm.org/cgi/reprint/63/3/1002?view=long&pmid=16535534.

It was necessary to revisit these studies, as increasing number of research groups have been using one or more of these methods to demonstrate the resustability of VBNC cells. It should noted that these methods were extensively revisited by several leading authors around the word and already proved unsuitable for resuscitating dormant or VBNC cells (see ref [1], [2], [3], [6] and references cited therein). However, these studies were unable to propose a method for resuscitating VBNC cells. This has also clearly implied the amount of experimental work involved. Hence, it is established in our study that use of mathematical models simplifies our efforts and provides highly sustainable scientific findings through cost-effective, rapid and robust approaches. In addition, these models lay a strong foundation for the successful resuscitation of VBNC cells.

2.1. *Implications of mathematical model on in-situ resuscitation*

In the method of in-situ resuscitation used by Nilsson and colleagues (1991) [5, 11] and Coutard and colleagues (2007) [4] to resuscitate VBNC cells of *Vibrio vulnificus* or *V. parahaemolyticus,* cells grown in artificial cultures were transferred to external media. In these studies stress was induced by lowering the incubation temperature to 4 or 5°C. A major portion of the inoculated cells lost culturability at this temperature. When the residual culturable cell count reached approximately <100 cells/L (colony forming units) in these batch cultures, the overall physiological state of the population was described as VBNC state. As implied by the methods described in these research articles, cultures were not supplied with substrates or nutrients to growth supporting levels during the temperature stress. Three boundary conditions were derived from these studies: (i) growth promoting substrate was zero or insufficient in the stressor medium to support bacterial growth, (ii) no cell growth and/or death occurred and (iii) two physiological states were reported that include one upon entry into VBNC state (loss of culturability) and another upon completion of resuscitation (regain culturability) (see Fig. 1 in ref. [5]; Fig. 2[C]&[D] in ref. [4]). Therefore, the dynamics of cells entering VBNC state in these microcosms can be described as,

$$\frac{dN_{vbnc}}{dt} = r_e[N_{vo} - N_{vbnc}] - -Eq.(10)$$

r_e = specific VBNC formation rate. N_{vbnc} = cell count of VBNC cells (these studies described VBNC cell count as the difference between cell count estimated by culture independent and culture dependent methods). The level of N_{vbnc} constantly increased and reached a plateau ($N_{vbnc}(p)$). Note Eq. (10) uses N_{vo} as a parameter (its numerical value remains constant i.e. original viable cell count at t = 0) and N_{vbnc} as a variable (its numerical value varies with time). When the culture entered VBNC state,

$$\frac{dN_{vbnc}}{dt} = 0 - -Eq.(11)$$

Therefore, as per the material balance shown in the above equations together with the experimental conditions described in these studies, at this state the numerical value of cell count of N_{vo} was nearly equal to the numerical value of cell count of N_{vbnc}. Note that when cells of *Vibrio sp.* was subjected to temperature stress, cells begin to lose culturability. Hence N_{vbnc} was a variable and which has increased with time and reached a plateau when the entire culture entered VBNC state. At this state it remains constant which we designate as

$N_{vbnc}(p)$. When the temperature was raised, it was shown that all the VBNC cells regained culturability. This process was termed as in situ resuscitation. Resuscitation of VBNC cells during temperature up-shift can be written as:

$$\frac{dN_R}{dt} = r_{ex}[N_{vbnc}(p) - N_R] - -Eq.(12)$$

N_R = resuscitated cell count; r_{ex} = specific resuscitation rate constant.

Note Eq. (12) uses $N_{vbnc}(p)$ as a parameter (its numerical value remains constant i.e VBNC count at the start of resuscitation process when t =0) and N_R as a variable (its numerical value varies with time). A second state of plateau was attained upon completion of resuscitation (See. Fig. 1 in ref. [5]; Fig. 2[C]&[D] in ref. [4]). At this point the level of resuscitated cells is designated as $N_R(p)$. Therefore, Eqn (12) can be written as

$$\frac{dN_R}{dt} = 0 - -Eq.(13)$$

Therefore, as per the material balance shown in the above equations together with the boundary conditions described in these studies, at this state the numerical value of cell count of N_{v0} should be at least equal to the numerical value of cell count of $N_R(p)$. Based on the assessment of cell counts shown in the in-situ resuscitation together with the above equations, whether a given process is bacterial growth or resuscitation can be inferred using the following simple tools.

(i) If VBNC cells regain culturability by the process of resuscitation, the resulting resuscitated cell count should at least reveal (to be initially qualified as a non-growth process):

$$N_R(p) = N_{vo} - -Eq.(14)$$

(ii) For example, when cells subjected to environmental stress, if a portion of a population experiences cell death while the remaining is transformed into VBNC cells, then (non-growth process):

$$N_R(p) < N_{vo} - -Eq.(15)$$

(iii) If what has been shown as resuscitation is actually due the re-growth of culturable cells (growth process) utilizing externally supplied substrate(s), the best index of this process will be implied by:

$$N_R(p) > N_{vo} - -Eq.(16)$$

All three categories are diagrammatically represented in Figure 1. Therefore, these models have helped us to rationally unveil these bioprocess and demonstrated that these processes of in-situ resuscitation is actually "bacterial growth".

3. Conclusion

The objective of these mathematical models was not to disprove the method of in-situ resuscitation or allied techniques such as substrate dependent resuscitation of VBNC cells. This has already been extensively done in an innumerable number of studies in the past (see ref [1], [2], [3], [4], [6] and references cited therein). However, these models would strongly help in designing robust resuscitation techniques in the future especially in temperature stressed cultures. Moreover, use of these models would significantly cut down the experimental load associated with these studies. In addition, they would also help in rationally differentiate mixed processes of growth and resuscitation.

References

1. G. Bogosian, and EV. Bourneuf. *EMBO Report,* **9**, 770-774, (2001).
2. A. Aertsen, and CW Michiels. *Crit. Rev. Microbiol.* **4**, 263-73, (2004).
3. F. Coutard, P. Crassous, M. Droguet, E. Gobin, RR. Colwell, M. Pommepuy, and D. Hervio-Heath, *ISME J*, **1**, 111–120, (2007).
4. F. Coutard, S. Lozach, M. Pommepuy, and D. Hervio-Heath, *Appl. Environ. Microbiol.* **73(16)**, 5183-5189, (2007).
5. L. Nilsson, JD. Oliver, and S. Kjelleberg. *J. Bacteriol.* **173(16)**, 5054-5059, (1991).
6. J. Ravel, IT. Knight, CE. Monahan, RT. Hill, and RR. Colwell, *Microbiology,* **141(Pt 2)**, 377-83, (1995).
7. R. Sachidanandham, Y. Al-Shayji, N. Al-Awadhi, and KY. Gin. *Biotechnol Bioeng.* **91(7)**, 838-847, (2005).
8. R. Sachidanandham, and G. Kalaiselvi, *Bioprocess Biosyst Eng.* **19(4)**, 243-245, (1997).

9. R. Sachidanandham, IAR. Kumar, MRV. Krishnan, and K. Jayaraman, *Bioprocess Biosyst Eng.* **21**, 319-322, (1999).
10. B. Smith, and JD. Oliver, *Appl. Environ. Microbiol.* **72(2)**, 1445–1451, (2006).
11. MD. Whitesides, and JD. Oliver, *Appl. Environ. Microbiol.* **63(3)**:1002-1005, (1997).

ENHANCEMENT OF CU(II) ADSORPTION CAPACITY BY SURFACE-IONIC IMPRINTED BIOSORBENT

A. SHOFIYANI[†], NARSITO, S. J. SANTOSA and S. NOEGROHATI
Department of Pharmacy, University of Gadjah Mada,
Sekip Utara Yogyakarta, 55281, Indonesia

Ion-imprinting technique based on microalgae biomass surface was applied to the preparation of surface ionic-imprinted biosorbent (SIIB) for Cu(II) adsorption. The biosorbent was prepared by coating Cu(II)-imprinted chitosan hydrogel onto biomass surface followed by cross-linking procedure. Desorption of the imprinted-ions left a specific template for Cu(II) binding sites. Batch adsorption was conducted to study the characteristics adsorption of Cu(II) onto SIIB, and the maximum capacity was calculated to be 53.1918 mg/g, higher than that on IIB (51.5469 mg/g). Desorption metal ion-imprinted on SIIB was more effective (96-100%) than that on IIB (60-70%), which offers promising applicative prospects in the wastewater treatment.

1. Introduction

Chitosan, poly (N-acetyl-D-glucos-amine), is deacetylated product of chitin, the second most abundant polysaccharides in nature. Due to its large content of $-NH_2$ and –OH groups that can chelate with many heavy metal ions such as Cu(II), Pb(II), Zn(II), and Ni(II), chitosan had been widely applied for remove heavy metal ions in wastewater. The disadvantage of using chitosan for industrial wastewater treatment is that chitosan dissolves in dilute acids. This can be avoided by crosslink chitosan with many kinds of bi/poly-functional agent chitosan e.g. glutaraldehyde 1, epichlorohydrin 1, and ethylene glycol diglycidyl ether 2. Crosslinking procedure can improve the stability of chitosan in acid media but the capacity significantly decreased because most of $-NH_2$ active sites consumed by crosslinking agent.

Using the molecular imprinting methods introduced by Ekberg and Mosbach, Tan *et al.* 3 prepared imprinted chitosan resin to adsorb Ni(II) ions. Molecular imprinting is an inexpensive technique for preparing adsorbent which have a specific space for molecular recognition, so the adsorption selectivity of desired molecule is enhanced. For metal ions, molecular imprinting can be interpreted as ionic imprinting exactly i.e. preparing such template for metal ions recognition. By this methods, active sites of chitosan was firstly shielded with metal ions before crosslinked so the stability of this resin increased without losing of its adsorption capacity.

Su *et al.* 4 developed this technique to synthesis of surface molecular imprinted chitosan adsorbent by using biomass of *Penicillium chrysogenum* as core materials. Surface molecular imprinting is one of the impotant types of molecular imprinting. Surface molecular imprinted adsorbent (SMIA) not only posseses high selectivity but also avoids problem with mass transfers 6.

The adsorption capacity of the Ni(II) (imprinted ions) on SMIA prepared by Su *et al.* [7] reported to be increased 40-50% compared with non-SMIA. According to Su and co workers, SMIA combines the advantages of the chitosan and mycelium adsorbent giving low cost, good ability and stability that allows to reuse the adsorbent for up to 15 cycles 7. However, the role of biomass onto SMIA is still un-explored yet and along literatures which have been reviewed, there is no research that using microalgae as core material for preparing surface ionic imprinted bio-sorbent (SIIB).

In this paper, we studied the use of microalgae *Chlorella sp* as support materials for chitosan, and use the SIIB prepared for selective Cu(II) adsorption.

2. Materials and Methods

2.1. *Materials*

Chitosan (from crab shells 90% DD) was purchased from Sigma-Aldrich. $CuCl_2.2H_2O$, EDTA and all other chemicals were of analytical grade.

[†]Work partially supported by postgraduate student grant of The D.G Higher Education, Ministry of National Education Indonesia.

Biomass of *Chlorella sp* collected from Lampung Bay, Bandar Lampung, Indonesia, washed thoroughly with water and deionized water to neutral pH, dried till constant weight then sized by 80-mesh sieve for further use.

The surface morphological image was obtained from Scanning Electron Microscopy (SEM) (JEOL T-330A) after 300 μ gold plating at accelerating voltage of 15 kV, 0.5A. The characteristic of functional groups among adsorbents was canfirmed by Fourier-Transform infrared (FTIR) (Shimadzu FTIR Prestige-21).

2.2. Preparation of surface imprinted biosorbent

SIIB was prepared according to the reported procedure 4,6 and our improved work. Firstly, 0.5 g chitosan (dry weight) was dissolved in 10 mL dilute acetic acid solution of 2,5% (v/v). Then a 5000 mg/L of 10 mL $CuCl_2$ was added into the chitosan solution and stirred for 60 minutes to get an equilibrium adsorption of Cu(II) ions. 2.0 g (dry weight) biomass of *Chlorella* then added into the above solution and stirred for 30 minutes. Amount of 0,5 mL epichlorohydrin was added into the mixture, stirred and allowed to react for 6 hours at room temperature.

The Cu(II) ions on the chitosan surface was desorbed by treating with 0.1M EDTA solution for 5 hours in a rotational shaker at room temperature. Regeneration was carried out by dipping the biosorbent into 0.1M NaOH solution for 1 hour. Subsequently, the suspensions filtered using vacuum filtration and then converted into pellets. The pellets was washed several times with deionized water until the washings were neutral, then crushed, and dried in a oven at 600C. The dry biosorbent was sized by an 80-mesh sieved and stored at a sealed bottle for further use.

2.3. Adsorption experiments

Batch adsorption studies were conducted by mixing 0.05 g (dry weight) adsorbent with 10 mL metal solution at pH 5.5-5.7 in a 100 mL erlenmeyer flasks, shaking at 120 rpm for 3 hour using a rotational shaker at room temperature. The relation between varying Cu(II) initial concentration (0-1000 mg/L) and its adsorption equilibrium data for Cu(II)-SIIB and non-SIIB was studied.

3. Results and Discussions

3.1. Characteristics of FT-IR spectra

To ascertain the various groups of different biosorbents, FT-IR spectra were obtained from biomass, chitosan, Cu(II)-ionic imprinted biosorbent (IIB) and Cu(II)-surface ionic imprinted biosorbent (SIIB). The FT-IR spectra is given in figure 1.

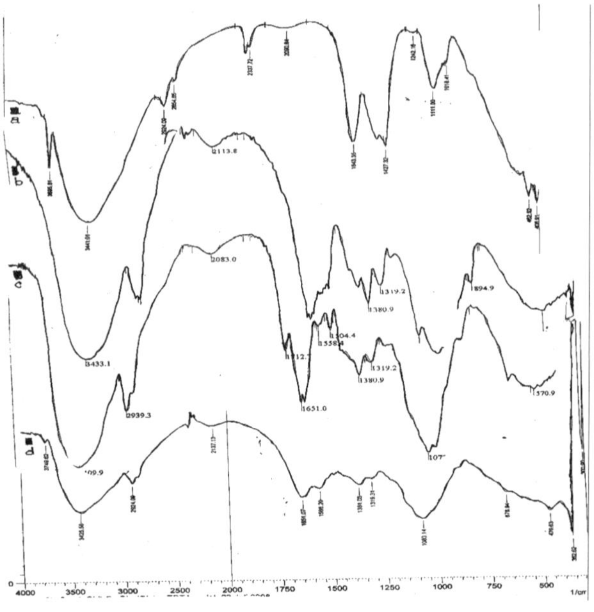

Figure 1. FT-IR spectra of (a) biomass, (b) chitosan, (c) IIB, and (d) SIIB.

It was obvious that the most characteristics peaks are at 3100-3400 cm^{-1} confirming the presence of overlapping –OH and $–NH_2$ stretching vibrations. The characteristic peaks at around 2925 cm^{-1} corresponding to C-H stretching of methyl or methylene groups of chitosan and biomass, and the bands around 1650 cm^{-1} is attributed to N-H bending of primary amine groups.

An interesting observations is the considerable shift in the positions and intensity of $–NH_2$ and –OH peaks of imprinted biosorbents. It was confirmed that $–NH_2$ and –OH of chitosan and biomass act as binding sites for metal ions adsorption. The study by Inoue *et al.* 8 confirmed that amino and hiyroxyl groups of chitosan can form chelates with Cu(II) ions by releasing hydrogen ions as shown in figure 2.

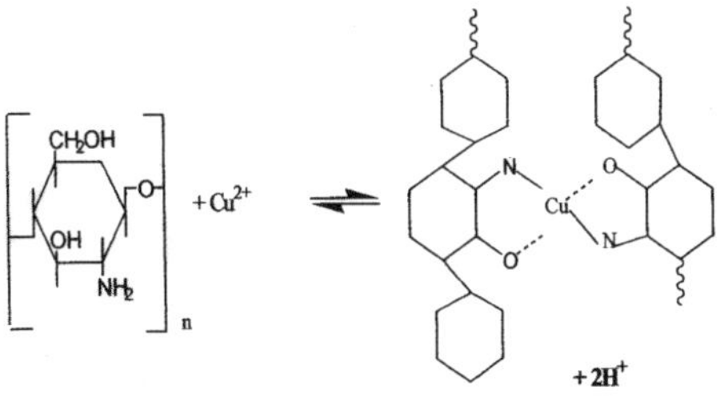

Figure 2. Cu(II) chelating ability of chitosan.

The SEM images of the surface of Chitosan, IIB and SIIB as shown in figure 3 display a rough structure on biosorbent surface.

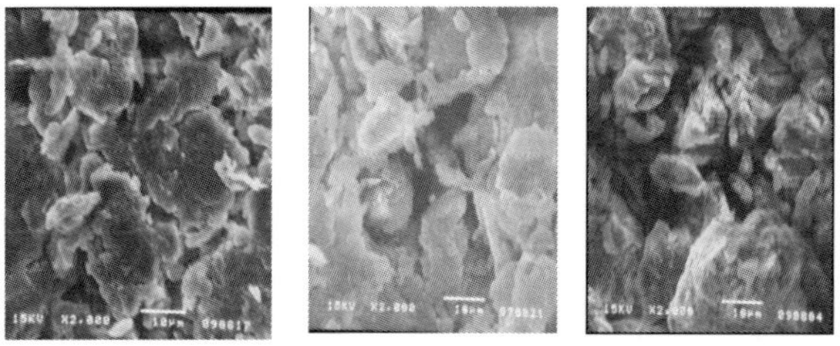

Figure 3. SEM image of chitosan, non-SIIB and SIIB.

3.2. Adsorptive characteristics of SIIB

The adsorption of Cu(II) on IIB and SIIB was carried out with 0.05 g biosorbents. Figure 4 shows the experimental equilibrium isotherm for adsorption of Cu(II) on IIB and SIIB, respectively.

The isotherm on both biosorbents were similar in shape. It can be seen that adsorption has a rapid increase in first concentration because it still much opened active sites so Cu(II) ions reached the sites effectively. The relatively constant state seen after concentration of 500 mg/L, indicating the equilibrium of adsorption and desorption process.

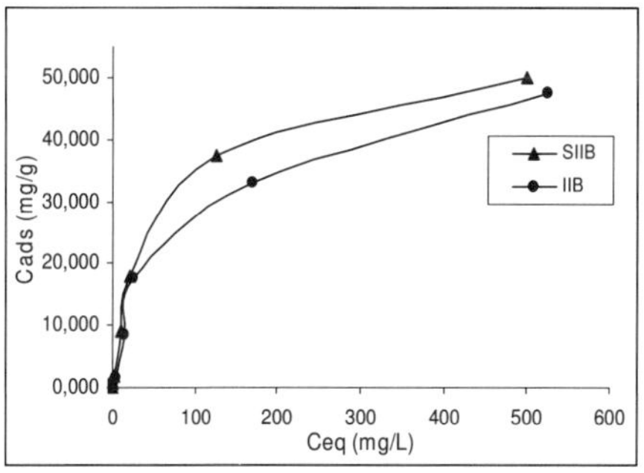

Figure 4. Equilibrium isotherm adsorption of Cu(II) on IIB and SIIB.

The adsorption behavior of the IIB and SIIB can be described by the Langmuir adsorption equation, as follow:

$$\frac{C_e}{Q_e} = \frac{C_e}{Q_{\max}} + \frac{1}{b\,Q_{\max}} \qquad (1)$$

Where C_e is the equilibrium or final concentration of Cu(II) (mg/L), Q_e is the adsorption capacity of Cu(II) adsorbed per unit weight of adsorbent (mg/g), Q_{max} is the theoretical maximum adsorption capacity (mg/g) and b is the Langmuir adsorption equilibrium constant (L/mg).

A linearized plot of C_e/Q_e versus C_e gives Q_{max} and b. The calculated results showed in table 1.

Table 1. Langmuir isotherm constant for Cu(II) adsorption.

Adsorbent	Q (mg/g)	*b* (L/mg)	R^2
SIIB	53.1918	0.0131	0.9883
IIB	51.5469	0.0219	0.9625

From the table, it can be seen that Langmuir equation fits well for Cu(II) adsorption both on the IIB and SIIB (R^2>0,95) under the concentration range studied. From the experimental results, the adsorption capacities for Cu(II) on SIIB and IIB were calculated to be 53.1918 mg/g and 51.5469 mg/g, respectively. The adsorption capacity of the SIIB developed in the presence study are about 20 percent higher than that of SMIA developed by Su *et al.* [5]. Su and co-workers synthesized the SMIA based on chitosan and biomass of *Penicllium chrysogenum* with Ni(II) as metal ions template. It had been found that adsorption capacity of Ni(II) to SMIA prepared is only 84% compared to chitosan imprinted resin.

It is considered that the higher adsorption capacity of SIIB toward IIB is attributed to the role of biomass. The presence of biomass as surface support for chitosan enable to expose more active sites of polymer that allows the enhancement adsorption ability of biosorbent. Another reason is that the functional groups of algae biomass may also took parts in the Cu(II) adsorption.

When the ratio of chitosan:metal ions concentration and chitosan:crosslinking agent was controlled exactly, most of active sites were chelated by metal ions imprinted (on the formation of biosorbents template) and only a few of it was used for crosslinking reaction. So the adsorption capacity still high without losing its stability in acid media.

3.3. Desorption

In order to prepare such as metal ions template on SIIB, the imprinted metal ions bound in previous step of preparation must be desorbed for further use. In this study, we used 0.1M EDTA solution as desorption agent by onetime batch method.

As comparation, percent desorption of Cu(II) ions-imprinted to SIIB was higher than that to IIB, i.e. about 96-100% for SIIB and 60-70% for IIB. As known, morphology of the adsorbents played an important role for its adsorption-desorption performance, due to material density and mass-diffusion resistance. Surface ionic-imprinting may direct the adsorption to take place on the surface of biosorbent, so ion-imprinted more easily removed by desorption.

This experimental results confirmed that surface ionic imprinted biosorbent offers advantage in matter of ease of imprinted-metals desorption. Therefore, the SIIB has potentially applied for removal of heavy metal ions in wastewater treatment.

4. Conclusions

The surface ionic imprinted biosorbent based on chitosan and biomass of microalgae *Chlorella sp* can be prepared by simple and rapid methods. The adsorption capacity of Cu(II) ions onto SIIB was higher than that on IIB, considered as biomass contribution to open more active sites of biosorbent. Metal ions imprinted on SIIB could be desorbed more effectively than that on IIB which offers good practical prospects in the wastewater treatment.

5. References

1. T.Y. Hsien and G.L. Rorrer, "Heterogenous Cross-Linking of Chitosan Gel Beads: Kinetics, Modeling, and Influence on Cadmium Ion Adsorption Capacity," *Ind. Eng. Chem. Res.*, **36**, 3631-3638 (1997).
2. Y. Baba, Y. Kawano, and H. Hirakawa, "Highly Selective Adsorption Resin: Preparation of Chitosan Derivatives Containing 2-Pyridiyl-methyl, 2-Thienylmethyl, and 3-(methylthio) propil Groups and Their Selective Adsorption of Precious Metals, *Bull. Chem. Soc. Japan*, **69**, 1955-1960 (1996).
3. W.S.W. Ngah, A.B. Ghani, and S. Kamari, "Adsorption Behavior of Fe(II) and Fe(III) Ions in Aqueous Solutions on Chitosan and Cross-Linked Chitosan Beads, *Bioresour. Technol,* **96**, 443-450 (2005)
4. T. Tan, X. He, and W. Du, "Adsorption Behaviour of Metal Ions on Imprinted Chitosan resin," *J.Chem.Technol. Biotechnol.*, **76**, pp 191-195 (2001).
5. H.J. Su, Z. Wang, and T.W. Tan, "Adsorption of Ni^{2+} on The Surface of Molecularly Imprinted Adsorbent from *Penicillium Chysogenum* Mycellium, *Biotechnol. Lett*, **25**, pp. 949-953 (2003).
6. Chang, *et al.*, Solid-Phase Extraction of iron(III) with an Ion-Imprinted Functionalized Silica Gel Sorbent Prepared by a Surface Imprinting Technique, *Talanta*, **71**, 38-43 (2007).
7. H.J. Su, Z.X. Wang, and T.W. Tan, Preparation of Surface Molecular Imprinted Adsorbent for Ni Based on *Penicillium chysogenum*, *J. Chem. technol. Biotechnol*, **4** (80), 439-444 (2005).
8. H.J. Su, Y. Li, and T.W. Tan, "Biosorption of Ni^{2+} by the Surface Molecular Imprinting Adsorbent, *Process. Biochem*, **41**, 1422-1426 (2006).
9. K. Inoue *et al.*, "Selectivity Series in the Adsorption of Metal Ions on a Resin Prepared by Crosslinking Copper(II) Complexed Chitosan, *Chem. Letters*, 1281-1284 (1988).

ON THE EFFICIENCY OF DISCRETE PARTICLE MODELING OF GAS-SOLID FLUIDIZED BED REACTOR*

C. L. WU†
Engineering College, Guangdong Ocean University, Zhanjiang 534088, China

A.S. BERROUK AND K. NANDAKUMAR
Chemical Engineering Program, The Petroleum Institute, P.O. Box 2533, Abu Dhabi, United Arab Emirates

A robust three-dimensional discrete particle model (3D DPM) for unstructured meshes is presented to model the gas-solid flows in fluidized beds. Inter-particle interactions are taken into account through a hard-sphere approach. A multifaceted numerical strategy is developed to enhance the computational efficiency. Simulation results demonstrate the ability of the model to capture many important characteristics such as spouting, particle clustering and core-annulus flow structures in gas-solid fluidization systems at specified operating conditions.

1. Introduction

Gas-solid fluidized beds are widely used in the chemical and petrochemical industries for a large variety of processes. The ongoing research efforts to develop numerical models for such complex systems are significant. Due to the multi-scale character of these systems, several models at different scale levels have been formulated. The discrete particle model (DPM) is a meso-scale (particle-scale) modeling technique between direct numerical simulation (DNS) and two-fluid model (TFM) for dense two-phase flows. It is deemed that DPM has a great potential for studies on dense two-phase flow problems, and for applications in chemical engineering field. In this approach, the particle motion is described in a Lagrangian framework by directly solving the Newtonian kinetic equations of each individual particle. The particulate-phase constitutive relations are not required because the particle-particle interactions are modeled through a two-variant collision strategy; the hard-sphere variant or the soft-sphere variant.

The DPM has become a versatile tool since the pioneering works that focused on its development and validation [1-3]. Although a lot of progress has been made on the numerical aspects of DPM, many studies were based on the two-dimensional (2D) model and only structured regular grids were adopted to discretize the computational domain. Despite the relatively high computational cost of 3D DPM compared to 2D DPM, it appears to be the only realistic way to simulate many industrial processes taking place in complex geometries. From a theoretical point of view, the 2D model for instance cannot characterize the radial hydrodynamics in cylindrical Circulating Fluidized Bed (CFB) risers because the flow in a riser is not axis-symmetric in most of the cases. Therefore, 3D DPM implemented in conjunction with unstructured grid should be developed and extended to incorporate necessary details to model real-world chemical engineering problems.

The high computational cost linked to 3D DPM has limited its use to applications in a laboratory scale with the number of particles typically less than 100,000. This shortcoming can be overcome by designing a computationally-efficient multifaceted numerical strategy. It is deemed that the model accuracy and its implementation efficiency depend strongly but not solely on: (1) 3D void fraction computation, (2) inter-phase coupling, and (3) collision event handling through proper data structures.

2. Governing equations

The governing equations of the gas phase mainly follow the volume-averaged form of the two-fluid model presented by Gidaspow [4]. The particles' motion obeys Newtonian second law of motion and particle collisions

*This work is partly supported by a grant from Guangdong Natural Science Foundation.
†Corresponding author. E-mail address: chunliangwu@gmail.com.

are described by a three-parameter hard-sphere model [5,6]. Interactions between the two phases are accounted for through the momentum exchange term.

2.1. *Gas phase hydrodynamics*

The conservation of the gas mass is balanced by the convective mass fluxes without inter-phase mass transfer:

$$\frac{\partial}{\partial t}(\varepsilon\rho_g)+\nabla\cdot(\varepsilon\rho_g\mathbf{u}_g)=0 \tag{1}$$

where ρ_g is the density of the gas phase, ε is the void fraction and $\mathbf{u_g}$ is the gas velocity.The momentum balance equations are similar to the single-phase momentum equations

$$\frac{\partial}{\partial t}(\varepsilon\rho_g\mathbf{u}_g)+\nabla\cdot(\varepsilon\rho_g\mathbf{u}_g\mathbf{u}_g)=-\varepsilon\nabla p+\nabla\cdot(\varepsilon\bar{\bar{\boldsymbol{\tau}}}_g)+\mathbf{S}_g+\varepsilon\rho_g\mathbf{g} \tag{2}$$

where p is the static pressure and $\bar{\bar{\boldsymbol{\tau}}}_g$ the stress tensor. The source term $\mathbf{S_g}$ due to particle-fluid drag interaction was suggested by Goldschmidt et al. [6]:

$$\mathbf{S}_g=-\frac{1}{V}\int_V\sum_{k=1}^{N_P}\frac{V_{p,k}\beta}{1-\varepsilon}(\mathbf{u}_g-\mathbf{u}_{p,k})\delta(\mathbf{x}-\mathbf{x}_{p,k})dV \tag{3}$$

where V is the integral volume, V_p is the particle volume, N_p is the particle number, $\mathbf{u_p}$ is the particle velocity and the δ function ensures that the reaction force acts as a point force at the position of the particle. β is the generic drag coefficient, which is a function of the particle Reynolds number and the void fraction.

2.2. *Dispersed phase hydrodynamics*

Since the gas inertia is much smaller than that of the solid particles, only the drag and the pressure gradient forces are taken into account in the particle momentum equation, which is written in a Lagrangian framework and reads as follows

$$m_p\frac{d\mathbf{u}_p}{dt}=-V_p\nabla p+\frac{V_p\beta}{1-\varepsilon}(\mathbf{u}_g-\mathbf{u}_p)+m_p\mathbf{g} \tag{4}$$

where m is the particle masse.

In the hard-sphere approach, collisions among dispersed particles are assumed binary and quasi-instantaneous and particle contact occurs at one point. During a collision only impulse forces are considered. By introducing the coefficient of normal restitution e_n, the coefficient of friction μ_f and the coefficient of tangential restitution e_t, the particle energy dissipated is calculated and the total impulse can be calculated according to the particle positions and velocities before collision. Thus the post-collision velocities can be computed [2, 5].

3. Multi-faceted numerical strategy

Due to the weak gas-phase inertia compared to that of the dispersed phase, the complex gas-particle system is often divided in two subsystems or processes that are solved separately at each simulation time step. In the first process, the two-phase interactions are taken into account while the particles are assumed fixed in space. Only the momentum exchange between the two phases is accounted for in this first step. The gas-phase governing equations are solved using a finite volume method together with the particle momentum equation. In the second process, all possible collisions between particles are detected and handled one by one in a chronological order. To ensure an efficient implementation of the discrete particle model used to simulate dense gas-particle systems on unstructured grids, a many-sided numerical strategy is designed. This strategy is based on (i) an accurate calculation of the void fraction (ii) a strong implicit coupling of the two-phase interaction and (iii) an adequate data structure designed to obtain a complexity *O(1)* time per collision regarding the event handling of particle collisions.

3.1. *Voidage calculation*

An analytical method to calculate the void fraction for general 3D unstructured meshes is proposed herein. It is developed for cells of different shapes (wedged, tetrahedral, or hexahedral) and for different particle-cell intersections (node, edge or face). Four cases are identified for which the volume V_n is accurately computed (see Fig.4 in [7]). Obviously if V_n is known, then we get $V_i=\frac{4}{3}\pi r_p^2-\sum V_n$. Herein V_0 and V_n ($n=1$, 2...) are the volumes of the cell C_0 hosting the centre of the particle and the ones of the other cells C_n that share the particle, respectively. For case 1 and 2, V_n can be calculated easily.

Case 3: one edge (or more than one) of the cell C_0 passes through the sphere, so the particle is split into several parts by all the faces sharing this edge. To analyze this case, we specify this edge as VV', the particle centre O as the origin of the local cylindrical coordinate system, the line parallel to VV' as the z-axis, and the line normal to the edge VV' as the x-axis (see Fig. 1a). The plane at $z=0$ is shown in Fig. 1b and the x-z plane in Fig. 1c. The volume of the sphere cut by the two faces (*face 1* and *face 2*) of the cell C_n can be calculated by:

$$V_n(l,\theta_1,\theta_2)=V(l,\theta_2)-V(l,\theta_1)$$

$$V(l,\theta)=2\int_0^H S(l,\theta,z)dz \qquad (0\leq\theta\leq\frac{\pi}{2},\ 0<l<r_p)$$

$$V(l,\theta)=\pi s^2\left(r_p-\frac{s}{3}\right)-2\int_0^H S(l,\pi-\theta,z)dz \ (\frac{\pi}{2}<\theta\leq\pi,0<l<r_p)$$

Here $s=r_p-l\sin\theta$. The upper limit of the above integrals is $H=\sqrt{r_p^2-l^2}$ $(0<l<r_p)$.

In the above definition of V_n, l is the distance from the particle centre to *VV'*, θ_1 and θ_2 are the angles from the x-axis to face 1 and face 2, respectively. $S(l,\theta,z_0)$ represents the area of the pseudo sector *VMP* at the height $z=z_0$. Details about the calculation of the integrals can be found in Ref.[7].

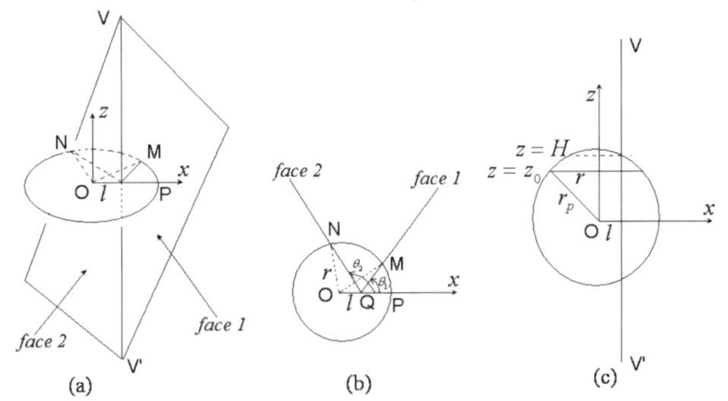

Fig. 1. Accurate calculation of porosity on 3D unstructured mesh (case 3) (a) perspective view; (b) plane at z=0 and (c) x-z plane.

Once the length l and the angle θ are calculated according to the positions of the particle and the cell faces, the above integrals yield the sphere volume encompassed by the *x-z* plane and the face. Then the volume of a sphere segment cut by any two faces can be calculated.

Case 4: a node or vertex *V* that is common to the cell C_0 and its neighbouring cells C_n, is occupied by a particle. The calculation of the volume V_n of the particle shared by C_n is performed as follows. First the vertex *V* is identified by checking all the vertices of the cell C_0. The intersection points of each edge (sharing *V*) with the particle surface are calculated. For cell C_n, these intersection points are denoted by *P*, *Q* and *R*, since there are only three edges between the three faces (denoted by f_1, f_2, f_3) for a tetrahedral, wedged or hexahedral cell that are intersected by the particle (see Fig.6 in [7]). Then the particle volume V_k encompassed by the face *PQR* and f_k $(k=1,2,3)$ can be calculated by the approach described above in case 3 for the sphere volume surrounded by any two faces. Finally we obtain:

$$V_n=V_t+\left(V_{sg}-\sum_{k=1}^{3}V_k\right)$$

$$V_t=\frac{1}{6}\left|(PR\times PQ)\times PV\right|$$

Where V_{sg} is the volume of the sphere segment split by the face *PQR*.

3.2. *Implicit two-phase coupling*

The semi-implicit scheme is used to discretize the particle momentum equation (Eq.4). The nonlinear drag coefficient is evaluated explicitly, which results in a system of linear equations. When only buoyancy and two-phase drag forces are taken into account, the particle velocity is calculated as follows:

$$m_p\frac{\mathbf{u}_p^{n+1}-\mathbf{u}_p^n}{\Delta t}=-V_p(\nabla p)_p^{n+1}+\frac{V_p\beta_p^n}{1-\varepsilon_p^{n+1}}(\mathbf{u}_{g,p}^{n+1}-\mathbf{u}_p^{n+1})+m_p\mathbf{g} \tag{5}$$

where the Eulerian properties such as void fraction ε , static pressure gradient ∇p and gas phase velocity $\mathbf{u}_g$ are mapped to the particle position through gradient interpolations [5].

For the momentum source term due to forces exerted by the dispersed phase on the gas phase, the integration of Eq.(3) is taken over the grid cell:

$$\mathbf{S}_g=-\frac{1}{\Delta V}\sum_{k=1}^{NPC}\frac{V_{p,k}\beta_{p,k}^n}{1-\varepsilon^{n+1}}(\mathbf{u}_g^{n+1}-\mathbf{u}_{p,k}^{n+1}) \tag{6}$$

where *NPC* is the number of particles associated with the cell, and $V_{p,k}$ are the corresponding fractional volumes. This formula implies that a particle may be acted on and also react to the fluid in several cells.

The above coupling scheme shows a conservation of the momentum exchange due to the two-phase drag interactions. This owes to the combined facts: (i) the nonlinear drag force is linearized and the drag coefficient is evaluated explicitly; (ii) both the particle and gas velocities are calculated in the same time layer. At the beginning of a time step, the drag coefficient for each particle is first calculated along with the gradient of the void fraction. Then, in each iteration for solving the gas pressure-velocity equations, the gradients of the involved Eulerian variables are calculated, and the linearized particle momentum equations are solved. The iteration stops only when the solution of both the gas mass and momentum equations converges. The particle drag force exerted on the gas is weighted by the particle fractional volume, which has a smooth effect on the source forces and helps numerical stability. The solution of the particle velocity and the gas field is synchronized, resulting in a strongly implicit two-phase coupling.

3.3. *Data structure for collision events*

In the hard sphere model, the identification of particle collision events in a numerical step consumes a large amount of CPU time. The collision events must be sorted by the collision time and thus form a sequence for computing the collision dynamics one after another. The collision time of any two particles is defined as the time interval from the beginning of a time step to the time of their collision and is calculated from the initial positions and velocities of them [2,5]. Because the particle positions and velocities are changed continuously, element insertions into or deletions from the collision sequence are frequent in the computation. If all the potential collision events are maintained only in one global list, operations on its elements will take more time because of its length.

The so-called local minimum algorithm[8] was applied for handling the collision sequence[5]. A global linked list, in which no more than one collision event for every particle is identified by the minimum collision time, together with an individual list for all the collisions involving a particular particle, is introduced (see Fig.7b in [5]). Using the individual particle lists, invalid collisions related to any particle can be deleted without any searching of the long global list. Potential collision events can be inserted into the global list more efficiently because the length of the global list has been limited to the number of particles simulated. That is, the operation complexity in time is decreased to $O(N_P)$. If a complete binary tree data structure is applied for the global list, the operation complexity will be decreased to $O(\log N_p)$. Since the global list is operated very frequently (twice for each collision), the above storage strategy is still computationally burdensome with increasing number of particles.

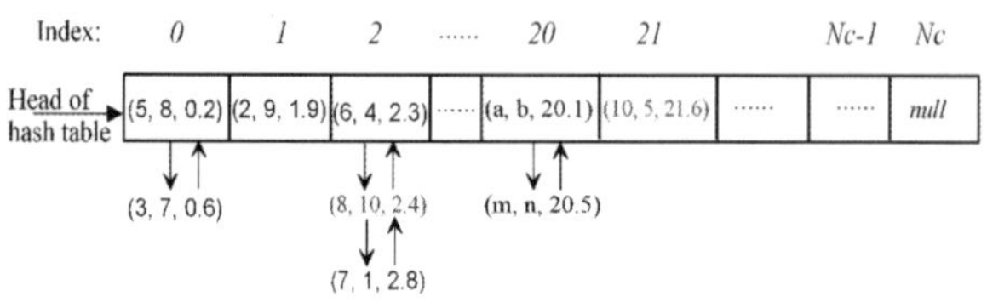

Fig. 2. Logical view of a chained hash table. The collision time is normalized by Δt_c.

Here a chained collision events table is introduced for the storage of the elements in the global list, which is illustrated in Fig.2. The key of the collision events table is the collision time t_c and the event function maps the key to a fixed address or index I in the table. A simple event function is defined as

$$I(t_c) = INT(t_c / \triangle t_c),\ \triangle t_c = \triangle t / NC$$

where the function $INT(x)$ gives the integral part of x and NC is the number of real collisions in a time step $\triangle t$. The collision time t_c of any two particles is defined as the time interval from the beginning of the time step to the time of their collision. If the particles collide successively with a uniform time interval, the above event function maps only one element to a table address. When inserting or deleting a collision event, its address is determined by the event function at once and the computational complexity in time is reduced to $O(1)$.

Since NC cannot be determined before the collisions happen, its value is taken as that in the last time step. Considering the fact that particles rarely collide with a uniform interval, it is possible that the above function maps more than one element to an identical address of the collision events table. This problem can be resolved by chaining, that is, the collision events table is treated as an array of linked lists, as shown in Fig.2.

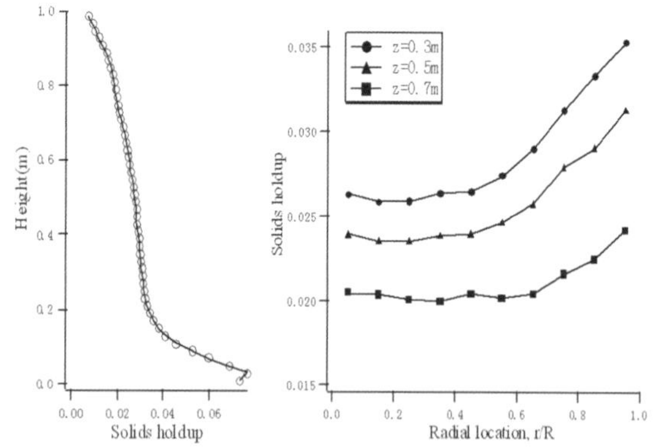

Fig. 3. Axial and radial profiles of particle holdup.

4. Test cases

A laboratory scale riser with a diameter of 0.1m and a height of 1m is simulated by the present 3D model. Particle properties and numerical parameters are given in Table 1. At the beginning, the particles are distributed uniformly inside the cylindrical column with an average voidage of 0.982. The simulation is run for 12s real time. To reduce the influence of the initial conditions, time-averaged results are obtained for the last 6 seconds of the simulation. The CPU time is about 150 hours for one simulation run. Fig.3 shows the time-averaged axial solids hold-up and the radial profiles of the particle volume fraction at different bed levels. The radial particle distribution is characterized by the core-annulus flow structure with dense solids flowing downward near the wall and a more dilute core flowing upward in the riser centre. The numerical model also captures the transient state of the heterogeneous structure such as particle clusters or swarms near the riser wall, as shown in Fig.4. Two particle circulations exist below and above the swarm (see Fig.4b), which lead to a converged flow regime at the interface. The shear movement of particles in the interface elongates the swarm and forms the particle sheet. The particle sheet disappears and splits into discrete particles as the particle shear movement continues. The cluster shape and size are comparable to the observations reported by Horio and Kuroki [9].

Table 1. CFB simulation parameters.

Riser diameter	0.1 m
Riser height	1.0 m
Superficial gas velocity	5.46 m/s
Particle diameter	0.8 mm
Particle density	2520 kg/m^3
Grid number	17,440
Particle number	521,000
Restitution coefficients e_n, e_t	0.97, 0.35
Friction coefficient	0.1
Time step $\triangle t$	0.0001s

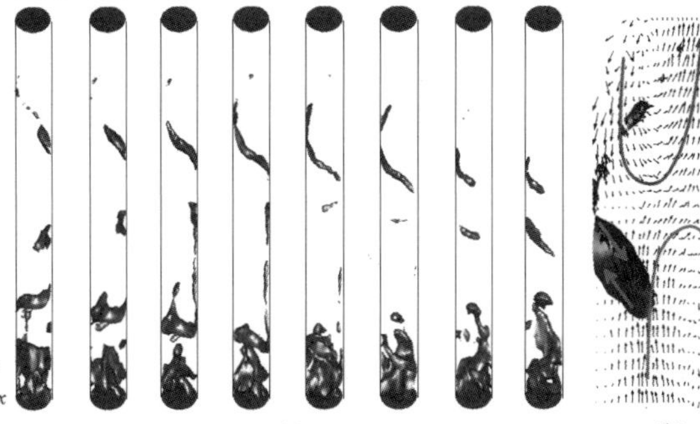

Fig. 4. Snapshots of heterogeneous structures in CFB riser.

We also performed a 3D discrete particle simulation of the quasi-2D spouted bed using the proposed model. The bed dimensions and simulation parameters are the same as those of the experiment in [10]. Fig.5 shows the flow pattern of particles at different points of time. The numerical simulation captures the main features of the spouted bed: the annulus, the spout and the fountain flow regions are distinguished very clearly. The transient granular flow is not steady but appears to be periodic. Also, a denser neck appears near the inlet, flows upward and finally disappears at the end of the spout. That means the particles do not move individually but as a cluster in the spouting region. While the particle cluster in the neck moves upward, it progressively grows bigger and denser with more particles entrained from the annulus. When it reaches the spout end, the spout is almost 'choked'. After that the cluster is injected into the fountain and the particles fall down to the annulus. The above flow behavior agrees well with the experimental observations [10]. The predicted period is about 0.11s, which is slightly less than the experimental one (0.15s).

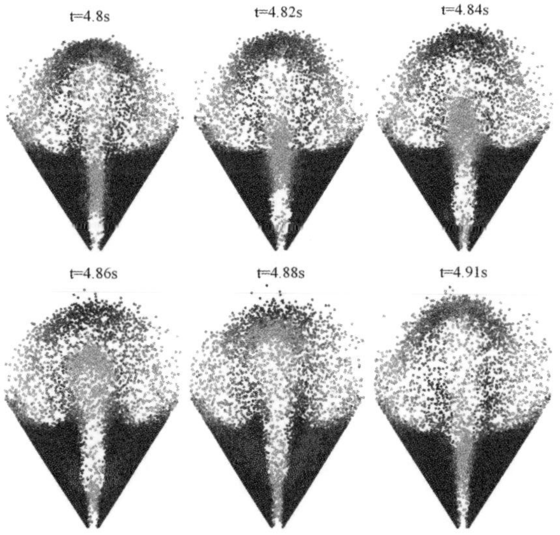

Fig. 5. Snapshots of particle distribution in spouted bed.

5. Concluding remarks

A robust and efficient three-dimensional discrete particle model based on unstructured grids was developed with the aim to simulate gas-solid systems with large number of particles. The numerical results show that the present model can capture many important features of the gas-solid flow systems, such as the core-annulus structure and cluster in a CFB riser, and the spout-annulus structure in a spouted bed.

References

1. Y. Tsuji, T. Kawaguchi, T. Tanaka. Discrete particle simulation of two-dimensional fluidized bed. *Powder. Technol.* 77, 79 (1993)
2. B.P.B. Hoomans, J.A.M. Kuipers, W.J. Briels, W.P.M. van Swaaij. Discrete particle simulation of bubble and slug formation in a two-dimensional gas-fluidised bed:a hard-sphere approach. *Chem. Eng. Sci.* 51, 99(1996)
3. J. Ouyang, J. Li. Discrete simulations of heterogeneous structure and dynamic behavior in gas–solid fluidization. *Chem. Eng. Sci.* 54, 5427(1999)
4. D. Gidaspow. *Multiphase Flow and Fluidization–Continuum and Kinetic Theory Descriptions.* Academic Press, San Diego (1994)
5. C.L. Wu, J.M. Zhan, Y.S. Li, K.S. Lam. Dense particulate flow model on unstructured mesh. *Chem. Eng. Sci.* 61, 5726(2006)
6. M.J.V. Goldschmidt, R. Beetstra, J.A.M. Kuipers. Hydrodynamic modelling of dense gas-fluidised beds: comparison and validation of 3D discrete particle and continuum models. *Powder Technol.* 142 , 23(2004)
7. C.L. Wu, J.M. Zhan, Y.S. Li, K.S. Lam, A.S. Berrouk. Accurate void fraction calculation for three-dimensional discrete particle model on unstructured mesh. *Chem. Eng. Sci.* 64,1260(2009)
8. M. Marin, D. Risso, P. Cordero. Efficient algorithms for many-body hardparticle molecular dynamics, *J. Comput. Phys.* 109, 306(1993)
9. M. Horio, H. Kuroki. Three-dimensional flow visualization of dilutely dispersedsolids in bubbling and circulating fluidized beds. *Chem. Eng. Sci.* 49, 2413(1994)
10. X.L. Zhao, S.Q. Li, G.Q. Liu, Q. Yao, J.S. Marshall. DEM simulation of the particle dynamics in two-dimensional spouted beds. *Powder Technol.* 184, 205(2008)

IMPACT FACTORS OF COMMUNITY HEALTH AND WELFARE: APPLICATION OF FAHP

KUEI-YANG WU
Department of Architecture, National United University, No. 1 Lien-Da, Kung-Ching Li, Miaoli 36003, Taiwan

HONG-DI HUANG
Department of Architecture, National United University, No. 1 Lien-Da, Kung-Ching Li, Miaoli 36003, Taiwan

Sustainable development is currently a hot issue that has progressed from global thinking to local action in places all around the world. This study applied the Fuzzy Analytic Hierarchy Process (FAHP) to construct a new three hierarchical expert group decision evaluation model for assessing the health and welfare (CHW) management strategy of sustainable development in Miaoli City, Taiwan. The result showed that the creation of a community health program (CCH) is the most important factor influencing CHW management strategy. The sub-factors of organizational operation flexibility (OOF), spontaneous sense of community (SSC), creating a community health (CCH) program, local self-sufficiency (LS), and average income of everybody (AIE) do not directly affect human life and health. However, they are important sub-factors in the CHW assessment, and would affect the sustainable management strategy assessment. Therefore, the results of the analysis of the global weights were 0.15~0.09. In addition, the global weights of the community and child care (CCC), included sufficient employment opportunities (PSEO), a security and disaster prevention system (SDPS), state of poor people (SPP), importance of regional links (IRL), and health education program (HEP), which were 0.09~0.05. Even though, these sub-factors are very low, they would affect the whole assessment of the CHW.

Keywords: Community health and welfare, FAHP, sustainable development.

1. Introduction

The concept of CHW is a real problem with which every community must cope carefully. As a result, the implementation of CHW has become a very important problem that modern society has to face seriously; this comprises sustainable development, environment, human health and welfare. Besides, there are several additional phenomena related to CHW, such as cancers and chronic diseases which have aroused widespread public concerns. In addition, due to the variety and complexity of factors that should be evaluated when developing the infrastructure and various appraisal systems for the community, the development of a new CHW evaluation system is of extreme importance [1]. Thus, this study aimed to develop a decision-making assessment model for the sustainable development with a specific-defined concept of CHW. In the present industrialized and urbanized society, this phenomenon will exacerbate the climate, human lifestyle, and the environmental quality. In turn, the style of human living is a critical factor affecting the conditions of human health, environment and sustainable development [2].

At any time, the condition of CHW (good or bad) would consist of various problems and multiple standards. Such a complex problem involves the effective integration of social information, the economy, environment, etc., so that all relevant experts, scholars, economists, environmentalists, and politicians will be able to draw up a set of community management strategies [3]. Many prior studies attempted to develop an assessment method for CHW because the information derived from the relevant assessment results would facilitate the quality decision-making of the management department and the government. Moreover, the quality of a community comprises very complex factors, which not only refer to the tangible environment, but also to providing the people with opportunities to meet their needs and desires within the human living conditions [4]. The problems result from human behaviors in daily life and affect human health as well as CHW due, for example, to the pollution of air, water and the earth [5].

For application in the environmental domain, Wang et al. [6] adopted AHP together with a fuzzy mathematics method in order to provide a compatibility model of decision-making suitable for the environment domain. Actually, FAHP is a kind of operational method with integrated applications, whereby this operational method would turn a complex problem into a simple decision-making process in hierarchies, while combining with the fuzzy logic to obtain the relative weight for various undefined indicators [7].

2. Fuzzy Analytic Hierarchy Process (FAHP)

2.1. *Analytic hierarchy process (AHP)*

Developed in 1970 by Saaty [3], AHP is an impersonal, systematic and scientific analysis method. AHP not only can resolve complicated decision-making problems, but also be used to resolve particular problems under an uncertain and unclear situation. AHP utilizes the Normal Scale to perform the pairwise comparison for factors. Generally speaking, AHP adopts the five elements within Saaty's method to calculate the complicated problems and perform the pairwise comparisons among them; the five elements of this method are listed in Table 1 [8].

Table 1. Pair-wise comparison scale for AHP preferences.

Numerical rating preferences	Verbal judgments of
1	Equal importance
3	Little importance
5	Common importance
7	Certain importance
9	Absolute importance

2.2. *Fuzzy set theory*

Fuzzy Set Theory was developed in 1965 by Zadeh; it could resolve unclear, uncertain, imprecise situations and problems. Fuzzy Set Theory can be used as a modeling tool within an uncertain and complex system that is difficult to define accurately [9].

3. Methodology

This study applied the FAHP analysis technique in the CHW management strategy for the study case of the Maoli community. The proposed FAHP model was used to analyze the CHW in the study case of the Maoli community, whereas this study is primarily composed as follows [10]:

1. Identify the factors and sub-factors to be used in the relevant models;
2. Construct the structure of AHP model based on the goal, factors and sub-factors; then, list the AHP model structure with the goal, factors or sub-factors, respectively at the first, second or the third level;
3. Utilize the Fuzzy Theory to make the FAHP questionnaires; then, ask six experts and scholars to complete the FAHP questionnaires;
4. Compare the varied FAHP questionnaires as well as determine the varied opinions provided by the said six experts and scholars regarding the relative importance of fuzzy number and relative membership function of the fuzzy judgment matrix and weight vector;
5. Calculate the local weights of factors and sub-factors by using the pairwise comparison;
6. Summarize the global weights of varied sub-factors and resolve the complicated decision-making problems in the CHW.

4. Result and discussion

The results showed that the three factors are in the FAHP structure (Fig. 1). In the structure, the goal is community health and welfare (CHW) and the factors are: create a community health (CCH) program, community income (CI), community organizations and awareness (COA). The state of poor people (SPP), security and disaster prevention system (SDPS), organizational operation flexibility (OOF), local self-sufficiency (LS), importance of regional links (IRL), spontaneous sense of community (SSC), health education program (HEP), community and child care (CCC), create a community health (CCAH) program, average income of everybody (AIE), and provide sufficient employment opportunities (PSEO) are the sub-factors. This research employed the fuzzy theory of trigonometry. In the initial stage of the questionnaire preparation, experts and scholars were asked to choose the important value of the fuzzy value in regard to the maximum value, median value and minimum value, respectively. Their opinions were calculated in relation to the fuzzy value of different important intensity. The result is shown in Table 2.

According to the researcher's experience, using the arithmetic mean to integrate the experts and scholars' opinions into the consistency index (C.I.) and consistency ratio (C.R.) of the value will be higher; thus, the result would be inconsistent. On the other hand, if a geometric mean is used for our calculations, the result would be consistency. Thus, this research integrated the experts and scholars' opinions with the geometric mean method to integrate the opinion consistency. The FAHP could calculate every factor and sub-factor's weight value. The results are as shown below:

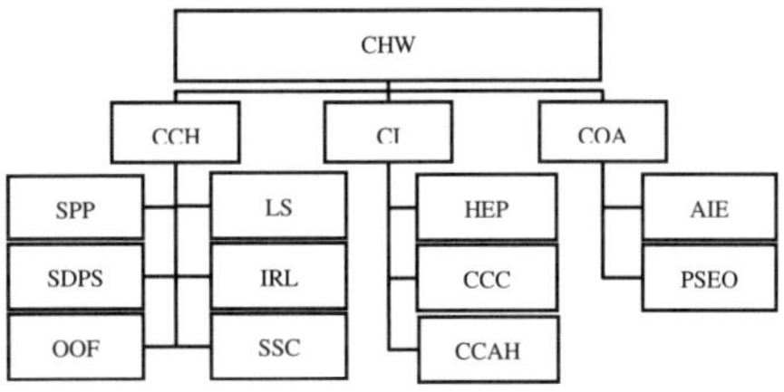

Fig. 1. Hierarchy structure of the fuzzy analytic hierarchy process.

Table 2. Definitions of the relative importance for fuzzy number and relative membership function for the fuzzy judgment matrix and weight vector.

Item	Meaning	Absolute important	Certain important	Common important	Little important	Equal important
1	MF	(7,8,9)	(6,7,8)	(4,5,6)	(2,3,4)	(1,2,3)
	RI	8	7	5	3	2
2	MF	(8,9,9)	(6,7,8)	(4,5,6)	(2,3,4)	(1,1,2)
	RI	8.66	7	5	3	1.33
3	MF	(7,8,9)	(5,6,7)	(3,4,7)	(2,3,4)	(1,2,3)
	RI	8	6	4.66	3	2
4	MF	(8,9,)	(6,7,8)	(4,5,6)	(2,3,4)	(,1,2)
	RI	8.66	7	5	3	1.33
5	MF	(7,8,9)	(6,7,8)	(4,5,6)	(2,3,4)	(1,2,3)
	RI	8	7	5	3	2
6	MF	(7,8,9)	(6,7,8)	(4,5,6)	(2,3,4)	(1,2,3)
	RI	8	7	5	3	2

MF: Membership function, RI: Relative importance.

The influence of the factors in the CHW management strategy of the community is shown in Fig. 2. The local weight of the CCH is 0.5770, and the local weight is higher than the local weight of the CI (local weight: 0.2480). In addition, the local weight of the CI is higher than the local weight of the COA (local weight: 0.1750). Therefore, the factor of the CCH is the most important in regard to influencing the CHW management strategy of the community; the lack of a complete CCW management system would seriously influence human health and sustainable development.

In Fig. 2, the sub-factors of the CCH in the step were calculated. The local weight of the OOF is 0.2567, and the local weight is higher than the local weight of the SSC (local weight: 0.1935), the local weight of the LS (local weight: 0.1745), the local weight of the SDPS (local weight: 0.1460), and the local weight of the SPP (local weight: 0.1171); then the sub-factor of the IRL is lowest (local weight: 0.1121). Therefore, the OOF is a very important sub-factor within the factor of the CCH.

In Fig. 2, the CI is the second most important factor among these factors; the sub-factors of the CI in the step are calculated. The local weight of the CCAH is 0.4322, and the local weight is higher than the local weight of the CCC (local weight: 0.3504), and the local weight of the HEP (local weight: 0.2174). The result shows that the sub-factor of the CCAH is a very important sub-factor in the factor of the CI.

The Fig. 2 shows that the COA is the third most important factor among these factors; the sub-factors of the COA in the step are calculated. The local weight of the AIE is 0.5184, and the local weight is higher than the local weight of the PSEO (local weight: 0.4816). Therefore, the sub-factor of the AIE is a very important sub-factor within the factor of the COA.

The global weights of all of the sub-factors in the FAHP structure are shown in Fig. 3; the global weight of the OOF is 0.1481 and it's the most important sub-factor of the step. The global weights of the SSC, CCAH, LS, and AIE are 0.1117, 0.1072, 0.1007, and 0.0907, respectively. In the community sustainable management strategy assessment, the sub-factors of the SSC, CCAH, LS, and AIE do not directly affect human life and health. But they are important sub-factors in the CHW assessment, and would affect the sustainable management strategy assessment. Therefore, the analysis of the global weights are at 0.15~0.09.

In addition, the sub-factors of the CCC, PSEO, SDPS, SPP, IRL, and HEP are 0.0869, 0.0843, 0.0842, 0.0676, 0.0647, 0.0539, respectively. The global weights of the CCC, PSEO, SDPS, SPP, IRL, HEP are at 0.09~0.05. Although these sub-factors are very low, they would affect the whole assessment of the CHW.

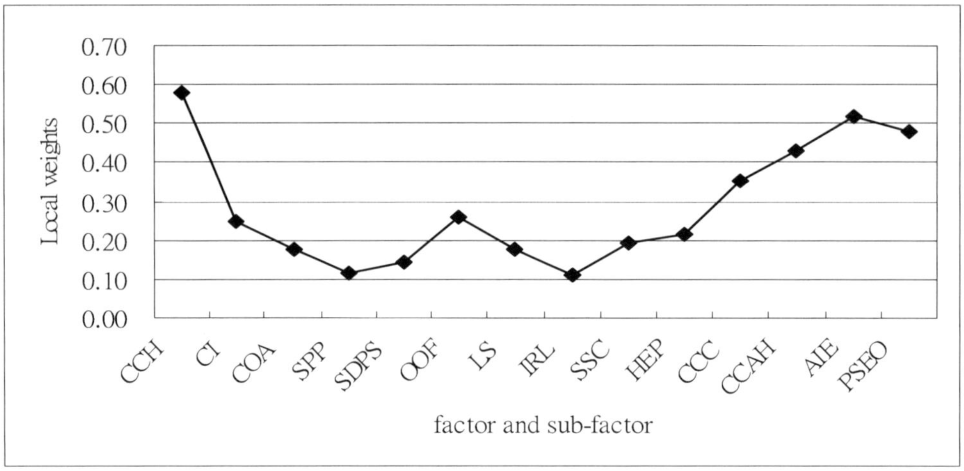

Fig. 2. Local weights and pairwise comparison matrix of factors.

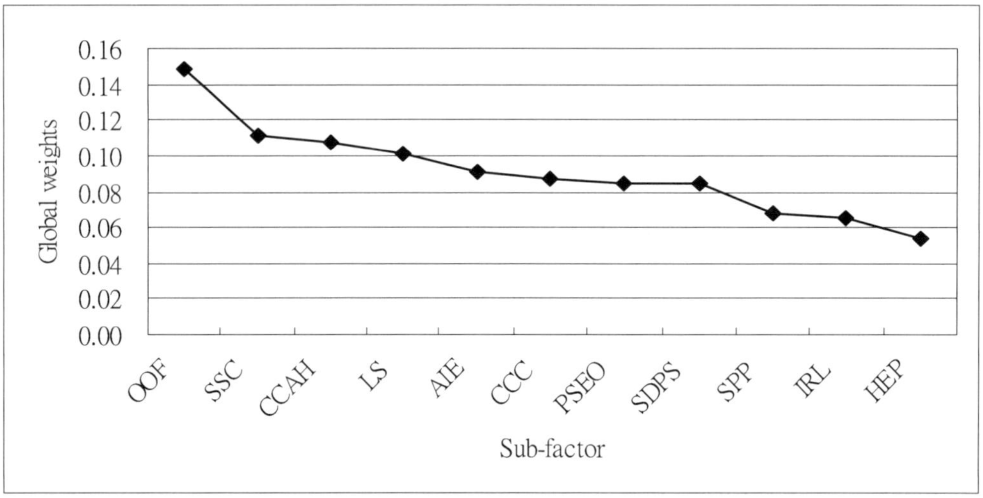

Fig. 3. Computed global weights for sub-factors.

5. Conclusion

This study applied the FAHP for the decision-making strategy in the CHW as below:

The factor of the CCH is the most important in influencing the CHW management strategy of the community. The sub-factors of the OOF, SSC, CCAH, LS, and AIE do not directly affect human life and health, but they are important sub-factors in the CHW assessment, and would affect the sustainable management strategy assessment. Therefore, the analysis of the global weights are at 0.15~0.09.

In addition, the sub-factors of the CCC, PSEO, SDPS, SPP, IRL, and HEP are at 0.09~0.05; even though these sub-factors are very low, they would affect the whole assessment of the CHW.

Acknowledgments

The authors thank NSC (National Science Council) 96-2415-H-239-002-SS3, Taiwan, for financially supporting this research.

References

1. Yueh-Hsia Chiua, Li-Sheng Chen, Chang-Chuan Chan, Der-Ming Liou, Shiao-Chi Wu, Hsu-Sung Kuo, Hong-Jen Chang, Tony Hsiu-Hsi Chen, Health

information system for community-based multiple screening in Keelung, Taiwan (Keelung Community-based Integrated Screening No. 3), International Journal of Medical Informatics, 75, 369-383, 2006.

2. Carol Molinari, Melissa Ahern and Michael Hendryx, The relationship of community quality to the health of women and men, Soc. Sci. Med. 47(8), pp. 1113-1120, 1998.
3. Saaty TL. The analytic hierarchy process. New York: McGraw-Hill, 1980.
4. Tara Smith, Maurice Nelischer, Nathan Perkins, Quality of an urban community: a framework for understanding the relationship between quality and physical form, Landscape and Urban Planning 39, 229-241, 1997.
5. Chen, LX., Zhu, WQ., Zhou, X., Zhou, ZJ., Characteristics of the Heat Island Effect in Shanghai and Its Possible Mechanism, Atmospheric Science 20(6), pp. 991-1001, 2003.
6. Wang, BM., Wang, LJ., Vital process of concrete based on fuzzy analytic hierarchy process and evaluation model of compatibility with environment. Environmental Ecology and Technology of Concrete Key Engineering Matterials 302-303, pp. 55-66, 2006.
7. Lee, WB., Lau, Henry, Liu, ZZ., and Tam, S. A Fuzzy Analytic Hierarchy Process Approach in Modular Product Design, Expert Systems 18(1), pp. 32-42, 2001.
8. Saaty TL. Fundamentals of decision making and priority theory. Pittsburg: RWS Publications, 2000.
9. Fong-Gong Wu, Ying-Jye Lee, Ming-Chyuan Lin, Using the fuzzyanaly tic hierarchyprocess on optimum spatial allocation, International Journal of Industrial Ergonomics 33, 553–569, 2004.
10. Metin Dagˇdeviren, Ihsan Yuksel, Developing a fuzzy analytic hierarchy process (AHP) model for behavior-based safety management, Information Sciences 178, 1717–1733, 2008.

HYDROTALCITE-LIKE SYNTHESIS USING MAGNESIUM FROM BRINE WATER*

EDDY HERALDY†
Department of Chemistry, Faculty of Mathematics and Natural Sciences, Gadjah Mada University, PO Box BLS 21, Sekip Utara, Yogyakarta 55281, Indonesia

Department of Chemistry, Faculty of Mathematics and Natural Sciences, Sebelas Maret University, Jl. Ir. Sutami 36A Kentingan, Surakarta 57126, Indonesia

TRIYONO, SRI JUARI SANTOSA, KARNA WIJAYA, DIAN PRASASTI, GITA SAVITRI, DANIK HERMAWAN, R. IWA and EMANUEL H. SURYO
Department of Chemistry, Faculty of Mathematics and Natural Sciences, Gadjah Mada University, PO Box BLS 21, Sekip Utara, Yogyakarta 55281, Indonesia

Synthesis of hydrotalcite-like (HTlc) from artificial brine water has been done. In preparation samples, 0.1 M Na2CO3 was added to calcium free artificial brine water containing AlCl3 with Mg/Al molar ratio of 2.0, 2.5 and 3.0 respectively until a pH 10.5 was obtained. The solution was then continuously stirred for 1 hour at 70°C. Characterization of solid crystal hydrotalcite-like was done by using FT-IR spectrophotometer and XRD.

Keywords: Hydrotalcite-like, artificial brine water, Na2CO3, Mg/Al molar ratio.

1. Introduction

Hydrotalcite (HT) or hydrotalcite-like (HTlc) is a layered double hydroxides (LDH) with anion exchange property. HTlc have a great variety of uses in many applications as an anionic exchange, catalyst, photocatalyst or precursor for the preparation of inorganic materials (Kovanda et al., 2005). Therefore, there have been numerous studies on adsorption capacity and the catalytic activity of HTlc and several studies on their use for the preservation of water environment (Kameda et al., 2000). HTlc is usually produced by adding Na2CO3, solution or NaOH and Na2CO3, solutions to a mixed solution of MgCl2, and AICl3, (Miyata, 1980) but the production cost is quite high. There are few studies on the production of HTlc on a commercial scale production.

The aim of the present study is to examine the feasibility of a new inexpensive method to prepare HTlc using artificial brine water, a waste from desalination process.

2. Experimental

2.1. *Materials*

Magnesium Chloride hexahydrate (MgCl2.6H2O), sodium chloride (NaCl), calcium chloride dihydrate (CaCl2.2H2O), potassium chloride (KCl), sodium carbonate (Na2CO3), sodium bicarbonate (NaHCO3), and aluminum chloride hexahydrate (AlCl3.6H2O) were used for the make artificial brine water and synthesis of hydrotalcite in p.a. grade. The redistilled water was used during the synthesis.

2.2. *Synthesis of Hydrotalcites by Coprecipitation Method*

2.2.1. *Starting solution*

(i) Artificial brine water contained 0.710M NaCI, 0.411M MgCI2.6H2O, 0.056M CaCI2.2H2O and 0.019M KCl. (ii) Filtrate was obtained by adding a mixed 0.1MNa2CO3-0.2MNaHCO3, solution to (i) until the 1.2 equivalent of calcium ion in (i) and stirring for 1h at 95°C.

*This work is supported by BPPS program of the Higher Education General Directorate of Indonesian
†Work partially supported by doctors grant of the Higher Education General Directorate of Indonesian

2.2.2. *Preparation of Mg/Al HTlc*

One thousand milliliter of 0.1OM Na2CO3 solution was added slowly to 50 mL of (ii) containing AICI3.6H2O, with an initial Mg/AI molar ratio of 2.0; 2.5 and 3.0 at constant pH 10.5 and then the solution was stirred for 1 h at 70°C. The product was centrifuged to recover the white solid at a speed of 2800 rpm for 15 min. The wet cake was washed with redistilled water until free of ion Chloride (AgNO3 test) and dried overnight with oven. The structure of materials were characterized by X-ray diffraction (XRD; MPX3, MAC SCIENCE) and Fourier transform infrared spectroscopy (FT-IR; HORIBA).

3. Results and Discussion

Calcium ion should be removed from brine water as CaCO3 since it is an impurity that coprecipitates with HTlc when Na2CO3 solution is added to brine water containing AlC13 (Figure 1).

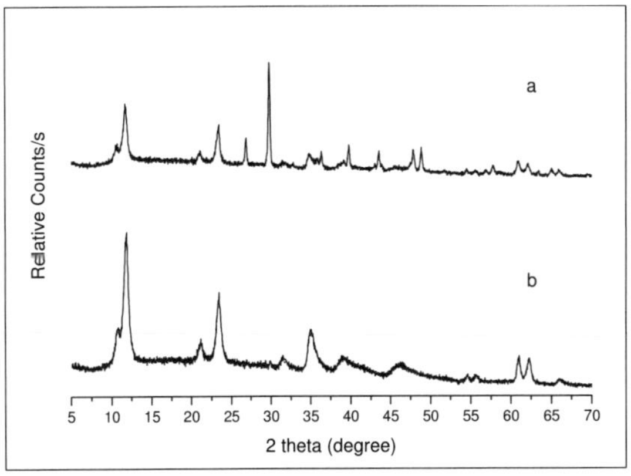

Figure 1. XRD patterns of Mg/Al HTlc with Ca^{2+} impurities (a) and no Ca^{2+} impurities (b).

When a mixed Na_2CO_3-$NaHCO_3$ solution was added to (i) to prepare (ii), Ca^{2+} was selectively removed from (i). Figure 2 shows the XRD pattern of product precipitated from (ii) containing $AlCl_3$ with an initial Mg/Al molar ratio of 2.0; 2.5 and 3.0 respectively at 70°C.

XRD patterns of HTlc prepared with different Mg/Al molar ratio are shown in Figure 2 where all sample have the same patterns that means the sample are hydrotalcite-like compounds.

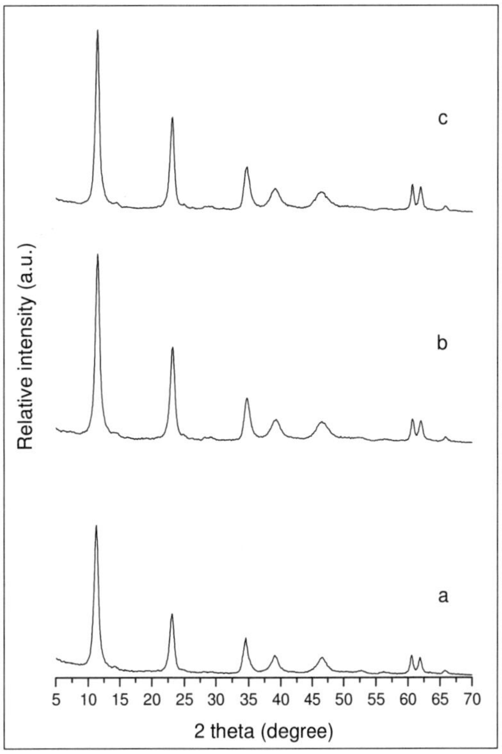

Figure 2. XRD patterns of Mg/Al HTlc with Mg/Al molar ratios of (a) 2.0, (b) 2.5 and (c) 3.0.

The three strong peaks at 2θ values of about 11.3°, 23.1° and 34.6° are characteristic of a layered structure and correspond to the (00*l*) reflections. The corresponding values of d_{003}, d_{006} and d_{009} are listed in Table 1. It was thought that CO_3^{2-} was intercalated into the precipitated HTlc since (*003*) reflection of HTlc intercalated CO_3^{2-} is around 7.83 Å. According to Kloprogge, the value of 7.83Å is characteristic for the (*003*) reflection of carbonate-bearing hydrotalcite (Kloprogge et al., 2002) and the good one is Mg/Al molar ratio 2. The diffraction peak near 61.0° corresponds to the (*110*) crystal plane. It can be seen in Figure 2 that the diffraction peaks are sharp, indicative of relatively well formed crystalline layered structures.

FT-IR spectra of the HTlc samples are shown in Figure 3.

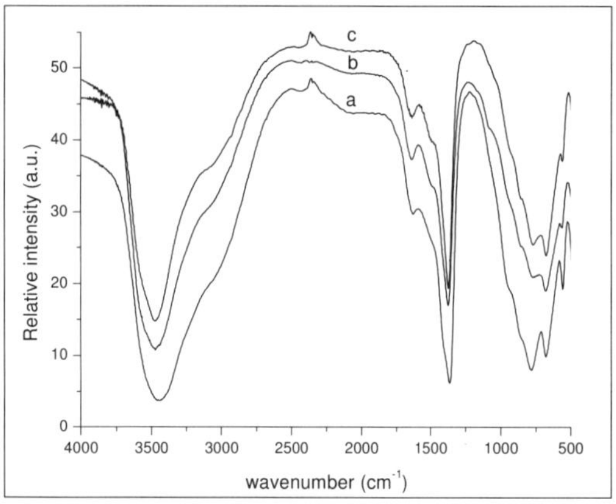

Figure 3. FT-IR spectra of HTlc with Mg/Al molar ratios of (a) 2.0, (b) 2.5 and (c) 3.0.

The broad peak around 3565 cm^{-1} can be ascribed to the stretching of OH groups attached to Al and Mg in the layers. The shoulder at 3046 cm^{-1} is due to hydrogen bonding of interlayer water with interlayer CO_3^{2-} anions. The number of interlayer CO_3^{2-} anions decreases with increasing Mg/Al ratio and this shoulder consequently becomes weaker and is not observed at high Mg/Al ratios. The bending vibration of interlayer water is found at 1637 cm^{-1}; this peak moves to lower wavenumber with a slight decrease in intensity with decreasing Mg/Al ratio. The antisymmetric vibration of CO_3^{2-} appears at 1384 cm^{-1} for high values of Mg/Al ratio and moves to 1361 cm^{-1} when the charge density of the layers increases and leads to stronger interaction between the layers and CO_3^{2-} ions. The peak around 644 cm^{-1} is attributed to the overlap of the antisymmetric vibration of CO_3^{2-} and Mg-O stretching vibrations, and its intensity increases with increasing Mg/Al ratio. The bands at 947, 790, and 554 cm^{-1} can be assigned to Al-O stretching modes, and their intensity decreases with decreasing Al content.

Table 1. XRD data of d_{00l} of Mg/Al HTlc with different Mg/Al molar ratios.

Mg/Al molar ratio	mol 2.0	mol 2.5	mol 3.0	HT Standard
d_{003}(Å)	7.838	7.688	7.688	7.840
d_{006}(Å)	3.844	3.824	3.837	3.900
d_{009}(Å)	2.590	2.579	2.577	2.600

4. Conclusion

As prepared the other HT or HTlc, this sample exhibit the (003) reflection around 7.83 Å, typical for Mg/Al HTlc with carbonate interlayer calcium ion

Acknowledgments

The authors are grateful to Higher Education General Directorate for Research Grants and Prof Sogo Shimazu, Chiba University, Japan for obtaining both XRD and IR data.

References

1. F. Kovanda, D. Kolousek, Z. Cılova and V. Hulınsky, *Appl. Clay Sci.* **28**. 101 (2005).
2. T. Kameda, T. Yoshioka, M. Uchida and A. Okuwaki, *Mol. Cryst. Liq. Cryst.* **341**, 407 (2000).
3. J.T. Kloprogge, D. Wharton, L. Hickey and R. L. Frost, *Am. Mineral.* **87**, 623 (2002)
4. S. Miyata, *Clays Clay Miner.* **28** (1), 50 (1980).

DEVELOPMENT OF POLYSULFONE-CARBON MOLECULAR SIEVES MIXED MATRIX MEMBRANES FOR CO_2 REMOVAL FROM NATURAL GAS

AMELIA SUYONO WIRYOATMOJO, HILMI MUKHTAR and ZAKARIA MAN

Chemical Engineering Department, Universiti Teknologi PETRONAS, Bandar Seri Iskandar, 31750 Tronoh, Perak Darul Ridzuan, Malaysia

Highly selective carbon molecular sieves (CMS) were dispersed in polysulfone (PSU) matrix with various loading concentration (10, 20, and 30 wt.%). A thorough investigation of membrane morphology, mechanical, and thermal properties are discussed. Separation performance of the membranes was tested by using CO_2 and CH_4 gas, respectively, in five different feed pressures within the range of 2-10 bar. The presence of CMS increases the permeability of CO_2 and decreases the permeability of CH_4. In correspondence, the ideal selectivity of CO_2/CH_4 is enhanced.

1. Introduction

Gas separation through selective polymeric membrane is one of the fastest growing branches of membrane technology [1]. A number of advantages such as low capital and operational costs, enhanced weight and space efficiency, ease of operation, and environmentally friendly are offered by membrane separation [2, 3]. However, polymer membranes seem to have reached its permeability and selectivity tradeoff limit, as identified by Robeson [4].

In effort to exceed its permeability-selectivity limitation, highly selective molecular sieves were incorporated in the polymer matrix. Mixed matrix membranes (MMMs) have the potential to achieve membrane with higher permeability and selectivity due to the narrow pore distribution inherent by the molecular sieve particles, and in the same time hindered their costly process and fragility by using cost-effective and flexible polymer as the continuous matrix [1, 5]. The successful key of this implementation lies on both, the selection of polymer matrix and molecular sieve, and the elimination of interfacial defects [6].

2. Background

Investigation of mixed matrix membrane was first reported by Paul and Kemp in 1970s when they discovered a delayed diffusion time lag effect for CO_2 and CH_4 by adding zeolite into rubbery polydimethyl siloxane (PDMS) matrix [7]. Jia et al. [8] reported that there were a slight improvement in O_2/N_2 and CO_2/CH_4 selectivity when silicalite particles were incorporated in PDMS matrix. At their highest loading, 70 wt.% silicalite in PDMS, the O_2/N_2 selectivity increase from 2.14 to 2.94 and CO_2/CH_4 selectivity increase from 3.42 to 8.86 at 30 °C. Even though it is interesting, these result still relatively low compared to the available neat glassy polymer membranes. Duval et al. [9] investigated the effect of the introduction of both zeolites (silicate-1, 13X and KY) and CMS having various pore size and surface properties, in several polymeric matrix (PDMS, EPDM, PCP, NBR 45 and 50). The results showed that zeolites improved the separation properties of poorly selective rubbery polymer towards a mixture of CO_2/CH_4. However, they found that mixed matrix membranes prepared from commercial CMS showing no improvement due to the dead-end nature of the material.

As described above, early works in mixed matrix membrane fabrication predominantly used rubbery polymers as the continuous phase. Industrial applications in gas separation tend to use glassy polymer because of their mechanical properties and better size-dependent separation characteristics compared to rubbery polymer [10, 11]. When used as the continuous phase, glassy polymers tend to form macroscopic voids which lead to no selectivity enhancement due to the poor polymer-filler adhesion. In they work, Vankelecom et al. [12] found that polyimide-zeolite appear to have poor polymer-filler adhesion compared to PDMS-zeolite membrane. They postulated that the high chain rigidity of polyimides resulting in voids in the mixed matrix membranes.

In previous works, zeolites were predominantly used as the molecular sieve particles, the homogeneous pore size become one of the most interesting factor of zeolite usage. On the other hand, most research identified poor polymer-filler adhesion becomes the hardest challenge in forming successful film formation.

In the view of this situation, carbon molecular sieve (CMS) were viewed as an alternative molecular sieve material, Vu et al. [5] had successfully formed Matrimid 5218-CMS (laboratory prepared) mixed matrix membranes with good polymer-filler adhesion. CMS particles appear to have good affinity to glassy polymers with minimal preparation and casting modifications.

Polysulfone is a high performance engineering thermoplastic with resistance to degradation, good gas permeability and selectivity values, low cost and high critical pressure of plasticization [13]. To achieve a good polymer-filler adhesion, glassy polymers with relatively low chain rigidity, such as polysulfone, were preferred than the rigid chain polymer to act as the continuous phase.

In the present work, commercial CMS particles were incorporated within the polysulfone matrix in various loading (10, 20, and 30 wt.%). The resulting membranes were characterized in term of its morphology, mechanical strength, thermal, and gas separation properties. CO_2/CH_4 system was used to characterize the permeation properties of the membranes due to the importance of its commercial use.

3. Experimental Works

3.1. *Materials*

Dense homogeneous and mixed matrix membranes were prepared from Polysulfone (PSU) Udel® P-1800, Solvay Advanced Polymer. Dicholoromethane (DCM, Merck®), with boiling point of 40 °C, were used as the solvent. Carbon molecular sieves (CMS), Shirasagi MSC-3K, purchased from Japan Enviro Chemical (Takeda) were used as the inorganic filler.

Prior to disperse in the polymer matrix, CMS pellets were crushed into a fine particles using Rocklabs® pressure grinder and dried at 250 °C for at least 12 h. CMS particles resulting have size distribution of 0.964±0.588 μm as verified by Scirocco 2000 particle size analyzer from Malvern Instruments Ltd., with density 2.1664 g/cc as verified by Ultrapycnometer 1000.

3.2. *Preparation of Homogeneous Dense Membranes*

A dense homogeneous polymeric membrane has the simplest morphology, therefore it is the simplest and most suitable structure to investigate the chemical and structure modification of the membranes [14]. Membranes were prepared by solvent-evaporation method. Casting solution were prepared by dissolving 15 wt.% of PSU in DCM and stirred for 24 h. The solution was drawn by a casting knife with 400 μm thickness on a flat and dry glass plate. A glass cover was placed covering the glass plate immediately after casting process to slower the evaporation rate of the solvent. After 24 h of evaporation, the dry film was carefully peeled-off from the glass surface. The membrane film was then dried in an oven at 110 °C for 24 h to remove the residual solvent.

3.3. *Preparation of Mixed Matrix Membranes*

Mixed matrix membranes were prepared by introduces a desired amount of CMS into DCM. The mixture were sonicated (Transsonic Digital S, Elma®) in 20 Hz for 10 min to broke CMS aggregates and enhance the homogeneity of CMS dispersion in solution. The CMS particles were then primed by adding approximately 10 wt.% of total polymer used. It is suggested that priming technique increase the compatibility of the CMS with the polymer by introducing a thin layer of the polymer onto the CMS surface, and to minimize the aggregation of CMS particles at high loadings [5]. After sizing the CMS particles, the remaining amount of polymer was introduced to the solution and the solution was stirred for 24 h. The casting process is similar with the formation of dense homogeneous membrane. The solvent was evaporated for 24 h and after the membranes peeled from the glass surface, it was dried in an oven for 24 h at 110 °C.

3.4. *Characterization of Membrane Films*

Membrane morphology determined using scanning electron microscope (SEM), LEO 1430 VP. Membrane mechanical properties were determined by following standard tensile test method from ASTM D-882-02. The sample area is 100 mm × 7mm, and minimum five samples were tested. The tests were carried by using universal testing machine LR 5K from Lloyd Instruments. Dynamic mechanical thermal analysis (DMTA) measurements were carried out on a Mettler-Toledo DMTA 861 in Universiti Sains Malaysia (USM). All experiments were performed with a 1 Hz frequency, 0.1% strain, and 2 K/min heating rate.

3.5. *Gas Permeation Test*

Gas permeation measurements were performed using gas membrane permeation unit. Pure CO_2 and CH_4 were employed as tests gases. A 14.53 cm^2 circular membrane area was mounted into membrane module. Before experiment start, the system was vacuumed to remove any gases or vapors sorbed into the membrane. Permeation tests of membranes were performed by introducing the desired single gas at the desired pressure (2, 4, 6, 8, and 10 bar) in ambient temperature. The permeation rate was measured at steady state condition using a soap bubble flow meter. The permeability coefficient (P_i) of each gas defined by the expression:

$$P_i = \frac{J_i \ell}{\Delta p_i} \qquad (1)$$

4. Results and Discussions

4.1. *Morphology*

Cross-sectional view of the resulting membranes shows in Figure 1. All of the membranes formed a dense membrane structure due to the solvent evaporation casting technique employed. In mixed matrix membranes fabrication, it can be seen that CMS particles were distributed homogeneously throughout the polysulfone matrix. Although in bigger size of CMS particles interface gap of polymer-filler still appear. The size of the gap is said to be very small so that it can still act as sieve for the gas molecules that passing through the membranes. It is showed by the selectivity of the membrane which remains high while increasing CMS loading within the matrix.

4.2. *Mechanical and Thermal Properties*

The mechanical and thermal properties of membrane films were affected by the incorporation of CMS particles within the polymer matrix. Data of membrane's mechanical properties are presented in Table 1. Young's modulus and maximum tensile stress appeared to be increase with higher CMS loading within polymer matrix. In opposite of that, strain at maximum tensile stress and toughness shows decreasing trend.

Young's modulus was determined from the elastic region of stress-strain curve. The value of Young's modulus is related to the rigidity of the material [15] since its value was determined from the slope of linear response of strain to a certain stress applied. Incorporation of rigid inorganic particles to polymer matrix increases the rigidity of the overall material. Higher loading of CMS within polymer matrix results in more rigid material as shown by increment of Young's modulus with higher CMS loading.

Maximum tensile stress appears to have the same behavior as Young's modulus, more loading of CMS was observed that the material break at maximum tensile stress. Strain at break decreased for all MMM with higher CMS loading. Inherent brittleness of CMS adding brittle characteristic to the hybrid material (MMM), at the same time decreases its ability to elongate when the material was stretched. This result meets an agreement with the value of Young's modulus. Introduction of CMS particles in polymer matrix cause the descreasing on stress and strain curve, hence the energy required to break the material needed is less, shown by the decreasing of toughness value.

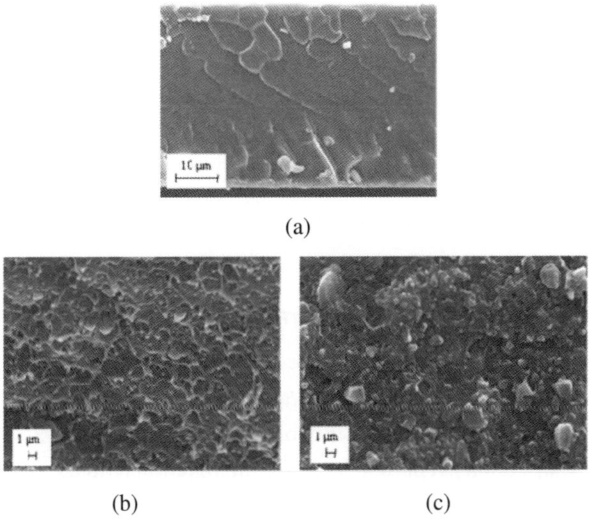

Figure 1. Cross-sectional micrographs of dense polysulfone membrane (a) and mixed matrix membranes in various CMS loading (20 wt.% (b); 30 wt.% (c).

Table 1. The Effect of CMS loading to the Young's modulus, tensile stress, strain and toughness of the membrane films.

	CMS loading (wt.%)		
PSU	10 wt.%	20 wt.%	30 wt.%
Young's modulus (MPa)			
1786.3	1868.3	2258.7	2449.5
Maximum tensile stress (MPa)			
33.7	34.4	35.2	36.4
Strain at maximum tensile stress (%)			
0.057	0.031	0.021	0.021
Toughness (J/m^3)			
0.5810	0.553	0.513	0.423

Glass transition temperature (T_g) of membranes were determined from tan δ peak, as depicted in Figure 2. Lower glass transition temperature observed as increased the CMS loading. Shifting to lower T_g is attributed to the local heterogeneity in the mixed matrix membranes with well dispersed of CMS particles. Similar result also founded by Kim et al. [16] in their work in SWNT-PSU system. They postulated that T_g shifting is possible that the favorable molecular interactions between filler particles and the polymer molecules create an interfacial zone of polymer segments with distributions in mobility [16].

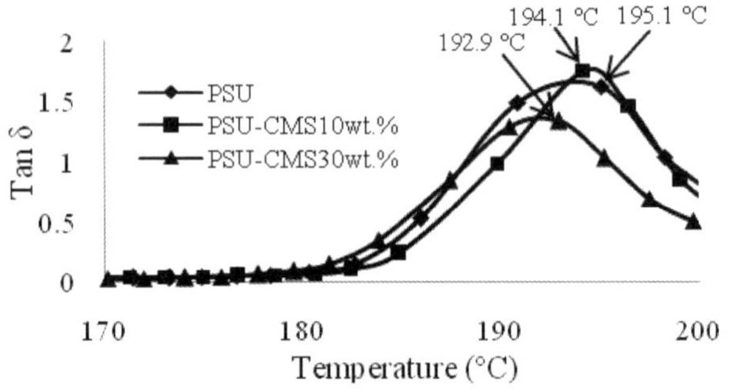

Figure 2. Tan δ graph from DMTA result of pure polymer and mixed matrix membranes.

4.3. *Gas Permeation Properties*

Gas permeation properties of the membranes were evaluated by subjecting membranes to pure gas of CO_2 and CH_4, respectively, in 5 different feed pressure (2, 4, 6, 8, 10 bar). The results of the permeation studies were plotted in Figure 3 and 4. Mixed matrix membranes show interesting permeation properties. As desired, the addition of CMS was able to increase the fast gas (CO_2) permeability and the CO_2/CH_4 selectivity, while slow gas (CH_4) permeability showing an inverted trend. It observed that there was a simultaneous increment of gas permeability and selectivity by increasing the percentage of CMS loading within the matrix.

The highest increment of CO_2 permeability and CO_2/CH_4 ideal selectivity were observed in low feed pressure (2 bar) with the highest CMS loading. P_{CO_2} P_{CO_2} were enhanced from 5.0 (pure PSU) to 6.84 Barrer (30 wt.% CMS loading), showing an increment of 27%. While P_{CO_4} P_{CH_4} are decreased by 117.39% from 1.49 (pure PSU) to 0.46 (30 wt.% CMS loading) Barrer. Consequently, the ideal selectivity of CO_2/CH_4 from 3.35 (pure PSU) to 14.87 (30 wt. CMS loading), representing increment of 77.48%. Meanwhile in high feed pressure (10 bar) mixed matrix membrane with 30 wt.% addition of CMS was able to have ideal selectivity of CO_2/CH_4 as much as 52.5 Barrer, increasing by 32.91% compared to the neat PSU membrane.

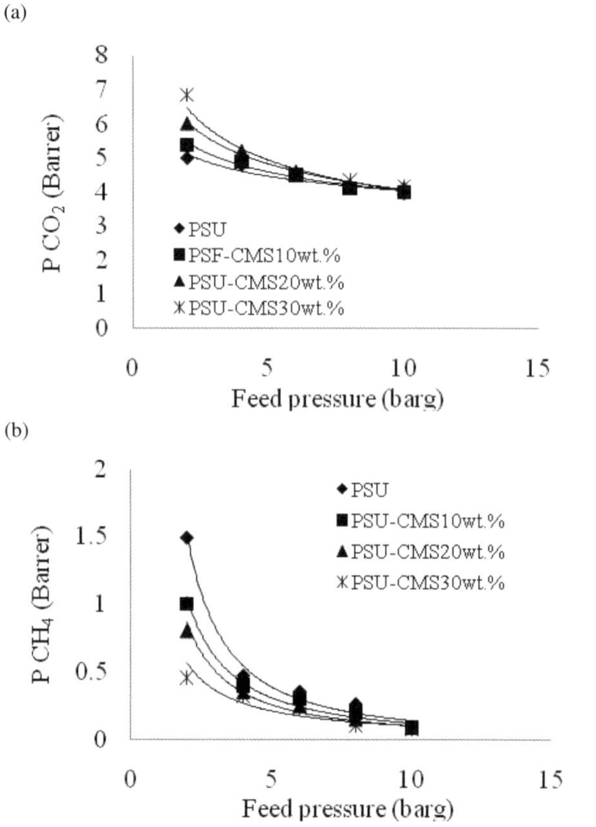

Figure 3. CO_2 (a) and CH_4 (b) permeability for pure polysulfone membrane and mixed matrix membranes with various CMS loadings (10, 20, 30 wt.%).

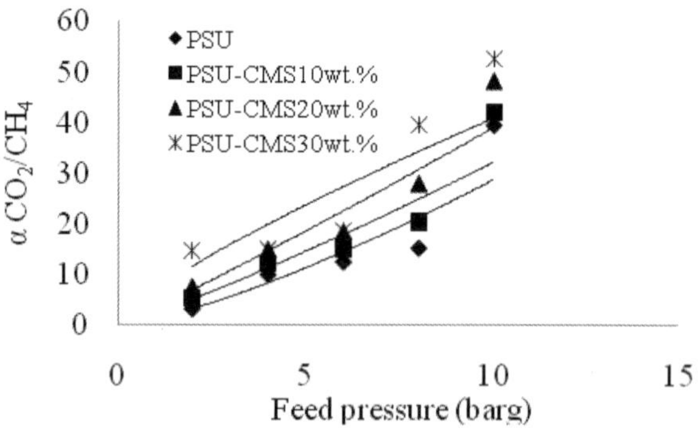

Figure 4. Ideal selectivity of CO2/CH4 for pure polysulfone membrane and mixed matrix membranes with various CMS loadings (10, 20, 30 wt.%).

5. Conclusions

CMS particles had been successfully incorporated within the polysulfone matrix. Scanning electron micrographs of the membranes shows a homogeneous distribution of CMS in the polymer matrix. The existence of filler particles reinforced the mechanical properties of the polymer membranes. The rigidity of the material was increased as shown by the increment of Young's modulus value. Tensile stress showing the same trend as Young's modulus. The inherent brittleness of CMS decreases the ability of the overall material to elongate when it's stretched, as shown by the decreasing of strain at break. The energy required to break the material needed is less, shown by the decreasing of toughness value. From dynamic mechanical analysis, lower glass transition temperature observed as increased the CMS loading. Shifting to lower T_g is attributed to the local heterogeneity in the mixed matrix membranes with well dispersed of CMS particles. Addition of CMS particles are indeed gave interesting permeation properties. The CO_2 permeability is improved as much as 27% in 30 wt.% CMS loading, while the permeability of CH_4 is decreased by 117.39% on the same CMS loading. In correspondence, CO_2/CH_4 ideal selectivity is increased as much as 77.48% in addition of 30 wt.% CMS within the polysulfone matrix.

Acknowledgment

Authors gratefully acknowledge the financial support granted by Universiti teknologi Petronas under Short Term Internal Research Funding (STIRF) no 22/08.09.

References

1. T.-S. Chung, L.Y. Jiang, and S. K. Y. Li, "Mixed matrix membranes (MMMs) compromising organic polymers with dispersed inorganic fillers for gas separation," Prog. Polym. Sci., vol. 32, p. 483, 2007.
2. D. Dortmundt and K. Doshi, "Recent developments in CO2 removal membrane technology," UOP LLC, Des Plaines1999.
3. M. Iqbal, Z. Man, H. Mukhtar, and B. K. Dutta, "Solvent effect on morphology and CO2/CH4 sepration performance of asymmetric polycarbonate membranes," J. Membr. Sci., vol. 318, p. 167, 2008.
4. L.M. Robeson, "Correlation of separation factor versus permeability for polymeric membranes," J. Membr. Sci., vol. 62, p. 165, 1991.
5. D.Q. Vu, W.J. Koros, and S. J. Miller, "Mixed matrix membranes using carbon molecular sieves: I. Preparation and experimental results," J. Membr. Sci., vol. 211, p. 311, 2003.
6. M. Anson, J. Marchese, E. Garis, N. Ochoa, and C. Pagliero, "ABS copolymer-activated carbon mixed matrix membranes for CO2/CH4 separation," J. Membr. Sci., vol. 243, p. 19, 2004.
7. C.M. Zimmerman, A. Singh, and W. J. Koros, "Tailoring mixed matrix composite membranes for gas separations," J. Membr. Sci., vol. 137, p. 145, 1997.
8. M. Jia, K.V. Peinemann, and R.D. Behling, "Molecular sieving effect of the zeolite-filled silicone rubber membranes in gas permeation," J. Membr. Sci., vol. 57, pp. 289-296, 1991.
9. J.-M. Duval, B. Folkers, M.H.V. Mulder, G. Desgrandchamps, and C. A. Smolders, "Adsorbent filled membranes for gas separation. Part 1. Improvement of the gas separation properties of polymeric membranes by incorporation of microporous adsorbents," J. Membr. Sci., vol. 80, p. 189, 1993.
10. T.T. Moore, "Effect of materials, processing, and operating conditions on the morphology and gas transport properties of mixed matrix membranes," vol. PhD: The University of Texas at Austin, 2004.
11. D. Shekhawat, D.R. Luebke, and H. W. Pennline, "A review of carbon dioxide selective membranes: a topical report," National Energy Technology Laboratory, United States Department of Energy, Pittsburgh2003.
12. I.V.J. Vankelecom, E. Merckx, M. Luts, and J. B. Uytterhoeven, "Incorporation of zeolites in polyimide membranes," J. Phys. Chem., vol. 99, p. 13187, 1995.
13. G.C. Kapantaidakis and G.H. Koops, "High flux polyethersulfone–polyimide blend hollow fiber membranes for gas separation," J. Membr. Sci., vol. 204, pp. 153-171, 2002.
14. P. Hacarlioglu, L. Toppare, and L. Yilmaz, "Effect of preparation parameters on the performance of dense homogeneous polycarbonate gas separation membranes," J. Appl. Polym. Sci., vol. 90, pp. 776-785, 2003.
15. C. Torras, F. Ferrando, J. Paltakari, and R. Garcia-Valls, "Performance, morphology, and tensile characterization of activated carbon composite membranes for the synthesis of enzyme membrane reactors," J. Membr. Sci., vol. 282, pp. 149-161, 2006.
16. S. Kim, L. Chen, J.K. Johnson, and E. Marand, "Polysulfone functionalized carbon nanotube mixed matrix membranes for gas separation: Theory and experiment," J. Membr. Sci., vol. 294, p. 147, 2007.

LIFE CYCLE ASSESSMENT OF WASTE CO-PROCESSING IN CHINESE CEMENT PRODUCTION

MICHAEL E. BOESCH and STEFANIE HELLWEG

ETH Zurich, Institute of Environmental Engineering, 8093 Zurich, Switzerland

China is the largest cement producer worldwide. The cement industry is very energy and raw material intensive and a large emitter of CO2. At the same time, China's waste management system still has many deficiencies, and the waste is an enormous source of pollution. Waste used as alternative fuel and raw material (AFR) in cement kilns can reduce greenhouse gas emissions, save primary resources, and provide a waste treatment service. A large variety of wastes can be co-processed if certain rules are followed and process adaptations are made in order to avoid negative impacts on operation and emissions. In this study, the Life Cycle Assessment (LCA) methodology has been applied to identify the effects of waste co-processing in cement production on greenhouse gas emissions. With substitution rates of 40% for fuels (20% plastics, 10% tires, 10% sugarcane leaves) and 5% for raw materials (blast furnace slag), a reduction in greenhouse gas emissions of 89 kg CO2-equivalents can be achieved per ton cement. From a waste management perspective, AFR co-processing may achieve even larger emission reductions, since the co-processed wastes need not be landfilled or incinerated.

Keywords: Alternative fuels and raw materials (AFR), cement industry, Life Cycle Assessment (LCA), greenhouse gas emissions, waste treatment.

1. Introduction

1.1. Cement production in China

With 47% of global production, China has the biggest cement industry in the world [1]. A significant share of the production is based on inefficient technology and on fossil fuels [1]. Cement is a blend of ground clinker from the cement kiln and additional mineral components. Clinker production is the most energy intensive process step in cement manufacturing. Energy intensity in Chinese cement industry will substantially decrease in the next decades due to national regulations [1]. Between 1997 and 2008, the share of clinker production from outdated vertical shaft kilns declined from 80% to about 30%, while new plants are equipped with efficient suspension preheater kilns [1, 2]. Despite these efforts to modernize production infrastructure, virtually no improvements have been made regarding the processed resources. In 2005, fuel share was 94% coal and 6% oil [1], while in 2008, oil was no longer used as fuel [2].

1.2. Waste management in China

The quantity of municipal, industrial and hazardous waste is increasing rapidly with the economic development in China [3]. Significant improvements have been made in the waste management sector in the last ten years, and a substantial increase in sanitary landfill and incinerator capacity has been implemented or is projected [4]. However, large waste fractions are still dumped to the environment, and most landfills lack landfill gas collection systems, leading to substantial emissions of the greenhouse gas methane [3].

1.3. Co-processing of alternative fuels and raw materials

Co-processing of alternative fuels and raw materials (AFR), i.e. the substitution of primary resources with wastes, has become a common procedure in various countries around the world since it has been introduced in the 1970s [5]. Fuel substitution rates of more than 70% are achieved in regions with optimized waste supply, specific waste preparation platforms, and adapted kiln systems [6]. For China, a fuel substitution potential of 40% has been estimated [7]. The substitution rate of traditional raw materials with waste is around 15% [8]. A wide range of municipal, industrial and hazardous wastes are feasible to be utilized as AFR in cement kilns (table 1).

Table 1: Examples of wastes suited and not suited for co-processing in cement kilns [5].

Waste material suited for co-processing
• Agricultural and non-agricultural biomass wastes • Waste tires, rubbers, plastics • Sewage sludge, paper sludge • Waste oil, solvents, paints, detergents • Slags, ashes, sands, dusts
Waste material not suited for co-processing
• Unsorted municipal waste • Electronic waste • Infectious and biologically active medical wastes • Radioactive waste • Explosives

To avoid negative effects on cement quality and the generation of toxic emissions when co-processing AFR, knock-out criteria regarding contamination must be followed [5]. For hazardous waste, regulations may require test burns for the verification of the destruction and removal efficiency of principal organic hazardous compounds [5]. Successful and sustainable implementation of waste co-processing requires the consideration of legal, technical, and social aspects. Examples of these include regulatory permissions, coordinated waste collection and preparation activities, adjustments of cement plant components (e.g. feeding systems), occupational health and safety measures, as well as information campaigns for local communities and other stakeholders. In Europe, Japan, US and various other countries, AFR co-processing is applied for economic and environmental reasons in the cement industry and has become an integrated option in waste management systems [9]. Despite this extensive and long-term experience with AFR co-processing around the world, only three of the approximately 5000 cement plants in China co-processed waste fuels in 2004 [10].

1.4. Life cycle assessment

Life cycle assessment [11] determines the environmental impacts of products, processes or services, through production, usage, and disposal. Cradle-to-gate assessments encompass the impacts of the production system and the upstream supply chain of all resources and operating materials. In the present study, only preparation is included for the assessment of wastes, but not their production. The impact of waste production is to be attributed to the industry responsible for its generation, and not to the one that assumes its treatment or disposal [12]. The LCA method has been previously applied to assess cement production in general [13] and for comparisons of waste treatment options (e.g. [14]). This cradle-to-gate assessment focuses on greenhouse gas (GHG) emissions. Supply chain emissions are from an LCA database [15]. Carbon dioxide from biogenic fuels is considered neutral because the carbon released during combustion is taken out of the atmosphere by the species during the growth phase [16].

2. Case study

The production of 1 ton of Portland cement is analyzed, which is composed of 95% clinker and 5% gypsum [17]. Chinese average electricity demand for the kiln system and cement grinding is 118 kWh [1]. The assessed cement kiln is a suspension preheater with precalciner and requires 3000 MJ per ton of clinker. This is the state-of-the-art production system which replaces most of the outdated vertical shaft kiln in China [1]. It is assumed that cement quality is not affected by AFR co-processing.

The analyzed wastes are plastics, tires, sugarcane leaves, and blast furnace slag. Plastics, tires and sugarcane leaves are alternative fuels and substitute coal. Blast furnace slag is an alternative raw material and substitutes limestone. The effect on greenhouse gas emissions is assessed for each waste and for a mix of AFR. The maximal substitution rate (table 2) is applied for single waste substitution. For the AFR mix, a substitution rate of 20%, 10%, 10%, and 5% is assumed for plastics, tires, sugarcane leaves and blast furnace slag, respectively.

Table 2: Characteristics of wastes for co-processing.

Fuels, AFR	Feasible substitution rate (% heat) [9]	CO_2 emission factor (kg CO_2 / GJ) [18]	Carbon neutral [16]
Coal	-	96	N
Plastics	n.a.[1]	75	N
Tires	<20	85	N
Sugar-cane l.	20	110	Y
BF slag	15	-	-[2]

[1] A substitution rate of 20% is assumed for plastics.
[2] Blast furnace slag is a calcined material and does not emit calcination CO_2 when processed.

3. Results

Reductions in greenhouse gases emissions are achieved with all assessed wastes. Co-processing of plastics, tires and sugarcane leaves reduce fossil CO_2 emissions at the plant. Plastics and tires lead to reduced CO_2 emissions since they contain less fossil carbon per unit of energy than coal (table 2). CO_2 from sugarcane leaves is neutral because it's a biogenic waste. Further greenhouse gas emissions are saved in the resource supply chain, since the decrease in coal consumption reduces the requirement for coal production and preparation. Blast furnace slag reduces CO_2 from the calcination process in the cement kiln ($CaCO_3$ -> $CaO + CO_2$) since it is a calcined material in contrast to the substituted traditional resource limestone. The assessed AFR mix (20% plastics, 10% tires, 10% sugarcane leaves, 5% blast furnace slag) results in emission reductions of 89 kg CO_2-equivalents per ton of cement, which corresponds to a relative reduction of 8% of total greenhouse gas emissions (fig. 1).

4. Discussion

Co-processing of AFR can achieve substantial environmental benefits in clinker and cement production regarding greenhouse gases. The substitution rate of 40% for fuels and 5% for raw materials in this case study is not the limit, as successful co-processing of substantially higher waste fractions in various regions

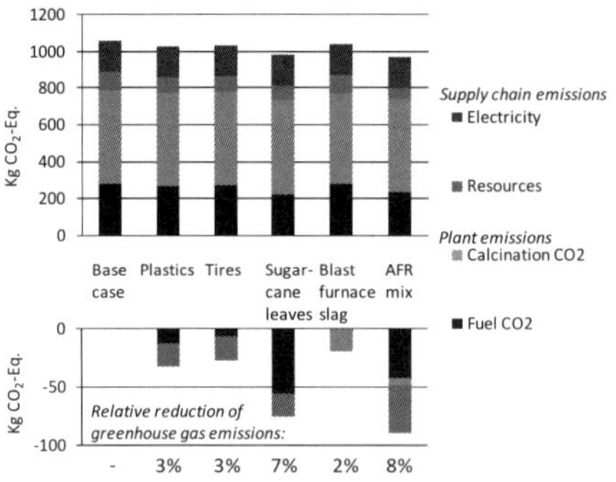

Fig. 1: Greenhouse gas emissions of the production of 1 ton cement with different AFR. In the base case, no AFR is co-processed. Top: Cumulative greenhouse gas emissions. Bottom: Absolute and relative difference to base case.

worldwide demonstrates (e.g. [6]). However, AFR co-processing requires investments to adapt preparation and production processes. Further, waste characteristics may incur production losses and increase process heat requirements. From a waste management point of view, overall benefits of co-processing may be substantially larger than the emission savings in cement production if alternative treatments such as dumping, landfilling or incineration are taken into consideration. However, to quantify these indirect effects, knowledge of the technical standard of the respective treatment system is required. For instance, modern landfills feature landfill gas collection systems and incinerators recover the waste heat by producing electricity and/or district heating with varying efficiencies, which substantially reduce the impact of traditional waste treatment.

Regarding sustainable cement production, kiln system and operation have a large influence on energy requirement and hence on greenhouse gas emissions, which may exceed benefits from AFR co-processing [9]. Furthermore, a reduction of the clinker content results in substantial reductions of the environmental impact per unit cement. Cement types with reduced clinker contents (e.g. CEM II: ≥65%, CEM III: ≥20% clinker) are specified in the European cement standard [17].

For a comprehensive assessment of the environmental effect of AFR co-processing, further environmental subjects such as acidification, ozone depletion and human health etc. should be assessed, in addition to climate change. The influence on these types of impact depends strongly on the kiln type, operation, the availability of flue gas treatment systems, and on characteristics of both primary resources and co-processed wastes.

5. Conclusion

In China, both the cement industry and the waste management sector undergo a rapid development. The integration of AFR co-processing in cement production and its implementation in waste management strategies can lead to substantial reductions in greenhouse gas emissions and provide a valuable waste treatment service. Regarding waste management, the availability, capacity, technological standard and cost effectiveness of the existing and planned waste treatment industries should be compared to co-processing in cement kilns, with regard to environmental impact, to identify optimal waste strategies.

References

1. IEA Tracking Industrial Energy Efficiency and CO2 Emissions; International Energy Agency: Paris, 2007.
2. Wang, H., College of Architecture and Environment, Sichuan University Chengdu, China. Personal Communication. 13.6.2009.
3. Huang, Q.; Wang, Q.; Dong, L.; Xi, B.; Zhou, B., The current situation of solid waste management in China. Journal of Material Cycles and Waste Management 2006, 8, (1), 63-69.
4. Delvoie, C.; Plessis-Fraissar, M. Waste Management in China: Issues and Recommendations; World Bank: 2005.
5. Ziegler, D.; Schimpf, W.; Dubach, B.; Degré, J.; Mutz, D. Guidelines on Co-processing Waste Materials in Cement Production. www.coprocem.com/Documents/guideline_coprocem_v06-06.pdf
6. Chandelle, J. Waste Co-Processing in Cement Plants - The European Experience; Brussels, 2008.
7. SINTEF Solid help for the Chinese climate. http://www.sintef.no/Home/Press-Room/Research-News/Solid-help-for-the-Chinese-climate/
8. Price, L.; Galitsky, C. Opportunities for Improving Energy and Environmental Performance of China's Cement Kilns; Ernest Orlando Lawrence Berkeley National Laboratory: Berkeley, 2006.
9. Murray, A.; Price, L. Use of Alternative Fuels in Cement Manufacture: Analysis of Fuel Characteristics and Feasibility for Use in the Chinese Cement Sector; Ernest Orlando Lawrence Berkeley National Laboratory: Berkeley, 2008.
10. Worrell, E.; Galitsky, C.; Price, L. Energy Efficiency Improvement Opportunities for the Cement Industry; Ernest Orlando Lawrence Berkeley: Berkeley, 2008.
11. ISO 14040: 2006 International Standard ISO 14040: Environmental Management - Life Cycle Assessment - Principles and Framework; International Organisation for Standardization: Geneva, 2006.
12. Doka, G. Life Cycle Inventories of Waste Treatment Services - Part III, ecoinvent report Nr. 13; Swiss Center for Life Cycle Inventories: Dübendorf, Switzerland, 2003.
13. Josa, A.; Aguado, A.; Heino, A.; Byars, E.; Cardim, A., Comparative Analysis of available Life Cycle Inventories of Cement in the EU. Cement and Concrete Research 2004, 34, (8), 1313-1320.
14. Winkler, J.; Bilitewski, B., Comparative evaluation of life cycle assessment models for solid waste management. Waste Management 2007, 27, (8), 1021-1031.
15. ecoinvent-center, ecoinvent database v2.0. In Swiss Centre for Life Cycle Inventories: Dübendorf, Switzerland, 2007.
16. IPCC Guidelines for National Greenhouse Gas Inventories, Vol 5 Waste; Intergovernmental Panel on Climate Change: 2006.
17. DIN DIN EN 197-1: Cement - Part 1: Composition, specifications and conformity criteria for common cements; 2004.
18. WBCSD CO2 Emissions Inventory Protocol, Version
19. www.wbcsd.org/web/publications/cement-tf1.xls

Session 6

OPTIMAL SIZING, ECONOMIC STUDY AND RELIABILITY OF DIFFERENT CONFIGURATIONS OF PHOTOVOLTAIC SYSTEMS

AMAL A. HASSAN, FATEN H. FAHMY and ABD EL-SHAFY A. NAFEH
Electronics Research Institute, National Research Center Building, Cairo, Egypt

MOHAMED A. EL-SAYED
Centre of Energy Studies, University of Trinidad & Tobago

This paper represents a comparative study between different configurations of a photovoltaic (PV) system. The objective of this study is to select the optimal system configuration for supplying the electrical loads of an emergency health clinic according to their energy requirements. These considered systems are stand alone PV system, and grid connected PV system. A computer program is developed to determine the optimal size, cost and reliability of each system based on selecting different values of loss of power supply probability (LPSP) for the different configurations of the PV systems. For each value of LPSP the different size combinations of solar array and battery is calculated. This method aims at finding the PV system configuration which meets the desired system reliability requirement with the minimum total cost over the life time of the system. To get optimum design of PV systems, it is needed to collect load data, meteorological data (solar radiation and temperature) for the site under consideration and data for the investment cost of the system. The Kharga Oasis located at The Western Desert of Egypt is selected as the site used for the case study.

1. Introduction

Conventional methods of generating electricity can produce pollutants such as carbon dioxide, the main gas responsible for global warming. The only resource needed to power a solar cell is sunlight; through the photovoltaic effect the energy contained in the sun light can be converted directly to electrical energy. Photovoltaic systems represent a silent, safe, not pollutant and renewable source of electrical energy [1]. This paper provides a design of PV system for supplying the electrical loads of an emergency health clinic with electrical energy including a comparative study between different configurations of a photovoltaic (PV) system to select the optimal system configuration.

Most of the emergency health clinics require an environmentally clean power supply. At the same time, the electrical load of the health clinics is critical and requires continuity of the power supply. In Egypt, there are many new projects such as those carried out in New Valley in the western desert of Egypt which will be accompanied by a new population [2]. From this point of view, it is important to introduce a model for an emergency health clinic powered by using photovoltaic system. The clinic provides services such as normal examination and emergency medical care.

2. PV System Sizing

Sizing of PV system is an important part of its design and it starts by estimating the energy consumption of the electrical loads of the clinic and the available solar energy radiation of the selected location. The sizing procedure is as follows [3]:

2.1. *Sizing procedure*

Due to intermittent nature of solar radiation, the power system reliability analysis has been considered as an important step in any system design process and must be reflected in the sizing calculations. It is then highly desirable that the procedure gives as a final result, the cost effective combination of the array size and storage capacity, subject to a desired system reliability level. In this study, the reliability of the system is expressed in terms of loss of power supply probability (*LPSP*). The LPSP is defined as the ratio of all energy deficits (LPS) to the total load demand during the considered period. This can be shown as Eq. (1) [3, 4]:

$$LPSP = \sum LPS(t) \Big/ \sum E_{Load}(t) \tag{1}$$

Where *LPS(t)* represents the deficit energy for hour t. it occurs when the total energy available for a time interval and the energy stored in batteries at the beginning of that interval are insufficient to satisfy the load demand for that interval. At any time the energy deficit can be calculated as follows [4]:

$$LPS(t) = (P_{Load}(t) - P_{tot}(t))\Delta t \tag{2}$$

Where, $P_{tot}(t)$ is given as follows [5, 6]:

$$P_{tot}(t)\Delta t = (P_{PV}(t)\Delta t + C_{bat}(t-1) - C_{bat\,min}) \times \eta_{inv} \tag{3}$$

Where, $P_{PV}(t)$ is the total generated power by the PV array at time t, $C_{bat}(t\text{-}1)$ is the available battery capacity (Wh) at time $(t\text{-}1)$, η_{inv} is the inverter efficiency and C_{batmin} is the minimum allowable storage capacity and it can calculated as follows [4]:

$$C_{bat\,min} = DOD \times C_{batn} \tag{4}$$

Where, DOD is the allowable depth of discharge for the battery bank and C_{batn} is the rated capacity of the battery bank. For different values of LPSP, the PV array size versus battery size is calculated and the optimal solution which minimizes the total system cost and ensures the high reliability is selected as shown below in Fig. 1 which represents the flow chart of the sizing procedure.

2.2. *PV array modeling*

Electrical modeling of suggested PV system is represented in the following equations [5]:

$$I_{PV} = N_p I_L - N_p I_{os}(exp(^{qV_{pv}}/_{(BKTN_S)}) - 1) \tag{5}$$

$$I_L = [I_{s.c} + K_I(T_c - 28)] * \frac{H}{1000} \tag{6}$$

$$P_{PV} = I_{PV} \times V_{pv} \tag{7}$$

Where, I_{PV} is the PV array output current (A), V_{pv} is the PV array output voltage (V), I_L is the light generated current (A), I_{os} is the cell reverse saturation current (A), B is the ideality factors (1.92), K is the Boltzmann's constant (1.38x10-23 joule/°K), q is the electronic charge (1.602x10-19 coulomb), T_c is the cell temperature (°C), T is the cell temperature (°K), K_I is the short-circuit current temperature coefficient (0.0017 A / °C), H is the cell illumination (W/m^2), $I_{s.c}$ is the cell short-circuit current at 28 °C and 1000 W/m^2 (4.8 A for SP75 module).

2.3. *PV array sizing*

The number of modules in series (N_S) is determined by dividing the designed system voltage V_{System} (usually determined by the battery bank or the inverter) and the nominal module voltage V_{Module} at standard test conditions (STC) (STC at 1000 W/m^2 and 28 °C) [6].

$$N_S = \frac{V_{System}}{V_{Module}} \tag{8}$$

The designed array output ($P_{PVarray}$) is determined by multiplying the number of strings in parallel N_P, the number of series modules N_S and the selected module output $P_{Modulemax}$ as given in Eq. (9) [6].

$$P_{PV\,array} = N_P \times N_S \times P_{Module\,max} \tag{9}$$

2.4. *Battery sizing*

The storage battery can be calculated using the following equation [2]:

$$Battery Storage = \frac{N_c \times E_L}{DOD \times \eta_{Out}} \tag{10}$$

Where, N_c is the number of storage days, for stand alone system it will be taken as the continuous cloudy days in the selected region (4 days), while for grid connected PV system the number of storage days N_c is 1 day for the expected grid failure. E_L is the load energy in Wh, and η_{out} is the balance of system efficiency and equals to 0.765 [2].

3. Economical Model

The price of the PV system and its installation are important factors in the economics of PV systems. For the present PV system, the life cycle cost (LCC) will be estimated as Eq. (12) [7, 8]:

$$LCC = IC + OM + R - S \tag{11}$$

Where, IC is the initial capital cost, OM is the operation and maintenance cost, R is the repair and replacement cost and S is the salvage value. In this work, all the costs are converted to present worth value as given below [7, 8]:

$$P_F = \frac{F}{(1+i)^N} \quad (12)$$

$$P_A = \frac{A[(1+i)^N - 1]}{i(1+i)^N} \quad (13)$$

Where, P_F and P_A are the present worth for the sum of money for the single payment and annual payment respectively, F is the single payment, A is the annual payment, i is the interest rate, N is the project life time.

For grid connected PV systems, the total cost of the system (TC) calculated as follows [9, 10]:

$$TC = LCC + E_p \times P_p - E_s \times P_s \quad (14)$$

Where, $E_p \times P_p$ is the present worth of the total cost of energy purchased from the electric grid (E_p) and $E_s \times P_s$ is the present worth of the total cost of energy sold to the electric utility grid (E_s). P_p and P_s are the prices of the energy bought from and sold to the utility respectively.

4. Programs Input Data

4.1. *Radiation data*

To get optimum design of PV system, it is important to collect meteorological data (solar radiation and temperature) for the site under consideration. Table 1 shows the monthly average values of global solar radiation over The Kharga Oasis [11]. It is clear from the figure that solar energy incident on the region is very high especially during summer months. The annual average global solar radiation is 6.303 KWh/m2/day.

Table 1: The monthly average values of daily global solar radiation (kwh/m2/day) in The Kharga Oasis.

Month	Daily Radiation in Kwh/m^2/day
January	4.43
February	5.43
March	6.38
April	7.71
May	7.66
June	8.02
July	7.9
August	7.45
September	6.7
October	5.64
November	4.68
December	4.13

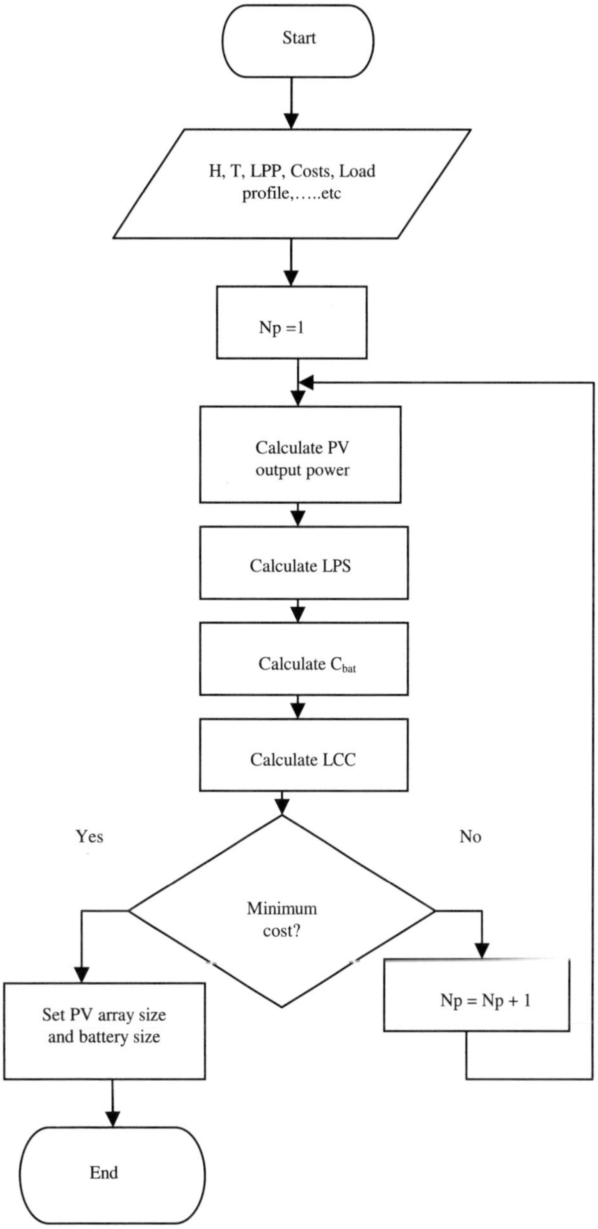

Fig. 1: Flow chart of the sizing procedure.

4.2. *Load data*

The emergency health clinic is suggested to operate two intervals during the day. The daily load profiles were determined by calculating the power demand (KWh/day) for all load types in the clinic during the four seasons. Figure 2 shows a typical daily load profiles during the four seasons. It is shown that, summer has the highest energy consumption (16.44 KWh/day) compared to the other seasons. The average annual load energy is 15.16 KWh/day.

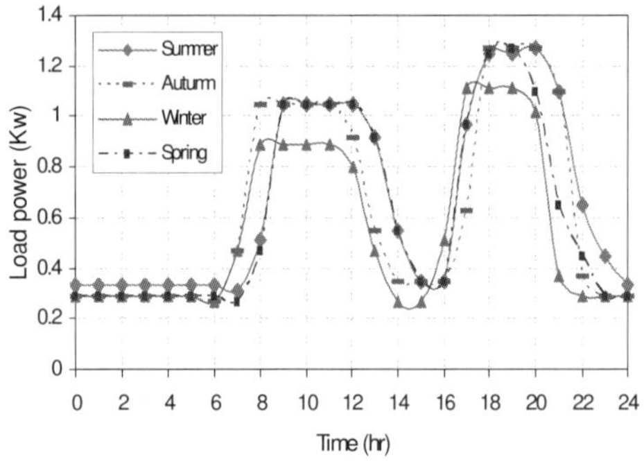

Fig. 2: Daily load profiles during the year.

4.3. *Economics and constraints*

The installation, replacement and maintenance cost of all components of photovoltaic system are shown in Table 2 [12]. The life time of PV arrays is 25 years and typical value for the interest rate is 8%. The cost of buying power from the grid (power price) is 0.05 \$/KWh. The price that the utility pays for power sold to the grid (sell back rate) is 0.05 \$/KWh [10]. The two prices are taken the same based on net metering principle that is, small renewable energy producers below a certain power limit typically (30 to 50 KW) have the same rate for both buying and selling electric power [13]. For the battery commercially available models such as Hoppecke 8 OPzS 800 battery (12 V, 800 Ah, and 9.6 KWh) are considered in the model. The battery life time is taken 5 years [10].

Table 2: Cost of different components of PV system.

Rated capacity	Initial cost	Replacement cost	Operation and maintenance cost
PV	5000 \$/KW	4000 \$/KW	0 \$/year
Inverter	800 \$/KW	750 \$/KW	40 \$/year
Battery	150 \$/KWh	150 \$/KWh	10 \$/year

5. PV System Sizing Results

Figures 3 and 4 shows the configuration of the system meeting different LPSP values for grid connected PV system and stand alone PV system. One point with the minimum total cost value occurs in each curve which means the best configuration for a certain LPSP; this is illustrated in Fig. 5 and Fig 6. This point is considered as the optimal one which meets the system reliability requirement with the lowest total cost. Table 3 and Table 4 summarize the best configuration and the cost for each LPSP value.

For grid connected PV system, the desired LPSP is 0.00001 which means the reliability of the system is 0.99999. The best configuration at that LPSP as shown in Table 3 is 11 PV parallel strings with 2 modules connected in series and the number of batteries is 4 batteries each 800 Ah, 12 volt. The total cost of the system at that point is 32,520 \$.

For stand alone PV system the best configuration at the desired LPSP = 0.0089 which indicates a reliability of .9911 as presented in Table 4, the number of PV parallel strings is 20 with 2 modules connected in series. The battery bank at that point consists of 12 batteries each 800 Ah, 12 volt. The total cost of the system is 76,048\$.

For grid connected PV system, the battery bank capacity reduced due to the existence of the power grid. The battery bank capacity then calculated to feed the electrical loads when the PV output power is smaller than the load power in case of failure of the electric grid. The period of absence of the electric grid for the site under consideration is selected to be 24 hrs.

For the stand alone system the total system cost is large due to using increased number of PV modules and the battery bank capacity is larger than that in grid connected PV system. In stand alone PV system, the battery is selected to cover all the electrical loads during night and during periods when the load required power is greater than the PV output power. This proves that the battery bank is heavy in the calculation of system costs.

It is clear that from results that the grid connected PV system is more economic and reliable than stand alone PV system. So the grid connected PV system will be selected to feed the electrical loads of the emergency health clinic.

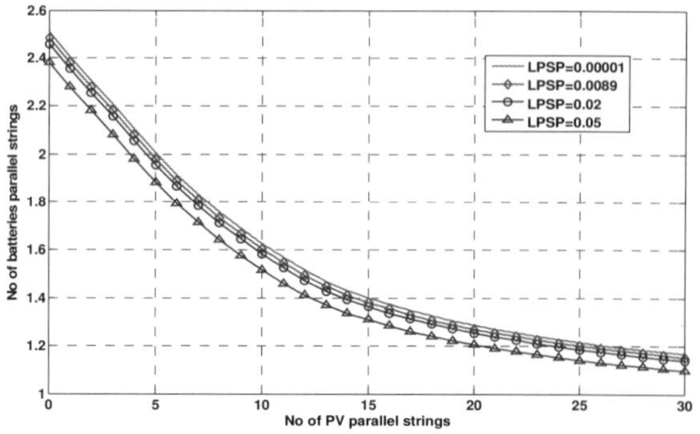

Fig. 3: Combination of PV and battery for different values of LPSP for grid connected PV system.

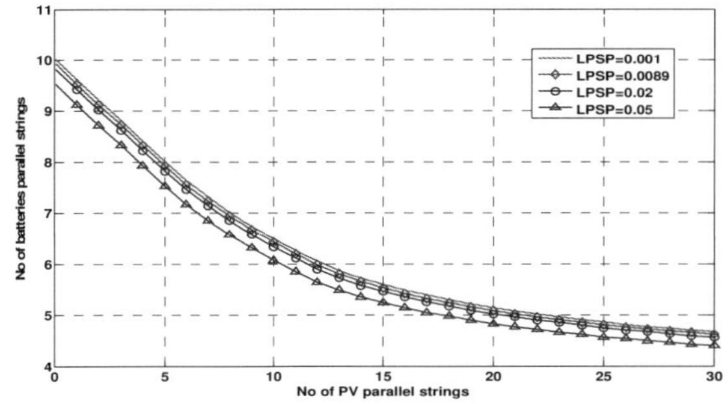

Fig. 4: Combination of PV and battery for different values of LPSP for stand alone PV system.

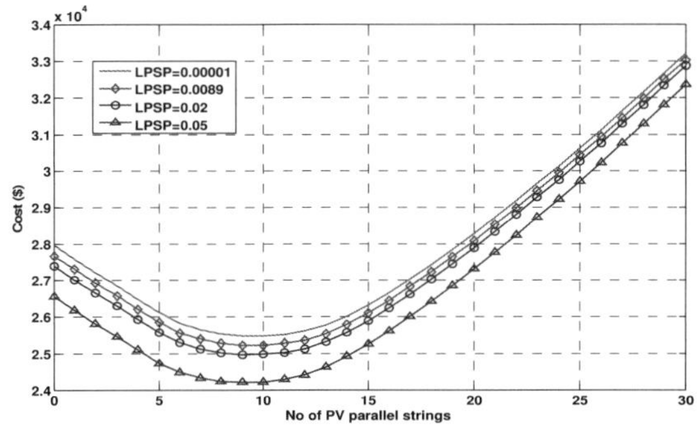

Fig. 5: System configurations and cost for different values of LPSP for grid connected PV system.

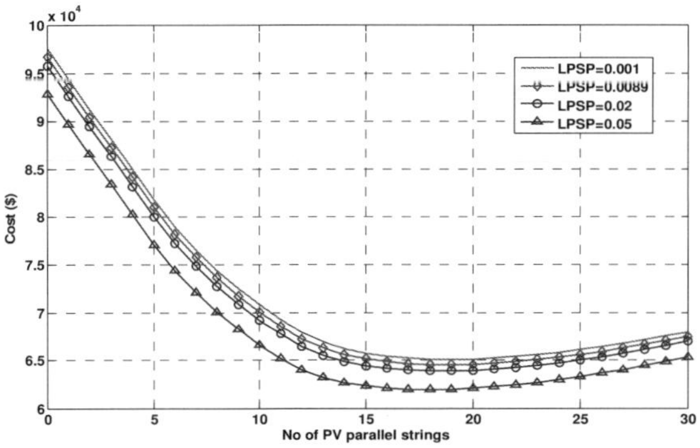

Fig. 6: System configurations and cost for different values of LPSP for stand alone PV system.

Table 3: Sizing results of grid connected PV system.

LPSP	No. of PV parallel strings	No. of batteries	Total cost ($)
0.00001	11	4	32,520
0.02	11	4	32,520
0.05	10	4	31,460

Table 4: Sizing results of stand alone PV system.

LPSP	No. of PV parallel strings	No. of batteries	Total cost ($)
0.0089	20	12	76,048
0.02	20	10	66,380
0.05	19	10	65,650

6. Conclusion

In this paper, a design of a PV system for supplying the electrical load of an emergency health clinic in the Kharga Oasis was presented. A computer program was developed to determine the optimal size, cost and reliability of different configurations of PV systems. The results show that PV systems sizing depend on load data, solar radiation, investment cost of the system and the prices of power sold to or bought from the grid.

For grid connected PV system, the best configuration is 1.65 KW PV array and 4 batteries (each 800 Ah, 12 V) connected to the electric grid with a reliability of 0.99999 and the total cost of the system is 32,520 $. The stand alone PV system configuration is 3 KW and 12 batteries with a reliability of 0.9911 and a system total cost of 76,048 $.

The results proved that the grid connected PV system is more economic and more reliable than stand alone PV system. So the grid connected PV system will be selected to feed the electrical loads of the emergency health clinic. PV systems also are renewable and environmental friendly power sources that contribute in reducing the green house gas emissions.

References

1. D. Cruz Martins, R. Demonti and R. Ruther, "Analysis
2. of utility interactive photovoltaic generation system using a single power static inverter", Photovoltaic Specialists Conference. Conference Record of the Twenty-Eighth IEEE, 2000, 1719 – 1722.
3. G.E. Ahmad, "Photovoltaic-powered rural zone family house in Egypt", Renewable Energy, 26, 2002, 379–390.
4. Tomas Markvart, "Solar electricity", John Wiley & Sons Ltd, USA, 1994.

5. S. Diaf, G. Notton, M. Belhamel, M. Haddadi, A. Louche, "Design and techno-economical optimization for hybrid PV/wind system under various meteorological conditions", Applied Energy, 85, 2008, 968-987.
6. Eftichios Koutroulis and Kostas Kalaitzaiis, "Development of a microcontroller-based, photovoltaic maximum power point tracking control system", IEEE Transactions On Power Electronics, 16(1), January 2001, 46-54.
7. Friedrich Sick and Thomas Erge, "Photovoltaics in buildings - a design handbook for architects and engineers", Freiburg, Germany, 1996.
8. France Lasnier and Tony Gan Ang, "Photovoltaic engineering handbook", Bangkok, Thailand, 1990.
9. Ali Naci Celik, "Present status of photovoltaic energy in Turkey and life cycle techno-economic analysis of a grid-connected photovoltaic-house", Renewable and Sustainable Energy Reviews, 10, 2006, 370–387.
10. Jose L. Bernal Agustin, Rodolfo Dufo Lopez, "Economical and environmental analysis of grid connected photovoltaic systems in Spain", Renewable Energy, 31, 2006, 1107-1128.http://www.nrel.gov/homer/
11. Egyptian solar radiation atlas, Cairo, Egypt, 1998.
12. Athanasia A. Lazou, Anastassios D. Papatsoris, "The economics of photovoltaic stand-alone residential house holds: A case study for various European and Mediterranean locations", Solar Energy Materials & Solar Cells, 62, 2000, 411-427.
13. G. G. Merino, D. Jones, L.E. Stetson, "Performance of a grid connected photovoltaic system using actual and kriged hourly solar radiation", Transactions of the ASAE, 43, April 2000, 1011-1018.

ENVIRONMENTAL ASSESSMENT OF WASTEWATER POLLUTION IN AL-QASSIM INDUSTRIAL CITY

T. I. Sabry

Associate Professor of Sanitary Engineering, Ain Shams University, Cairo, Egypt. tsabry@consultant.com

S. M. ElKholy

Associate Professor of Geotechnical Engineering, National Water Research Center, Cairo, Egypt. selkholy@excite.com

I. S. Al-Salamah

Associate Professor of Hydrology Engineering, Qassim University, Qassim, Saudi Arabia. alsalamah90@hotmail.com

A new approach for a continuous monitoring program is used to assess the status of the wastewater in the sewage network in accordance with the environmental standards and to identify the industrial contamination agents that may exist in the sewage and their possible sources. Moreover, an investigation program is used to assess the potential contamination of the surrounding soil and groundwater due to leakage from the sewage ponds. Five wastewater paths (streams) were selected in the sewage network to trace and follow the sources of the pollutants by collecting samples from selected locations (manholes) in these paths. In order to examine the existence of polluted industrial wastewater in the sewage network, the pollutants concentrations were compared with their extreme values in Buraidah domestic sewage and with their allowable values in the Saudi Executive register (SER) standard for treated wastewater and reuse (Ministry of water and electricity). The study showed the existence of polluted industrial wastewater in the sewage network. However, except for TDS values in the Northern pond and NH3-N values in the Southern pond, the pollutants concentrations entering the ponds have rarely exceeded the law limits due to the dilution factor with other low pollutants concentrations from the low-polluted factories. The study also showed that there is some seepage of the wastewater into the surrounding soil and groundwater. This seepage resulted in an increase in the concentrations of the pollutants agents compared to the typical values found in other sites within the City of Buraidah.

Keywords: Industrial wastewater, wastewater pollution assessment, sewage network, soil and groundwater pollution.

1. Introduction and Literature Review

1.1. *Wastewater contamination*

Industrial technology usually involves the use of a huge amount of chemicals to keep modern life humming. Therefore, factories dump more than 220 million pounds of toxic chemicals into surface water each year, relying on dilution to render them harmless. However, pollutants still accumulate in bottom sediments or become part of the food chain, sometimes causing serious threats to human health later on (12).

Most industries produce some wet waste, although recent trends have been to minimize such production or recycle such waste within the production process. Minimizing wastewater and emissions, and using raw materials more efficiently are some of the gains several industries have seen through using good housekeeping management. However, many industries remain dependent on processes that produce large amount of wastewater (4).

The generated wastewater contaminants from some sources of pollutions are introduced in the following sections: (1), (2), (3), (4), (8)

Food industry. Wastewater generated from agricultural and food industries have distinctive characteristics that set it apart from common municipal wastewater managed by public or private wastewater treatment plants. It is biodegradable and nontoxic, but it has high concentrations of biochemical oxygen demand (BOD) and suspended solids (SS).

Complex organic chemicals industry. A wide range of industries use complex organic chemicals such as pesticides, pharmaceuticals, paints and dyes, petro-chemicals, detergents, plastics etc. Wastewaters can be contaminated by feed-stock materials, by-products, product material in soluble or particulate form, washing and cleaning agents, solvents and added value products such as plasticisers. Many authors have review articles in the wastewater treatment and recycling in the textile processing industry (5), (7). Rott et al (7) gave some examples of applied and effective end-of-pipe-steps in the textile processing industry.

Water treatment. Water treatment for the production of drinking water is likely to be found in any industrial cities. Many industries have a need to treat water to obtain very high quality water for demanding purposes. Water treatment produces organic and mineral sludge from filtration and sedimentation processes.

Fertilizer Manufacturing. The expected main constituents of fertilizers industry wastewater are: Nitric, sulfuric, phosphoric acid, anhydrous ammonia, organics, ammonium nitrate, sulfate, phosphate, chloride, and Urea.

Disposal of wastewater from such industrial plants is a difficult and costly process. Nowadays, most industries have onsite facilities to treat their wastewater so that the pollutant concentrations in the treated wastewater comply with the local and/or national regulations (4), (8).

1.2. *Soil contamination*

It is defined as the presence of man-made chemicals or other alteration in the natural soil environment. This type of contamination typically arises from the rupture of underground storage tanks, application of pesticides, and percolation of contaminated surface water to subsurface strata, leaching of wastes from landfills or direct discharge of industrial wastes to the soil. The most common chemicals involved are petroleum hydrocarbons, solvents, pesticides, lead and other heavy metals. The occurrence of this phenomenon is correlated with the degree of industrialization and intensity of chemical usage. The concern over soil contamination stems primarily from health risks, both of direct contact and from secondary contamination of water supplies.

Contaminants in the soil can hurt plants when they attempt to grow in contaminated soil and take up the contamination through their roots. Contaminants in the soil can adversely impact the health of animals and humans when they ingest, inhale, or touch contaminated soil, or when they eat plants or animals that have themselves been affected by soil contamination. Animals ingest and come into contact with contaminants when they burrow in contaminated soil. Humans ingest and come into contact with contaminants when they play in contaminated soil or dig in the soil as part of their work. Certain contaminants, when they contact our skin, are absorbed into our bodies. When contaminants are attached to small surface soil particles they can become airborne as dust and can be inhale.

2. Research Problem and Objectives

The industrial city of Al-Qassim, which lies at the Southern entrance of Buraidah City, has been at work for more than 25 years. The city comprises industries such as aluminum products, food industries, fertilizing, metal plating, pharmaceutical manufacturing, and desalinated water treatment, which may produce potential hazards to the surrounding environment (air, soil, and water resources).

The sewage system in the city relies on collecting sewage into two large ponds, with total area of 6 hectares, and treating it by the biodegradation action of aerobic/ anaerobic bacteria. These ponds are known as "stabilization ponds". However, the common practice now is to use the ponds as evaporation ponds without sewage outlet. Bad smell and colored waste water have frequently been detected in the stabilization ponds, especially the Southern one. Those are primary evidence of the existing of the contaminated industrial wastewater in these ponds.

The sides and bed of the ponds are lined with polyethylene sheets in order to protect the soil and groundwater from getting contaminated through seepage of sewage. However, since their construction, none of the stabilization ponds, the soil, or the groundwater has ever been examined for potential wastewater seepage which shows the necessity for the current study. Wastewater seepage or wastewater overflow are expected since there is no sewage outlet from these ponds and evaporation rate might not match the inlet flow rate.

The objectives of the current research can be summarized as follows:

1. Identifying the industrial contamination elements that could exist in the sewage and their possible sources.
2. Examining the status of the wastewater in the sewage network in accordance with the environmental standards.

3. Investigating the possible leakage from the ponds into the surrounding soil and groundwater.
4. Remediation program of industrial pollutants.

3. Methodology

3.1. *Sampling locations and analysis*

The existing sewage networks are divided into two parts: Northern part and Southern part, and each part is connected to one of the stabilization ponds.

Grab water samples are collected from critical locations (manholes) in the network and numbered from downstream to upstream of the network. In the numbering system, the letters "N" and "S" stand for Northern and Southern networks, respectively. The selected sampling locations in the network and the type of the existing heavily polluted factories that may have contributed on the wastewater pollution are illustrated in Figures (1) and (2).

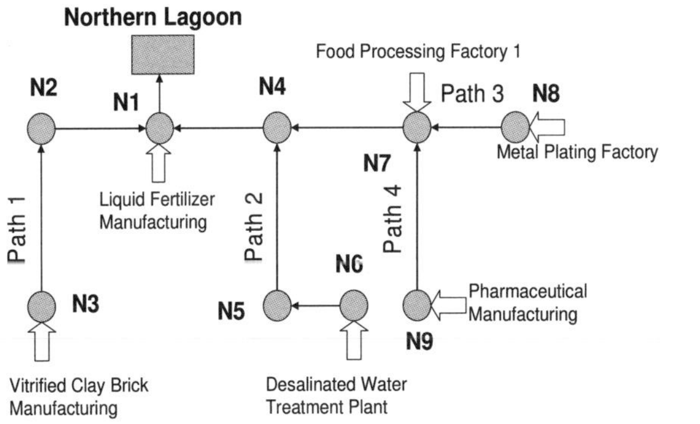

Figure (1): Sewage network for the Northern part of the industrial city.

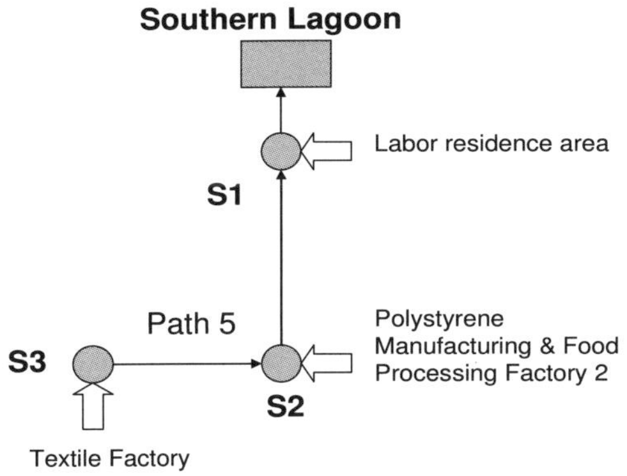

Figure (2): Sewage network for the Southern part of the industrial city.

Table (1): Factories that have polluted wastewater and their disposed location.

Factory Type	**Inlet Location**
Pharmaceutical Manufacturing	N9
Metal Plating	N8
Food Processing 1	N7
Desalinated Water Treatment Plant	N6
Vitrified Clay Brick Manufacturing	N3
Liquid Fertilizer Manufacturing	N1
Textile	S3
Polystyrene Manufacturing	S2
Food Processing 2	S2

Table (1) shows the type of these factories and their wastewater inlet locations in the sewage network.

Four streams in the Northern sewage pipelines and one stream in the Southern sewage pipelines were selected to trace and follow the sources of the pollutants throughout these streams as shown in the Figures (1) and (2). The five paths (streams) are as follows:

Path 1: N3, N2, N1; Path 2: N6, N5, N4, N1; Path 3: N8, N7, N4, N1; Path 4: N9, N7, N4, N1; and Path 5: S3, S2, S1

To identify the existence of industrial wastewater pollution in the sewage network, the pollutants concentrations were compared to their extreme values in the Buraidah domestic sewage. The data of the Buraidah domestic sewage characteristics (BSC) were collected from the raw sewage that enters the Buraidah wastewater treatment plant during the last 5 years.

Pollutants concentrations were also compared to the maximum allowable limits for the disposal of raw wastewater in the Saudi Executive register (SER). For the industrial wastewater, SER asks the factories to have wastewater pretreatment before disposing their wastewater into the municipal sewer lines. Table (2) shows the SER standards for the industrial wastewater that may be disposed into municipal sewer lines.

Table (2): Some of the parameters allowable limit for the raw wastewater that may be disposed into municipal sewer lines according to SER.

Parameters	Max. Allowable Limit (mg/l)
Total suspended solids (mg/l), TSS	600
pH	6-9
Biochemical oxygen demands (mg/l), BOD	500
Chemical oxygen demands (mg/l), COD	1000
Ammonia (mg/l), NH3-N	80
Phosphate (mg/l), PO4	25

Biochemical oxygen demand (BOD), Chemical oxygen demand (COD), Total suspended solids (TSS), Total Dissolved Solids (TDS), Ammonia (NH3-N), Nitrate (NO2), and orthophosphates (PO4) contaminants concentrations were determined according to the *Standard Methods for the Examination of Water and Wastewater (APHA, 1995)* (6).

Although some industries can produce other contaminants, such as heavy metals in metal plating industry and fluoride and silicate in fertilizer manufacturing, in addition to that mentioned in the above paragraph, the above-chosen parameters for analysis considered the minimum requirements for detecting the contamination for all types of the existing industries.

3.2. *Soil and groundwater investigations*

For the current environmental study, the data collected for soil and groundwater samples constitute a small subset of the possible values. In order to assess the existence and the extent of any soil/groundwater contamination due to leakage from the stabilization ponds of the sewage network, an exploratory drilling program was implemented. The program comprised four exploratory boreholes which have been drilled to a maximum depth of 15.0m each at the vicinity of both Northern and Southern ponds. Groundwater level was carefully observed for each borehole. Soil and groundwater samples from boreholes have been carefully collected at pre-set intervals under the close supervision of field geologist. The collected samples were subjected to chemical analyses to detect the existence of any pollutants.

Moreover, two surface water samples were collected from overflow spots that occurred outside the boundary wall of the industrial city near the southern pond were thoroughly analyzed.

4. Results and Discussion

☆ ***Results of Waste Water Analysis***

For more than 6 months of monitoring and analysis, 3 to 5 samples were collected from each location at the most critical locations in the sewer pipeline and 1 to 2 samples were collected from each location at the less critical locations in the sewer pipeline. These samples were analyzed for determining the required parameters.

The pollutants concentrations at the five streams are compared to their maximum limits in the Buraidah Sewage Characteristics (BSC), table (3), and in the Saudi Executive register (SER) standards for treated wastewater and reuse (Ministry of water and electricity), table (2). SER does not include limiting value for the TDS concentration in Table (2). However, it mentions the maximum limit for TDS of the treated sewage to be used in the irrigation, which is 2500 mg/l. Since TDS concentration is nearly not influenced by the conventional treatment process, TDS of 2500 mg/l was used in the study as the maximum permissible value for the TDS in the industrial wastewater.

Table (3): Some of parameters values for the raw sewage entering Buraidah wastewater treatment plant.

Parameters	Values
Total suspended solids (mg/l), TSS	130-340
pH	6.5-8.7
Biochemical oxygen demands (mg/l), BOD	120-190
Chemical oxygen demands (mg/l), COD	240-450
Ammonia (mg/l), NH3-N	14-30
Phosphate (mg/l), PO4	15-35

COD concentrations were measured at selected locations in the sewage network. The results of the average COD concentrations are presented in Table (4):

Table (4): Values of the COD concentrations at the selected locations.

Location	*N1*	*N4*	*N6*	*N8*	*S1*	*S2*	*S3*
COD, mg/l	830	757	1980	430	939	1800	1930

As shown in Table (4), all COD values are high and much higher than that of BSC (maximum 450 mg/l) except for location N8. However, only locations N6, S2, and S3 have COD values violating the limits that mentioned in the SER (maximum 1000 mg/l). The average COD values at both ends of the networks, at N1 and S1, are close but less than the SER limit. This can be attributed to the dilution factor that happened due to mixing of the human excrete with the industrial wastewater especially for S1 where large labor compound sewage collector is connected to stream 5 just before S1 location.

The ratios of BOD to COD in the selected locations ranged between 0.21 and 0.32 which are much less than the typical values for domestic sewage in Buraidah city which is around 0.5. This indicates that the raw industrial wastewaters of some factories are disposed directly into the sewage network without treatment. This value is only close to that of the domestic sewage at location N8.

The results of each of the five streams are discussed in the following sections:

Path 1:

Some TDS concentrations at N1 are much higher than the SER allowable limit. The TDS values for all other locations upstream N1 are less than or slightly higher than the SER allowable limit. This indicates that the potential source of this high TDS values is the nearby liquid fertilizer manufacturing company.

The TSS values through stream 1 are always less than the law limit but sometimes higher than that of the value of BSC especially at N3 and N2 locations (581 and 546 mg/l, respectively). These rather high TSS values at N3 and N2 locations can be originated from the clay particles that are disposed from the vitrified clay brick manufacturing company into the sewer line.

The low pH value, which is 3.2, at N2 location in only one sample, is perhaps due to the disposal of acidic solutions that probably used in small factories that are connected to path 1. Except this, pH values are always within the normal limits.

Path 2:

At N6 location, the BOD, COD, and TDS values are higher than that of SER allowable limits. The high TDS value is mostly due to the disposal of the brine water (refused salts) from the desalinated water treatment plant into the sewer line. However, the TDS values at N5 location are always less the 2500 mg/l due to the dilution with other low-TDS wastewater.

Although the TSS values at N4 location with an average value of 845 mg/l are higher than the SER limit, the TSS values at locations upstream of N4 in paths 2, 3, 4 are less than SER limit. The wastewater of small glass manufacturing company, which is located close to N4, can be the reason for the high TSS values at N4.

Path 3:

Almost there is no indication of polluted industrial wastewater in the sewer path 3 up to N4 location. The BOD and COD concentrations, which are 270 mg/l and 430 mg/l, at N8 location is less than SER limit and they are close to the average concentrations of BOD and COD in the BSC. The very high BOD value (1066 mg/l) at N7 location can be resulted from the wastewater of the food processing factory 1. At location N8, there is no evidence of any polluted wastewater (solids or organics) coming from metal plating company.

Path 4:

As in path 3, there is no evidence of polluted industrial wastewater coming from the pharmaceutical manufacturing company (solids or organics). However, it should be taking into account that the results of the TSS and TDS concentrations at N9 location, which are 520 mg/l and 2394 mg/l, are close to, but less than, the SER limits. Heavy metals and COD concentrations analysis for the Pharmaceutical Manufacturing wastewater need to be done to ensure this result. Only one sample was taken from N9 location because the manhole at this location is always dry whenever samples were collected.

Path 5:

Contradictory to the Northern sewage network, the wastewater in the southern sewage network seemed to be more polluted, although the number and size of the factories into this part are much less than the Northern part. This can be attributed to that wastewater in the southern sewage network imposes to higher organic loads than that of Northern sewage network.

Some results of TSS values such as 2530 mg/l, COD values such as 1930 mg/l, and pH values such as 5.25 are higher than SER limit at S3 location. This indicates that Textile factory is discharging wastewater that is not complying with the SER standards. Also at S2 location, some values of BOD such as 507 mg/l, COD such as 1800 mg/l, and pH such as 4.6 do not comply with the SER standards. The source of these pollutants can result from the textile factory and/or from the food processing-2 company as the BOD "the main source of pollution in the food processing" value is higher than the allowable limit. There is no evidence that polystyrene manufacturing company is polluting the sewage network.

Although, this part of sewage network is heavily polluted at some locations, the wastewater at the outlet network location at S1 is not polluted, except for NH3-N value which was 371 mg/l, and their pollutant concentrations are less than the SER limits. This can be attributed to the dilution that occurs with the domestic wastewater of the labor compound and from non-polluted wastewater of the small factories that exist along with the sewage stream.

☆ ***Results of the Overflow Water Samples from the Ponds:***

The overflow water samples were analyzed to determine the chemical indices, and the results are summarized in Table (5).

Table (5): Chemical indices of the two overflow water samples.

Parameter	*Value For Sample # 1*	*Value For Sample # 2*	Max. Allowable Limit (mg/l)*
pH – at 25^0C	7.42	7.37	6- 8.4
Electrical Conductivity (micro ohms/cm)	3350	5140	--
Total Dissolved Salts (mg/l)	2155	3245	--
Total Alkalinity (M) as $CaCO_3$ (mg/l)	132	128	--
BOD_5 (mg/l)	62	54	40
COD (mg/l)	155	135	--
Ammonium (mg/l)	0.6	0.44	5
Sodium (mg/l)	258	447	--
Potassium (mg/l)	5	6	--
Chlorides (mg/l)	412	690	--
Sulfates (mg/l)	785	1185	--
Bicarbonate (mg/l)	162	156	--
Nitrate (mg/l)	22	27	10
Total Phosphorus (mg/l)	3.27	5.16	--
Total Plate Count (/100ml)	900	346	--
Total Coliform Bacteria (/100ml)	208	52	1000

The results show agreement with the results of the analysis of the sewage water samples collected from the wastewater inlets to ponds. The ponds overflow incidents show the threats of soil contamination in case the storage capacity of the ponds are exceeded.

☆ ***Results of Soil and Groundwater Analysis***

The results of the field and lab tests revealed that the soil profile comprises the following:

Northern Pond

Soil profile: a surface soil layer, followed by intercalated layers of fine sand and silty sand, and then a moderately weathered limestone layer. The ground water level settled at an approximate level of 5.5 m from ground surface.

Southern Pond

Soil profile: a surface soil layer, followed by a layer of weathered limestone extended to the end of the borehole. The ground water level settled at an approximate level of 3.0 m from ground surface.

Chemical Analysis:

Various parameters have been taken into account for the detailed chemical analysis of the boreholes water and soil samples. Some of the results are presented in Tables (6) and (7).

Table 6: Chemical indices of borehole water samples in mg/l*.

NO.	*Parameters*	*B.H.# 01 WS-01*	*B.H.# 02 WS-01*	*B.H.# 03 WS-02*	B.H.# 04 WS-01
01	pH – Value	7.24	7.33	7.38	7.32
02	Electrical Conductivity (micro ohms/cm)	3570	3875	3075	4235
03	Total Dissolved Salt	2320	2520	2000	2750
04	Chlorides	365	508	346	426
05	Sulfates	795	1207	845	1310
06	BOD	Not Detected	Not Detected	Not Detected	Not Detected
07	COD	Not Detected	Not Detected	Not Detected	Not Detected
08	Total Phosphorus	0.05	0.07	0.06	0.08
09	Ammonium	Not Detected	Not Detected	Not Detected	Not Detected
10	Nitrate	8	7	8	8

*SER standard limits are similar to that in table 5

Table 7: Results of chemical analysis of borehole soil samples in mg/l*.

NO.	*Parameters*	*B.H.# 01*	*B.H.# 02*	*B.H.# 03*	B.H.# 04
	Depth of sample (m)	2.0- 3.0	1.5	9.0-10.5	4.5- 6.0
01	pH- Value	7.23	7.26	7.24	7.31
02	Total Soluble Salts	1625	2150	1890	2235
03	Chlorides	221	290	260	330
04	Sulfates	635	690	650	785
05	Total Phosphorus	2.5	2.7	2.5	3.4
06	Ammonium	Not Detect.	Not Detect.	Not Detect.	Not Detect.
07	Nitrate	2.2	2.6	3.3	3.1

*SER standard limits are similar to that in table 5

The results of the chemical analysis show that, for the tested spots, the contaminations concentration in the soil are within the normal figures in this region and do not show evidence of contamination until now. This is attributed to the effectiveness of the geotextile sheet that is used to protect against possible seepage from the stabilization ponds.

5. Conclusion and Recommendation

The results of investigating the status of contamination in wastewater of the Industrial City of Buraidah show evidence of contamination with industrial wastewater at some locations that are not in comply with the Saudi laws and regulations. Although, almost all wastewater pollutants values in the final outlet to ponds complied with the Saudi law, this shall not give the right to non-compliance industries to dispose polluted wastewater in the sewage network. These factories shall change their conditions and make their disposed wastewater comply with the law.

From this study, the source of the polluted wastewaters in the sewage networks are expected to be from the liquid fertilizer manufacturing, textile manufacturing, and food processing industry.

For those non-compliance industries with the Saudi law, Best Management Practices (BMPs) can be implemented to reduce contaminants from entering sewage network (source controls) or to treat their wastewater before disposing it into the sewage network (treatment controls). In general, source controls are more effective in reducing pollutant levels in sewage network, and are therefore, preferred by EPA and other regulatory agencies (9), (10), (11).

The results of testing the soil and groundwater in the region don't show serious evidence of contamination at the investigated spots of the stabilization ponds. However, the threat of overflow from the stabilization ponds into the surrounding areas of the industrial city must be thoroughly observed and monitored in order to safe guard the environment against possible increase in the sewage collected in the ponds beyond the storage capacity.

Acknowledgement

This research work is a research grant with title “Environmental Study for the wastewater contamination in Al-Qassim industrial city” which is supported by scientific research unit in Qassim University, Saudi Arabia.

References

1. Al Goodman, Minimizing Waste and Reducing Costs in the Food and Beverage Industry, Industrial Wastewater Volume 7, Number 1, owned by Water Environment Federation (WEF), February/March (2008).
2. American Petroleum Institute (API), Management of Water Discharges: Design and Operations of Oil-Water Separators, 1st Edition, American Petroleum Institute, (February 1990).
3. M. R. Beychok, Aqueous Wastes from Petroleum and Petrochemical Plants, 1st Edition, John Wiley & Sons. LCCN 67019834, (1967).
4. Industrial wastewater treatment from Wikipedia, the free encyclopedia. http://en.wikipedia.org/wiki/Wastewater
5. Philippe C. Vandevivere, Roberto Bianchi, and Willy Verstraete, Review: Treatment and reuse of wastewater from the textile wet-processing industry: Review of emerging technologies. Journal of Chemical Technology & Biotechnology, volume 72 Issue 4, Pages 289 – 302, (1999).
6. Standard Methods for the Examination of Water and Wastewater, (APHA, 19th Edition, 1995), Available at www.standardmethods.org.
7. U. Rott, and R. Minke, Overview of wastewater treatment and recycling in the textile processing industry. Water science and technology, ISSN 0273-1223 , CODEN WSTED4, (1998).
8. G. Tchobanoglous, F.L. Burton, and H.D. Stensel, Wastewater Engineering (Treatment Disposal Reuse)/Metcalf & Eddy, Inc., 4th Edition, McGraw-Hill Book Company. ISBN 0-07-041878-0, (2003).
9. U.S. Environmental Protection Agency." Code of Federal Regulations, 40 CFR 122.2.
10. U.S. Environmental Protection Agency. Washington, D.C. "Guidance Manual for Developing Best Management Practices (BMP)." Chapter 3. October 1993. Document No. EPA-833-B-93-004.
11. U.S. Environmental Protection Agency. EPA, Office of Water Regulations and Standards. Preliminary Data Summary for the Pharmaceutical Manufacturing Point Source Category, EPA 440/1-89/084, U.S. Environmental Protection Agency, Washington, D.C. September 1989.
12. Woods hole oceanographic institution. Industrial pollution, http:// www.whoi.edu/page.do?pid=1246

CEMENT/ACTIVATED-CARBON SOLIDIFICATION/STABILIZATION TREATMENT OF NITROBENZENE*

ZHAOJI SU, JIANGUO LIU, YIYING JIN, YONGFENG NIE

Department of Environmental Science and Engineering, Tsinghua University, Beijing 100084, China

The feasibility of using cement/activated-carbon Solidification/stabilization technology to treat nitrobenzene was estimated in this study. It was found that the using of activated carbon (AC) could reduce both the nitrobenzene (NB) leaching ratio (LR) in the leaching tests and the NB mass volatilization during the mixing and curing process. And the solidified/stabilized product had better to be disposed in a hazardous waste landfill to achieve a less leaching out amount. The increasing AC adding amount resulted in the continue reduction of LR, but did not continuously reduce the NB mass volatilization. The adding dosage of AC needs future study.

1. Introduction

Cement-based Solidification/stabilization (S/S) is an important technology in hazardous waste management. It is widely applied to treat hazardous waste with heavy metals. But for hazardous wastes with organic contaminant, especial with volatile organic contaminant, it is difficult to apply the conventional cement-based S/S technology. But using of proper additives, such as activated carbon (AC) and organoclays, could improve the S/S efficiency for the organic contaminant. [1-3]

In this study, nitrobenzene (NB), whose vapor pressure is 0.15 mm Hg (20°C), is chosen to be solidified/stabilized by ordinary Portland cement. It is found that the organic compounds which are volatile, water soluble, and with a pKa value below 12 are likely to readily leach from cement based binder systems. [2] According to such hypothesis, NB is quite a challenge to be treated using cement alone. Faschan carried out a research using organoclays blending with the cement to reduce the NB volatilization, and found that the organoclays was highly effective and could reduce the volatilization caused during cement-based S/S process of NB to 1/6 of its original values. [4] But organoclay was quite different from type to type. So here a more common additive, AC was chose with cement together to treat NB.

The feasibility of applying cement and activated-carbon to treat NB was assessed in this research. The leaching tests based on three disposal hypothesis of the solidified/stabilized product were applied to estimate the S/S efficiency. And the NB volatilization was measured during the mixing and curing processes.

Paragraphs should have its first line indented by about 0.25 inch except where the paragraph is preceded by a heading and the abstract should be indented on both sides by 0.25 inch from the main body of the text.

2. Experiment

Major headings should be typeset in boldface with the first letter of important words capitalized.

2.1. *Sample preparation*

Specimens were prepared by using ordinary Portland cement (OPC, 42.5R) and activated carbon to immobilize NB. The NB was spiked into the OPC to achieve a NB content of 3000 ppm and 5000 ppm, representing low NB content and high NB content specimen respectively. Then a calculated amount of 2 wt% or 5 wt% activated carbon (AC) with respect to weight of OPC was added. A water/cement ratio of 0.35 was used in this study. After mixing with a mechanical blender, the fresh paste was put into cylindrical moulds. After cured for 7 d or 28 d under the natural conditions in the lab, the specimens were tested for leaching. During the curing process, the temperature in the lab varied from 9.3°C to 23.9°C; and the relative humidity from 16% to 76%. The specimens prepared were shown in Table 1.

*This work is supported by a grant from the National High Technology Research and Development Program of China (863 Program) (No. 2006AA06Z370).

Table 1. Specimens prepared.

Specimen	NB content	AC adding amount (wt%)
A-1	3000 ppm	0
A-2		2
A-3		5
B-4	5000 ppm	0
B-5		2
B-6		5

2.2. *Leaching test*

'Co-disposal scenario' and 'acid-precipitation scenario' were taken into account based on the specific disposal conditions of the solidified/stabilized products to estimate the leaching behavior of NB from the specimens. The former one described the scenario that the solidified/stabilized products were co-disposal with municipal solid wastes (MSW) in a MSW landfill, where the environment was slightly to moderately acidic during the active stage of waste decomposition; and the later one assumed that the solidified/stabilized products were disposed in a hazardous waste landfill, and the contaminants could reasonably leach out under the acid rain/snow caused by the heavy industry and coal utilization. Hence, acetate buffer solution and 'synthetic acid water' were used as leachant 1# and 2# for 'co-disposal scenario' and 'acid-precipitation scenario' respectively. To reflect the real conditions of the two scenarios in China, the preparation of leachant 1# and 2 # were referred to the Chinese regulation standard method, Solid waste – extraction procedure for leaching toxicity - Acetic acid buffer solution method (HJ/T 300-2007) and Solid waste – extraction procedure for leaching toxicity - Sulphuric acid & nitric acid method (HJ/T 299-2007), respectively. In addition, deionized water was used as leachant 3# for comparison. The characteristics of the leachants were summarized in Table 2.

The main parameters of the leaching test were given below:

- Maximum Particle Size: 9.5mm
- Ambient temperature: 23±2°C
- Liquid to solid ratio (L/S): 10:1.
- Duration of test: 18±2h.
- Agitation method: end over end (30±2 rpm).

Following the leaching test, the samples were filtered through a 0.45 μm cellulose nitrate membrane filter. The filtrate was extracted with carbon disulfide. And then the extractions were merged and dried with anhydrous sodium sulfate. The measurement of the NB was carried out using a gas chromatograph (HP5890, Agilent Technologies) with hydrogen flame ionization detector (FID) and a capillary column HP-5 (30 m × 0.32 mm × 0.25 μm).

2.3. *Volatilization measurement*

To estimate the NB volatilization amount during the mixing process, the samples were mixed using a blender machine which was placed in a steel box with a volume of 260 L, and the fresh paste was put into a cylindrical moulds and cured under the natural conditions in a 3 L brown glass bottle which was named curing bottle. The cylindrical mould had a diameter of 4.2 cm and a height of 10 cm, and the top side of the moulds was exposed to air. Each container had one inlet and one outlet for air sampling.

The sampling and analysis procedure referred to the Air quality-Determination of nitrobenzene (mononitro-and dinitro-compound)-Reduction by zine-N-(1-Naphthyl) ethylene diamine dihydrochloride

Table 2. Characteristics of the leachants.

Leachant	Scenario	pH	Composition
1#	Co-disposal scenario	4.93±0.05	Buffered acetic acid: add 5.7 mL glacial CH_3COOH to 500 mL of deionized water, add 64.3 mL of 1N NaOH, and dilute to a volume of 1 L.
2#	Acid-precipitation scenario	3.20±0.05	Synthetic acid water: slowly add the diluted 2:1 H_2SO_4:HNO_3 acid solution a couple of drops at a time to the deionized water until the pH reaches the desired level.
3#	Comparison scenario	7	Deionized water

spectrophotometric method (GB/T 15501-1995). This standard method is used to determine NB compounds from industrial waste gas and ambient air. Schematic representation of the air sampling system was shown in Figure 1.

During the analysis process, NB could be reduced into aniline with catalyst copper salt and through the diazotization with N-(1-Naphthyl) ethylene diamine dihydrochloride, and then formed a stable purple complex which could be determined by spectrophotometry. The analyze using a UV-VIS spectrophotometer (UV757CRT, Shanghai Jingyan Technology Co., Ltd.) at a wavelength of 550 nm.

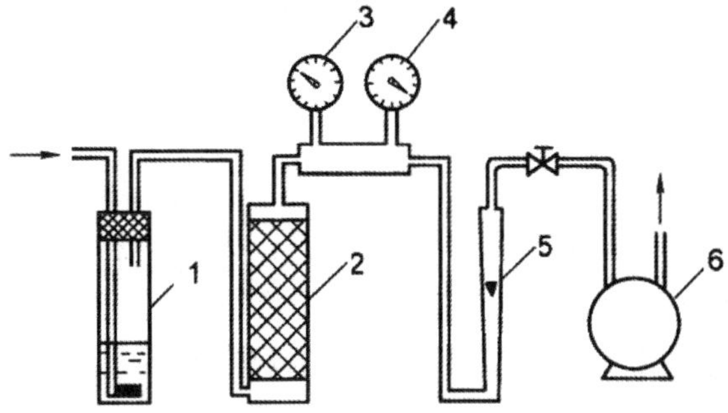

1- absorber with porous glass plate; 2- dryer; 3- temperature meter; 4- pressure meter; 5- flow meter; 6- sampling pump.

Figure 1. Schematic representation of the air sampling system.

3. Results and discussion

3.1. *Comparison of the S/S efficiency in leaching test*

The leaching ratio (LR) was introduced to estimate the S/S efficiency, which could be calculated as follows:

$$LR(\%) = \frac{\text{The amount leaching}}{\text{The original amount}} \times 100\% \quad (1)$$

The LR calculating results were showed in Figure 2.

3.1.1. Effect of NB content

In this study, two NB concentrations of 3000 ppm and 5000 ppm were used. As shown in Figure 2, the specimen with higher NB content had larger LR under the same conditions of AC adding amount and leachant used in the leaching test.

When cement was used to solidified/stabilized organic compounds, the hydration of cement was found to be interfered with the organic compound depending on both the organic compound existing in the system and its concentration.[5] Normally, the extent of hydration of the cement-based solidified/stabilized specimen was very important to achieve a discontinue pore structure in order to limit both water and ion exchanges between the specimen and the environment so that the leaching of the contaminant would be reduced as well. And when the AC was added, the hydration product was expected to coat and encapsulate a fraction of the carbon particulate, which might also limit the accessibility of the contaminants to the environment [6]. The presenting of NB did cause detrimental effect on the cement hydration; and as the NB content increased, the detrimental effect was enhanced, resulting in a larger LR. The affect of NB on the hydration of cement still need a further study.

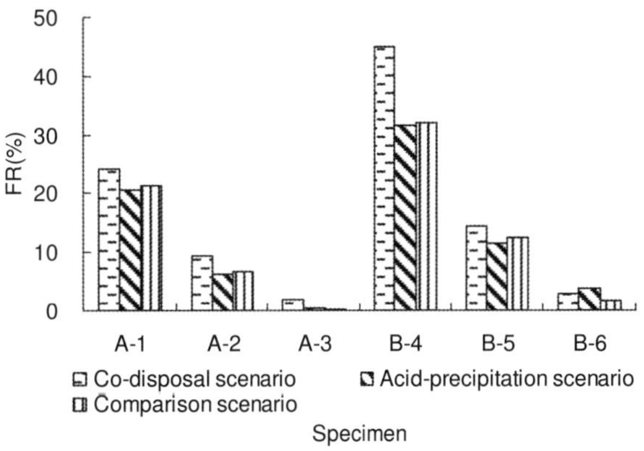

Figure 2. Leaching ratio of specimens (curing time=28 d) in leaching test.

3.1.2. Effect of activated carbon

Without AC, the LR of NB could reach as high as 24.1% and 44.9% for low NB content specimen and high NB content specimen respectively. When AC was added, the LRs were significantly reduced. When comparing with the LRs of the specimens without AC, the LRs were less than 40% of that value as 2 wt% AC was added; while less than 13% of that value as 5 wt% AC was added.

As a result, AC could be used as adsorbent for the S/S of NB. Without counting the NB volatilization in, the adding of AC could significantly improve the S/S efficiency of NB in the leaching test. Actually, the NB were continuously volatilizing during the mixing and curing processes in the S/S procedure, which was an important NB release source that could not be ignored. And the NB volatilization would be discussed later.

3.1.3. Effect of leachant

As mentioned before, the leachant chosen here was based on the disposal hypothesis of the solidified/stabilized product, and buffered acetic acid, synthetic acid water and deionized water were used to simulate co-disposal scenario, acid-precipitation scenario and comparison scenario respectively. For specimens without AC and with 2 wt% AC, the LR decreased in the following order: co-disposal scenario > comparison scenario > acid-precipitation scenario. Although such order changed when 5 wt% AC was added, the LR of all the related specimens was very low, no more than 4%. This phenomenon indicated that the solidified/stabilized product had better to be disposed in a hazardous waste landfill in order to achieve a less leaching out of NB.

3.2. *NB volatilization and NB volatilization controlling*

As showed in Figure 3 and Figure 4, the adding of AC could reduce the NB volatilization during the mixing and curing processes. During the mixing process, when the AC adding amount increased from 2 wt% to 5 wt%, a continuous NB volatilization reduction was not found, unlike what was happened in the leaching test. And adding 2 wt% AC could reduce the NB volatilization to 34% and 26% of its original values when no AC was added for the specimens with 3000 ppm NB and 5000 ppm NB respectively.

During the curing process, most NB volatilization occurred during the first 3 days. And then the NB volatilization of the specimens without AC increased slowly, while the NB volatilization of the specimens with AC was almost undetectable. After 28 d curing, for specimens without AC, the NB mass volatilization could reach about 0.29‰, which was higher than that during the mixing curing process. When AC was added, the NB mass volatilization was reduced to no more than 21% of that. For specimens with 3000 ppm NB, when the AC adding amount increased from 2 wt% to 5 wt%, the NB mass volatilization increased from 0.02‰ to 0.06‰.

In fact, NB was generally adsorbed or encapsulated in the product matrix in the specimen. The adding of AC might enhance the NB adsorbing, but also increased the total porosity of the specimen which could cause an easy volatilization of NB. This might partly explain the phenomenon happened during the mixing and curing processes, where the NB mass volatilization increased as the AC adding amount increased.

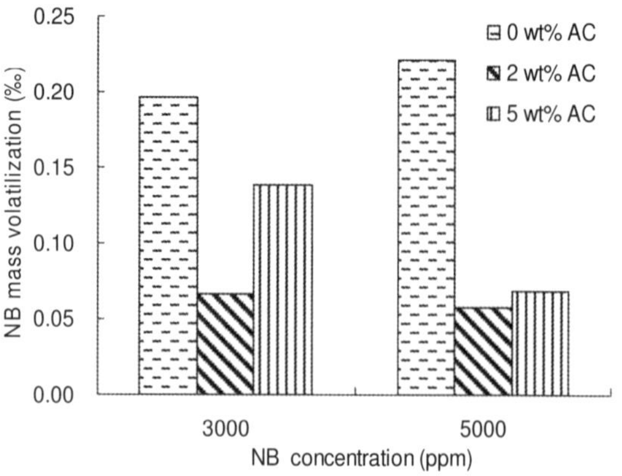

Figure 3. The NB volatilization during mixing process.

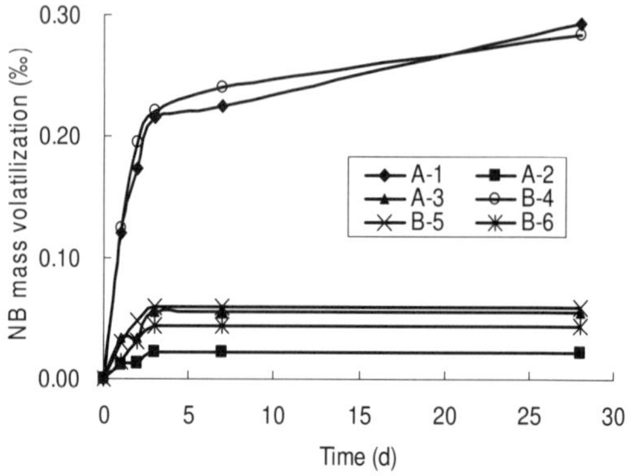

Figure 4. The NB volatilization during curing process.

4. Conclusion

The using of AC could reduce both the NB LR in the leaching tests and the NB mass volatilization during the mixing and curing process. And the solidified/stabilized product had better to be disposed in a hazardous waste landfill to achieve a less leaching out amount. When the AC adding amount increased from 2 wt% to 5 wt%, the LR was continuously decreased, but the NB mass volatilization did not follow such rule. As the adding of AC did largely increase the cost, the dosage of AC adding to the specimen needs future studies to obtain better economic efficiency.

References

1. J. R. Conner, Van Nostrand Reinhold, New York, 1990.

2. B. D. Bone, L. H. Barnard, D. I. Boardman, et al., Environment Agency, Bristol, Science report SC980003/SR2, 2004.
3. S. Paria, P. K. Yuet, Environ. Rev. **14**, 217~255. (2006).
4. A. Faschan, *Hazard. Waste Hazard. Mater.* **9**, 345-354 (1992).
5. S. Trussell, R. D. Spence, Waste Manage. **14**, 507-519 (1994).
6. H. A. Arafat, V. M. Hebatpuria, H. S. Rho, et al., J. Hazard. Mater. **70**, 139-156 (1999).

MICROSTATE SEGMENTATION OF NON-MUSICIAN BRAIN IN SIMULTANEOUSLY AUDIO-VISUAL PERCEPTION

WICHIAN SITTIPRAPAPORN
College of Music, Mahidol University,
Salaya, Nakhonpathom, Thailand

Associations between picture and music sound form the basis of reading. Learning the correspondences between them is a crucial step in sight reading acquisition. This study was designed to investigate whether task-related processing of audio and visual features was independent or task-related processing in one modality might influence the processing of the other. The present study employed simultaneous audio-visual stimulus in the oddball paradigm to re-examine the effects of attention on audio, visual and audio-visual perception in the non-musician brain. Electroencephalographic (EEG) was recorded from 28 normal participants. Participants were presented with pitch (auditory: A), picture (visual: V), and simultaneous pitch and picture (audio-visual: AV). The result showed that the Korean participants showed larger mean interaction amplitude and longer in time than the Chinese participants. This reveals that experience influences the early cortical automatic processing of pitch contour. These results suggest that efficient processing of associations between pictures and sounds relies on neural mechanisms similar to those naturally evolved for integrating audiovisual perception.

Keywords: Brain, audiovisual perception, cognition, event-related potential (ERP), microstate.

1. Introduction

Theoretically, the human central auditory system has a remarkable ability to establish memory traces for invariant features in the acoustic environment such as music, acoustic and speech sounds in order to correct the interpretation of natural acoustic sound heard. Even when no conscious attention is paid to the surrounding sound, changes in their regularity can cause the listener to redirect his or her attention toward the sounds (1). Event-related potential (ERP) recordings have bought new insight to the neuronal events behind auditory change detection in audition. ERPs components reflect the conscious detection of a physical, semantic, or syntactic deviation from the expected sounds (1). The ERP recordings thus allow one to probe the neural processes preceding the involvement of the attentional mechanisms. For instances, ERPs have been recorded that reflect memory traces representing sounds composed of several simultaneous or successive tonal elements (2-4).

It is well-known that auditory signals can be differentiated by a variety of factors, including spectral and temporal information. It is important to recognize that music, acoustic and speech sounds differ in the way they exploit spectral and temporal cues. The present work focuses on three related issues: (1) how is the brain's acoustic sound/music processing represented in brain electric potential fields? (2) how the acoustic sound/music are with different pitch contours distinction embedded in this processing? and (3) which implications do these findings have for theories of the human brain processing, respectively. The aim of the present study thus was to study the human brain's audio-visual integration mechanisms for pictures, i.e., for stimuli that have been previously associated through learning. This study explored the effects of the relatedness of picture and acoustic sound/music. The goal of the present study thus was designed to re-examine the effects of attention in audio (acoustic), visual and audio-visual dimensions by employing an oddball paradigm of ERP recording.

2. Materials and Methods

2.1. *Subjects*

Fourteen right-handed native speakers of Chinese, and fourteen right-handed native speakers of Korean, participated in the ERP experiment. None of the participants had more than three years of formal musical training and none had any musical training within the past five years. After a complete description of the intended study, written informed consent was obtained.

2.2. *Stimuli and Procedure*

Stimuli consisted of a set of three synthesized acoustic sounds that were distinguished minimally by pitch contour: [T1]; [T2]; [T3] and presented in an oddball paradigm of ERP recording. The duration of the stimuli was 300 ms. All three synthesized acoustic sounds had physiological features similar to Chinese pitch/tone. The audio-visual experiment included four stimuli: congruent [T1] (acoustic sound [T1] + visual [T1]), congruent [T2] (acoustic sound [T2] + visual [T2]), incongruent [T1] (acoustic sound [T1] + visual [T2]) and congruent [T4] (acoustic sound [T4] + visual [T4]). The auditory and visual experiments included only the audio (acoustic) and the visual parts of these stimuli, respectively. The stimulus onset asynchrony was 1300 ms. Stimulus sequences consisted of frequent ($P = 0.60$) congruent [T1] stimuli and congruent ($P = 0.15$) and incongruent ($P = 0.15$) [T2] stimuli. Congruent [T4] stimuli were presented as target ($P = 0.10$) to be able to check that subjects were attending the stimuli (see Fig. 1). The stimulus sequences presentation was controlled by the stimulus system (STIM2, Neurosoft, Inc. Sterling, USA). While visual stimuli were presented on the computer screen, audio (acoustic) sound stimuli were delivered binaurally to the participants through plastic tubes and earpieces. Sound density was adjusted to be 85 dB above the participant's hearing threshold. Subjects sat in an electrically shielded and soundproofed room with the response buttons under their hands. The participant had to press the button on the response pad when the target was presented and ignore any other types of stimuli.

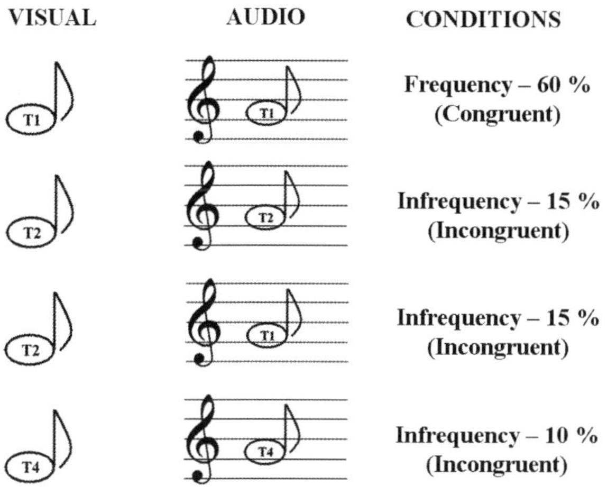

Figure 1. Illustration of the experimental conditions.

2.3. *Electroencephalographic Data Processing*

EEG data were collected in an electrically and acoustically shield room. EEG was recorded with a Quick-Cap equipped with 64 channels according to the international 10-20 system using Scan system (Scan 4.3, Neurosoft, Inc. Sterling, USA). Reference electrode was at mastoids. The signals were bandpass filtered at 0.05-100 Hz and digitized at 1000 Hz. The impedance of the electrode was below 5 kΩ. Eye movements were monitored with two EOG electrodes. Four electrodes monitored horizontal and vertical eye movements for off-line artifact rejection. EEG was segmented into 1000 ms epochs, including the 100 ms pre-stimulus period. The average waveforms obtained from the standard, deviant and target stimuli were digitally filtered by a 0.1-15 Hz band-pass filter and finally baseline-corrected. Cross-modal interaction was investigated by subtracting the ERPs to the auditory (A) and the visual (V) stimuli alone from the ERP to the combined audio-visual (AV) stimuli (i.e. interaction = AV - (A+V).

2.4. *Statistical Analysis*

Statistical analysis was performed on the GFP area of 21 electrodes sites within the time range of difference waveform of cross-modal interaction (100-250 ms). Two groups consisted of Chinese and Korean separately and two conditions were meaningful and meaningless. Five sites were prefrontal line, frontal line, central line, parietal line, and occipital line, respectively. ERP was analyzed with three-way ANOVAs with a repeated measure (group X condition X electrode site). Four electrodes sites such as prefrontal line (Fp1, Fpz, Fp2), frontal line (F7, F3, Fz, F4, F8), central line (T7, C3, Cz, C4, T8), parietal midline (P7, P3, Pz, P4, P8), and occipital line (O1, Oz, O2) sites were used.

3. Results

A three-way repeated measures ANOVA conducted on the mean amplitude yielded main effects of group ($F_{1,26} = 7.31$, $p = 0.01$) and condition ($F_{1,26} = 9.63$, $p < 0.01$), and an interaction effect between group and condition ($F_{1,26} = 5.40$, $p = 0.04$). No main effect of electrode location was found ($F_{1,26} = 1.44$, $p = 0.282$). Korean participants showed a larger mean amplitude relative to Chinese participants for both condition ($F_{1,26} = 9.73$, $p < 0.01$). Global Map Dissimilarity was computed between all maps of the mean map sequence of the Korean (k) and Chinese (c) (Fig. 2 and 3, right), and between all maps of the mean map sequence of the Chinese with themselves (Fig. 2 and 3, left), resulting in what is called now "landscape correlation matrices".

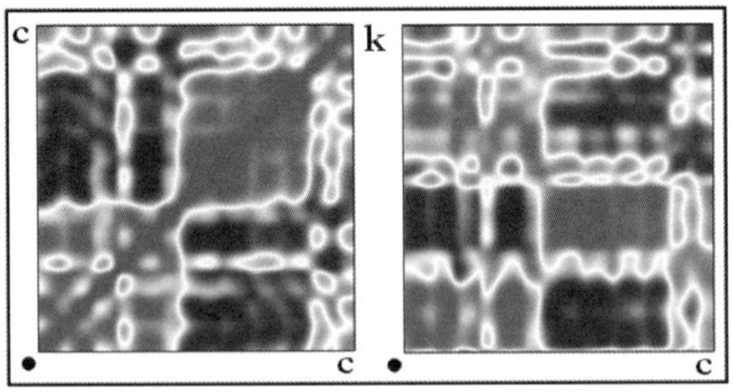

Figure 2. Matrices of correlations between map landscapes of meaningful stimuli. Original is the left lower corner (dot); time runs up and to the right; thus, maps of equal latency are compared along the diagonal from the lower left corner to the upper right corner. Left graph: mean ERP maps of the Chinese vs themselves; Right graph: mean ERP maps of the Chinese (horizontal) vs. those of the Korean (vertical).

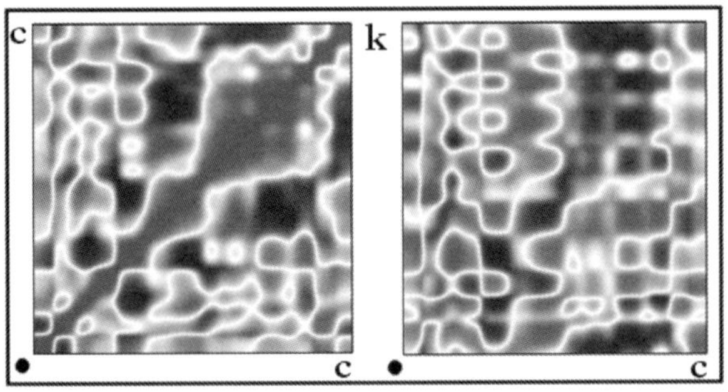

Figure 3. Matrices of correlations between map landscapes of meaningless (B) stimuli. See more information in figure 2.

4. Discussion

The present study was able to find evidence of both non-phonetic and phonetic audio-visual interactions in the ERPs to the same AV stimuli. The differences in the sum of the ERPs to the unimodal A and V stimuli and in ERPs to AV stimuli indicated interactions presumably based on temporal of the A and V components of the AV stimuli. These interactions appeared to be similar for both Chinese and Korean groups. In addition, the differences in the ERPs to the meaningful and meaningless of AV stimuli probably reflect multisensory interactions in phonetic processing. When acoustic and visual phonemes were meaningful, they formed a natural multisensory interaction stimulus. The results of the reaction-time experiment are clearly shown. When subjects categorized acoustic pitch contours, incongruent visual pitch prolonged reaction times ($P < 0.01$) and congruent visual pitch ($P < 0.05$) shortened the reaction times. When subjects categorized visual pitch, incongruent acoustic pitch prolonged reaction times ($P < 0.01$), but the congruent ones did not affect reaction times. These results demonstrate that phonetic-level interactions between acoustic and visual affect identification speed of acoustic/visual perception. The phonetically incongruent audio-visual pitch contours were recognized more slowly than the congruent ones. The comparison of ERPs to the Chinese and Korean in audio-visual pitch contours perception revealed significant differences at latencies of 106 and 116 ms after the sound onset. Both interactions could be explained by modulated activity in parieto-occipital and temporal regions. Importantly, the non-phonetic interactions, which were probably generated in the sensory-specific regions, were earlier than the first phonetic interaction. All phonetic interactions were probably generated in the high-level multisensory regions. The results suggest that sensory-specific and multisensory cortices are involved in audio-visual processing at separate latencies and that they are sensitive to different features of audio-visual stimuli.

The present results collaborate previous findings showing that visual stimuli have access to the early levels of auditory processing hierarchy (5-10) and support the auditory integration models. Integration of auditory and visual non-phonetic information is primarily based on temporal and spatial coincidence of the stimuli. These mechanisms are important in audio-visual integration of speech as well. However, seeing and hearing speech also provide phonetic information. Therefore, both general and speech-specific multisensory mechanisms might be important in audio-visual perception of speech (5, 9). Some comments on the rationale for analyzing the sources of the AV - (A+V) difference wave may be discussed. This difference wave may have a non-zero value whenever the bimodal AV stimulus elicits a different amount or a different configuration of neural activity from the linear sum of the activities elicited by the unimodal A and V stimuli. This differential neural activity could result from an enhancement (or suppression) of activity within one (or both) of the unimodal source configuration, or it could arise from a different source not activated by either unimodal stimulus alone. In any case, this differential neural activity due to cross-modal interaction is, in principle, localizable to a particular brain region or regions. Previous studies have made inferences about the localization of such interactive neural events on the basis of their associated scalp fields manifest in the AV - (A+V) difference wave (11,12). In the present study, scalp topographical evidence was supplemented by map landscape in order to estimate the locations of the contributing sources.

5. Conclusion

The current study demonstrates that the audio-visual interaction is an indicator for investigating the automatic processing of acoustic sound perception. This finding gives support for the view that both sensory-specific and heteromodal cortices are involved in the audio-visual integration process at separate latencies and are sensitive to different features of audio-visual stimuli.

References

1. M. Tervaniemi, M. Rytkönen, E. Schröger, R. Ilmoniemi and R. Näätänen, Learning & Memory 8, 295-300 (2001).
2. E. Schröger, M. Tervaniemi, C. Wolff and R. Näätänen, Brain Res. Cogn. Brain Res. 4, 145-148 (1996).
3. C. Alain, D.L. Woods and K.H. Ogawa, Brain indices of automatic pattern processing. NeuroReport 6, 140-144, 1994.
4. K. Alho, M. Tervaniemi, M. Huotilainen, J. Lavikainen, H. Tiitinen, R.J. Ilmoniemi, J. Knuutila and R. Näätänen, Psychophysiology 33, 396-375 (1996).
5. V. Klucharev, R., Mttonen and M. Sams, Brain Res. Org. Brain Res. 18, 65-75 (2003).
6. R. Mottonen, M. Schurmann and M. Sams, Neurosci. Lett. 363, 112-5 (2004).
7. M. Sams, R. Aulanko, M. Hämäläinen, R. Haru, O.V. Lounasmaa and S-T. Lu, Neurosci. Lett. 27, 141-145 (1991).
8. J. Besle, A. Fort, C. Delpuech and M.H. Giard, Euro. J. Neurosci. 20, 2225-34 (2004).
9. G.A. Calvert and T. Thesen, J. Phys. Paris. 98, 191-205 (2004).
10. V. van Wassenhove, K.W. Grant and D. Poeppel, Proc. Natl. Acad. Sci. U.S.A. 102, 118-6 (2005).
11. J.J. Foxe, I.A. Morocz, B.A. Murry, B.A. Higgins, D.C. Javitt and C.F. Schoeder, Cog. Brain Res., 10, 77-83 (2000).
12. M.H. Giard and F. Peronmet, J. Cog. Neurosci. 1, 473-90 (1999).

AN EXPERIMENTAL STUDY ON THE EFFECTS OF EGR AND EQUIVALENCE RATIO ON CO AND SOOT EMISSIONS OF DUAL FUEL HCCI ENGINE

M. R. KALATEH and M. GHAZIKHANI

Department of Mechanical Engineering, Ferdowsi University of Mashhad, P.O. Box No. 917751111, Mashhad, Iran

HCCI engines improvements are challenged with high CO emissions. In this study, effects of EGR and equivalence ratio on CO and soot emissions of a dual fuel HCCI engine are investigated. The experiments were conducted on a variable compression ratio (VCR) single-cylinder research engine with compression ratio of 17.5:1. Premixed gasoline is provided by a carburetor connected to intake manifold and equipped with a screw to adjust premixed air-fuel ratio, and diesel fuel is injected directly into the cylinder through an injector at pressure of 250 bars. Intake charge temperature was increased up to 110-115°C by using an electrical heater. The higher advanced injection timing (35 BTDC) was used to initiate the gasoline auto-ignition in the HCCI dual fuel engine. The results show that increasing EGR rate increases CO emissions due to dilution effect which reduces combustion temperature, also addition of EGR increases soot emission as a result of decreases the inlet air and its oxygen, which create rich points in the cylinder mixture. Results also show that the CO emission decreases by increasing the equivalence ratio due to formation of more OH radicals in the cylinder and increasing the overall reactivity.

Keywords: Dual fuel HCCI engine, EGR, equivalence ratio, CO and soot emissions.

1. Introduction

With the rapid development of world economy, the problems of energy resource crisis and environment pollution become more and more serious. As one of the main energy consumers and the main source of environment pollution, the automobile receives comprehensive attentions of worldwide researchers [1]. In the history of research and development of IC engines, two engine concepts have played dominant roles which are SI (gasoline) engine and CI (diesel) engine. In conventional stoichiometric charge SI engines, a spark plug ignites the air-fuel mixture in the cylinder, creating high local temperatures and resulting in high NO_X emissions. In CI engines, after taking air into the cylinder and compressing it, the start of combustion is controlled by the injection of fuel into the hot and high-pressure air. This system creates a combustion pattern that produces a high temperature combustion zone and a fuel-rich zone, which yield NO_X and PM emissions, respectively. Many studies are focused on the reduction of both NO_X and PM from these engines. However, due to the combustion mechanism, it is difficult to reduce both NO_X and PM simultaneously [2-4]. Respecting the Euro IV emission norms in 2005, possibilities as the catalytic oxidation, the NO_X traps and the particulate traps can be used, in other words: post-treatment. However, for the future Euro emission norms, the restrictions are more severe and another solution has to be found. The world-wide fuel consumption and exhaust emissions can realistically be reduced if an alternative for the IC engine is developed with characteristics that are significantly better than those of present engines. Concerning the emission reduction during the combustion process, Homogeneous Charge Compression Ignition (HCCI) promises to be a good solution to respect these future Euro norms [5, 6]. HCCI has been researched for some thirty years. It was first identified by Noguchi et al. and Onishi et al. as a method to reduce emissions and fuel consumption of two-stroke engines at part-load conditions [7]. The main concepts of HCCI are breathing premixed air/fuel mixture, as in conventional spark ignition (SI) engines, and ignition without a spark plug, as in conventional compression ignition (CI) engines [8, 9]. In HCCI engine the fuel and air are premixed to form a homogeneous mixture before the compression stroke. As a result, the mixture ignites throughout the bulk without discernable flame propagation due to occurrence of auto-ignition at various locations in the combustion chamber (multi-point ignition), which may cause extremely high rates of heat release, and consequently, high rates of pressurization [10-13]. HCCI combustion is the process in which a homogeneous mixture is auto-ignited by the compression from the piston motion, so the fuel chemical kinetics plays a dominate role during the whole

Nomenclature	
HCCI	Homogeneous Charge Compression Ignition
EGR	Exhaust Gas Recirculation
BTDC	Before Top Dead Center
DI	Direct injection
CI	Compression Ignition
SI	Spark Ignition
IC	Internal Combustion
r_p	Premixed ratio
h_{up}	Heating value of premixed fuel (kJ/kg)
$\dot{m}_p$	Mass flow rate of premixed gasoline (kg/s)
$\dot{m}_d$	Mass flow rate of injected fuel (kg/s)
h_{ud}	Heating value of diesel fuel (kJ/kg)
$\dot{m}_{aEGR}$	Mass flow rate of intake air with EGR (kg/s)
$\dot{m}_{EGR}$	Mass flow rate of EGR gases (kg/s)
ϕ	Equivalence ratio
MON	Measured octane number
$(A/F)_s$	Stoichiometric air-fuel ratio
UHC	Unburned hydrocarbon
CO	Carbon monoxide
CO_2	Carbon dioxide
NO_X	Oxides of nitrogen
PM	Particulate matter

combustion process, which means that HCCI ignition is determined by the charge mixture composition and its time–temperature history. Several parameters affect the quality of the HCCI combustion and the ignition delay: mixture homogeneity, inlet charge temperature, fuel composition, equivalence ratio, coolant temperature, internal and external EGR, engine speed and kinetics of the fuel oxidation at lower temperatures [14-18]. Since the HCCI combustion is lean burn and occurs without flame propagation, it provides much lower combustion temperature than that of conventional SI and CI engines. As a result, HCCI combustion produces very low levels of NOx and particulate matter (PM) while maintaining high thermal efficiency. However, greater amounts of hydrocarbon (HC) and carbon monoxide (CO) emissions are released in HCCI engine relative to conventional SI and CI engines [8, 19-21].

Although stable HCCI operation and its substantial benefits have been demonstrated at selected steady state conditions, several technical barriers must be overcome before HCCI can be widely applied to IC engines. Control of auto-ignition process over different engine operating conditions, achieving cold start, the expansion of operating range and meeting emission standards are challenges. Several potential control methods have been proposed so far, the most effective including EGR, variable compression ratio (VCR) mechanisms or variable valve timing (VVT) to change the effective compression ratio and the amount of hot residual gas respectively [4, 8, 11]. Now, it is generally accepted that the EGR technique is an efficient way to control HCCI combustion. HCCI combustion can occur in internal combustion engines by varying the inlet air temperature and exhaust gas recirculation (EGR) fraction over a range of equivalence ratios [15, 22].

EGR is widely used as the main method to depress the NO_X emission from diesel engines. Currently, EGR is also used as the basic method to control the ignition timing and burn rate of HCCI combustion [15]. EGR consists of many gaseous chemical species, which includes the main components of burned gases, CO_2, H_2O, N_2 and O_2, partial burned gases such as CO, particular matters, HCs and high temperature combustion products NO_X. Different species has different heat capacity and chemical reactivity, therefore has different effect towards ignition timing and heat release rate of HCCI combustion [23]. The application of EGR on HCCI combustion engine has a number of effects on the combustion process and emissions. The effects of EGR on HCCI combustion that have been investigated are: increase in intake charge temperature (heating effect), reduction of oxygen concentration (dilution effect), increase in specific heat of the mixture (heat-capacity effect), chemical interactions involving the CO_2 and H_2O species of the recycled burned gases (chemical effect; this influences not only the overall kinetics, but also can change a specific reaction path [24, 25]) and stratification of the recycled burned gases (stratification effect). The dilution and heat capacity effects are responsible for reducing the heat-release rates and extending the combustion duration. The heating effect is mainly responsible for the advance of auto-ignition timing, and the residuals-stratification effect facilitates HCCI combustion. Reactive species, present in the residuals, facilitate auto-ignition [10, 15, 23].

The objective of this study is to investigate the effects of EGR and equivalence ratio on CO and soot emissions of dual fuel HCCI engine using premixed gasoline.

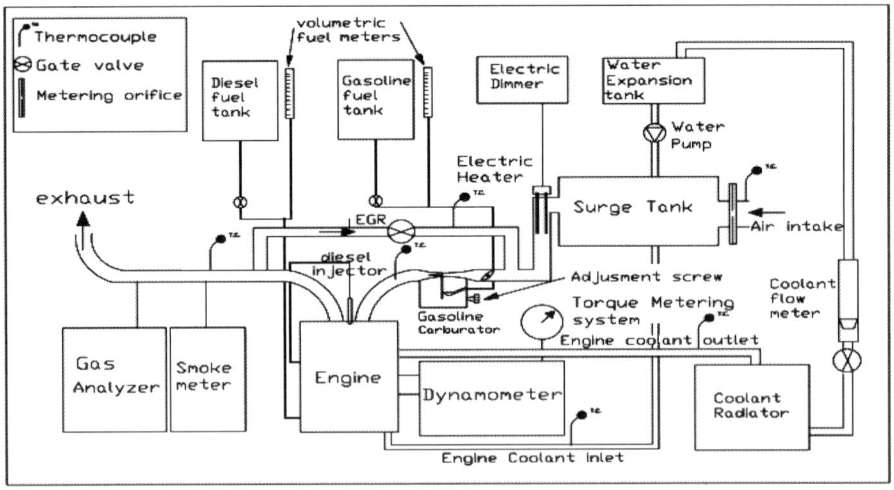

Figure 1: Schematic diagram of experimental apparatus.

2. Experimental setup and procedure

2.1. *Experimental setup*

In this study all experiments were conducted on a four stroke VCR single-cylinder naturally-aspired research engine with a displacement volume of 582 cm^3, which the test rig is TD43 model equipped by Techquipment Co. The engine has a bowel type piston with a bowel diameter of 41.5 mm. The base engine can operate as SI engine with the compression ratio in the range of 7 to 11, and also can convert to diesel type (spark plug is replaced by an injector) in the range of 14 to 18 for the compression ratio. Premixed gasoline is introduced to the inlet manifold by means of a carburetor which is mounted on 180 mm from intake valve upstream and equipped with a fuel adjustment needle screw. Diesel fuel is injected into the cylinder through an injector. The specifications of the test engine are listed in the table1.

The experimental apparatus is composed of electrical dynamometer, AVL-415 smoke meter, RE 205 Plint exhaust analyzer, which measures HC (as C_6), K-type thermocouples, 220 V single-phase 2Kw intake air heater which placed in the air flow stream, external EGR system, adjustable coolant system and air mass flow meter (surge tank and orifice system). Fig. 1 shows the schematic diagram of the experimental apparatus.

Table 1: Engine specifications.

Engine	Four stroke
Number of cylinders	1
Compression ratio	17.5:1
Displacement volume	582 cm^3
Bore	82 mm
Stroke	95 mm
Injection mode	DI
Number of injection holes	4
Injection pressure (bar)	250
Gasoline fueling	Carburetor

2.2. *Experimental procedure*

In this work, the diesel engine was firstly started at idle position in the compression ratio of 17.5:1. Then intake charge temperature was adjusted on 115 ^{0}C. After a few minutes to achieve steady condition at constant cooling temperature of 50 ^{0}C, premixed gasoline was introduced to engine and torque increased gradually. Then by using an external EGR system different rates of EGR were introduced to get HCCI auto-ignition close to TDC. Table 3 shows the specifications of the fuels used in dual fuel HCCI engine. It should be noted that using early injection (35BTDC) in HCCI engine in compare with conventional CI engine was the necessity of initiating the auto-ignition of dual fuel HCCI engine.

To show the effect of fuels in dual fuel HCCI engine, the premixed ratio (r_p) is defined as a ratio of premixed fuel energy (Q_p) to total energy (Q_t). It can be obtained from the following equation [1]:

$$r_p = \frac{\dot{Q}_p}{\dot{Q}_t} = \frac{\dot{m}_p h_{up}}{\dot{m}_p h_{up} + \dot{m}_d h_{ud}} \tag{1}$$

In the equation (1), $\dot{m}_P$ represents mass flow rate of premixed gasoline, $\dot{m}_d$ is mass flow rate of injected fuel, hup is heating value of premixed fuel and hud is heating value of diesel fuel. Therefore, rp = 1 corresponds to single fuel HCCI combustion and rp = 0 corresponds to typical CIDI combustion.

EGR rate also is calculated as follow [25]:

$$EGR(\%) = \frac{\dot{m}_{EGR}}{\dot{m}_{EGR} + \dot{m}_{aEGR}} \times 100 \tag{2}$$

Where, $\dot{m}_{aEGR}$ is mass flow rate of intake air with EGR, and $\dot{m}_{EGR}$ is mass flow rate of EGR.

Different EGR rates were applied in each test by using a simple computer code. The code can estimate the orifice pressure drop at specific EGR rate in term of engine speed, ambient conditions and intake air properties at orifice.

The electrical dynamometer which is connected directly to the engine plays the role of rotating initiator in starting mode, and when running the engine, consumes the power output for generating electricity. The stator of dynamometer can rotate freely around its shaft and as a result of engine torque during power generation; it is forced out from the horizontal equilibrium position. Using a Newton-meter on a known-length beam, for bringing the dynamometer back to the horizontal, the engine torque is measured.

In this study the intake charge temperature was increased up to 115°C, which helps the fuel to overcome its activation energy and improves the pre-ignition chemical reactions. Altering the coolant temperature at the 40 to 70°C, HCCI-DI combustion showed better results at 50°C. Therefore, coolant temperature was maintained 50°C throughout all the tests.

Engine tests were carried out at the operating conditions shown in Table 2, trying to modify every one of the main parameters affecting the oxidation kinetics of any fuel, such as the engine speed, injection timing and the charge composition (quantified by the EGR percentage). All these parameter combinations provide different levels of pressure and temperature in the cylinder. As it can be observed in Table 2, HCCI conditions were achieved by an early start of injection (SOI) diesel fuel, which promote the auto-ignition of gasoline premixed fuel due to heat releases by diesel fuel before HCCI combustion of premixed fuel. Having reached HCCI combustion, emissions and other data were recorded.

Table 2: Test conditions.

Speed (rpm)	1200-1700
Intake charge temperature (°C)	110-115
Coolant temperature (°C)	50
EGR rate (based on mass flow rate of intake air)	0-15%
Injection timing	35 BTDC
Premixed ratio (r_p)	0-1

Table 3: Fuel specifications [19].

Fuel type	Gasoline	Diesel fuel
MON	87	-
Cetane number	-	54
Higher Heating Value (kJ/kg)	47300	46100
Lower Heating Value (kJ/kg)	44000	43200
Heat of vaporization (kJ/kg), at 1 atm, 25°C	305	270
Density (kg/m^3)	720	780
(A/F)s	14.6	14.5

3. Results

3.1. *Soot emission*

Fig. 2 shows the effects of EGR and engine speed on soot emission of dual fuel HCCI engine. As it can be seen, for any EGR rates, soot emission increases by increasing the engine speed due to reduction of mixture homogeneity and incomplete combustion. Also, as the engine speed is raised, due to insufficient time for re-burning the soot, which is produced in diesel fuel combustion process, during the HCCI combustion of premixed fuel (gasoline), soot emission increases according to Fig. 2.

In these experiments due to usage of diesel fuel as the pilot fuel, scrutiny of the production of soot emission in dual fuel HCCI engine is substantial. Soot emission in dual fuel HCCI engine is produced in the first combustion step (diesel fuel combustion), and then some portion of it is burned in the HCCI combustion step of gasoline premixed fuel. The measured soot emission in the exhaust is a part of produced solid carbon in the combustion process of diesel fuel which cannot be burned in HCCI combustion of premixed fuel.

The comparison of soot emission between with EGR and without EGR in Fig. 2 shows that, at the presence of the EGR, soot emission increases due to reduction of inlet air and its oxygen which create more rich points in the cylinder mixture.

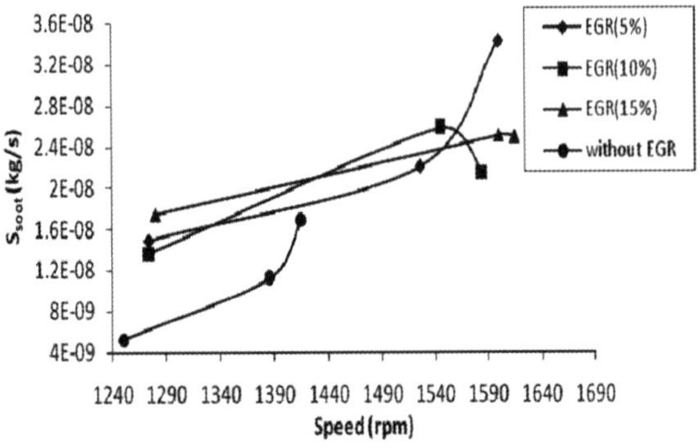

Figure 2: Effects of EGR and engine speed on soot emission of dual fuel HCCI engine.

3.2. *CO emissions*

Variation of CO emission of dual fuel HCCI engine with equivalence ratio and engine speed for different EGR rates are shown in Figure 3-6.

CO emissions variations as a function of equivalent ratio and engine speed for 5 and 10 percentages of EGR are shown in Figures 3 and 4, respectively. According to these Figures, by increasing the engine speed, there is insufficient time for formation a completely homogenous mixture, therefore CO emission increases at first due to incomplete combustion. However, gradual increase in equivalation ratio as the engine speed rises, CO emissions decrease due to high reactivity created by producing more radicals (as OH) at higher equivalence ratio.

In Fig. 5, it is observed that for 15% EGR despite the rise in engine speed, CO emission decreases because of increase the equivalence ratio which raises the reactivity.

CO emissions for without EGR is shown in Fig. 6, in which it is observed that in high engine speed, regarding to the decrease in equivalent ratio that reduces the reactivity, CO emission increases.

Moreover, in general, CO formation in a HCCI engine can be explained as follows:

Chemical kinetics is essential to the formation and consumption of CO, and the equivalence ratio is a crucial parameter of CO emission. Also, overall reactivity is the key phenomenon in at HCCI engine for explains most of the results. This implies that the emissions are not independent of the auto-ignition timing. Furthermore, a certain minimum equivalence ratio is need in order for the system to have enough energy to convert CO into CO_2. For any other cases, however, a higher equivalence ratio will only increase the CO in the emission.

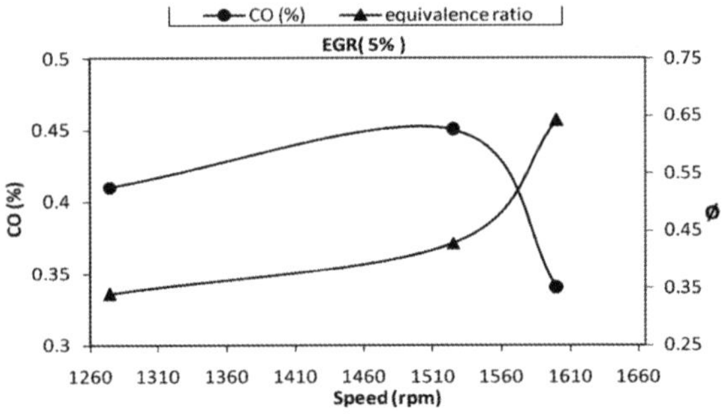

Figure 3: Variation of CO emission with equivalence ratio and engine speed at 5% EGR.

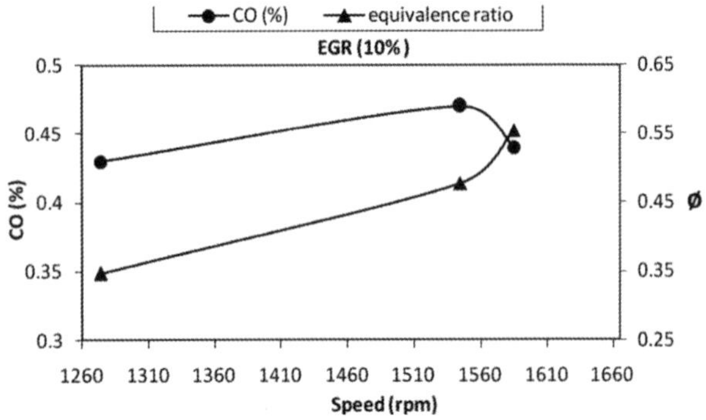

Figure 4: Variation of CO emission with equivalence ratio and engine speed at 10% EGR.

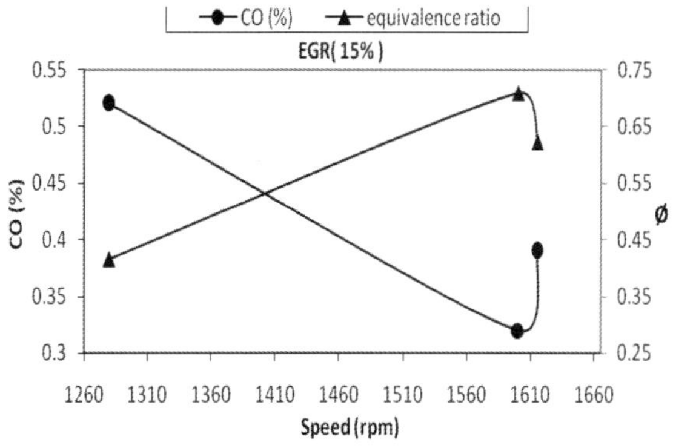

Figure 5: Variation of CO emission with equivalence ratio and engine speed at 15% EGR.

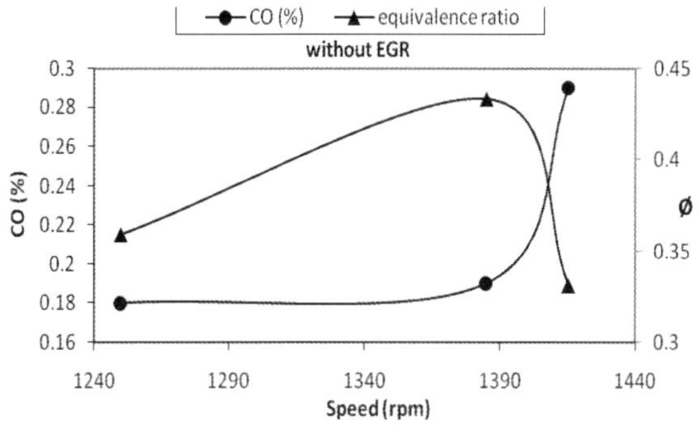

Figure 6: Variation of CO emission with equivalence ratio and engine speed for without EGR.

3.3. *Effects of EGR and equivalence ratio on soot and CO emissions and operating range of dual fuel HCCI engine*

With scrutiny and analysis of CO and soot emissions of dual fuel HCCI engine in Figures 2-6, the general results can be explained as follow:

In a HCCI engine, a higher overall reactivity would be expressed by a decreasing ignition delay. As the OH radicals increases, combustion delay decreases which results in rise in overall reactivity. Generally, the effect of the chemical species in EGR played a very important role in influencing the amount of OH radicals that are present in the cylinder. Since, at higher equivalence ratios more OH radicals are formed, the effect of chemical species of EGR in the producing of OH radicals is negligible at higher equivalence ratios. Accordingly, the main effect of EGR at high equivalence ratios is dilution of in-cylinder mixture and reduction of maximum combustion temperature. Decrease in maximum combustion temperature, results in knock to occur in higher engine speeds and equivalent ratios. Therefore, EGR can extend the operating range of HCCI engine to higher speeds and equivalence ratios. This can be noticed in the Figures 2-6. It can explain why the engine operating range is limited to lower engine speeds and equivalence ratios with no EGR.

As could be seen in Fig. 2, addition of EGR leads to decrease the inlet air and its oxygen, which create rich points in the in-cylinder mixture and results in increase the soot emission. Also, as it can be observed in Fig. 3-6, CO emission decreases with increasing the equivalent ratio due to formation of more OH radicals which increase the overall reactivity.

4. Conclusions

In this study, the effects of EGR and equivalence ratios on CO and soot emissions of dual fuel HCCI engine were investigated. Results can be summarized as:

1) Engine speed rise for any EGR rate results in increase the CO and soot emissions due to reduction of mixture homogeneity and incomplete combustion.

2) EGR rise increases the soot emission due to decreases the inlet air and its oxygen, which create rich points in the cylinder.

3) As the equivalence ratio increases CO emission decreases due to formation of more OH radicals in the cylinder and increasing the overall reactivity.

4) Addition of EGR can extend the operating range of dual fuel HCCI engine to higher engine speeds and equivalence ratios due to dilution effect and reduction of maximum combustion temperature.

References

1. W. Zeng, M. Xie, M. Jia, Numerical investigation on the application of catalytic combustion to HCCI engines, Chemical Engineering Journal 127 (2007) 81–93.
2. M. Canakci, An experimental study for the effects of boost pressure on the performance and exhaust emissions of a DI-HCCI gasoline engine, Fuel 87 (2008) 1503–1514.
3. Sh. Tanaka, F. Ayala, J. C. Keck, J. B. Heywood, Two-stage ignition in HCCI combustion and HCCI control by fuels and additives, Combustion and Flame 132 (2003) 219–239.

4. S. Soylu, Examination of combustion characteristics and phasing strategies of a natural gas HCCI engine, Energy Conversion and Management 46 (2005) 101–119.
5. H. Machrafi, S. Cavadias, J. Amouroux, The development and experimental validation of a reduced ternary kinetic mechanism for the auto-ignition at HCCI conditions, proposing a global reaction path for ternary gasoline surrogates, Fuel processing technology 90 (2009) 247–263.
6. H. Machrafi, K. Lombaert, S. Cavadias, P. Guibert, J. Amouroux, Reduced chemical reaction mechanisms: experimental and HCCI modelling investigations of autoignition processes of iso-octane in internal combustion engines, Fuel 84 (2005) 2330–2340.
7. J. A. Gaynor, R. Fleck, R. J. Kee, R. G. Kenny, G. Cathcart, a study of efficiency and emissions for a 4-stroke SI and a CAI engine with EEGR and light boost, International Journal of SAE, 32-0042, 2006.
8. K. Yeom, J. Jang, Ch. Bae, Homogeneous charge compression ignition of LPG and gasoline using variable valve timing in an engine, Fuel 86 (2007) 494–503.
9. M. Yao, Z. Chen, Z. Zheng, B. Zhang, Y. Xing, Study on the controlling strategies of homogeneous charge compression ignition combustion with fuel of dimethyl ether and methanol, Fuel 85 (2006) 2046–2056.
10. A. C. Alkidas, Combustion advancements in gasoline engines, Energy Conversion and Management 48 (2007) 2751–2761.
11. D. Ganesh, G. Nagarajan, M. M. Ibrahim, Study of performance, combustion and emission characteristics of diesel homogeneous charge compression ignition (HCCI) combustion with external mixture formation, Fuel 87 (2008) 3497–3503.
12. X. C. Lü, W. Chen, Z. Huang, A fundamental study on the control of the HCCI combustion and emissions by fuel design concept combined with controllable EGR. Part 1. The basic characteristics of HCCI combustion, International Journal of Fuel, 84, 1074–1083, 2005.
13. D. S. Kim, Ch. S. Lee, improved emission characteristics of HCCI engine by various premixed fuels and cooled EGR, International Journal of Fuel, 85, 695–704, 2006.
14. M. Jia, M. Xie, A chemical kinetics model of iso-octane oxidation for HCCI engines, Fuel 85 (2006) 2593–2604.
15. X. C. Lü, W. Chen, Z. Huang, a fundamental study on the control of the HCCI combustion and emissions by fuel design concept combined with controllable EGR. Part 2. Effect of operating conditions and EGR on HCCI combustion, International Journal of Fuel, 84, 1084–1092, 2005.
16. H. Machrafi, S. Cavadias, An experimental and numerical analysis of the influence of the inlet temperature, equivalence ratio and compression ratio on the HCCI auto-ignition process of Primary Reference Fuels in an engine, Fuel processing technology 89 (2008) 1218–1226.
17. H. Machrafi, S. Cavadias, Three-stage autoignition of gasoline in an HCCI engine: An experimental and chemical kinetic modeling investigation, Combustion and Flame 155 (2008) 557–570.
18. J. Andrae, D. Johansson, P. Bjrnbom, P. Risberg, G. Kalghatgi, Co-oxidation in the auto-ignition of primary reference fuels and n-heptane/toluene blends, Combustion and Flame 140 (2005) 267–286.
19. J. Ma, X. Lü, L. Ji, Z. Huang, an experimental study of HCCI-DI combustion and emissions in a diesel engine with dual fuel, International Journal of Thermal Sciences, 47, 1235–1242, 2008.
20. H. Machrafi, S. Cavadias, J. Amouroux., a parametric study on the emissions from an HCCI alternative combustion engine resulting from the auto-ignition of primary reference fuels, Applied Energy 85 (2008) 755–764.
21. M. Sjöberg, J. E. Dec, An investigation into lowest acceptable combustion temperatures for hydrocarbon fuels in HCCI engines, Proceedings of the Combustion Institute 30 (2005) 2719–2726.
22. L. Shi, Y. Cui, K. Deng, H. Peng, Y. Chen., study of low emission homogeneous charge compression ignition (HCCI) engine using combined internal and external exhaust gas recirculation (EGR), Energy 31 (2006) 2665–2676.
23. R. Chen, N. Milovanovic, A computational study into the effect of exhaust gas recycling on homogeneous charge compression ignition combustion in internal combustion engines fuelled with methane, International Journal of Thermal Sciences 41 (2002) 805–813.
24. H. Machrafi, S. Cavadias, Ph. Guibert, "An experimental and numerical investigation on the influence of external gas recirculation on the HCCI autoignition process in an engine: Thermal, diluting, and chemical effects," International Journal of Combustion and Flame, 2008.
25. H. Machrafi, "Experimental validation of a kinetic multi-component mechanism in a wide HCCI engine operating range for mixtures of n-heptane, iso-octane and toluene: Influence of EGR parameters," Energy Conversion and Management 49, 2008, 2956–2965.
26. J. B. Heywood, "Fundamentals of Internal Combustion engine," McGraw Hill, New York, 1988.

BIODEGADATION OF PHARMACEUTICAL ORGANIC POLLUTANTS BY PHOTOSYNTHETIC BACTERIA*

MADUKASI, E. I. and CHUNHUA HE

State Key Lab of Urban Water Resource and Environment,
Harbin Institute of Technology, 202 Haihe Road, Harbin 150090, PR China

Suspended growth photo bioreactor (PBR) was used to treat pharmaceutical wastewater by a wild strain purple non sulfur photosynthetic bacterium (PNSB) isolated from the soil; the isolate was identified as being closely allied to Rhodobacter sphaeroides. External illumination was with 2 (40W) fluorescent compact light sources on both sides. Optimum growth of the isolate was obtained after enrichment of the pharmaceutical wastewater with 0.5% $(NH_4)_2SO_4$ and 0.1% yeast extract under micro aerobic optimum light (6000lx) condition at 5d retention, maximum dry cell weight (DCW) and COD% removal were 880 mg/L and 80%. Analysis of the culture after treatment of the enriched and non-enriched wastewater showed the crude protein content of the biomass to be 54.6% and 42.4% respectively.

1. Introduction

Characteristics of the pharmaceutical industries are the diversity of their process operations, which gives rise to a wide variation in the liquid wastes. In most cases, these effluents contain little or no biodegradable organic matters and the pollutant loads in terms of biological oxygen demand (BOD) may be negligible hence higher COD than BOD (Bitton, 2005). Most substances produced in a pharmaceutical industry are structurally complex organic chemicals that are resistant to biological degradation (Ren et al, 2008). Conventional biological treatment methods are usually inappropriate for the treatment of pharmaceutical wastewaters because of the negligible BOD content of the wastewater. For instance, the wastewater for this study has a BOD of 146mg/L and COD of about 8000mg/L, resulting in the BOD/COD ratio of about 0.02. Hence the need to try other methods of biological treatment such as the application of photosynthesis bacteria in degrading pollutants found in industrial wastewater. Photosynthetic bacteria is a cost effective system for treating wastewater, this is because PSB does not only produce quality effluent, but produces substances of commercial interest such as single cell protein (SCP), biopolymers, and antimicrobial agents, etc., (Kantachote et al., 2005). Purple non sulfur photosynthetic bacteria are metabolically the most versatile among all prokaryotes; an aerobically photoautotrophic and photo heterotrophic in the light or micro aerobic-light condition (Holt et al., 1994). It has been utilized by some researchers to treat various sources of wastewaters such as, Concentrated latex wastewater (Choorit et al., 2002), and Latex rubber sheet wastewater (Kantachote et al., 2005). The aims of this study are to isolate, identify and test the ability of wild strain purple non sulfur photosynthetic bacteria as well as to determine the optimum conditions for this isolate to be most effective in treating pharmaceutical wastewater. The harvested biomass from the treatment will be evaluated to ascertain its nutritional value.

2. Material and Methods

2.1. *Bacterium culture, isolation and identification*

The bacterium was isolated with Sistrom minimal medium (RCVBN) consisting of (in g/l) DL-malate: 4. MgSO4: 0.12. (NH4)2SO4: 1. CaCl2: 0.075. KH2PO3: 0.3. Na2 EDTA: 0.020. VB1: 0.001. Nicotinic acid: 0.001. Biotin: 0.015. Trace element solution, 1ml. The trace element solution contained (in mg/l) Zn $SO_4.7H_2O$: 20. $MnCl_2.4H_2O$: 6. H_3BO_3: 60. $CoCl_2.6H_2O$: 40. $CuCl_2.2H_2O$: 2. Na_2MoO_4: 6. 2g of soil was added to 95ml sterile medium and incubated an aerobically at 28-35°C for about 96hrs under illumination with a 100W incandescent lamp.

The isolate was characterized using both morphological and physiological properties and

*This work is supported by the State Key Laboratory of Urban Water Resource and Environment (2008QN01, 2008DX01).

according to Bergey's Manual of Systematic Bacteriology vol.3 (Imhoff and Trüper, 1989) also Bergey's Manual of Determinative Bacteriology 9th ed. (Holt et al., 1994). Bacteriochlorophylls were isolated, characterized, and identified as described by Oelze (Okubo, et al, 2006). Cellular fatty acids contents were determined by using the MIDI procedure (MIDI, Inc, New York Del). Identification of the isolates were based on a comparison of fatty acid profiles using the fatty acid profiles in tryptic soy broth agar anaerobe database.

2.2. *Wastewater*

Wastewater from a pharmaceutical company located in Harbin, China was collected severally at different times from the company's holding tank; each collection was analyzed for COD, BOD, Suspended solids (SS), and pH etc. Representative wastewater sample was used as the growth medium.

2.3. *Optimization of wastewater composition and treatment conditions*

Ammonium or nitrate was added to the sterilized wastewater (SW) medium at varied concentrations of 0, 0.1, 0.2, 0.3 and 0.4%. Ammonium being the best nitrogen source was tested again at higher concentrations of 0 0.5, 0.75, 0.125 and 0.15% respectively. Effect of added yeast extract (YE), SW medium supplemented with optimal concentration (0.5%) of ammonium was used as the base control. Yeast extract is a complex natural material with high vitamin B levels; it was tested for its effect on growth at the following concentrations; 0, 0.01, 0.05, 0.1, 0.2, 0.3%. All the cultivations were grown 3 consecutive times for 72 hrs under micro aerobic-light conditions.

Effect of light variation, the effect of light intensity was investigated by growing the isolate in sterilized optimized wastewater (SOW); i.e. the wastewater was supplemented with 0.5% $(NH_4)_2SO_4$ and 0.1% yeast extract, grown under light intensities of 500, 1000, 2500, 3500, 4000, 6000 and 8000 lumens respectively. In all these, the temperature variation was same (20-31°C).

2.4. *Photo bioreactor*

A test-tube of 100ml with cellophane tapered edge to prevent volatilization was used as a bioreactor. Three treatments were set as follows; sterilized optimized wastewater (SOW) as a control without inoculums; SOW plus 10% inoculums; and sterilize wastewater (SW) plus 10% inoculums. Its temperature was between 20-31°C. Illumination was with 40W compact fluorescent lamps at the desired lumens on both sides of the bioreactor. Cell growth, pH and COD etc were monitored at the end of each retention time of 3, 5 and 7 days of cultivation respectively. Micro aerobic condition was achieved by creating a head space in the PBR, the cultivation condition was static-light exposure. The dissolved oxygen (DO) levels of all treatment were between 0.7-1.63mg/L.

2.5. *Analytical methods*

The incident light intensity was measured by using a TES-1330A digital light meter. Organic compounds in the wastewater were identified qualitatively using Gas Chromatography Mass Spectrometry (GC-MS). The bacterial cell concentration was determined by optical density at 660nm (i.e.OD660) using a UV-VIS spectrophotometer. The OD660 value was converted to DCW concentration via proper calibration, (where 1.0 OD660 approximately equals 0.6g dry cell/L.). Crude protein was expressed as total nitrogen (%N) multiplied by 6.25. Biomass productivity was calculated from the total volume of wastewater used and the total weight of biomass produced in the process.

3. Results and Discussion

3.1. *Identification of the bacterium*

The isolate was a non motile, Gram negative, oval rod, 0.25μm wide and 1.50μm long and reproducing by binary fission, internal photosynthesis membrane appeared as lamella. By using GC-FAME (Gas chromatography of fatty acid methyl ester) analysis the purple non sulfur photosynthetic bacteria isolate was identified as being closely allied to R. sphaeroides strains with similarity indices of 0.85 ± 0.05. This also identifies the isolate as Rhodobacter sphaeroides [Choorit et al., 2002] and it's named Rhodobacter spheroids z08. After growth under anaerobic – photo heterotrophic conditions cell suspensions were red and the absorption spectra of living cells suspension showed maxima at 370, 782, 800, 827, and 852 as shown in Figure1. The two main peaks at 800 and 850nm, which are allied to bacteriochlorophyll a, & b as well as carotenoid, are characteristics of purple non-sulfur bacteria (Madigan et al., 2000).

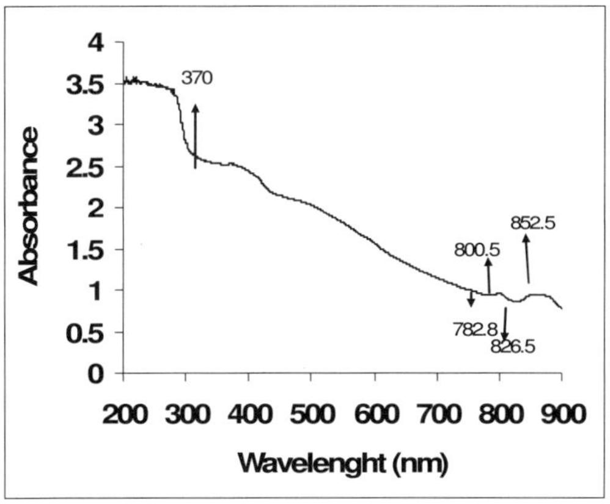

Fig. 1: In vivo spectrum of the (Pure culture).

The wastewater has a COD of 8400mg/l and a BOD of 145mg/l. The organic pollutants identified in the wastewater are shown in Table1.

Table 1: List of organic compounds tentatively identified in the pharmaceutical wastewater.

Peak No.	Name of compound	$t_{R\ (min)}$	Molecular ion peak (m/z)	Base peak (m/z)
1	Benzenamine	15.082	121	120
2	Benzothiazole	24.818	181	181
3	1,2-Benzenedicarboxylic	31.315	278	149
	acid	34.082	101	44
4	1-Butanamine	36.262	149	44
5	Benzenepropanamine	38.066	265	44
6	Desmethyldoxepin	38.388	99	44
7	Methylpent-4-	39.414	219	44
	enylamine	40.517	283	44
8	Nitro-L-arginine	41.760	254	57
9	1-Octaecanamine	43.144	380	57
10	Octadecane	44.702	394	57
11	Heptacosane			
12	Octacosane			

According to Table1, the wastewater is embedded with recalcitrant organic compounds which are resistant to biodegradation. These identifications were made by GC-MS and supported by GC-FTIR where 2 peaks were evident at 1304 cm^{-1} and 1600-1300 cm^{-1}, suggesting the presence of aliphatic amino salts and aromatic compounds respectively. It's high in chemical oxygen demand than biological oxygen demand, hence low value of BOD to COD ratio (0.02). This depicts that the wastewater contains high level of non biodegradable organic substances [Bitton, 2005] and difficult to treat with conventional biological treatment methods. These organic compounds found in the pharmaceutical wastewater could be vital part of everyday life, but most of them are of environmental concern because of their toxicity, persistence, and tendency for bioaccumulation [Sapana et al., 2008].

3.2. *Wastewater optimization*

The test water sample has a TN value of about 530 mg/L, about 0.05% nitrogen by the weight of the BOD. This did not conform to the biological wastewater treatment ideology of carbon to nitrogen ratio of 5% nitrogen by the weight of the BOD. Hence the need for additional nutrients in the form of either $(NH_4)_2SO_4$ or $NaNO_3$ as shown in fig.2 a , b.

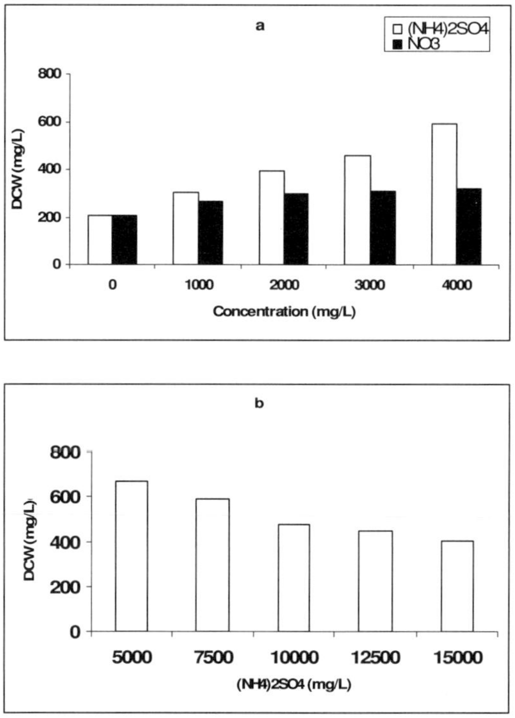

Fig. 2: Effect of nitrogen sources and their concentrations on the growth of the PNSB in pharmaceutical wastewater with micro aerobic-light conditions for 72 hr, (a) Source of nitrogen. (b) Added (NH4)2SO4 on bacterial growth.

Fig. 1a shows that (NH4) 2SO4 produced the highest dry cell weight (592.3 mg/L) at 400 mg/L against NaNO3 with (318.2 mg/L) at the same concentration. Ammonium being the best nutrient was tried further at higher concentrations (Fig. 1b), it gave maximum growth of 672.0 mg/L DCW at 5000 mg/L while further increment retards the growth. Addition of yeast extract on ammonium supplemented wastewater further increased the cell growth of the PNSB from

OD660 1.03 – 1.502 in the pharmaceutical wastewater test medium. It recorded 0.997 g/l DCW at 1.0g/l concentration. Therefore, in order to treat the pharmaceutical wastewater, additional nutrients need to be added to the wastewater to stimulate increase in growth of the microbes. This cultivation was carried out over 96hrs under micro aerobic-light conditions. High concentration had no significant

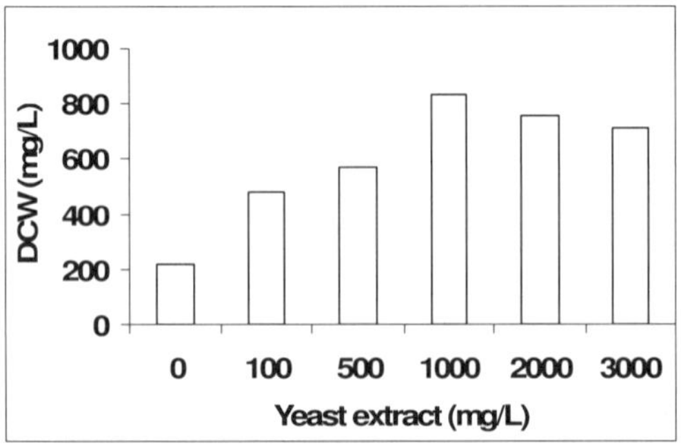

Fig. 3: Effect of added Yeast extract and its concentration on the growth of PNSB in pharmaceutical wastewater. With (NH4)2SO4 = 5 g/L as base control.

Although yeast extract contain many B vitamins, it may also act as a source of protein and its use depends on the economic evaluation of the process. The incident light intensity was varied between 500 to 8000 lx. The best growth was observed at 6000 lx (data not shown) with a COD reduction of about 75% (initial COD value of 2500 to final value of 604 mg/L respectively) and the cell growth was 820.5 mg/L DCW. This finding could be attributed to the fact that the PNSB thrives better in an intense light.

Figure 3 shows the experiment carried out to infer the effective retention time for the treatment. The isolate grows up to 377.4 mg/L in sterilized non optimized wastewater (SW) in 5 days while the optimized wastewater (SOW) recorded a higher growth of 688.3 mg/L at the same 5 days retention (Fig. 3a). Also Figure 3b shows that the COD reduction was higher in SOW at 5d than SW (64.8 and 43.0% respectively), although at 7 d, the COD and dry cell weight were slightly higher than at 5 d, but 5 d was chosen as the optimal retention time for the treatment for economic reasons. The SW with lower COD reduction could be attributed to the inhibition of the cell activity by the recalcitrant pollutants that the wastewater is embedded with (Table 1), while the sharp reduction of COD in the SOW with inoculums shows that the bacteria activity was increased with additional nutrient hence able to metabolize most of the pollutants faster before the onset of inhibition.

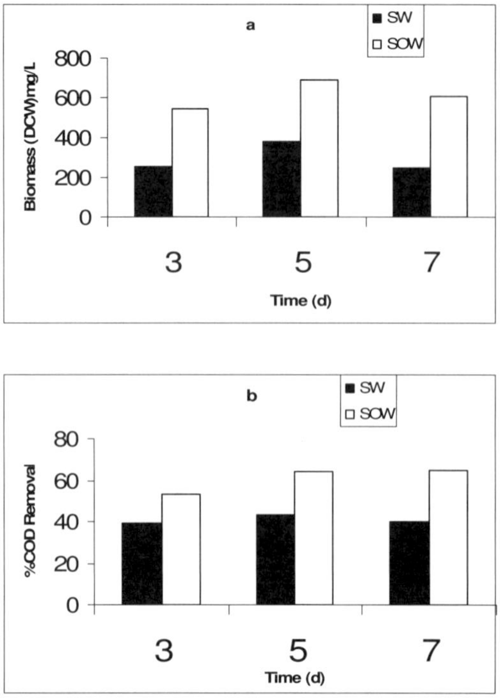

Fig. 4: Cell growth & COD removal as a function of time (3500 lx, initial COD = 2500 mg/L; (a) Biomass, mg/L (b) COD removal, (%).

The conditions chosen for the treatment of the pharmaceutical wastewater were therefore as follows: supplemented wastewater was with 0.5% ammonium sulphate and 0.1% yeast extract with a final pH of 6.6-7.0, incubation without shaking at 20-30°C for 5 days with fluorescent compact incident light intensity of 6000 lx. The choice of fluorescent compact lamp for illumination of the reactor was as a result that greater proportion of its power is converted to usable light while smaller proportion is converted to heat as against incandescent lamp which is heat driven and energy draining lamp. Although the initial cost of fluorescent lamp is higher than the incandescent lamp but fluorescent's longer life reduces lamp replacement cost.

Specific growth rate was calculated as 0.015 h^{-1} and 0.011 h^{-1} for SOW and SW respectively. While the process productivity based on the five days optimal growth is 0.082 g/L .day with a biomass yield of 12.8% for the optimized wastewater based on the chemical oxygen demand.The GC-MS fragmentation analyses of the pharmaceutical wastewater after treatment of the optimized wastewater (Table 2) indicate that the PNSB was able to either ameliorate or transform those recalcitrant and xenobiotic materials found in the initial wastewater into acceptable organic compounds. This shows that the PNSB used for this treatment is candidate for bioremediation.

Table 2: Lists of organic compounds tentatively identified in treated optimized pharmaceutical wastewater.

Peak No.	Name of compound	$t_{R\ (min)}$	Molecular ion peak (m/z)	Base peak (m/z)
1	4-methoxybenzene-1,2-diol.	10.504	140	140
2	Naphthalene	13.431	128	128
3	2-Butenoic acid	15.091	144	43
4	1-2Benzene-dicarboxylic acid.	21.946	148	142

The protein content of the culture after treatment in both the supplemented and non-supplemented wastewater is well above the required protein content for it to be acceptable as a protein source for either fish feed or animal feed. However, the process productivity was low (0.082 g/L.day). Photosynthetic bacteria typically have lower process productivity than heterotrophic bacteria because the cell densities achieved are low. As for the yield, normal heterotrophs growing with aerobic growth conditions, about 50% of organic carbon can be converted into biomass, whereas with anaerobic conditions only 5% is converted into biomass (Kantachote et al., 2005). For this micro aerobic condition the yield is about 12% which is much favorable than complete anaerobic condition. Also in this present study, the bacteria did not thrive well in a complete anaerobic light condition, for instance, after about 2weeks cultivation an aerobically, its optical density was ($OD_{660} = 0.442 \pm 0.001$). Whereas at micro aerobic light, it recorded greater optical density values in less than 2weeks of cultivation ($OD_{660} = 1.234 \pm 0.002$). This suggested that the isolated PNSB could thrive better in micro aerobic-light than an aerobic light condition.

References

1. Bitton, G. (2005). Wastewater Microbiology, 3rd Ed. pp 55-72, 213-222, 502-518. Wiley-Liss Pub, N.Y.
2. Choorit, W., P. Thanahoset, J. Thongpradistha, K. Sasaki, and N. Noparatnaraporn (2002). Identification and cultivation of photosynthetic bacteria in wastewater from a concentrated latex processing factory. Biotechnology Letters. 24: 1055-1058.
3. Holt, J.G., N.R. Krieg, P.H.A. Sneath, J.T. Staley, and S.T. Williams (1994). Bergey's Manual of Determinative Bacteriology. 9th ed., pp787. The Williams Wilkins Co., Baltimore.
4. Imhoff, J.F., and H.G. Trüper (1989) Purple nonsulfur bacteria. In: J.T. Staley ed. Bergey's Manual of Systematic Bacteriology. Baltimore, Williams & Wilkins. 3: 1658-1682.
5. Kantachote, D., T. Salwa, and U. Kamontam (2005). The potential use of anoxygenic photosynthetic bacteria for treating latex rubber sheet wastewater. Electronic Journal of Biotechnology. (*on-line*) 8:3. Valparaiso dic.
6. Okubo, Y., F. Hiroyuki, and H. Akira (2006). Characterization of phototrophic purple non-sulfur Bacteria forming colored microbial mats in a swine wastewater Ditch. Applied & Environmental Microbiology 12: 6225-6233.
7. Ren, N., Z. Chen, A. Wang, A, Z.P. Zhang, and S. Yue (2008). A novel application of TPAD-MBR system to the pilot treatment of chemical synthesis-based pharmaceutical wastewater. Water Research 42: 3385-3392.
8. Sapana, S., B. Sujata, T. Amruta, K. Kalal, U.D. Phalgune, and N.R. Deshpande (2008). GC-MS Study of hydrocarbons - A Renewable Biofuel with high calorific value from aerial roots of ficus Benghalensis Linn. Electronic Journal of Environmental, Agricultural and Food chemistry. 7: 3.

WHEAT RESPONSE TO NITROGEN AND POST-ANTHESIS WATER DEFICIT

A. BAHRANI
Ph.D. Student, Islamic Azad University, Science & Research branch, Khuzestan, Iran

H. HEIDARI SHARIF ABAD
Professor, Seed and Plant Improvement Institute, Karaj, Iran

Z. TAHMASEBI SARVESTANI
Associate Professor, Tarbiat Modares University, Tehran, Iran

G. H. MOAFPORIAN
Assistant Professor, Agricultural Researches Organization, Zarghan, Iran

A. AYENEHBAND
Assistant Professor, Shahid Charan University, Ahwaz, Iran

Relationship between post anthesis water deficit and nitrogen on wheat (*Triticum aestivum* L.) in Southern of Iran are not well defined. Therefore, we established an experiment on a clay loam at Shiraz Agricultural Researches Station, between 2006 and 2008. Three wheat cultivars were grown under post anthesis water deficit and three N fertilizers. Two-year average data showed that spike number per square meter and kernel per spike was similar at both nonstressed and post anthesis water deficit treatments. Kernel weight was reduced by water stress under 0 and 80 kg N ha^{-1}, but increased under 160 Kg N ha^{-1}. Under water deficit grain yield was reduced by 25%. Grain yield of the crop that no received N being 15% greater than for that receiving 80 kg N ha^{-1}. Chamran produced the highest grain yield, which was 19% higher than that of Shiraz and Marvdasht. In water deficit treatment, straw yields increased with 80 kg ha^{-1} increment of N but further increments of N had little effect. Under post anthesis water deficit DM remobilization efficiency increased with 80 kg ha^{-1} increment of N by 29%, but further increments of N decreased it. The contribution of DM remobilization to the grain ranged from 7 to 23% of grain dry weight and under water deficit was 78% more than nonstress treatment. It is concluded that there is great potential to increase winter wheat yield by properly managing nitrogen fertilization in this region. Stored carbohydrate represented an important buffer for yield production when stress occurred during grain filling.

1. Introduction

In Mediterranean climate region of Iran, rainfall decreases and evaporation and temperature increases in spring, when wheat enters the grain filling stage (Ahmadi and Baker, 2001; Fathi, 2005). Consequently, wheat crops often experience water deficit during grain filling, thereby limiting grain yield (Spiertz et al., 2006; Barnabas et al., 2008; Bluem, 1998; Nicolas and Turner, 1993).

Irrigation water and fertilizers are two vital but costly inputs in irrigated farming. Their economic use demands that maximum yield should be obtained per unit application of these inputs. The importance of N fertilization in increasing wheat production has been well documented (Zhang and Shanguan, 2006; Fan et al., 2005; Bahrani, and Tahmesebi 2005), but still it is difficult to determine the quantities to apply under water stress condition. It seems that in arid and semi-arid regions, heredity of drought tolerance should be include in studies investigating the interaction between N efficiency and water management.

Grain growth in wheat depends on C from three sources:current assimilation, remobilization of preanthesis assimilates stored in the stem and other plant parts, and retranslocation of assimilates stored temporarily in the stem after anthesis. In evaluating the reduction in grain yield arising from postanthesis water deficits, it is necessary to identify which of these sources of C is limiting grain filling. Postanthesis water deficits are known to reduce C assimilation and hence the availability of current assimilates for grain filling, but are not considered to affect the translocation of C to the grain (Nicolass et al., 1985). Water deficits during grain filing increase the proportion of stored assimilates relative to current assimilates in the grain (Palta et al., 1994; Yang et al., 2000; Pheloung and Siddique 1991), but whether this reflects a larger actual mobilization of stored assimilates rather than simply a reduction in current assimilation is not known (Turner and Begg, 1981).

There is little information on which to base fertilizer recommendations for wheat grown under postanthesis water deficit in this region. Also wheat production in northern Iran suffers from a continental hot, dry wind usually at the end of the growing season. If crop maturation is unfavorably delayed, such wind may dehydrate the wheat rapidly and lead to reduce grain wheat. In this paper we evaluated (i) the interacting effects of N fertilization and postanthesis water deficit on grain yield and yield components of three wheat cultivars; and (ii) remobilization and contribution of DM to grain yield in this condition.

2. Materials and Methods

The studies were conducted on a clay loam at Shiraz Agricultural Researches Station, for 2 yr (2006-2008). The experiment was randomized block, split-split plot design with four replications. Main plots, Irrigation treatment, were I1 (nonstressed) and I2 (Post anthesis water deficit with 65% FC). Sub plots, fertilizer treatment, were N at rates of 0, 80, and 160 kg ha^{-1}. Sub-sub plots, Cultivars, were Shiraz, Marvdasht, and Chamran. Plots were sown on 11 November 2007 and on 14 November 2008 with a cone seeder, and were 8 m long and 1.5 m wide, with 6 rows 0.2 m apart. Irrigation of each mainplot was measured volumetrically by field calibrated gypsum block. Six gypsum blocks were install in each replication randomly. Measurements were made after anthesis, before each irrigation, at harvest, and at other selected times during the growing period. And change in soil moisture was measured weekly to a depth of 30 cm by it. The irrigation system was operated so that runoff did not occur. Above-ground DM production at heading in both years was measured by cutting at ground level 0.3 m^2 quadrants per plot. Immediately prior to harvest, spike no. m^2 was determined by averaging three counts of 1-m sections of row within each plot. The number of kernel per spikes was determined from 20 spikes taken at random from a 1 m section of each plot and counted with an electronic seed counter. And average kernel weight was determined by weighing 250 kernels randomly drawn from the bulk grain sample from each plot. The center four rows (of 6 rows) of each plot were then harvested for grain yield, and converted to grain yield per hectares. Harvest indexes (HI) were calculated using yield from the square meter samples. Twenty main stems that headed on same day were tagged for each treatment. As the tagged main stems of each cultivar reached anthesis, 10 plants in each plot were removed. At maturity, 10 additional tagged plants were removed. Anthesis was scored when anthers in the central florets of 50% of the spikes in plant had dehisced and maturity when almost all the spikes in plot showed complete loss of green color. Anthesis dated of the cultivars differed by about 10 d both years. Samples were dried in a forced-air oven at 70°C for at least 48 h. The various parameters referring to DM that are discussed will be evaluated as fallows:

1-DM remobilization efficiency (%)=(DM remobilization / DM at anthesis) ×100 2- Contribution of dry mater remobilization to grain (%)=(DM remobilization/grain yield) × 10

Data were analyzed by analysis of variance (SAS Institute 1982).When significant differences were found (*P=0.05*), Duncan's multiple range test (DMRT) were calculated.

3. Results and Discussion

Table 1. Results of analysis of variance combined across years, Irrigation, fertilizer and wheat cultivars to both years.

Sources	df	Spikes no. m^{-2}	Kernel spike^{-1} no.	Kernel weight mg	Grain yield kg ha^{-1}
Year (Y)	1	**	**	**	*
Irrigation(I)	1	NS	NS	**	**
Y×I	1	NS	NS	*	NS
Nitrogen(N)	2	*	*	*	*
Y×N	2	NS	NS	NS	NS
I×N	2	*	NS	NS	NS
Y×I×N	2	NS	NS	NS	NS
Cultivar (C)	2	**	**	**	**
Y×C	2	NS	*	**	*
I×C	2	NS	NS	*	*
N×C	4	*	NS	NS	NS
Y×I×C	2	NS	NS	*	*
Y×N×C	4	NS	NS	NS	NS
I×N×C	4	NS	NS	NS	NS
Y×I×N×C	4	NS	NS	NS	NS
CV, %		14	15	10	13

Continue of Table 1.

Sources	df	Straw yield kg ha^{-1}	Harvest index (%)	DMRE (%)	Contribution of DMR to grain (%)
Year (Y)	1	**	**	*	NS
Irrigation(I)	1	*	*	**	**
Y×I	1	NS	NS	NS	NS
Nitrogen(N)	2	**	*	*	**
Y×N	2	NS	NS	**	**
I×N	2	NS	NS	NS	**
Y×I×N	2	NS	NS	NS	*
Cultivar (C)	2	*	*	*	*
Y×C	2	NS	NS	NS	*
I×C	2	NS	NS	NS	*
N×C	4	NS	NS	NS	NS
Y×I×C	2	NS	NS	*	NS
Y×N×C	4	NS	NS	NS	NS
I×N×C	4	NS	NS	*	*
Y×I×N×C	4	NS	NS	NS	NS
CV, %		16	17	15	18

*,** Significant at 0.05 and 0.01 prpbability level, respectively. NS = nonsignificant at $P > 0.05$.

Total spike number per square meter was similar at both nonstress and water deficit treatments, indicating that the water deficit treatment did not affected tiller per plant (Table 1). Application of N affected the number of spike per m^{-2} from 718 m^{-2} in wheat receiving 0-N to 789 m^{-2} in wheat receiving 80 kg N ha^{-1} and 743 m^{-2} in wheat receiving 160 kg N ha^{-1} (mean of 2 yr). The crop that received 80 kg N ha^{-1} produced 10% more ears m^{-2} than the crop that received 0-N resulting in grain yield. Among the three wheat cultivars, Chamran produced the highest spike per at either N level, and the doubling of N fertilization increased grain yield by 1130 kg ha^{-1}.

Grain number per plant and in each individual shoot was similar at both nonstress and WD treatments, indicating that the WD treatment did not affected floret abortion. The crop that received 80 kg N ha^{-1} produced 10% more grains per ear than the crop received 0-N resulting in grain yield. On the adequately fertilized treatments, limited irrigation treatment reduced seed number in both years. Marvdasht produced 11% and 25% more grain per spike than Shiraz and Chamran, respectively (Table 2).

Kernel weight was reduced by water stress under 0 and 80 kg N, but increased under 160 kg N with compared with their respective nonstressed treatment in both experiment. Possibly because only the kernel weight, rather than the spike number or kernel number per spike, was influenced by water deficit during grain filling. Nitrogen up to 0 kg/ha also increased grain dry weight per ear. Kernel weight was greatest for Chamran and least for Marvdasht. The significant effects were on treatment I-2, where seed weights were decreased with increasing increments on N through 160 kg ha^{-1}.

Under water deficit grain yield was reduced by 25%. The reduction resulted from a reduction in grain size, but not in grain number. Grain yield increased in response to applied N, with the grain yield of the crop that no received N being 15% greater than for that receiving 80 kg N ha^{-1} (Table 2). Average winter wheat yields ranged from 5226 to 5953 kg grain ha^{-1} for the control and 80 kg N ha^{-1} rate of the first experiment and from 4052 to 4642 kg grain ha^{-1} for the control and 80 kg N ha^{-1} rate of the second experiment (Table 2). In 2006-2007 the maximum grain yield of three varieties were attained with 80 and 160 kg N ha^{-1}, and ranged from 5569 to 6199 kg ha^{-1} Marvdasht produced the lowest grain yield at either N level and the doubling of N fertilization increased grain yield by 1130 kg ha^{-1}. Grain yield reductions for Shiraz was obtained with up to 80 kg N ha^{-1}. This reduction was caused by decrease in number of spike per meter square without compensation by mean weight per grain. However, Increasing in grain yield beyond 80 kg N ha^{-1} was marginal. This dose not implies that there was no response to the application of higher rates of N at nonstressed. For the I-2 treatment, water becomes more limiting than fertility, and there was no yield response to applied N. The yearly variations were associated with weather, which influenced growth and maturity of the crop.

Straw DM responded positively to increasingly higher rates of N. The three cultivars responded similarly to higher rates of N. With the I-2 treatment, straw yields increased with 80 kg ha^{-1} increment of N but further increments of N had little effect. Straw yields on treatment I-2 were 13 and 15% lower than those on treatment I-1 in 2006-2007 and 2007-2008, respectively.

Stress during grain filling reduced harvest index by reducing grain yields after most of the vegetative growth had been completed. Harvest index varied with N treatments in both years (Table 2). These results could be explained by the separate effects of N treatments on the yields of grain and straw DM at maturity. Marvdasht, which had the lowest grain yields also, had the lowest HI (8-14%). Furthermore, since Marvdasht did not always produce the lowest amount of TDM, its low yields were due, at least in part, to a low HI. In 2007-2008 harvest index was greatest for Chamran and least for Marvdasht.

Water deficit markedly affected DM remobilization efficiency. Under water deficit DM remobilization efficiency increased by 29% (Table 2). With the I-2 treatment, DM remobilization efficiency increased with 80 kg ha^{-1} increment of N but further increments of N decreased it). Chamran showed higher remobilization efficiency (25 and 18%) in 2006-2007 and 2007-2008, respectively. One explanation for differences in remobilization efficiency is that during grain filling period, the plant retains an amount of DM at anthesis that is essential for survival and various biological functions, while the remainder is available for remobilization. It appears that DM retained depends on cultivars and prevailing growth conditions, although genetic variability in DM remobilization has been reported (Austin et al., 1977; Davidson and Birch, 1978).

The contribution of DM remobilization to grain ranged from 7 to 23% of grain dry weight (Table 2). Similar values were reported for barley by Gallagher et al. (1975, 1976), who concluded that pre-anthesis storage of carbohydrates is very important for grain yields in barley and wheat. Higher proportions (6-73%)

Table 2. Average number of spikes, kernel per spike, kernel weight, grain yield (GY), straw yield (SY), harvest index (HI), DM remobilization efficiency (DMRE), and contribution of DM to the grain of three wheat cultivars under three N and two water regime in 2006-2008.

2006-2007 year

Treatments	Spikes no. m^{-2}	Kernel Spike^{-1} no.	Kernel Wt. mg	GY Kg ha^{-1}	SY Kg ha^{-1}	HI (%)	DMRE (%)	DMR to grain (%)
Irrigation (I) †								
I1	709	55	31	6684	10654	39	19	7
I2	682	55	26	4922	9465	34	25	14
Avg.	699	55	29	5803	10059	37	22	11
LSD (0.05)	NS	NS	5	1286	1500	6.3	3.9	1.8
Nitrogen (N) Kg ha^{-1}								
0	648	50	31	5226	8361	35	19	10
80	756	55	35	5953	11280	36	22	11
160	683	56	36	5831	10536	40	23	13
Avg	696	54	34	5670	10059	37	22	11
LSD (0.05)	92	7.1	5.8	1296	1468	5.9	3.2	1.7
Cultivar(C)								
Shiraz	626	59	33	5841	9842	38	23	11
Marvdasht	644	60	31	5387	10330	35	19	10
Chamran	816	47	38	6182	9951	38	25	13
Avg.	695	55	34	5803	10041	37	22	11
LSD (0.05)	94	6.9	5.9	1301	1398	6.1	3.3	1.7

2007-2008 year

Treatments	Spikes no. m^{-2}	Kernel Spike^{-1} no.	Kernel weight mg	GY Kg ha^{-1}	SY Kg ha^{-1}	HI (%)	DMRE (%)	DMR to grain (%)
Irrigation(I) †								
I1	818	49	37	5109	12121	36	15	8
I2	831	47	29	3897	10247	31	19	15
Avg.	825	48	33	4503	11184	34	17	12
LSD (0.05)	NS	NS	5.9	1056	1612	5.3	2.1	1.9
Nitrogen(N) Kg ha^{-1}								
0	789	46	32	4052	11325	32 b	18	16
80	822	49	34	4642	14781	35 a	20	11
160	804	49	37	4614	13448	36 a	13	8
Avg	805	48	34	4436	13185	34	17	12
LSD (0.05)	106	6.8	6.1	1162	1697	5.7	2.4	1.7
Cultivar(C)								
Shiraz	788	46	35	4257	12991	36	15	12
Marvdasht	802	57	29	4927	13445	30	16	14
Chamran	924	41	36	5824	13118	35	18	8.5
Avg.	838	48	33	5002	13184	34	17	11.5
LSD (0.05)	104	6.9	4.5	1165	1683	5.6	2.4	1.5

of yield provided by pre anthesis were calculated by other investigators (Austin et al., 1977; Bidinger et al., 1977). Contribution of DM remobilization to the grain under water deficit was 78% more than nonstressed (Table 2). In our experiments plants fertilizing well showed a lower contribution of stored DM than unfertilized plants (Table 2). Chamran was higher in contribution of DM to the grain than Shiraz and Marvdasht.

Water stress at grain filling period induces early senescence, reduces photosynthesis and shortens the grain filling period (Kobata et al., 1992; Palta et al., 1994; Zhang et al., 1998; Souza et al., 1997). Such responses would be resulted to reduce in kernel weight, straw yield and grain yield. We found, however, that the effect of water stress at this time could greatly promote the remobilization of previously restored carbon reserved in plant. Early senescence induced by water stress can improve grain filling when senescence is unfavorably delayed has great significance to wheat production to Iran. The grate parts of Iran are limited by adverse weather conditions. Continental hot and dry wind (temperature > 30ºC, relative humidity <30%, and wind speed >10 km hr) is common in early June to July in this area and can dehydrate wheat before it matures. Water stress may accelerate senescence so that the wheat can mature before the adverse condition occurs. The heavy use of N in Iran results in unfavorably delayed senescence, leading to a low kernel weight. Early senescence induced by water stress could increase the rate of grain filling and improve kernel weight in this case. Nitrogen fertilizer is one of the major input costs in production of winter wheat in Iran. Our soil test indicated that the experimental field was low in available soil N. It was not surprising that the response of winter wheat to N application was dramatic under nitrogen application. The effect of high levels of soil N on vegetative growth, grain yield and yield components of wheat are well documented (Ayoub, et al., 1994; Donaldson, et al., 2001 and Patil, et al., 1995). Grain yield was highly related to spike per square meter, showing that the number of spike per square meter is the essential component determining the final grain yield in wheat. Strong relationships between grain yield and number of spike per square meter in wheat have been reported by other researchers (Fisher et al., 1977; Shanahan et al., 1984; Entz and fowler 1991; Gan and Stobbe 1996). When a high amount of N was applied, kernel weight was increased under water deficit. The obvious explanation for such a result is that, when N was heavily used, delayed senescence led to a slow grain filling and a poor remobilization and partitioning of assimilates into the grain. Some lodging- resistance wheat cultivars bred recently in Iran have low harvest index and poor grain filling because they stay green for too long and remobilize assimilates poorly to the grains. Water deficit may induced these cultivars to mobilized more presorted assimilates to grain.

References

1. Bahrani, A. and Z.A. Tahmasebi. 2005. Effect of rate and time of nitrogen fertilizer application on quantitative and qualitative traits, nitrogen and DM remobilization efficiency in bred and durum wheat. Iranian Journal of Agricultural Science. 36:5:1263-1271.
2. Barnabas B., K. Jager & A. Feher, 2008. The effect of drought and heat stress on reproductive Processes in cereals. Plant, Cell & Env. 31:11-38. Bidinger, F.R., R.B. Musgrave, and R.A. Fischer. 1977. Contribution of stored pre-anthesis assimilates to grain yield in wheat and barley. Nature. 270:431-43.
3. Davidson, J.L. and J.M. Brich. 1978. Response of standard Australian and Mexican wheat to temperature and water stress. Aust. J. Agric. Res. 29:1091-1106. Entz, M.H and D.B. Fowler. 1991. Agronomoc performance of winter versus spring wheat. Agron. J. 83:527-532.
4. Fan X. L., D Jiang , Dai T, Jing Q, Cao W. 2005. Effects of nitrogen supply on flag leaf photosynthesis and grain starch accumulation of wheat from its anthesis to maturity under drought or water logging. The journal of applied ecology. 16(10):1883-1888.
5. Fan X. L., Y. K. Li., 2001. Effect of drought stress and drought tolerance heredity on nitrogen efficiency of winter wheat . Plant Nutrition. 92:62-63
6. Fathi, GH. 2005. Effect of drought and nitrogen on remobilization of dry mater and nitrogen in six wheat cultivars. Iranian Journal of Agricultural Science. 36:5:1093-1101.
7. Fischer, R.A., I. Aguular, and D.R., Laing. 1977. Post-anthesis sink size in a high yielding dwarf wheat: yield response to grain number. Aust. J. Agric. Res. 28:165-175.

8. Gallagher, J.N., P.V. Biscoe, and R.K. Scott. 1975. Barley and its environment. V. Stability of grain wheat. J. Appl. Ecol. 12:319-336.
9. Gallagher, J.N., P.V. Bisco, and B. Hunter. 1976. Effect of drought on grain growth. Nature. 264:541-542.
10. Gan, Y.T., and E.H. Stobbe. 1996. Seedling vigor and grain yield of "Roblin" wheat affected by seed size. Agron. J. 88:456-460.
11. Kobata, T. J.A. Palta, and N.C. Turner 1992. Rate of development of post anthesis water deficits and grain Filling of spring wheat. Crop Sci. 32:1238-1242. 30-Nicolas, M.E., and Turner, N.C. 1993. Use of chemical desiccants and senescing agents to select wheat lines maintaining stable grain size during Post-Anthesis drought. Field Crop Res. 31:155-171.
12. Nicolas, M.E., H. Lambers, R.J. Simmpson, and M.J. Dalling. 1985. Effect of drought on metabolism and partitioning of carbon in two wheat varieties differing in drought-tolerance. Ann. Bot. 55:727-747. Palta, J.A. T. Kobata, N.C. Turner, and I.R. Fillery.1994. Remobilization of carbon and nitrogen in wheat as influenced by post-anthesis water deficits. Crop Sci. 34:118-124.
13. Patil, D.M., M.S., Mahjan, E., Patil, and S.M. Jawale. 1995. Response of wheat varieties to different levels of nitrogen. Field Crop Abstract. Vol: 48, P. 268, No. 2266. Pheloung, P.C. and K.H.M. Siddique.1991. Contribution of stem DM to grain yield in wheat cultivar. Aust. J. Plant Physiol. 18:53-64. Shanaban, J.F., D.H. Smith, and J.R., Welch. 1984. An analysis of post-anthesis sink-limited winter wheat grain yields under various environments. Agron. J. 76:611-615.
14. Souza, P.I.D., D.B. Egli, and W.P. Bruening. 1997. Water stress during seed filling and leaf senescence in soybean. Agron. J. 89:807-812.
15. Spiertz J.H.J., R.J. Hamer, H.Y. Xu, C. Primo-Martin, C. Don and P.E.L. Van derPutten, 2006. Heat stress in wheat; effects on grain weight and quality within genotypes. Europ. J. Agr. 25: 89-95.
16. Turner, N.C., and J.E. Begg. 1981. Plant-water relations and adaptation to stress. Plant Soil.58:97-131. Zhang, J. X. Sui, B. Li, B. Su, J. Li, and D. Zhoo.1998. An improve water use efficiency for winter wheat grown under reduced irrigation. Field Crop Res. 59:91-98.
17. Zhang X, Shanguan Z. 2006. Effects of nitrogen fertilization on leaf photosynthesis and respiration of different drought-resistance winter wheat varieties. Journal of Applied Ecology 17(11):2064-9.
18. Yang J, J. Zhang Z, Huang Q, Zhu, and L. Wang. 2000. Remobilization of Carbon reserves Is Improved by Controlled Soil-Drying during Grain Filling of Wheat Crop Science. 40: 1645-1655.

BIOSORPTION OF LEAD (II) AND CADMIUM (II) FROM AQUEOUS SOLUTIONS BY PRE-TREATED NONLIVING ALGA PADINA PAVONIA

M. NIKAZAR

Amirkabir University of Technology (Center of Excellency for Petrochemical Engineering)
No. 424, Hafez Ave., Tehran, Iran

MAJDIAN and F. VAHABZADEH

Amirkabir University of Technology (Center of Excellency for Petrochemical Engineering)
No. 424, Hafez Ave., Tehran, Iran

This paper describes the adsorption of heavy metal ions (Lead and Cadmium) from aqueous solutions by marine alga Padina pavonia. Metal sorption yields were modified using different kinds of pretreatment reagents: HCl, CaCl2, formaldehyde and NaOH. NaOH improved the maximum sorption capacity of the biomass. The biosorption kinetics was found to be fast, with 90% of adsorption within 30 min.

1. Introduction

Heavy metals released by a number of industrial processes are some of the major pollutants in marine, ground, industrial, and even treated wastewaters. Lead is widely used in many important industrial applications, such as storage battery manufacturing, printing, pigments, fuels, photographic materials, and explosive manufacturing, but it is also one of the three most toxic heavy metals [1]. The major sources of cadmium release into the environment by waste streams are electroplating, smelting, alloy manufacturing, pigments, plastic, battery, mining and refining processes [2]. Its presence in drinking water above the permissible limit (5 mg ml−1) may cause adverse health effects such as anemia, encephalopathy, hepatitis, and nephritic syndrome [3].

The biosorption of heavy metals from aqueous solutions can be considered as an alternative technology in the industrial wastewater treatment. It is based on the ability of biological materials to accumulate heavy metals from wastewater by either metabolically mediated or physico-chemical pathways of uptake [4]. This innovative depurative process uses biomaterials, such as marine algae [5] or wastes coming from industrial [6] and biological processes [7], which are either low-cost or abundant in nature.

In this work, the use of *P. pavonia* as a biosorbent for Pb^{2+} and Cd^{2+} ions from aqueous solutions was studied. The effect of pretreatment with different materials, biosoption equilibrium time and the appearance of disturbing ion in multicomponent system was studied in a batch system.

The experimental data have been analyzed by the adsorption isotherms, kinetics, equilibrium and thermodynamic parameters.

2. Materials and Methods

2.1. *Preparation of algae and metal solutions*

P. pavonia, a brown alga, obtained from the Persian Gulf on the coast of Qeshm Iran was extensively washed with distilled water and sundried on the beach and ovendried in an oven at 50°C overnight. The dried biomass was ground in a laboratory blender and sorted by sieving using the standard test sieves. The batch of the biomass with a particle size of 0.25-0.354 mm was selected for subsequent pretreatments. Pretreatment of the biomass was carried out as follows: a sample of 10 g of biomass was treated with a 0.1N NAOH solution (1000 ml) for 12 hrs under slow stirring. After 12 hrs of contact, the biomass was rinsed with deionized water in the same volume many times until a stable wash solution pH was reached. The biomass was then dried in an oven at 50°C overnight. So the prepared biomass was stored for later use.

Analytical grade reagents were used in all experiments. Stock single and multi metal solutions (1000 mg/L) were prepared by dissolving metal salts (cadmium nitrate [$Cd(NO_3)_2$_6H2O], lead nitrate [$Pb(NO_3)_2$], Merck, Darmstadt, Germany) in acidic distilled water. All working solutions were prepared by diluting these stock solutions with distilled water.

2.2. *Determination of metal concentration*

Single metal concentrations in the relevant samples were determined by an atomic absorption spectrophotometer (SpectrAA-300, Varian ABS, Tempe, Arizona) [8].

The biosorption of the metal ions (q) in the sorption system was calculated using the mass balance

$$q = V(C_i - C_e)/W \quad (1)$$

where V is the solution volume, W is the amount of the biomass, and C_i and C_e are the initial and final (or equilibrium) metal concentrations, respectively[9].

2.3. *Batch adsorption experiments*

Preliminary experiments were performed to determine optimum conditions. For this purpose, 100 mg of each dried biomass was added to 100 ml of solution with a known concentration and initial pH 5.2(±.1) for Lead and 5.9(±.1) for Cadmium in 250 ml Erlenmeyer flasks. The flasks were placed on a shaker shaking constantly at 150 rpm, at 25°C. The pH of the solutions during the contact period was adjusted using a small amount of 0.1M HNO_3 or 0.1M NaOH as required. Samples were periodically withdrawn and the solutions were separated from the biomass by filtration through the filter papers (Whatman No. 40 Ashless). After each time interval (0, 5, 15,...360 min), the algal biomass in one of the flasks was taken for Cd and Pb analysis as previously described.

The equilibrium data were obtained in similar conditions. The initial metal concentration ranged from 50 to 1000 mg l^{-1}.

To determine the adsorption characteristics of a selected metal in the binary metal mixtures, the initial concentrations of the selected metal were varied over a range of 50-850 mg/L while the competing metal-ion concentration in each biosorption medium was held constant at 50-850mg/L. To fully observe any competitive effect, the equilibrium experiments were conducted for binary metal mixtures with the equivalent concentration of each metal ion in the range of 50-850 mg/L.

The pretreatment of the biomass was achieved by the immersion of 1 g of the biomass in 100mL of pretreatment solution (NaOH, $CaCl_2$, Formaldehyde, HCl 0.2M) at constant stirring (50 rpm) for 18 h.

3. Non-linear Regression Analysis

All the model parameters were evaluated by a non-linear regression using the MATLAB® software. The optimization procedure requires an error function to be defined in order to be able to evaluate the fit of the equation to the experimental data. Apart from the correlation coefficient (R^2), the residual root mean square error (RMSE) and the sum of error squared (SSE) were also used to measure the goodness-of-fit.

$$RMSE = \sqrt{\frac{1}{n-2}\sum_{i=1}^{n}(q_{\exp i} - q_{cali})^2} \quad (2)$$

$$SSE = \sqrt{\frac{1}{n}\sum_{i=1}^{n}(q_{\exp i} - q_{cali})^2} \quad (3)$$

where the subscripts "exp" and "cal" are the experimental and calculated values of q, respectively and N is the number of measurements [10], [11].

4. Results and Discussion

4.1. *Effect of biomass pretreatments on metal removal*

Inorder to compare the effect of different pretreated on biosoption using the Longmuir model. Langmuir proposed a theory to describe the adsorption of gas molecules onto metal surfaces. The Langmuir adsorption isotherm has been successfully applied to many real sorption processes. The linearized Langmuir isotherm is represented by following equation:

$$C_e/q_e = 1/q_{max}b + Ce/q_{max} \quad (4)$$

where C_e is the concentration at equilibrium (mg/L), q_e the adsorption capacity in equilibrium (mg/g), b the Langmuir adsorption constant (L/mg), and q_{max} is the significant adsorption capacity (mg/g). To determine if the adsorption process is favorable or unfavorable, for the Langmuir type adsorption process, isotherm can be classified by a term 'R_L', a dimensionless constant separation factor, which is defined as below [12]:

$$R_L = 1/(1 + bC_0) \quad (5)$$

These experimental data were mathematically represented by linearized Langmuir sorption models (Figs. 1 and 2 and Table 1).

The literature data point out that the addition of calcium to alginate, whose main component is alginic acid, produces a cooperative association between chains (preferentially of α-L-guluronic acid and β-D-mannuronic acid), giving rise to a new molecular structure with large cavities occupied by calcium ions. This treatment produces a cross-linking reaction of the alginate polymeric chains. Also, this calcium retained by alginate plays an important role during ion exchange [13].

Formaldehyde treatment also produces chemical cross-linking between adjacent functional groups, mainly hydroxyl groups. However, this does not increase the cadmium sorption capacity as shown in our study. Additionally, cross-linking between polymeric chains by formaldehyde provokes the reduction of the biomass surface area, inhibiting the effective interaction between the biomass and metal cations in the solution [14].

Biomass conditioning with hydrochloric acid did not improve the sorption capacity either. It is accepted that acid washing produces the displacement of light metals by protons on the surface of the biomass, though this does not mean an increase in the sorbent capacity of the biomass. The acid treatment can dissolve polysaccharide compounds of the external cell wall of some fungi, creating additional bonding sites (amino groups). However, for several algae, such as *Padina*, the treatment can provoke the dissolution of alginates or other constituents of the cell wall matrix, and the subsequent decrease of sorption capacity. It has also been reported that the treatment with protons can alter the electronegativity of the cell wall of *Padina*, producing a decrease in the sorption capacity [15].

Generally alkali treatments decrease protonation. This treatment substitutes sodium inons on functional groups, increasing the electrostatic attraction by positive metal cations and facilitating ion exchange. The pretreatment of the biomass with NaOH was highly effective for the sorption metals.

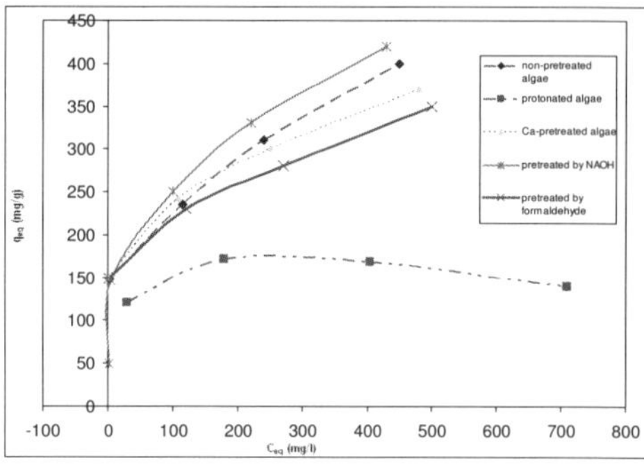

Fig. 1. Effect of different treatment on the biosorption of Pb (II).

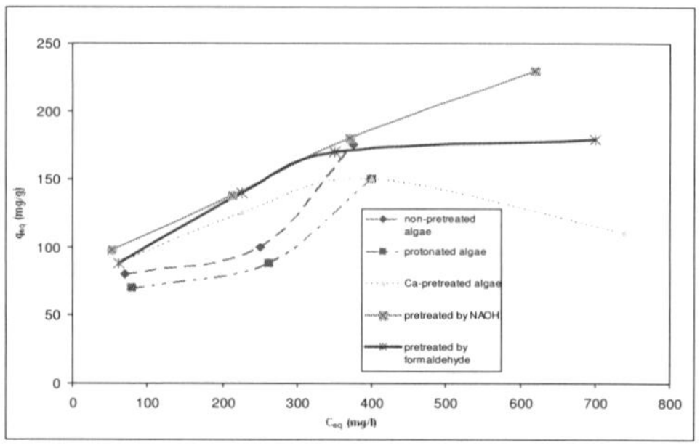

Fig. 2. Effect of different treatment on the biosorption of Cd (II).

4.2 . *Effect of contact time*

The kinetic profiles of Lead and cadmium adsorption by P. pavonia biomass are shown in Fig. 3. A two-stage kinetic behavior is evident: a very rapid initial sorption over a few minutes, followed by a long period of much slower uptake (especially for *Padina*–metal systems). The equilibrium time needed for the different metal–biomass systems ranged from about 1 to 3 h. In general, about 90% of the total metal ion sorption was achieved within 5 min for cadmuim and 23 min for lead. Therefore, in subsequent equilibrium experiments, 3 h was deemed more than sufficient to establish equilibrium.

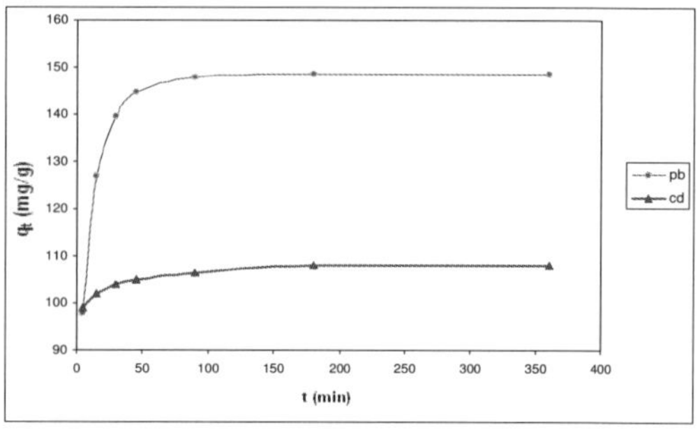

Fig. 3. Effect of contact time for the biosoption Lead and Cadmium.

Table 1. Langmuir isotherm constants for the biosorption of cadmium after different pretreatments of the biomass.

	Pb (II)			Cd (II)		
	q_{max}^*	b^*	R^2	q_{max}^*	b^*	R^2
Non-pretreated algae	316.45	.066	.985	222.26	.0056	.91
Protonate algae	145.77	-.15	.99	198.41	.005	.89
Ca-pretreated algae	303.95	.09	.993	179.86	.0083	.9
Pretreated by NaOH	415.45	.15	.99	297.9	.009	.97
Pretreated by formaldehyde	282.96	.095	.994	264.2	.006	.9

*b: L/g, q_{max}: mg/g

5. Conclusion

The study indicated that the biomass of the *Padina pavonia* algae particularly its Na-pretreated form could be used as an efficient biosorbent material for the treatment of metal (Lead and Cadmium) ions-bearing waste water streams.

The pretreatment with different materials for the biomass and treatment with NaOH were highly effective for the sorption of Lead and Cadmium. The kinetics of adsorption by this biomass was rapid with 90% of the total adsorption occurring within the first 30 min.

The sorption of the metal ions in the binary system was investigated in terms of metal-ion removal and equilibrium isotherms. Experimental data fitted the competitive Langmuir model. The sorption of the metal ions was reduced by the presence of coion, with the inhibitory effect increasing as the concentration of the co-ion increased. Lead manifested the higher inhibitory effect on the biosorption of other metal ions, followed by cadmium.

Acknowledgment

The authors express their thanks to Ms Samadian for editing the article.

References

1. R. Jalali, H. Ghafourian, Y. Asef, S.J. Davarpanah, S. Sepehr, Removal and recovery of lead using non living biomass of marine algae, J. Hazard. Mater. B92 253–262 (2002).
2. M. Tsezos, Biosorption of metals. The experience accumulated and the outlook for technology development, Hydrometallurgy 59, 241–243 (2001).
3. W. Lo, H. Chua, K.-H. Lam, S.-P. Bi, A comparative investigation on the biosorption of lead by filamentous fungal biomass, Chemosphere 39, 2723–2736 (1999).
4. E. Fourest, J.-C. Roux, Heavy metal biosorption by fungal mycelia by-products: mechanism and influence of pH, Appl. Microbiol. Biotechnol. 37, 399–403 (1992).
5. S. Schiewer, B. Volesky, Modeling of the proton–metal ion exchange in biosorption, Environ. Sci. Technol. 29, 3049–3058 (1995).
6. M.J. Gallagher, H. Moo-Young, The use of paper-mill wastes as a sorbent material for heavy metals in aqueous solution, in: Proceedings of 14th International Conference on Solid Waste Technol Technology Management IC.1/1–IC.1/8, Philadelphia, USA, 1–4 November, (1998).
7. W.C. Leung, M.F. Wong, H. Chua, W. Lo, P.H.F. Yu, C.K. Leung, Removal and recovery of heavy metals by bacteria isolated from activated sludge treating industrial effluents and municipal wastewaters, Wat. Sci. Technol. 41, 233–240 (2000).
8. Ofer Raize,Yerachmiel Argaman, Shmuel Yannai, "Mechanisms of Biosorption of Different Heavy Metals by Brown Marine Macroalgae", J. Biotechnology and Bioeng,VOL. 87, NO. 4, AUGUST 20, (2004).
9. Ping Xin Sheng, Yen-Peng Ting and J. Paul Chen, "Biosorption of Heavy Metal Ions (Pb, Cu, and Cd) from Aqueous Solutions by the Marine Alga Sargassum sp. in Single- and Multiple-Metal Systems", J. Ind. Eng. Chem. Res., 46, 2438-2444, (2007).
10. K.V.Kumar, S. Sivanesan, Pseudo-second order kinetic models for safranin onto rice husk: Comparison of linear and non-linear regression analysis, Process Biochem. 41, 1198–1202 (2006).
11. S.C. Tsai, K.W. Juang, Comparison of linear and non-linear forms of isotherm models for strontium sorption on a sodium bentonite, J. Radioanal. Nucl. Chem. 243, 741–746 (2000).
12. T.W. Weber, R.K. Chakkravorti, Am. Inst. Chem. Eng. J. 20, 228 (1974).
13. MM, Figueira, B. Volesky, VST Ciminelli and FA, Roddick Biosorption of metals in brown seaweed biomass. Water Res 34:196–204 (2000).
14. RS Bai and TE Abraham, Studies on enhancement of Cr(VI)Nbiosorption by chemically modified biomass of Rhizopus nigricans. Water Res 36:1224–1236 (2002).
15. Davis TA, Volesky B andMucci A, A review of the biochemistry of heavy metal biosorption by brown algae. Water Res 37:4311–4330 (2003).

SONOCATALYTIC DEGRADATION OF VARIOUS DYES IN WASTEWATER OVER HEAT-TREATED TIO_2 CATALYSTS AND ASSISTED BY HYDROGEN PEROXIDE*

AHMAD ZUHAIRI ABDULLAH† and PANG YEAN LING

School of Chemical Engineering, Universiti Sains Malaysia, Engineering Campus, 14300 Nibong Tebal, Penang, Malaysia

†Tel: +604-599 6411, Fax: +604-594 1013, E-mail: chzuhairi@eng.usm.my

The ambient temperature sonocatalytic degradation of congo red, methyl orange, and methylene blue by heat-treated titanium dioxide (TiO_2) catalyst to high temperature at an initial concentration of between 10-50 mg/l, catalyst loading of between 1.0-3.0 mg/l and hydrogen peroxide (H_2O_2) concentration up to 600 mg/l at 20 kHz and 50 W has been investigated. Proper amount of rutile phase of TiO_2 form on anatase had better sonocatalytic effect that the pure anatase. The degradation of organic dye depends on the chemical structure and molecular weight. With the highest molecular weight and highest reactivity with hydroxyl (·OH), peroxyl (HO_2•) and superoxide (O_2•-) radicals, congo red exhibit higher degradation rate. The highest degradation efficiency of congo red occurred at the initial concentration of 10 mg/l, 1.5 g/L of catalyst loading and 450 mg/L of H2O2 to give the efficiency of above 45% in 75 min.

1. Introduction

Application of ultrasounds in water with frequency range 18 kHz-10 MHz can result in acoustic cavitation, namely, the formation, growth and collapse of cavity bubbles that entrapped dissolved gasses/vapors surrounding water [1, 2]. Subsequently, many hot spots with local extremely high temperature (up to 5,000 K) and high pressure (up to 1,000 atm) are generated and then induce the dissociation of water (Equation (1)), oxygen molecules (Equation (4)) hydrogen peroxide (Equation (8)) [3, 4].

$$H_2O \xrightarrow[TiO_2]{)))} \cdot OH^\bullet + {}^\bullet H \quad (1)$$

$${}^\bullet OH + \cdot H^\bullet \rightarrow H_2O \quad (2)$$

$$2{}^\bullet OH \rightarrow H_2O_2 \quad (3)$$

$$O_2 \xrightarrow[TiO_2]{)))} 2{}^\bullet O \quad (4)$$

$${}^\bullet O + H_2O \rightarrow \cdot 2\,{}^\bullet OH \quad (5)$$

$$O_2 + \cdot H^\bullet \rightarrow \cdot O_2H \quad (6)$$

$${}^\bullet O_2H + {}^\bullet O_2H \rightarrow H_2O_2 + O_2 \quad (7)$$

$$H_2O_2 \xrightarrow[TiO_2]{)))} 2\,{}^\bullet OH \quad (8)$$

In this study, the TiO_2 powder was adopted as sonocatalyst as its advantages such as inexpensive, nontoxic, biologically stable and reusable [5, 6]. With the presence of a heterogeneous catalyst, it can increase the rate of formation and growth of cavitation bubbles which will then increase the rate of formation ${}^\bullet OH$ radicals [7]. The aim this study is to investigate effects of sonocatalytic degradation on organic dyes and different heat-treated TiO_2 catalysts. Other factors such as initial dye concentration, TiO_2 catalyst loading and H_2O_2 concentration were examined as well.

2. Experimental

2.1. *Materials and Apparatus*

The TiO_2 powders (R & M Marketing, Essex, U.K.) were adopted as sonocatalysts with and without heat-treatments. H_2O_2 30% (Merk Bchuchardt OHG) was bought to be used as a promoter for the sonocatalytic degradation process. Congo red, methyl orange and methylene blue powders (R & M Marketing, Essex, U.K.) were used as model compounds for confirming the sonocatalytic activity of heat-treated TiO_2 powders.

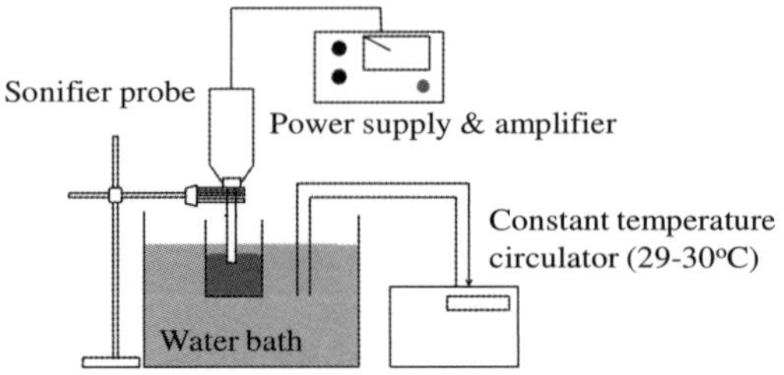

Figure 1. Schematic illustration of sonocatalytic degradation experiment apparatus.

*This work is supported by USM's Research University grant.

A general schematic diagram of experimental scheme is as illustrated in Figure 1. Digital ultrasonic apparatus with probe type sonifier (Branson, model 250) was adopted to irradiate organic dyes and operated at a fix frequency 20 kHz. The output power was 50 W. UV-vis spectrometer (Perkin-Elmer Company, USA) was used to study the degradation process of organic dyes.

2.2. *Process study*

2 sets of 15 g of TiO_2 powders were weighted and heat-treated at 800, 1,000°C for 2 hours and 4 hours respectively in a muffle furnace. Subsequently, the heat-treated TiO_2 powders were added into the prepared 100 mL of dye solution. Then, the beaker was put inside water bath to regulate the temperature at 29-30°C. Next, the sonifier probe was inserted at the center of the beaker to provide ultrasonic irradiation. The experimental conditions adopted throughout the whole experiments were: initial dyes concentration of 20 mg/L, TiO_2 powders of 2g/L and neutral initial pH unless otherwise specified. The liquid samples were centrifuged a few times to separate the TiO_2 powders. After that, the liquid sample's concentration was measured by means of UV-vis spectrometer to estimate the degradation efficiency of the dye solution. The efficiency was calculated as:

$$\text{Degradation efficiency} = \frac{C_o - C_t}{C_o} \times 100\%$$

where, C_o = initial dye concentration
C_t = dye concentration at time t (min)

3. Results and Discussions

3.1. *Effect of different chemical structures of dye and heat-treated catalysts*

The degradation efficiencies for three types of dyes after 75 minutes of ultrasonic irradiation are shown in Figure 2. The dyes degradation efficiency generally followed the order as below:

Congo red > methylene blue > methyl orange

Similar result has been observed by Lachheb *et al.* [8] for the photocatalytic degradation by using anatase of TiO_2. They found that the degradation efficiency decreased from congo red to methylene blue, then crocein orange G where could be attributed that this order might due to the decreasing molecular weight of dyes. Azo bonds or azobenzene (chromophore for congo red and methyl orange) has the higher reactivity coefficient with ·OH radical than dimethylamine group (auxochromes for methyl blue and methyl orange) which was reported to accelerate the rate of degradation.

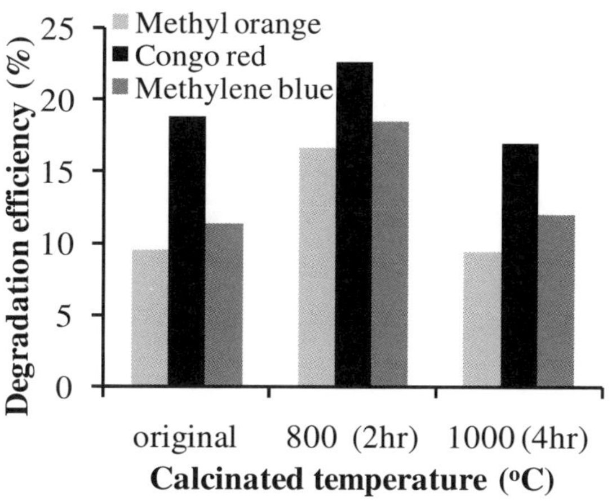

Figure 2. Effect of different dyes chemical structure and heat-treated catalysts (power = 50 W, frequency = 20 kHz, temperature = 30°C, initial dye concentration = 20 mg/L).

The reactivity coefficient of azobenzene, dimethylamine, and benzene ring are shown as below:

$$k^{\cdot OH}_{azobenzene} = 2 \times 10^{10} Lmol^{-1}s^{-1} \quad (9)$$

$$k^{\cdot OH}_{dimethylamine} = 1.4 \times 10^{10} Lmol^{-1}s^{-1} \quad (10)$$

$$k^{\cdot OH}_{benzene} = 1.7 \times 10^{10} Lmol^{-1}s^{-1} \quad (11)$$

Thus, congo red which contained 2 azo groups would possess higher reactivity with $^{\bullet}OH$ radicals, followed by methylene blue and methyl orange. By considering all the rate of reactivity to hydroxyl ($^{\bullet}OH$), peroxyl ($HO_2^{\bullet}$) and super peroxide radicals ($O_2^{\bullet -}$) with respect to the compound in dye structure, it can be found that the rate of degradation in congo red was more rapid, followed by methylene blue and methyl orange. From the Figure 3, it is found that the highest degradation efficiency is obtained when the heat-treated temperature is 800°C for 2 hours. This may be due to the formation of rutile phase with appropriate area and thickness on the surface of anatase phase that has the ability to inhibit the recombination of e^--h^+ pairs.

3.2. *Effect of initial dye concentration*

Results from Figure 3 show that the degradation efficiency decreased with the increasing of initial dye concentration. There are two possible reasons. Firstly, highly initial concentration of dye solution might cause excessive adsorption of the congo red molecules on the surface of TiO_2 powder until it was fully covered the TiO_2 powders. Secondly, when increasing the initial dye concentration, the mutual screens between congo red molecules and TiO_2 particles increased as well to negatively affect the efficiency. These two phenomena would prohibit TiO_2 particles from absorbing heat and energy resulting from the acoustic cavitation and also disturbed the transmission of ultrasound. Thus, this would limit the •OH radical that could attack certain dye molecules, thus leading to a decrease in the degradation efficiency.

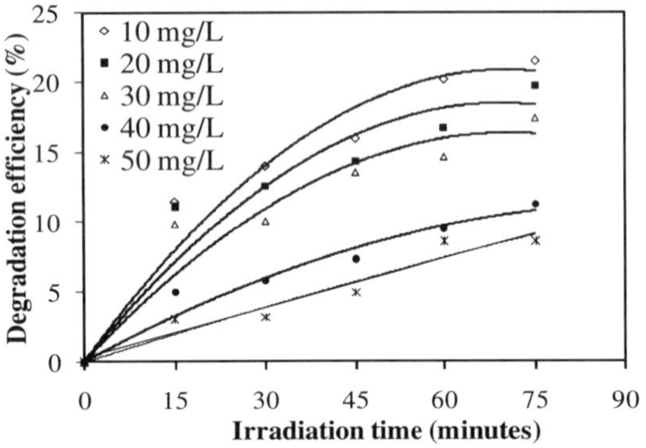

Figure 3. Effect of initial dye concentration on sonocatalytic degradation of congo red (power = 50 W, frequency = 20 kHz, temperature = 30°C).

3.3. *Effect of TiO_2 loading*

Figure 4 shows a series of experiments done by varying addition amount of TiO_2 starting from 1.0 g/L to 3.0 g/L in order to obtain the optimum addition amount of TiO_2 catalyst. Results show that the degradation efficiency of congo red rose sharply when the addition amount of TiO_2 increased from 1.0 g/L to 1.5 g/L. It began to decline after the added amount was increased to 2.0 g/L and further.

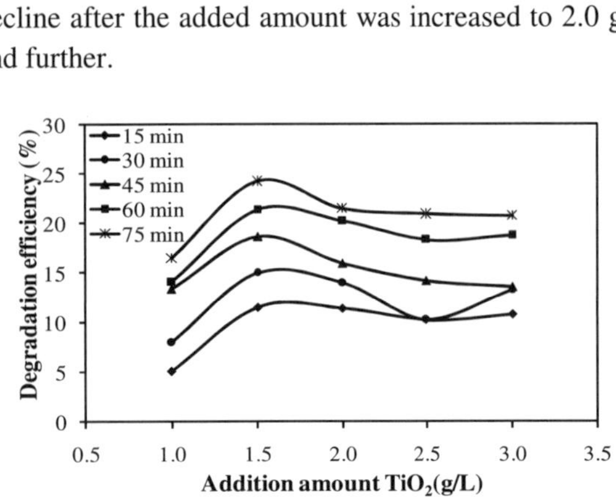

Figure 4. Effect of addition amount TiO_2 on sonocatalytic degradation of congo red (power = 50 W, frequency = 20 kHz, temperature = 30°C, initial dye concentration = 10 mg/L).

This phenomenon can be explained by the fact that the excessive TiO_2 particles caused TiO_2 particles to agglomerate. This would result the decreasing number of surface active sites at TiO_2. In addition, mutual screens among TiO_2 particles increased that would make the congo red molecules to receive extra shielding effect. Then, it would prohibit TiO_2 particles from absorbing heat and energy resulting from the acoustic cavitation [9]. Consequently, it might result in a reduction in the generation of exciting TiO_2 and the •OH radicals. Only small chances that dye molecules encountered •OH radicals produced by ultrasound. However, such a phenomenon was not observed in this study.

3.4. *Effect of H_2O_2 addition*

The addition of H_2O_2 was done for the purpose of initiating and promoting the free radical reaction. As seen in Figure 5, in the concentration range studied, the amount of H_2O_2 added has positive contribution on degradation of congo red compared with that with the absence of H_2O_2. Its degradation rose as the added amount was increased slowly from 75 mg/L to 450 mg/L. Further increase in the concentration of H_2O_2 from 450 mg/L to 600 mg/L, would cause a decrease in the degradation of congo red.

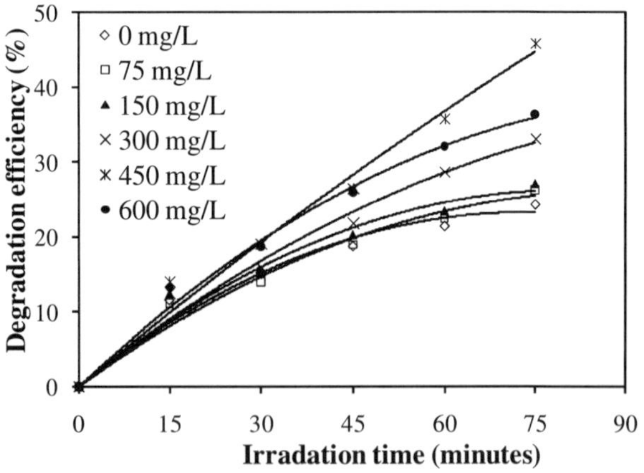

Figure 5. Effect of addition amount H_2O_2 on the sonocatalytic degradation of congo red (power = 50 W, frequency = 20 kHz, temperature = 30°C, initial dye concentration = 10 mg/L, TiO_2 amount = 1.5 g/L).

This could be attributed to the fact that in the presence of high concentration of H_2O_2, H_2O_2 became a scavenger and •OH radicals that would result in a slower degradation of the dye molecules. As well known, •OH and H$^\bullet$ radicals were generated by the dissociation of water in a solution and H_2O_2 could be scavenging the $^\bullet OH$ radicals produced. Below equations show that H_2O_2 acts as a hydroxyl scavenger (reaction 12) and as a hydroxyl source (reaction 13) under the sonochemical conditions [10]:

$$H_2O_2 + {}^{\cdot}OH \rightarrow H_2O + HO_2^{\cdot} \quad (7)$$

$$H_2O_2 + H^{\cdot} \rightarrow {}^{\cdot}OH + H_2O \quad (8)$$

4. Conclusion

In present study, it was found that TiO_2 with a proper amount rutile phase form on anatase TiO_2 had better effect on the sonocatalytic activity than that of pure anatase. The degradation of organic dyes (congo red, methylene blue and methyl orange) were demonstrated to depend on their chemical structures. With the highest molecular weight and highest rate of reactivity with $\bullet OH$, $HO_2^\bullet$, and $O_2^{\bullet -}$ radicals, congo red exhibited the highest sonocatalytic degradation efficiency. The highest sonocatalytic degradation efficiency for congo red (45% within 75 minutes) occurred at the initial concentration of dye of 10 mg/L, the added of TiO_2 amount of 1.5 g/L, the added H_2O_2 concentration of 450 mg/L neutral pH.

References

1. N. H. Ince, and G. T. Guyer, Ultrason. 42, 591 (2004).
2. J. Wang, Z. J. Pan, Z. H. Zhang, X. D. Zhang, F. Y. Wen, T. Ma, Y. F. Jiang, L. Wang, L. Xu, and P. L. Kang, Ultrason. Sonochem. 13 493 (2006).
3. M. Abbasi, and N. A. Razzaghi, J. Hazard. Mater. 153, 942 (2008).
4. Y. G. Adewuyi, Environ. Sci. Technol. 39, 8557 (2005).
5. J. Wang, Y. F. Jiang, Z. H. Zhang, G. Zhang, T. Ma, and W. Sun, Desalination 216, 196 (2007).
6. H. H. Huang, Ph.D dissertation, Graduate Institude of Environmental Engineering, National Central University, 2006.
7. N. Shimizu, C. Ogino, M. F. Dadjour, and T. Murata, Ultrason. Sonochem. 14, 184 (2007).
8. H. Lachheb, E. Puzenat, A. Houas, M. Ksibi, E. Elaloui, C. Guillard, J. M. Herrmann, Appl. Catal. B 39, 75 (2002).
9. J. Wang, T. Ma, Z. H. Zhang, X. D. Zhang, J. F. Jiang, D. B. Dong, P. Zhang, and Y. Li, *J. Hazard. Mater.* **137**, 972 (2006).
10. Claudiominero, Mircolucchiari, Davidevione, and Valtermaurino, *Environ. Sci. Technol.* **39**, 8936 (2005).

TRANSPORTATION OF DANGEROUS SUBSTANCES: A DECISIONAL SUPPORT SYSTEM FOR RISK ANALYSIS

VINCENZO TORRETTA

Insubria University, Department Environment-Health-Safety, via GB Vico, 46
Varese, 21100, Italy, tel. +39 0332 218782, e-mail: vincenzo.torretta@uninsubria.it

A decision support system is proposed to quantify risk due to transport of dangerous substances: road, rail, pipeline and inland waterways. The aim of the work is the evaluation of:

- population global risk (n. subject /year) considering the exposition to the potential relevant incident;
- individual risk (yearly probability/person), with the hypothesis that the single person would be exposed during all the dangerous event duration without any protection.

It allows the comparison of risk distribution in different solutions of dangerous substances transportation, permitting the user to decide what course of action to follow with the aim to transport dangerous substances minimizing the risk distribution on the territory.

1. Introduction

Potential user of this instrument can be decision maker in Public Authority for territorial planning. Mathematical approach has been derived from a previous software named "Transit" (GWBasic language application), that could quantify effects due to the dispersion of hydrofluoric acid from rail unit. For the current application, relevant modifications have been done to the algorithms of Transit.

A risk assessment analysis for a specific transportation may includes:

- description of the transport stream (number of yearly loaded transport unit per substances or category);
- description of transport units;
- description of the transport routes;
- description of n. of accidents and level of traffic in order to determine accident frequencies;
- description of ignition sources;
- properties of transported substances;
- terrain classification of the surroundings of transport route (urbanized, agricultural area, etc);
- meteorological data;
- living and working population in the surrounding area of transport route (density of population).

Mathematical has been geared out not considering tunnel and environmental effects. In order to evaluate risk about transport and using Geographical Information System potentiality, a management system could be adopted for creating a flexible instrument for planning at regional scale, thus combining state of art guidelines and needs of the project. The scale of the study area needs a practical instrument for managing a large number of input data. The model allows to estimate the impact of new roads or different ways for the trasport, modify transported quantity, consider potential restrictions on time/routes or tract deviations: in general it allows to decide what course of action to follow with the aim to transport dangerous substances minimizing the risk distribution on the territory.

2. Main characteristic of the software

The model and maps have been integrated each other. The result is a GIS software where, over Technical Regional Map, can be done a simulation for a specific road/rail/pipeline tract. In case of new data the system can be easily updating. The main characteristics of the software are:

- it's has been developed in accessible standard language and is compatible with the most common used operating systems;
- it's equipped by user interface for selection of roads, rail and pipeline streams;
- the results will have geo-references into Technical Regional Map.

User can select the area of interest and move and operate with a friendly pan operator. The software automatically shows data associated to the selected tract, in particular:

- λ = incident frequency of the selected tract (number of accidents yearly per km);
- L = length of the tract (km);
- NT = hazardous traffic flow (number of dangerous substances transported daily);
- A = impact area (km^2);
- DW = population in the surrounding associated to selected tract.

Table 1. Structure of the data vector inherent to the examined tract.

NI	accident frequency on the selected tract (number of accidents yearly per km)
d_k	population in the surrounding associated to selected tract
N	traffic flow (vehicle per day)
NT_{1k}	n. of hazardous vehicles per year, referring to substance n.1
NT_{2k}	n. of hazardous vehicles per year, referring to substance n.2
......	n. of hazardous vehicles per year, referring to substance n...
NT_{54k}	n. of hazardous vehicles per year, referring to substance n.54

First of all, 54 substances (6 for each transport class) have been considered, in particular hydrofluoric acid, ammonia, gasoline, benzene, liquefied oxygen, ethylene dioxide, methanol, sulphuric acid, acetylene, considering also the corresponding potential accident scenario: flash fire, pool fire, UVCE, toxic release, bleve, jet fire. The choise of the substances and scenario has been carryed out following this priority criteria selection: substances which present an elevated flux (using classification of CNIT - Conto Nazionale delle Infrastrutture e dei Trasporti) and substances which present an elevated magnitudo (historical analysis).

3. Criteria for modeling

First step corresponds to the identification of the transport stream with characteristic of the hazardous flux. The most important parameters are λ (accident frequency), L (length of the tract) and NT (hazardous traffic daily flow). Data are visualized with a GIS software; user can select a road tract and automatically the software provide to identify critical parameters (number of dangerous flux). The quantified risk analysis has been done considering different type of accident scenario, also taking into account that one substance can lead to one more accident (i. e fire and toxic release). The evaluation of physical effects on human health is calculated using an off-line software (TNO Effects), that can quantify the distance, in meters, of the threshold considered limits from the transport road line.

4. Estimation of damage area

4.1. *Critical event*

Typical scenario considered are:

- dispersion (release of toxic, flammable o dangerous for human health without fire);
- pool fire (flammable liquid);
- flash fire (vapor cloud fire without explosive effects);
- UVCE (unconfined vapour cloud explosion);
- BLEVE (boiling liquid expanding vapour explosion);
- jet-fire (flammable pressurized gas).

It has been made a distinction between the following transport classes:

- flammable liquid transported in atmospheric tank unit;
- flammable or toxic gas compressed or liquefied;
- compressed cryogenic flammable or toxic;
- compressed flammable gas.

4.2. *Identification of threshold limit for possible damage area*

Human and environmental effects due to an accident directly depend on exposition time to the source damage, such as toxic concentration and exposition to thermal flux related to a fire. It's necessary to establish some threshold limit for potential damage area.

Overcoming limits can lead to serious consequence for health. A valid reference for threshold limit can be done by D.M. LL.PP 9/5/2001 "*Basic risk assessment for territorial planning for areas interested by major risk activities*" (see Table 2). This Italian law defines three territorial classes in relation to possible damage to human health due to major activities, defining for different scenario the threshold limit:

- first territorial planning zone - certain impact (elevated death probability - the first zone is closed to accident place and is characterized by very high damage over public health, typically death);

- second territorial planning zone – damage area (not reversible – this zone, more external than the first one, is characterized by not reversible damage for human health);
- third territorial planning zone (reversible - not serious damage for population).

Table 2. Limits established by D.M. LL.PP. May 2001.

Incident scenario	Death	Not reversible	Reversible
Fire (kW/m^2)	12,5	5	3
Flashfire	LFL	0,5 LFL	0,1 LFL
Explosion (bar)	0,3	0,07	0,03
Toxic dispersion	LC_{50}	IDLH	LOC

With:

- LFL = Lower Flammable Limit;
- LC_{50}(h, 30 min) = Concentration of toxic substance that causes death for 50% of exposed people in 30 minutes;
- I.D.L.H. = Immediately Dangerous to Life or Health;
- LOC = Level Of Concern.

For each scenario the software gives back three lines corresponding to three different territorial planning zones with progressive distance from the road axis.

For every considered substance it is necessary to study all the possible scenario, identifying, for every scenario, the relative three thresholds of damage. In such way it results possible to calculate the worst scenario, with the maximum extension of the area of potential damage.

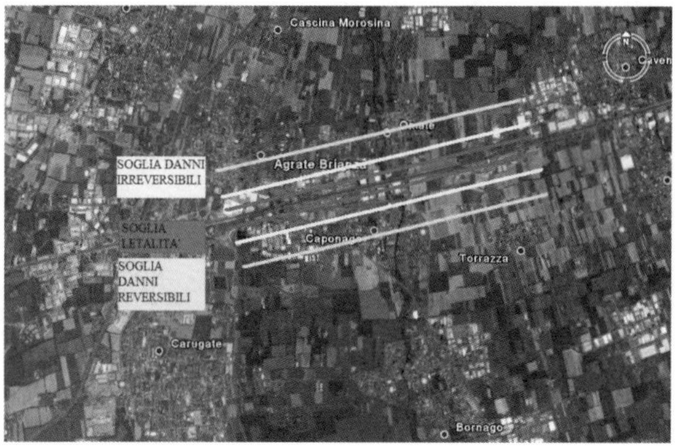

Figure 1. Example of the three threshold limit for road transport.

The TNOs charts furnish the necessary distances of damage from the road and it allows to appraise the theoretical risk in the involved areas, considering the parameters on which the evaluation of the consequences depends on:

- typology of transported substance;
- typology of accident scenario;
- typology of territory surrounding;
- threshold of damage (reversible damages, irreversible damages, lethality);
- loss of containment classes (pinhole, rupture, tank collapse);
- atmospheric conditions.

5. Mathematical model

The parameters for quantifying risk are:

- λ = accident frequency of the selected tract (n. of accidents yearly per km);
- L = length of the tract (km);
- NT = hazardous traffic flow (daily quantity of dangerous transported substances);
- PK = probability of release by transport unit;
- PS = probability of ignition of flammable substances;
- PJ = probability of certain meteorological conditions;
- PW = probability of terrain classification of the surrounding area of transport route (urban, rural, industrial);
- A = impact area (km^2);
- DW = population in the surrounding associated to selected tract;
- FUG = human shelter factor for escaping and protection.

$\underline{\lambda}$: GIS data bank created for this project, provides to give, for every selected tract, the number of the accident yearly per km. The hypothesis adopted for road transport is that λ is the initial frequency of serious damage to vessel per unit distance (km). The initial accident frequency is strongly depending on local factors. For that reason each kilometre of the road system has been characterized by a specific initial frequency. By the way, rail and pipeline streams have been characterized with a constant accident rate per km given by the network owners.

L: software allows to identify on the Technical Map an accurate length of the transport system between two points. Minimum unit is 1 km.

NT: in the selected road tract, a very important parameter is the number of passages in a unit time of dangerous substances. A specific study has been carried out to determine in all the investigated area, the flow of dangerous substances. The identification of those flows in rail and pipeline streams has been done with an elevated degree of details. In fact in those system initial and final point of the transported quantity is simply identified. On the contrary in road transport more specific hypothesis have been done. Despite its analysis complexity, a GIS data bank about hazardous material has been constructed concerning nine transport classes, defined by ADR, European agreement for the transportation of dangerous goods. The high number of the transported materials makes necessary a selection of certain dangerous substances.

PK: the probability to have a substance release following an accident, can be: total lose of product caused by a tank breakdown (hypothesis corresponding to K=3), release of 50% of product (K=2) and release of 10% of product (K=1). K=0 is assumed in the hypothesis that no release follows as consequence of an accident. For instance, the probability adopted for pressurised road tank is (%):

- P (K=0) = 99.00
- P (K=1) = 0.90
- P (K=2) = 0,07
- P (K=3) = 0,03

PS: the events that have been considered are: toxic exposure, pool fire, jet fire, flash fire. For a given accident and for given quantity released, a probability of ignition may be adopted for flammable substances. This data has been quantified with simple event trees.

PJ: represents probability of typical meteorological conditions in the area, in particular Pasquill Stability Classes F and D.

PW: a classification of the surroundings has been done for all transport networks. The GIS data bank, referring to the location examined, gives three values of probability, i.e. 30% rural territory, 30% industrial area and 40 % urban area.

A: the physical effects of the considered events are calculated and A represents the damage area extension.

DW: for all transport networks the number of people that could be exposed to the accident is calculated. A specific level of population density (taking into account the contribution of schools, hospitals, public offices) corresponds to every kilometre.

FUG: in order to estimate physical effects to human health, some shelter/flight factor have been considered, i.e. the escaping possibility.

Frequency F (event/year) and Impact I (number of injured people) is calculated using these following expressions:

$$F\left(J, s, K, W, S\right) = \lambda \cdot L \cdot NTs \cdot PK \cdot PS \cdot PJ \cdot PW \quad (1)$$

$$I\left(So, s, K, W, S\right) = \lambda\left(So, J, s, K\right) \cdot DW\left(W\right) \cdot FAT\left(W\right) \cdot \left(1 - FUG\left(1\right)\right) \quad (2)$$

Using a simple function for combining F and I, the Population Risk (people affected yearly), is:

$$R\left(I, J, s, K, S, W\right) = F \cdot I \quad (3)$$

6. Risk parameter dynamic analysis

The quantification of the risk connected to the transport of dangerous substances, needs the evaluation of various inherent parameters among the others the network characteristics (traffic, accident rate, etc.), the systems of transport, the surrounding urbanization and population density and so on. The parameters above listed can suffer variations and it means a dynamic analysis of data base feeding the software. For this reason the model allows:

1. updating of the used dates referred to the network characteristics;
2. infrastructural interventions (i.e. new roads realization)
3. changes of use of the transport vectors (i.e. moving on railroad a part of dangerous substances transported on the road or on other transport systems);
4. regulation of the access for determined line roads and/or limitations of schedule;
5. change of the run covering the same way.

7. Conclusion

The methodology chosen allows to quantify risk related to transportation of dangerous materials. In this paper a simplified method for Individual and Population risks evaluation has been proposed. Information data (regarding network characteristics, flows, etc.) can be automatically and easily updated by the user (i.e. Public Authority). Finally, the approach allows to:

- identify areas with high or critical level of risks;
- develop possible planning instrument for Public Authority for the evaluation of the impact of new roads, expansion of existing networks, etc.
- give to a decision maker a simple instrument in order to evaluate the reduction of the risks, and determine possible and effective interventions, such as restrictions on time/routes or way deviations.

References

1. Istituto Nazionale di Statistica ISTAT (2006).
2. Provincia di Milano, *Programma di Previsione e Prevenzione dei Risch*, pp 1-50 (2005).
3. V. Torretta, *Analisi di rischio di incidenti rilevanti,* Sistemi Editoriali (2006).
4. TNO Safety Software, Effects version 7.5 - Modelling the effects of accidental release of hazardous materials, User manual, The Hague, The Netherlands (2000).

SULFATION OF CASTOR OIL WITH DEVELOPMENT OF EMULSIFIERS

ABDUL BASET M. ELGELLAL[†]
Chemistry Department, Faculty of Science, 7th of October University, Misurata, Libya

IMAD H. AL-WAHAIB
Chemistry Department, Faculty of Science, 7th of October University, Misurata, Libya

The work is related to the field of preparation of emulsifiers for the use in asphalt emulsions, and formulating of solution type cooling-lubricating fluids. Thus, castor oil was reacted with gaseous sulfur trioxide at 25°C using carbon tetrachloride as a solvent. Emulsification properties of the sulfurized castor oil were studied in relation to the content of combined SO_3. The highest emulsification tendency was attained at approximate 8 percentages by weight of SO_3 content corresponding to the sulfation of one hydroxide group in castor oil. email: Baset1950@com

1. Introduction

1.1. *Aromatic hydrocarbons, petroleum oil and vegetable oil*

Sulfonation of aromatic hydrocarbons, petroleum oil and vegetable oil has great industrial importance, from which useful surfactants material could be produced. Such surfactant could be used as detergents, emulsifying agents, softening agents, demulsifying agents etc. The sulfonation of aromatic hydrocarbons is used to induce the sulfonyl group (SO_3H) while the sulfation of aliphatic organic molecules such as alcohols and olefins produce sulfuric acid esters. The sulfated castor oil (Turkey red oil) is one of the most effective surfactants.

The sulfated castor oil is used in fabric manufacture as a softening agent. This sulfated oil is classically prepared, in early 1970s, by gradual addition of sulfuric acid to castor oil under harsh condition of high temperature and long reaction time in addition to producing side products.[1-2]

Recently, the sulphated castor oil is prepared by using diluted gaseous sulfur trioxide.[1] (diluted with 4% V in N_2) under mild condition (i.g., at 45-50°C for 20-25 min).The resulted sulphated castor oil should be Highly soluble in water, making it good surfactant and/or good emulsifying agent and stable under acidic conditions.

The experimental data showed that the optimum reaction temperature was 45-50°C.[1] Lowering the temperature below this range leads to increasing the viscosity of the reaction mixture resulting poor interaction between the gas and liquid phases. However, raising the reaction temperature higher than 50°C could allow some oxidation reactions to take place producing sulphur dioxide.

It has been proved that the sulfation reaction of the castor oil at <50°C occurs by a reaction between gaseous SO_3 and the hydroxyl groups of esters of ricinolic acid.[1]

There have been a number of studies focusing on the sulfation of fatty acids using SO_3.[3-5]

According to these studies, the esters of the higher sulfonyl fatty acids. Based on that the sulfation of the castor oil takes place at relatively high temperature (>60°C) producing free sulfonates without the formation of sulfated glycerides.[6] The sulfated castor oil has wide industrial uses.

It is used in leather industry as the aqueous filling material for leather contains about 3-4% by weight of the saponifiied sulfated castor oil.[7]

The sulfated castor oil along with the coconut oil are used as a hydrating material.[8]It is also used in processing leather in which 20% by volume of ethanol is added to an aqueous solution of the sulfated castor oil to form an emulsifier for so aching chrome-treated leather in order to enhance its flexibility

In the presence of 6% by weight sulfated castor oil, the hydrating and flexibility properties of the mixture have been improved.[9] The sulfated castor oil has been used to purify the asphaltic sand from the asphalt.[10]

In addition, it has been involved with the lithium lipids for improving their thermal properties.[11] Further

[†]Work partially supported by chemistry department, faculty of science, 7th of October University.

more, the sulfated castor oil has been utilized with other materials to form cooling-lubricating emulsifiers that are employed in cutting and polishing the glass. [12]

Other types of lubricating emulsifiers have been synthesized from the sulfated castor oil and other substances concentration ranging between 30 to 50%.[13]

2. Experimental

2.1. *Materials*

Pure castor oil (B.P) was used ricinolic acid (C12------OH, 9C) was prepared by saponification pure castor oil using 20 % sodium hydroxide solution at 90- 95°C with stirring for 2 hr. the resulting mixture was acidified by dilute sulphuric acid. The resulted ricinolic acid was watched with hot water and dried.

2.2. *Sulfation*

Two methods have been used for sulfation: in the first method. Oleum containing 30% SO_3 by weight was used. The oleum was added slowly over 30 minutes with stirring at certain temperature. In the second method, the SO_3 diluted with nitrogen (approximate 4% by volume) was used. The excess SO_3 of was liberated by heating the oleum and passing current of nitrogen. The sulfating agent (SO_3 diluted with nitrogen) was passed through the gas distributor placed in bottom of reactor that contains oil. This was within ±2°C by heating the reactor. By using sulfation in this way, the unreacted residual gas was expelled at the end of reaction of reaction by heating to 70°C.

The reaction depth was determined by measuring the content of reacted SO_3 gas and also by measuring the acidity of the resulted sulfated oil. Hexane and CCl4 were used as solvents.

2.3. *Method of test and extraction*

2.3.1. *Acidity of sulfated oil*

The acidity of sulfated oil was determined by titration with standard solution of KOH. The sulfated castor oil (1g) was dissolved in mixture ($100cm^3$) of toluene and iso propyl alcohol in ration of 2:1 V/V at high temperature.

$$\text{Acidity (mg KOH/g)} = \frac{V \times N \times 56}{W}$$

V = volume of KOH (ml)
N = normality of KOH solution
W = weight of the sample (gm)

2.4. *Emulsification properties*

The emulsification properties of sulfated castor oil and its derivatives were studied by forming of asphaltic emulsions. The asphaltic emulsions were prepared by mixing equal volumes of water, light asphalt and toluene in the presence of emulsifying agent.

The stability of the resulted emulsions was monitored over 24 h. The amount of separated water gave indication to the stability of the emulsion. Thin asphalt, and toluene and in presence of certain ration of emulsifier, the stability of emulsion with time has been followed.

Over night separated water ration was standard for stability of emulsion. The basic model makes the following assumptions:

3. Results and discussion

3.1. *Sulfation of castor oil using SO_3 and studying the properties of the resulting emulsion*

3.1.1. *The relationship between the depth of sulfation and acidity of the sulfated castor oil*

The sulfation was carried out at 45 C using different ratios of sulphur trioxide and the castor oil after which the acidity of the products was determined.

The theoretical values of the acidity of the sulfated oil were calculated and compared with the experimental data.

The content of SO_3 in the product when one hydroxyl group was sulfated is:

$$\frac{Mwt.of SO3 \times 100}{Mwt.0f Castoroil} = \frac{80 \times 100}{932} = 8.6\%$$

This means that the molecular weight of the product is 1012 and the acidity of product is 55mg KOHgm. The experimental and theoretical correlation (Figure 1), Show the sulfation of castor oil without observing any side products. Moreover, the experimental results are in accordance with the theoretical calculation with error percentage of 3.5%.

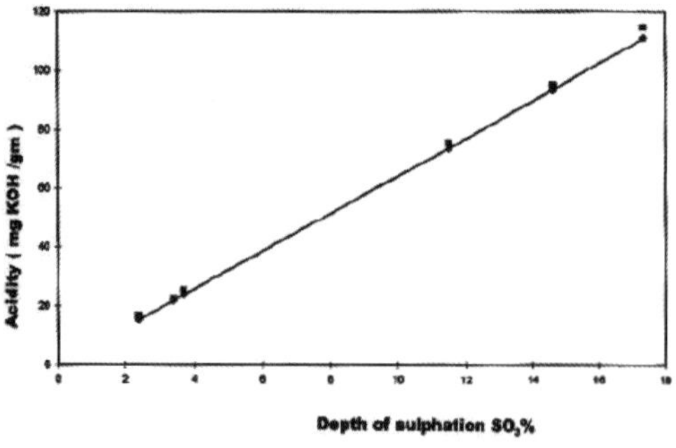

Figure 1: The relationship between sulphation's depths and the acidity of the product (sulphation was carried out at 45°C).

3.1.2. *Optimization of the sulfation of the castor oil*

A range of chlorinated hydrocarbons and other low-boiling hydrocarbon were used as solvent in the sulfation process. These solvents are in reactive towards sulfur trioxide. In addition, such solvents are easy to remove by evaporation where the sulfation is complete.

The ration of solvent to castor oil was (2:1 w/w). The temperature of the sulfation reaction was reduced to 25°C due to the use of such low-boiling solvents. The rate of the sulfation reaction in carbon tetrachloride was found to be much faster than the reaction rate when the sulfation was carried out on neat castor oil or in hexane. The rate of the sulfation, in carbon tetrachloride, increased by 2.4 times (Table 1).

Table 1: The effect of solvents on the rate of the sulfation of castor oil using SO_3.

Solvent	Temp. of Sulfation °C	*Reaction Rate
No solvent	50	19
Hexane	25	05
Carbon tetrachloride	25	46

*(Increase of acidity/h) (mg KOH/gr/h)

3.1.3. *Study of the emulsification properties of sulfated castor oil*

The sulfated castor oil with different levels of sulfation was emulsified with petroleum asphalt in which the content of the emulsifying agent (sulfated caster oil) was 2% by weight. The stability of the resulting emulsions was evaluated according to the content of SO_3 in the sulfated castor oil (Figure 2). The pure caster oil showed on emulsification properties, however, the sulfation of one hydroxide group in the oil (8-10, by wt of SO_3), improved dramatically the emulsification properties. Sulfation of a second hydroxyl group decreased the emulsification ability of the product.

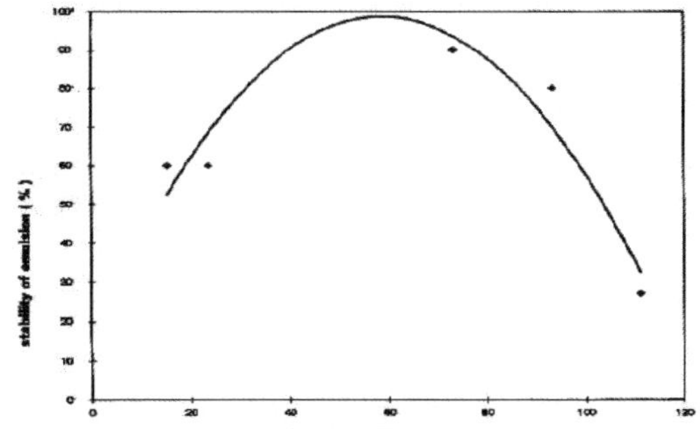

Figure 2: Changing the emulsification properties of the sulphated oil by changing the acidity.

3.2. *Study of emulsification properties of the sulfated ricinolic acid and its salts*

3.2.1. *The sulfated ricinolic acid and its salts as emulsifiers*

Ricinolic acid showed no ability to product emulsion with petroleum asphalt. This was observed when ricinolic acid was combined with water and petroleum asphalt (diluted with toluene) in ration of (1: 1: 1 w/w) with concentration between 0.5 and 2.0%. An immediate separation of the layers was occurred.

However, when the ricinolic acid was sulfated, using diluted SO_3 (diluted with N_2) in which different levels of sulfation were obtained, the resulting products showed emulsification properties. The most stable emulsion resulted from the use of sulfated ricinolic acid containing 2.3% of SO3. The emulsification property decreased as the SO3 content increased (Figure 3).

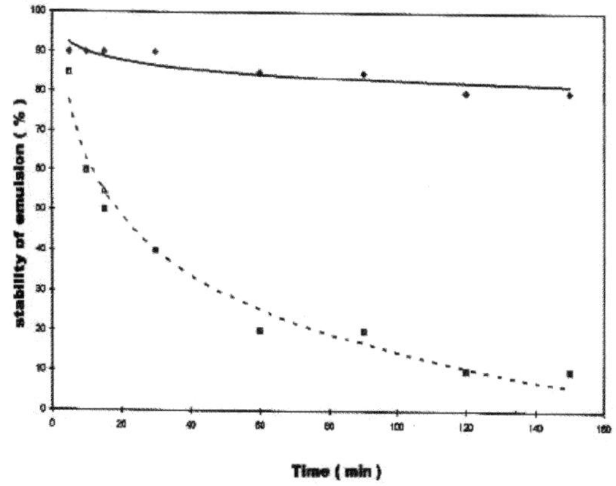

Figure 3: Stability of asphalitic emulsions of the sulphated ricinolic acid (concentration of emulsifying agent is 2.0% by weight).

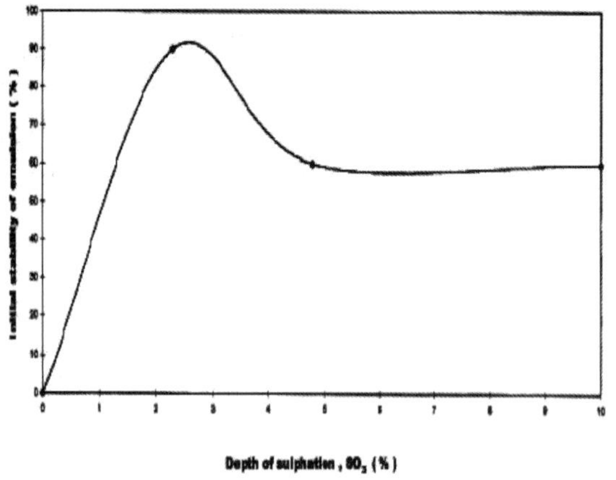

Figure 4: The relationship between the initial stability of asphaltic emulsions and the depths of the sulphation of ricinolic acid (concentration of ricinolic acid is 2.0% by weight).

The sodium salt of ricinoleic acid showed no emulsification properties. However, partial emulsification properties were seen when the sodium salt of sulfated ricinolic acid was used (Figure 5).

The emulsification properties of the latter case decreased as the content of SO_3 increased which means that the sodium salt of the sulfated ricinolic acid became a detergent rather than emulsifying agent.

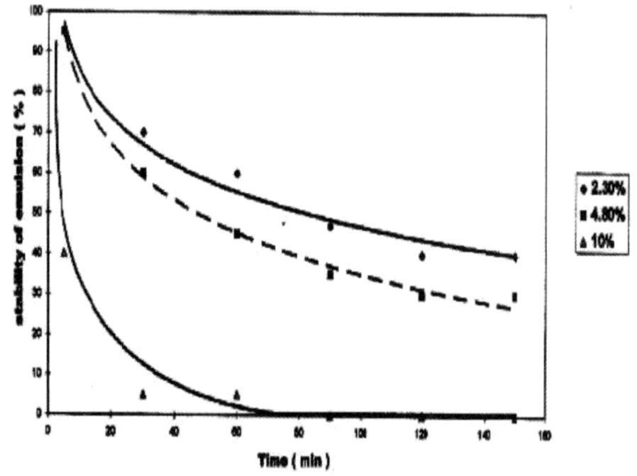

Figure 5: The stability of asphaltic emulsion that contain sodium salt of sulphated ricinolic acid (concentration of emulsifying agent 2.0% by weight).

3.2.2. *The formation of emulsifying agent by the partial neutralization of sulfated ricinolic acid*

The possibility of preparing some emulsions by partial neutralization of high molecular weight carboxylic acids has been mentioned in the literature. This technique leads to the formation of emulsions consisting of free carboxylic acid and their salts (e.g sodium salts) in the same time. Based on that, an attempt was made to form an asphaltic emulsion containing an emulsifying agent prepared by the partial neutralization of sulfated ricinolic acid. The emulsifying agent was prepared by the sulfation of ricinolic acid up to 2.3% (by weight) of SO_3 content. Partial neutralization of the resulting sulfated acid was carried out using a concentrated solution of sodium hydroxide to afford 25% neutralization. This allowed a good level of solubility in water. An asphaltic emulsion was prepared from water, toluene and liquid petroleum asphalt in ration of 1:1:1 (by weight) in the presence of 2.0% (by weight) of the emulsifying agent. This emulsion showed high level of stability. For comparison, the partial neutralization of ricinolic acid resulted to no emulsification properties.

Conclusion

1. The sulfation of castor oil with gaseous sulfur trioxide was studied by using carbon tetrachloride as solvent, therefore, the rate of the reaction was very fast and the sulfation process was carried out low temperature.
2. The properties of emulsified sulfated castor oil changes according to the depth of sulfation with the enhancement of the concentration of sulfur trioxide in emulsifier up to (8%w). this percentage lead to sulfation of one hydroxyl group in caster oil and the ability of the emulsifier increases, whereas if it is more than this percentage the ability of the emulsifier decrease and lead to formation of bisulfonic acids.
3. Stable tar emulsion was formed by using prepared emulsifier through molecular neutralization of sulfated ricinolic acid. This emulsifier contains free sulfuric acids and their salts.

Acknowledgement

My Thanks goes to the chemistry department, Science faculty, 7th October University.

References

1. F. J. Kremers, JAOCS, 48, No. 7, 314 (1971).
2. E. E. Gilbert, Sulfonation and Related Reactions, Interscience Publishers, P. 363(1965).
3. Japan Paten 7802, 419 (1978); Chem. Abs., 88:172294.f

4. Japan Patent 77. 156, 817 (1977); Chem. Abs., 88:152044.
5. B. L. Kapur, J. M. Solomon and B. R. Bluestein, JAOCS, 55, No. 6, 549 (1978).
6. B. M. Milwidsky and S. Holtzman, Soap, Cosmet., Chem. Spec., 48, No. 6, 446 (1972); Chem. Abs., 77:116345.
7. USSR Patent 977, 496 (1982); Chem. Abs., 98:74236.
8. K. J. Kedlaya, K. Vijayalakshmi, S. C. Sumathi and V. V. Rao, Leather Sci., 20, No. 6, 199(1973); Chem. Abs., 80:97359.
9. S. Alberto and V. Victor D; Jaco La Plata, No. 1, 63 (1973); Chem. Abs., 80:49296.
10. U. S. Patene 3, 951, 778 (1976); Chem. Abs., 85:49177.
11. P. N. Vdovichenko, V. A. Dagaev and A. I. Petlitskii; Neftepererab. Neftekhim (Kiev) 17, 74 (1979); Chem. Abs., 92:8603.
12. Czech. Patent 254, 462 (1988); Chem. Abs., 110:196190.
13. Rom. Patent 84, 989 (1985); Chem. Abs., 104:150747.

Session 7

INVESTIGATION OF FISH SPECIES BIODIVERSITY IN HARAZ RIVER

ISSA PIRI
Department of Agriculture, Payame Noor University (PNU), Zahedan, Iran

IMAN HOMAYOON NEZHAD
Department of Environment, Payame Noor University (PNU), Zahedan, Iran

PARIA AMIRIAN
Department of Environment, Payame Noor University (PNU), Zahedan, Iran

Fish diversity of Haraz River has studied from November 1999 to August 2000. This river is situated in The Southern Caspian basin of Iranian Mazandaran Province. Five Stations Were Selected along The River. The Stations Were Selected in The Fallowing Positions: One station in Upstream, 1450 meters above sea level. Two station at a gravel mining site 350 meters asl. Three station Above the Hezar Sangar dam at 220 meters asl. Four station Just below The Hezar Sangar dam, 190 meters asl. And five station at The Estuary with Elevation of – 26 above Sea Level (asl). Altogether, 522 specimens were Collected and studied. These belonged to 20 species and 9 families. The names of species and families are cited. Ciprinids comprised 67.2 percent of the species. Seventy percent of species were resident and 30 percent anadromous. Rain bow trout (oncorhynchus mykiss), Gold fish (carassius auratus), (Liza saliens) and Stickle back (Gasterosteus aculeatus) and Gambusia (Gambusia holbrooki) were introduced species.

1. Introduction

Iran is located in an area which is very much important with regard to animal geographical field. We know little about the fauna of Iranian fish, yet this information is interesting and is worth to pay attention to. Studies about fish communities in the rivers show that none-subsistence parameters such as temperature, the speed of water circulation, and the type of river bed have a vital role in plenty of kinds of fish. So sort enrichment, variety enrichment, and fish productions aren't the same in different parts of the river [1]. Therefore, for a better management and in accordance with fishery goals and objectives, also environmental purposes and factors of fish communities a river can be divided in to different segments. Usually each segment has been called based on the name of the majority of the kind of fish in that particular part of section. Consequently many rivers like those in Europe have been divided. However these things have not been done for Iranian rivers, and if something have been done that's very little. Not alone irregular and immethodical interference of human kind on rivers (construction of dams, agricultural taken from the river. Third and fourth stations were development, construction of centers near the rivers to increase fish reproduction, bringing none native sorts of fish to these areas, and taking sand and fine from the bank and river bed) caused a lot of damages to all kinds of native Iranian fish. Haraz River in mazandaran province in the north part of Iran is among of those protected rivers by Iran environment organization which contains lots of kinds of native fish. The main goals and aims of this research are to identify the kinds of fish, study the species of fish in different parts of the river, and also study of ecological niche overlapping, Also food selection in order to recognize the effects of rainbow trout on red mole trout and yet study of human interferences on the rivers (construction of dams, taking sand and fine from the bank and river bed and construction of centers near the rivers to increase fish reproduction).

2. Materials and methods

To start the job, at first five stations were selected. First station was assigned at the upper of the river (trout area), second station was assigned at the area which sand is appointed at the area which hezarsangar dam of Amol is located between them. Finally the fifth station was chosen at the mouth of the river. Catching the fish was done by salick net (10 meters in diameter) sampling was taken during the four seasons. At each station for each day of working, six hours of catching fish was considered. Catching the insects living in water was done by surber sampler with dimension of 30 by 30 centimeters. The fish and other insects caught by the mentioned style were fixed in formalin (10%) and transferred to the laboratory of Tarbiat Modarres University. Identification of the fish was based on the family origin, genus and species. The content of the stomach of trout fish and counting the number of insects living in the water, to find out the effects of rainbow trout on red mole trout, food selection comes as:

1) Species niche overlapping

We used the Pianka formula for Species niche overlapping [Putman, 1996]

$$\alpha_{j,k} = \frac{\sum P_{i,j} \cdot P_{i,k}}{[(\sum P_{i,j}^2)(\sum P_{i,k}^2)]^{\frac{1}{2}}}$$

Pi,j is the food material in the stomach of the first fish in percent and Pi,k is the food material In the stomach of the second fish in percent.

$$\text{Food selection} = \frac{\text{The food stuff in the stomach of the animal in percent}}{\text{Food stuff in environment in percent}}$$

If the found number is greater than one, it means the animal has selected that food. If it's equal or less than one, then it means, the animal hasn't selected that particular food or its indifferent forward that food.

3. Results

Biodiversity of fish: among those fish kippor fish presents 9 genus and 11 species is the most various. Salmon fish presents 2 genus and 2 species and other kinds of fish present one genus and species. From the existence species, 70% is river inhabitant and 30% is immigrant (from sea to river). (Table 1)

Table 1. Biodiversity of fish in Haraz River.

Species	) Exotic ( ) Native (	Immigrant from the sea to the river	Inhabitant in the river
(Salmonidae):			
Salmo trutta fario	N	-	+
Oncorhynchus mykiss	E	-	+
(Cyprinidae):			
Capoeta capoeta	N	-	+
Barbus mursa	N	-	+
Barbus lacerta	N	-	+
Alburnoides bipunctatus	N	-	+
Leuciscus cephalus	N	-	+
Rhodeus sericeus	N	-	+
Carassius auratus	E	-	+
Rutilus frissi kutum	N	+	-
Chalcalburnus chalcoides	N	+	-
Vimba vimba	N	+	-
Rutilus rutilus caspicus	N	+	-
(Gobiidae):			
Neogobius fluviatilis	N	-	+
(Gasterosteidae):			
Gasterosteus aculeatus	E	–	+
(Petromyzontidae):			
Caspiomyzon wagneri	N	+	-
(Mugilidae):			
Liza saliens	E	+	-
(Cobitidae):			
Cobitis taenia	N	-	+

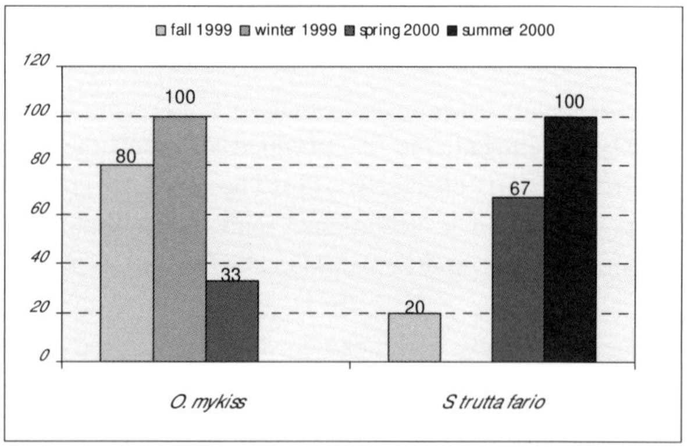

Fig. 1. The plenty of fish species at the first station of Haraz River in percent.

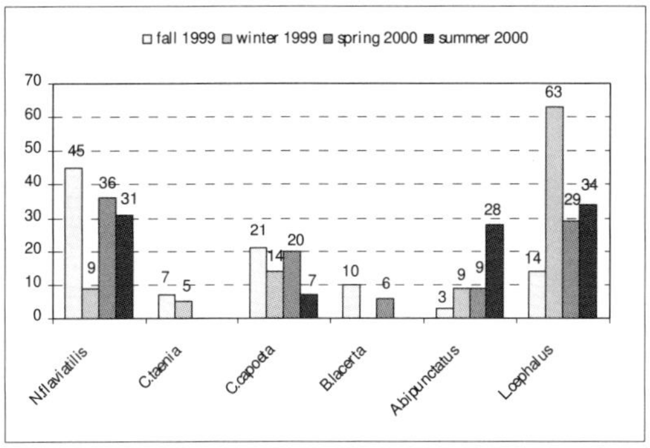

Fig. 2. The plenty of fish species at the second station of Haraz River in percent.

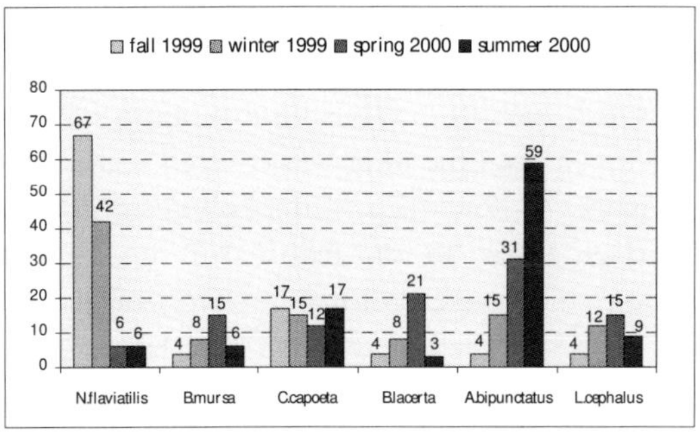

Fig. 3. The plenty of fish species at the third station of Haraz River in percent.

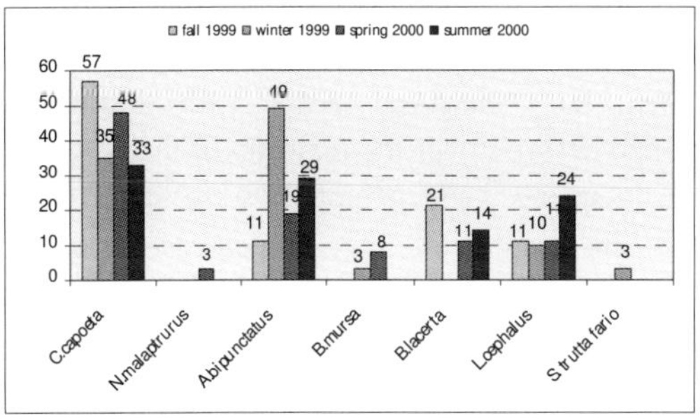

Fig. 4. The plenty of fish species at the fourth station of Haraz River in percent.

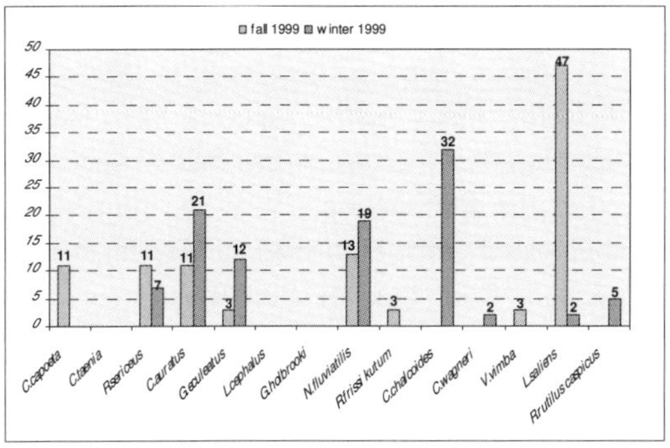

Fig. 5. The plenty of fish species at the fifth station of Haraz River in percent (fall & winter, 1999).

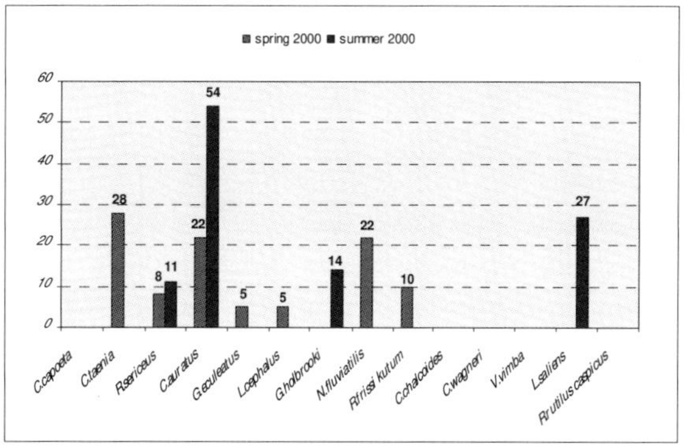

Fig. 6. The plenty of fish species at the fifth station of Haraz River in percent (spring & summer, 2000).

Species niche overlapping & food selection: to determine the effects of rainbow trout as a none native species over the red mole trout as a native species at the first station, consider fig. 6. Since in the two seasons of fall and spring two species of rainbow trout and red mole trout were observed together, the calculation was done just for these two seasons. (Table 2) In the summer only red mole trout and in the winter just rainbow trout were caught. So the rate of coexistence wasn't calculated. The food selection was calculated only in the summer. (Table 3) Therefore, considering Tables 4, 5 the rate of coexistence and food selection are as follow:

Table 2. Species niche overlapping of rainbow trout & red mole trout.

Fall of 1999	Spring of 2000	Season
0.88	0.7	Overlapping

Table 3. Food selection for red mole trout, summer of 2000.

Trichoptera	Ephemeroptera	Simulidae	Chironomidae	Gammaridae
1.9	1.09	0.58	0.78	0.16

As you see, the found numbers for Ephemeroptera and Trichoptera is greater than one. Therefore we may conclude that, they have selected this food stuff. But for Chironomidae, Simulidae and Gammaridae, since it's less than one, we could tell that the red mole trout a select or with regard to its one food stuff.

Table 4. The amount of stomach content of red mole trout at the first station in percent.

	Spring of 2000	Summer of 2000	Fall of 1999	Winter of 1999
Ephomeroptera	4.4	51.9	87	-
Trichoptera	53.9	44.3	11.5	-
Chironomidae	0.6	1.9	0.5	-
Simulidae	-	1.4	0.5	-
Gammaridae	0.6	0.5	0.5	-
Plecoptera	40.5	-	-	-

Table 5. The amount of stomach content of rainbow trout at the first station in percent.

	Spring of 2000	Summer of 2000	Fall of 1999	Winter of 1999
Ephemeroptera	1.1	-	45.9	17.4
Trichoptera	11.1	-	31.2	58.4
Gammaridae	-	-	5.8	5.6
Diptera	-	-	0.3	0.6
Simulidae	-	-	1.1	15.5
Chironomidae	1.1	-	7.2	1.9
Tipulidae	-	-	2.5	0.6
Coleoptera	-	-	0.6	-
Odonata	-	-	0.3	-
Nematod	-	-	3.1	-
Annelida	-	-	1.1	-
Gastropoda	-	-	0.3	-
Plecoptera	86.7	-	-	-

4. Discussion and conclusion

Considering the results, it's obvious that the river mouth (fifth station) where many immigrant fish entered for reproduction purposes has more variety compare to the other parts of the river.(fig 5 & 6) In fall season from 8 caught fish species, L.saliens with 47% had the most share and G.aculatus ، V.vimba ، R . frissi kutum with only 3% had the least share. Yet in the winter the highest share belonged to C.chalcoides with 32% and the lowest share belonged to C.wagneri and L.saliens with just 2%. The important point is that C.chalcoides wasn't seen during other seasons, so we can conclude that, this immigrant species has got the highest accumulation in this season. This species usually has been seen in the middle parts of the river whereas the river bed is covered with rocks and pretty small pieces of stones [2]. In this research, the catching of fish was done only at the mouth of the river where its bed is covered with sandy naterials. Also L.saliens wasn't observed in the spring either. However, it was seen again in the summer season (27%). Therefore it can be concluded that this kind of fish are present at the mouth of the river in the falls and summers. In the winter season 8 species were identified. In the spring season the majority belongs to C.taenia (27%) and the least belonged to L.cephalus and G.aculatus (5%). C.taenia in the seasons of fall and winter was also observed at the second station. (Fig 2) At the second station sand and small pieces of stones and mud are taken from the bed river as well as the bank river, and the river has been divided into some branches. In one branch of the river that its bed is covered with sand, mud and small pieces of stone, this type of fish has been observed, and other types of fish such as N.fluviatilis ، B.mursa ، B.lacerta and ... which need the river bed covered with rock and bigger pieces of stones [2]. Weren't observed in this part of the river. It can be said that if this trend or pattern continues, could have destructive effects on the native fish of the river. These native fish have had a vital role on the overall condition of the river and its animals. Of course at the second station because of the division of the river and immethodical withdraw of sand and mud, ecological conditions of the river have been changed too much. C.taenia usually lives in the lower parts of the river where the bed river is sandy and mudy [2]. C.taenia wasn't observed at the fifth station and as mentioned before it was the majority of the fish in spring season. In spring 7 species totally were identified in the summer the number of Species decreased rapidly and reached to 4 species. In this season C.auratus which is a none native type presents the majority (54%). Of course this kind of fish has been seen in the other catching seasons. The least belonged to G.holbrooki (8%). This sort of fish at first was brought to Iran in order to fight with malaria fly. At the present time a none-native species may be found in most lakes and rivers and water bodies of the country. This type is also a food competitor for many native species [2]. Gambuzia wasn't seen in other seasons. Altogether 14 species were identified at the fifth station. Third and fourth stations which are located in the both sides of Hezarsangar Amol deviation dam from type of Bed River, the speed of water circulation and ... are very much alike. (Fig 3 & 4) Five species C.capoeta ، A.bipunctatus ، B.lacerta ، B.mursa and L.cephalus were observed in both stations. But N.fluviatilis which in every four season was observed at the third station wasn't observed at the 4th station. Interesting to mention that fish cow in the two seasons of fall and winter was the top fish to be caught at the third station (47% & 67%). Now considering the fact that the ecological conditions of the two stations have been very similar and even the same species have been seen in the two stations, why the fish cow hasn't been seen at the 4th station? Is the deviation dam an obstacle for the fish cow so that it can not enter the area or maybe there are of her reasons that can find the answer to this puzzle? It seems as if more research must be done in this area. Another thing is that the existence of S.trutta fario at the 4th station. In the winter just 3% of this kind of fish was caught. Mean while the height of water in this area is 190 meters. In fact this is a plain station. Can it be concluded that the living conditions for this type of fish in its main living place (the upper parts of the river) have been inappropriate or somehow improper? As it was mentioned in the results part coexistence between rainbow trout and red mole trout at the first station is a lot. (Fig 1) This rate in the fall of 1999 was 88% and in spring of 2000 was lowered to 70%. On the other hand by considering the calculation done in summer of 2000, it was found out that the red mole trout is a selector and selects its own food. Also observation with regard to some characteristics of rain bow trout and red mole trout indicates that rain bow trout doesn't allow the red mole trout to get closer to its living place and makes a distance with it [3]. Yet we must pay attention to the fact that rain bow trout is not very sensitive against changes in water temperature and the present oxygen in the water but red mole trout in fact is very sensitive against these changes and also against food stuff and its changes [3]. So by realizing all of these facts, its possible that in the long run the workshops used for raising and reproducing of the rain bow trout, where the fish after escaping from them and entering the river, could have a destructive effect on the native species of the river (red mole trout) and consequently caused them to be destroyed. In the two seasons of fall and winter the most fish that was caught at the first station belonged to rain bow trout (80% & 100%). In winter and at the fourth station red mole trout was observed but at the first station where it's the original living place of this kind of fish, it wasn't seen. Also the entrance of rain bow trout to the Caspian sea via haraz river for Salmo trutta capius that according to Kad and Abdoli (1995) has been ranked EN(endangered) in the IUCN rankings will also cause a problem in the future [4]. At the third station six species were caught during the fall season and it happened again during other seasons, the caught species have almost the same conditions. Of course it has to be mentioned that during the seasons of fall and winter, the majority belonged to N.fluviatilis and during the seasons of spring and summer the majority belonged to A.bipunctatus. Interesting to consider that N.fluviatilis during the spring and A.bipunctatus during the fall have had the least numbers compare to the other types of fish? At the second station the majorityof fish that were caught belonged to N.fluviatilis and L.cephalus and the lowest number of fish that were caught belonged to C.taenia and B.lacerta. Considering the point that 20 species from overall 9 families during 4 seasons of the year by salick net were caught, one can conclude that haraz river is very rich for having different species of fish compare too many other rivers in the country. For instance during one year research when sampling was done each month by electro shocker machine and salick net, in the two rivers of Chalous and Sardabroud, totally 17 species from 5 families were identified [2]. Also another research which was done from Farvardin Thruoght Aban of 1998 on Helleh River in the southern part of iran, 14 species from overall 10 families were identified [5].

References

1. Abdoli, A., 1994, the Study of Population of Fish in Sardabroud and Chalous River in Mazandaran Province, Msc Thesis, Tehran University.
2. Abdoli, A., 1999, the Fish of Inside Water in Iran, Nature and Wild Life Museum.
3. Vosoughy, Gh., Mostajir, B., 1994, The Fresh Water Fishes, Tehran University.
4. Riyahy Bakhtiary, Y., Moghaddasi, D., 2001, Rain Bow Trout in Haraz River, Water Magazine, No 30.
5. Jamalpour, M., 1998, Biodiversity of Fish Species in Helleh River (An Ecological Approach), Msc Thesis, Tarbiat Modarres University.

INVESTIGATION OF HAZARDOUS WASTE, A CASE STUDY OF FARS PROVINCE IN IRAN

IMAN HOMAYOON NEZHAD

Department of Environment, Payame Noor University (PNU), Zahedan, Iran

PARIA AMIRIAN

Department of Environment, Payame Noor University (PNU), Zahedan, Iran

ISSA PIRI

Department of Agriculture, Payame Noor University (PNU), Zahedan, Iran

Hazardous wastes are applied to the wastes that due to having features such as toxicity, corrosivity, ignitability, reactivity or other similar characteristics jeopardize man's health and environment or somehow bring about this possibility. Hazardous wastes are produced from several industrial and commercial sources, agricultural or even domestic activities. The risks of this kind of materials are very different according to the type of hazardous wastes and environmental conditions. Different effects including acute, chronic, short-term and long-term are to be expected. Economic and health consideration initiated a study aiming to identify hazardous wastes in Fars province that has an important role in the developing industry of Iran. In this study the necessary data are gathered through questionnaire, interviewing the authorities of the units and referring to the available documents. The information encompasses the type and amount of waste, method of temporary storage, period of waste discharge, method of final disposal and the condition of recovery and reuse of waste. Also for storing and analyzing the data, "waste database" has been made. The results attained from analyzing the information indicate the production of 1049042 tones of industrial wastes, 13701 tones of hospital wastes and 135563 liters agricultural wastes in the current year (2006-2007), the waste production ratio of different industrial groups is obtained as 31% by chemical industries, 16% by oil and coke industries and 53% by other industries. Furthermore, 10915 tones of wastes (in conformance with appendix no 1 of Basel convention) categorized as hazardous wastes. In the context of the mentioned hazardous wastes, y11, y35, y34, y12 and y13 (Basel codes) has reported as the major part of this categorization. All of aspects of waste management especially hazardous wastes analyzed and their results demonstrated.

1. Introduction

The Hazards of hazardous wastes for human and the environment is increasing due to increase of production rate, and different types of chemical reactions. The sources of hazardous wastes are different (industrial, household, commercial, hospital, and etc) [1]. However the major sources are industrial, agricultural and hospital activities [1]. Fars is one of the provinces with appropriate conditions and status which enables it to handle a lot of industrial and other kinds of activities related to hazardous wastes and are supposed to handle more in the near future. Due to population increase, urban development and industrials development in this province, especially in recent decades, the hazardous wastes become a real challenge for the province authorities. It is most urgent to protect the environment and to avoid the unpleasant happenings (which have already occurred in other countries) by controlling the hazardous wastes. It is worth to mention that at Fars province, there have been some events with regard to the hazardous wastes which mostly have been underestimated because of the conservative policy of the responsible people at the province. As an example in 1978 the hazardous wastes of a canned food factory caused the death of hundreds of thousand of fish and other aquatic animals. As it was mentioned, the management of hazardous wastes even in developed countries such as the U.S, the U.K and Japan doesn't have a long history and its not more than 4 decades old [2]. What other countries can learn from the researches and the experiences of these countries (U.S, the U.K and Japan) is that, the control and prevention cost of hazardous wastes is between 10 to 100 times less than the money that must be paid to clean up the environment because of a case of hazardous wastes [3]. So common sense tells us, that the past and present experiences should be used in order to save the environment as well as to develop and expand industrial activities at the same time. By realizing this fact, the first step to make a correct management system in the area of hazardous wastes is to identify and then classify the hazardous wastes [4] & [5]. In each management system, the statistics and input information play an important role which can be used most in the future planning [6]. So by considering the fact that, there is no proper system in gathering, storing, management and processing the information in this field and the lack of an appropriate action relating to the hazardous wastes in Fars province, the importance of this research is evident more than ever. We hope that, the results of this research would be helpful to the experts and the people with key responsibilities in Fars environment protection agency and other responsible organizations in the field of hazardous wastes.

2. Materials and Methods

During this research from 2850 industrial units in Fars province which are active in 22 industrial groups, 800 industrial units with possible hazardous wastes residue have been selected as samples for future evaluation and research. To select and screen these samples many factors were considered. In fact those units were selected that had productivity or mining nature and repeatedly in articles, reference books and international and accredited organizations such as U.S environmental protection agency (EPA) and RCRA (Resource Conservation & Recovery Act) and Basel convention have been emphasized [7], [8] & [9]. After selecting the samples by using proper questionnaire, educational meetings, and interview with critics and top managers at the industrial units (licensed by the department of environmental protection agency of Fars province) the necessary information was collected. After that, in order to save and process the information, software called Microsoft ACCESS XP was used and industrial wastes database was formed. To monitor the hospitals and clinical centers, with help of the department of hygiene of Shiraz Medical Science University, the departments of social security and other centers appropriate questionnaires were sent to all clinical centers. Right after collecting the information, to save and better process of information, hospital wastes database was formed. For agricultural specific wastes, all the necessary information from organizations and departments such as the department of plants protection and storage management of Fars, Fars supportive services company and the agriculture department of Fars were gathered and processed [7].

3. Results

The results of analyzing this information including the uncontrolled industrial sewage indicate that 1049042 tons of wastes are produced each year. About 10915 tons of coded wastes (expressed in enclose 1 Basel convention) under the title of hazardous wastes is classified, the production of this amount of wastes comes from 17 industrial groups among 22 industrial groups in Fars province. From these 17 industrial groups, oil and coke industries (4694 tons in each year), new metal productions (3115 tons in a year), chemical industries(790 tons in a year), radio, TV and mass media communications (824 tons in a year) and textile (612 tons in each year) have got the most share in producing hazardous wastes in Fars province each year respectively and the industrial groups of office machines and accounting (0.1 tons in a year), clothing, fabric and processing fur coat (0.13 tons in a year) have got the lowest share, respectively (in producing the hazardous wastes) (Table 1).

Table 1. Producing of hazardous wastes in each industrial groups (Fars Province, 2007).

Industrial Group	ISIC code	Hazardous waste (ton/year)	percent
Food & drinks products	15	97.27	1%
Textile	17	611.31	6%
Clothes & process of fur	18	0.13	0%
Tannery, leather, bag, shoes & suitcase	19	299.32	3%
Paper & paper products	21	18.85	0%
Publication, print, copy	22	21.33	0%
Products of oil	23	4693.75	43%
Material & chemical products	24	789.50	7%
Plastic & rubber products	25	24.00	0%
Mineral none metal products	26	27.78	0%
Basic metal production	27	248.54	2%
New metal products	28	3115.16	29%
Machinery & equipment	29	83.67	1%
Office machines & accounting	30	0.08	0%
Electric machines & appliances	31	59.64	1%
Radio, TV & mass communication	32	824.13	8%
Other means of transportation	35	0.36	0%
Total		10915.00	100%

Also 19 code series from Basel codes have been registered in these wastes. Codes y11 (4621 tons a year), y35 (2069 tons a year), y34 (1283 tons a year), y12 (864 tons a year), y13 (653 tons a year) have got the highest share respectively and codes y17 (0.1 tons a year), y33 (2 tons a year) have got the lowest share. (Table 2)

From all physical conditions observed among wastes in Fars province, physical conditions of liquids (5593 tons per year) and semi-solid (564 tons per year) have got the highest share between hazardous wastes respectively in Fars province. The physical condition of sludge (137 tons per year) has got the lowest share. (Figure 1)

Among the characteristics in Fars province hazardous wastes, the characteristics of inflammability and toxicity together with Basel characteristic code (H3 & H12) at a rate of 54% and characteristic of corrosivity with Basel characteristic code (H8) at a rate of 31% have got the highest share. (Figure 2)

How to discharge hazardous wastes, it has to be said that about 2773 tons (25%) of these wastes are discharged in the nature with no control at all. 7% is buried and only 1% is burned. 7381 tons (67%) are classified under other methods that 33% of this is discharged in sewage system network, 21% is filtered or refined in factories. 10% is buried in abandoned mines, 10% is sold and 1% is reused in the respected unit. (Figure 3 and Table 3)

About the responsibility of discharging the wastes, nearly 7897 tons (73%) is discharged by the respected unit and 2974 tons (27%) is discharged by the private sector. With regard to recycle and salvage of these wastes, we must mention only 813.5 tons (7%) is recycled and reused and for the remaining 10101 tons (93%) no recycling or salvaging is done. According to the recent assessment with regard to hazardous wastes, total weight of hospital wastes is nearly 13701 tons a year which 6728 tons a year (49%) of this figure belongs to ordinary wastes (semi-household), and 6973 tons a year (51%) belongs to infectious wastes. (It has been noted that the infectious wastes has received the code y_1 and characteristic code H6.2 in Basel classification. Clinical centers and semi-hospital sections produce almost 93 tons of constant and advent materials in Fars province yearly. The physical state of this material is liquid with corrosive characteristic which has received Basel especial code H_8 and Basel cod Y_1. Based on the list of expired poisons in Fars province obtained from the department of agriculture of Fars, the entire expired poisons at this province is equal to 22405 liters, Neiriz town has got the most (2618 liters) and pasargad town has got the least (50 liters) of the above amount. Also according to the list of expired poisons at the storage of protection management of plants in Fars, the total weight of expired poisons has been announced as 16657 liters. Tetradifon poison has got the most of it (5000 liter), and fitrotion poison has got the minimum (4500 liters). The list of expired agricultural poisons at Fars province supportive service company indicates 965000.4 liters. The most belonged to micronized sulphur (20168) liters and Asian sulphur (2000 liters). Overall the total amount of expired agricultural poisons in Fars province is equal to about 135563 liters. According to the department of protection of plants in Fars, in this province nothing is done to deal with expired poisons, they just keep them in the storage and eventually are sent to the ghazvin plant (Shimi Keshavarz) and be destroyed by this company.

Table 2. Classification of hazardous wastes based on Basel classification (Fars Province, 2007).

Basel code	Code definition	Hazardous wastes (ton/year)	percentage
Y_{11}	Waste tar coming from refining, filtration & distillation	4621.25	42%
Y_{12}	Wastes related to production of formulation, application of ink, colors, pigments alcoholic, lacquer, coloring materials, waxes, oil	863.99	8%
Y_{13}	Wastes related to production of formulation, application of resins, sedate, latex, glues & softness	653.29	6%
Y_{17}	Wastes coming from the metal & plastic covering	0.10	0%
Y_{19}	Metal carbons (reaction of metals with co)	21.16	0%
Y_{21}	Chrome relations (6 capacity)	65.73	1%
Y_{21},Y_{22},Y_{23},Y_{31}	Chrome reactions, copper reactions, zinc reactions, lead & its reactions	10.00	0%
Y_{22}	copper reactions	34.64	0%
Y_{22},Y_{23}	copper reactions & zinc reactions	20.01	0%
Y_{23}	zinc reactions	390.50	4%
Y_{29}	Mercury & its reactions	152.26	1%
Y_{33}	Minerals cyanides	1.80	0%
Y_{34}	Acid liquids or solid acids	1282.68	12%
Y_{34},Y_{21}	Acid liquids or solid acids, chrome reactions (6 capacity)	1.29	0%
Y_{34},Y_{35}	Acid liquids or solid acids & alkaline liquids or alkalis in the state of solid	50.00	0%
Y_{35}	alkaline liquids or alkalis in the state of solid	2069.16	19%
Y_{36}	Powder or fiber of asbestos	5.00	0%
Y_7	Waste materials coming from refining & plating of metals (including cyanides)	200.75	2%
Y_8	Wastes from mineral oils which aren't suitable for usual usage	471.24	4%
Total		10915.00	100%

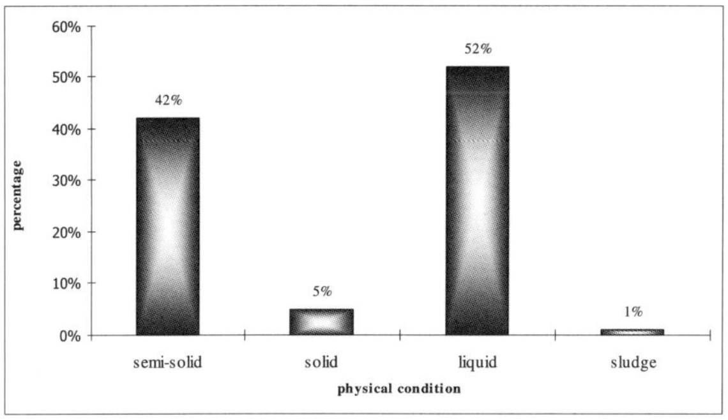

Figure 1. The Share of Physical conditions for hazardous wastes (Fars Province, 2007).

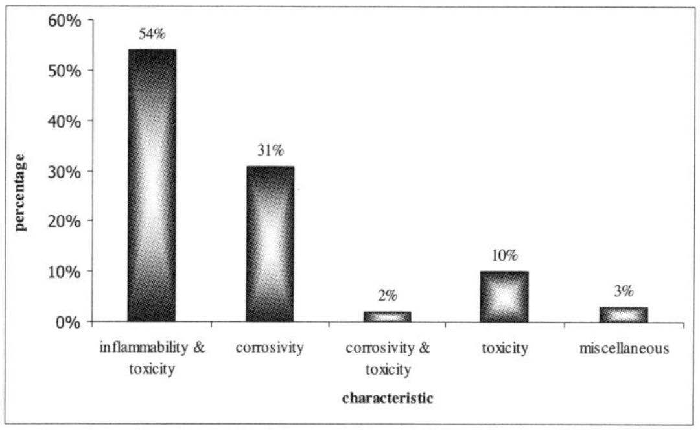

Figure 2. The share of different characteristics of hazardous wastes (Fars Province, 2007).

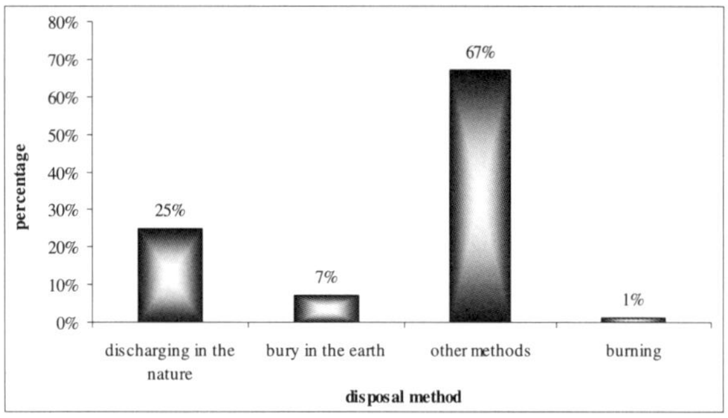

Figure 3. The share of final disposal methods for hazardous wastes (Fars Province, 2007).

Table 3. Other methods of disposal for hazardous wastes (Fars Province, 2007).

other methods of disposal	(ton/year)	percent
Probably will be thrown away	741.96	10%
Is sold	761.70	10%
Discharging in the well	88.96	1%
Discharging in the city sewage system	1.91	0%
Discharging in the canals around the factory	2380.52	33%
Filtering in the factory	50.96	1%
Discharging in the factory sewage system network	1554.76	21%
It being reused in the some unit	51.82	1%
To be buried in abandoned mines	734.19	10%
Un known	951.20	13%
Total	7318.03	100%

4. Discussion and Conclusion

During this research around 10915 tones of encoded wastes (registered in enclose no 1 Basel convention) produced by sampled units were checked. These hazardous wastes belong to 17 industrial groups out of 22 industrial groups in Fars province. Among the 17 industrial groups in the province, industrial groups of coal and oil industries (43%), new metal products (29%) have the highest share in production of hazardous wastes. So these industries are more important than other industries in the field of hazardous wastes, the mentioned industries should be considered as the first priorities. The wastes coming from these industries and factories include all kinds of acids, used soda, ooze and sludge containing heavy metals, oily wastes, acid tars, cyanide and other hazardous and poisonous wastes. These wastes must be discharged under a proper management and minimizing the risk of this danger for human and the environment. Also 19 code series among Basel codes have been registered for these wastes. Codes y11 (42%), y35 (19%) and y34 (12%) have got the highest shares. Therefore for choosing a certain management method, these wastes must be realized as the first priorities. From all physical conditions among the wastes in Fars province, liquid physical condition (52%) and semi-solid (42%) have the first and second rank. So any decisions with regard to selecting the methods, collecting technologies, transportation, filtration, refining , final discharging, recycling and salvaging must be considered and matched with regard to these physical conditions. Among the present characteristics in hazardous wastes in the province, inflammability and toxicity with characteristic Basel cod (H3-H12) together at a rate of 54% and the characteristic of corrosivity with characteristic Basel code (H8) at a rate of 31% have got the highest shares. Paying attention to the fact that the above characteristics are much more important than the others, at the stage of managing the hazardous wastes and final disposal a great care must be taken to neutralize the methods of inflammability, toxicity and corrosivity. With regard to the method or style of discharging the hazardous wastes we have to confess that not only there is no certain or particular organization that would handle the responsibility, but also, no scientific or logical methods is used by the units which are the major producers of these dangerous wastes. Therefore it is essential that some managing actions be done very soon. Only 7% of these wastes are recycled and reused but for the remaining 93% the story is different and unfortunately nothing is done about it. In general it is for sure that, the share of recycling and salvage with regard to the hazardous wastes is very little or basically nothing. How ever, it's obvious that recycling and salvage is one of the key and main solutions against the problem of hazardous wastes. It's worthy and deserving to accelerate and encourage all types of investigation, inquiry and research to decrease and deduct and reduce danger of hazardous wastes in the environment. All in all it is necessary to adopt methods, collecting technology, transportation, distillation, final disposal, recycling and ... infectious wastes, constant and advent materials at clinical units and expired agricultural poisons in the province, appropriately and properly. Generally speaking the way that is dealt with hazardous wastes and their final discharge is not satisfactory at all. These wastes based on the presented list by the Basel convention are with no doubt dangerous and require special and fast attention.

References

1. Lilja, R., Liukkonen, S., Industrial hazardous wastes in Finland – trends related to the waste prevention goal, Journal of Cleaner Production, Volume 16, Issue 3, Pages 343-349.(2008).
2. Musee, N., Lorenzen, L., Aldrich, C., New methodology for hazardous waste classification using fuzzy set theory: Part I. Knowledge acquisition, Journal of Hazardous Materials, In Press, Corrected Proof, Available online 13 November 2007.
3. Silva, m.a.r., Mater, l., Small hazardous waste generators in developing countries: use of stabilization/solidification process as an economic tool for metal wastewater treatment and appropriate sludge disposal, Journal of Hazardous Materials, Volume 147, Issue 3, 25 August 2007, Pages 986-990.
4. LaGrega, d., Buckinham,p.l., Evans, j.c., Hazardous Waste Management, McGraw-Hill, New York.(1999).
5. Mackay, d., McCarty, l.s., MacLeod, m., On the validity of classifying chemicals for persistence, bioaccumulation, toxicity, and potential for long-range transport, Environ. Toxicol. Chem. 20 (2001), pp. 1491–1498.(2001).
6. Mittal, a., Gupta, v.k, "et al", Process development for the batch and bulk removal and recovery of a hazardous, water-soluble azo dye (Metanil Yellow) by adsorption over waste materials (Bottom Ash and De-Oiled Soya), Journal of Hazardous Materials, In Press, Corrected Proof, Available online 20 June 2007.
7. USEPA 2006. Inventory update rule, Office of Prevention, Pesticides, and Toxic Substances; http://www.epa.gov/oppt/iur/ (accessed on 19.09.2006).
8. Wilke, B., Riepert, F., Koch, C., Kühne, T., Ecotoxicological characterization of hazardous wastes, In Press, Corrected Proof, Available online 9 November 2007.
9. Mar Babin, m., Cañas, i., Tarazona, j.v., An in vitro approach for the ecotoxicity testing of toxic and hazardous wastes, Toxicology Letters, Volume 172, Supplement 1, 7 October 2007, Pages S157-S158.(2007).

NANO-TUNGSTEN CARBIDE PREPARED FROM PALM KERNEL SHELL FOR CATALYTIC DECOMPOSITION OF HYDRAZINE

ABDUL RAHIM YACOB*

Chemistry Department, Science Faculty,
Universiti Teknologi Malaysia, 81310, UTM Skudai, Johor, Malaysia

SITI ZUBAIDAH HANAPI, VICINIVARRI INDERAN and RATNA SARI DEWI DASRIL

Physical Laboratory, Chemistry Department, Science Faculty,
Universiti Teknologi Malaysia, 81310, UTM Skudai, Johor, Malaysia

Tungsten carbide (WC) is known to have similar activity as the noble metals in catalytic reaction. WC has attracted interest for research in heterogeneous catalysis. In this study, nano tungsten carbide is prepared by mechanical alloying using activated carbon prepared from palm kernel shell. High surface area activated carbon was prepared by chemical and physical activation for comparison. The activated carbon and the WC prepared catalyst were chemical characterized using FTIR, XRD, SEM and BET single point respectively. Finally, the mechanical alloyed nano WC, prepared using chemical activated carbon was chosen for the catalytic decomposition reaction of hydrazine. The prepared nano WC indicates positive catalytic decomposition of hydrazine.

1. Introduction

The incentive to find substitute materials to replace platinum group metals are growing rapidly with the number and amplitude of industrial processes require their use. They are widely used in several reactions of industrial importance such as hydrotreatments, hidrocracking of heavy fractions from oil, dehydration of alcohols, metathesis, and isomerization of olefins [1-2].

Transition metal carbides, especially tungsten carbides, received a great interest since their catalytic properties, similar to Pt-like noble metals [1-3]. These metal carbides possess significantly modified physical and chemical properties compared to the parent metals due to the incorporation of carbon into the metal lattice [4-6].

Previously a number of studies carried out about tungsten catalysts supported on alumina by different techniques [5-10]. But all the results show that the surface molecular structure of supported tungsten catalyst is quite complex and depend on the metal oxide content and the presence of humidity [6].

In the other hand, there are very few studies of tungsten catalysts supported on carbon materials despite their importance as supports for catalysts [1, 4].

For this purpose, in this study tungsten catalyst with different metal content were prepared by using tungsten hexacarbonyl, which were deposited on activated carbon prepared in laboratory from palm kernel shell (PKS).

To generate economic utilization for this cheap and abundant solid waste, palm kernel shell is proposed as a prospective starting material for the preparation of activated carbon because of its relatively high fixed-carbon content (about 18 wt%), low ash content (less than 1.0 wt%) and the presence of *A*, pre-exponential or frequency factor (s^{-1}) of inherent porous structures [7]. In fact, a large number of naturally occurring carbonaceous solid wastes, such as coconut shell, almond shell and grape seed, walnut shell, sawdust and olive stones have been successfully converted into activated carbon [7-9].

Several methods to prepare WC are proposed in the literature, [10] and the most used involve reaction of the solid precursor, either as metal or oxide, with gas

*This work is supported by Universiti Teknologi Malaysia and the Ministry of Science, Technology and Innovation Malaysia for the financial support through FRGS Vote 78088.

phase containing the reducing or carburizing agents. In general this methods yield WC samples with a small surface area and polymeric carbon deposited on their surface.

Hydrazine is a liquid fuel used as thrusters in rockets and propellants. The reactions involved catalytic decomposition of hydrazine to nitrogen and ammonia. The catalytic decomposition of hydrazine used commercial catalyst of Ir/Al_2O_3, which is rare and expensive. Rodrigues *et al* (3), has demonstrated the ability of molybdenum carbide (Mo_2C) to show similarity reaction for hydrazine. Thus, with WC is known to have similar activity as the noble metals and have similar chemical and physical properties as Mo_2C, it is suggested that WC be used for hydrazine decomposition reaction.

$$N_2H_4\ N_2 + 2\ H_4 \longrightarrow \quad (1)$$

$$3\ N_2H_4\ N_2 + 4\ NH_3 \longrightarrow \quad (2)$$

The main objective of this study is to prepare and characterize high surface area carbon prepared from waste palm kernel shells. The best-prepared carbon will be selected for mechanical alloying to form tungsten carbide for catalytic decomposition reaction of hydrazine.

2. Materials and Methods

2.1. *Chemicals*

Tungsten hexacarbonyl powder (99%) and commercial activated carbon (99.9%) with molecular weight of 183.85 and 12.01 gmol^{-1} respectively were purchased from Scharlau Chemise S.A. Zinc chloride ($ZnCl_2$) and sodium carbonate (Na_2CO_3) from MERCK was use along with concentrated hydrochloric acid (HCl), 37% w/w purchased from J.T. Baker. Nitrogen gas and Helium used was supplied by Malaysian Oxygen Bhd. The palm kernel shells were obtained from Kulai palm oil mill.

2.2. *Preparation of Activated Carbon*

For preparation of carbon, 100 g of $ZnCl_2$ used as a chemical activating agent was added to 100 g of palm kernel shells in a 1L beaker. 100 mL of concentrated HCl (37% w/w) and 10 mL deionized water were added to the beaker. The mixture was stirred and heated at 100-110°C until it is homogeneous. The treated char was kept in oven overnight for dehydration purpose. The sample was then heated at 800°C for four hours. The activated sample was neutralized with diluted Na_2CO_3 solution and washed with deionized water. The washing process was repeated until the pH approximately 7. Sample was then dehydrated in oven at 100°C. The sample obtained from this activation was labeled CZ-AC. For comparison, chemical activation using potassium carbonate K_2CO_3, physical activated in vacuum of 10^{-3} mbar and H_3PO_4 was also prepared and labeled as CK-AC, P-AC and HP-AC respectively. Activation by H_3PO_4 was slightly different compared to others especially in activation part. After impregnation of PKS in 50% H_3PO_4 for 24 hours, the mixture undergone dehydration process in furnace at 120ºC for 1 hour. The temperature was increased to 500ºC and hold at that temperature for another 2 hours under flow of inert gas. After completed, the char was rinsed with hot distilled water until the pH reach approximately 7 and dried in oven at 100ºC.

2.3. *Preparation of Tungsten Catalyst*

Tungsten hexacarbonyl powder was mixed with the appropriate amount of activated carbon in a stainless steel container, which sealed under argon atmosphere. This closed container mechanically rotated for about 2 hours at a rotation speed 400 rpm at a temperature close to 70°C to force the carbonyl to sublime evenly onto the support. After cooling to the room temperature, the supported catalysts were sieved in order to remove the finest particles produced during the preparation. Two catalysts containing 6.1% and 15.1% W were prepared by following this procedure. Latter the catalysts were heated either in Helium flow or under high vacuum at 950°C for 4 hours.

2.4. *Characterization of Samples*

Fourier Transform Infrared (FTIR) analysis was carried out using Shimadzu 8300 spectrometer with wave number ranging from 400 to 4000cm^{-1}. Single point BET surface area was measured at temperature of 77 K using Micromeritics Pulse Chemisorb 2705. XRD analysis was carried out using Bruker X-ray powder diffractometer with a CuKa radiation over a 2-theta range from 20-90 degrees at a step width of 0.05 and step time of 1 sec. The identification of peaks is based on JCPDS-ICDD database (PDF-2 files) incorporated in the software.

3. Results and Discussion

3.1. *Fourier Transform Infrared Spectroscopy*

Figure 1 shows the FTIR spectra of commercial activated carbon, raw palm kernel shell, $ZnCl_2$ activated carbon CZ-AC, K_2CO_3 activated carbon CK-AC, H_3PO_4 activated carbon HP-AC and physical activated carbon P-AC respectively. The commercial activated carbon FTIR spectrum was used as a reference for comparison with other synthesized activated carbon. By interpreting the infrared transmittance spectrums, the functional group in the sample can be determined.

Based on Figure 1, only raw PKS shows the most complicated and apparent spectrum. A strong and broad adsorption peak appeared at 3415.7 cm^{-1}, corresponding to the stretching of O–H functional group and this indicates the presence of bonded hydroxide in the raw sample. There was another peak observed at 2920.0 cm^{-1} corresponding to the C – H sp^3 stretching. A strong C=O peak could be observed at 1735 cm^{-1}. This sample also shows two important absorption peaks at 1247.8 and 1037.6 cm^{-1} which represents the stretching of C – O functional group.

Unlike the FTIR spectrum showed by raw PKS, other spectrums illustrate less absorption peaks. Although the samples were prepared via various activation methods, there was a similarity in the adsorption patterns. Basically all the samples showed a weak broad peak around 3350 – 3500 cm^{-1}, which indicates the presence of hydroxide in the samples. It is suggested that the samples traps moisture on its surface during sample handling due to their adsorptive nature.

From the samples spectra obtained, clearly, most of the absorption peaks of functional groups were diminished. During the carbonization and activation process, the functional groups as observed from the raw PKS spectrum were evaporated as volatile materials when heat was applied to the samples. The trends of samples spectra almost similar with the commercial activated carbon also proved that the activation process has taken place successfully.

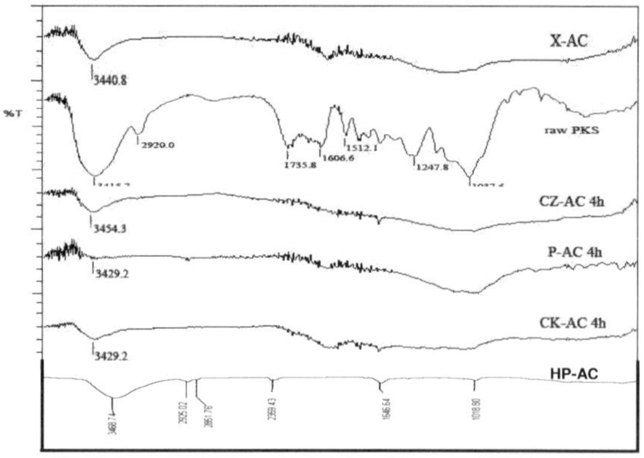

Figure 1. FTIR spectra for (a) commercial activated carbon (b) raw palm kernel shell, (c) CZ-AC, (d) P-AC, (e) CK-AC and (d) HP-AC.

3.2. *Single Point BET Surface Area*

From Figure 2, raw palm kernel shell gives only 0.8673 m^2g^{-1} of BET surface area, while commercial activated carbon has surface area of 1068 m^2g^{-1}. From the graph, chemical activation using $ZnCl_2$ gave comparatively high surface area of 1100 m^2g^{-1} compared to chemical activation using K_2CO_3, H_3PO_4 or physical activation.

This result suggests that the chemical activating agent used, $ZnCl_2$, has contributed to the higher surface area. The used of $ZnCl_2$ as the activating agent most probably resulted in more new pores being created and the existing pores widened to give activated carbon with higher BET surface area [20]. This is further supported by SEM micrograph of chemical activated (a) CZ-AC, (b) CK-AC and (c) HP-AC respectively (Fig. 3).

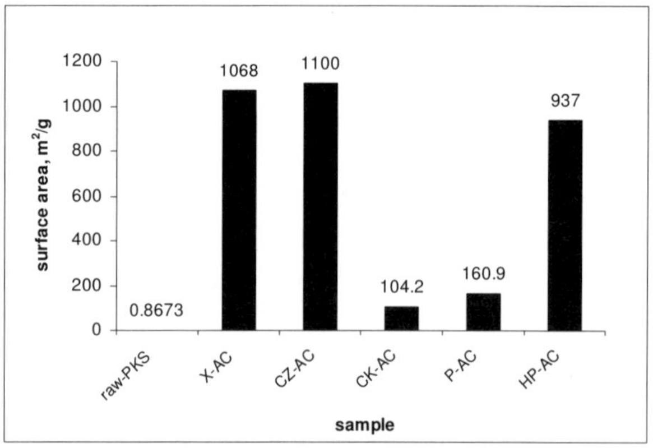

Figure 2. Single Point BET surface area of raw palm kernel shell, commercial activated carbon, CZ-AC , CK-AC, P-AC and HP-AC.

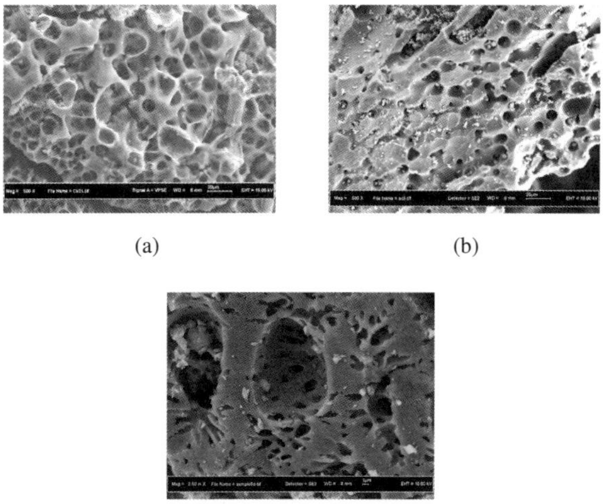

(a) (b)

(c)

Figure 3. SEM micrograph of (a) CZ-AC, (b) CK-AC and (c) HP-AC.

3.3. *XRD Analysis*

The chemical state and dispersion state of the metallic phase of the prepared catalysts were studied by XRD. Figure 4 (a) and (b) present the XRD diffraction patterns of the prepared mechanical alloyed WC using chemical activated carbon CZ-AC with equi-atomic ratio of 0.01. For comparison XRD for pure tungsten was recorded and as shown in figure 4 (a). While, the main diffraction peaks of mechanical alloyed WC are 31.596, 35.617 and 48.252 respectively figure 4 (b). By applying Scherer's Equation for the prepared WC, it was calculated that the particle size was ca. 0.25 nm. This further support that the mechanical alloyed WC was of nano size [10].

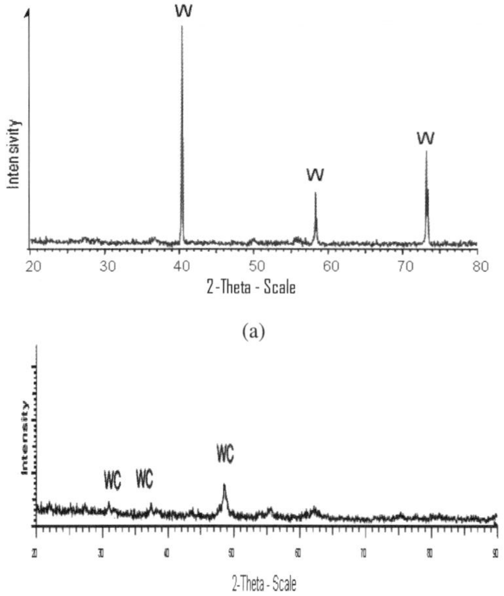

(a)

(b)

Figure 4. XRD diffraction patterns of prepared WC catalysts after mechanical alloyed.

3.4. *Catalytic Decomposition of Hydrazine*

In this study, the catalytic decomposition of hydrazine was carried out in a micro reactor whereby the hydrazine was passed over the tungsten carbide catalyst at room temperature and the composition of the products were analyzed using gas chromatography with time recorded in minutes.

Figure 5 is the product composition detected by GC, which comprises of nitrogen and ammonia only. This reaction further support the two reactions product discussed earlier. It was also observed that the prepared WC could be used for a long period of time.

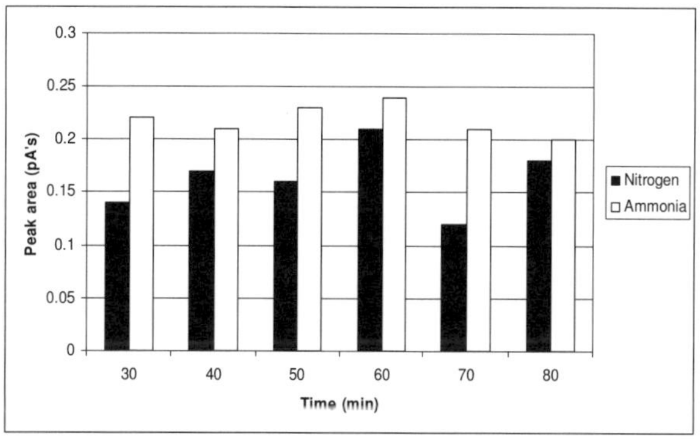

Figure 5. The product composition of nitrogen and ammonia detected using GC.

4. Conclusion

In this study, chemical activated high surface area carbon; using $ZnCl_2$ is successful prepared from waste palm kernel shells. The best carbon prepared was use for mechanical alloying to form tungsten carbide. The prepared WC further indicates positive catalytic decomposition of hydrazine. This study further supports the need for an alternative and cheaper catalytic material for the decomposition reaction.

References

1. Yacob A.R. Hanapi S.Z. inderan V, "Nano tungsten carbide supported on carbon from palm kernel shell in remediation of chlorofluorocarbon" Journal IEEE Explore 2.6, 556-563, 2009.

2. Delannoy L., Giraudon J.M. Granger P. Leclercq L. and Leclercq G. "Group VI transition metal carbides as alternative in the hydrochlorination of chlorofluorocarbon" Catalysis today 59, 231-240, 2000.
3. Rodrigues, J.A. Cruz, G.M. Buglia, G. Boudart, M. and Djega Mariadassaou "Nitride and carbide of molybdenum and tungsten as substitutes of iridium for the catalysis used for space communication", Catalysis Letter, 45, 1-3, 1997.
4. Alvarez-Merino M. A., Arrasco-Marin F., Fierro J. L. G, and Moreno-Castilla C., "Tungsten catalyst supported on activated carbon 1 Preparation and Characterization after their heat treatment in inert atmosphere" Journal of Catalysis 192, 363-373, 2000.
5. Levy, R.B., and Boudart, M., "Platinum like of tungsten carbide in surface catalysis" Science 181, 547-549, 1973.
6. Perez-Cadenas A.F., Moreno-Castilla C., Maldonado-Hodar F.J., and Fierro J.L.G., "Tungsten oxide catalyst supported on activated carbon: Effect of tungsten precursor and treatment on dispersion, distribution, and surface acidity of catalyst", Journal of Catalysis 217, 30-37, 2001.
7. Yacob A.R., Majid Z.A., Dasril R. S. D. and Inderan V., "Comparison of Various Sources of High Surface Area Carbon prepared by Different Types of Activation" The Malaysian Journal of Analytical Sciences, Vol.12 No.1, 2008.
8. Aik Chong Lua and Jia Guo, "Preparation and characterization of chars from oil palm waste" Carbon Vol. 36; No. 11, 1663-1670, 1998.
9. Alvim Ferraz M.C.M, and Gaspar, "Micropore size distribution of activated carbon impregnated after carbonization", J. of Porous Materials, 10, 47-55, 2003.
10. Kim J.C and Kim B.K. "Synthesis of nano sized tungsten carbide powder by chemical vapor condensation process" Scripta Materialia, 50, 969-972, 2004.

METHANE REFORMING IN A SMALL CHANNEL REACTOR

MOHAMED AL NAKOUA
Chemical and Petroleum Engineering Department, United Arab Emirates University, Al Ain, Abu Dhabi, U.A.E.

MUFTAH EL-NAAS
Chemical and Petroleum Engineering Department, United Arab Emirates University, Al Ain, Abu Dhabi, U.A.E.

In this study, the performance of methane reforming reactions in a small channel reactor has been investigated. Carbon deposition is rapid in the cases of partial oxidation of methane (POM) and CO_2 reforming separately. Our results showed that when POM was carried out simultaneously with steam and/or CO_2 reforming reactions in small channels coated by thin film catalyst, carbon deposition on the catalyst was drastically reduced. The results revealed that methane-air and steam and/or CO_2 mixtures can efficiently be converted into synthesis gas 'syngas' at temperatures higher than 800°C. It has been also shown that the stability and coking resistance of Ni/Al_2O_3 catalyst are increased by addition of Ba, Cr and La_2O_3.

1. Introduction

The channel reactor concept offers major benefits compared with conventional packed bed, slurry, or tubular reactors. The first is a major reduction in size – process intensification. Employing the catalyst as a thin layer (<100 μm) coated on the channel surface reduces mass and heat transfer restrictions compared with pellet catalysts and can improve the effectiveness factor. It is reported that the "functioning layer" of catalyst pellets in a conventional reforming tube is only about 50 μm thick [1]. The vital advantage of channel reactor concept would be that thin catalyst films would have much less resistance to transport processes than traditional catalyst pellets. Hence catalyst effectiveness factors would be near unity.

There have been several recent studies of the application of channel reactors; steam reforming of methane to produce hydrogen as the endothermic reaction:

$$CH_4 + H_2O \leftrightarrow CO + 3H_2 \quad \Delta H^0_{298k} = +206.1 \text{ kJ/mol} \quad (1)$$

$$CO + H_2O \leftrightarrow CO_2 + H_2 \quad \Delta H^0_{298\,k} = -41.0 \text{ kJ/mol} \quad (2)$$

The steam reforming of methane, Reaction (1) produces a synthesis gas 'syngas' having a H_2/CO ratio of 3 or greater if accompanied by the Water Gas Shift Reaction (2). Such a gas composition is useful for processes requiring a H_2-rich feed such as NH_3 synthesis or iron ore reduction.

Alternative syngas compositions can be obtained by processes such as CO_2 reforming; Reaction (3) or partial oxidation of methane (POM); Reaction (4), producing H_2/CO ratio of 1 and 2, respectively.

$$CH_4 + CO_2 \leftrightarrow 2CO + 2H_2 \quad \Delta H^0_{298k} = +247\text{kJ/mol} \quad (3)$$

$$CH_4 + \tfrac{1}{2}O_2 \leftrightarrow CO + 2H_2 \quad \Delta H^0_{298k} = -45\text{kJ/mol} \quad (4)$$

However, there is an increased risk of carbon being deposited on the catalyst under CO_2 reforming and POM circumstances as compared to steam reforming conditions. Steam reforming is broadly industrialized as a method of synthesis gas production, which is a highly energy and capital-intensive process. As it supplies syngas with $H_2/CO > 3$, an energy efficient route for the production of syngas with a H_2/CO ratio of about 2, which is desirable for methanol and Fischer-Tropsch synthesis process, is essential [2].

In recent years, a few studies have been reported on simultaneous POM with steam and/or CO_2 reforming [3-5]. The process occurs in a most energy efficient and safe manner by the coupling of the exothermic POM and endothermic steam and/or CO_2 reforming, Reaction (5):

$$CH_4 + O_2 + (H_2O \text{ and/or } CO_2) \rightarrow CO + H_2 \quad (5)$$

The reaction can be mildly exothermic, thermo neutral and mildly endothermic by manipulating the reaction temperature and/or concentration of O_2 in the feed and consequently it requires little or no external energy [6]. It is an interesting reaction because of the consumption of both reactive CO_2 and CH_4, which are critical for the green house effect. This is a particularly attractive target for the application of channel reactors.

The objective of this study was to examine the performance of simultaneous partial oxidation of methane with steam and/or CO_2 reforming reactions in a small channel reactor coated by thin layers of catalyst via a sol gel method.

2. Experimental Methods

2.1. *Channel reactor*

The reactor consists of two 316-type stainless steel blocks with external dimensions 185 mm x 25 mm x 15 mm each.

The upper block had a groove, depth 1mm, width 5 mm and length 160 mm. 2 mm ID holes were drilled in both groove ends as reactor inlet and outlet, see Figure 1.

The groove depth of the lower block is 2 mm, giving a total reactor channel surface area of 25 cm^2. Five thermocouple holes were drilled in the lower block, with their tips 0.5 mm from the channel, see Figure 2.

A 1 mm thickness copper shim was grooved and placed between the blocks as a gasket. This made the channel 2 mm deep.

The reactor assembled with two clamps 15 mm thick, 16 holes were drilled in each clamp for the bolts and nuts, see Figure 3.

The reactor was placed axially in a muffle furnace, which also preheated the feed to the reaction temperature. Both furnace ends were sealed by insulation.

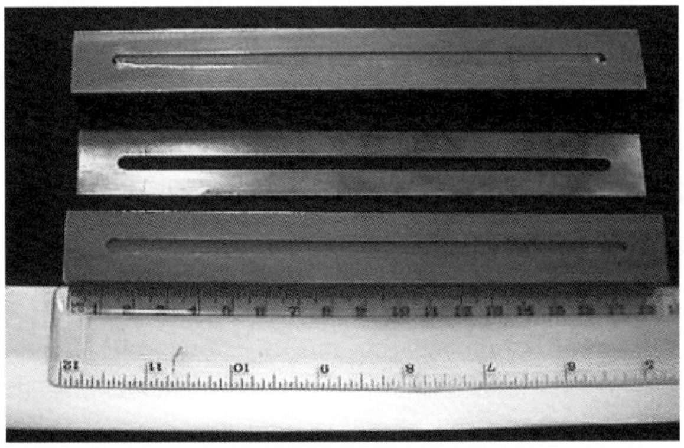

Figure 1. Two stainless steel blocks and copper shim.

Figure 2. Five thermocouple holes were drilled in the lower block and copper shim between blocks.

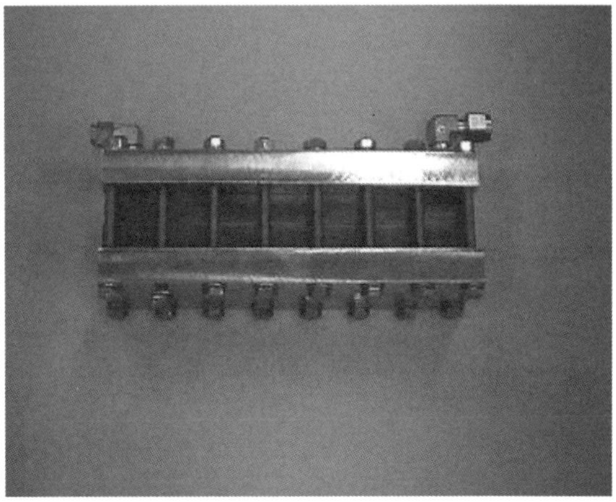

Figure 3. Assembled channel reactor.

2.2. *Catalysts*

Catalysts were prepared by dispersing 'Disperal' alumina supplied by Sasol [7] in 1% nitric acid and adding the appropriate nickel and other additives as nitrates to give composition of calcined catalysts in wt%; Ni(49%)/Al_2O_3(51%) (catalyst A) and Ni(33%)-Cr(5.6%)-Ba(11%)/La_2O_3(19%)-Al_2O_3(31.4) (catalyst B), respectively.

The reactor was coated with catalyst by forcing the sol-gel through the channel, blowing out the surplus and air-drying.

Multiple coats were used to build up the desired catalyst weight and the catalyst then calcined in air at 500°C (this temperature gave highest surface area).

The reactor's blocks were light enough to be weighed to +/-1 mg, giving accuracy in the mass of catalyst of +/- 1% or better.

2.3. *Apparatus and experimental procedure*

Air was used as feed instead of pure oxygen. The methane reforming reactions apparatus, experimental setup and procedure were presented in details in [8].

3. Results and Discussion

3.1. *Coupling of POM and steam*

The catalyst preparation method outlined above produced catalysts that adhered well to stainless steel (type 316). The result of the catalyst B activity test is shown in Figure 4.

Three coats of sol-gel gave a catalyst loading of 6.87 mg/cm^2 after calcination.

The catalyst film gave over 95% methane and 100% oxygen conversion at 830°C furnace temperature and atmospheric pressure. The methane conversion was decreased to 87% as feed ratios O_2/CH_4 and H_2O/CH_4 decreased hence oxygen conversion was steady ≈ 100%.

The methane flow rate was 0.5, water 0.5 and air 1.61 mol/h.

The produced H_2/CO ratio was varied between ≈ 2 to 2.5, as shown in Figure 5.

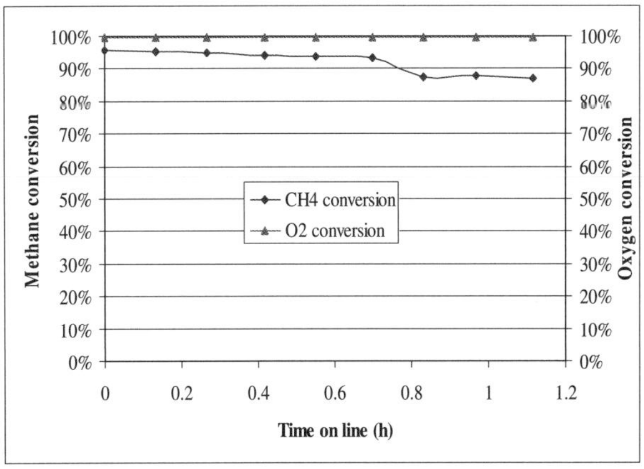

Figure 4. Methane, oxygen conversion, vs. time on line over catalyst B at 830°C furnace temperature, catalyst loading 6.87 mg/cm^2.

The results showed that methane, air and steam mixtures can efficiently be converted into synthesis gas at temperatures higher than 800°C.

The results of POM experimental work in literature [8] showed that the coking rate was high and varied from 0.044 to 1.1 g carbon/g cat. h. However, addition of steam to methane partial oxidation reduced the coking rate to 0.017 g carbon/g cat. h.

Other reports concluded that no coke was formed when H_2O added to POM, [9].

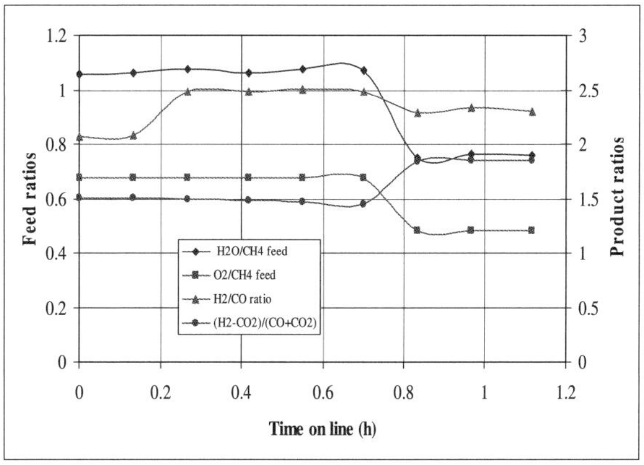

Figure 5. Feed, product ratio vs. time on line over catalyst B at 830°C furnace temperature, catalyst loading 6.87 mg/cm^2.

3.2. *Coupling of POM and CO_2*

The results showed that catalyst B gave methane and oxygen conversion 75 and 100% respectively at 788°C furnace temperature and atmospheric pressure.

The CO_2 conversion was changed as the feed ratios O_2/CH_4 and CO_2/CH_4 changed hence H_2/CO ratio was varied between ≈ 1.2 to 1.55, as shown in Figure 6 and Figure 7.

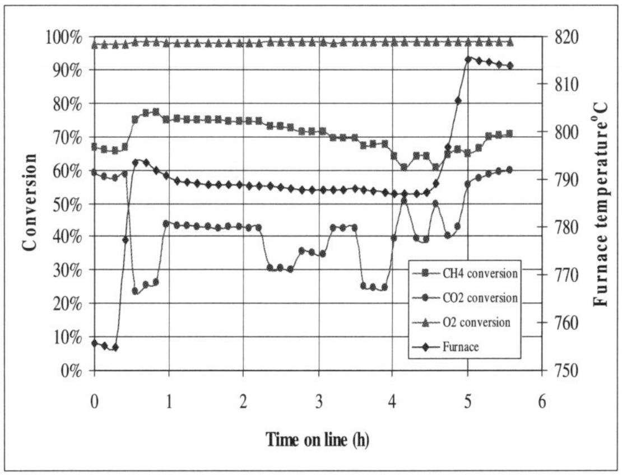

Figure 6. Conversion, furnace temperature vs. time on line over catalyst B at atmospheric pressure, catalyst loading 8.63 mg/cm^2.

The literatures concluded that the addition of CO_2 to the partial oxidation of methane reaction not only alters the H_2/CO ratio of the products, but also causes the temperature of the hot spots in the catalyst bed to reduce significantly [10].

Other reports found that presence of O_2 strongly diminishes the catalyst tendency to carbon lay down; therefore, it is worth considering the addition of O_2 in the feeding mixture of CO_2 reforming, as way of improving the stability of catalysts [11].

In our work, no rapid coke lay down was observed when POM combined with CO_2 reforming compared to CO_2 reforming and POM separately.

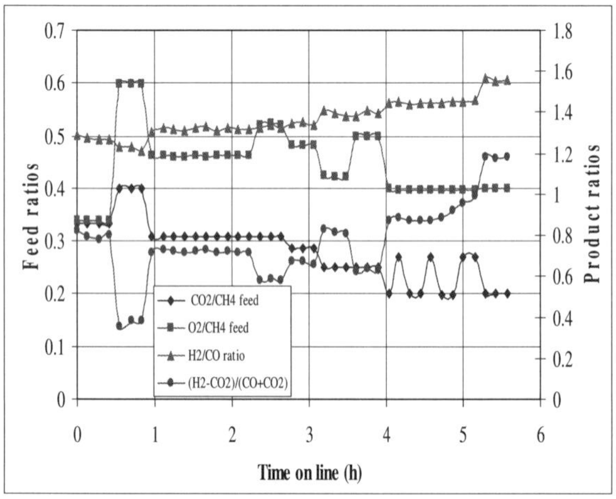

Figure 7. Feed, product ratios vs. time on line over catalyst B at atmospheric pressure, catalyst loading 8.63 mg/cm^2.

3.3. *Simultaneous POM, steam and CO_2*

The results showed that the catalyst B gave from 75 to 86% methane conversion and from 42 to 62% CO_2 conversion at 830°C furnace temperature and atmospheric pressure hence the conversion changed with the feed ratios change as shown in Figure 8 and Figure 9.

H_2/CO ratio ≈ 2 was achieved by manipulating the feed ratio and the reaction temperature.

The reaction can be mildly exothermic, thermo neutral and mildly endothermic by manipulating the reaction temperature and/or concentration of O_2 in the feed.

Oxygen conversion was always ≈ 100% for all mixtures, at any residence time, pressure and catalyst activity. The supported Ni catalysts are found to yield synthesis gas via the so called indirect partial oxidation mechanism that was reported in literature [12]. The oxygen first consumed in the full oxidation of methane to carbon dioxide and water, Reaction (6):

$$CH_4 + O_2 \rightarrow CO_2 + 2H_2O \quad \Delta H^0_{298k} = -802 kJ/mol \quad (6)$$

Consequently, the remaining methane is converted to synthesis gas by reforming with fed, produced water and carbon dioxide, Reactions (1, 2 and 3). Although high conversions to synthesis gas can be obtained in this way the primary full oxidation is believed to be tricky in excess of oxygen. It could be all the oxygen was consumed by Reaction (6).

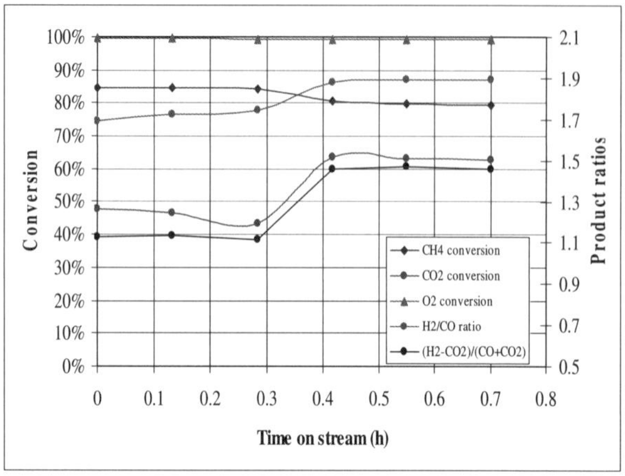

Figure 8. Conversion, product ratios vs. time on stream over catalyst B at 830°C and atmospheric pressure, catalyst loading 6.87 mg/cm^2.

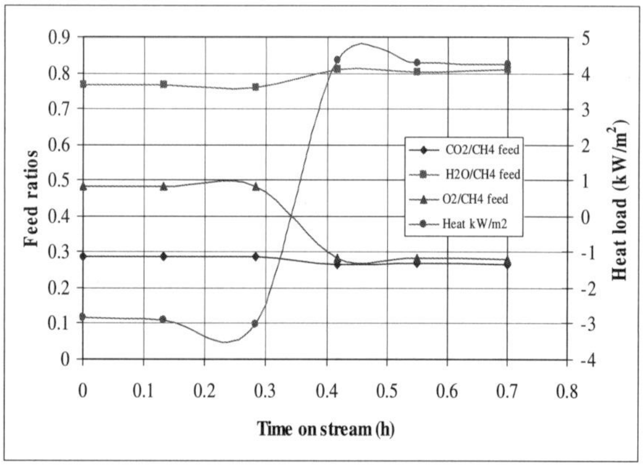

Figure 9. Feed ratios, heat load vs. time on stream over catalyst B at 830°C and atmospheric pressure, catalyst loading 6.87 mg/cm^2.

4. Conclusions

- Methane, air, steam carbon dioxide mixtures can efficiently be converted into synthesis gas at temperatures higher than 800°C. Ni-based catalysts exhibit very high activity for these processes.

- Carbon deposition rate was reduced by coupling of POM with steam and/or CO_2 reforming reactions and addition of Ba, Cr and La_2O_3 to Ni/Al_2O_3 catalyst.
- The concept of a channel reactor to produce synthesis gas 'syngas' for Fischer-Tropsch and methanol synthesis is feasible and has potential advantages over conventional types of reactor.
- Lower axial temperature gradient along the reactor was achieved compared to endothermic and exothermic reactions separately.
- The pressure drop along the whole length of the reactor was negligible.
- The catalyst effectiveness factor was about 1 and the values of activation energy indicated the reactions did not have any diffusion limitations.
- It has been demonstrated that sol gel method facilitates catalyst preparation.

References

1. M. Twigg, *Catalyst Handbook 2nd edition*, Wolfe Publishing, London (1989).
2. G. Froment, *J. of Mol. Cat. A: Chem.* **163**,147-156 (2000).
3. D. Qin, J. Lapszewicz, and X. Jiang, *J. of Cal.* **159**, 140-149 (1996).
4. Y. Matsuo, Y. Yoshinaga, Y. Sekine, K. Tomishige and K. Fujimoto, *Cat. Today*. **63**, 439-445 (2000).
5. K. Tomishige, Y. Matsuo, Y. Sekine and K. Fujimoto, *Cat. Comm.* **2**, 11-15 (2001).
6. V. Choudhary, B. Uphade, and A. Mamman, *Mic. and Mes. Mats*. **23**, 61-66 (1998).
7. http://www.sasoltechdata.com/tds/DISPERAL_DISPAL.pdf
8. M. Nakoua, Ph. D. thesis, Sch. of Chem. Eng. and Adv. Mats. U. of Newcastle upon Tyne, Newcastle upon Tyne, UK (2004).
9. U. Olsbye, O. Moen, A. Slagtern and I. M. Dahl, *App. Cat. A: Gen.* **228**, 289-303 (2002).
10. S. Liu, G. Xiong, H. Dong, and W. Yang, *App. Cat. A: Gen.* **202**, 141-146 (2000).
11. N. Nichio, M. Casella, G. Santori, E. N. Ponzi and O. A. Ferretti, *Cat. Today*. **62**, 231-240 (2000).
12. A. Steghuis, J. van Ommen and J. Lercher, *Cat. Today*. **46**, 91-97 (1998).

PALM DATE SEEDS AND ITS APPLICATION AS ACTIVATED CARBON — A REVIEW

ANIYANKUNJU.CM
CALEDONIAN COLLEGE OF ENGINEERING, MUSCAT, OMAN, P.B 2322, CPC 111

S. FEROZ
CALEDONIAN COLLEGE OF ENGINEERING, MUSCAT, OMAN, P.B 2322, CPC 111

Agricultural wastes like dates stones, corncob, coconuts shells, nuts shells and stones, oil palm stones and shells, apple pulp, chickpea husks, rice straw, lignin and palm tree branches are considered as a cheap and renewable sources of raw materials for activated carbon (AC) production. The impregnation ratio (R), carbonization time (Ct) and, carbonization temperature (CT) is the most influential factors in preparation of activated carbon from the agricultural wastes. This paper presents a complete review of activated carbon from palm date seeds and its application in treatment processes.

Keywords: Activated carbon, palm dates, impregnation ratio, carbonization time, carbonization temperature.

1. Introduction

Activated carbon is useful in treatment because it acts as an adsorbent, and can effectively remove in-organic and organic pollutants by the process commonly known as adsorption. Adsorption is a separation process in which certain components of a fluid phase are held to the surface of a solid adsorbent. Most adsorbents are highly porous materials, and adsorption takes place initially on the walls of the pores and then at specific sites inside the pores. It is a removal process where particles are bound to adsorbent particle surface by either chemical or physical attraction. Activated carbons are the most popular adsorbents used for the removal of toxic substances from water. This could be related to their extended surface area, high adsorption capacity, microporous structure and special surface reactivity. However, the adsorptive capacities of activated carbons depend mainly on the precursor nature, the operating conditions of adsorption such as temperature, pH and the nature of the adsorbate.

Activated carbons could be produced by either physical (thermal) activation or chemical activation. Thermal activation involves carbonization or calcination of the raw materials at elevated temperatures (500-900°C) in an inert atmosphere followed by mild oxidation (gasification) of the substance with steam, air and/or carbon dioxide at high temperatures (800-1000°C). Simultaneous carbonization and activation can also be obtained chemically by impregnation with dehydrating agents such as phosphoric acid and zinc chloride.

Adsorption is found to be widely used technique for the removal of toxic metal ions from wastewater. Many reports [1-27] have been appeared in literature review on the production of low cost adsorbents using cheaper and readily available materials. Activated carbon has versatility and wide range of applications and it has been proven to be an effective adsorbent for the removal of a wide variety of organic and inorganic pollutants from different media. Activated carbon is classified based on its size and shape into four types: powder, granular, fibrous, and clothe and each type of them has its specific application as well as inherent advantages and disadvantages in waste water treatment. Therefore, production of low-cost activated carbon becomes a great goal of many researchers.

In recent years, the importance of agricultural waste as a cheap and renewable source of raw materials for activated carbon production is becoming more significant. The most common agricultural wastes used in the production of AC are date stones, corncob, coconuts shells, nuts shells and stones, oil palm stones and shells, apple pulp, chickpea husks, rice straw, lignin and palm tree branches. This paper presents a detail review on the production of activated carbon from palm date seeds/stones and its application in treatment process.

2. Composition of date pits and date carbon

Date pits, also called pips, stones, kernels, or seeds are produced in packing plants or in industrial date are available in large quantities when pitted dates processing plants. An approximate composition of date pits and date pit carbon is given in the Table1 & Table 2 respectively.

Table 1. Typical Composition of Date Pits.

Parameter	Date Pits
Moisture	0.0%
Volatiles	8.8%
Ash	4.0%
Apparent Spec. Gravity	0.67%
Real Spec. Gravity	1.36%
Porosity	51%
Ion-Absorption	1.8%mg aeq iodine/g carbon

Table 2. Typical Composition of Date Pit Carbon.

Parameter	Date Pits
Moisture	5-10%
Protein	5-7%
Oil	7-10%
Crude fiber	10-20%
Carbohydrate	55-65%
Ash	1-2%

Myhara Mark et al., [18] investigated the chemical composition of two Omani date varieties, Fard and Khalas. Composition analysis showed that Khalas contained more glucose (44.69±0.04%) than Fard (43.55±0.70%) and less non-starch polysaccharide (NSP) in Khalas (3.95±0.04%) than Fard (4.44±0.03%). The seeds of two date palm (Phoenix dactylifera L) cultivars, Deglet Nour and Allig, from the Degach region – Tunisia, were analyzed for their main chemical composition by Souhail et al [9]. The following values (on a dry-weight basis) were obtained for Deglet Nour and Allig cultivars, respectively: protein 5.56 and 5.17%, oil 10.19 and 12.67%, Ash 1.15 and 1.12% and total carbohydrate 83.1 and 81.0%. Their studies also included the properties of oil extracted from date pits using gas-liquid chromatography and the data revealed that the major unsaturated fatty acid was oleic acid (41.3-47.7%), while the main saturated fatty acid was lauric acid (17.8%) for the Deglet Nour cultivar and palmitic acid for the Allig cultivar (15.0%). Capric, myristic, myristoleic, palmitoleic, stearic, linoleic and linolenic acids were also found and their results showed that date seed oil could be used in cosmetic, pharmaceutical and food products.

Chemical composition and nutritional quality of date seeds were analyzed by considering the seeds of Ruzeiz and Sifri date cultivars by Sawaya et al. [7]. The results showed that the seeds on average found to contain 6.5% protein, 10.4% fat, 22.0% fiber, 1.1% ash, and 60.0% carbohydrates on a dry weight basis. Mineral analysis showed higher concentration of potassium followed by phosphours, magnesium, calcium and sodium. Oleic (44.25%), lauric (17.35%), myristic (11.45%), palmitic (10.30%) and linoleic (8.45%) were the major fatty acids in date seed oil.

2.1. *Activated carbon from palm date seeds*

Two distinct methods are used for the production of activated carbon, namely, chemical activation (activation with mineral salts) and physical activation (activation using oxidizing agents such as steam or CO2). In chemical activation, the precursor is impregnated with an activating agent such as zinc chloride or phosphoric acid followed by carbonization in an inert atmosphere at temperatures ranging from 400 to 800oC and this is followed by a leaching process to remove the activator. Chemical activation is carried out if the raw material is wood or peat and it has some distinct advantages over physical activation that include higher yield, simplicity, and lower temperature of activation and good development of the porous structure. Production of activated carbon involves the following steps: pretreatment of the date's stones/seeds, impregnation of the dates stones/seeds with the activator, carbonization of the impregnated stones/seeds and finally the removal of activator.

Girgis et al., [12] investigated the porosity development in activated carbons obtained from date pits under chemical activation with phosphoric acid. To facilitate effective crushing and grinding, they dried washed-clean whole date pits in an air oven at 80°C for 48 h and then sixty grams of the crushed pits (0.5–2.0

mm in size) were mixed with 100 cm3 of pre diluted H_3PO_4 (85 wt. %) to varying concentrations in the range 30–70 vol. %. The slurry was left overnight at room temperature, and then transferred to a stainless steel reactor (diameter of 40 mm and length of 60 cm) with narrow ports of 10 mm diameter at both ends, which vents the exhaust gases and liquid condensates. The reactor is then admitted into an electric tube furnace and heated at a slow rate to allow free evolution of volatiles, up to the hold temperatures 300, 500, or 700°C. Soaking at each final temperature was conducted for 2 h. The cooled mass was then subjected to thorough washing with hot water until the washings were neutral in effect, then dried at 110°C for 24 h, and finally kept in tightly closed bottles. Texture characteristics of the products were determined by adsorption of N_2 at 77K, as well as iodine, phenol, and methylene blue values. They found that carbons obtained at 300oC were very poorly porous, although with anomalously high capacity for the uptake of probe molecules from solution and this was attributed to the contribution of unsaturated chemical bonds initiated in the partially decomposed material. Carbons obtained at 500 and 700°C are good to excellent adsorbents and attain best developed porosity at 700°C, contrary to the earlier well-established temperature of 500°C recommended for treatment of agricultural precursors. They also found that phosphoric acid inflict physical and chemical modifications on the botanical structure by penetration, particle swelling, and partial dissolution of the biomass, bond cleavage and reformation of new polymeric structures resistant to thermal decomposition. Yahia [6] studied the effect of preparation conditions on the yield and quality of activated carbon produced from dates stones using zinc chloride as an activator. The optimum conditions for activated carbon production was evaluated based on the determination of various adsorption parameters of methylene blue (MB) and phenol. Using MB as an adsorbate it was found that the optimum ratio of activation to dates stones (R) is 2.0. The maximum unit capacity for MB was 148 mg/g at carbonization time (Ct) =1 hour and carbonization temperature (CT) = 600°C and corresponding yield was 32%. Milder conditions (Ct = 0.5 hr and CT = 500°C) produced activated carbon with maximum adsorption capacity of 286.3 mg/g with a yield of 31%. Using phenol as an adsorbate, it was found that the optimum value of R is 0.5. The maximum unit capacity for phenol was 19 mg/g at Ct = 3 hours and CT = 600°C and activated carbon yield of 47.3%. On the other hand, milder conditions (Ct = 0.5 hr and CT = 500°C) were required to produce activated carbon with maximum capacity for phenol removal of 75.0 mg/g and yield of 64.6%

Haimour and Emeish [5] investigated the utilization of date stone wastes for production of activated carbon by chemical activation with phosphoric acid using a fluidized bed reactor. The effects of the activation time, activation temperature, impregnation ratio, and particle size on the yield and the adsorptive capacity towards iodine were studied. The yield and quality of the activated carbon prepared by using H_3PO_4 were compared with that prepared from date stones using same equipment, and under similar conditions using $ZnCl_2$ as an oxidizing agent. The maximum value of the iodine number of the activated carbon produced using H_3PO_4 in this work was about 495. The iodine number for the produced activated carbon was higher when phosphoric acid was used, compared to that when zinc chloride was as impregnation reagent; however, the yield obtained when H_3PO_4 was used was lower than the yield $ZnCl_2$ was used. The iodine number increases significantly with increasing the activation temperature. By increasing the impregnation ratio at same temperature, the iodine number decreased sharply and an oscillation is noticed for all cases but it was clearer at 800°C. The average variation of the iodine number for the whole range of particle size used in the work is ± 10%.

Ahmed El Nemr [1] in his studies developed activated carbon by adding dried date palm seed (2.0 kg) in a small portion to H_2SO_4 (98%) kept soaking for 5 h followed by boiling for 20 h in a fume hood. After cooling in ice bath, the reaction mixture was poured onto cold water (5 lt) and filtered. The obtained carbon was dried in an open oven at 120°C for 24h and then immersed in 5%$NaHCO_3$ (4 lt) to remove any remaining acid and after that filter the carbon. The produced carbon was then washed with distilled water until pH of the activated carbon reached 6, dried at 150°C for 24 h and then hand sieved to the particle size ≤0.100mm and kept in a glass bottle for further usage in experimental studies.

3.0 Adsorption on Date Palm Seeds (Activated Carbon):

Activated carbon is perhaps one of the most widely used adsorbents. Over the last few decades, adsorption systems involving activated carbon have gained importance in purification and separation processes on an industrial scale. It is now considered as one of the best available technologies in removing both organic and inorganic trace contaminants. Its use can be broken down to discrete areas such as effluent treatment, potable water treatment, solvent recovery, air treatment, decolorizing, metal ore processing, and many more general domestic applications. Activated carbon owes their distinguished properties to the extensive surface area and internal porosity, as well as their developed surface structure. The inherent nature of the precursor, or starting material, as well as the methods and conditions employed for carbon synthesis, determine the final pore size distribution and the adsorption properties of the activated carbon.

The adsorption kinetics and isotherms of methylene blue (MB) on raw and thermally activated date pits, an agricultural solid waste, were investigated by Fawzi et al., [10]. The effects of activation temperature, solution pH, adsorbent particle size, and solution salinity on methylene blue removal were studied. Dye removal increased with an increase in solution pH but decreased with an increase in solution temperature. Raw date pits were effective in removing MB from aqueous solutions and are a promising adsorbent material. Unexpectedly, it was found that physical activation of date pits reduced the adsorption capacity. The measured adsorption kinetics was well described by pseudo-second order kinetic model. Furthermore, the Langmuir isotherm model fitted the experimental equilibrium data.

Myhara Mark et al., [18] investigated the chemical composition and water sorption isotherms of two Omani date varieties, Fard and Khalas. Moisture sorption isotherms conducted at three temperatures displayed a crossing effect due to the dissolution of crystalline sugars at higher temperature and moisture contents. Isosteric heat of sorption, for both varieties, varied from 9.4 to -1.6 kJ/mol as the moisture content changed from 5.0 to 40.0%. Differences in moisture sorption behavior of the two varieties were attributed to compositional differences.

S.M. Mane et al., [8] investigated the adsorption of phenol on date seed carbon to assess its possible use as an adsorbent for the processing of phenolic waste water. The influence of various factors such as initial concentration, agitation speed, adsorbent dosage and temperature on the adsorption capacity has been studied. The percentage removal of phenol was observed to increase with an increase in initial concentration of phenol and decreases with an increase in temperature thereby indicating the exothermic nature of the reaction. Kinetic data was obtained by using a pseudo-second-order equation to understand the reaction mechanism and also the Langmuir and Freundlich equations were employed to interpret the adsorption isothermal data.

The use of activated carbon developed from date palm seed wastes, generated in the jam industry, for removing toxic chromium from aqueous solution has been investigated by El Nermr et al., [1]. The activated carbon has been achieved from the date palm seed by dehydrating methods using concentrated sulfuric acid. The batch experiments were conducted to determine the adsorption capacity of the biomass. The effects of initial concentration, pH, contact time, and concentration of date palm seed carbon have been studied at room temperature. A strong dependence of adsorption capacity on pH was observed, the capacity increases as pH value decreases and the optimum pH value is observed to be 1.0. Kinetics and adsorption equilibrium were studied at different sorbent doses. The adsorption process was fast and equilibrium was reached within 180 min. The kinetic data were analyzed using various kinetic models pseudo first order equation, pseudo second order equation, Elovich equation and intraparticle diffusion equation – and the equilibrium data were tested using several isotherm models, Langmuir, Freundlich, Koble-Corrigan, Redlich-Peterson, Tempkin, Dubinin-Radushkevich and generalized isotherm equations and the same was reported.

Yakubu et al., [3] has derived activated carbon from date seed by treatment with zinc chloride and with steam, and sieved into 250, 500 and 1000 nm particle sizes. They investigated experimentally that reduction of zinc, nickel and COD from tannery effluent was found to decrease with increase in particle size, and the carbon activated by ZnCl2/steam recorded the highest waste removal. The ZnCl2 only activated carbon removed Ni (88%), Zn (90%) and COD (91%); the ZnCl2/steam activation treatment removed Ni (91%), Zn (99%) and COD (99%). The adsorption mechanism has been modeled by Nernst and Freundlich isotherms with the partition coefficient observed to decrease with temperature and it is lower in tannery coefficients. The intensity of adsorption defined by Freundlich isotherm is greater than 1 in all the cases, which indicated that the process is favorable. Decolorisation of 85% to 97% was recorded for the textile effluent, and 75 to 90% for the tannery effluent, in all cases decreasing with temperature increase. It was also found out that the $ZnCl_2$/steam activation treatment is better for the textile effluent and the $ZnCl_2$ activation is better for the tannery effluent.

3. Conclusions

There has been an intense interest in using activated carbon derived from cheap agricultural waste. Date seed can be considered as potential raw material for preparation of activated carbon and subsequent use for the removal of pollutants from aqueous solutions. A review on date palm seeds as activated carbon and its application in treatment process is presented.

References

1. Ahmed El Nermr, Azza Khaled, Ola Abdelwahab and Amany El-Sikaily, "Treatment of waste water containing toxic chromium using new activated carbon developed from date palm seed", J Hazard Mater, vol 152, no 1, 2008, p 263-275.
2. Zoulikha Merzougui and Fatima Addoun, "Effect of oxidant treatment of date pit activated carbons application to treatment of waters", Desalination, vol 222, 2008, p 394-403.
3. MK Yakubu, MS Gumel, and Abdullahi, "Use of activated carbon from date seeds to treat textile and tannery effluents", African Journal of Science and Technology (AJST), Science and Engineering Series, vol 9, no 1, 2008, p 39-49.
4. B Abdelrahman, Fadhil and M Mohammed Deyab, "Conversion of some fruit stones and shells into activated carbons", The Arabian Journal for Science and Engineering, vol 33, no 2A, 2008, p 175-184.
5. NM Haimour and S Emeish, "Utilization of date stones for production of activated carbon using phosphoric acid", Waste Management, vol 26, no 6, 2006, p 651-660.
6. A Yahia, "Activated carbon from date's stone by ZnCl2 activation", JKAU. Eng. Sci, vol 17, no 2, 2006; p 75-100.
7. WN Sawaya, JK Khalil and WJ Safi, "Chemical composition and nutritional quality of date seeds", Journal of Food Science, vol 49, no 2, 2006, p 617-619.
8. SM Mane, AK Vanjara, MR Sawant, "Removal of Phenol from waste water using date seed carbon", Journal of the Chinese Chemical Society, vol 52, 2005, p 1117-1122.
9. Souhail Besbes, Christophe Blecker, Claude Deroanne, Nour-Eddine Drira and Hamadi Attia, "Date seeds: chemical composition and characteristic profiles of the lipid fraction", Food Chemistry, vol 84, 2004, p 577-584.
10. Fawzi Banat, Samer Al Asheh, and Leema Al Makhadmeh, "Evaluation of the use raw and activated date pits as potential adsorbents for dye containing waters", Process Biochemistry, vol 39, 2003, p 193-202.
11. F Garcia, A Alonso and J Tascon, "Pyrolysis of apple pulp: chemical activation with phosphoric acid", J Anal Pyrol vol 63, 2002, p 283-301.
12. B Gergis, A Abdel-Nasser and El-Hendawy, "Porosity development in activated carbons obtained from date pits under chemical activation with phosphoric acid", Microporous Mesoporous Mater, vol 52, 2002, 105-117.
13. W Lafi, "Production of activated carbon from acrons and olive seeds", Biomass Bioenergy", vol 20, 2001, p 57-62.
14. AC Lua and J Guo, "Preparation and characterization of activated carbons from oil palm stones for gas-phase adsorption", Colloids Surf, vol 179, 2001, 151-162.
15. R Juang, F Wu and R Tseng, "Mechanism of adsorption of dyes and phenols from water using activated carbons prepared from plum kernels", J Colloid Interf Sci, vol 227, 2000, p 437-44.
16. J Guo and A Lua, "Preparation of activated carbons from oil-palm-stone chars by microwave-induced carbon dioxide activation", Carbon vol 38, 2000, p 1985-93.
17. L Khalil, "Porosity characteristics of chars derived from different ligocellulosic materials", Adsorp Sci Technol, vol 17, 1999, p 729-39.

18. M Robert Myhara, S Mark Taylor, A Bogdan Slominski and Ismail Al Bulushi, "Moisture sorption isotherms and chemical composition of Omani dates", Journal of Food Engineering, vol 37, no 4, 1998, p 471-479.
19. Y Magdy and A Daifullah, "Adsorption of a basic dye from aqueous solutions onto sugar-industry-mud in two modes of operations", Waste Management, vol 18, 1998, p 219-226.
20. Y Ho and G McKay, "Kinetic models for the sorption of dye from aqueous solution by wood", Trans IChem, vol 76, 1998, p 183-191.
21. M Nassar and Y Magdy, "Removal of different basic dyes from aqueous solutions by adsorption on palm-fruit bunch particles" Chem Eng J, vol 66, 1997, p 223-226.
22. C Namasivayam, N Muniasamy, K Gayathri, M Rani and K Ranganathan, "Removal of dyes from aqueous solutions by cellulosic waste orange peel", Bioresour Technol, vol 57, 1996, p 37-43.
23. C Namasivayam and R Yamuna, "Adsorption of direct red 12 B by biogas residual slurry, Equilibrium and rate processes", Environ Pollution, vol 89, 1995, p 1-7.
24. M Ahmed and R Ram, "Removal of basic dye from waste water using silica as adsorbent", Environ Pollution, vol 77, 1992, p 79-86.
25. C Kirubakaran, S Krishnaiah and S Seshdri, "Experimental study of the production of activated carbon from coconut shells in a fluidized bed reactor", Ind Eng Chem Res, vol 30, 1991, p 2411-2416.
26. S Khare, K Panday, R Srivastava and V Singh, "Removal of Victoria blue from aqueous solution by fly ash", J Chem Technol Biotechnol, vol 38, 1987, p 99-104.
27. KK Panday, G Parsed and VN Singh, "Copper (II) removal from aqueous solutions by fly ash", Water Res, vol 19, 1985, p 769-873.

THE USE OF LOW GRADE DIATOMITE TO RECOVER REPEATINGLY USED PALM OIL

LADDA MEESUK
Department of Chemistry, Faculty of Science, Kasetsart University, 50 Pahonyotin Road, Bangkok 10903, Thailand

PORNSAWAT WATTHANAKUL
Department of Earth Science, Faculty of Science, Kasetsart University, 50 Pahonyotin Road, Bangkok 10903, Thailand

CHERDSUK ATTHA-AROON
The Gem and Jewelry Institute of Thailand, Rama IV Road, Bangkok 10400, Thailand

This research involved the use of diatomite from Northern Thailand, a low silica content diatomite, to bleach dark color in palm oil which has been used several times. Diatomite could be used to bleach such oil. The optimum condition was to use 8% w/v of diatomite and heat at 80-85°C for 60 min. The red color in oil decreased most, at 61.1% of untreated, used oil. The bleaching mechanism could be explained by chemical and physical adsorptions. Diatomite also had ability to adsorb a sample of toxic compound, benzo(a)pyrene which was added in oil.

1. Introduction

Diatomite or diatomaceous earth is a pale colored, soft, light weight rock composed principally of the silica microfossils of aquatic unicellular algae known as diatoms.

Diatomite has wide variety shapes of diatoms, typical size is 10-200 μm. Major components in diatomite are silica (SiO2) and alumina (Al2O3), which is about 70 and 11-18%, respectively. Minor components are Fe, K, Mg, Na, K and Ca. Diatomite has highly porous structure, low density and high surface area which results in a number of applications such as filter-aid in processing of liquid foodstuffs and chemical fluids [1,2] sorption of inorganic and organic chemicals [3-7] although, the article of Shawabkeh and Tutunji [6] was commented by Lin et al., 2006 [8], as well as an adsorbent for oil spills [9]. Erdogan et al., 1996 [10] reported the use of diatomite and other minerals to remove color and turbidity in sugar beet juice and they recommended diatomite to use in the sugar processing. This report is the first time that diatomite is used to bleach dark colour in palm oil.

Palm oil composes of triglycerides as major components, minor components are diglycerides and monoglycerides, other constituents are fatty acids and non-glyceride compounds. The ratio of saturated and unsaturated fatty acids is about 1:1. Among edible oils, such as corn oil, soybean oil and palm oil (palm olein), it was found that palm oil is the most appropriate for frying of food at about 180°C, because there are smaller amount of linolenic and linoleic acids, which are easily oxidized, in palm oil than in other oils. Furthermore, an antioxidant, tocopherol is higher in palm oil. Foods which are fried with palm oil can be kept longer without rancidity.

The frying process involves many reactions such as oxidation, isomerization, polymerization, pyrolysis condensation and hydrolysis.

These reactions can produce more than 400 mpounds, about 220 compounds are volatile and the rest will remain as contaminants in oils. Some of these compounds are coloured and mostly toxic [11], for examples oxidation and some pyrolysis condensations produce red compounds, while peroxides and aldehydes are yellow. The red compounds can be hydroxyalkyl quinone which is an oxidized product of alpha-tocopherol. Some polymerization products cause viscosity of the used oil. Some products are toxic polycyclic aromatic hydrocarbons (PAHs) such as benzo(a)pyrene, pyrene and fluoranthene. The repeating used of oils will produce larger amount and also enhance toxicity of these compounds, actions of

*This work is supported by Kasetsart University Research and Development Institute (KURDI).

vitamins and enzymes will be destroyed, mutation can occur as well as irritation of digestive systems. Waste, used oils are produced at an enormous amount each year, some of these are reused. The reuse of these oils will be more profitable if some toxic compounds are minimized, this is the main purpose of this work.

2. Experimental

2.1. *Preparation of diatomite samples*

Diatomite samples were obtained from Lampang, Northern Thailand. The samples were ground manually, washed with distilled water, dried at 110°C and ground again to get smaller sizes before use (an average particle size was 6.88 μm, determined by using mastersizer model S. ver. 2.19, Malvern Instruments)

2.2. *Bleaching of used palm oil*

The repeatingly used palm oil is the palm oil that has been used several times to fry "patonggo" or starch fingers and obtained from a market in Bangkok area.

Used palm oil 100 ml in a 250 ml beaker was heated and stirred with magnetic stirrer, when the temperature reached 60°C, known amount of diatomite was gradually added with stirring. The amounts of diatomite were varied from 1 to 10% by weight of oil, each by 1% interval. The bleaching temperature was varied from 70-75, 75-80, 80-85, 85-90 and 90-95°C. The bleaching time was 15, 30, 45 and 60 min. After bleaching, the mixture was set aside to cool to room temperature, then diatomite was filtered off with Whatman filter paper (no. 42). The colour of oil before and after bleaching were measured by Lovibond Tintometer (Pardero Model E), a 5¼ inches cell was used.

2.3. *Analyses of diatomite*

Chemical and physical properties of the diatomite samples were analyzed before and after used by X-Ray Fluorescence Spectrometer (Philips PW 1480), Scanning Electron Microscope (LEO 1450 VP) and Fourier Transform Infrared Spectrometer (Perkin Elmer System 2000).

2.4. *Characterization of the repeatingly used palm oil*

The repeatingly used palm oils were characterized before and after removing dark colour by fatty acid compositions. (Gas Chromatography, Perkin Elmer model Auto System XL). To determine fatty acids, American Oil Chemist's Society (AOCS) standard method was used. Peroxide and acid values of oil were also determined. Peroxide value of oil was determined by iodometric method. The peroxide value indicates the degree of oxidation reaction in oil, the oil that had been used repeatedly will be expected to have higher peroxide value than unused oil. The acid value of oil is the milligram of potassium hydroxide which is equivalent to free fatty acid in 1 g of oil. The acid value of the palm oil samples was measured by using The Association of Official Analytical Chemist (AOAC) standard method.

2.5. *Adsorption of a toxic compound added in oil*

Adsorption of a toxic compound in oil was studied by UV-visible spectrophotometer (Jasco 7800). Benzo(a)pyrene, a PAH, was used as a representative of a toxic compound.

The adsorption of benzo(a)pyrene was studied by adding a known amount of benzo(a)pyrene (25 ppm) in unused oil and measured the absorption spectrum in ultraviolet-visible range, the maximum absorption was found at 363 nm. Diatomite was then added (8% w/v) and the mixture was stirred at 80-85°C for 30 min. The absorption intensities at 363 nm of unused oil with benzo(a)pyrene and unused oil with benzo(a)pyrene after treated with diatomite were measured and compared. The amounts of benzo(a)pyrene adsorbed by diatomite were determined from a calibration graph.

3. Results and discussion

Chemical compositions of the diatomite samples which were used in this work were analyzed by X-ray Fluorescence Spectrometer and found that it contained only 63.6% SiO2 which is a low quality diatomite in terms of SiO2 content. The high quality diatomite should contain more than 88% of total SiO2 content [5].

The measurement of oil colours by Lovibond Tintometer is shown in Figure 1 (the optimum condition is shown). The optimum condition was obtained when 8% of diatomite was used at bleaching temperature 80-85°C and bleaching time 60 min. It was found that the red colour decreased most, 61.1% of the original colour.

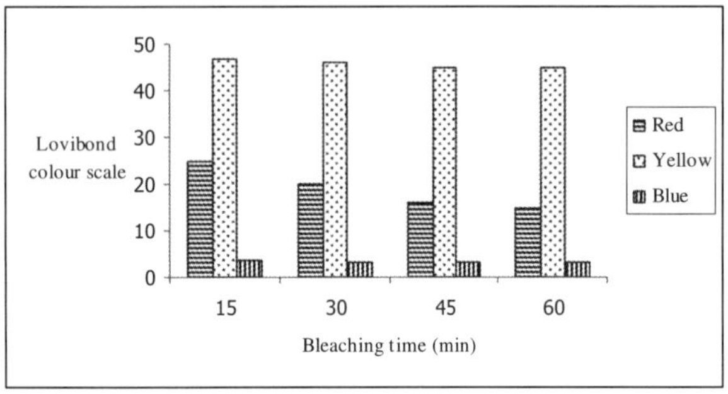

Figure 1. Oil colours in Lovibond scale and bleaching time, using 8% diatomite at bleaching temperature 80-85°C.

The fatty acid compositions of the used palm oil before and after treated with diatomite were not different which meant that diatomite could bleach the dark colour in oil without affecting the fatty acid composition in oil. The peroxide value and acid value were decreased after the used oil samples were treated with diatomite i.e. they were decreased from 8.5 to 4.25 and 1.24 to 0.82, respectively.

FT-IR spectra of diatomite before and after use are shown in Figure 2(a) and (b). As can be seen, the vibrations of diatomite structural bands such as SiO–H stretching of the silanol group at 3626 and 3445 cm-1 shifted to lower frequencies at 3619 and 3439 cm-1, respectively. This could be caused by hydrogen bonding between H of the silanol group, SiO–H at the surface of diatomite and O of coloured compounds in oil. The Si–O stretching (out of plane and in plane) around 1100 cm-1 exhibited some shifts resulted from hydrogen bondings at the surface of diatomite. In the IR spectrum of diatomite after use (Figure 2(b) some additional bands due to organic functional groups such as C–H, C=O and C=C vibrations at 2925, 1745 and 2361 cm-1, respectively are found. This reveals that diatomite had adsorbed some organic compounds which caused dark colour in the oil. This can be said that the colour bleaching was both chemical and physical adsorptions.

Scanning Electron Micrographs of diatomite before and after use are shown in Figure 3(a) and (b), respectively. As can be seen, diatomite before use has a porous surface but after use the pores at the surface are less seen because diatomite had adsorbed coloured compounds into the pores (diatomite sample has been washed thoroughly after used). This confirms the adsorption of coloured copounds on the surface of diatomite.

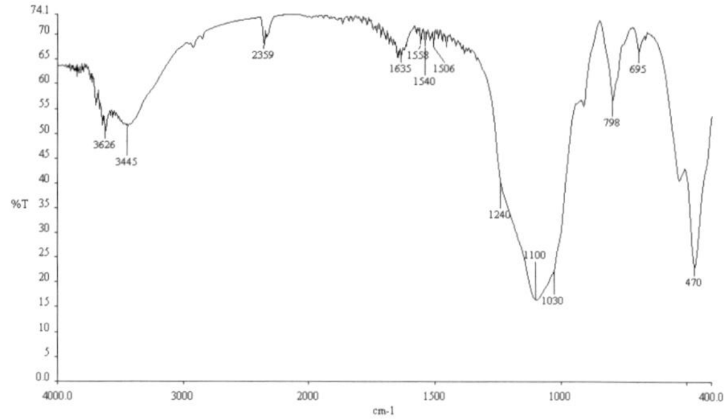

(a)

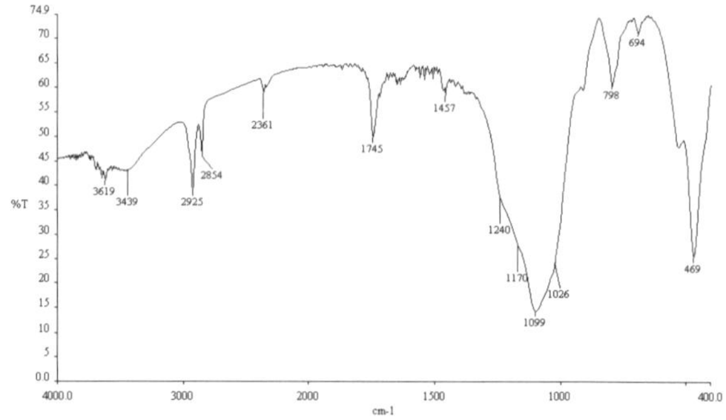

(b)

Figure 2. FT-IR spectra of diatomite before (a) and after use (b).

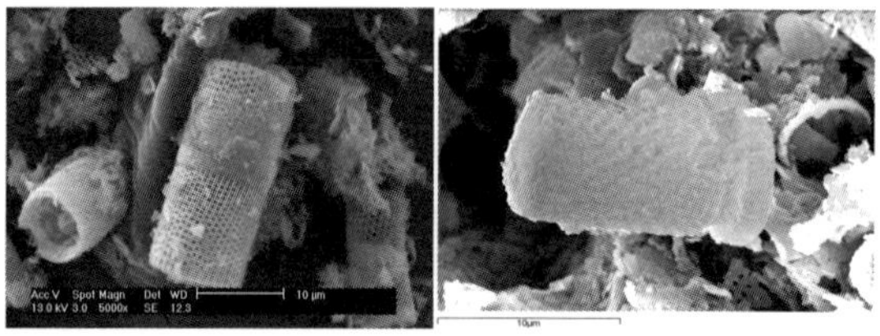

(a) (b)

Figure 3. Scanning electron micrographs of diatomite before (a) and after use (b), at magnification x5,000.

The study of adsorption of benzo(a)pyrene which was added in unused oil was found that the absorption of 25 ppm benzo(a)pyrene at 363 nm was equal to absorbance 2.4534. After the oil was stirred with 8% diatomite for 30 min, the absorbance decreased to 0.6680, from the calibration curve this absorption was equal to 6.52 ppm of benzo(a)pyrene. This means that 18.48 ppm or 73.92% of benzo(a)pyrene has been adsorbed by diatomite. The absorption spectra of benzo(a)pyrene in oil before and after treated with diatomite are shown in Figure 4.

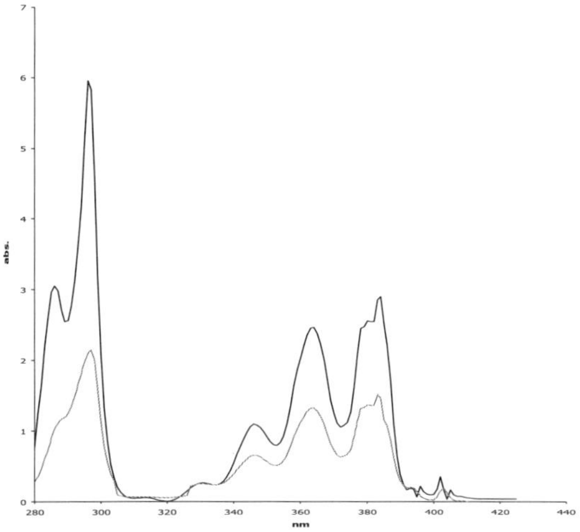

Figure 4. Absorption spectra of benzo(a)pyrene 25 ppm in oil before (upper spectrum) and after adsorption (lower spectrum).

4. Conclusion

A low silica content diatomite can be used to bleach dark colour in repeatingly used palm oil. The red colour decreased 61.1% of the untreated, used oil. The optimum condition was obtained when 8% w/v of diatomite was used to bleach the oil at 80-85°C for 60 min. The results from IR spectra and SEM indicated that the bleaching was both chemical and physical adsorptions. Diatomite could also adsorb 73.92% of benzo(a)pyrene added in oil.

Acknowledgement

The authors would like to thank Ms. Sontaya Phothong for the technical assistance.

References

1. S. D. J. Inglethorpe, "Diatomite." Technical Report WG/92/39, British Gelogical Survey, Nottingham, 37 p., 1992.
2. S. Martinovic, M. Vlahovic, T. Boljanac and L. Pavlovic, Int. J. Miner. Process. 80, 255-260, 2006.
3. S. Akyüz, T. Akyüz and N. M. Ozer, J. Mol. Struct. 565, 493-496, 2001.
4. P. Huttenloch, K. E. Roehl and K. Czurda, Environ. Sci. Technol. 35, 4260-4264, 2001.
5. G. G. Martirosyan, A. G. Manukyan, E. B. Ovsepyan and K. A. Kostanyan, Russ. J. App. Chem. 76(4), 531-535, 2003.
6. R. A. Shawabkeh and M. F. Tutunji, App. Clay Sci. 24(1-2), 111-120, 2003.
7. M. A. M. Khraisheh, Y. S. Al-degs and W. A. M. Mcminn, Chem. Eng. J. 99, 177-184, 2004.
8. J. X. Lin, S. L. Zhan, M. H. Fang and H. Yang, Appl. Clay Sci. 11, 55-67, 2006.
9. J. F. Lemonas, Am. Ceram. Soc. Bull. 76(6), 92-95, 1977.
10. B. Erdogan, S. Demirci and Y. Akay, Appl. Clay Sci. 11, 55-67, 1996.
11. W. L. Clark and G. W. Serbia, Food Technol. 45(2), 84-94, 1991.

INVESTIGATION OF GRAZING EFFECTS ON COMPOSITION AND VEGETATION DIVERSITY

SEYED AKBAR JAVADI
Department of Range Management, Islamic Azad University, Science and Research Branch, Tehran, Iran

EHSAN ZANDI ESFEHAN
Department of Range Management, Islamic Azad University, Science and Research Branch, Tehran, Iran

The study was performed at Lar watershed, 84 Km far from Tehran, to investigate the grazing effect on composition and vegetation diversity. To achieve this goal, the vegetation of inside and outside the enclosure was compared to each other. The number of the quadrates was calculated on the basis of the statistic method and then fifteen 1m2 quadrates were established along with five transects inside and outside the enclosure. The results of vegetation cover sampling indicates that the percentage of grasses and forbs inside the enclosure is greater than that outside the enclosure while the percentage of shrubs is more outside the enclosure. The species of palatable class I and II also showed the maximum canopy cover percentage inside the enclosure while the species of class III had the maximum canopy cover percentage outside the enclosure. The vegetation diversity inside and outside the enclosure were compared through Shanon-Winer and Hill Indices. The results showed that the vegetation diversity inside the enclosure (3.9-15) had been more than that outside the enclosure (3.8-14.7) respectively.

1. Introduction

Composition and diversity of vegetation are of important parameters to determine the health and resistance of rangeland ecosystems. Inappropriate management in using of rangeland ecosystems results in vegetation destruction and decline of composition and diversity in rangelands. The ultimate goal of natural resources management is to preserve the diversity in range and forest ecosystems (9). Diversity has been constituted of two elements namely species richness and evennes (7) and high richness is considered as one of the important attributes in each community or ecosystem by many ecologists (10).

Due to the importance of vegetation composition and vegetation diversity in range management studies, several researchers have considered these parameters in different methods especially to determine the effect of grazing on the mentioned factors. Amrollahi and Zare (2001) studied the diversity of plant species in three different grazing intensity and concluded that diversity and evennes of no grazing treatment had been more than that of mediocre or heavy grazing treatments. Mesdaghi (2000) compared the indices of species diversity under different uses and noted that Hill indices (N_{0}) and Pielou indices ($E1$) had been the best indices to determine the richness and evennes respectively. He also concluded that the uniformity in critical area had been higher and this is due to the existence of rare species in reference and key areas and invasive species in critical areas.

Gharadaghi (1999) stated that overgrazing would result in destruction of palatable species and intensive reduction of the species of class II. Ghanbarian (2001) noted that species of class I and class II had been dominant in the enclosure area while species of class III are dominant in under grazing areas.

Johnston et al. (1971) showed the effects of long term grazing and founded that the dominant vegetation had been different under light and heavy grazing and the amount of forbs and shrubs would increase with increased grazing intensity. Yavari (2001) indicated that the range type of the enclosure area (*Festuca ovina*) had differed from the range type of open grazing (*Phlomis sp*). Mosavi (2001) concluded that the enclosure would improve the vegetation composition and increase the palatable species.

Bowns and Bagley (1986) studied at mountainous rangelands of Utah and noted that the total amount of species and forbs in the Enclosure area had been higher than that of the areas under grazing. Dormaar et al. (1989) noted that heavy grazing would decrease the amount of desirable species and consequently would lead to the regression of the grassland. Willms et al. (2002) stated that the grazing factor had no effect on presence and richness of the species but affected their abundance.

2. Materials and Methods

The Study area is located at Lar watershed, 84 Km far from Tehran, between 51°, 32′ and 52°, 4′ E longitude and from 35°, 48′, 40″ to 36°, 4′, 35″ N latitude. The area is mountainous and annual average rainfall level is 410 mm. Snowfall is predominant in the area.

Considering the density and dimension of the canopy cover in the study area, 15, 1m^2 quadrates were separately established on five transects inside and outside the enclosure area. The percentage, palatability class and vegetation form of the species were recorded inside the quadrates.

Shannon – Wiener (H') and Hill (N_1) indices were used to calculate and to compare the species diversity. Shannon – Wiener (H') index which is used widely in ecology as follows:

$$H' = -\sum_{i=1}^{S}\left[\left(\frac{n_i}{n}\right)\ln\left(\frac{n_i}{n}\right)\right] \quad (1)$$

And Hill (N_1) index is defined as follows:

$$N_1 = e^{H'} \quad (2)$$

Where:

S: number of species, N: total number of the individuals in the sample, n_i: canopy cover percentage. Calculations were done on the basis of software published by Ludwig and Reynolds (1988).

3. Results and Discussion

The results of the research are presented in table (1).

The results show that enclosure in the study area leads to the increase of canopy cover percentage of grasses and forbs. These species are of high palatability hence first of all will be grazed by livestock outside the enclosure. They will also be weakened and replaced by shrubs and unpalatable species due to the overgrazing and premature grazing during the time. Our results are similar to the results of Bowns (1986), Willms et.al (1990) Ghelichnia(1996) and Gharedaghi (1999).

Table 1. Results of the vegetation sampling.

Area	enclosure area	Grazing area
Canopy cover percentage of grasses	49	12.9
Canopy cover percentage of shrubs	7.3	25.5
Canopy cover percentage of forbs	38.9	17.6
Percentage of litter	3	4.9
Percentage of bare soil	1.3	33
Percentage of gravel and stone	0.5	6.1
Percentage of class I species	52.5	7.5
Percentage of class II species	11	2
Percentage of class III species	30.8	30
Percentage of none palatable species	1	16.5
total Percentage of vegetation cover	95.3	56

With a view to palatability, Negative changes have also occurred in vegetation composition. Considering the results, the percentage of class I species like *Agropyron trichophorom,* Dactylis *glomerata and Astragalus* inside the enclosure area is higher than that outside the enclosure while species of class III such as *Cosinia sp, Thymus sp and Poa bulbosa* are more outside the enclosure. Our findings also reveal that the amount of legumes inside the enclosure is (20.4%) more than that outside the enclosure (5.6%) and consequently it may increase the amount of nitrogen and soil fertility inside the enclosure.

The values of Shanon and Hill indices are 3.9 and 15 for inside the enclosure respectively. For outside the enclosure the values of the mentioned indices are 3.8 and 14.7 respectively. The results indicate that vegetation diversity will increase inside the enclosure due to the opportunity given to range species to grow, while many species will be eliminated outside the enclosure because of heavy and continuous grazing.

The results of evennes considered by Sampson index shows that inside the enclosure it is higher than that outside the enclosure and it may be referred to the presence of rare species in vegetation composition of the enclosure. On the basis of the results, overgrazing and premature grazing leads to the change of the vegetation composition and diversity.

To have a permanent use of rangeland and to preserve its sustainability, it is necessary to suppress overgrazing with appropriate grazing systems which results in improvement of vegetation composition and diversity.

Table 2. Dominant specie inside and outside the enclosure.

Grazing area	Enclosure area
Agropyron trichophorum	*Agropyron trichophorum*
Ag. repens	*Ag. repens*
Bromus tectorum	*Ag.elongatum*
Poa bulbosa	*Ag.pectiniformis*
Alyssum sp	*Bromus tomentllus*
Achillea sp	*Dactylis glumerata*
Astragalus sp	*Alyssum sp*
Artemisia aucheri	*Achillea sp*
Thymus sp	*Astragalus sp*
Acantholimon sp	*Artemisia aucheri*
Eryngium sp	*Thymus sp*

References

1. Amrollahi, A. and Zare, A. Investigation on species diversity in three different grazing areas. Jangal and Marta Journal of Iran.No.48:44-47. (2001).
2. Bowns, E.J. and Bagley, F.C. Vegetation Responses to long-term sheep grazing on mountain Ranges, J. Range Manage, 39:431-434. (1986).
3. Dormaar, J.F., Smoliak, S. and Willms, W.D. Vegetation and Soil responses to short-duration grazing on fescue grasslands, J. Range Manage, 42:252-256. (1989).
4. Ghelichnia, H. Comparison of vegetation among the key, critical and reference area in Gholestan park. Pjohesh and Sazandeghi Journal of Iran. No.30:72-75. (1996).
5. Gharadaghi, H. and Adeli, J. Comparison of plant composition in enclosed and grazing area of Rodshor rangeland. Jangal and Marta Journal of Iran. No.43:28-44. (1999).
6. Ghanbarian, Gh. Investigation of enclose effect on plant composition. Second national conference proceeding of range and range management in Iran. (2001).
7. Johnston, A., Dormaar, J. F. and Smoliak, S. Long-term grazing effects on fescue grassland soils, J. Range Manage, 24:185-188. (1971).
8. Ludwig, A. J. and Reynolds F.J., Statistical Ecology a primer on methods and computing. By john Wiley and sons, Inc. (1988).
9. Mesdaghi, M. and Sadeghnejad, M. Indices of species diversity comparison under three different uses in semi step grassland. Journal of agriculture and natural resources science, Ghorgan, Iran. No.3:63-76. (2000).
10. Mesdaghi, M. vegetation description and analysis. Jahad, pub. Mashhad, Iran. 287p. (2001).
11. Mosavi, S. Enclose effects on soil and vegetation trend in Semnan rangeland. Second national conference proceeding of range and range management in Iran. (2001).
12. Willms, W.D., Dormaar, J.F., Adams, B.W. and Douwes, H.E. Response of mixed prairie to protection from grazing. J. Range Manage. 55:210-216. (2002).
13. Yavari, A. Investigation on vegetation dynamic under different management. Second national conference proceeding of range and range management in Iran. (2001).

CHEMICAL ACTIVATION OF MANGO SEED COAT (*MANGIFERA INDICA L*) — PREPARATION AND CHARACTERIZATION

B. G. PRAKASH KUMAR and B. B. GULYANI
Chemical Engineering Department, BITS-PILANI, Dubai
P.Box 345055, UAE

B. VIJAIKUMAR and LIMA ROSE MIRANDA
Chemical Engineering Department, A. C. College of Technology, Anna University
Chennai - 600025, India

The present study deals with the preparation of activated carbon using mango seed coat (MSC) using various activating agents (HCl, HNO_3, $ZnCl_2$, H_3PO_4, NaOH, and KOH). The influence of activation temperature (300 to 600°C) of the prepared activated carbon on the iodine number, methylene blue number, yield and surface area were investigated. Iodine number was the maximum at 400°C for all the activating agents except for HCl while for $ZnCl_2$, it was at 500°C. Methylene blue number was highest for H_3PO_4 and KOH activation, for others it was negligible. Results showed that in all cases, on increasing the activation temperature the yield decreased. Specific surface area of the activated carbon was maximum at about 1114 m^2g^{-1} at 400°C with impregnation ratio of 75% at an activation time of 1hr.

1. Introduction

Activated carbon has been widely used in many fields, and can be produced from a variety of carbonaceous source materials. Various materials are used to produce activated carbon, some of the most commonly used being agriculture wastes such as coconut shell [1], waste biomass [2], jackfruit peel [3], guava leaf powder [4], coconut flowers [5], nut shells [6], etc. Chemicals, which are frequently used for activation, are $ZnCl_2$, H_3PO_4, $ZnCl_2$, KOH and NaOH [2 and 3].

India accounts for 66% of worldwide mango production. There are nearly 1060 mango varieties in India. Of these, only about 20 varieties are grown on commercial scale [7]. So a large tonnage of bio-waste results from mango plantation. To enhance the value of this bio-waste, preparation of activated carbon has been attempted in the present study using various activating agents. Also, the feasibility of preparation of the activated carbon from mango seed coat (*Mangifera Indica L*) was investigated by chemical activation using agents such as HCl, HNO_3, $ZnCl_2$, H_3PO_4, NaOH, and KOH.

1.1. *Experimental materials and methods*

The mango seed coat (MSC) was used as the precursor for the preparation of activated carbon in this study. These agricultural solid wastes appear to be suitable materials for high quality activated carbon because of their inherent high solid density and relatively high fixed-carbon but low ash content. A sample of 20 g was soaked in various chemical agents (50% vol HCl, 50% vol. HNO_3, 1N $ZnCl_2$, 50% vol H_3PO_4, 1 N NaOH, 1N KOH) in an amount sufficient to cover the raw material completely and left overnight at room temperature. The impregnated solid was heated from 300 to 600°C (10°C min^{-1}) for 1 h under self-generating atmosphere. In all experiments, the heating rate was kept constant. After cooling, the product was washed sequentially several times with hot distilled water. The cleaned samples were dried at 110°C for 24 hr. The samples were weighed in order to know the percentage yield. The prepared carbons were stored in the self sealing covers for characterization.

1.2. *Porous characterization of the sample*

The activated carbon products were characterized by measuring their iodine number (IN), methylene blue number (MBN), and Brunauer–Emmett–Teller (BET SA) surface area. The iodine number (ASTM Designation D4607- 86) and methylene blue number (BIS 877-1977) of activated carbons (mg g^{-1}) was measured at 30°C based on the standard test method which indicate the development of micropores and of mesopores respectively [8]. BET surface area was determined using standard adsorption of N_2 gas at 77 K, using Quantochrome, NOVO 4200.

2. Results and Discussion

2.1. *Effect of the activating agents*

The chemical used as the activating agent greatly influences the IN (iodine number) i.e. micropore development of the AC (activated carbon), as can be seen from Figure -1.

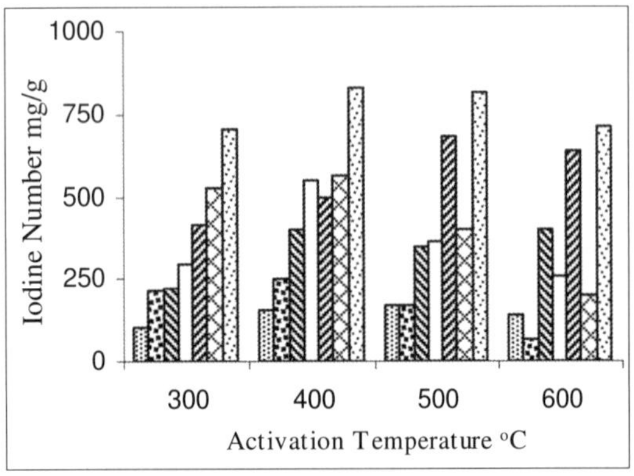

Legends for Figure 1.

HCl Raw HNO_3 NaOH

$ZnCl_2$ KOH H_3PO_4

Figure 1. Effect of activation temperature on IN.

The steady decrease in IN of raw MSC (mango seed coat) as well as NaOH and KOH shows the destruction and shrinkage of pores with an increase in temperature after 400°C. For HCl and HNO_3 the activation temperature does not influence the IN up to a great extent. For $ZnCl_2$, IN (and thereby micropore development) is highest at 500°C. However, the highest IN is that of H_3PO_4 at 400°C.

The effect of activation temperature on the MBN (methylene blue number) is characterized in Figure-2. MBN is highest for H_3PO_4 as the activating agent and second highest for KOH activation at 400°C with 1 hour activation time at 1N concentration for each. An increase in MBN signifies the initial phase of mesopores. MBN is at maximum for all the activating agents at 400°C except for NaOH (for which it is 500°C). It is seen clearly from the graphs that changing the activation temperature mainly influences the MBN for H_3PO_4 and KOH compared to all the other activating agents. With steam activation of mango pit at 750°C, MBN obtained only 31.04 mg g^{-1} (9).

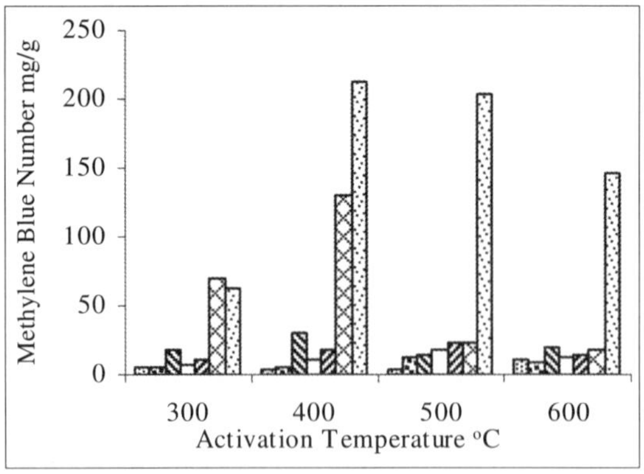

Legends for Figure 2.

HCl Raw HNO_3 NaOH

$ZnCl_2$ KOH H_3PO_4

Figure 2. Effect of activation temperature on MBN (methylene blue number).

The Figure-3 shows that the percentage yield decreases on increasing activation temperature for the various chemical activating agents. The yield is at a maximum for H_3PO_4 at 300°C followed by $ZnCl_2$ and KOH for whom it is equal at 300°C. Though it is expected that yield will increase due to the swelling or widening of pores with an increase in temperature, the temperature increase might be causing a shrinking effect in the mesopores and micropores resulting in a decrease in the percentage yield due to a high value of burn-off for most of the activating agents.

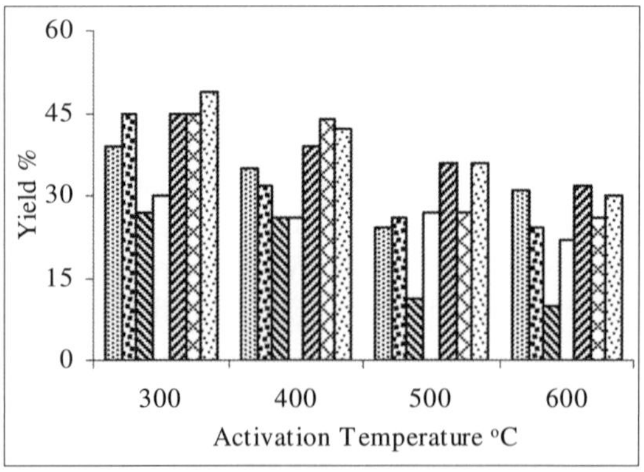

Legends for Figure 3.

HCl Raw HNO_3 NaOH

$ZnCl_2$ KOH H_3PO_4

Figure 3. Effect of activation temperature on yield.

From above figures, H_3PO_4 has the highest MBN and IN at 400°C, which suggests that the highest amount of pore development obtained using H_3PO_4 as the chemical activating agent.

2.2. *Effect of H_3PO_4 impregnation ratios*

It was observed from the above results that H_3PO_4 activating agent gives best results. The experiments were conducted to observe the influence of impregnation ratios (IR) of H_3PO_4 from 0 to 100. From the Figure-4 it is observed clearly that IN has a peak value at 75 IR, MBN has at 50 IR. It is known that as the concentration of H_3PO_4 increases the resistance to burn-off increases. It can also be seen from the Figure-4 that yield increases as IR increases. From the above results it is evident that micropore generation is highest at 75% IR.

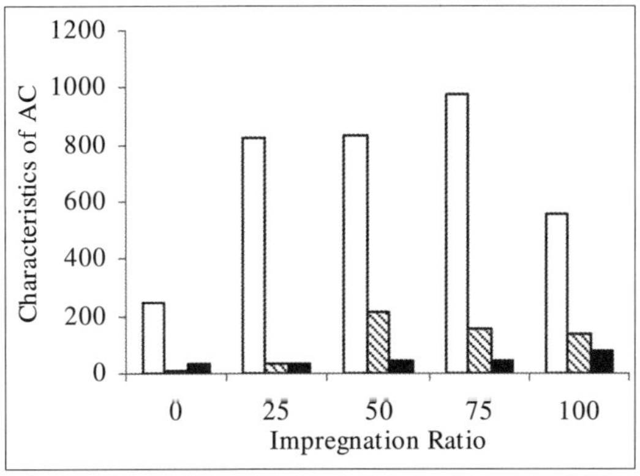

Legends for Figure 4.

IN MBN Yield %

Figure 4. Effect of IR on characteristics.

2.3. *Effect of activation temperature*

The activation temperature plays an important role in determining the pore development, surface area and percentage yield of the AC. Preparation of AC with temperature ranging from 300°C to 600°C were studied (Figure-5). The BET surface area, IN as well as methylene blue number were highest for an activation temperature at 400°C suggesting the highest possible micropore and mesopore development at this temperature. A further increase in the activation temperature results in the destruction of pores possibly due to shrinking core behavior. While increasing the temperature from 300°C to 600°C there is a predictable decrease in the yield due to a high degree of burn-off.

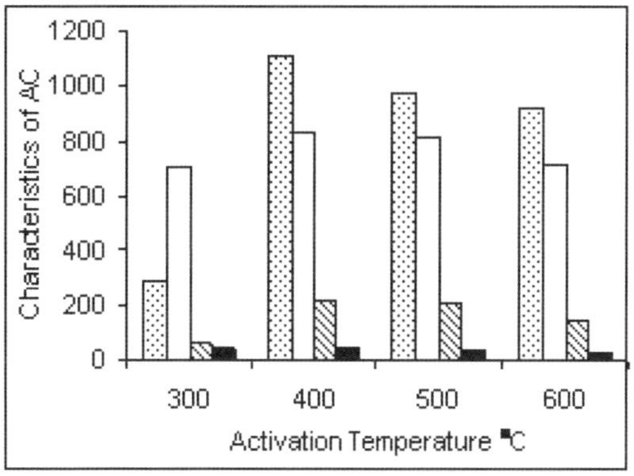

Legends for Figure 5

BET SA IN MBN

MVN Yield %

Figure 5. Effect of activation temperature on characteristics.

2.4. *Effect of activation time*

Activation time has a primary influence on the pore development process. A high iodine number for an activation time of 1 hr indicates the opening of restricted pores in the carbon skeleton (Figure-6). But as the activation time increases, the iodine number decreases which means that the presence of micropores decreases. The methylene blue number increases till 2 hr activation time but later decreases indicating that the initial phase of mesopores and pore widening process of the initially formed micropores took place up to 2 hr after which it declined. This decline in methylene blue number from 2 hr of activation to 3 hr may be the result of pore shrinkage of well developed mesopores back into the initial phase of mesopore-development. The slight increase in the methyl violet number (MVN) indicates the presence of well developed mesopores at the end of 2 hr but further increase in activation time hardly makes any changes in the methyl violet number.

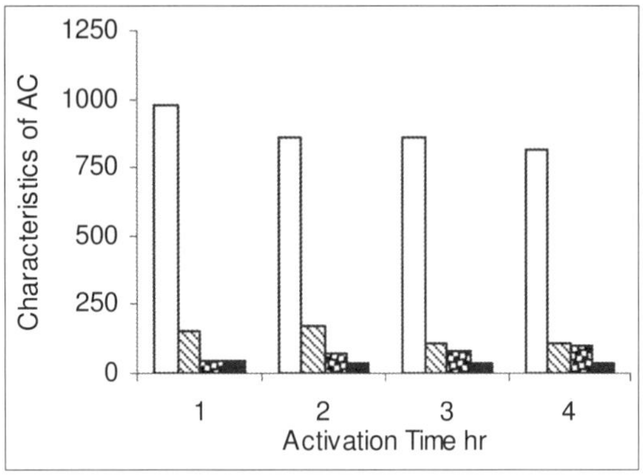

Legends for Figure 6.

IN MBN

MVN Yield %

Figure 6. Effect of activation time on characteristics.

2.5. *Surface characteristics*

To visually ensure the process of pore development, scanning electron microscope (SEM) images were taken for samples of activated carbon prepared by the chemical activation of mango seed coat using 75% H_3PO_4 for various activation time (Figure-7).

The images were taken at 1000x and 2000x magnifications. At 1hr of activation time, the process of breaking of organic material to create large number of new micropores takes place. The image having an activation time of 2 hr shows the presence of well developed mesopores. However, the activation time of 3 hr and 4 hr indicates the destruction of pores through the disintegration of the carbon skeleton.

3. Results and Discussion

The development of micropores and mesopores in the activated carbon prepared from mango seed coat using H_3PO_4 as the chemical activating agent with impregnation ratio 75% is at its highest at a temperature of 400°C and an activation time of 1 hr resulting in a BET surface area of 1114m^2 g^{-1}.

The present study on preparation of AC from mango seed coat using H_3PO_4 as the chemical activating agent shows that it would add value to these agricultural commodities as effective adsorbents. Henceforth mango seed coat can be viewed as good precursor for AC preparation.

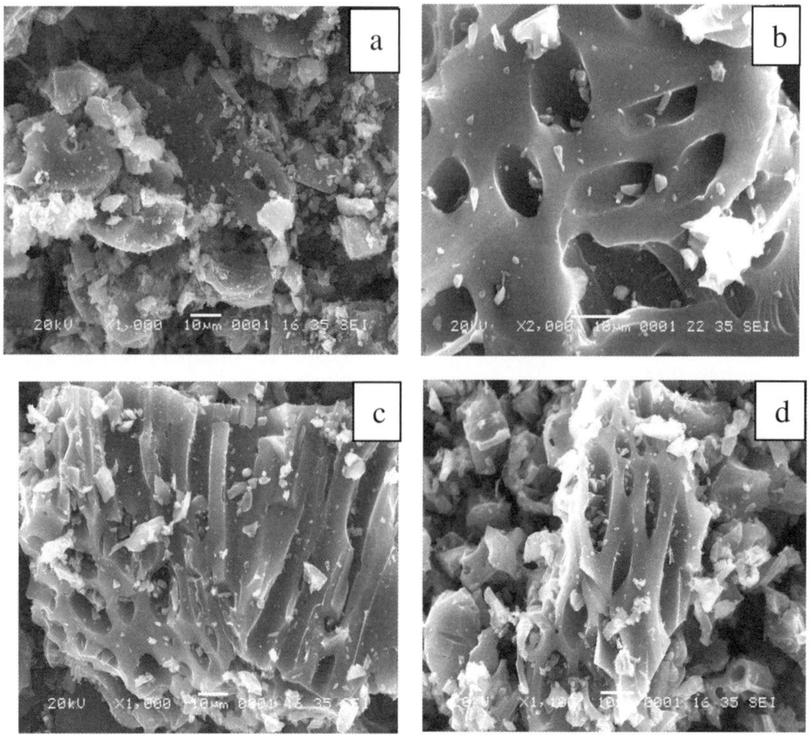

Figure 7. Surface characterization at 75% H3PO4 (a - at 1 hr, b - at 2 hr, c - at 3 hr and d - at 4 hr).

References

1. C. J. Kirubakaran, K Krishnaiah, S. K. Seshadri, *Ind. Eng. Chem. Res.* **30**, 2411 (1991).
2. Turgay Taya, Suat Ucarb, Selhan Karagöz, *Journal of Hazardous Materials*. **165**, 481(2009).
3. Devarly Prahas, Y. Kartika, N. Indraswati, S. Ismadji, *Chemical Engineering Journal*, **140**, 32 (2008).
4. V. Ponnusami, R. Madhuram, V. Krithika, S.N. Srivastava, *Chemical Engineering Journal*, **140**, 609 (2008).
5. J. Herbert Asirvatham, Nikhar Gargieya, Paradkar Manali Sunil, B. G. Prakash Kumar, and Lima Rose Miranda, *Carbon letters*, **9**, 203 (2008).
6. Amir Fouladi Tajar, Tahereh Kaghazchi, Mansooreh Soleimani, *Journal of Hazardous Materials,* **165**, 1159 (2009).
7. Kostermans, J. M. Bompard, "The Mangoes." Academic Press, London, UK, 1993.
8. http://www.idsenviroment.com/enviroment/europe/Carbochem/Activated_Carbon/19_0/s_supplier_2.html
9. M. P. Elizalde-Gonźalez, V. Herńandez-Montoya, *Biochemical Engineering Journal*, **36**, 230 (2007).

EFFECT OF PROCESS VARIABLES ON MICROWAVE, CONVECTIVE, MICROWAVE-CONVECTIVE AND SUN DRYING OF PLASTER OF PARIS

MAHESH GANESAPILLAI†

Department of Biotechnology, Sathyabama University, Old Mahabalipuram Road, Chennai, 600 119, India

LIMA ROSE MIRANDA

Department of Chemical Engineeering, A.C. Tech, Anna University, Sardar Patel Road, Chennai, 600 025, India

Batch-type experimental dryer with an online weighing mechanism was developed for determining the drying behavior of casted and tempered plaster of paris. Drying rates of plaster slabs were experimentally determined by microwave, convective, combined microwave–convective and natural open air sun drying systems at three levels of microwave output power (180, 360 and 540 W), drying air temperature (40, 50 and 60°C), different combinations of microwave power and temperature (180W/60°C, 360/50°C and 540/40°C) and sample geometries (cylindrical, rectangle and square faced cuboids) for studying the drying kinetics with respect to varied initial moisture content and thicknesses of plaster. The type of dryer, microwave output power and sample thickness had a strong effect, whereas sample geometry and temperature did not have much impact on the drying rate. Microwave drying was investigated as potential method for drying.

1. Introduction

Plaster of paris ($CaSO_4 \cdot 1/2H_2O$), is obtained by calcining gypsum or calcium sulfate dehydrate ($CaSO_4 \cdot 2H_2O$). Reciprocally, when hemihydrate grains are mixed with water, entangled gypsum needles crystallize, as a result of setting, the final product becomes rigid. The recyclable character of gypsum and its excellent insulating properties has given rise to renewed interest in this environmental material [1]. Plaster is used as fire resistance on residential and other structures, cast into various shapes including sheets, sticks, and molds. The conditions of moisture removal can be changed to adjust the porosity of the hemihydrate [2]. Several drying methods are commercially available and the selection of the optimal method for drying is determined by the quality requirements, raw material characteristics, and economic factors [3]. Advances in the drying technology have increasingly stimulated over the last decades in contrast to the traditional convective drying with hot air drying. The use of microwaves as a source of energy is rapidly growing and has already been successfully exploited to industrial and domestic applications. The objectives of the present work were to dry plaster with different levels of drying in microwave, convective, combined microwave convective and natural open air sun drying. To study the rate of moisture removal as drying rate with respect to the effect of microwave output power, temperature and combined effects of microwave power and temperature on drying kinetics.

2. Methods and Materials

2.1. *Sample Preparation*

All tests were made on cylindrical (CY), square (SQ) and rectangular (RC) faced cuboids molded in different initial moisture content (IMC) by mixing commercially available high purity plaster of paris (SRL India Ltd. Mumbai, India.) with small quantities of de-ionized water in the ratio 3:2. Slip casting was undertaken at customary ambient temperature, the freshly molded samples were initially tempered to improve the rigidity and stability of the material [4].

2.2. *Drying*

Microwave (MD), convective (CD) and combined microwave convective drying (MCD) tests were performed in a domestic microwave oven SAMSUNG C-103F model. The experimental assembly of natural open air sun drying (SD) was located on the open top

†Work completely supported by Defense Research and Development Organization, Government of India.

floor of a building to have a good view of the sky direct radiation. The temperature of plaster bed on the tray, air temperature, relative humidity just above plaster bed surface of drying sample were measured at 15 min interval during the experiments. A Kanemay 8004 model thermo-hygrometer was used to measure relative humidity.

2.3. *Experimental Procedure*

Drying of plaster was set off with seven different IMC (60, 65, 70, 75, 80, 85 and 90% (w.b.)), three sample geometries (SQ, RC and CY) and continued for a period of time with constant drying conditions. A digital balance of 0.01 g precision (Precisa XB 320M, Switzerland) was used for determination of mass. During SD, the initial weight of the samples were noted and then exposed to normal day light at ambient conditions for thirty hours. The list of experiments conducted along with the range of variables was covered as given in Table I and Table II.

Table 1. List of experiments conducted with plaster samples of 0.015 m thickness.

S. No	IMC (%)	Shape	Type of Dryer			
			MD (W)	CD (°C)	MCD (W/°C)	SD (°C)
1 2 3 4 5 6 7	60 65 70 75 80 85 90	SQ RC CY	180 360 540	40 50 60	180/60 360 /50 540 /40	35
Total Experiments			63	63	63	21

Table 2. List of experiments conducted with plaster samples of varied thickness.

S. No	Thick (m)	IMC (%)	Shape	Type of Dryer		
				MD (W)	CD (°C)	MCD (W/°C)
1 2 3 4	0.012 0.013 0.014 0.015	70 80 90	SQ RC CY	180 360 540	40 50 60	180/60 360/50 540/40
Total Experiments				36	36	36

All experiments were performed in triplicate, the average values of these replications were used for further analysis of drying parameters like drying rate, moisture content at different drying times. In all drying methodologies, the initially dried samples were then placed in a hot air oven set at 105°C until bone dry, finally the weight of the sample was noted. As such, a total sum of 318 experiments was conducted, which includes samples of uniform and varied thicknesses.

3. Results and Discussions

3.1. *Effect of Drying Methodology*

The drying curves for plaster under MD, CD, MCD and SD conditions are shown in Figure 1. Plaster samples with 90% IMC dried up to 10.83% within 640 s at 180 W for a sample thickness of 0.015 m, whereas the same reached 31.99% in 320 s at 540 W with MD. Convection drying curves followed a same trend for all temperature ranges. As expected; there was an acceleration of the drying process due to the increase of air temperature. Treatments of 180 W at 60°C required about 20 min to obtain plaster with final moisture content of 19.57% (w.b.) for SQ cuboids with 90% IMC, whereas those of 360W/50°C and 540W/40°C reached a final moisture content of only 30.02 and 29.67% in the stipulated time period of 20 min. The positive interaction between microwave and hot air treatments on plaster is possibly due to a direct heat generation inside plaster, coupled with the convective heat transfer at the surface by hot air [5] and [6]. However, the MW powers dominated the shortening in drying time rather than the hot air temperatures. SD solely depends for their operation entirely on solar-energy, where air is circulated through the sample by the result of wind pressure. When the IMC remained at 90% (w.b.), the samples were dried to a maximum of 76.84 to 81.06% (w.b.) in 30 hours for all three geometries.

3.2. *Effect of Microwave Power*

The effect of changing the microwave output power on dehydration characteristics of 0.015 m plaster samples were shown in Fig. 1(a). The results indicated that mass transfer is rapid with minimum MWP (180W), because more heat is generated within the sample, creating a larger vapor pressure differential between the centre and the surface of products. The observed initial acceleration in drying may be caused by an opening of the physical structure allowing rapid evaporation and transport of water [7]. However higher MWP (above540 W) has also been associated with physical damages to the products. The moisture content of the material was very high during the initial phase of drying, which resulted in higher absorption of MWP due to the higher moisture diffusion. As the drying progressed the loss of moisture in the product caused a decrease in the absorption of MWP and resulted in a falling drying rate. Maximum value of moisture removal from the materials at 180 W MWP was obtained during 80 to 180 s of the drying period. About 60% of the drying was completed by 480 s at 180 W, whereas the same reached in 320 s with 360 and 540 W. The drying rates increased with the decrease in power level.

3.3. *Effect of Temperature*

The dependence on temperature was greater at the beginning of the process and almost negligible after saturation. Drying plaster samples at 60°C decreased the drying time compared to 40 and 50°C, to reach the target moisture content of 32.13% (Fig. 1(b)) for SQ samples. Plaster dried at 60°C for 8 h or more had significantly lower moisture content than other temperatures. It has been reported that almost all of the drying takes place in falling rate period [8]. However, Mowlah et al (1983) [9] have found both constant and falling rate periods for fruit samples. Though casehardening was encountered with hot-air dried products, an increase in air temperature decreased drying time for all samples.

3.4. *Effect of Initial Moisture Content*

Since the solvent used was water (dipolar), they absorb microwave instantly and rapidly. During all the experiments in MD, it was observed that the drying rate was higher for samples with higher IMC, which is impossible to achieve by other conventional drying methods. In the air drying process, drying rate decreased with decrease in IMCs, which might be attributed to slower moisture migration rate from interior to surface and evaporation rate from surface to air caused by lower sample moistures. These results are in agreement with the previous observation [10, 11]. The time taken to reach a final moisture content of 10.83% (w.b.) from 90% for MD was 660 s at 180 W MWP for a sample thickness of 0.015 m for RC cuboids, whereas CD samples reached 56.54% (w.b.) in 8 h at 60°C. MCD of plaster at 180 W and 60°C required about 20 min to obtain a final moisture content of 15.37% (w.b.) for a RC cuboids sample with 90% IMC (Fig. 1(c)), whereas it took 30 hrs to dry the same sample in natural open air sun drying to reach a final moisture content of 51.39%.

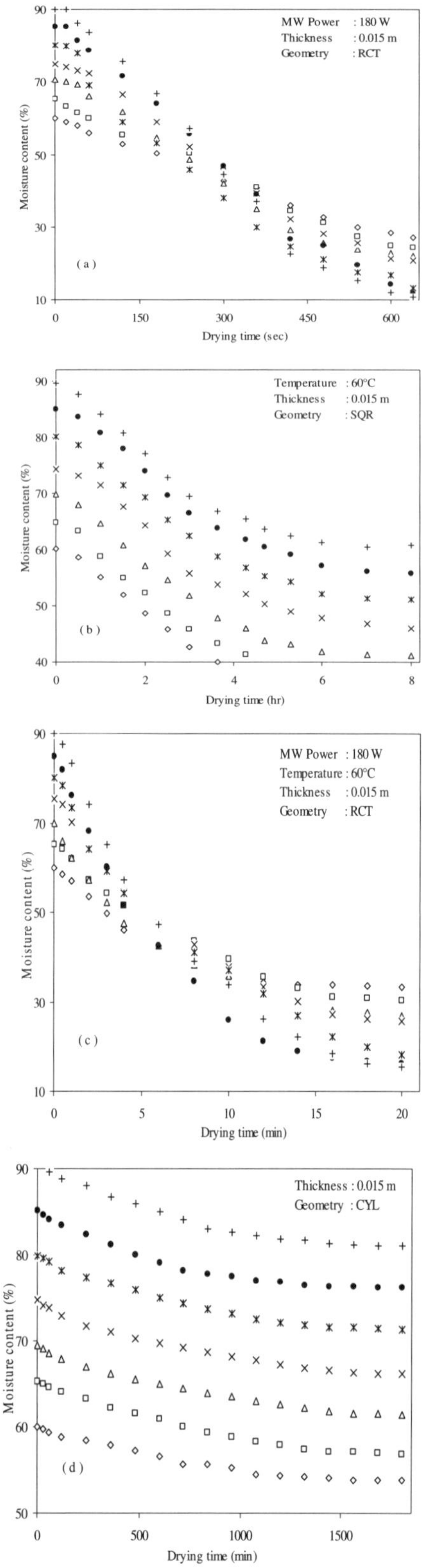

Fig. 1 (a)-(d) Drying curves of plaster at various IMC, (a) Microwave drying, (b) Convective drying, (c) Combined microwave convective drying, (d) Natural open air sun drying (◊, 60% : □, 65% : Δ, 70% : x, 75%: *, 80%: •, 85%: +, 90%).

3.5. *Effect of Geometry*

As reported by several authors [12, 13] no significant changes were observed for plaster samples. SD process reduced the moisture content of plaster from 60% to moisture content of 51.67% (w.b.) for SQ cuboids in 1800 min, when exposed at sunlight at a temperature of 35°C (Fig.1(d)). Whereas in RC and CY cuboids, final moisture content reached 51.39% and 53.80% (w.b.) respectively for the same duration. It was observed that the drying rate was high for RC cuboids due to the comparatively larger surface, further the drying rates were identical with other geometries.

3.6. *Effect of Thickness*

3.6.1. *Microwave drying kinetics*

Plasters having 70% IMC dried to a final moisture content of 24.06, 25.17, 26.22, and 27.92,% within 640 s at MWP level of 180 W for sample thicknesses of 0.012, 0.013, 0.014 and 0.015 m respectively, at pressure level of 15 kPa for SQ cuboids (Fig. 2a). Whereas, final moisture content of 11.36, 12.08, 12.98 and 13.71% was obtained for the same geometry in 640 s with 90% IMC at 180W. SQ cuboids dried up to 13.56, 14.98, 16.17 and 17.22% for 0.012, 0.13, 0.14 m and 0.015 m thicknesses respectively in 640 s. The drying rate was indirectly proportional to sample thickness.

Faster drying time was achieved to remove the moisture content and it may be due to the rapid diffusion process. Thus, the MD times were around 120% shorter at 180 W than the corresponding convective air drying times for all sample thicknesses. These results are in agreement with the previous observation of different food drying [14].

3.6.2. *Convective drying kinetics*

It has been observed that the maximum moisture removal occurred in sample with a thickness of 0.012 m (final moisture content – 4.89%) in a steady drying time of 900 min when the IMC was maintained at 70% (w.b.). Whereas the IMC increased from 70 to 80% (w.b.) the final moisture content touched 51.01, 52.80, 54.11 and 55.66% (w.b.) in 720 min (Fig. 2b), for sample thicknesses of 0.012, 0.013, 0.014 and 0.015 m at consistent temperature (50°C), for RC cuboids. Whereas the temperature was raised to 60°C the samples responded faster drying rate, the final moisture content reached 51.22, 52.89, 55.02 and 56.54% from 90% (w.b.) final moisture content in 480 min for 0.015 m samples (RC).

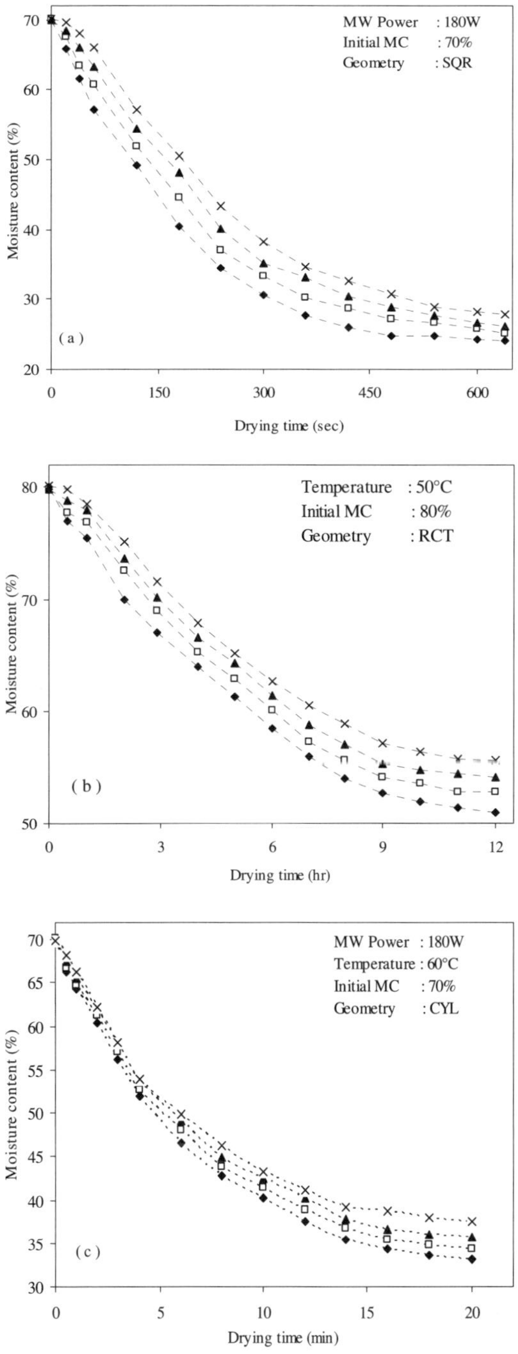

Fig. 2. (a)-(c) Drying curves of plaster at varied sample thickness, (a) Microwave drying, (b) Convective drying, (c) Combined microwave convective drying, (♦, 0.012 m: □, 0.013 m: ▲, 0.014 m: x, 0.015 m).

3.6.3. *Combined microwave-convective kinetics*

The sample thickness provides an important driving force to overcome all mass transfer changes within the sample, thereby increasing the drying rate. The final moisture content reached 33.17, 34.43, 35.66 and 37.49% (w.b.) for CY samples of 0.012, 0.013, 0.014 and 0.015 m thickness respectively, consuming 1200 s at 180W/60°C (MWP/Temperature) (Fig. 2c). RC and SQ samples followed the same trend. During MD the dielectric properties, particularly dielectric loss of molded plaster may change affecting microwave absorbed in the sample. Feng et al (2001) [15] reported that a decrease in moisture content resulted in a decrease in dielectric constant and dielectric loss of dehydrated components. Therefore MCD yielded higher drying rate than that of the conventional hot air drying. This phenomenon also agreed with the findings by Ruız et al (2003) [16].

4. Conclusions

The kinetics of microwave, convective, combined microwave convective and natural open air sun drying of plaster of paris can be successfully described with all process conditions, although microwave drying and sample thickness had significant effect, whereas sample geometry and temperature did not have much impact on the drying rate. A significant difference was observed in drying times, where the shortest time was obtained using microwave drying. The longest drying time was observed for natural open air sun drying. Drying rate and process duration depend on microwave output power, initial moisture content and sample thickness of sample. Higher rates and shorter drying times were achieved at lesser microwave output power. The results obtained imply that microwave drying is the most appropriate method for drying plaster, with regard to drying rate and process duration.

References

1. G. Dumazer, V. Narayan, A. Smith and A. Lemarchand, J. Phys. Chem. C, **113**, 1189, (2009).
2. G. Mahesh, I. Regupathi and T. Murugesan, Dry. Tech. **26**, 963, (2008).
3. A. Reyes, C. Campos and R. Vega, Dry. Tech. **24**, 1469, (2006).
4. S.G. Walde, K. Balaswamy, V. Velu and R.G. Rao, J. of Food Eng., **55**, 271, (2002).
5. T.N. Tulasidas, G.S.V Raghavan and E.R. Norris, ASAE, **36**, 1861, (2002).
6. H. Feng and J. Tang, Dry. Tech.**18**, 1183, (2000).
7. A.E Kostaropoulos and G.D Saravacos, Food Sci., **37**, 1043, (1995).
8. P.S.Madamba, R.H. Driscoll and K.A. Buckle, J. of Food Eng., **29**, 75, (1996).
9. G. Mowlah, K. Takano, I. Kamoi and T. Obara, Lebensmittel Wissenshaft und Technologie, **16**,103, (1983).
10. C. Ertekin and O. Yaldiz, J. of Food Eng., **63**, 349, (2004).
11. T. Gunhan, V. Demir, E. Hancioglu and A. Hepbasti, Energy Conser. and Mgt., **46**, 1667, (2005).
12. D.G. Prabhanjan, H.S. Ramaswamy and G.V.S. Raghavan, J. of Food Eng., **25**, 283, (2005).
13. J. Wang. and Y.S. Xiong, J. of Food Eng., **68**, 505, (2005).
14. H. Sabarez, W.E. Price, P.J. Back and L.A Woolf, Food Chem., **60**, 371, (1997).
15. H. Feng, J. Tang, R.P. Cavalieri and O.A Plumb, AIChEJ., **74**, 1499, (2001).
16. C.G. Ruíz, M.J. Martínez, P. Fito and A. Chiralt, Innovative Food Sci. and Emerging Technologies, **4**, 203, (2003).

RESPONSE SURFACE METHODOLOGY STUDY ON REMOVAL OF HUMIC ACID FROM AQUEOUS SOLUTIONS USING ANIONIC CLAY HYDROTALCITE*

YAMIN YASIN†

Chemistry Department, International Education Centre, Universiti Teknologi MARA, Kampus Seksyen 17 40200, Shah Alam, Selangor, Malaysia

SITI MARIAM SUMARI AND ABD HAFIZ ABD MALEK

Department of Environmental Technology, Faculty of Applied Sciences, Universiti Teknologi MARA 40450, Shah Alam, Selangor, Malaysia

Removal of humic acid from aqueous solutions using anionic clay hydrotalcite was carried out. Response Surface Methodology (RSM) based on a five-level, four variable central composite rotatable design (CCRD) was used to evaluate the interactive effect of various parameters. The parameters were: time (2-6 h); temperature (30-70°C); amount of dosage (100-300 mg) and initial concentration of humic acid solution (50-100 mg/l). Simultaneously increasing adsorption time, temperature, adsorbent dose and initial concentration of humic acid increased the percentage of humic acid removal from aqueous solution. The optimum conditions derived via RSM for the removal of humic acid were adsorption time of 4.0 h, temperature 50°C, adsorbent dose of 200 mg and initial concentration of 75 mg/l.

1. Introduction

The presence of humic acid has been a problem in the water industry and in environmental purification such as soil remediation. The contamination of humic substances has been known to induce a deterioration of adsorbents and a prevention of adsorption onto the adsorbents. There are various methods of removing humic substances using conventional methods including coagulation and flocculation, oxidation or ozonation and membrane separation [1]. However, these methods are not widely used due to their high cost and economic disadvantage. In contrast, an adsorption technique is by far the most versatile and widely used and has proven successful in removing humic substances.

Hydrotalcite is a class of anionic clays with high anion exchange capacities, are effective adsorbents for removal of a variety of anionic pollutants. The chemical composition of hydrotalcite can be described by the formula $[M^{2+}_{1-x}\ M^{3+}_{x}\ (OH)_2]^{x+}(A^{n-})_{x/n}.mH_2O$ (1) where M^{2+} and M^{3+} are metal cations, for example Mg^{2+} and Al^{3+}, that occupy octahedral sites in the hydroxide layers, A^{n-} is an exchangeable anion, and x is the ratio $M^{3+}/(M^{2+} + M^{3+})$. Carbonates are the interlayer anions in the naturally occurring mineral hydroxides, which is a member of this class of materials. The hydrotalcite's anion exchange ability, large surface area and regeneration ability ensures that this adsorbent can be effectively utilized in wastewater purification. Response surface methodology (RSM) is an effective statistical technique for the investigation of optimization process. RSM is a collection of mathematical and statistical techniques useful for developing, improving and optimizing the processes and can be used to evaluate the relative significance of several affecting factors even in the presence of complex interactions [2]. The main objective of RSM is to determine the optimum operational conditions of the process or to determine a region that satisfies the operating specifications [3]. RSM comprising a five level-four-factor central composite rotatable design (CCRD) was used in our work to evaluate the interactive effects and to obtain the optimum conditions for humic acid removal from aqueous solution using anion clay hydrotalcite as an adsorbent.

*This work is supported by etc, etc.

†Work partially supported by grant 2-4570.5 of the Swiss National Science Foundation.

2. Materials and Methods

2.1. *Preparation and characterization of hydrotalcite*

Hydrotalcite was prepared from magnesium sulphate hydrate [$Al_2(SO_4)_3.16H_2O$], aluminium sulphate hydrate [$Mg(SO_4).7H_2O$] and sodium carbonate at moderate conditions following Reichle's procedures (Reichle 1998). X-ray diffraction (XRD) pattern of the sample powder was recorded on a Siemen diffractometer D500 with Ni filtered, CuKα radiation at 40Kv and 20 mA. The sample was mounted on a glass slides and scans at 2°-65° 2θ/min at 0.003° steps. The basal spacing was determined via powder technique.

2.2. *Experimental design*

A five-level-four-factor CCRD was employed in this study, requiring 21 experiments [3]. The fractional factorial design consisted of 8 factorial points, 8 axial points and 5 center points. The variable and their levels selected for the removal of humic acid were: time (2-6 h); temperature (30-70°C); amount of dosage (100-300 mg) and initial concentration of humic acid solution (50-100 mg/l). The data obtained were fitted to second order polynomial equation (1) as shown below:

$$Y = \beta_0 + \sum_{i=1}^{4} \beta_i x_i + \sum_{i=1}^{4} \beta_{ii} x_i^2 + \sum_{i=1}^{3} \sum_{j=I+1}^{4} \beta_{ij} x_{ij} \quad (1)$$

where Y is percentage of humic acid removal; β_0, β_i, β_{ii}, β_{ij} are constant coefficients and x_i the uncoded independent variables.

2.3. *Batch adsorption and analysis*

Humic acid purchased from Merck was used for the adsorption experiment. Stock solutions of humic acid were prepared in deionized water and were diluted according to the working concentration. Humic acid concentration was measured using Lambda 20 uv-visible spectrophotometer. 0.25 g of synthesized compounds was put into 25 ml of 50-100 mg/l solutions of the humic acid solutions. Twenty five milliliters of humic acid solution in 250 ml conical flask was contacted with 0.25 g hydrotalcite and was agitated using water batch shaker operating at 120 rpm. Experiments were performed according to the central composite rotatable design (CCRD) as given in Table 1. The response was expressed as percentage of humic acid removal, calculated as [$(C_0 - C_t) / (C_0)$] x 100.

2.4. *Data analysis*

The data from the experiments performed are analyzed using design expert 6.06 version and then interpreted. Three main analytical steps; analysis of variance (ANOVA), a regression analysis and the plotting of response surface were performed to establish an optimum condition for the removal of humic acid.

Table 1. Central composite rotatable quadratic polynomial model, experimental data, actual and predicted values for five-level-four-factor response surface analysis.

Exp.	Time (h)	Conc. (mg/l)	Temp. (°C)	Dos. (mg)	Actual (%)	Predicted (%)
1	4	75	50	200	70	70.5
2	6	50	30	300	70	65.5
3	6	50	70	300	35	35.5
4	4	32.9	50	200	78	73.5
5	4	75	50	200	35	30.5
6	4	75	50	200	27	27.5
7	2	100	70	300	79	79.5
8	2	50	70	100	29	24.5
9	4	75	50	200	20	22.9
10	2	100	30	300	78	80.8
11	6	100	70	100	15	17.9
12	4	75	83.6	200	80	82.9
13	4	75	16.4	200	55	63.9
14	4	75	50	360	75	71.8
15	2	50	30	100	35	37.9
16	7.5	75	50	200	79	81.9
17	4	75	50	30	75	73.5
18	6	100	30	100	75	73.5
19	4	117	50	200	75	73.5
20	4	75	50	200	75	73.5
21	0.5	75	50	200	74	73.5

3. Results and Discussion

3.1. *Characterization of hydrotalcite*

The XRD pattern of original hydrotalcite synthesized at different ratio are presented in Figure 1. As shown in the Figure, original hydrotalcite indicate fairly good crystallinity, with d-spacing 7.9Å which is demonstrated general features of hydrotalcite. The d-spacing show the characteristic values for trigonal structures with symmetrical peaks assigned to the (003) and (006) planes, respectively. The interlayer spacing of the sample corresponding to the 006 plane was found to be 4.5 Å. Hydrotalcite at ratio 4 was subsequently chosen as adsorbent for removal of humic acid in this study.

3.2. *Model fitting and ANOVA*

Experimental data for removal of humic acid using anion clay hydrotalcite are given in Table 1. The predicted values were obtained from model fitting technique using the software design expert version 6.06 and were seen to be sufficiently correlated to the observed values. Fitting of the data to various models (linear, two factorial, quadratic and cubic) and their subsequent ANOVA showed that removal of humic acid from aqueous solution using hydrotalcite were mostly suited with quadratic polynomial model. From the design expert, the quadratic polynomial model is given as Eq. (2) below:

$$\text{Removal (\%)} = 73.46 + 17.24\,A + 19.32\,B + 2.39\,C + 13.08\,D - 7.64\,A^2 - 8.17\,B^2 - 1.98\,C^2 - 4.81\,D^2 \quad (2)$$

where A is the time; B the concentration; C the temperature and D is the amount of adsorbent dosage. The computed model *F-value* of 19.80 was higher than tabular value of F = 12.97, implying the model are significant at 1% confidence level. The model also showed a very low value of pure error of 0.8 which indicate good reproducibility of the data obtained. The high coefficient of determination (R^2 = 0.9788) and a very small P-value (0.0001) from the analysis of ANOVA showed that, quadratic polynomial was highly significant and sufficient to represent the actual relationship between the response (% removal) and the significant variables. The variables of time, concentration and dosage of hydrotalcite are the most significant in adsorption of humic acid from aqueous solution using hydrotalcite. This result indicated that time, concentration and dosage of hydrotalcite were the important variables.

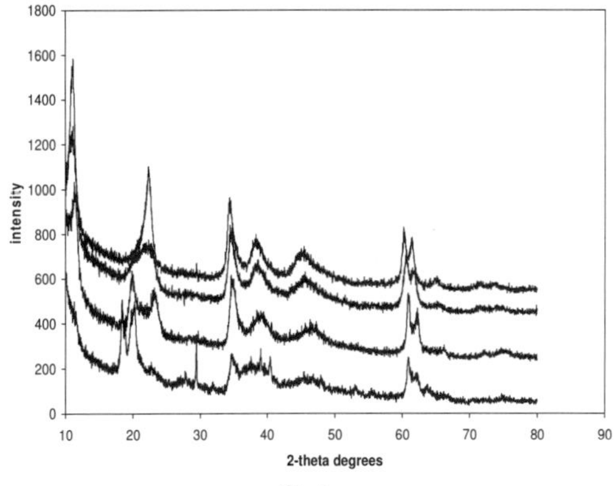

Figure 1. XRD pattern of synthesized hydrotalcite at different preparation ratio.

3.3. *Response surface plot*

The quadratic polynomial equation was then used to facilitate plotting of response surfaces. Two parameters were plotted at any one time on the x_1 and x_2 axes, respectively, with the other two remaining parameters set at their centre points values (coded level:0). Figure 2 shows the profile of time and initial concentration of humic acid solution on the removal of humic acid using anionic clay hydrotalcite. As expected, increase in both time and initial concentration of humic acid solution will increase in percentage removal of humic acid from aqueous solution. This indicates that higher the contact time between humic acid and hydrotalcite, higher the removal efficiency till the equilibrium time is reached.

The same trend of adsorption was for dyes removal on novel adsorbent [2]. As shown in the Figure, humic acid removal is increasing with increasing humic acid concentration. The adsorption is fast at early stages and will approach equilibrium after 4 hours. Adsorption with high concentration of humic acid and longer contact time between humic acid and hydrotalcite favored maximal percentage removal of humic acid.

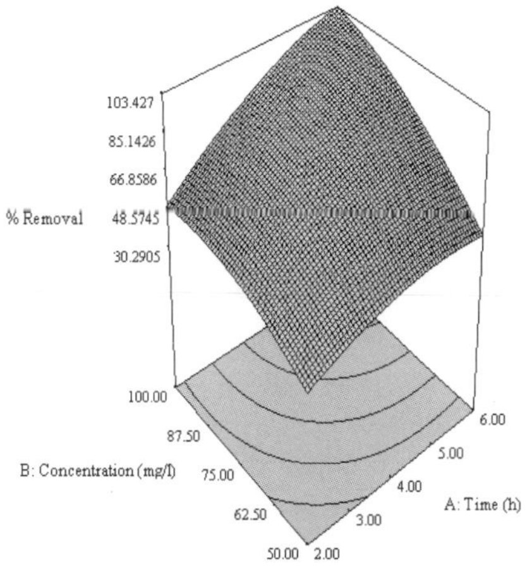

Figure 2. Response surface plot showing the effect of time and concentration and their mutual effect on removal of humic acid using anionic clay hydrotalcite.

Response surface predicted for interaction of time and temperature of adsorption is illustrated in Fig. 3. The effects of temperature can be appointed to its effects on adsorbate solubility as well as its direct influences on the adsorption using hydrotalcite. On increasing reaction temperature, adsorbate solubility is improved reducing mass transfer limitations and making the adsorbate more available to the hydrotalcite. Higher reaction temperatures also promote collisions between molecules to result in accelerated rates of reaction. As shown in the Figure, adsorption with highest temperature and highest contact time favored maximal percentage adsorption. The increase of efficiency in humic acid removal with the increase of the temperature suggested that anion-exchange mechanism may occur during the adsorption.

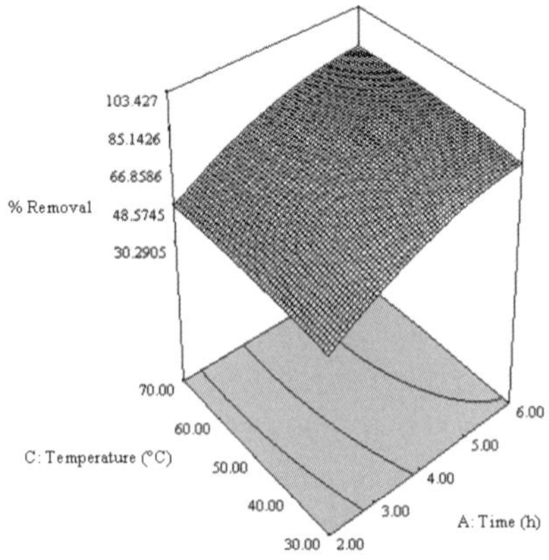

Figure 3. Response surface plot showing the effect of time and temperature and their mutual effect on removal of humic acid using anionic clay hydrotalcite.

Figure 4 depicts the response surface plot showing the effect of time and amount of adsorbent dosage on the removal of humic acid using hydrotalcite. As shown in Figure, at any given amount of adsorbent dosage from 100 mg to 300 mg, and increased contact time of adsorbent and humic acid solution led to higher percentage of removal. As expected, adsorption rate increase with increase in adsorbent dosage. This is due to the fact that, increase in adsorbent dosage increase area available for adsorption.

Figure 5 and Figure 6 depict the response surface plots as a function of concentration versus temperature and concentration versus amount of adsorbent dosage respectively. As expected, adsorption with highest concentration of humic acid and highest temperature favored maximum percentage of adsorption. Figure 6 represents the effect of varying concentration and amount of dosage of hydrotalcite. At low amount of dosage and low concentration of humic acid, the percentage removal of humic acid was lower. High percentage of removal was recorded when high amount of hydrotalcite and high concentration of humic acid were used. Presence of larger amount of adsorbate generally increases the availability of adsorbate to adsorbent. This relationship holds when there are no limiting factors such as presence of activators or inhibitors or mass transfer effect.

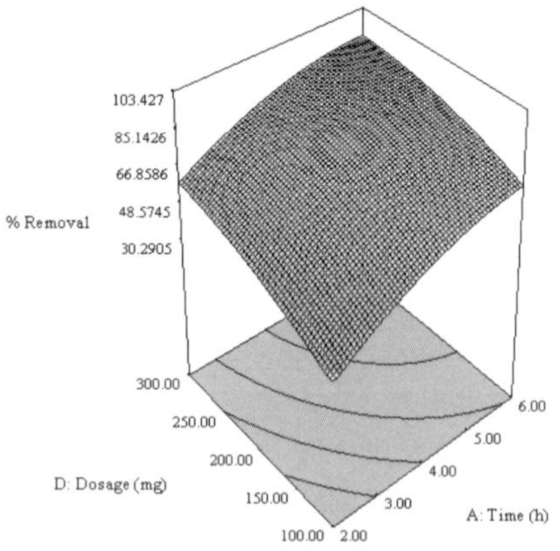

Figure 4. Response surface plot showing the effect of time and dosage and their mutual effect on removal of humic acid using anionic clay hydrotalcite.

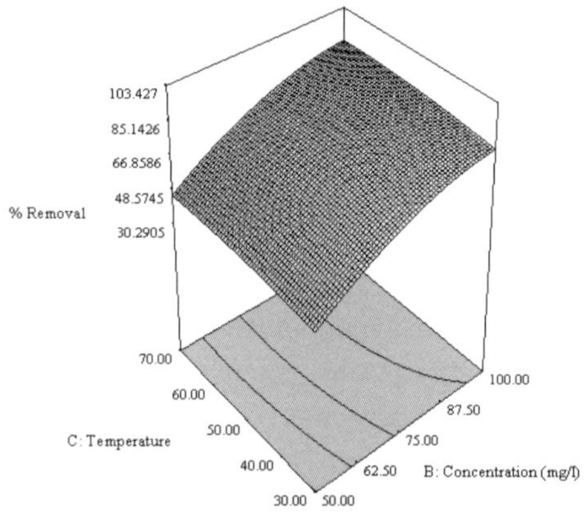

Figure 5. Response surface plot showing the effect of concentration and temperature and their mutual effect on removal of humic acid using anionic clay hydrotalcite.

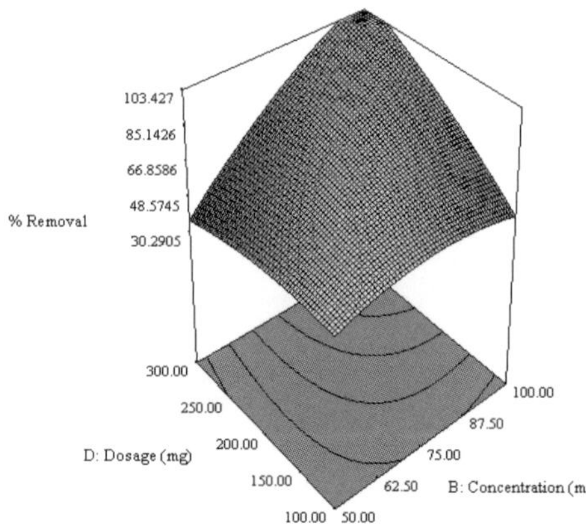

Figure 6. Response surface plot showing the effect of concentration and dosage and their mutual effect on removal of humic acid using anionic clay hydrotalcite.

3.4. *Optimization of adsorption*

Within the experimental range studied, optimum conditions for removal of humic acid using anionic clay hydrotalcite were predicted using the optimization function of the Design Expert. These are presented in Table 2 along with their predicted and actual values. The analysis indicated that maximum percentage removal of humic acid was at adsorption time of 4.0 h, temperature 50°C, adsorbent dose of 200 mg and initial concentration of 75 ppm. The response behavior of time and initial concentration of humic acid (Figure 3) were followed while holding the other reaction parameter constant at suggested point. Therefore the reaction conditions (adsorption time 4.0 h, temperature 50°C, adsorbent dose 200 mg and initial concentration 75 ppm) were recommended as optimal for removal of humic acid using anionic clay hydrotalcite. Comparison of predicted and experimental values revealed good correspondence between them, implying that empirical models derived from the RSM could be used to adequately describe the relationship between the factors and response in removal of humic acid using anionic clay hydrotalcite.

Table 2: Solutions of optimum conditions

Exp.	Time (h)	Conc. (mg/l)	Temp. (°C)	Dos. (mg)	Actual (%)	Predicted (%)
1	4	75	50	200	75.2	73.5
2	2	50	30	100	63.2	66.9
3	6	100	70	300	82.4	79.9

4. Conclusion

The adequacy of the predicted model was examined by additional independent experiments under the suggested optimal removal conditions. From the data obtained, the observed values were statistically near the predicted values and the generated model adequately predicted the percentage removal. Thus, the removal of humic acid using anionic clay Hydrotalcite was successfully developed by fractional factorial design of RSM.

References

1. C. P. Schutless and C. P. Huang, *Soil Sci. Soc Am J,* **55**, (1991)
2. K. Ravikumar, S. Krishnan, S. Ramalingam and K. Balu, *Dyes pigments* **72**, 66 (2007).
3. R. H. Myers and D. C. Montgomery. Response Surface Methodology. *Second ed. Wiley*. (2001)

Session 8

MOISTURE DEPENDENT GEOMETRICAL PROPERTIES OF SUNFLOWER SEED, AZARGOL VARIETY AS A CASE STUDY

R. KHODABAKHSHIAN

Department of Agricultural Machinery, Ferdowsi University of Mashhad, P.O. Box 1163, Mashhad, Iran

B. EMADI

Department of Agricultural Machinery, Ferdowsi University of Mashhad, P.O. Box 1163, Mashhad, Iran

M. H. ABBASPOUR FARD

Department of Agricultural Machinery, Ferdowsi University of Mashhad, P.O. Box 1163, Mashhad, Iran

M. H. SAIEDIRAD

Technical and Engineering Center of Agricultural Ministry, Khorasan Razavi, Mashhad, Iran

The knowledge of geometrical properties such as shape and size is essential for the design of processing equipment. Some geometrical properties of sunflower seed, Azargol variety as a case study, was investigated at various moisture content levels (3–14% d.b.) in three categories of big, average and small sizes. The ranges of length, width and thickness were 13.31–15.97, 5.01–7.50 and 3.11–4.38 mm, respectively. Also, the ranges of arithmetic, geometric and equivalent diameter were 5.93–8.07, 7.15–9.29 and 6.04–8.26 mm, respectively. The sphericity and surface area varied from 0.44 to 0.53 and 110.47 to 204.6 mm², respectively.

1. Introduction

Sunflower seed (*Helianthus annuus L.*) is an important oilseed crop because it contains a large amount of highly nutritious oil (Shukla et al., 1992). About 35 different varieties of sunflower are cultivated in Iran (Iranian Agricultural ministry's Reports, 2005). In this country, sunflower seeds are cultivated and consumed in two different ways either directly as nut or as cooking oil. The geometrical properties of sunflower seed, like those of other grains and seeds are essential for the design of processing equipment including separating, handling, sorting, sizing, drying, and dehulling. These properties affect the conveying characteristics of solid material by air or water and cooling and heating loads of food materials (Sahay and Singh, 1994). Many researchers have determined the geometrical properties of different types of grains and seeds such as soybean (Deshpande, et al., 1993), oil bean seed (Oje and Ugbor, 1991), neem nuts (Visvanathan, et al., 1996), karingda seed (Suthar and Das, 1996), cumin seed (Singh and Goswami, 1996), sunflower seeds (Gupta and Das, 1997), chick pea seeds (Konak et al., 2002).

Very limited published results on the detailed geometrical properties of sunflower seed and its dependency on functional parameters are available. Some physical properties of sunflower seeds (Modern variety) as a function of moisture content have been reported by Gupta and Das (1997). Arithmetic mean diameter, geometric mean diameter, surface area and volume as function of size and variety have not been investigated yet. They indicated that the length of the seed was closely related to its thickness and width while there was less association between mass of seed and its dimensions. Also, they revealed that the length of the kernel is closely related to its width but not with either thickness or mass. Isik and Izil (2007) have also investigated some moisture-dependent properties of sunflower for only the Turkey variety of sunflower seed. They have revealed an increase in geometric mean diameter when moisture content varies from 10.06 to 27.06%. The current study was conducted to investigate the effect of moisture content, size and variety on geometrical properties of Azargol variety as one of the most common Iranian varieties of sunflower seed. The properties including size, geometric mean diameter, arithmetic mean diameter, equivalent diameter, sphericity and surface area were determined at various moisture contents ranged 3–14% d.b.

2. Materials and Methods

For this study, Azargol variety was obtained from a region of Khorasan Razavi province, Iran, during autumn season in 2008. A portion of seeds equal to twenty kilograms was selected and transported to the lab. The sunflower seeds were manually cleaned to get rid of foreign matters, broken and immature seeds. The initial moisture of seeds was obtained 7.2% d.b. using the standard hot air oven method with a temperature setting of 105 ± 1°C for 24 h (Gupta and Das, 1997). To study the effect of seed size on geometrical properties, the seeds were graded into three size categories (small, medium and large) using 5.5, 6.5 and 8 mesh sieves. All the geometrical properties of the seeds were measured for three moisture contents in the range of 3 to 14% (d.b.).

To get the seeds with the desired moisture contents, sub-samples of seeds from each category (small, medium and large), each weighing 0.5 kg, were drawn form the bulk sample and dried (by putting them in the oven at 75°C for 2 h) or adding calculated quantity of water to the seeds then, pre-determined quantity of water was added to the sub-samples and they were kept in double-layered low-density polyethylene bags of 90μm thickness, sealed and stored at low temperature (5°C in a refrigerator) to avoid the growth of microorganisms and allowing to uniformity of moisture distribution. Before starting the tests, the required quantities of seed was taken out of the fridge and allowed to warm with room temperature for approximately 2 h (Joshi et al, 1993; Gupta and Das, 1997).

To determine size and shape of the seed for revealing the relations among length, width and thickness with each other in, totally 50 seeds of each sub-samples were randomly selected and labeled for easy identification. Finally, three main dimensions namely length, *L*, width, *W*, and thickness, *T*, of seed were carefully measured using a digital vernier caliper (Diamond, China) with an accuracy of ±0.02 mm. The average diameter of seed was computed using geometric mean, arithmetic mean and equivalent diameters of the three axial dimensions. The geometric mean diameter, D_g, arithmetic mean diameter, D_a, and equivalent diameter, D_e, were calculated using the following equations (Mohsnin, 1986):

$$D_g = (L.D.T)^{1/3} \quad (1)$$

$$D_a = \frac{(L+W+T)}{3} \quad (2)$$

$$D_e = [L.\frac{(W+T)^2}{4}]^{1/3} \quad (3)$$

The criteria used to describe the shape of the sunflower seed was the sphericity. The sphericity, φ, of seed was determined using following formula (Mohsnin, 1986):

$$\varphi = \frac{(L.D.T)^{1/3}}{L} \quad (4)$$

The surface area of the seeds was determined by similarity with a sphere of the same geometric mean diameter, using the following expressions (Altuntas et al. 2005):

$$S = \pi.D_g^2 \quad (5)$$

The tests were carried out with five replications for each level of moisture content, categories and the average values reported. Average, minimum, maximum, standard deviations, correlation coefficients of dimensions and regression equations were computed using Microsoft Excel software (2003).

3. Results and Discussions

All main dimensions (length, width and thickness) for each size of seed have increased linearly with increase in moisture content (Table 1). This indicates that during the moisture absorption process, sunflower seed will expand in length, width and height. Deshpande et al. (1993) reported similar results for soybean. For the studied range of moisture content (3-14% d.b.), the total average expansion was maximum along the seed length and minimum along its thickness. Deshpande et al. (1993) found that the expansion rate of soybean seeds to be the largest along their thickness in comparison with their other two principal axes. The ranges of length, width and thickness of sunflower seed were 13.31–15.97, 5.01–7.50 and 3.11–4.38 mm, respectively. Gupta and Das (1997) reported the ranges of 8.92–9.52

mm for the length of Modern variety of sunflower seed. In addition, they also reported the ranges of 3.92–5.12 mm for width and thickness of Modern variety of sunflower. In the other study, the average length, width and thickness of sunflower seed were indicated 7.79, 7.12 and 4.18 mm at a moisture content of 10.06% d.b., respectively by Isik and Izil (2007).

Table 1. Variation of dimensions, parenthesis shows standard deviation.

Dimensions	Size	Moisture content (% d.b.)		
		3	7	14
Length	Big	14.48 (0.90)	15.13 (1.13)	15.97 (0.89)
	Average	13.87 (1.14)	14.27 (0.96)	15.18 (1.02)
	Small	13.31 (0.98)	13.79 (0.77)	14.12 (0.89)
Width	Big	7.24 (0.29)	7.35 (0.44)	7.50 (0.45)
	Average	6.32 (0.23)	6.34 (0.39)	6.50 (0.38)
	Small	5.01 (0.38)	5.31 (0.43)	5.52 (0.35)
Thickness	Big	4.26 (0.31)	4.28 (0.38)	4.38 (0.40)
	Average	3.80 (0.28)	3.83 (0.37)	4.12 (0.29)
	Small	3.11 (0.38)	3.39 (0.46)	3.59 (0.29)

The equivalent, geometric and arithmetic mean diameters of the seed increased linearly with increase in moisture content (Fig. 1).

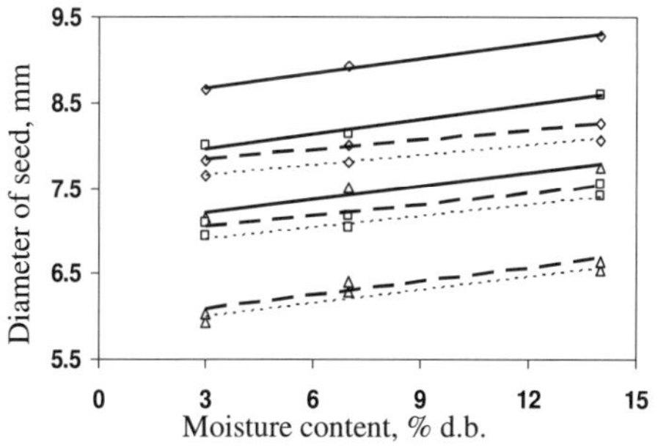

Figure 1. Effect of moisture content and size on diameter of sunflower seed (◊, big; □, average; Δ, small; full, dash and dot lines show arithmetic, equivalent, and geometric diameter, respectively).

The average of geometric mean diameter in all sizes was higher than those reported for cumin seed (Saiedirad et al. 2008). The minimum and maximum of equivalent diameter for sunflower seed were 6.04 and 8.26, respectively. In addition, the ranges of arithmetic mean and geometric mean diameters of sunflower seed were 7.15–9.29 and 5.93–8.07 mm, respectively. Gupta and Das (1997) reported that the values of 4.72 mm for equivalent diameter of sunflower seed. Also, the variation of geometric mean diameter for sunflower seed was indicated 6.15–7.93 mm in the moisture content ranged from 10.06–27.06% by Isik and Izil (2007).

Table 2 shows the regression models and coefficients of determination (R^2) achieved by fitting the dimension values for sunflower seed as a function of moisture content and size. In all cases, very high correlation was found between diameters and moisture content.

Table 2. Regression models and coefficients of determination achieved for diameters.

Size	Equation of diameter	R^2
Big	$D_a = 0.0568Mc + 8.503$	0.99
Average	$D_a = 0.0556Mc + 7.805$	0.98
Small	$D_a = 0.0524Mc + 7.048$	0.94
Big	$D_g = 0.0389Mc + 7.719$	0.99
Average	$D_g = 0.0423Mc + 6.929$	0.96
Small	$D_g = 0.0523Mc + 5.938$	0.93
Big	$D_e = 0.0389Mc + 7.529$	0.99
Average	$D_e = 0.0440Mc + 6.774$	0.97
Small	$D_e = 0.0532Mc + 5.826$	0.93

The average of sphericity variation with moisture content and size are shown in Table 3. The range of variation for sphericity was 0.44–0.53. Deshpande et al. (1993) found that the sphericity of soybeans to increase linearly with moisture content up to 25%. The average of sphericity for sunflower seed was reported 0.57 by Gupta and Das (1997). Isik and Izil (2007) indicated an increase in sphericity (0.79 to 0.84) with increase in moisture content.

Table 3. The sphericity of sunflower seed.

Size	Moisture content (% d.b.)		
	3	7	14
Big	0.53	0.52	0.5
Average	0.5	0.49	0.49
Small	0.44	0.45	0.46

The variation range of surface area for Azargol variety of sunflower seed was 110.47–204.6 mm² (Fig. 2). The surface area of sunflower seed increased linearly with increasing moisture content in all categories. Deshpande et al. (1993) found a similar result working with soybeans. In contrary, Hsu et al. (1991) revealed that the surface area of pistachios to decrease with absorbing moisture. The values of surface area of the studied sunflower seed also increased with increasing in size.

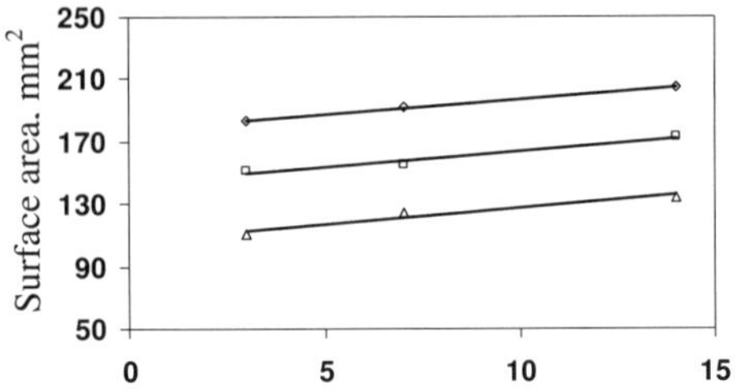

Figure 2. Effect of moisture content and size on surface area of sunflower seed (◊, big; □, average; Δ, small).

The regression relationships between surface area and moisture content for each size of the studied variety of sunflower seed are represented in Table 4.

Table 4. Regression models and coefficients of determination achieved for spherecity.

Size	Surface area	R^2
Big	S=1.92 Mc + 177.83	0.99
Average	S=1.98 Mc + 143.79	0.96
Small	S=2.09 Mc + 106.32	0.94

4. Conclusions

1. All the main dimensions (length, width and thickness) for each size of seed increased linearly with the increase of moisture content.
2. For the studied range of moisture content (3–14% d.b.), the total average expansion was maximum along the seed length and minimum along its thickness.
3. Coefficients of determination between dimensions and moisture content showed very high correlation between them in all cases.
4. Equivalent, geometric and arithmetic mean diameters of the seeds increased linearly with an increase in moisture content.
5. Except some cases, the sphericity of the seed in studied variety increased with an increase in size and moisture content.
6. The surface area of seed increased linearly with increasing moisture content in all categories.

References

1. B.D. Shukla, P.K. Srivastava, and R.K. Gupta, "Oilseed Processing Technology," Central Institute of Agricultural Engineering Publications, Bhopal, India (1992).
2. Iranian Agricultural Ministry's Reports, The bureau of statistics, Ministry of Agriculture, Tehran, Iran (2005).
3. K.O. Sahay and K.K. Singh, "Unit operations in agricultural processing," New Delhi: Vikas Publishing House Pvt. Ltd. (1994).
4. S.D. Deshpande, S. Bal and T.P. Ojha, "Physical properties of soybean," J. Agric. Eng. Res., 56, 89-98 (1993).
5. K. Oje and E.C. Ugbor, "Some physical properties of oil bean seed," J. Agric. Eng. Res., 50, 305-313 (1991).
6. R. Visvanathan, P.T. Palanisamy, L. Gothandapani and V.V. Sreenarayanan, "Physical properties of neem nut," J. Agric. Eng. Res., 63, 19-26 (1996).
7. S.H. Suthar and S.K. Das, "Some physical properties of karingda (*Citrullus lanatus* (*Thumb*) *Mansf*) seeds," J. Agric. Eng. Res., 65, 15-22 (1996).
8. K.K. Singh and T.K. Goswami, "Physical properties of cumin seed," J. Agric. Eng. Res., 64, 93-98 (1996).
9. R.K. Gupta and S.K. Das, "Physical properties of sunflowerseeds," J. Agric. Eng. Res., 66, 1-8, 1997.
10. M. Konak, K.C. Arman and C. Aydin, "Physical properties of chick pea seeds," Biosystems Eng., 82, 73-78 (2002).
11. E. Isik and N. Izli, "Physical properties of sunflower seeds," J. Agric. Eng. Res., 8:677-686 (2007).
12. N.N. Mohsenin, "Physical Properties of Plant and Animal Materials," 2nd Revised and Updated Edition.Gordon and Breach Science Publishers. New York (1986).
13. E. Altuntas, E. Ozgoz, and O.F. Taser, "Some physical properties of fenugreek (*Trigonella foenum-graceum* L.) seeds," J. Food .Eng. Res., 71:37–43 (2005).
14. M.H. Saiedirad, A. Tabatabaeefar and Borghei, "Effect of moisture content, seed size, loading rate and seed orientation on force and energy required for fracturing cumin seed under quasi-static loading," J. Food Eng. Res., 86:565–572 (2008).
15. R.H. Hsu, J.D. Mannapperuma and R.P. Singh, "Physical and thermal properties of pistachios," Journal of Agricultural Engineering Research. 49:311–321 (1991).

STUDIES ON THE EXTRACELLULAR TANNASE FROM NEWLY ISOLATED *BACILLUS THURANGIENCES* BN2

PRASANNA D. BELUR

Department of Chemical Eng. National Institute of Technology Karnataka, Surathkal, Srinivasanagar, Mangalore, Karnataka State-575 025, India

GOPAL MUGERAYA

Department of Chemical Eng. National Institute of Technology Karnataka, Surathkal, Srinivasanagar, Mangalore, Karnataka State-575 025, India

SUBBALAXMI. S.

Department of Chemical Eng. National Institute of Technology Karnataka, Surathkal, Srinivasanagar, Mangalore, Karnataka State-575 025, India

Nine tannase producing bacteria have been isolated from rotting tea, rotting coffee pulp and tannin rich forest soil. All the isolates were tested for extracellular and cell-associated tannase activity. Five isolates showed both extracellular and cell-associated tannase activity, three showed only cell-associated tannase activity and one isolate showed only extracellular activity. The isolate showing highest extracellular activity was identified as *Bacillus thuringiensis.* This isolate could consume 62.3% of gallic acid in 24 hours when cultured in medium having gallic acid as the sole carbon source. Growth studies indicate that enzyme is inducible and growth associated. Full factorial analysis was carried out to identify the best combination of carbon and nitrogen supplements. Maximum extracellular enzyme production of 0.0084 U/ml was observed. The pH and temperature optima of the enzyme were found to be 5.0 and 30°C.

1. Introduction

Tannin acyl hydrolase (E.C. 3.1.1.20), commonly known as "Tannase" catalyses the hydrolysis of ester and depside bonds in hydrolysable tannins [1]. Tannin protects the vulnerable parts of plants from microbial attack by inactivating the invasive microbial extracellular enzymes. But many bacteria have developed the ability to grow in the presence of tannins. The mechanisms by which bacteria can overcome inhibition include tannin modification, degradation, dissociation of tannin-substrate complexes, tannin inactivation by high affinity binders, membrane modification/repair and metal ion sequestration. Many microorganisms evolved the ability to produce tannase to nullify the inhibitory effect of tannins. Tannase hydrolyses tannins thereby reduce the concentration and also produce the products, some of which are used as energy source [2].

The major applications of this enzyme are in the production of Gallic acid, instantaneous tea and acorn wine. Gallic acid is used for the manufacture of an antimalarial drug Trimethoprim and in the synthesis of esters, such as Propyl gallate, used as anti-oxidant in food industry. The enzyme is also used as clarifying agent in some wines, juices of fruits and refreshing drinks with coffee flavour. The enzyme has potential uses in the treatment of tannery effluents and pre-treatment of tannin containing animal feed [3].

Many Fungi, Yeasts and Bacteria are known to produce tannase. Tannase is now known to be a ubiquitous enzyme of the microbial world. Number of investigations revealed that *Aspergillus* spp. are the most studied and efficient producers of tannase. About 13 fungi and 3 yeast have been reported to produce tannase [2]. Tannase is commercialized by Biocon (India), Kikkoman (Japan), ASA Specilaenzyme GmbH (Germany) and JFC GmbH (Germany) [3].

Report on tannase from bacterial origin is scanty in literature. Deschamps et al., [4] isolated nine isolates belongs to genera *Bacillus, Staphylococcus* and klebsiella by enrichment culture technique, using tannic acid as the sole carbon source. *Bacillus lichiniformis* [5], *Bacillus cereus* [6] have been isolated by enrichment culture technique using tannic acid as the sole carbon source. Ayed and Hamdi [7] studied *Lactobacillus plantarum* and reported significant enzyme production.

Lekha and Lonsane [8] have reported both cell-bound and extra-cellular activity of *Aspergillus niger.*

Similarly Seth and Chand [9] reported intra-cellular activity in case of *Aspergillus awamori*. There are no reports of cell-associated tannase activity in case of Bacteria. In this communication we are reporting the cell-associated tannase activity of bacterial isolates for the first time.

2. Materials and Methods

2.1. Isolation and identification of tannase producing bacteria

Samples were collected from effluent pit of a tea processing factory and coffee curing house. Samples were also collected from top soil of western ghats rich in tannin. All the samples were collected in fresh glass vials. About 1g of sample is added to 100 ml of Nutrient broth, incubated in Incubator shaker for 48 hours at 30°C, 150 rpm. 1 ml of turbid broth is inoculated into selection medium. The selection medium is having Commercial tannic acid as the sole carbon source and Casein hydrolysate as the supplemented complex nitrogen source. The medium is having following composition (w/v): Tannic acid, 0.5 %; NH_4NO_3, 0.3%; Casein hydrolysate, 0.2%; KH_2PO_4, 0.05%; K_2HPO_4, 0.05%; $MgSO_4$. $7H_2O$, 0.05%; salt solution III, 1ml. The salt solution III is having composition, Cobalt sulfate -1.75 mg, Manganeous chloride-1.15 mg, Ammonium molybdate 2.95 mg, Nickel sulfate- 9mg, Zinc sulfate 37 mg, in 100 ml distilled water. 100 ppm of cycloheximide was used to suppress the fungal growth. Tannic acid was filter sterilized and added after the sterilization of other media ingredients. The salt solution was separately prepared and steam sterilized, which was added after adding tannic acid. Initial pH was 5.5. The cultures showing substantial growth were transferred to fresh medium. After three more transfers, turbid broth was diluted appropriately and spread plated in Nutrient agar plates. Individual discrete colonies were again inoculated into selection medium, incubated for 24 hours. Nine bacterial isolates designated as KKO, KKD, KKL, BN2, AC6, BN1, DT222, BA3 and TB. The isolates were preserved at 4°C on Nutrient agar slants for further work.

Identification of one selected strain was made after studying various morphological, cultural and physiological characteristics according to *Bergey's Manual of Systematic Bacteriology* [10]. Later it was confirmed by 16S rDNA sequence analysis.

2.2. Preparation of inoculum

Inoculum was prepared by growing the organism in 50 ml sterile medium (composition same as isolation medium) in 100 ml Erlenmeyer flask for 24 hours in an incubator shaker at 30°C, 150 rpm.

2.3. Production of tannase enzyme

Organism was cultured in 250 ml Erlenmeyer flasks containing 100 ml liquid medium (pH 5.5). The production medium composition is as follows(w/v): $K_2H PO_4$, 0.7%; $MgSO_4.7H_2O$, 0.04%; $(NH_4)_2 SO_4$, 0.4%; $FeSO_4.7H_2O$, 0.002%; Yeast Extract, 0.1%; Tannic acid, 0.5%; Salt Solution III- 1ml. The salt solution III is having composition Cobalt sulfate -1.75 mg, Manganeous chloride-1.15 mg, Ammonium molybdate 2.95 mg, Nickel sulfate- 9mg, Zinc sulfate 37 mg, in 100 ml distilled water. A 1% inoculum was added and incubated at 30°C in incubator shaker at 150 rpm. The broth was centrifuged (5500xg, 15 min), supernatant and pellet were assayed periodically for tannase activity.

2.4. Effect of gallic acid

All the isolates were cultured in a medium very similar to the production medium, but the tannic acid was replaced with 0.44% (w/v) gallic acid. The broth was centrifuged (5500 x g, 15 min), supernatant and pellet were assayed periodically for tannase activity. Gallic acid concentration in the supernatant was determined periodically.

2.5. Assay of tannase and gallic acid

Tannase activity was estimated by reaction with tannic acid as substrate followed by the reaction with methanolic rhodanine as described elsewhere [19]. The quantity of gallic acid released during hydrolysis of tannic acid represents the tannase activity. The sensitivity of the assay was 36 x 10^{-6}U $ml^{-1.}$ While measuring the cell-associated activity, sediment of 1 ml of broth was taken after centrifugation and suspended in 1 ml acetate buffer (0.1 m, ph 5.0). Dry biomass weight of 1 ml broth was measured by drying the sediment at 105°C for 36 hours. Using this data, activity was calculated for one liter broth.

One unit of enzyme activity is defined as the micromole of gallic acid released per milliliter per minute.

Gallic acid concentration was measured making use of Lagemaat and Pyle [11] method. A standard curve was obtained by determining the absorbance of standard solutions of gallic acid with concentrations ranging from 0 to 0.20 g/l.

2.6. Effect of carbon and nitrogen supplements

Full factorial analysis was carried out to find the best combination of carbon & Nitrogen sources. Glucose, Fructose, Sucrose and Lactose (0.5% w/v) were chosen as carbon source supplement with tannic acid. Urea, Ammonium sulfate, Ammonium nitrate and sodium nitrate (0.4% w/v) were chosen as inorganic nitrogen sources along with Yeast extract in the above mentioned production medium.

3. Results and Discussion

3.1. Screening of the isolates

Tannins are the fourth most abundant plant constituent after cellulose, hemi cellulose and lignin. Tannins acts as deterrent against invasive microorganisms, insects and animals. The ability of tannins to inflict damage is mainly due to its ability to complex with extracellular and membrane bound proteins. The formation of protein-tannin complex possibly reduces the ability of microorganisms to scavenge and ingest nutrients. But several microorganisms evolved the ability to produce tannase, a glycoprotein [12] which does not get inactivated by tannins. In our quest to develop a bacterial system to produce tannase which degrade "hydrolysable tannins" we took up this work. We could isolate nine pure cultures, using conventional enrichment culture technique using tannic acid as the sole carbon source. Proficient growth in the production medium indicated that the organism is capable of growth in the presence of tannic acid as well as consuming tannic as carbon source for its metabolism. However when we measured the tannase activity of cell free supernatant, we could find tannase activity in only five isolates, this led us to carry out assay for cell-associated tannase activity. Eight isolates showed cell-associated activity, and five showed both cell-associated and extracellular activity. This indicates that perhaps cell-associated tannase activity is quite common in the microbial world. Possibly cell-associated activity provides the additional advantage for the organism when it is competing for the precious nutrients in the soil. There are several reports on cell-bound enzyme activity in case of bacteria [13, 14]. (Fig. 1).

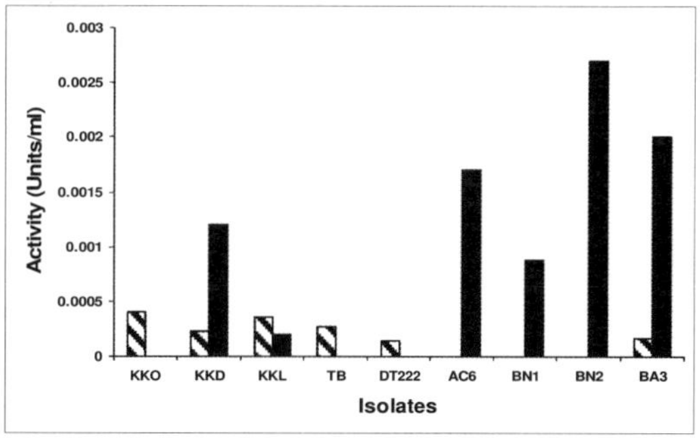

Fig. 1. Comparison of extracellular ▇ and cell-associated ▨ activities of all the isolates.

Tannic acid, a polygalloyal ester upon hydrolysis by tannase releases gallic acid and glucose molecules.

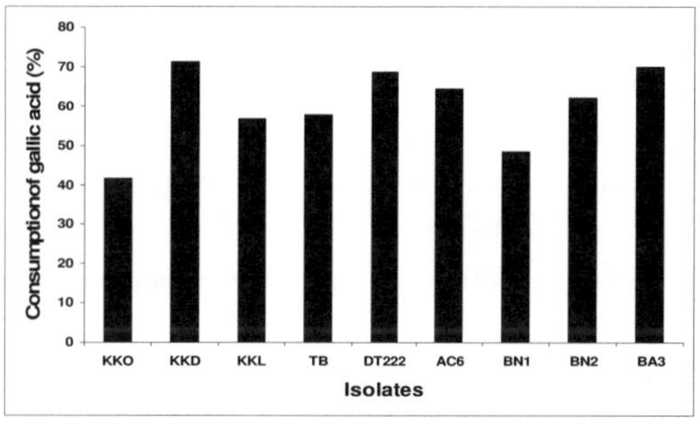

Fig. 2. Comparison of gallic acid consumption of isolates.

Rapid utilization of these energy rich molecules would help in the better growth and product formation. Hence, all the isolates were cultured in production medium having replaced tannic acid with gallic acid (0.44% (w/v). The gallic acid content of the broth was measured at the 24th hour. Among the isolates which showed extracellular activity, KKD, AC6, BN2, BA3 consumed more than 60% of gallic acid in 24 hours We found that the consumption was 41% to 71.2% in 24 hours. (Fig. 2). Based on these results we chose BN2, which showed highest extracellular tannase production (0.0027 U/ml) with reasonably good ability to consume gallic acid (62.3%).

3.2. Identification of organism

The selected isolate BN2 was identified based on Morphological, nutritional, physiological and biochemical characteristics (Table 1) as described in Bergey's Manual of Determinative Bacteriology [10].

Table 1. The morphological, physiological and Biochemical characteristics used to identify BN2.

Morphology	Casein hydrolysis
Cell shape	Argenine
Cell size	hydrolysis
Motility	Lipase production
Endospore	Oxidae production
Gram staining	Catalase
Cultural characteristics	production
Colony shape	Gelatinase
Optimum Temperature	production
Optimum pH	Utilization of
Growth on nutrient	Succinate
agar	Citrate
Growth on NaCl 2.5-	Acid production from
7.0%	Glucose
Growth at 5,20 and	Galactose
50°C	Sucrose
Biochemical properties	Fructose
Urease production	Trehalose
Voges-Proskauer	Arabinose
Nitrate reduction	Cellobiose
Indole	Inositol
Nitrate reduction	Mannitol
Esculin hydrolysis	Mannose
ONPG	Raffinose
Methyl red	Xylose

The culture was Gram positive, aerobic rod, endospore forming, motile, Oxidae negative and catalase positive. Spores were ellipsoidal and were located centrally. Other phenotypic characteristics were the ability to reduce nitrate, use of citrate, and production of acid from various carbohydrates, hydrolysis of casein, Tweeen 80, Phenyl alanine, Egg-yolk, gelatin and Esculin. It was also positive for Voges-Proskauer and Methyl red test. Based on these results, it was presumptively identified as *Bacillus thuringiensis.* However to confirm, the partial nucleotide sequence (884 bp) were determined and compared with those available in public databases. Identity of BN2 was determined based on the 16S rDNA sequence similarity of > 99% with that of the prototype sequence in the GenBank. The sequence was submitted to GenBank and accession number was assigned as GQ249351.

3.3. Growth studies

The isolate BN2 was cultured in the production medium. The cell free broth was analyzed for extracellular tannase yield periodically. The biomass content of the broth was measured in terms of turbidity at 620 nm. The data shows that enzyme production is possibly growth associated, reaches peak activity in the stationary phase. The stationary phase lasts for 9 hours, during which peak enzyme production was recorded (Fig. 3).

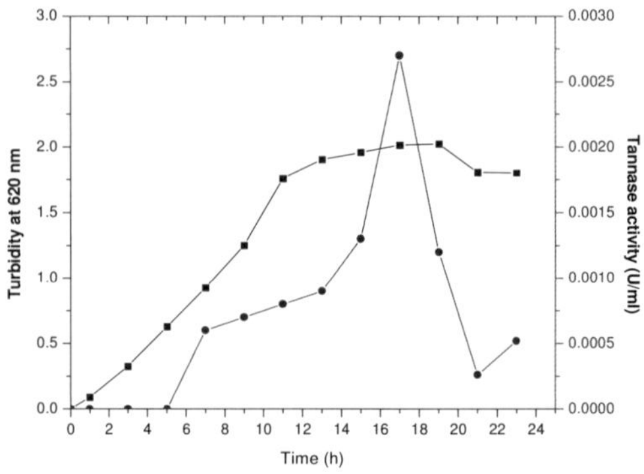

Fig. 3. Time course of biomass productivity (■) and Tannase activity (●).

The growth studies suggest that maximum production occurs in stationary phase of growth, which is in concurrence with the findings of Mondal et al., [15]. Where as Mondal et al., [5] have reported the peak activity at the exponential phase (21st hour) in case of *Bacillus lichiniformis* and there are other reports of bacteria showing peak activity at the initial phase of growth [4,16].

Replacing tannic acid with methyl gallate, Gallic acid and Glucose in the production medium failed to produce tannase in agreement with several studies [4,16,5].

3.4. Effect of initial pH on enzyme production

The production medium having initial pH of 6.6 to 8.0 was taken. The medium with the initial pH of 7.2 gave highest enzyme production at 16 hours (Fig. 4).

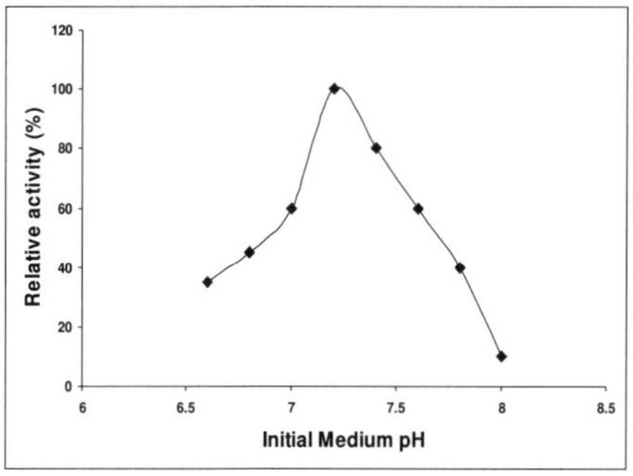

Fig. 4. Effect of initial medium pH on Tannase production. 100% relative activity is 0.006 U/ml.

3.5. Effect of pH and tTemperature on tannase activity

The results indicated that the extracellular tannase enzyme produced by BN2 has been active over a pH range of 4.0 to 8.5.Showed optimum pH at 5.0 (Fig. 5).

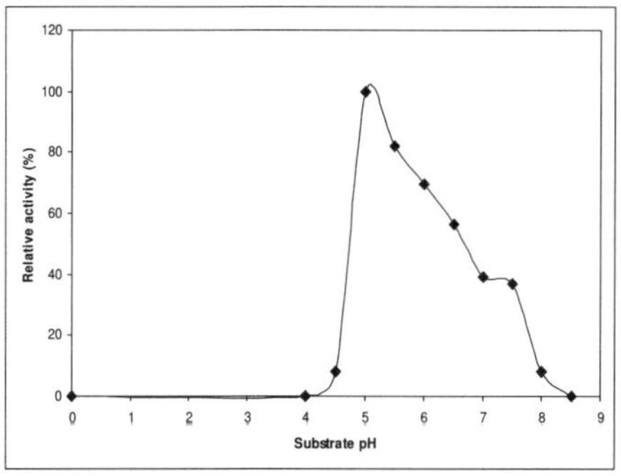

Fig. 5. Effect of pH on Tannase activity, 100% relative activity is 0.007 U/ml.

Enzyme remained active from 15-60°C and optimum activity was found at 30°C (Fig. 6). It may be due to the fact that the strain was isolated from decomposing plant material which is acidic in nature.

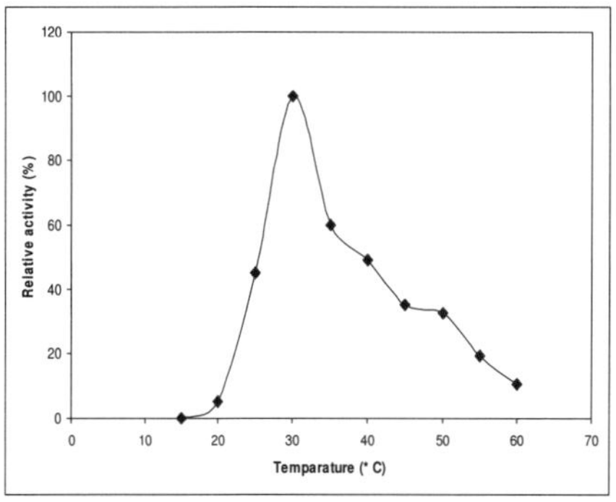

Fig. 6. Effect of temperature on Tannase activity. 100% relative activity is 0.0067 U/ml.

3.6. Effect of carbon and nitrogen supplements

Urea with Sucrose combination gave highest enzyme production of 0.0084 Units/ml after 12 hours of growth (Fig. 7).

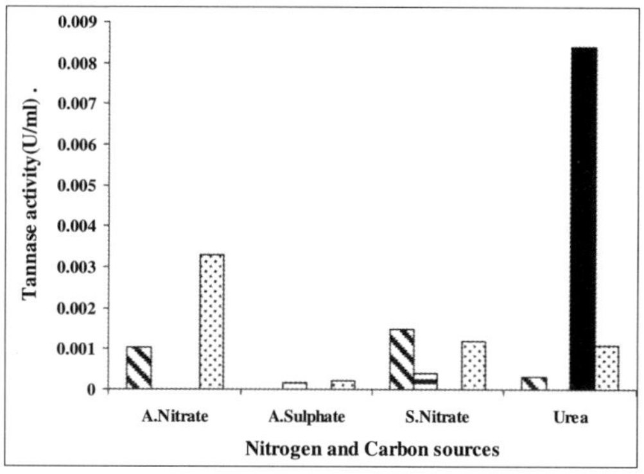

Fig. 7. Effect of carbon and Nitrogen supplements on enzyme production. Glucose, Fructose, Sucrose and Lactose are plotted against nitrogen sources.

Peak tannase production was achieved in 12 hours compared to 16 hours in the medium having no carbon and nitrogen supplements. This could be explained by the fact that availability of easily metabolizable carbon and nitrogen sources will help in the build up of critical biomass required to show tannase activity.

4. Conclusion

Nine tannic acid (hydrolysable tannin) degrading bacteria were isolated, tested for cell-associated and extracellular tannase activities. Results suggest that cell-associated tannase activity is quite common in nature among tannin degrading bacteria. Based on extracellular activity and gallic acid consumption ability, strain BN2 chosen for further studies. Based on Physiological, Morphological and biochemical studies it is been identified presumptively as *Bacillus thuringiensis*. Identity of BN2 was confirmed based on the 16S rDNA sequence similarity of > 99% with that of the prototype sequence in the GenBank. The sequence was submitted to GenBank and accession number was assigned as GQ249351. Growth studies in shake flask trials suggest that tannase production is inducible and peak production takes place in the stationary phase. Medium having pH 7.2 gave peak enzyme production. The pH and temperature optima of the enzyme are 5 and 30°C respectively. Urea and sucrose supplementation in the fermentation medium resulted in 0.0084 U/ml extracellular activity.

Acknowledgments

We wish to thank Dr. Yogesh S. Shouche, National Centre for Cell Sciences (NCCS) Pune for carrying out the 16S rDNA analysis.

References

1. E.Haslam and J.E.Stangroom. *Biochemistry Journal,* **99**, 28-31(1965)
2. Tej.K.Bhat, Bhupinder Singh and O.P.Sharma *Biodegradation,* **9**, 343-357(1998)
3. R. Belmares, J.C.Conttreras-Esquivel, R.Rodriguez-Herrera, A.R.Coronel, C.N.Aguilar *Lebensm-Wiss. U.-Technol*, **37**, 857-864(2004)
4. Alain M.Deschamps, Gulten Otuk, and Jean-Michel Lebeault. *J.Fermentation Technology.* **61**, 55-59(1983).
5. Keshab C.Mondal;Rintu Banarjee & Bikas R.Pati, *Biotechnology letters*, 0, 767-767(2000)
6. K.C.Mondal, D.Banarjee, M.Jana, B.R.Pati, *Analytical Biochemistry* , **295**, 168-171(2001)
7. Lamia Ayed & Moktar Hamdi, *Biotechnology Letters,* **24**, 1763-1765(2002)
8. P.K.Lekha and B.K.Lonsane, *Process Biochemistry,* **29**, 497-503(1994)
9. Malini Seth and S.Chand, *Process Biochemistry* **36**, 39-44(2000)
10. Bergey's manual of Determinative Bacteriology, ninth edition, Lippincott Williums & Wilkins. Baltimore(1994)
11. J.Van de Lagemaat, D.L.Pyle, *Chemical Engineering Journal,* **84,** 115-123(2001)
12. Barthomeuf. C., Regerat. F., Pourrat. H. *J Ferment Bioeng* 77, 320–323(1994).
13. F.V.Pereira-Meirelles, Maria.H.M, Rocha-Leao, Geraldo L, Anna Jr. , *Biotechnology Letters,* **22**, 71-75(2000)
14. S.Sinsuwan, S.Rodtong, J.Yongsawatdigul, *LWT-Food science and Technology*, 41:2166-2174(2008)
15. Keshab C.Mondal and et.al, *Gen.Appl.Microbiol,* **47**, 263-267(2001)
16. Kumar. R., Singh M *J.Agric. Food Chem.* **32**, 447-453(1984)

BIOPHYSICAL PROPERTIES OF SOME PERSIAN HONEYS

HAMID TAVAKOLIPOUR

Department of Food Engineering, Azad University-Sabzevar Branch (IAUS), Sabzevar, Khorasan Razavi Province, Iran

AHMAD KALBASI ASHTARI

Department of Food Engineering, Faculty of Agriculture, Tehran University Karaj, Iran

The rheological properties of two varities of Persian honeys were investigated over temperature range of 10-30°C. The honeys had a moisture content of 16.8-17.2% and exhibited Newtonian behaviour over shear rate range of 0-100 s^{-1}. The viscosity varied between 6.7 and 150 Pa.s according to honey type and temperature. The temperature dependence of viscosity was described using the Arrhenius model. The values of activation energy of honey samples are 104.5 and 109.7 kJ/mol for thyme and shahjahan variety, respectively.

Keywords: Honey; rheological properties; Newtonian; Arrhenius model.

1. Introduction

Honey is a natural food product having a high nutritional and medicinal value. It is an aqueous dispersion of carbohydrates (mono, di and polysaccharides), colloidally dispersed macromolecules of protein, organic acids and many biologically active compounds [1,2]. Honey varieties with different floral origins are produced in various parts of Iran. Rosa, thyme, astragalus, trifolium and medicago are common varieties. Iran with annual production 42000 tons in 2007 and with approximately 3.5 million hives, ranks eleventh in honey production among the honey producing countries. Study of honey rheology should be helpfull in process design, quality control, shelf life assessments and product development. The apparent viscosity of honey depends on several factors, most important are water content, temperature and chemical composition. In general, the moisture content of different origin honeys may be in the range of 13-29% [2]. A 1% change in water content has the same effect on viscosity as a 3.5°C change in temperature [3]. Apart from composition, honeys are highly sensitive to temperature.

The Arrhenius model is widely used for describing of temperature dependence of viscosity but other models like Williams-Landel-Ferry (WLF) have proved useful for the viscosity-temperature relationship of honey varities [3,4]. Although Iran is an important honey producer in the world but there are a few documents on the physio-chemical characteristics [5] but a report on the rheological properties of Persian honeys is lacking in the literature. This research focuses on studying the rheological properties of two types of Persian honey. A rotational viscometer was used to measuring the rheological properties of honey samples within temperature range 10-30°C.

2. Materials and Methods

Two common Persian honey varities named thyme from north and shahjahan from Northeast of Iran were obtained from a local processor. An Abbe refractometer (model 2WAJ, China) was used to determine of moisture content of honey samples at 20°C.

An Anton Paar DV-3P digital rotational viscometer fitted with small sample adapter with flow jacket for thermostatting the sample (Anton Paar GmbH, Austria) was used for viscosity determination over shear rate range of 0-100 s^{-1}. A water bath (LAUDR E200, Windaus, Germany) connected to viscometer was used for maintain constant temperature over range of 10-30°C during experiments. Because the air bubbles and crystals affected to viscosity determination, the samples preheated to 55°C in a water bath for 1h to dissolve any crystal nuclei which may have been present in the sample. The preheated honey samples were then stored in an incubator (Memmert, Germany) at 30°C for overnight to ensure complete removal of air bubbles.

To minimize recrystalization, the honey samples were then stored frozen at –18°C until analysis. Before each test, honey samples were removed from frozen storage, heated and mixed to ensure homogeneity. To regulating water content at 21 and 25%, subsamples (50 ml) of each honey variety were diluted with distilled water.

3. Mathematics

The Arrhenius model was used for temperature dependence of viscosity:

$$\eta = \eta_\infty \exp(E_a/RT) \qquad (1)$$

Where η apparent viscosity (Pa.s), η_∞ is the pre-exponential factor (value of viscosity as T approaches infinity) in Pa.s, T absolute temperature (K), E_a activation energy (kJ/mol) and R gas constant (8.314×10^{-3} kJ/mol.K).

A nonlinear regression technique that uses the Levenberg-Marquardt method to solve nonlinear equations was performed for the Arrhenius model [6].

To determine deviations between the experimental and calculated viscosity values, the mean absolute percentage error (P) was used for goodness of fit evaluation:

$$P = \frac{100}{n}\sum_{i=1}^{n}\frac{|Y_{ei} - Y_{ci}|}{Y_{ci}} \qquad (2)$$

Where Y_{ei} is experimental viscosity, Y_{ci} is calculated viscosity using models and n is number of experimental points.

4. Results and Discussion

Figures 1 and 2 illustrates apparent viscosity of thyme and shahjahan varities as a function of shear rate at different temperatures. The apparent viscosity of two varieties remained nearly constant with increasing shear rate, indicating Newtonian flow behaviour ($\tau = \eta\dot{\gamma}$) for the honeys. This result consistent with other researches on different varities such as Turkish honeys [2] and Polish honeys [3], but some honey varities such as a type of Spanish honey named Galician [7] had non-Newtonian behaviour. Figures showes viscosity values in the range of 6.7-120 Pa.s for thyme and 7.5-150 Pa.s for shahjahan variety at study temperatures.

Also, as shown in Figures 1&2, the viscosity of honey samples decreased with increasing temperature. This is expected because these honeys are Newtonian liquids and highly sensitive to temperature variation. The Arrhenius parameters can be found in Table 1.

Table 1. Arrhenius parameters of persian honeys.

Honey	Arrhenius model			
	η_∞ (Pa s)	E_a (kJ/mol)	P(%)	R^2
shahjahan	8.20×10^{-19}	109.7	8.8	0.989
thyme	5.35×10^{-18}	104.5	12.1	0.984

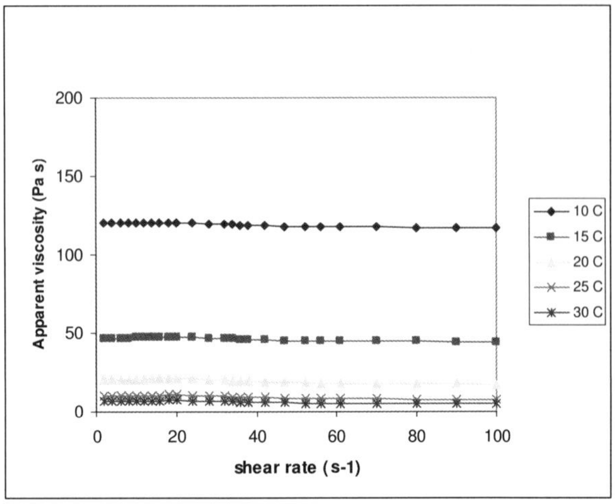

Fig. 1. Viscosity curve of thyme honey at different temperatures.

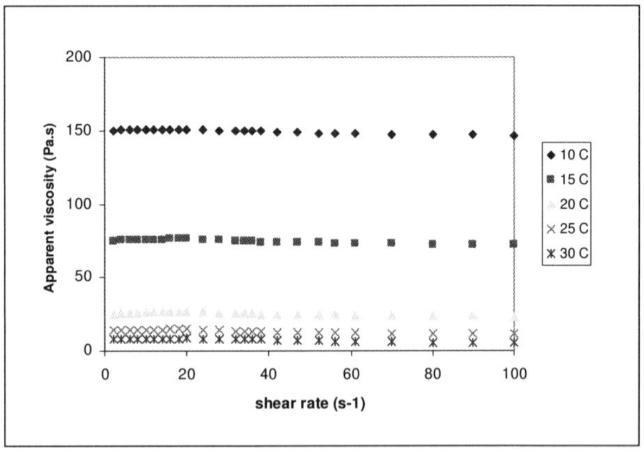

Fig. 2. Viscosity curve of shahjahan honey at different temperatures.

The mean absolute percentage error (P) for the honey samples was 12.1% for thyme and 8.8% for shahjahan variety, the first a little above the upper limit 10% and the second is lower than 10%. Therefore Arrhenius equation is relatively satisfactory for describing of the honey samples viscosity as function of temperature.

As shown in Table 1, the activation energy which reflecting the sensivity of honey to temperature is 104.5 for thyme and 109.7 kJ/mol for shahjahan honey. This values very similar to the results reported by Juszczak and Fortuna (2006) for Polish honeys but it is different from Turkish honeys with 63.4-78.5 kJ/mol [2].

5. Conclusion

The honey varieties investigated in this study showed Newtonian flow behaviour. The Arrhenius model has been shown to be suitable for modeling the rheological behaviour of Persian honeys within temperature range of 10-30 C.

References

1. S. Zaitoun, K. Al-Majeed Ghazwi, Al-Malah and B. Abujdayil, International Journal of Food Properties, 4, 139(2001).
2. A. Kayacier and S. Karamen, Journal of Texture Studies, 39, 17(2007).
3. L. Juszczak and T. Fortuna, Journal of Food Engineering, 75, 43(2006).
4. P.A. Sopade, P. Halley, B. Bhandari, B. D'Arcy, C. Doebler and N. Caffin, Journal of Food Engineering, 56, 67(2002).
5. Gh. R. Jahed Khaniki and A. Kamkar, Iranian Journal of Food Science and Technology, 1, 35(2005).
6. M.B. Cutlip and M. Shacham, "Problem solving in chemical and biochemical engineering with POLYMATH, Excel, and MATLAB, 2nd edition," Prentice Hall International Series, 705(2008).
7. D. Gomez-Diaz, J.M. Navaza and L.C. Quintáns-Riveiro, European Food Research Technology, 222, 439(2005).

RISK ASSESSMENT OF FLUORIDE IN INDIAN CONTEXT

VEENA CHAUDHARY
Department of Chemistry, C.S.S.S. PG College, Machhra, Meerut, UP, India

MUKESH KUMAR
S.V.B.P. University of Agriculture & Technology, Meerut, UP, India

After highlighting first case of fluoride (F) contamination in 1937, till then yet government fails to develop such a technique which would cheap and affordable to all. F contamination is increasing and spreading among different states of India and causative factors includes geogenic, phosphatic fertilizers, overexploitation, low nutritional status, lack of awareness and industrial effluents. Information, education and communication based project would be helpful to mitigate F contamination in drinking water.

Keywords: Drinking water, fluoride, geogenic.

1. Introduction

India, the seventh largest, second most populous country in the world, is plagued with numerous water quality problems of geogenic origin and fluorosis stands first among them.

It has been estimated that in Indian population of 18,197,000 are affected with dental fluorosis and 7,889,000 are affected with skeletal fluorosis.[1] However, as a surprising paradox incidence of dental, skeletal and crippling skeletal fluorosis was reported in India with average F concentration as low as 0.5, 0.7 & 2.8 ppm, respectively. Fluorosis turns out to be the most widespread in india, affecting more than 66 million people including 6 million children under 14 years age. In early 1930s fluorosis was reported only in 4 states of India, 1986 it was 13, in 1992 it was 15, in 2002 it was 17 and in 2006 it was 20 and spreading in rest of the states.[2]

The primary causative factor in endemic fluorosis has been unequivocally identified as the F intake through drinking water. So, the ever increasing F levels in drinking water pose the greatest threat to human health as evidenced by the endemic of fluorosis in India. The intensity of natural fluoride in groundwater ranges from 0.5 to 48 ppm or even more.[3]

Due to the severity of impacts due to excess F in groundwater, the WHO permissible limit of F in India has been reduced from 1.5 to 1.0 ppm in 1998.People suffering from fluorosis afraid to declare their pain due to impending daughter's marriages, because affected mostly people are below the age of 30 years.

2. Causative factors

2.1. *Geogenic origin*

As water has been identified as one of the potential source of F intake, groundwater is the major and preferred source of drinking water in rural as well as urban areas, particularly in the developing countries like India because disinfection treatment is often not required it is estimated that in India 85% of domestic needs in rural areas and 50% in urban areas are met by ground water.

In the central basaltic geological formation of India, over exploitation of finite ground water resources is forcing government authorities to explore deeper (previously un-suryed) aquifer. Many of these aquifers are located in older strata, which contain high levels of chemical minerals such as fluorite or hornblende. These pollutants when leached in groundwater sources have significant health effects.

2.2. *Fertilizer application*

Vast area of land around the India has F added on a yearly basis with the application of phosphatic fertilizers. F found as an impurity within these fertilizers. Consumption of DAP and SSP has risen many times in India. To meeting the demand, India imports rock phosphate and phosphoric acid. In 2006, import of rock phosphate was 5.3 Mte of world total 29.7 and import of phosphoric acid was 2.6 Mte of world total 4.8 Mte. India is the major importer of both rock phosphate and phosphoric acid.[4]

2.3. *Over exploitation*

India is projected to be one of the worst affected by physical water scarcity in the world by 2025, with fresh water availability per person per year of 1000 m^3. The demand for water is increasing dramatically and is leading to conflicts due to the rapid growth of population. 85% of ground water extracted is used for irrigation purposes and 15% for industrial and domestic purposes. Reciprocally as much as 70 to 80 % of India's agricultural output may be ground water depended.

The average level of ground water development in India is 32%, although some states have exploited their resources to a much greater extent (94% in Punjab, 84% in Haryana, 60% in Tamil Nadu, 64% in Lakshdeep, 51% in Rajasthan),[5] result into continuous declining of water table.

2.4. *Low nutritional status*

Fluorosis, in general is found to have a decisive relation to the economic status and poverty, as it is too severe and more chronic among the rural and semi urban Indian habitations afflicted with calcium, vitamin and nutritional deficiencies. Poor nutrition makes people more prone to fluorosis.

Protein supplementation has a beneficial effect on reducing the F toxicity[6] and calcium status modifies the type of bone changes seen in fluorosis.[7] Over 90% of the persons affected with severe skeletal fluorosis, bone disease and deformities belong to the low socioeconomic group, where calcium intake status is low.[8]

2.5. *Lack of awareness*

The failure of such drinking water projects is attributable mainly to a lack of efficiency and inadequate adaptation to local conditions. There are many reasons for this: the population is not sufficiently aware of the issue, treatment methods are too complicated, culturally unacceptable and water quality problems are often overshadowed by the daily struggle for survival.

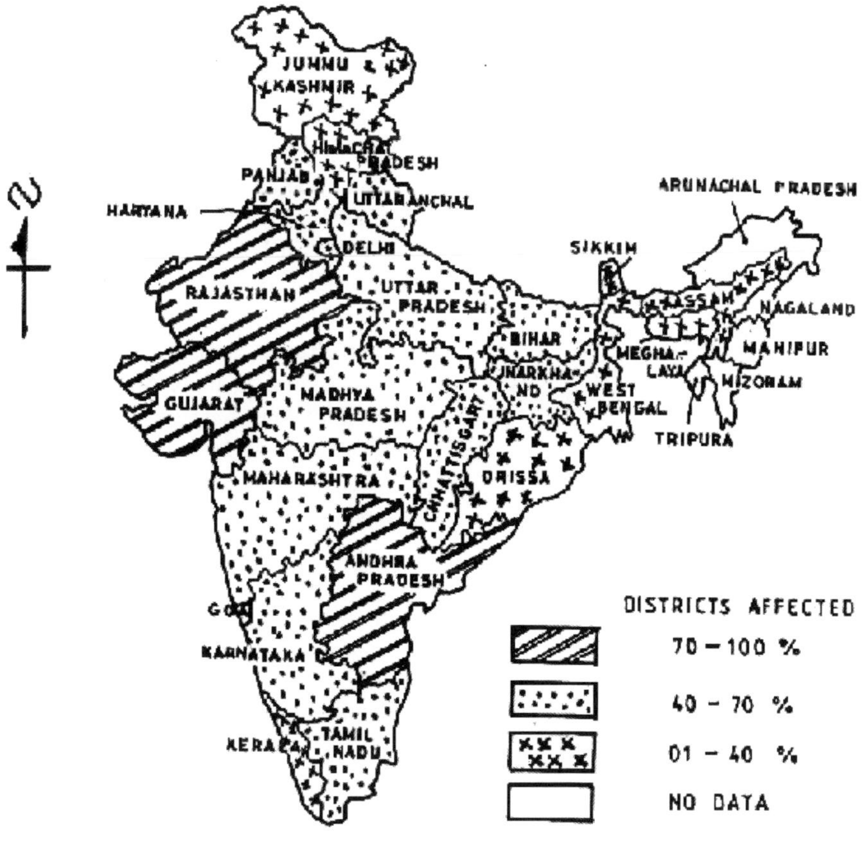

Figure 1. Fluoride map of India.

2.6. Other factors

The largest use of calcium (or sodium) fluoride is in steel making. Fluorite used as a flux in the smelting of Nickel, Copper, Gold and Silver, as rodenticides and as catalyst for certain organic reactions. Fluorapatite are used industrially for the preparation of phosphoric acid, fertilizer and mineral feeds. Cryolite is used as an insecticide and in molten forms, as an electrolyte in the production of Aluminium from Alumina. Fluoride and its compounds are used in producing Uranium (UF_6) for nuclear industry, high temperature plastics, air conditioning and refrigeration. Hydrogen fluoride (HF) is extensively used for etching the glass of light bulbs etc. The fluorinated sulphur (SF_6) is causing a slow but discernible warming of the earth's climate. All effluents leach f into the groundwater.

3. Mitigation programmes

To mitigate F contaminants, there is a need for a holistic health management approach. In this connection, a novel concept of integrated fluorosis mitigation (IFM) has been developed and implemented jointly by UNICEF and NEERI (National Environment Engineering Research Institute, India) to address the growing problems of fluorosis.

Rajiv Gandhi National Drinking Water Mission, Jawaharlal Nehru National Urban Renewal Mission, Urban Infrastructure Development Scheme for small and medium towns, Swajal Dhara Programme, Bharat Nirman and National Rural Drinking water Quality Monitoring and Surveillance Programme are some of the major affords made by government to provide clean drinking water for all by 2009. The 73rd and 74th constitutional amendments have developed the water supply responsibility to Panchayati Raj institutions/ local bodies.

Many NGO's of India working in this direction by information, education and communication (IEC) concept. IEC is an important component of any development programme particularly related to water and environmental sanitation.

4. Conclusions

F in ground water is increasing due to declining of water level, heavy use of phosphatic fertilizers and industrial effluents. Till the 10th plan, an estimated total of Rs 1,105 billion spent on providing safe drinking water since the first five year plan was launched in 1951. Despite such expenditure lack of safe and secure drinking water continuous to be a major hurdle and a national economic burden.

Already poverty prone families are unable to afford any water treatment system. For F, a range of filters from Nalgonda technique to reverse osmosis have come and gone but the problem still remains. This persistence of condition has led to severe distrust among people for any new intervensions, therefore making such efforts even more challenging.

Public counseling and developing public motivation may be needed to bring in a long term effect for reducing the menace of fluorosis.

References

1. National Environmental Engineering Research Institute (India), UNICEF, Regional Medical Research Centre for Tribals (India), Public Health Engineering Department, Integrated Fluorosis Mitigation, Guidance manual.
2. B. S. Ayoob, A. K. Gupta, Critical Reviews in Environmental Sciences and Technology. 36. 433 (2006).
3. A. K. Susheela, A Treatise on Fluorosis, revised 2nd ed. New Delhi, India. Fluorosis Research and Rural Development Foundation (2003).
4. Fertilizer Statistics, 53rd. The fertilizer Association of India, New Delhi. (2007-08).
5. Central Statistical Organization. "Compendium of Environment Statistics" New Delhi: Department of Statistics. Ministry of Planning and Programme implementation, Government of India (1997).
6. N. J. Chinoy, M. Dipti, XXIIIrd ISFR Conference abstracts. Fluoride 33 (1), (2000).
7. K.A.V.R. Krishnamachari, Progress in food and nutrition sciences 10 (3), 279 (1986).
8. S.P.S. Teotia, M. Teotia, D.P. Singh, V. Anand, C. V. Singh, N. P. Tomar Singh, Fluoride: 17 (1) 14 (1984).

STUDIES ON FLUIDIZATION CHARACTERISTICS FOR BIO-CHEMICAL PROCESSES

K. S. K. RAO PATNAIK*

Department of Chemical Engineering, University College of Technology (A), Osmania University, Hyderabad – 500 007, India Fax No: 91- 40 – 27098472 Email: ksk_raopatnaik@yahoo.co.in

P. VIJAY KUMAR

AVK Bio-Technologies, Hyderabad, India

K. SRIHARSHA

Department of Mechanical Engineering, Chaitanya Bharathi Institute of Technology (OU), Gandipet, Hyderabad – 500 075, India

Fluid Bed Reactors have been widely used in Chemical, Bio-Chemical and Petro-Chemical industries. The advantages of Fluid Bed Bio-Reactors (FBBR) are mainly Food, Pharmaceutical and Biological waste treatment sectors. In a fluidized Bio-Reactor, the particles can be much smaller and fine. During fluidization operation, the bed expands to accommodate microbial growth. The high surface area for biomass to grow results in high heat and mass transfer rates. This leads to isothermal and uniform mixing of particles tends to high concentration of active bio-mass per unit volume of reactor. This process also eliminates pollutants and toxins which slow down the process. The problem of cell washout that aggravates operation of continuous, stirred tank fermenters is less likely in a fluidized bed, indeed it is impossible as long as the superficial liquid velocity is kept below the settling velocity of the solid particles. Aerobic Reactors require oxygen and depending upon this they can be classified as two phase or three phase FBBR. In the present investigation the general fluidization characteristics for Bio-Chemical processes was studied and compared with the literature. Three - phase fluidization is an operation involving a bed of suspended particles in gas and liquid media. This occurs due to a net drag force of gas and liquid in the flow direction on the particles. Gas-Liquid-Solid Fluidization is considered in the present work as operating bed of fluidized solid particles simultaneously with concurrent upward flow of a continuous liquid phase and a gaseous phase dispersed in the forms of bubbles. In this paper mainly the experimental work of the author on FBBR & distributor design are discussed and estimated the Fluidization parameters namely minimum fluidization Velocity, Free Settling Velocity, Bed Expansion and Bed Stratification for calculating bio-film formation. Hence, the process of bio-film formation, its characteristics and properties need to be well understood to design and operate efficiently a FBBR and also important conclusions are drawn.

1. Introduction

In a fluidized bed bio-reactor can be classified by P. Ghosh et al. (1996) into 3 categories. In the first case the solid phase contains the enzyme immobilized on its surface or entrapped inside it and the immobilized enzyme catalyses the bio chemical reaction.

The liquid phase contains the substrate to be converted to product. In the second type of FBBR, the solid phase is a microbial floc which are kept in fluidized state by the upward flow of the medium. In these lower type reactors washout of flocs is minimized by having an expanded sections and an internal settler at top of the reactor. Due to this expanded sections superficial liquid velocity at the top of the rector decreases and hence the flocs settle back into the reactor without getting washed out. Also increase in floc diameter due to its growth increases its settling velocity and this helps in retaining it the bio-rector. In the third type, the microorganism grows as a film on media particles. The media particles usually clay or sand or other inert materials offer very high surface area for the biomass to grow as bio film. In this type a supported film FBBR higher upward flow velocity of the fluid can be employed compared to microbial floc bio rectors because the inert solid material increases the composite density of bio particles. However, during operation due

*Professor of Chemical Engineering, to whom correspondence should be addressed.

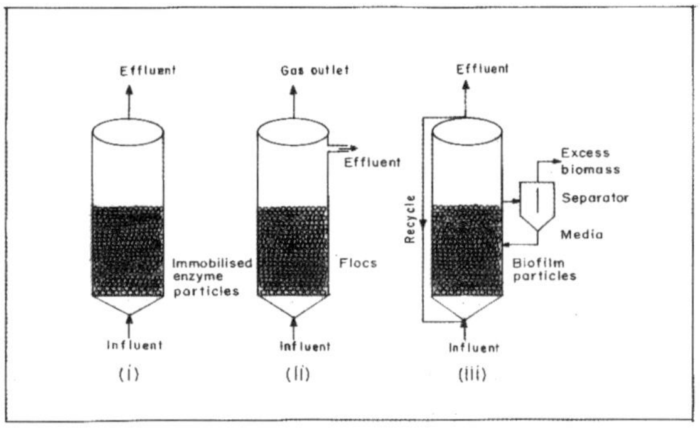

Figure 1.Type of fluidized bed bioreactors: (a) Immobilized enzyme reactor; (b) Tower reactor; (c) Biofilm reactor. (P. Ghosh et al. 1996).

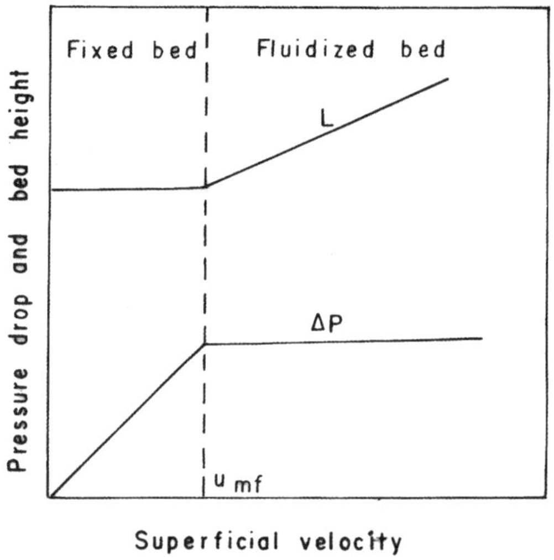

Figure 2. General Fluidization Characteristics.

to microbial growth, overall density of bio particle decreases with time which usually leads to washout of the over - grown bio particles. Under these circumstances these over-grown bio particles are intentionally washed out of the bio rector and by using a mechanical device, the biomass is separated from the inert media and the inert media returned to the bio rector where as the separated bio mass is washed as excess sludge. FBBR can be classified as aerobic or anaerobic reactors, aerobic reactors require oxygen and depending upon this they can be classified as two phase or three phase FBBR. In a three phase FBBR, bed of solids is fluidized by the co-current flow of a liquid and a gas. The liquid forms a continuous phase and the gas is a dispersed bubble phase.

2. Materials and methods

2.1. *Characteristics of fluidization*

Once the bed is fluidized, the pressure drop across the bed remains constant but the bed height continues to increase with increasing flow (Figure 2). Moreover, biochemical reactions are slow and require high residence time. This control the linear velocity of the fluid to lower ranges, where it sufficient to fluidize the particle and prevents egglomeration. Generally in the supported film FBBR, the superficial velocity is always maintained slightly above the minimum fluidization velocity or media used. This indirectly creates shear on the bio particles and hence allows the development of bio-film on the inert particle.

2.2. *Experimental set-up*

The experimental set up for investigating the fluidization characteristics for bio chemical process has been presented as a line diagram in Figure 3. The Test fluidization column consists of copper tube of 120 cm length and having a diameter of 10.16 cm. The column was fitted with a calming section having the same diameter as that of the test column and a distributor sandwitched between the column and calming section by means of flanges. The test column was heated externally by means of a kanthal resistance coil (6 kilowatt rate & 18 gauge) wound on the column and controlled through a voltage regulator. The 18 gauge kanthal wire having a total capacity of 6 k.w was wound with equal spacing over six slotted asbestos r. The strip were placed vertically over the test column, so as to keep the coil away from the column. The input and output terminals on the column winding were so selected as to divide the coil in the sections with a provision to yield 3 kw energy in each section. This is also provides uniform distribution of the kanthal energy generated.

Three copper constantan thermocouples fixed on the test column by brazing them in the grooves etched on the surface , at 3 different places, so as to measure at 3 wall temperature check the uniformly of the wall temperature. These thermocouples were connected to a Honey-Well Brown strip chart recorder with ¼ sec pen speed through a selector switch. The reference point in all case was ice point, provided by placing the ice in thermoflask.

The distributor was constructed by brazing 2 copper plates of 0.32 cm thick on either side of a 1.9 cm long copper tube of 10.16 cm diameter to form a chamber having the same diameter as that of the test column. After drilling holes in the distributor section uniformly 0.32 cm copper tube of 2.54 cm length were inserted and brazed to allow the fluid from the calming section to pass through to the test column without entering into the distributor section. Holes of diameter 0.06 mm were drilled on the upper plate of the distributor in between the spacing left after fixing of the 2.5 cm long copper tubes. 4 pots were provided at the side of the distributor chamber for admission of air into the test column. The arrangement enables independent of each fluid directly to the test column.

A brass wire screen of 100 mesh was spread over the distributor so as to prevent the particals weeping in to the calming section through the water inlets.

An overflow chamber 15.24 cm diameter was provided at top of the test column a wire mesh cylinder was used to prevent any solid paricals coming with exit stream from a test column. The out coming fluid collected in the chamber and led to a storage time through a 5.08 cm outlet tube.

A Beckman thermometer was placed the test column just at the termination of the heating section for measuring the outlet fluid temperature in the test column.

A glass column having the same diameter and length as the test column was provided to observe the fluid bed conditions existing in the test column. The glass column has a similar distributor and a calming section as the test column.

An overflow chamber of 10.16 cm diameter and 50 cm long, made of copper was fitted on the top of the glass column to allow for the variations in the bed height during the adjustment of flow rates of the fluids. The overflow chamber has a provision to introduce solids in the glass column. The outgoing fluids forms the glass column were let to the storage tank.

3. Results and discussion

Minimum Fluidization velocity (U_{mf}) has been calculated by making a force balance across the fluidized bed. For biological flocs and supported bio-film, the factors namely ρ_S, ϵ_{mf} and Φ_s varies with shape of the flocs, diameter of the flocs and nature of the bio-particals are observed. The minimum fluidization velocities are compared with the general equations $N_{Re}<1$ and $N_{Re}>10^3$. Particals are found to have large deviations.

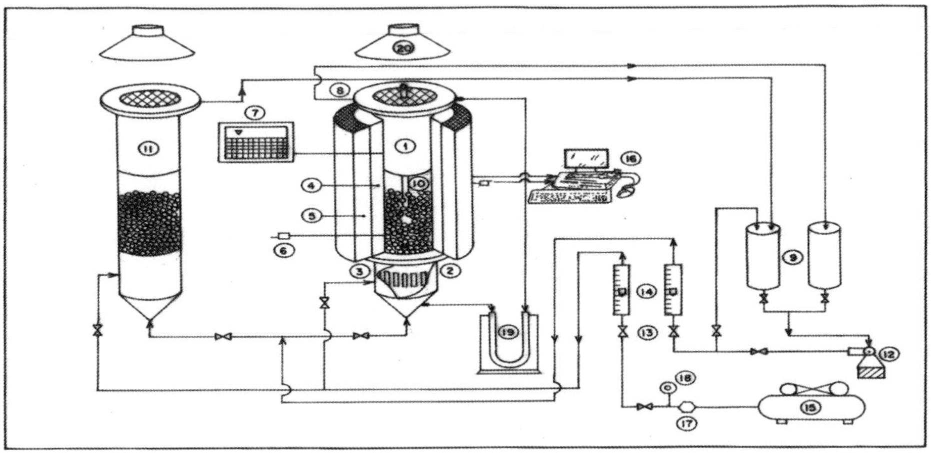

1.	Fluidized bed	8.	Chamber	15.	Air compressor
2.	Calming section	9.	Storage Tank	16.	Computer
3.	Distributor	10.	Probe/Thermometer	17.	Regulator
4.	Electric heaters	11.	Glass column	18.	Pressure gauge
5.	Insulation layer	12.	Pump	19.	Manometer
6.	Thermocouples	13.	Bypass valves	20.	Solid port
7.	Recorder	14.	Rotameter		

Figure 3. A Schematic Diagram of the experimental setup.

To correlate the authors data with Cleasby and Bauman (1977) empirical equation with ± 5% deviations.

$$U_{mf} = \frac{0.00381\, dp^{1.82} g^{1.88} [\rho_l(\rho_P - \rho_l)]^{0.94}}{\mu^{0.88}} \quad (1)$$

It is also suggested that fluidized beds can be operated for biochemical processes at velocities much higher than the minimum fluidization velocity but is limited by the terminal velocity of the particals. For non-sperical particals the terminal or free settling velocity was reported by Kunii and Levenspiel (1969)) from the experimental correlation of the dimensionless group

$C_D\ N_{Re,p}$ vs $N_{Re,p}$,. In FBBR, the shape of bioparicals are not rigid and also non-sperical. This may result in more C_D values than the predicted by the conventional equations. Hence, the authors are campared with the modified C_D in the Hermanowicz and Ganczarcyzk (1983) equation.

$$C_D = \frac{17.1}{N_{Re,\ p}^{0.47}} \quad (2)$$

It is observed that the microbial flocs and supported bio-film, the terminal or free settling velocity of the biopartical changes with time. This **is attributed for due to merely biomass growth. The** authors calculated the $N_{Re,p}$ lies in between 1.0 to 400, where most particals of behaviour is observed.

In searching for alternate and appropriate method under these circumstances, the authors compared with Andrews equation (1982) and successful results are reported. The terminal velocity at any time to the terminal velocity of the solid as follows.

$$\frac{u_t}{u_{ts}} = \frac{(1+\beta x)^{1/1.5}}{(1+x)^{1/3}} \quad (3)$$

Where

$$\beta = \frac{\rho_b - \rho_1}{\rho_p - \rho_1} \quad (4)$$

This equation is plotted in figure 4.

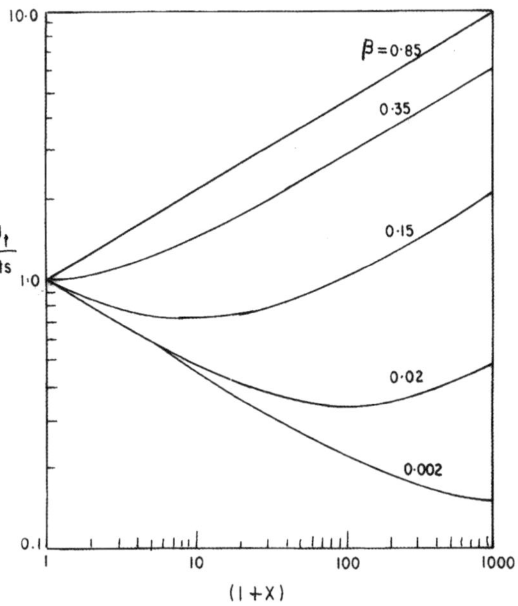

Figure 4. Bed Stratification Characteristics.

It is concluded that for heavy support particals (β«1), growth of biomass decreases the terminal settling velocity. This result in bed stratification. However, for flocs (β=1) growth always increases settling velocity.

Table 1. The following range of parameters are covered in the present authors data.

Particle sizes	0.17, 0.20, 0.25, 0.36, 0.46, 0.50, 0.66mm
Bed material	Fine river sand
Fluid mass	Water, aq-glycerol (5% to 40% wt %), Air
Superficial liquid velocity	$0.30 < u_L < 5$ cm/sec
Superficial gas velocity	$0 < u_G < 3$ cm/sec
Air pressure	0.65 kg/cm^2 (gauge)
Diameter of the fluidized column	10.16 cm
Heating length of the fluidized column	100 cm
Viscosity of aqueous glycerol	$0.01 < \mu_L < 0.0355$ pa.s
Solid hold up	$0.10 < \varepsilon_s < 0.3$
Liquid hold up	$0.16 < \varepsilon_L < 0.95$
Gas hold up$0.08 < \varepsilon_G < 0.65$	

4. Conclusions

In case of biological flocs and supported biofilms, since the factors sphericity, minimum bed porosity and density of solid changes with diameter, shape and composite nature of the bioparticle, the minimum fluidization velocities are often found to have large deviations.

Fluidized beds can be operated at velocities much higher than the minimum fluidization velocity but is limited by the terminal or free fall velocity of the particles.

Microbial flocs and supported biofilm, the terminal settling velocity of the bioparticle changes with time due to biomass growth.

It can be seen for heavy support particles ($\beta<<1$), growth of biomass decreases the terminal settling velocity. However, for flocs ($\beta=1$), growth always increases settling velocity.

References

1. Andrews. G.F (1982) Fluidized bed fermenters. A steady state analysis. BioTech mol, Bio Engg., vol.24, pp. 2013 – 2030
2. Bryers, J.D. and Characklis, W.G. (1982) Processes governing primary biofilm formation, Biotech mol., Bio engg., vol pp. 245 –2477
3. Characklis, W.G. (1981) Fowling bio-film development. A process analysis. Biotech mol., Bio Engg., vol – 23, pp. 1923–1960
4. Cleasby, J.L and Bauman, E.R. (1977) Backwash of granular filters used in waste water filtration, Environmental Protection Technology Series, EPA-60012-77-016
5. Ghosh, P, Ramachandran, K.B. and Praveen, V.V. (1996) Application of fluidization principles for bio processes, volume 8, Gordon and Breach publishers, pp. 303–338.
6. Hermanowicz, S.W and Ganczarcyzk, J.J (1983). Some fluidization Characteristics of biological beds, Biotechnol. Bioeng., vol 25, pp1321-1330
7. Hoehn, R.C. and Ray. A.D. (1973) Effects of thickness on bacterial film. J. water pollution control fed., vol. 45, pp. 2302–2320
8. Kunii, D. and Levenspiel, O. (1969) Fluidization Engineering, John Wiley & Sons, New York.
9. Shieh, W.K. and Keenan. J.D. (1986) Fluidization bed biofilm reactor for waste water treatment. Adv. Bio chem. Engg., vol. 33–167.
10. Wen, C.Y. and Yu Y.H. (1966) Mechanics of fluidization in fluid particle technology edited by S. Lee, Chem. Eng. Prog. Ser., vol. 52, pp. 100–111.

TOXICITY EVALUATION OF DIFFERENT URBAN SOLID WASTE COMPOST ON GERMINATION AND GROWTH PARAMETERS OF LEPIDIUM SATIVUM L

ALI REZA ASTARAEI

Associate Prof. of Soil Science Department, College of Agriculture, Ferdowsi University of Mashhad

A laboratory experiment was conducted to study the effect of water extracts of urban solid waste compost (USWC) of different seasons on germination and growth parameters of Lepidium sativum L. Four USWC produced in spring, summer, autumn and winter were selected .Their water extracts of 100, 30, 10, 3, and 0% concentrations was prepared by diluting the original water extract of 100% concentration as five treatments with three replications and analyzed statistically as a completely randomized design. Results showed that maximum germination percentage was noted in 0% (control) and minimum in 100% concentrations. Highest radicle length was noted in 0% and 3% and lowest in 100% concentrations. Similar results were noted in plumule length. Result of different USWC water extract concentrations showed that water extract concentration of 10% resulted in 42 and 55% reductions in germination percentage and germination growth rate, respectively, compared to 100% concentration. Whereas, 3% USWC water extract concentration resulted in radicle and plumule lengths decrease by 72 and 59%, compared to 100% concentration. Our results indicated that different seasonal USWC prepared were not processed completely and as they are not fully matured, they contain high soluble organic compounds, imposing high toxicity during germination and early seedling growth of Lepidium sativum L.

Keywords: USWC, water extract concentrations, Lepidium sativum L.

1. Introduction

Seventy percent of the dry lands, which constitute 25% of world agricultural land, are used for agriculture. Dry areas; provide habitat and livelihoods for more than one billion people.

The amount of Urban solid waste (USW) generated annually in Iran is increasing due to increasing rate of urbanization and industrialization as a result, its disposal is a major environmental problem. USW compost is being manufactured from organic fraction of the USW stream. Land application of USWC, allows for the recycling of nutrients, and produces a relatively low-cost product that can be used as a soil amendment in agriculture, horticulture, and land reclamation especially for countries with arid and semi arid climates such as Iran.

In general, composting of municipal solid waste has potential as a beneficial recycling tool.

Its safe use in agriculture, however, depends on the production of good quality compost, specifically, compost that is mature and sufficiently low in metals and salt content.

There are reports that the organic toxins are present in USW compost [7]. These compounds are highly chlorinated and known to be persistent in the environment [9].

Feedstocks that contribute organic pollutants included pesticides, household wastes such as oils and solvents, and paper products because of the printing ink. A survey of municipal and commercial wastes reported that under 0.5% by weight is hazardous and researchers indicated that phthalate esters are likely the most abundant xenobiotic present in USW compost [7]. The concentrations of dioxin/furans and polychlorinated biphenyls (PCB) were higher in mixed USW compost than source-separated USW compost [7].

The aim of the present work was to evaluate the toxicity of different urban solid waste compost on germination and growth parameters of *Lepidium sativum L.*

2. Materials and Methods

Four USWC produced in spring, summer, autumn and winter were examined for their toxicity on germination and growth parameters of *Lepidium sativum L.* Different USWC were dried at 70°C, each one with equal amount was mixed with distilled water (1:2.5) in mechanical shaker for 12 h., kept at room temperature of 25°C for 3 days, their water extracts were obtained and sterilized [1, 5]. Five treatments: T_1- 100%, T_2- 30%, T_3- 10%, T_4- 3% and T_5- 0%, were prepared by diluting the original water extracts of 100% concentration with distilled water and one control (0%), each treatment with three replications in a completely randomized design. Some physico-chemical properties of different USWC measured by standard methods are presented in table 1.

Table 1. Some physico-chemical properties of MSWC produced in different seasons of year.

Parameters	Range
OM (%)	20-26
Ash (%)	60-70
OC (%)	9-13
N (%)	0.7-1.1
Na (%)	0.4-0.7
K (%)	0.5-0.9
Moisture (%)	8-20
Saturation (%)	80-90
EC (1:2.5) (dS/m)	9-11.5
pH(1:2.5)	7-7.7
C/N	10-14

The 10 sterilized *Lepidium sativum L.* seeds were placed in between two filter paper in Petri dishes of 10 cm in diameter, 5 cc of each treatment was added to each Petri dish, and all Petri dishes were incubated at 20°C for 15 days. Germination count was performed daily and at the end of experiment, germination percentage, radicle and plumule lengths in Petri dishes were measured.

The data was analyzed statistically using MSTAT-C and mean values were compared by Duncan Multiple Range Test at 5% significance.

3. Results and Discussion

3.1. *Effect of different water extract concentrations on germination and growth parameters of Lepidium sativum L.*

Results of different water extract concentrations on seed germination of Lepidium sativum L. showed that control (0%) and 3% concentrations had highest and 30% and 100% had lowest germination percentage (table 2). But all treatments showed reduction in germination compared to control (0%), this reduction of germination, is attributed to increasing concentration of toxic soluble organic matter present in growth media [11, 12]. Plumule length of Lepidium sativum L. in control and 3% concentrations was maximum and in 30% and 100% was minimum (table 2).The reduction of Plumule length in 3%, 10%, 30% and 100% concentrations levels were 11%, 17.3%, 26.9%, 62.4%, respectively, compared to control. [4, 10].

Table 2. Effect of different water extract concentrations on germination percentage, radicle length, plumule length and germination growth rate of Lepidium ativum L.

Treatments	Germination (%)	Radicle length (cm)	Plumule length (cm)	Germination growth rate
T_1	55.57d	2.184 e	1.693 e	2.24 c
T_2	92.22 c	3.419 d	3.293 d	4.6 b
T_3	96.67 b	4.886 c	3.718 c	4.98 a
T_4	99.07 b	7.807 b	4.01 b	5.2 a
T_5	100 a	11.5 a	4.5 a	5.3 a

T_1 = 100%, T_2 = 30%, T_3 =10%, T_4 = 3%, T_5 = 0%
Means within a column followed by the same letter are not significantly different at $p = 0.05$.

Radicle length of *Lepidium sativum L.* in control and 3% concentrations was maximum and and in 30% and 100% was minimum (table 2). Comparing radicle length of all concentration levels to control, revealed that 3%, 10%, 30% and 100% concentrations showed reduction by 32%, 57.5%, 70% and 81%, respectively. The presence of soluble organic matter in growth media, imposes toxicity to seed, delays germination, resulted in phyto-toxicity, and highest negative impact is noted on radicle length, followed by plumule length of *Lepidium sativum L.*

Germination growth rate was highest in control (0%) and lowest in 100% concentration level. Treatments of 3% and 10% concentration levels were at a par and showed no significant differences compared to control, but their germination growth rates were significantly higher than 30% and 100% concentration levels.

Plumule length to radicle length ratio increased by increasing the concentration levels, indicating the highest toxicity impact on radicle length followed by plumule length (data not shown).

All composts with 100% concentration showed lowest germination percentage, plant plumule and radicle lengths (table 3) and minimum germination percentage, plumule and radicle lengths were noted in winter (C_4T_1), autumn (C_3T_1), winter composts of 100% concentration, respectively.

Table 3. Interaction effect of different seasonal compost produced and concentrations on germination percentage, Plant plumule and radicle lengths and germination growth rate (GGR) of Lepidium sativum L.

Compost	Conc (%)	Germination (%)	Plant plumule length (mm)	Plant radicle length (mm)	GGR
	T_1	80d	1.46h	0.4gh	3.9d
C_1	T_2	100a	3.57ef	1.67f	5.0a
	T_3	100a	4.17abcd	3.5e	5.2a
	T_4	100a	4.59a	8.17b	5.2a
	T_5	100a	4.5a	11.5a	5.27a
	T_1	66ef	1.5h	0.5gh	3.1e
C_2	T_2	90bc	2.44h	1.67f	4.7bc
	T_3	100a	3.03c	3.31e	5.2a
	T_4	95ab	4.1bcd	6.33c	4.9ab
	T_5	100a	4.5a	11.5a	5.27a
	T_1	70e	1.2h	0.2h	0.95f
C_3	T_2	85cd	3.0c	1.42f	4.2cd
	T_3	100a	3.03c	2.91e	5.2a
	T_4	100a	4.0 bcd	5.84c	5.2a
	T_5	100a	4.5a	11.5a	5.27a
	T_1	50g	1.65gh	0.0h	0.4g
C_4	T_2	78d	2.5g	1.0fg	3.9d
	T_3	80d	2.16fg	3.5e	4.1d
	T_4	100a	2.92cd	6.41c	5.2a
	T_5	100a	4.5a	11.5a	5.27a

C_1 = spring compost, C_2 = summer compost, C_3 = autumn compost, C_4 = winter compost,
T_1 = 100%, T_2 = 30%, T_3 =10%, T_4 = 3%, T_5 = 0% (control)
Values in the same column followed by the same letters are not significantly different at 0.05 level according to DMRT.

As mentioned earlier, all composts produced during different seasons showing reduction in measured plant growth parameters with their increasing water extract concentrations compared to control (T_5). Indicating, thereby, toxicity effects due to the release of soluble organic matter substances, produced because of improper processing of compost.

Similar results were reported by Zucconi et al. [11], Zucconi [12], Muthuchelian et al. [10], Astaraei and Ivani [2] and Astaraei and Sampietro

4. Conclusion

The results obtained, clearly, indicate the present of organic toxins in USWC produced in USWC Recycling industry of Mashhad city. Nearly, all seasonally, USWC produced showed their adverse effects on tested plant, indicating that the physical and chemical makeup of USW compost tends to shift with time and source and thus careful yearly monitoring of USW compost quality is required (6).

Standardizing the analytical procedures used to assess quality of the final product, as well as a standardized index of compost quality need to be developed for the industry.

This will allow easy monitoring of the quality of USW compost and allow comparison of products among different facilities. Year to year variation in the properties of compost from the same source prevents researchers from drawing conclusions and inhibits research and effective use of the material [8].

References

1. Agrawal, R.L., "Seed Technology," *Oxford and IBM Publishing Co. London*, pp. 515–564, 1982.
2. Astaraei, A .R. and R. Ivani, "Effect of organic sources as foliar spray and root media on nutrition of cowpea plant," *American-Euroasian J, Agric. & Enviro. Sci.* 3(3), 352–356, 2008.
3. Astaraei, A.R. and D.A. Sampietro, "Allelopathic effect of almond on cress and fenugreek," *Allelopathy Journal*. 22 (1), 239–244, 2008.

4. Baur, P., H. Marzouk, J. Schonherr, and T. Grayson, "Partition coefficient of active ingredients between plant cuticle and adjuvant as related to rates of foliar uptake," *J. Agric. Food Chem.*, 45, 3659–3665, 1997.
5. Fernandez, G., and M. Johnston, "Seed vigor testing in lentil, bean and chickpea," *Seed Sci. Technol.*, 23, 617–627, 1995.
6. Hicklenton, P., V. Rodd, P.R. Warman, "The effectiveness and consistency of source-separated municipal solid waste and bard composts as components of container growing media," *Sci. Hort.*, 91, 365–378, 2001.
7. Logan, T.J., C.L. Henry, J.L., Schnoor, M. Overcash, M., and D.C. McAvoy, "An assessment of health and environmental risks of trace elements and toxic organics in land-applied municipal solid waste compost," *Compost Sci. Util.*, 7(3), 38–53, 1999.
8. Mamo, M., C.Rosen, T. Halbach, "Nitrogen availability and leaching from soil amended with municipal solid waste compost," J. *Environ. Qual.,* 28, 1074–1082, 1999.
9. Muir, D.C.G., P.H. Howard, "Are there other persistent organic pollutants? A challenge for environmental chemists," *Environ. Sci. Technol.*, 40(23), 7157–7166, 2006.
10. Muthuchelian, K., D. Neri, F. Zucconi, "Influence of marine algae and humus application on growth, photosynthesis, and metabolic content in vegan saneness L." Mosonnmogyaro van., 185-192, 1996.
11. Zucconi, F., M. Fort, A. Monaco, M. Bertoldi, "Biological evaluation of compost maturity," *Biocycle,* 22(4): 27–29. 1981.
12. Zucconi, F., "Dechino del suolo, stanchezzadel terreno," Spazio verde padova, pp. 291, 1996.

PREPARATION OF PT-SN/AL$_2$O$_3$ CATALYST: MODELING OF PLATINUM AND TIN PRECURSOR ADSORPTION ONTO ALUMINA

MARYAM SAEEDIZAD, SAEED SAHEBDELFAR, ELNAZ SAMEI and MARYAM ZAMAN

Catalyst Research Group, Petrochemical Research and Technology Company, National Petrochemical Company, P.O. Box 14358, Tehran, Iran

The adsorption of hexachloroplatinic acid and tin chloride on γ-Al_2O_3 was studied. Pt-Sn/γ-Al_2O_3 catalyst samples were prepared by sequential impregnation of hexachloroplatinic acid and tin chloride on the support. The amount of solute adsorption was determined through the difference of the initial and final concentration of the solute. Among different adsorption models tested, the Langmuir adsorption model was found to predict satisfactorily the hexachloroplatinic acid and tin chloride adsorption data. The parameters of the models, i.e., adsorption constants were determined from separate experiments at different solution pH and surface loadings. Adsorption experiments were in the absence of diffusion resistances.

1. Introduction

Platinum-based catalysts are widely employed in chemical industries, e.g. for dehydrogenation of alkanes to olefins and catalytic reforming of naphtha.

Platinum is supported on high-surface area supports and is often promoted by other metal metals such as tin. Techniques for the preparation of such catalysts are based on impregnation of the support with solutions of hexachloroplatinic acid as a platinum precursor and tin chloride. The system can then be filtered, if excess solution has been employed, and the catalyst dried and further pretreated to transform the metal from its precursor state into its active form. Impregnation in the catalyst preparation is the first step that metal salt dissolved in aqueous solution are contacted with a porous alumina. During a contact time the metal adsorbs from an aqueous solution onto the high surface area support. Attempts to "transform the art of catalyst preparation into a science" have focused more closely on the process of adsorption.

Therefore, modeling of impregnation of the metals is beneficial both economically and from environmental point of view as it makes optimal use of the precursor possible.

Adsorption characteristics of systems being investigated include compounds of Pd on alumina [1], Pd on silica and alumina [2], Rh on alumina [3], Pt and citric acid on alumina [4].

The present study is devoted to modeling of Pt-Sn/γ-Al_2O_3 catalyst preparation by impregnation. Among different models, an adsorption model which predicts satisfactorily the hexachloroplatinic acid and tin chloride adsorption on the alumina is selected. The parameters of the model, i.e., adsorption constants and surface saturation coverage were determined from separate experiments.

2. Experimental

A known amount of commercial γ-alumina pellets provided by a European company (table 1) that was ball-milled to 30-60 µm and dried at 120°C over-night. Specific concentration of hexachloroplatinic acid or tin chloride was kept in a shaker during the impregnation period at a constant temperature of 30ºC. It was observed that the concentration of both precursor solutions change relatively little with time after approximately 30 min. The slurry was filtered and the filtrate concentration was determined by atomic absorption spectroscopy for platinum or tin concentration. The amount of solute adsorbed on γ-Al_2O_3 was determined through the difference in solution concentration before and after impregnation. In these way more than three sets of equilibrium adsorption data obtained at different pH and surface loading (m^2/lit) for hexachloroplatinic acid and tin chloride adsorption on alumina.

Table 1. Physical characteristics of γ-Al_2O_3 granules.

Diameters (mm)	1.8-2
Bulk density (g/cm^3)	0.5
Surface area (m^2/g)	130
Pore volume (ml/gr)	1.05

3. Equilibrium adsorption models

Adsorption isotherm models have been used to describe adsorption of catalyst precursors on oxide surfaces from aqueous solutions quantitatively. Adsorption isotherm models typically describe experiments within a limited range of conditions but are popular because they are simple to implement and require fewer parameter estimations. Table 2 introduces common adsorption isotherm models where K is the equilibrium adsorption constant.

C_e is the solution concentration; C_0 is the solution initial concentration. The quantity n_e is calculated from the difference between the initial and equilibrium solution concentration. By using a linear function or nonlinear algorithm the equilibrium adsorption constants K, and the surface saturation coverage, n_s, were obtained for each component.

Table 2. Common adsorption isotherm models [5].

Model	Expression
Temkin	$\frac{n_e}{n_s} = kln(KC_e)$
Freudlich	$n_e = n_s C_e^{1/k}$
Langmiur	$n_e = \frac{KC_e n_s}{1+K^*C_e}$

K*, K(lit/ mole), n_s(mol/ gr)
*dimensionless constant

4. Results and discussion

The results of data fitting for hexachloplatinic acid equilibrium adsorption on alumina are shown in Fig. 1. One observes that Freundlich isotherm could not describe the adsorption of hexachloroplatinic as good as Langmuir and Temkin isotherms. Langmuir model could describe adsorption at different conditions precisely. The equilibrium adsorption constants K and the surface saturation coverage, n_s, were obtained for Langmuir model (table 3). Surface saturation coverage and equilibrium adsorption constant obtained in the range of 23-58 μmole/m^2 and 216-672 lit/mol, respectively, at different pH and surface loadings.

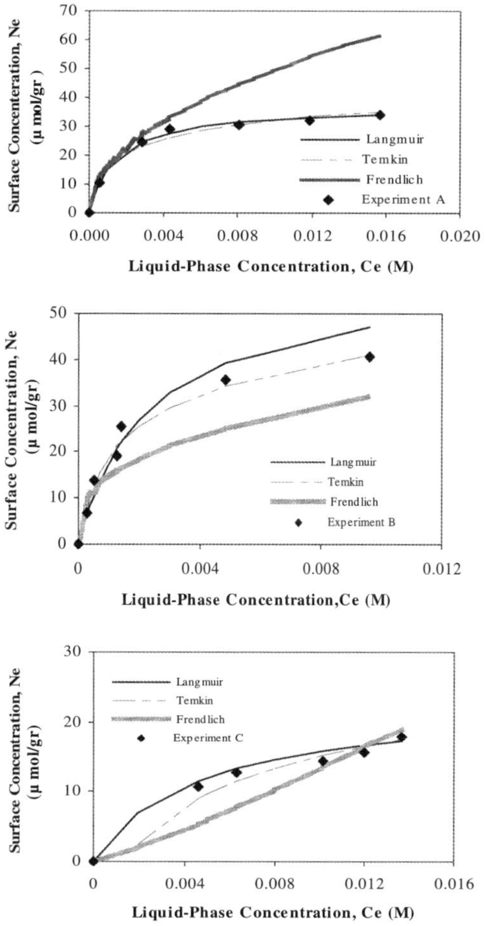

Fig. 1. Equilibrium adsorption of hexachloroplatinic acid on alumina.

Table 3. Equilibrium adsorption parameters of hexachloroplatinic acid on γ -Al_2O_3.
A(S/L=10000 m^2/lit, initial pH 1.5)
B(S/L=20000 m^2/lit, initial pH 1.5)
C(S/L=10000 m^2/lit, initial pH 1)

		A	B	C
Temkin	K	9666	6805	688
	c_1 n_s	7.023	9.77	7.8004
	R^2	0.9592	0.9871	0.9623
Freudlich	K	3.700	2.419	2.335
	b_1 n_s	109.9	290.7	107.3
	R^2	0.8922	0.9638	0.9548
Langmiur	K	672.5	430	216
	n_s	37.175	58.14	23.148
	R^2	0.9988	0.9508	0.9618

Deviation from the Langmuir behavior observed at higher surface loading could be associated with a decrease in availability of strongest binding sites, leading to decrease in apparent adsorption equilibrium that was observed elsewhere [6].

The solution initial pH could affect adsorption constants and surface saturation coverage and large values indicate strong adsorption. Solution pH after equilibrium adsorption of hexachloroplatinic acid on alumina increased as indicated in Fig. 2. Experiment A with initial pH 1.5 and final pH 2-4 showed larger adsorption equilibrium constant It can be explained by the increasing positive charge density at alumina surface at the optimum solution pH 3-4 and below the value of the IEPS of alumina, typically near pH 8, for hexachloroplatinic acid adsorption [7].

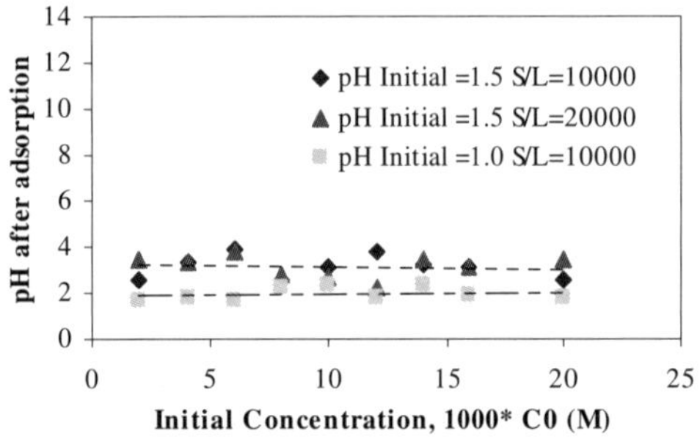

Fig. 2. Solution pH after equilibrium adsorption of hexachloroplatinic acid on alumina.

The results of data fitting for tin chloride equilibrium adsorption on alumina were shown in Fig. 3. It was seen that Freundlich and Temkin isotherm could not describe the adsorption of tin chloride as good as Langmuir isotherm.

For tin chloride, the surface saturation coverage was found to be 0.1-0.4 μmole/m2 and the equilibrium adsorption constant was 8.7-16 lit/mol (table 4). Higher surface saturation coverage was seen in higher surface loading and higher pH range. The decrease of surface saturation coverage with decreasing pH related to decrease of adsorption sites due to alumina dissolution. Apparent adsorption equilibrium constant decreased at higher surface loading with lower availability of strongest binding sites.

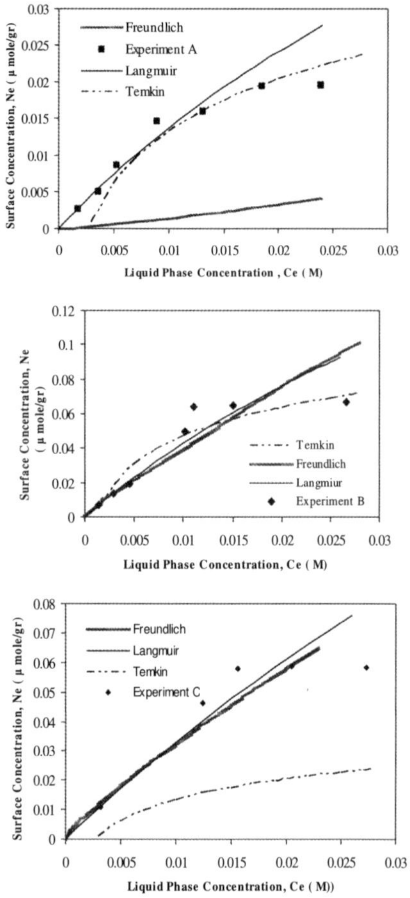

A(S/L=10000 m2/lit, Initial pH0.3)

B(S/L=10000 m2/lit, Initial pH 0.7)

C(S/L=20000 m2/lit, Initial pH 0.7)

Fig. 3. Equilibrium adsorption of tin chloride on alumina.

Table 4. Values of adsorption parameters of tin chloride on γ -Al_2O_3.

		A	B	C
Temkin	K	683.88	723.45	362.57
	kn_s	7.29 E-3	2.38E-2	1.03E-2
	R^2	0.978	0.938	0.905
Freudlich	K	1.2661	1.1535	1.1856
	kn_s	0.4675	2.2489	1.5722
	R^2	0.950	0.940	0.93
Langmiur	K	16	15.03	8.77
	n_s	0.1	0.33	0.41
	R^2	0.989	0.992	0.991

5. Conclusion

The Langmuir adsorption isotherm could predicts fairly the hexachloroplatinic acid and tin chloride adsorption on alumina. From the equilibrium, hexachloroplatinic acid was found have surface saturation coverage of 23-58 μmole/m^2 and equilibrium adsorption constant of 216-672 lit/ mol in surface loadings 10000-20000 m^2/lit and solution initial pH 1and 1.5. The equilibrium adsorption constants K and the surface saturation coverage, n_s decrease as the solution initial pH decrease below 1.5 and final pH solution decrease below 3. Deviation from the Langmuir behavior observed at higher surface loadings, associated with a decrease in availability of strongest binding sites, results in decrease in apparent adsorption equilibrium constant and surface saturation coverage.

For tin chloride, the surface saturation coverage was found to be 0.1-0.4 μmole/m^2 and the equilibrium adsorption constant was 8.7-16 lit/mol in surface loadings 10000-20000 m^2/lit and solution initial pH 0.3 and 0.7. The equilibrium adsorption constant decrease at higher pH range and lower surface loading. Surface saturation increase at higher pH with lower alumina dissolution.

References

1. Ch. Sivaraj, Cr. Contescu and J.A. Schwarz, J. Catal.132, pp 422-431, (1991).
2. J.A. Schwarz, C.T. Ugbor and R. Zhang, J. Catal.138, pp 38-54, (1992).
3. J.S. Hepburn, H.G.Stenger and C.E. Lyman, J. Catal.128, pp 34-47, (1991).
4. P. Papageorgiou, D. M. Price, A. Gavriilidis and A. Varma, J. Catal. 158, pp 439–451, (1996).
5. T. Horvath, Encyclopedia of Catalysis, Wiley, New York, vol. 3. pp 49-79.
6. M.M. Benjamin and J.O. Leckie, J. Coll. Interf. Sci. 79, pp 209-221, (1981).
7. Th. Mang, B. Breitscheidel, P. Polanek and H. Knozinger, Appl. Catal. A: General: 106, pp 239-258, (1993).

INFLUENCE OF BLADE ABSENCE ON PERFORMANCE OF DC TYPE LOW-NO_X BURNER FOR PULVERIZED-COAL FIRED BOILER*

ZHIQIANG WANG† and YONG DONG

School of Energy and Power Engineering, Shandong University
Jinan 250061, China

SHAOZENG SUN

Research Institute of Combustion Engineering, Harbin Institute of Technology
Harbin 150001, China

In the operation process of pulverized-coal fired boiler with DC Type Low-NO_x Burner (DCTLB), blades absence results in great changes of performance. According to the absence of five-level blades burner, the performance was studied on the experimental system of DCTLB. Results show that when one blade is damaged, the effect of the fifth blade on resistance coefficient, ratio of rich and lean air flux and concentration rate is least. Three key performance parameters of two blades absence can be widely adjusted. The case of three blades absence must be avoided in practice.

Keywords: Low-NO_x burner; resistance coefficient; ratio of rich and lean air flux; concentration rate.

1. Introduction

In a long time coming, Environmental pollution is a severe problem confronted by thermal power industry at home and abroad. With the rapid development of national economy and closely related energy industry, China environmental pollution has attracted wide attention. The technology of Low-NO_x Burner in pulverized-coal fired boiler is the effective low-NO_x measure. At low cost, it is very suitable for China national conditions [1].

Low-NO_x Burner can ensure the combustion efficiency of pulverized-coal fired boiler as well as the emission of NO_x, SO_2 and coal-dust pollution can be controlled into an allowable range [2]. Many researchers have accomplished lots of work in this respect, and developed a number of new type burners, mainly including DRB type Low-NO_x burner [3], PM type Low-NO_x burner [4], WR type Low-NO_x burner [5] and horizontal bias burner [6,7] and so on. In the operation process of pulverized-coal fired boiler with DC type Low-NO_x burner (DCTLB), blades absence result in great changes of performance parameters. According to the absence of five-level blades burner, much work has been done on experimental system of DCTLB. The results provide an important theoretical basis for the design and operation of pulverized-coal burner.

2. DC Type Low-NO_x Burner

2.1. *The operation of DCTLB*

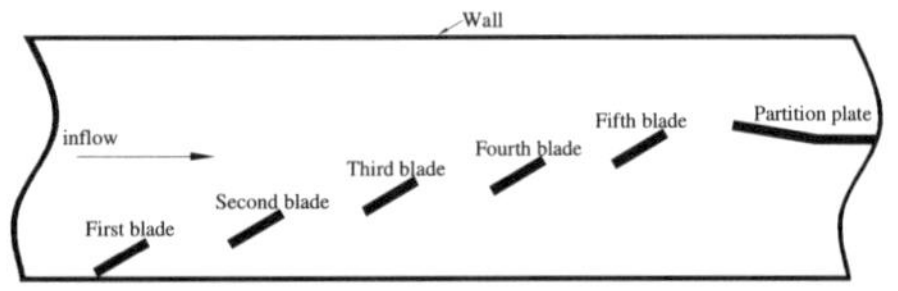

Fig. 1. Schematic diagram of DCTLB.

The operation of DCTLB is presented. The entrance of the burner is connected with the primary air duct. When the primary air carried pulverized-coal particles flows onto the blades of DCTLB, the inertial force difference between air and pulverized-coal particles makes the most particles entry into rich flow duct after the particle and blade collision, especially for large particles. Finally, the primary air is divided into two airflows, the high concentration pulverized-coal airflow and the low concentration pulverized-coal airflow based mainly on air. The operation schematic diagram of DCTLB is shown in Fig 1.

*This work is supported by the Special Funds for Major State Basic Research Projects of China (Contract No.: 2006CB200303).

†Work partially supported by Major Projects of State Power Corporation (Contract No.: KJ00-02-16-1).

2.2. *Key performance parameters*

The following parameters are mainly adopted for evaluating the performance of DCTLB.

1) Resistance coefficient

$$\xi = \Delta P / P_{d,in} \quad (1)$$

Where ΔP is total pressure difference between the inlet and outlet of DCTLB, and $P_{d,\,in}$ is dynamic pressure of the inlet.

Resistance coefficient is an important parameter in the engineering, especially for boiler retrofit. And the pulverized-coal feeding becomes more and more difficult as the resistance coefficient increases. According to the direct-fired system, the capacity of mill is also affected. Resistance coefficient reflects the resistance which the air and pulverized-coal should overcome to get though DCTLB. In general case, the resistance coefficient is the smaller the better.

2) Ratio of rich and lean air flux

$$R_Q = Q_n / Q_d \quad (2)$$

Where Q_n is rich airflow flux, and Q_d is lean airflow flux.

The ratio of rich and lean air flux is also a very important performance index of DCTLB, and has a very close relationship with concentration rate. It reflects homogeneous degree of airflow distribution between rich and lean air flux. The higher ratio can result in the large flux of rich flow duct, so that excessive airflow dilutes the concentrated effect of DCTLB. The structure of DCTLB should ensure the rich and lean air flux ratio appropriate.

3) Concentration rate

$$R_C = C_n / C_d \quad (3)$$

Where C_n is coal concentration of rich-flow duct, and C_d is coal concentration of lean-flow duct.

Concentration ratio reflects the effect of DCTLB on the degree of particles concentration. Rich-side pulverized-coal concentration which has influence on combustion is a major issue to be considered in engineering practice. Theory and engineering practice on pulverized-coal combustion technology of DCTLB shows that the higher concentration is not the better, but there is an optimum value. Therefore, the performance must be synthetically considered among resistance coefficient, rich and lean air flux ratio and concentration rate.

3. Experimental System

Experimental system is designed based on the theory of similarity and modeling, which includes geometric similarity, physical parameters similarity, boundary condition similarity and momentum ratio similarity and so on [8]. The test section is 780 mm in length, 200 mm in width and has five-level blades (the serial number of blades is from first to fifth along the flow direction). Schematic diagram of experimental system is shown in Fig. 2.

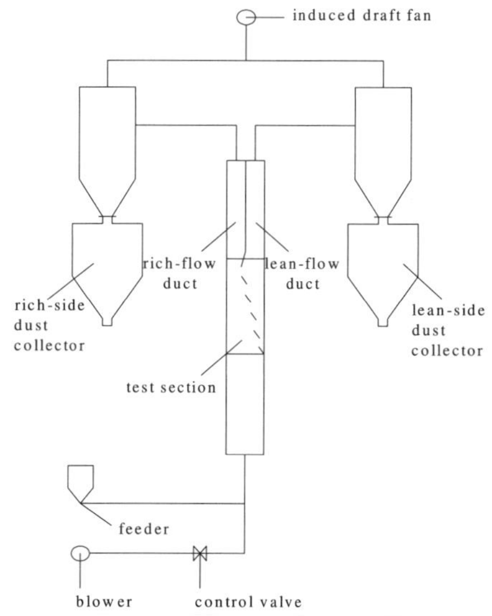

Fig. 2. Experimental system of DCTLB.

Glass bead is supplied by using the spiral type feeder, and then enters the vertically placed test section of DCTLB after adequately accelerated mixing. Due to the concentration efficiency of the septation baffle, the inflow carried glass bead is divided into rich airflow and lean airflow. Finally, the gas-solid flow enters the cyclone to recovery the solid particles.

4. Numerical Simulation

Process of gas-solid two phases flow is reproduced with numerical simulation of FLUENT software package. The turbulence model used is a standard k-ε model and the two phases flow model is a particle stochastic trajectory model. And the particle velocity is about 90 percent of the air. This is because there is the velocity slip between solid phase and gas phase. The particle concentration is well distributed at the inlet. The distribution of particle size normally follows Rosin-Rommler distribution. The outlet boundary is assumed as the fully developed outflow.

5. Results

5.1. *One blade absence*

The main reasons of the resistance in DCTLB are the fast attenuation of the airflow velocity after hitting the blade, the flow around the blade and the vortex energy loss generated in leeward side of the blade. Effect of one blade absence on resistance coefficient is shown in Fig. 3. The resistance coefficient is up to 2.3 when the blades are complete, but the resistance coefficient reduces to 1.95 owing to the one blade absence.

The experiment and simulation results show that the effect of the first blade absence on resistance loss is the greatest. The reason is that the inflow clings to the wall of DCTLB and the first blade has the most important influence on the inflow direction. However, the first blade absence reduces the resistance caused by the impact and the flow around the first blade, thus the airflow can access to the lean-side duct without any resistance. The influence of the fourth is only next to the first. The fifth is the least because the residual blades have already formed the complete concentrator structure.

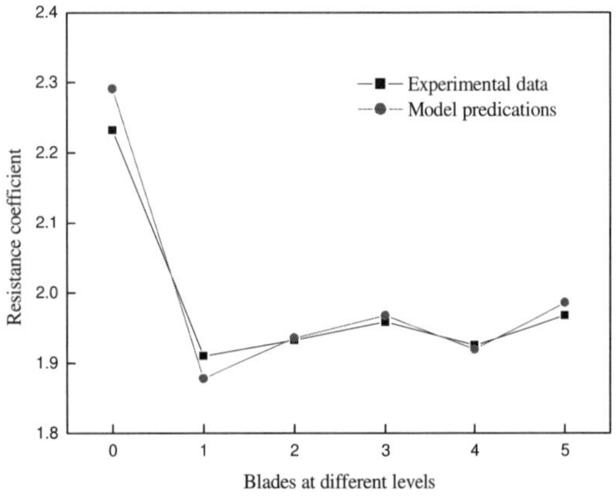

Fig. 3. Effect of one blade absence on resistance coefficient.

Effect of one blade absence on ratio of rich and lean air flux is shown in Fig. 4. The ratio is 0.78 when the blades are complete, but when one blade is damaged, the value is found to change greatly. The effect of the second and the third is greater (see Fig. 4).

As the residual blades form a relatively integrate structure of concentrator under the working condition of the first blade absence, the total rich and lean air flux ratio still remains higher values even if some airflow reaches directly into lean-side duct. The effect of the second is the greatest. This is because the inflow has been concentrated by the first blade, but the airflow velocity is not especially high and there is a flowing space between the first and the third, which greatly reduces the circle resistance. The most airflow enters into lean-side duct. The ratio of the third blade absence is higher than the second. The reason is that the airflow has the certain velocity and the stronger stiffness when the inflow has been concentrated by the first two blades. As Fig. 4 shows that the ratio of the fourth and the fifth is also higher.

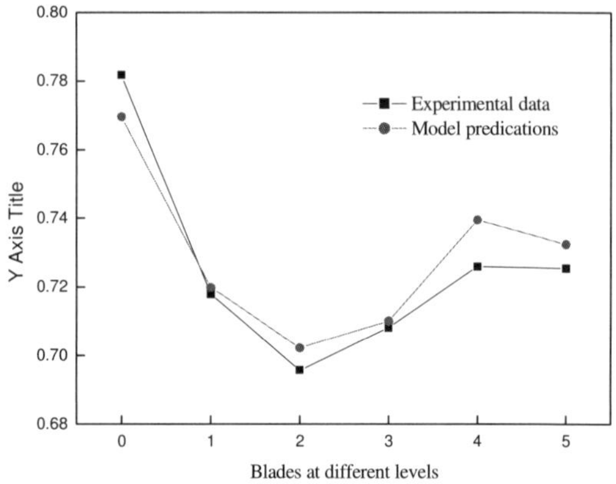

Fig. 4. Effect of one blade absence on ratio of rich and lean air flux.

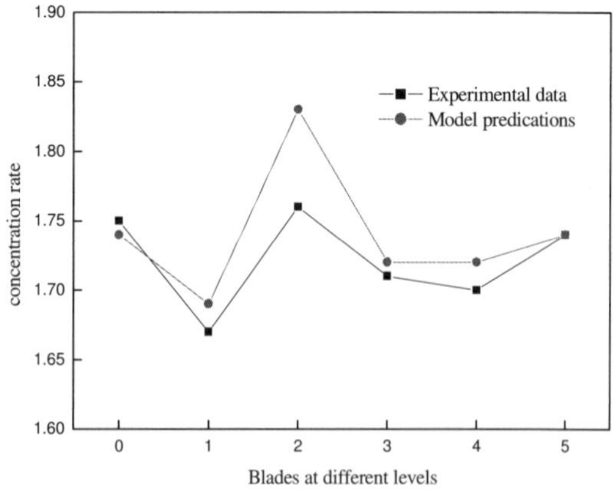

Fig. 5. Effect of one blade absence on concentration rate.

Effect of one blade absence on concentration rate is shown in Fig. 5. The concentration rate is not maximum and only 1.75 under the complete blades. But the maximum concentration rate has been obtained when the second blade is damaged. It is because that the second blade absence makes gas-solid separation higher and the ratio of rich and lean air flux least.

Pershing's results show that fuel NO_x is more than 90% in the combustion process of pulverized-coal. At the meantime, almost all of fuel N can rapidly decompose and the intermediate products of HCN, N, CN and NH_i are formed. These products react to form NO with oxygenated compounds (O, O_2 and OH), then further reacted to form N_2 with NO. Therefore, in order to reduce the NO emission, the important measure should be taken to limit fuel NO that is converted from fuel N, for example, reducing the content of oxygenated compounds and strengthening the reduction reaction of NO. Besides fuel NO, the gas pollutant also includes thermal NO and prompt NO, and they are formed in oxidation process of N_2 at a high temperature [9, 10].

The main reasons for reducing the NO emission are: the primary air carried pulverized-coal particle is divided into two airflows, the content of O_2 is least in rich-side duct and fuel NO can be greatly reduced because of enhanced reduction in fuel-rich combustion technology. The formation amount of thermal NO increases sharply with the increase of the temperature. But the excess air ratio is more than 1 in lean-side duct, which makes the combustion temperature and the oxidation rate decrease. Finally, the formation amount of thermal NO minimizes.

5.2. *Two blades absence*

Table 1. Results of two blades absence.

Case	Blades absence	Resistance coefficient	Rich and lean air flux ratio	Concentration rate
1-0	1st, 2nd	1.697	0.68	1.434
1-1	1st, 3rd	1.714	0.673	1.501
1-2	1st, 4th	1.738	0.697	1.461
1-3	1st, 5th	1.716	0.729	1.53
1-4	2nd, 3rd	1.638	0.664	1.514
1-5	2nd, 4th	1.705	0.68	1.55
1-6	2nd, 5th	1.687	0.705	1.645
1-7	3rd, 4th	1.716	0.675	1.491
1-8	3rd, 5th	1.711	0.691	1.625
1-9	4th, 5th	1.688	0.713	1.588

Table 1 presents experimental results of two blades absence in DCTLB. When the two blades are damaged, the variation range of resistance coefficient is relatively great. The difference between the maximum and minimum is 0.1. Effect of the second and other blade absence on resistance coefficient is greater, especially for Case 1-4. But the maximum resistance coefficient of Case 1-2 has less influence on DCTLB.

When the third and other blade are damaged, the ratio of rich and lean air flux is relatively small and that of Case 1-4 is minimum among them. Reasons for the greater ratio of Case 1-3 and Case 1-9 are that other three blades can form relatively complete concentration structure, and it ensures the better concentrated effect. According to two blades absence, concentration rate of Case 1-0 is minimum and that of Case 1-6 is maximum. In practice, the parameter can be adjusted in accordance with the laws.

5.3. *Three blades absence*

Table 2 presents experimental cases of three blades absence in DCTLB. Fig. 6 presents experimental results of three blades absence in DCTLB. Resistance coefficient of Case 2-5 is minimum. And the variation range of resistance coefficient of the first and other two blades absence is relatively great.

Table 2. Cases of three blades absence.

Case	Blades absence
2-1	1st, 2nd and 3rd
2-2	1st, 3rd and 5th
2-3	2nd, 3rd and 4th
2-4	2nd, 3rd and 5th
2-5	3rd, 4th and 5th

When three blades are damaged, the rich and lean air flux ratio of Case 2-1, Case 2-3 and Case 2-4 changes little and has a very small impact on the performance of DCTLB. The rich and lean air flux ratio of Case 2-2 is maximum. The rich and lean air flux ratio of Case 2-5 is minimum, so the concentration effect is very poor.

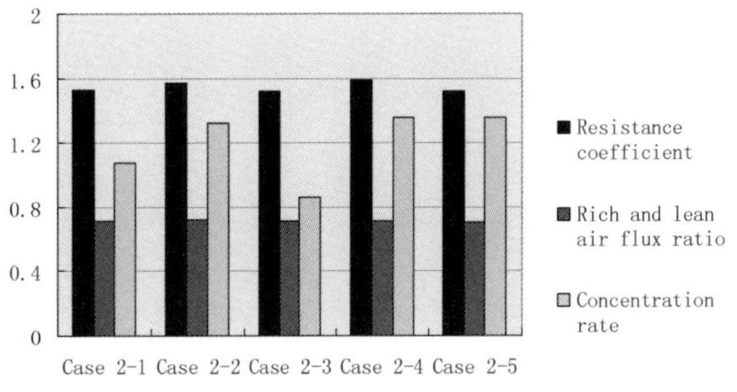

Fig. 6. Results of three blades absence.

According to three blades absence, the concentration ratio changes greatly. Concentration ratio of Case 2-3 has been less than 1. This shows that the concentration role is very poor. For other cases, the concentration rate is more than 1, but very small. In the evaluation of DCTLB performance, concentration rate is an important parameter. By comprehensive analysis, when the three blades are damaged, the blade group structure is seriously damaged, makes concentration rate very low and can not meet the actual requirements of the concentrator performance.

6. Conclusion

(1) The results of one blade absence show that the fifth blade has the least influence on three main performance parameters. The first has the most important influence.
(2) When two blades are damaged, to remove the intermediate two blades should be first considered if resistance coefficient is adjusted in large range. The maximum concentration rate is obtained in some cases that the second and the fifth are simultaneously damaged.
(3) Because three blades absence has great influence on the geometric structure of DCTLB, the case must be avoided in practice.

Acknowledgments

This work is sponsored by the Special Funds for Major State Basic Research Projects of China (Contract No.: 2006CB200303) and Major Projects of State Power Corporation (Contract No.: KJ00-02-16-1).

References

1. S. Glomsrod and T. Y. Wei, *Energy Policy* 33, 525 (2005).
2. P. A. He, Z. P. Zhao and Y. K. Qin. Machinery Industry Press, 198 (1987).
3. Q. S. Zheng, *Boiler Technolog* 5, 27 (1995).
4. J. H. Kuang, S. F. Ding and G. C. Jiao, *Power Engineering* 20, 611 (2000).
5. G. S. Liu, S. Y. Chen and Y. H. He, *Proceedings of the CSEE* 16, 54 (1996).
6. Y. k. Qin, Z. Q. Li and R. Sun, *Engineering Science* 4, 42 (2002).
7. X. F. Zhang and B. G. Tang, *Central China Electric Power* 17, 29 (2004).
8. Z. G. Li, National Defence Industry Press, 190 (1982).
9. D. W. Pershing and J. D. Wendlt, *Proceedings of 10th Int Symposium on Combustion*, 389 (1977).
10. Jinping Li, *Modern Electric Power* 17, 6 (2000).

CARBON CYCLE IN THE DRAINAGE AREA OF YALONG RIVER

HONGHAI KUANG†
School of Geography, Southwest University
Beibei, Chongqing, PR China 400715

MING SHENG KUANG
School of Geography, Southwest University
Beibei, Chongqing, PR China 400715

There are many human-projects in the area studied by the paper. By analyzing the water sample and rock sample from this area and combining the research result of local corrosion rate of carbonate rocks, the paper studies the local carbon cycle. It also analyzes CO_2 in the drainage area of YALONG river, calculates the CO_2 absorbed by the local carbonate by observing the corrosion rate of test block, uses chemical analysis method to calculate the CO_2 absorbed by YALONG river and draw a conclusion that ion of Na^+ and K^+ from feldspar have little influence on carbon, while that the ion of Ca^{2+} from carbonate and carbonate itself are the main resourcet in the local carbon. Moreover, the paper gets the proportion of ions and carbonate in Local total CO_2 and want to find a way to defined calculate the area CO_2.

1. Introduction

According to results of foreign researchers, various ions (Ca^{2+}, SO_4^{2-}, NO_3^- etc) which do not contain carbon have great influence on the formation of carbon cycle. In order to research the carbon cycle in the drainage area of YALONG river, with the way of researching practical points to entire area, this paper calculate local annual CO_2 moving amount by observing the corrosion rate of test block and then used chemical analysis method to analyze the rate of various ions during the process of carbon cycle. Paragraphs should have its first line indented by about 0.25 inch except where the paragraph is preceded by a heading and the abstract should be indented on both sides by 0.25 inch from the main body of the text.

2. Materials and methods

2.1. *Materials*

This paper chose 6 research places (baidiao, wali, qimulin, luning, lizhuang, bazhe), 35 observe points. The main samples collected are water samples. The collection of rock samples is getted by eastern China design institution.

2.2. *Methods*

This method is mainly based on the dissolution equation of carbonate：

$$CaCO_3+CO_2+H_2O \rightarrow Ca^{2+}+2HCO_3^-$$

the formula of getting carbon of CO_2：
$F=E \times S \times R \times 44/100$

F is the carbon of CO_2 and unit is 10^{10} g/a. E is the rate of the corrosion of rock and unit is mg/cm^2·a. S is the area of research region, we use the area of YALONG river and unit is Km^2. R is the purity of carbonate in the test piece rock, here we use the ideal value 1. E mainly counts from data researched in rock corrosion. The rate of corrosion has already be counted out, 4.01×10^{-2} mg/cm^2. Thus the total amount of CO_2 absorbed by local carbonate is 239958.4×10^8 mg, 23995840 kilogram, 23996 ton.

2.3. *Chemical analysis*

Because the areas researched in this paper have crossed a large span of latitudes, the elements such as the growth of plants and power of rock efflorescence are different and the distribution of the rock thus is distinct. Degree

†The paper is supported by the Research Fund for the Dectoral Program of Southwest University (No: SWUB2008055, 104220-20710913).

of mineralization of the surface water in the area is usually in 120-180 mg/L, while the content of other ions is under 10mg/L. Both belong to low fineralization kind. The surface water varies obvious with the change of temperature which has a large amount of water and high temperature in the raining period in summer, in contrast, small amount of water and low temperature in winter. The annual average temperature is between 7.5 to 15.8°C, varying form 9 to 20°C. Most part of it insinuate itself into underground through plumb infiltration and converge into YALONG river as drainage combining with corrosion water in the cranny, solution hole and covered way of carbonate strata, which the amount of drainage can be 60 to 70% of the water amount of corrosion, while a small part converge into YALONG river along with through flow of surface water and valley river. From it, by analysis of stone sample, we conclude that the ion of Ca^{2+} is mainly from limestone and can be described by equation as follow:

$CaCO_3+CO_2+H_2O \rightarrow Ca^{2+}+2HCO_3^-$
$CaSiO_3+CO_2+H_2O \rightarrow Ca^{2+}+2HCO_3^-+SiO_2$
$CaSiO_3+2CO_2 \rightarrow CaCO_3+SiO_2$

From this, we know that each ion of Ca^{2+} get one molecule of CO_2 in atmosphere; one molecule of SiO_2 get two molecule of CO_2 in atmosphere. By analysis of stone sample, we found that the ions of Na^+ and K^+ among positive ions are mainly from feldspar which can be described as follow equation:

$2NaAlSi_2O_8+2CO_2+11H_2O \rightarrow Al_2Si_2O_3(OH)_4+2Na^++2HCO_3^-+4H_4SiO_4$
$2KAlSi_2O_8+2CO_2+11H_2O \rightarrow Al_2Si_2O_3(OH)_4+2K^++2HCO_3^-+4H_4SiO_4$
$MgSiO_4+4CO_2+4H_2O \rightarrow 2Mg^{2+}+4HCO_3^-+H_4SiO_4$
$CaAl_2Si_2O_8+2CO_2+4H_2O \rightarrow 2Al(OH)_3+Ca^{2+}+2HCO_3^-+2SiO_2$
$2CaMg(CO_3)_2+Mg^{2+}+4SiO_2+H_2O \rightarrow Mg_3Si_4O_{10}(OH)_2+2CaCO_3^-+CO_2+CO$

From this, we know that each ion of Na^+ in water get one molecule of CO_2 in atmosphere; one ion of Mg^{2+} get two molecule of CO_2 in atmosphere; one ion of Me^{2+} get one molecule of CO_2 in atmosphere.

The local long period average flow is 1200 m^3/s; total flow is 1200×3600×24×365=37843200000 m^3. The average concentration of vary positive ions plus the local year flow, 37843200000 m^3, equals the practical number of every positive ion. Thus according to the formation that one ion of Na^+ get one molecule of CO_2, one Mg^{2+} get two molecule of CO_2, one Me^{2+}get one molecule of CO_2, we calculate the real absorbed number of CO_2 (44 as the amount of the molecule of CO_2), the equation as follow:

$[CO_2](Na^+)=[Na^+]\ (mmol \cdot l^{-1}) \times 37843200000 \times 44$
$[CO_2]\ (K^+)=[K^+]\ (mmol \cdot l^{-1}) \times 3\ 7843200000 \times 44$
$[CO_2]\ (Mg^{2+}=[Mg^{2+}]\ (mmol \cdot l^{-1}) \times 2 \times 37843200000 \times 44$
$[CO_2]\ (Ca^{2+})=[Ca^{2+}]\ (mmol \cdot l^{-1}) \times 37843200000 \times 44$

3. Conclusion

Followed methods mentioned above and added the local total amount of CO_2 is 27442444 kg, 27442 ton; Ions of Na^+, K^+ of the water sample from the area is mainly originated from feldspar .The content of Ca^{2+} ion which originated from carbonate and carbonate in local carbon cycle have the biggest influence on carbon sequestration; the permeability of strata in sample area coefficient is small, attributes to weak or ever weaker permeability of strata. Because the permeability of carbonate strata is very infirm in the area, no corrosion occurs in the internal of the whole strata of carbonate clod. Carbon cycle always occurs on the surface of the ground and in the cranny, solution hole and covered way of carbonate strata.

Therefore, the carbon cycle on the surface of the area is mainly affected by the function of surface corrosion; There is a ion of Ca^{2+} increase trend in the water sample, so does the local move amount of carbon; carbon cycle under the surface mainly occurs along secondary structure surface inter strata and form layers or strata, the moving amount of carbon is usually small and easily affected by the environment changes.

References

1. Mingsheng kuang, the research report of xiaojiang, southwest normal university press, 2003, P15-45.
2. Ecidi, the research report of jinping II, 2004

SYNTHESIS OF MESOPOROUS NANOCRYSTALLINE TIO_2 WITH POTATO STARCH AS PORE TEMPLATE AND ITS PHOTOCATALYTIC ACTIVITY

CANGGIH SETYA BUDHI

Chemistry Department, Universitas Gadjah Mada, Sekip Utara, Bulak Sumur Yogyakarta, 55281, Indonesia

INDRIANA KARTINI

Inorganic Chemistry Laboratory, Chemistry Department, Universitas Gadjah Mada, Sekip Utara, Bulak Sumur Yogyakarta, 55281, Indonesia

High porosity of nanocrystalline TiO_2 on mesoporous scale has been synthesized through sol-gel process by using potato starch as pore forming agent. The gel starch template was immersed for 4 days in the colloidal of TiO_2 nanoparticles which were prepared by hydrolyzing titanium (IV) tetraisopropoxide (TTIP) in ethanol at pH 1 for 24 hours. Mesoporous TiO_2 was obtained by template removing through combination of etanol-HCl extraction and calcination at 500 °C for 4h. XRD diffractogram reveals nanocrystalline anatase phase of TiO_2 with particle size up to 12.96 nm. The N_2 adsorption-desorption isotherm shows mesoporosity of TiO_2 powder with specific surface area up to 65.65 m^2/g. As the powders were intended to be used as dye substrate for dye-sensitized solar cells, the photocatalytic activity over I_3^- formation has been evaluated at 288 nm of UV region. It is resulted that potato starch templated-TiO_2 shows higher activity than those of Degussa P25 and nontemplated-TiO_2.

Keywords: Mesoporous TiO_2, potato starch, template, photocatalyst.

1. Introduction

Porous material with regular pore size and high porosity or specific surface area have been attracted significant interests due to their prospective applications. Particularly, mesoporous titania is fascinating for many application such as photocatalyst [1,2] and electrodes for wet solar cell [3]. In such applications, porous titania shows high photocatalytic activity because of its large surface area and nanocrystallinity. It has been widely known that photocatalytic is a kind of surface dependent-activity, that it depends on the microstructure and pore size of the photocatalyst [4,5]. Therefore, it is important to control the surface structure in order to increase the photocatalytic activity.

Many works have been reported on the preparation of porous titania via various routes such as sol-gel, hydrothermal, direct deposition from aqueous solution, etc. Amongst, the sol-gel method has been widely applied as a synthetic route for the preparation of new material, providing very homogenous samples, even at low synthesis temperature [6]. It has also been shown that TiO_2 in the anatase phase can be synthesized at room temperature without any previous or further thermal or hydrothermal treatment.

Antonelli and Ying in 1995 [7] have successfully synthesised mesoporous TiO_2 through templating agent method. They employed modified sol-gel reactions in the presence of alkylphosphate surfactant as template. Recently, investigations are directed into the use of block copolymers as template agent to generate the formation of mesoporous titania [8,9]. The use of neutral non-surfactant templating agent is supposed to be eco-environmental friendly route to prepare mesoporous materials. Other researchers have reported works on non-surfactant templated synthesis, such as synthesis route employing polyethyleneglycol (PEG) [10], dibenzoyl-L-tartaric acids [11] and urea [12] as pore templates. The resulted TiO_2 by these templates have shown high surface area and pore volume with narrow pore size distributions.

In this paper, work on potato starch monolith templated sol-gel method for nanocrystalline mesoporous titania synthesis is reported. The influence of template removal technique through combination of ethanol-acids extraction and calcination on porosity of the resulted TiO_2 is also described. It is shown that templated-TiO_2 powder after treated with combined techniques for template removal has higher photoactivity over the formation of I_3^- under UV light irradiation compared to the Degussa P25 and nontemplated-TiO_2 powders.

2. Experimental Section

2.1. Synthesis of potato starch-templated TiO_2

Potato starch template was prepared by diluting 4.5-15 g of starch granules to 50 mL of deionized water under stirring at room temperature [13]. The swelling granules can be obtained by heating the starch slurry at 95°C, then manual stirring was performed until gelatinization occurred. The gelatinized starch paste was kept at 95°C for 10 min without stirring. It was mold to the 2 cm x 2 cm x 2 cm dimension and left to age at room temperature for 15h, then freeze dried at 10°C for 3 days to initiate spongy gel template formation.

Typical mesoporous TiO_2 was then fabricated by the following method: Starch gel that has been dried for 2 days at room temperature was immersed in the TiO_2 colloidal solution (pH 1) and shaken gently for 4 days. TiO_2 colloidal was prepared through hydrolyzing 20 mL of TTIP (Aldrich 97%) in 140 mL of ethanol absolute under addition of 7 mL of 1M HCl. The TiO_2-loaded template (TiO_2-starch composite) was then removed from the colloidal solution, washed with deionized water and dried in air for 30 min. The starch template was removed by the extraction of TiO_2-starch composite in ethanol-HCl solution and subsequently calcined at 500°C for 4h to obtain mesoporous powders. The extraction process was taken place for 5h at 80°C.

2.2. Characterization

The crystal phase of TiO_2 powder was evaluated by X-ray powder diffraction (XRD, Philips) with Cu Kα radiation (λ = 1.5406 Å), and the crystallite size was estimated by Scherrer's formula [13].

The porosity of TiO_2 powders was examined using N_2 adsorption-desorption measurement at 77 K (NOVA 1200, Quantachrome) and the specific surface area was calculated using multi-point BET method using NovaWin2 version 2.2. The morphology of mesoporous TiO_2 nanocrystalline was examined using scanning electron microscope (JEOL JSM-6360 LA).

2.3. Photocatalytic activity

The photocatalytic activity was measured through the formation rate of I_3^- due to the oxidation photo-reaction of I^- to I_2 in excess I^- condition. A reaction system was set up by adding 50 mg of TiO_2 powder into 10 mL of 0.2 M KI aqueous solution then stirred and irradiated with UV light (15 W, with maximum emission at about 365 nm) at room temperature. After the irradiation of 45 min, the suspension was taken and centrifuged. The clear supernatant was diluted ten times then the concentration of liberated I_3^- was monitored by the absorbance at 288 nm, using UV-Vis Spectrophotometer (Shimadzu UV 2550). The molar extinction coefficient was determined to be 4.0 x 10^4 (cm^{-1} mol^{-1} L^1) [14]. For comparison, a photocatalytic reaction was carried out using Degussa P25 and nontemplated-TiO_2 powders.

3. Results and Discussion

A monolith templated sol-gel method for mesoporous titania fabrication has been generated using potato starch gel. Internal pore structure of starch facilitated porous titania formation. By this method, an eco-environmentally route on nanocrystalline mesoporous titania synthesis is envisaged.

3.1. Potato starch gel template and composite

Starch gel template was synthesized by adding starch granule into the water solvent under controlled stirring at room temperature. Starch is insoluble in cold water because of its crystalline structure. Free access of water is probably prevented owing to the formation of oxygen bridges between the nearly parallel crystals. Starch swells when treated with cold water and shrinks to its original size when dried to its previous water-content. Raising the water temperature up to 90 °C resulted in highly swollen starch and start gelatinizing to form paste of starch. Freezing the starch paste is the most important steps for preparing internal porous (macroporous) starch because the dehydration could take place during freezing to form three dimensional starch skeletons [13]. SEM images of Figure 1 show that potato starch gel framework consisted of intact 3-D continuous macroporous network with various pore sizes that were controlled by starch concentration. Starch gel morphology prepared with 9 wt% of starch concentration has approximately 8-150 μm pore size with 1-3 μm of wall thicknesses. Therefore, it is possible to incorporate various TiO_2 nanoparticles in the pores of potato starch gel.

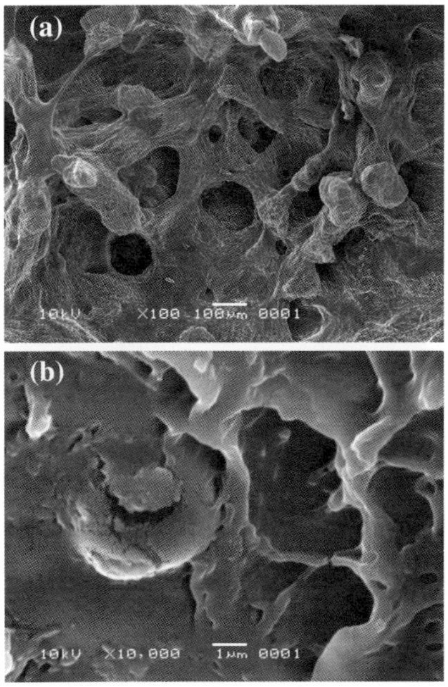

Figure 1. SEM images of potato starch sponge gel 9 wt% at (a) 100x magnification (scale bar 100 μm) and (b) 10,000x magnification (scale bar 1 μm).

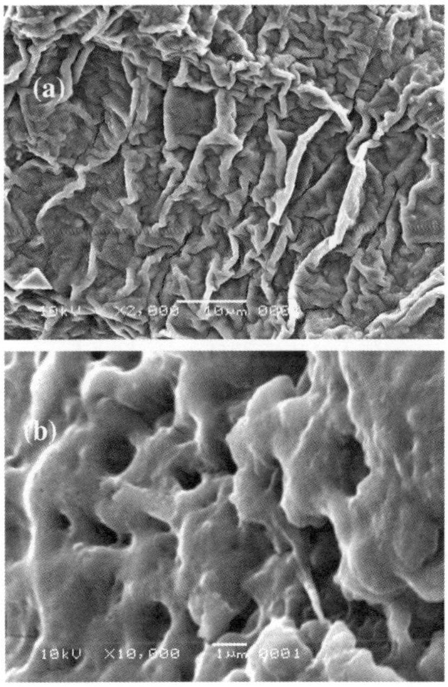

Figure 2. SEM images of TiO2-starch composite 9% of starch at (a) 2,000x magnification (scale bar 10 μm) and (b) 10,000x magnification (scale bar 1 μm).

Composites of TiO_2-starch were prepared by immersing 3-D skeleton of starch gel into white colloidal TiO_2 dispersions. TiO_2 nanoparticles were precipitated onto the surface of the starch framework to produce an intact composite. The TiO_2 loading are probably facilitated by electrostatic interactions between positive charge of TiO_2 nanoparticles at pH 1 and hydrated amylose of 3-D porous network. This interaction prevents the initial deposition of nanoparticles onto the external surface or in the near-surface regions of the pores, so that the colloids may penetrate deeply into the biopolymeric architecture [13]. SEM images of Figure 2 shows TiO_2 loading and precipitating on the starch gel surface that cause smaller opening and thicker wall compare to the initial porous starch gel. Image at higher magnification (Figure 2b) reveals that tightly TiO_2 layers attached on the starch surface.

3.2. *Potato starch templated-mesoporous TiO_2*

Mesoporous TiO_2 with nanocrystalline scaffolds was obtained by starch gel template removal, through combination of ethanol-HCl extraction at 80°C for 5h and calcination at 500°C for 4h. Template removal was also taken place by calcination at 600°C for 4h for comparison.

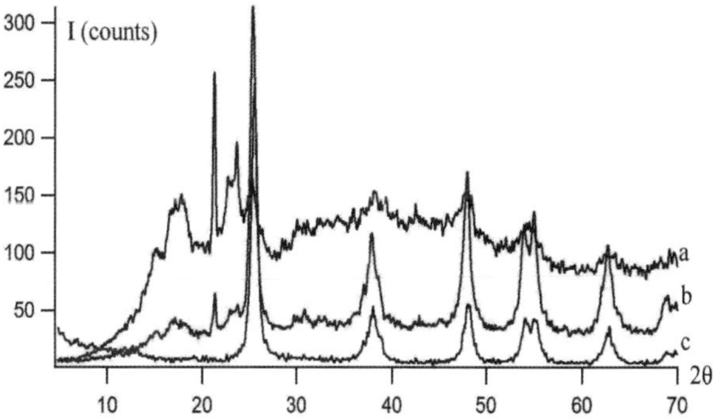

Figure 3. XRD pattern of composite TiO2-starch (a), mesoporous TiO2 after extraction-calcination (b), and calcined TiO2 at 600°C for 4h (c).

The crystalline phase of the resulted powder was examined using X-ray diffraction (XRD), which is shown in Figure 3. It is showed that, TiO_2-starch composite particles are amorphous when only extracted in ethanol-HCl without calcination as indicated in Figure 3a. Broad peaks on the 2θ 17.95°; 21.33° are related to the crystallite lattice of starch polymer, and 2θ of 25.22° corresponds to primarily anatase phase lattice of TiO_2. Calcination treatment at 500°C can improve the anatase crystallinity of mesoporous TiO_2 powders (Figure 3b). However, TiO_2 powders treated by calcination at 600°C without extraction have lower anatase crystalline intensity (Figure 3c). The crystallite size of TiO_2 particles after extraction and calcination is ~13 nm, which is smaller than both the resulted TiO_2 by calcination only (~24.86 nm) and the nontemplated-TiO_2 (~67.88 nm). It is clear that starch gel template may possibly limit the TiO_2 crystal growth.

Figure 4. SEM images of mesoporous TiO2 nanocrystalline at (a) 3,000x magnification and (b) 10,000x magnification.

SEM images in Figure 4 shows the morphology of mesoporous TiO_2 powder synthesized by our starch gel template method after the template removal using combined extraction and calcinations technique. This nano-architecture involves the chemical based construction and self assembly of organic-inorganic moieties with high-order structure controlled by starch gel templating. Hierarchically internal pores of resulted mesoporous TiO_2 is revealed by Figure 4a. Image at higher magnification (Figure 4b) shows nanocrystalline TiO_2 particles as building blocks in the mesoporous structures.

Nitrogen adsorption-desorption isotherms curve (Figure 5) of mesoporous TiO_2 prepared by starch gel 9 wt% indicates a type-IV isotherm curve with an H2 hysteresis loop according to IUPAC classification [15], indicating mesoporous structure of the TiO_2 powders. From BET analysis, the mesoporous TiO_2 from extraction-calcination treatment has specific surface area up to 65.65 m^2/g higher than TiO_2 with only calcination treatment (20.36 m^2/g) and non-templated TiO_2 (19.01 m^2/g). It means that extraction treatment has played significant roles in stabilizing the constructed pore by potato starch over gradually template removal steps.

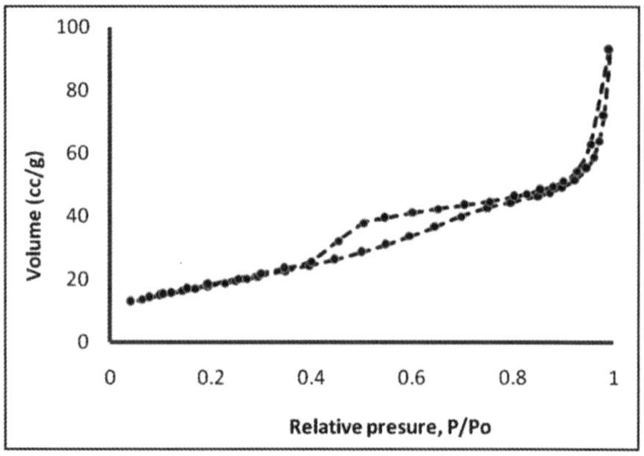

Figure 5. Nitrogen adsorption-desorption isotherms of the mesoporous TiO2.

3.3. *Photocatalytic activity*

Since we intends to apply the resulted TiO_2 for photoanode in dye-sensitized solar cells, the photocatalytic activity of the synthesized mesoporous TiO_2 through the formation of I_3^- due to the oxidation photoreaction of I^- to I_2 in excess I^- [14] are evaluated. P-25 (Degussa) and the only calcined TiO_2 were used for comparison. The results are displayed in Figure 6. It is clear that mesoporous TiO_2 prepared from extraction-calcination has highest activity on I_3^- forming due to its high specific surface area and nanocrystallinity features.

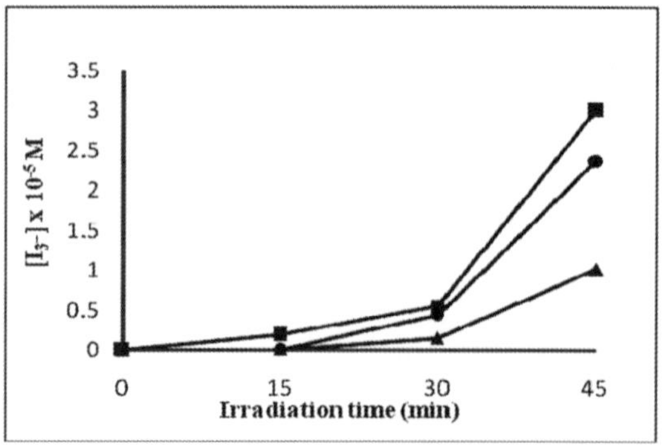

Figure 6. Photocatalytic activity (I3-) concentration) of the extraction-calcination mesoporous TiO2 (■); only calcined mesoporous TiO2 (●); and Degussa P25 (▲) under UV light irradiation.

The starch gel concentration effect during the pores formation of mesoporous TiO_2 was also studied through their photocatalytic activity on I_3^- forming. The results is depicted in Figure 7. It is clearly observed that mesoporous TiO_2 prepared from 15% starch gel has highest I_3^- formation up to 1.375 x 10^{-4} M, than TiO_2 synthesized with other starch gel concentrations.

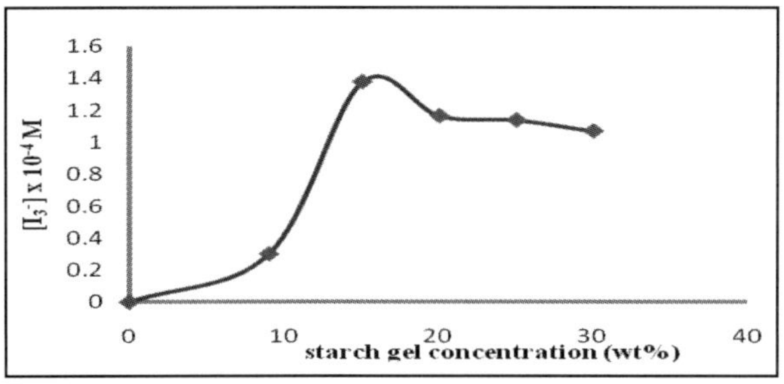

Figure 7. Photocatalytic activity (I3- concentration) of mesoporous TiO2 prepared from starch gel on 9; 15; 20; 25; and 30 wt% after 45 min UV light irradiation.

It can be seen that starch concentration influences the pores opening and presumably wall thickness of starch gel template. Therefore, its various internal pores may affect the TiO_2 nanoparticles infiltration during the immersion of colloidal TiO_2.

4. Conclusions

Highly porous and crystalline TiO_2 has been synthesized through potato starch templated method. It is also observed that template removal techniques play important role in stabilizing the created pores. High surface area TiO_2 has showed high photocatalytic activity over triiodide formation leading to potential substrate for dye loading in dye-sensitized solar cells.

Acknowledgements

The authors acknowledge financial support from Directorate of Higher Degree Education Indonesia under Competency Research Grant 2008-2009.

References

1. F. Victor, J. Stone, and R.J. Davis, *Chem. Mater.* **10**, 1468 (1998).
2. N. Negishi, K. Takeuchi, and T. Ibusuki, *Appl. Surf. Sci.* **417**, 121-122 (1997).
3. B. O'Regan and M. Gratzel, *Nature* **353**, 737 (1991).
4. H. Hirashima, C. Kojima, K. Kohama, and H. Imai, H., *J. Sol-Gel Sci. Technol.* **8**,843 (1997).
5. H. Hirashima, C. Kojima, K. Kohama, H. Imai, V. Balek, H. Hamada, and M. Inaba, *J. Non-Cryst. Solids* **225**, 153 (1998).
6. C.N.R. Rao and J. Gopalakrishnan, *New Direction in Solid State Chemistry*, 2nd Ed., Cambridge University press, Cambridge, **1997**.
7. D.M. Antonelli and J.Y. Ying, *Chem. Int. Ed. Engl.* **34**, 2014 (1995).
8. P. Yang, D. Zhao, D.I. Margolese, F.B. Chmelka, and G.D. Stucky, *Chem. Mater.* **11**, 2813-2826 (1999).
9. I. Kartini, D. Menzies, D. Blake, P. Meredith, J.C.D. da Costa, and G.Q. Lu, *J. Mater. Chem.* **14**(19), 2917 (2004).
10. M.M. Yusuf, H. Imai, and H. Hirashima, *J. Sol-Gel Sci. Technol.* **28**, 97-104 (2003).
11. Y. Wei, J. Xu., H. Dong, K.Y. Qiu, and S.A. Jansen-Varnum, *Chem. Mater.* **11**, 2023 (1999).
12. J. Zheng, J. Pang, K. Qiu, and Y. Wey, *Micropor. Mesopor. Mater.* **49**, 189-195 (2001).
13. M. Iwasaki, S.A. Davis, and S. Mann, *J. Sol-Gel Sci. Technol.* **32**, 101-102 (2004).
14. S. Pavaupree, J. Jitputti, S. Ngamsinlapasathian, Y. Suzuki, and S. Yoshikawa, *Structural, Photocatalytic Activity and Photovoltaic Properties of Mesoporous Anatase titania Nanopowders Prepared by Hydrothermal Method*, The 2nd Joint Internasional Conference on Sustainable Energy and Environment, November 21st-23rd, Bangkok (2006).
15. F. Rouquerol, J. Rouquerol, and K. Sing, Adsorption by Powders and Porous Solids: Principles, Methodology, and Applications; Academic Press, San Diego (1999).

Session 9

SURFACE WATER CHARACTERIZATION AT SEMENYIH DAM

NESAMALAR KANTASAMY
Faculty of Applied Sciences, MARA University of Technology (UiTM),
40450 Shah Alam, Selangor, Malaysia

SITI MARIAM SUMARI
Faculty of Applied Sciences, MARA University of Technology (UiTM),
40450 Shah Alam, Selangor, Malaysia

Semenyih Dam was the first investigation site as part of a national project aimed at assessing, monitoring and evaluating the state of acidification in the country. The objective of this investigation is to make preliminary and continuous assessment of the acidification status of an inland aquatic environment. In-situ measurements of pH, temperature and electrical conductivity were conducted while laboratory analysis included concentration measurements of major basic cations and acidic anions. The cations analyzed were Ca^{2+}, K^+, Mg^{2+}, Na^+ and NH_4^+ and the anions were Cl^-, SO_4^{2-}, NO_3^- and HCO_3^-. Basic pH, together with high acid neutralizing capacity (ANC) and alkalinity in both 2005 and 2008 indicate Semenyih Dam was capable and is still capable of neutralizing any acidification due to anthropogenic activities. This thus classifies Semenyih Dam as a non acid-sensitive or non-acidic inland aquatic ecosystem.

1. Introduction

Acidification effects on the environment began a long time ago, however, serious attention on this environmental issue only began in the late 1960s. Acid deposition is a process where acidic pollutants are deposited from the atmosphere to the earth's surface. There are two forms of acid deposition, wet and dry. Wet deposition refers to acidic rain, fog and snow. Dry deposition refers to acidic gases and particles.

Inland aquatic acidification is acidification effects on water bodies such as reservoirs, rivers, streams, lakes and other waterways. Acid deposition has many harmful ecological effects when pH of an aquatic system falls below 6 and especially 5. Inland aquatic acidification has profound impacts on both animals and plants and the effects can be both direct and indirect (Davies, 2005). Direct effects include toxic conditions while indirect effects can be lack or loss of suitable, acid-sensitive prey for food. Inland aquatic acidification effects have been observed in Northern Europe and North America in the 1970's when pH of lakes decreased and this resulted in declining fish populations (Street et al, 2001). According to Ebise (2004), numerous acidification studies in East Asia have been done on air, soil, vegetation and rainwater, but there are very few studies involving inland aquatic environments. However, this situation cannot be assumed to continue, as East Asian are experiencing countries rapid economic growth and urbanization which generate NO_x and SO_2, well-known acidic precursors of HNO_3 and H_2SO_4 (EANET, 2000).

As emphasized in Chapter 9 of Agenda 21 ((on transboundary atmospheric pollution) adopted at the 1992 United Nations Conference on Environment and Development (UNCED) in Rio de Janeiro, acid deposition is a problem that requires international cooperative efforts for its solution. Based on this realization, Malaysia through the Ministry of Natural Resources and Environment (NRE), the Meteorological Services Department (MMS) and UiTM have embarked on a national project aimed at assessing, monitoring and evaluating the state of acidification in the country and ultimately the possible effects on the environment. The objective of this study is to make preliminary and continuous assessment of the state of surface water acidification for an inland aquatic environment. Semenyih Dam has been chosen as the first urban inland aquatic site due to its accessibility and suitability. This study will contribute to a database for EANET (Acid Monitoring Network in East Asia) coordinated by Japan and of which Malaysia and other 12 East Asian countries are participants. Analysis from this study will form background information for other site selection criteria as well as expand future research direction for inland aquatic acidification in Malaysia.

2. Methodology

2.1. *Sampling Site*

Semenyih Dam located in Hulu Langat district, was built to ensure there is sufficient water at a downstream intake point for a water treatment plant. This treatment plant supplies a major portion of domestic water to Klang Valley which includes Federal Territory of Kuala Lumpur, and districts of Gombak, Petaling, Ulu Langat and Klang. Semenyih Dam is located 20 km from Bangi town and 25 km from Kajang town while Beranang, Seremban and Nilai are located within 25 km radius of this dam. All these towns have big and small industries, businesses and housing areas which are capable of contributing anthropogenic pollutants. According to Bhatia (2001), automotive, packaging, stainless steel, chemical and paint industries found in Beranang, Seremban, Kajang, Nilai and Bangi have greater potential to release acidic pollutants than other industries. Matori et al (2004) found sulfate and nitrate in dust samples in an air pollution study at Bangi. This study is also in agreement with a terrestrial study done by Rozali et al (2001) which showed presence of potential acidic pollutants, SO_4^{2-} and NO_3^- in street dusts of Kajang, Nilai, Bangi and Seremban.

2.2. *Sampling Points and Frequency*

4 sampling sites (M1, M2, M3 and M4) located along the middle of the dam were selected. Samples were collected approximately every three months according to EANET (2000) documents. Water samples were collected at 1 foot depth using a grab sampler. Sampling duration covered both rainy (late Oct to early March) and dry (early June to late September) seasons.

2.3. *Sampling Analysis*

The pH, temperature and electrical conductivity were measured in situ while the rest of the samples were brought to the laboratory in polyethylene bottles at 4^0C for analysis. Chemical analysis for anionic species (SO_4^{2-}, NO_3^-, Cl^-) and cationic species (K^+, Ca^{2+}, Na^+ and Mg^{2+}) were carried out using Metrohm 761 suppressed type ion chromatography. Analysis was done in duplicates. Quality check standard was run for every 10 samples. Alkalinity was determined by titration using 0.01 M H_2SO_4 to endpoint 4.8 using a pH meter. The Acid Neutralization Capacity (ANC) was calculated using a formula by Skjelkvale et al (2000) as the difference between concentration of strong base cations (SBC) and strong acidic anions (SAA).

3. Results

The various measurements and analytical results for the surface water are given in Table 1, Table 2, Fig 1 and Fig 2. Acid Neutralization Capacity (ANC) values for 2005 and 2008 are given in Table 3.

4. Discussion

4.1. *Physical Characteristics of Semenyih Dam for 2005 and 2008*

Table 1 shows the mean pH reading for Semenyih Dam varied between 6.42 to 8.12 with an average pH of 7.38 for 2005. In 2008, the mean pH varied between 6.79 to 7.44 with an average pH of 7.13 as indicated in Table 2. There seem to be no significant change in pH between rainy and dry seasons. Natural waters usually have pH values in the range of 4 to 9 and most are slightly basic due to presence of bicarbonates and carbonates of the alkali and alkaline earth elements. According to Elvington and Agreen (2004), lakes with a pH of less than 5.6 is considered acidic. Since Semenyih Dam has pH of more than 7.00 for both 2005 and 2008, this man-made lake is not considered to have an acidic status for 2005 and also for 2008. However, it should be noted that the average pH in 2008 has decreased to 7.13 compared to 2005 when it was 7.38. Data from 2006 which is not shown in this paper showed an overall average pH of 7.24 for Semenyih Dam. This indicates a disturbing decreasing pH trend from 2005 to 2008.

The mean temperature for Semenyih Dam varied between 24.04 to 31.19 with an average temperature of 27.17 for 2005. In 2008, the average temperature for Semenyih Dam was 27.78 with temperature variation between 25.1 and 30.3. Though temperature variations within each year for both 2005 and 2008 seem to be significant, the overall average temperature for both years did not show significant difference as shown in Table 1 and Table 2.

Electrical conductivity is affected by the presence of inorganic dissolved solids such as chlorides, nitrates and cations such as sodium and magnesium in water. According to EANET (2000), inland water bodies with low electrical conductivity are prone to acidification by acid deposition. Table 1 and Table 2 show that Semenyih Dam has an average electrical conductivity of 22.58 μS/cm for 2005 and 28.90 μS/cm for 2008.

Table 1. Mean, standard deviation and average for physical and chemical composition of surface water for Semenyih Dam (based on all sampling points) for 2005.

2005	Mth	EC (25^0C) µS/cm	pH	T (^0C)	Na^+ mg/L	K^+ mg/L	Mg^{2+} mg/L	NH_4^+ mg/L	Cl^- mg/L	NO_3^- mg/L	Alk meq/L	$nssSO_4^{2-}$ mg/L	$nssCa^{2+}$ mg/L	HCO_3^- mg/L
Mean	Jan	23.31	8.12	31.19	2.21	1.94	1.57	0.06	0.68	0.26	0.23	0.73	7.34	13.0
Sdev		1.00	0.08	0.31	0.03	0.02	0.02	0.01	0.20	0.01				
Mean	Mar	19.52	7.34	24.04	2.77	2.22	1.67	0.13	0.70	*5.53	0.23	0.74	7.32	13.0
Sdev		0.50	0.13	1,20	0.01	0.01	0.01	0.01	0.02	0.05				
Mean	July	23.15	7.62	25.14	3.37	3.24	2.80	0.15	0.79	0.11	0.23	0.66	4.23	13.0
Sdev		0.33	0.18	1.79	0.01	0.03	0.03	0.01	0.04	0.01				
Mean	Sept	24.32	6.42	28.30	2.17	1.84	0.65	nd	0.81	0.03	0.21	0.83	2.26	11.0
Sdev		0.40	0.11	0.97	0.01	0.03	0.10		0.01	0.02				
	Ave	**22.58**	**7.38**	**27.17**	**2.63**	**2.31**	**1.67**	**0.085**	**0.75**	**0.13**	**0.23**	**0.74**	**5.29**	**12.5**

*sample suspected contaminated nd: not detected nss: non sea salt

Table 2. Mean, standard deviation and average for physical and chemical composition of surface water for Semenyih Dam (based on all sampling points) for 2008.

2008	Mth	EC (25^0C) µS/cm	pH	T (^0C)	Na^+ mg/L	K^+ mg/L	Mg^{2+} mg/L	NH_4^+ mg/L	Cl^- mg/L	NO_3^- mg/L	Alk meq/L	$nssSO_4^{2-}$ mg/L	$nssCa^{2+}$ mg/L	HCO_3^- mg/L
Mean	Jan	37.60	6.79	25.1	2.80	2.00	0.59	0.08	0.90	0.40	0.30	0.50	1.99	15.0
Sdev		0.15	0.02	0.01	0.01	0.02	0.01	0.01	0.02	0.01				
Mean	March	18.00	7.44	30.3	2.50	2.30	0.80	0.07	0.80	0.10	0.34	0.47	2.51	17.0
Sdev		0.03	0.03	0.02	0.03	0.02	0.02	0.04	0.01	0.02				
Mean	July	30.60	7.31	30.2	2.72	1.84	0.63	0.02	1.58	0.71	0.23	1.21	1.88	13.0
Sdev		1.20	0.03	0.02	0.03	0.14	0.03	0. 13	0.33	0.02				
Mean	Sept	29.30	6.98	29.5	2.54	2.19	0.68	0.05	0.80	0.20	0.23	1.96	2.95	13.0
Sdev		1.03	0.02	0.01	0.02	0.02	0.12	0.01	0.34	0.01				
	Average	**28.90**	**7.13**	**27.78**	**2.64**	**2.08**	**0.67**	**0.05**	**1.02**	**0.35**	**0.27**	**1.03**	**2.33**	**14.5**

These values are considered high (above 20.00 µS/cm) so Semenyih Dam should have been less prone to acidification in 2005 and 2008.

Table 3. ANC values in meq/Lof Semenyih Dam for 2005 and 2006.

2005	Jan	Mar	July	Sept	Ave ANC
	0.47	*	0.63	0.27	**0.46**
008	0.28	0.32	0.23	0.30	**0.28**

*Contaminated sample

4.2. *Chemical Characteristics (Cations) of Semenyih Dam for 2005 and 2008*

Results of Table 1, Table 2 and Fig 1 show that the relative abundance of major cations in Semenyih Dam for 2005 followed the order: $nssCa^{2+} > Na^+ > K^+ > Mg^{2+} > NH_4^+$, while in the 2008 it was slightly different with this order: $Na^+ > nssCa^{2+} > K^+ > Mg^{2+} > NH_4^+$. According to Table 1 and Fig 1, the mean concentrations of Na^+, K^+, Mg^{2+}, Ca^{2+} and NH_4^+ were the lowest during the dry month Sept in 2005 and this pattern is also followed by Na^+ and Mg^{2+} ions in 2008. This probably indicates rainwater contributes some of these ions especially Na^+ and Mg^{2+} to the Semenyih Dam. This is in agreement with previous undocumented rainwater studies done in 2005 and 2006 at Semenyih Dam which showed the presence of Na^+, K^+, Mg^{2+} and Ca^{2+}. Soil studies at Semenyih Dam also indicated presence of Na^+, K^+, $Mg^{2+}$and Ca^{2+} and since some renovation work was being done at the jetty area late 2007 and 2008 to prepare for a royal visit, this could have disturbed the soil framewok and added these cations into the dam water. This may explain why the pattern of cation concentration was different for $Ca^{2+}$and K^+ in 2008 compared to 2005. For example, the nss Ca^{2+} concentration showed the highest mean in Sept 2008 while it had the lowest mean concentration in Sept 2005.

Overall, Na^+, K^+, Mg^{2+} and Ca^{2+} and NH_4^+ show decreased concentration in 2008 compared to 2005. Since K^+ and Na^+ are associated with soil dust and biomass burning, more stringent monitoring and checks lately on open burning especially during haze episodes may have decreased the presence of these ions in 2008. Another reason could be the increase in acidic anions such as Cl^-, NO_3^- and SO_4^{2-} in 2008 as shown in Fig 2. Base cations like Na^+, K^+, Mg^{2+} and Ca^{2+} act to neutralize these acidic anions so that a basic pH of the dam could be maintained.

4.3. *Chemical Characteristics (Anions) of Semenyih Dam for 2005 and 2008*

Results of Table 1, Table 2 and Fig 2 show that the relative abundance of major anions in Semenyih Dam for 2005 followed the order: $HCO_3^- > Cl^- > SO_4^{2-} > NO_3^-$ while in the 2008 it was slightly different with this order: $HCO_3^- > SO_4^{2-} > Cl^- > NO_3^-$. Concentration of all these anions show an upward trend in 2008 compared to 2005. The concentration of SO_4^{2-} is significantly higher than NO_3^- for both 2005 and 2008. The origin of SO_4^{2-} and NO_3^- can be traced back to combustion processes such as automobile combustion for NO_3^- and coal burning for SO_4^{2-}.

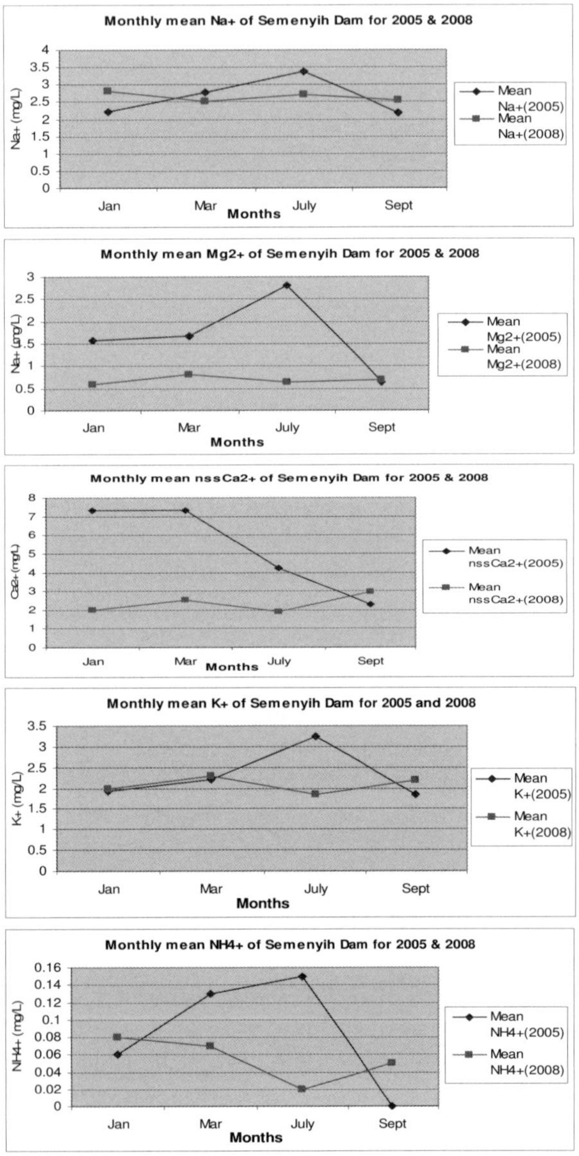

Fig. 1. Chemical characteristics (cations) of Semenyih Dam for 2005 and 2008.

The Cl^- ion is normally associated with seaspray and various anthropogenic activities. The Cl^- in Semenyih Dam probably originated from anthropogenic activities especially along the rivers supplying the dam. Alkalinity is a measure of the capacity of a water body to neutralize acids and is taken as a function of bicarbonates ions. The HCO_3^- represents the major form of alkalinity and therefore provides the natural buffering systems in water bodies. The HCO_3^- represents the largest ionic species in Semenyih Dam for both 2005 and 2008 and is about 10 to 12 times the concentration of the strong anions, SO_4^{2-} and NO_3^-. The high presence of HCO_3^- ions indicate high alkalinity values (more than 0.02 meq/L). Semenyih Dam has average alkalinity values between 0.23 to 0.27 meq/L. According to Elvington & Agren (2004), lakes with zero or less than 0.02 meq/L HCO_3^- are classified as acidic lakes.

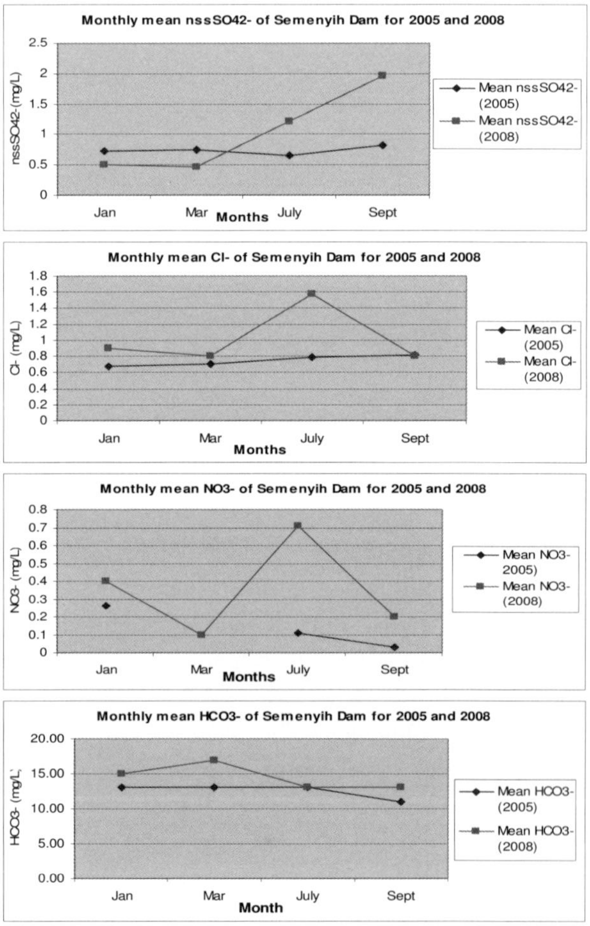

Fig. 2. Chemical characteristics (anions) of Semenyih Dam for 2005 and 2008.

5. Conclusion

In Semenyih Dam, the mean concentrations of the base cations such as K^+, Ca^{2+}, Na^+ and Mg^{2+} are more dominant in concentration than the acidic anions (NO_3^-, SO_4^{2-} and Cl^-) for both 2005 and 2008. This gives a high ANC (above 0.1 meq/L) value indicating Semenyih Dam has high neutralizing capacity. The high concentration of HCO_3^- in 2005 and 2008 also gives high alkalinity, which also indicates presence of a good buffering capacity of the dam water. High ANC, high alkalinity values together with basic pH in 2005 and 2008 indicate Semenyih Dam was capable and is still capable of neutralizing any form of acidification due to anthropogenic activities. This thus classifies Semenyih Dam as a non-sensitive or non-acidic inland aquatic ecosystem.

References

1. Acid Deposition Monitoring Network in East Asia, (EANET), (2000). Technical Documents for Monitoring on Inland Aquatic Environment in East Asia.
2. Bhatia, S.C. (2001). Environmental Pollution and Control in Chemical Process Industries. Khanna Publishers, Delhi, India.
3. Davies J.J.L, (2005). Trends in surface water chemistry of acidified UK freshwaters, pp 27 39.
4. Ebise, S (2004). Impact of Acidic Deposition on Inland Aquatic Environment, Department of Engineering, Faculty of Engineering, Setsunan University.
5. Elvington, P and Agren, C, (2004). Air and Environment – Chapter 3, Acidrain, org. The Swedish NGO Secretariat on Acid Rain.
6. Mohd Khalid Matori, Mohd Suhaimi Hamzah, Abd Khalid Wood, Shamsiah Abdul Rahman, Abdul Halim Baharuddin dan Ariffin Talib, (2004). Analisis Anion and Kation Menggunakan Teknik Ion Kromatografi Bagi Zarahan Terampai Udara Bagi Kawasan Bangi, Malaysian Journal of Analytical Sciences, Vol 8, No 1 (2004) 171-176.
7. Mohd Rozali Othman dan LimSun Hoo, (2001). Kajian beberapa parameter fizis dan kandungan kimia debu jalan raya di sekitar Kuala Lumpur, Kajang, Bandar Baru Bangi, Bangi Lama, Nilai dan Seremban. Malaysian Joutrnal of Analytical Scuiences, Vol 6, No 1 (2001): 162-172.
8. Streets, D.G., Tsai, N.Y., Akimoto, H and Oka, K., (2001). Trends in emissions of acidying species in Asia, Water, Air and Soil Pollution, 130: 187-192.
9. Skjelvale, B.L., Stoddard, J.L., Anderson, T., (2001), Trends in surface water acidification in Europe and North America (1989-1998), Water, Air and Soil Pollution, 130: 187-192.

BENCHMARKING HEAVY METAL CONTAMINATION FOR BROWNFIELD IN MALAYSIA

AIK HENG, LEE
Senior R & D Director, Institute Alam Sekitar Malaysia

HAMID NIKRAZ
Professor & Chairman, Department of Civil Engineering, Curtin University of Technology

YUNG TSE HUNG
Professor, Department of Civil Engineering, Cleveland State University

Malaysia has over 10,000 brownfield sites which have yet to be registered. In fact there is limited information and controls on soil and groundwater contamination from these brownfield sites. In this study, four brownfield sites were examined in terms of the soil and groundwater quality to benchmark the heavy metal contents to be used as indicators of soil and groundwater contamination and to derive suggested benchmarking reference for remedial action to be taken. It also described briefly the possible remediation techniques to clean up the heavy metals in brownfield sites and the need for registration of brownfield sites towards sustainable development.

Keywords: Brownfield, soil quality, groundwater quality, heavy metal contamination and remediation.

1. Introduction

Due to urbanization and increasing land values there is a dire need to ensure that potential land is strategized for development purposes. Potential of redeveloping brownfields has prompted the need to have proper management strategies that can identify, document and catologue these brownfields to assist decision makers and planning agencies to make appropriate decisions so that environmental risks can be minimized for brownfields to be redeveloped. Currently, in Malaysia there is no full listing and registration for brownfield sites and also there are no standards for soil and groundwater quality [1]. The brownfield sites may contain low contaminants such as debris to highly hazardous substance such as heavy metals, volatile organic substances, radioactive material and others that may contain in the ground due to man activities.

It is estimated there are over 10,000 brownfield sites exist in Malaysia. Some of these sites are developed, abandoned and some with potentials for redevelopment. Brownfields are analogue to contaminated sites that have been used for industrial, commercial, warehousing or other purposes that may be contaminated due to the presence of hazardous substances, pollutants or contaminants and usually existing in areas occupied by industrial activities, waste dumping, petrol and oil storage. The site of the brownfield sites varies from small premises to larger ones depending on the nature and type of activities at site.

The contaminants encountered at brownfields generally include heavy metals such as As, Cd, Co, Cr, Cu, Hg, Ni, Pb, Zn, VOC such as benzene, toluene and SVOC such as PAH and others.

Heavy metals cannot be chemically degraded and can affect all groups of organism and ecosystem processes [2-6] which are generally originating not only from industrial activities but also agriculture activities such as fertilizer application [7]. Heavy metals therefore are good indicators for contamination of brownfield sites.

The geology in Malaysia generally consists of granite, sedimentary rocks, shale, sandstone and limestone whereby heavy metals are also grounded in natural form. Heavy metals however can be introduced further to the land through fertilizer application, soil remediation and can cause contamination as well. On top of these activities, heavy metals are further introduced by other man activities namely industrial activities.

This study was therefore conducted with the aim to determine available heavy metals found in brownfields so as to benchmark the baseline with typical agriculture soil so relevant benchmarking references of heavy metal contamination can be established for remedial action to be taken.

2. Materials and method

2.1. *Soil and groundwater sampling*

Twenty soil samples obtained from four brownfield sites in Malaysia were evaluated to assess the range of heavy metals in the soil and groundwater at the brownfield sites. The four brownfield sites as depicted in Table 1 involve namely (i) Site 1 that is an industrial land in proximity to heavy manufacturing industries; (ii) Site 2 that is a site in proximity to motor vehicle repairs; (iii) Site 3 that is a site in proximity to high technology industrial activities; and (iv) Site 4 that is a site in close proximity to sanitary landfill.

Table 1. Description of brownfield sites.

Brownfield Site	Number of Samples	Type of Activity	Surrounding Land Use
Site 1	3	Heavy Industry	Heavy industrial activities
Site 2	9	Mechanical Workshop	Industrial, housing, warehousing
Site 3	2	High Technology Industry	High technology industrial activities
Site 4	6	Sanitary Landfill	Agriculture, landfill

Both test pits and borehole drilling were used in the sampling program. Soil samples were also taken from each test pit for classification of the soil profile. Soil samples collected during sampling were placed in pre-cleaned jars, sealed and labeled prior sending to laboratory for analytical purposes.

Groundwater wells were established in each test pit and left for 24 hours to ensure stability of the groundwater levels before measuring the groundwater levels. The monitoring wells were purged by removing one full well volumes of water. Groundwater samples were obtained with use of a bailer. The groundwater were then collected in 1000 ml bottles, labeled and sealed prior sending to laboratory for analytical purposes.

2.2. *Analysis*

The soil and groundwater samples were sent to laboratory for analysis. All samples were preserved according to the analytical requirement. The testing methods used are in accordance to APHA (American Public Health Association) 19th Edition 1995 as shown in Table 2.

Table 2. Testing methods and parameters tested.

Parameters Tested	Testing Method
As	APHA 3114C – Continuous Hydride Generation
Cd	APHA 3111B
Co	APHA 3111B
Cr	APHA 3111B
Cu	APHA 3111B
Hg	APHA 3112B – Cold Vapor Atomic Absorption Spectrometric
Ni	APHA 3111B
Pb	APHA 3111B
Zn	APHA 3111B

Note: APHA – American Public Health Association, 19th Edition 1995

The analytical results obtained from the soil and groundwater sampling programs were analyzed and compared to the selected references for soil and groundwater.

By way of this analogue, the significance of the heavy metal contamination in the soil and groundwater at the brownfield sites can be quantified with the relevant references.

3. Results and discussions

3.1. *Heavy metals in soil*

Table 3 provides a summary of the heavy metal concentrations found at the brownfield sites. The results were also compared with the soil investigation level published by the Ministry of Agriculture and the Dutch Reference Values. Fig. 1 provides a comparison of the heavy metal contents at the four brownfield sites.

Table 3. Summary of soil analysis (mg/kg).

Heavy Metals	Site 1	Site 2	Site 3	Site 4	Malaysia Reference *	Dutch Reference	
						S Value **	I Value ***
As	0.03	0.1	0.43-0.63	0.26-1.13	60	29	55
Cd	0.03-0.08	0.03	0.87-22.25	NA	0.3	0.8	12
Co	0.63-0.85	NA	1.35-2.56	NA	10	9	240
Cr	0.003	0.05-3.1	1.28-23.36	12.26-29.21	60	100	380
Cu	0.36-0.70	3.8-11.7	0.97-13.10	0.1-6.6	50	36	190
Hg	0.01	0.01	0.70	0.7	0.35	0.3	10
Ni	0.005	1.3-10.7	1.0-2.91	NA	45	35	210
Pb	0.26-0.29	12.1-46.5	0.10-10.56	3.31-11.29	65	85	530
Zn	NA	12.4-96.2	NA	NA	95	140	720

Note: Values indicate range detected at each site

* Soil investigation level published by Ministry of Agriculture Malaysia

** S Values (Target Values) – Below this value regard as being multi functional and land suitable for all usage

*** I Values (Intervention Values) - Above these values the soil is regard as a 'serious case of pollution' and some form of measures are to be adopted

NA – Not Available

NR – No Recommendation

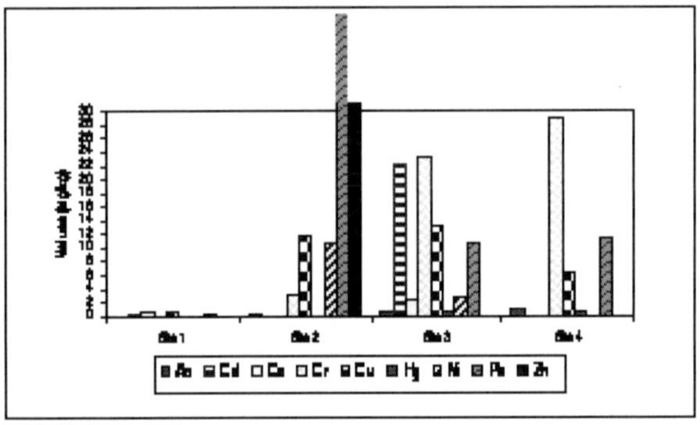

Fig. 1. Metal concentrations in the soil at brownfield sites.

As values ranging from 0.03 to 0.63 mg/kg were detected at the brownfield sites. The values were within the Malaysian reference as well as within the Dutch reference.

Cd values ranging from 0.03 to 22.25 mg/kg were detected at the brownfield sites. Cd values at site 1, site 2 and site 4 were within the Malaysian reference as well as the Dutch reference. Cd values at site 3 however exceeded these references.

Co values ranging from 0.63 to 2.56 mg/kg were detected at the brownfield sites. The values were within the Malaysian reference and the Dutch reference.

Cr values ranging from 0.003 to 29.21 mg/kg were detected at these sites which were within the two references used for comparison purposes.

Cu values ranging from 0.1 to 13.10 mg/kg were detected at these brownfield sites which were also within the references used for comparison purposes.

Hg values ranging from 0.01 to 0.7 mg/kg were detected at the brownfield sites. Hg values at site 3 and site 4 exceeded the Malaysian reference. Hg levels at these sites were also detected to be above the Dutch S Values and below the Dutch I Values for intervention purposes.

Ni values ranging from 0.005 to 10.7 mg/kg were detected at the brownfield sites which were within the Malaysian reference and the Dutch reference.

Pb values ranging from 0.10 to 46.5 mg/kg were detected at the brownfield sites which were within the Malaysian reference and the Dutch reference.

Lastly, Zn of 12.4 to 96.2 mg/kg were detected at site 2 which were detected to be above the Malaysian reference and the Dutch reference.

In the summary based on the analysis conducted, the heavy metal concentrations detected at the brownfield site can be used as a good indicator to assess the level of contamination of the brownfield sites. The study to compare the extend of the heavy metal contamination can be used to benchmark heavy metal contamination for remedial actions.

3.2. *Heavy metals in groundwater*

Table 4 provides a summary of the groundwater analysis at the four brownfield sites while Fig. 2 depicts the metal concentrations in the groundwater at the brownfield sites. Comparisons with the NDWQ (National Drinking Water Quality) and the Dutch reference values were also made.

Table 4. Summary of groundwater analysis (mg/l).

Heavy Metals	Site 1	Site 2	Site 3	Site 4	Malaysia Reference *	Dutch Reference	
						S Value **	I Value ***
As	0.3	0.01	0.001-0.005	0.001-0.006	0.05	0.01	0.06
Cd	0.2	0.003	0.006	0.003	0.005	0.0004	0.006
Co	0.2-3.67	NA	0.002	NA	NR	0.006	0.03
Cr	0.3	0.01	0.001-0.063	0.01-0.7	0.05	0.001	0.03
Cu	1.36-2.39	0.01	0.001	0.01-0.34	1	0.015	0.075
Hg	1	0.01	0.001	0.001-0.006	0.001	0.00005	0.0003
Ni	0.5	NA	0.001-0.003	NA	NR	0.015	0.003
Pb	2.0	0.03	0.148-0.196	0.09-0.21	0.1	0.015	0.075
Zn	NA	0.38-0.51	NA	NA	1.5	0.065	0.7

Note: Values indicate range detected at each site

* National Drinking Water Quality, Ministry of Health Malaysia
** S Values (Target Values) – Below this value regard as being multi functional and land suitable for all usage
*** I Values (Intervention Values) - Above these values the soil is regard as a 'serious case of pollution' and some form of measures are to be adopted
NA – Not Available
NR – No Recommendation

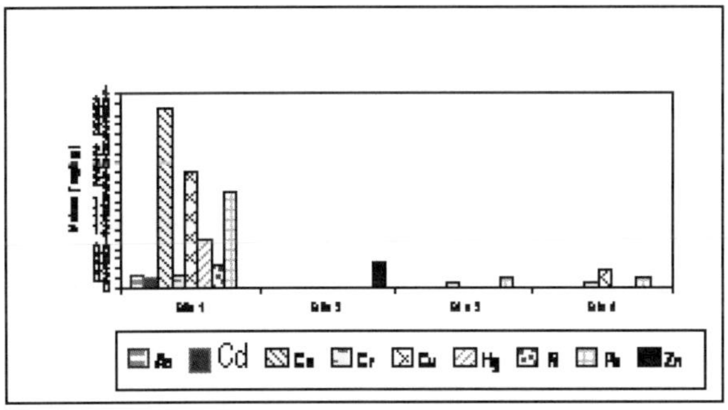

Fig. 2. Metal concentrations in groundwater at brownfield sites.

As values ranging from 0.001 to 0.3 mg/l were detected at the brownfield sites. As values at site 1 were above the NDWQ as well as above the Dutch S Values.

Cd values ranging from 0.003 to 0.2 mg/l were detected in the groundwater at these brownfield sites. Cd values at site 1 were detected to be above the selected reference while Cd values at site 3 were above the Dutch S Values.

Co values ranging from 0.002 to 3.67 mg/l were detected in the groundwater samples. Co values at site 1 were detected above the Dutch I Values.

Cr values ranging from 0.001 to 0.3 mg/l were detected in the groundwater samples. Cr values at site 1 were detected above the NDWQ and Dutch Reference while Cr values at site 2 were found to be above the Dutch S Values. At site 4, Cr values were above the NDWQ and Dutch I values.

Cu values ranging from 0.001 to 2.39 mg/l were detected in the groundwater samples. Cu values at site 1 were above the selected references while at site 3 Cu values were above the Dutch I for intervention purposes.

Hg values ranging from 0.001 to 1 mg/l were detected in the groundwater samples. Hg values at site 1 were detected above the selected while Hg values at site 2 and site 3 were above the Dutch S Values. Hg values at site 4 were above the Dutch S Values.

Ni values ranging from 0.001 to 0.5 mg/l were detected in the groundwater samples. Ni values at site 1 were above the Dutch I Values for intervention purposes.

Pb values ranging from 0.03 to 2.0 mg/l were detected in the groundwater samples. Site 1, site 2 and site 3 showed Pb values in the groundwater to be above the NDWQ and Dutch I values for intervention.

Zn values ranging from 0.38 to 0.51 mg/l were detected in the groundwater samples which were found to be above the Dutch S Values.

Thus, from the above analysis, it can be summarized that the heavy metals detected in the groundwater are good indicators to be used to assess the level of contamination in the brownfield sites.

3.3. Benchmarking reference of heavy metal contamination for brownfield

A suggested reference was derived from the study to provide a benchmark for remedial actions required to be taken as shown in Table 5.

Table 5. Suggested reference for benchmarking heavy metal contamination in brownfield for remedial action to be taken.

Heavy Metals	Value In Soil (mg/kg)	Value In Groundwater (mg/l)
As	5	0.05
Cd	10	0.005
Co	5	0.1
Cr	30	0.05
Cu	20	1.0
Hg	0.3	0.002
Ni	20	0.002
Pb	50	0.2
Zn	95	1.5

This provides as an alternative reference value for benchmarking soil and groundwater contamination in brownfields which had exposed to contaminants such as industrial pollutant or leaching of metal ions from nearby sources.

3.4. *Remediation for heavy metal contamination*

Various considerations are to be given after evaluating the contaminants found above the suggested reference derived. Nevertheless, a consistent and common framework is required to address the issues. For this purpose a hierarchy of remedial actions is proposed so that the brownfield site can be redeveloped for other purposes. The remedial techniques involve onsite treatment or offsite treatment as depicted in Fig. 3.

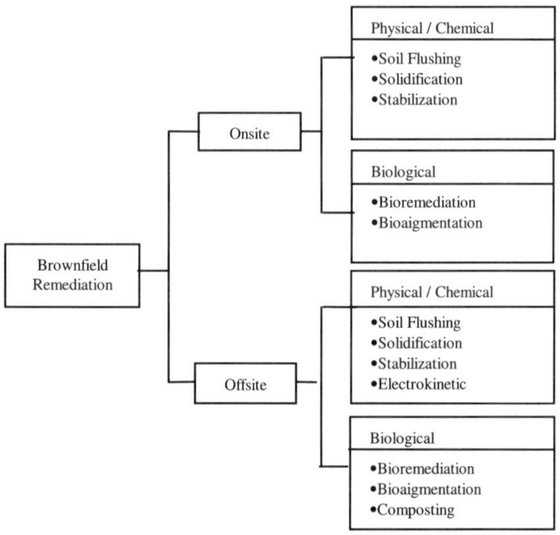

Fig. 3. Remediation for heavy metal contamination in brownfield.

The remediation techniques include physical chemical treatment involving soil flushing and solidification method applicable depending on the heavy metal contaminants found at site. Biological treatment involving bioremediation and bioaugmentation whereby heavy metals are removed by use of microorganism.

4. Conclusions

The study aims to derive reference for benchmarking heavy metals at brownfield sites which serves as a good indicator for heavy metal contamination for the brownfield sites.

Further studies however are to conducted for the brownfield sites as a move towards registration of the brownfield as an additional approach towards sustainable development in Malaysia.

References

1. Lee, A.H. (2001) The need for the registration of contaminated sites in Malaysia. Proceeding of 1st National Conference On Contaminated Land, XII, 1-12, Kuala Lumpur.
2. Salt, D.E., Blaylock, M., Kumar, N.P.B.A., Dushenkov, V., Ensley, B.D., Chet, I. and Raskin, I. (1995) Phytoremediation: a novel strategy for the removal of toxic metals from the environment using plants. Biotechnology 13, 468-474.
3. Babich, H. and Stotzky, G. (1985) Heavy metal toxicity to microbe-mediated ecologic processes: A review and potential application to regulatory policies. Environ. Res. 36, 111-137.
4. Baath, E. (1989) Effects of heavy metals in soil microbial processes and populations (a review). Water. Air Soil Pollut. 47, 335-379.
5. Griller, K.E., Witter, E. and McGrath, S.P. (1998) Toxicity of heavy metals to microorganisms and microbial processes in agricultural soils: a review. Soil Biol. Biochem. 30, 1389-1414.
6. Vangronsveld, J. and Cunningham, S.D. (1998) Introduction to the concepts. In: Metal-Contaminated Soils: In Situ Inactivation and Phytorestoration (Vangronsveld, J. and Cunningham, S., Eds.) pp. 1-15. Springer, New York.
7. Fauziah C.I., Zaharah A.R., Zauyah S., and Rahim A.R. (2001) Proposed heavy metals reference values for site assessment based on its distribution in agricultural soil of Peninsular Malaysia. Proceeding of 1st National Conference On Contaminated Land, I, 1-20, Kuala Lumpur.

ASSESSMENT OF HEAVY METALS IN SOIL, SEDIMENT, WATER AND SUGARCANE PLANT AROUND EWEKORO CEMENT FACTORY IN OGUN STATE, NIGERIA

OLUJIMI, O. OLANREWAJU

Department of Environmental and Occupational Studies, Faculty of Applied Sciences, Cape Peninsula University of Technology, South Africa. olujimiolusoji@yahoo.com; 209076151@cput.ac.za

OLUWADARE, IDOWU

Department of Science Laboratory Technology, Moshood Abiola Polytechnic, Abeokuta, Ogun State, Nigeria. idowuani@yahoo.co.uk

Soil, sediment, water and sugarcane plant stem around Ewekoro Cement Factory in Ogun state, South Western Nigeria were assessed for heavy metals (Cd, Pb and Zn) contamination, exchangeable bases and pH over a range of about half kilometer from the factory during the wet (harvest) season of 2008. Bioavailability of Zn (as the only heavy metals detected) in sugarcane plant stem was determined as well. The metal concentrations in topsoil ranged from 35.65 ± 0.29 to 28. 45 ± 1.60; 14.10 ± 0.44 to 9.65 ± 0.84; 2.43 ± 0.04 to 1.90 ± 0.08 mgKg-1 for Zn, Pb and Cd respectively while in subsoil; it ranged from 29.56 ± 0.50 to 18.53 ± 1.63; 8.21 ± 1.53 to 7.34 ± 0.67; 1.93 ± 0.03 to 1.19 ± 0.09 for Zn, Pb and Cd respectively. There is reasonable spatial correlation of Zn to Cd and Zn to Zn in soil and sugarcane (R2 = 0.699; R2 = 0.665) respectively.

Keywords: Cement factory, exchangeable bases, heavy metals, soil, water and sugarcane plant.

1. Introduction

Metals are elements that may occur naturally in rocks, soils and water (Bornman et al., 2007). Some metals at low doses are essential for growth and development of some plants (e.g. Cu, Zn, Cr, Ni) but at high doses may cause metabolic disorders and growth inhibition for plant species (Bornman et al., 2007; Chojnacki et al., 2005; Goyer, 1997). Metals pollution of environment can result from direct atmospheric deposition, geological weathering, coal combustion, and discharge of agricultural, municipal, residential or industrial waste (Oymak et al,. 2009; Demirak et al., 2006; Fatoki et al., 2004). Mining and milling operations of mineral deposits with the attendants wastes disposal, provide obvious sources of contamination of the environment (Yilmaz et al., 2007; Jung et al., 2001).

Pollution of aquatic environment by metals and their accumulation in fish and other aquatic life has posed so much environmental concern in recent years (Dural et al., 2007; Guo et al., 2006; Krishnamurti and Naidu, 2000). Build up of metals in sediment has significant implication for local communities as well as river quality (Demirak et al., 2006; Awofolu et al., 2005). Elevated levels of metals have been reported in different environmental and biological samples (Horng et al., 2009; Dural et al., 2007; Demirak et al., 2006; Ikem et al., 2003).

High concentrations of these metals in sediments, sludge, soils, water, groundwater, plants and biomagnifications in the food web through trophic transfer may have a negative effect on animals and human health as they are considered non-essential but rather toxic to body growth and development (Kazi et al., 2009; Oymak et al., 2009; Horng et al., 2009). Consumption of heavy metals through soil-crop-food chain has been proposed as the major source of human exposure to potentially toxic metals (Jing et al., 2009; Horng et al., 2009; Kazi et al., 2009). The endocrine disruptive properties of these metals have been demonstrated in different experimental work of Zeng et al., 2003; Choe et al., 2003; Handy, 2003 and Lafuente et al., 2000. The objectives of this study were to investigate soil, water and sediment for heavy metals contamination in the vicinity of Ewekoro (Nigeria) cement factory and to assess their bioavailability in sugarcane.

2. Materials and Methods

2.1. Study area

The study areas are Osun river and sugarcane plantation around Ewekoro cement factory in Ewekoro Local Government area of Ogun State, Nigeria. The Osun river is located in the Southern part of the factory. It flows through Alaguntan village and other rural settlement, all in Ewekoro Local government area of Ogun state, Nigeria. It extends to about 8km southwards with an average width of 80 meter. The river supplies the villagers water needs and also provides them with fishes, shrimps and other seafood. It is also used for solid waste disposal. The sugarcane plantation is located next to the factory and at the bark of Osun River and extends about 6 Km south west and east of the factory.

2.1.1 Sample collection

Triplicate samples of water and sediment were collected from eight sampling points along the Osun river. They were collected by scuba diving and manually pushing the corer into the sediment. The corer was stoppered at the bottom and carefully retrieved into the boat. The samples were placed in clean polythene bags and labeled. The pH and conductivity of the river were carried out at the point of collection. 5 ml of nitric acid was added to the water and stored in an ice bag during transportation. Sugarcane, topsoil and subsoil were collected from the same spot over a distance of half a kilometer from the factory (i.e. 0 m, 100 m, 300 m, and 500 m). The samples were collected in triplicates, stored in polythene bag, labeled and transported to laboratory for analysis.

2.2. Sediment sample treatment

The wet sediment samples collected were air dried for three days ensuring that contamination from surrounding was avoided by covering with mesh cloth. The sediments were crushed in a mortar using a pestle and sieved using a 2 mm sized sieve. Sample pH was obtained in triplicates using a mixture of 1: 2 soil to distilled water, stirred on mechanical shaker at 200 revolutions per hour for 30 minutes and allowed to settle for 5 minutes and pH was determined using a calibrated pH meter.

2.2.1 Heavy metals determination in sediments

1 g of finely sieved sediment sample was weighed into previously rinsed acidified conical flask and 5 ml sulphuric acid, 10 ml nitric acid and 5 ml perchloric acid were added. The mixture was heated at low heat on a hot plate until white fumes appear. Distilled water was added and the solution was filtered into 50 ml volumetric flask. The solution was made up to the 50 ml mark with distilled water. Blanks were also prepared using the same procedure. Heavy metal concentration was determined using Atomic Absorption Spectrophotometer (A Varian 220 model).

2.3. Soil sample treatment

Soil samples (top and sub-soils) were air dried for three days ensuring that contamination from surrounding was avoided by covering with mesh cloth to avoid contamination. The samples were crushed in a mortar using a pestle and sieved with 2 mm sieve size. Sample pH was determined in triplicates and the mean value calculated using a mixture of 1: 2 soil to distilled water, stirred on mechanical shaker at 200 revolutions per hour for 30 minutes, allowed to settle for 5 minutes and pH was determined using a calibrated pH meter. The organic carbon in the soil was determined using Walkey-Black method and particle size distribution was determined using hygrometer method (Juo, 1988).

Table 1. Sampling points for water and sediment showing various human activities.

Points	Area	Human activities
1	Alagutan Village I	Fishing, Drinking
2	Alagutan Village II	Fishing, Farming, Waste disposal
3	Ewekoro Village Head	Farming
4	Backyard	Solid waste disposal, Fishing
5	Ewekoro bridge	Farming
6	Ewekoro sugarcane farm I	Farming
7	Ewekoro sugarcane farm II	Farming
8	Ewekoro new plant 2	Farming
	Ewekoro water works	

2.3.1 Heavy metals determination in soil

1 g of finely sieved sediment sample was weighed into previously rinsed acidified conical flask and 5 ml sulphuric acid, 10 ml nitric acid and 5 ml perchloric acid. The mixture was then heated at low temperature on a hot plate until white fumes appear. Distilled water was added and the solution was filtered into 50 ml volumetric flask and then was made up to the 50 ml mark with distilled water. Blanks were also prepared using the same procedure. Heavy metal concentration was determined using Atomic Absorption Spectrophotometer.

2.4. Sugarcane treatment

2.4.1 Sugarcane digestion

10 ml of the liquid content was extracted and measured into a beaker and 10 ml of perchloric acid was added into the beaker and heated at 150°C until a clear fume was obtained. The digest was pour into a 50 ml standard volumetric flask and made up to the mark with distilled water. Ca, Mg, Zn, Pb and Cd were determined using Atomic Absorption Spectrophotometer while Na and K was determined using flame photometer.

2.4.2 Nitrogen and carbon

Nitrogen: 5 ml of the sample was measured into a volumetric flask and 5 ml of H_2SO_4 was added into it with selenium as a catalyst. It was digested at 350°C. This was then washed into 100 ml standard volumetric flask and made up to 100 ml mark with distilled water. Nitrogen was then determined using Atomic Absorption Spectrophotometer.

Carbon: 0.2 ml of the sample was measured and 10 ml of $K_2Cr_2O_7$ was added to oxidize the carbon. It was then acidifies with 10 ml H_2SO_4 and heated. It was then allowed to cool and washed into 50 ml standard volumetric flask which was made to the 50 ml mark with distilled water and read on spectrophotometer at 600 nm.

3. Result and Discussion

Metal concentrations in water, sediment, soil and sugarcane plant stem are presented in Tables 2-6, which include mean concentrations with samples concentrations of three heavy metals i.e. Cadmium (Cd), Zinc (Zn) and Lead (Pb) were determined.

3.1. Water

The metal pollution status of the river water is given in Table 6. Cd and Zn were not detected in the samples; however, Pb was detected to range between 0.10 and 0.17 mg/L at P8 and P1 respectively. Comparing the values with Nigeria Standard for Water Quality (FEPA, 1991), the water is accepted for Cd and Zn as unpolluted but polluted for Pb as it concentration is 17 times higher than the maximum acceptable limit of 0.01mg/L in

Table 2. Mean (±SD) heavy metals and pH in sediment (mg/kg).

Point/Metal	Cd	Pb	Zn	pH
P1	nd	1.78 ± 0.31	nd	7.80 ± 0.49
P2	nd	2.15 ± 0.43	3.00 ± 0.28	7.55 ± 0.37
P3	nd	1.40 ± 0.89	nd	7.60 ± 0.67
P4	nd	1.05 ± 0.08	2.00 ± 0.17	7.60 ± 0.51
P5	nd	1.55 ± 0.05	nd	8.00 ± 0.56
P6	nd	0.55 ± 0.01	1.00 ± 0.29	7.85 ± 0.23
P7	nd	0.70 ± 0.04	nd	8.05 ± 0.17
P8	nd	1.25 ± 0.09	nd	7.75 ± 0.34

nd = not detected
n = 3

Table 3. Mean (±SD) of exchangeable bases, phosphorus and heavy metals in sugarcane stem (mg/kg).

Distance (M)	Ca	Mg	Na	K	P	Zn	Pb	Cd
100	33.25 ± 2.00	60.78 ± 2.25	56.76 ± 1.94	55.90 ± 2.68	44.00 ± 1.69	3.05 ± 0.45	nd	nd
300	21.30 ± 1.90	53.65 ± 0.58	58.48 ± 0.88	58.48± 0.43	88.00 ± 3.05	2.98 ± 0.99	nd	nd
500	28.35 ± 1.15	61.40 ± 1.90	54.18 ± 0.99	51.60± 1.49	47.50 ± 1.90	2.60 ± 0.24	nd	nd
Control	20.08 ± 1.08	43.85 ± 2.74	46.85 ± 2.05	39.65 ± 5.27	38.32 ± 1.39	3.68 ± 0.68	nd	nd

nd = not detected
n = 3

Table 4. Mean (±SD) of pH, exchangeable bases, phosphorus and heavy metals in topsoil (mg/kg).

Distance	0	100	300	500	Control
pH (water)	7.40 ± 0.60	6.90 ± 0.35	6.70 ± 0.17	6.70 ± 0.15	7.20 ± 0.10
pH (KCl)	6.80 ± 0.06	6.60 ± 0.12	4.50 ± 0.10	4.20 ± 0.27	5.30 ± 0.29
Ca (Cmol/Kg)	45.43 ± 4.71	21.68 ± 1.68	19.12 ± 1.33	21.93 ± 1.22	4.56 ± 0.54
Mg (Cmol/Kg)	4.11 ± 0.19	10.89 ± 0.34	5.78 ± 0.19	7.38 ± 0.39	1.26 ± 0.08
Na (Cmol/Kg)	2.52 ± 0.09	0.37 ± 0.02	0.19 ± 0.02	0.47 ± 0.09	0.28 ± 0.04
K (Cmol/Kg)	2.52 ± 0.27	0.33 ± 0.05	0.16 ± 0.02	0.93 ± 0.07	0.28 ± 0.06
P (ppm)	29.00 ± 0.64	33.00 ± 1.01	27.00 ± 0.97	32.00 ± 0.69	14.14 ± 1.11
H^+ (Cmol/Kg)	0.10 ± 0.03	0.12 ± 0.01	0.14 ± 0.01	0.18 ± 0.02	0.07 ± 0.04
% Organic carbon	0.83 ± 0.07	3.54 ± 0.08	3.64 ± 0.19	2.28 ± 0.26	2.92 ± 0.45
Zn	35.65 ± 0.79	30.70 ± 1.04	28.05 ± 1.56	2.64 ± 0.49	2.64 ± 0.49
Pb	9.65 ± 0.84	11.38 ± 0.94	13.77 ± 0.20	14.10 ± 0.44	3.48 ± 0.29
Cd	2.43 ± 0.04	2.07 ± 0.08	1.90 ± 0.08	1.93 ± 0.03	0.18 ± 0.03
% Sand	83.20 ± 1.69	72.40 ± 0.55	64.30 ± 1.81	72.70 ± 0.10	64.50 ± 0.80
% Silt	12.30 ± 1.31	22.60 ± 0.64	24.60 ± 1.07	14.30 ± 0.59	20.20 ± 1.16
% Clay	4.50 ± 1.58	5.00 ± 0.38	11.10 ± 1.52	13.00 ± 2.07	15.30 ± 1.48

Soil textural class is sandy loam.
n = 3

Table 5. Mean (±SD) of pH, exchangeable bases, phosphorus and heavy metals in subsoil (mg/kg).

Distance	0	100	300	500	Control
pH (water)	75.20 ± 0.31	5.10 ± 0.43	4.80 ± 0.12	5.15 ± 0.27	6.20 ± 0.14
pH (KCl)	4.20 ± 0.87	4.25 ± 0.43	4.00 ± 0.11	4.20 ± 0.93	5.30 ± 0.29
Ca (Cmol/Kg)	1.11 ± 0.09	0.92 ± 0.07	0.85 ± 0.03	0.75 ± 0.43	1.39 ± 0.54
Mg (Cmol/Kg)	0.86 ± 0.01	0.79 ± 0.04	0.73 ± 0.05	0.69 ± 0.21	1.01 ± 0.07
Na (Cmol/Kg)	0.62 ± 0.02	0.57 ± 0.23	0.51 ± 0.21	0.48 ± 0.03	0.75 ± 0.04
K (Cmol/Kg)	0.25 ± 0.01	0.19 ± 0.10	0.18 ± 0.02	0.17 ± 0.01	0.34 ± 0.01
P (ppm)	1.05 ± 0.03	2.10 ± 0.05	1.60 ± 0.40	2.99 ± 0.11	7.35 ± 1.63
H^+ (Cmol/Kg)	0.08 ± 0.02	0.10 ± 0.01	0.13 ± 0.03	0.15 ± 0.01	0.04 ± 0.01
% Organic carbon	0.50 ± 0.04	2.43 ± 0.89	2.69 ± 1.10	2.24 ± 1.12	1.76 ± 0.24
Zn	29.56 ± 0.15	26.44 ± 1.21	21.60 ± 0.98	18.53 ± 1.63	28.80 ± 1.23
Pb	7.34 ± 0.67	7.69 ± 0.89	8.21 ± 1.53	7.35 ± 0.69	1.92 ± 0.80
Cd	1.93 ± 0.03	1.47 ± 0.78	1.41 ± 0.89	1.19 ± 0.09	0.59 ± 0.03
% Sand	81.30 ± 2.57	70.60 ± 3.55	60.60 ± 5.76	74.60 ± 4.65	68.60 ± 4.78
% Silt	15.20 ± 1.32	22.20 ± 1.67	24.20 ± 2.46	20.20 ± 2.33	22.20 ± 3.57
% Clay	4.50 ± 0.45	7.20 ± 0.89	15.20 ± 3.23	5.20 ± 1.23	9.20 ± 2.7

Soil textural class is sandy loam.
n = 3

Table 6. Showing water physicochemical properties and heavy metal content.

Distance	P1	P2	P3	P4	P5	P6	P7	P8
pH at 27°C	6.60	6.40	7.20	7.30	6.00	6.30	6.40	6.60
Conductivity (μScm^{-1})	210	220	170	160	100	120	200	200
Alkalinity (ppm)	45	55	25	30	30	35	130	125
Hardness (ppm)	123	134	108	106	66	74	252	266
Chloride (ppm)	56	48	34	30	12	12	28	26
Total Dissolved Solid	0.25	0.28	0.19	0.18	0.09	0.16	0.44	0.41
(ppm)	0.28	0.32	0.32	0.40	0.14	0.28	1.16	1.18
Total Suspended Solid	0.03	0.07	0.06	0.05	0.051	0.06	0.06	0.06
(ppm)	8.27	11.22	9.43	10.56	9.60	8.77	9.56	9.08
Phosphate (ppm)	8.65	9.60	78.80	74.96	117.24	100.60	13.44	16.80
Nitrate (ppm)	nd	nd	nd	nd	nd	nd	nd	nd
Sulphate (ppm)	nd	nd	nd	nd	nd	nd	nd	nd
Cd (mg/L)	0.17	0.10	0.12	0.10	0.10	0.16	0.11	0.10
Zn (mg/L)								
Pb (mg/L)								

n = 3
nd = not detected

drinking water according to Codex Allimentarus Commission. Other water parameters may be acceptable for irrigation since pH and speciation patterns determines bioavailability of metal to plant.

3.2. Heavy metal concentration in soils and sediment

Aqua regia digestion extract between 70 to 90% of the total content of trace elements, thus several workers have used the extraction method to determine heavy metals in soil, sediment and sludge (Demirak et al., 2006; Jung et al., 2001). Table 2, Table 4 and Table 5 shows the level of metals in the sediment, topsoil and subsoil respectively. The mean sediment pH at each location is presented in Table 2. The pH ranged from 7.55 to 8.05 showing to be slightly alkaline. Cadmium was below detection limit while Zn was detected at locations P2, P4 and P6 in sediment. This could be attributed to waste disposal and use of animal waste to improve crop yield by people living in Alagutan village. Pb in sediment ranged from 2.15 ± 0.43 to 0.55 ± 0.01 with highest concentration at P2. This can be due to dry deposition from the atmosphere and the ability of sediment to serve as sorption medium for metals and organic compounds (Xue and Xu, 2006; Alatriste-Mondrageon et al., 2003; Li et al., 2006; Yuan et al., 2002).

The metal concentrations in topsoil ranged from 35.65 ± 0.29 to 28. 45 ± 1.60; 14.10 ± 0.44 to 9.65 ± 0.84; 2.43 ± 0.04 to 1.90 ± 0.08 $mgKg^{-1}$ for Zn, Pb and Cd respectively while in subsoil; it ranged from 29.56 ± 0.50 to 18.53 ± 1.63; 8.21 ± 1.53 to 7.34 ± 0.67; 1.93 ± 0.03 to 1.19 ± 0.09 for Zn, Pb and Cd respectively. Zn concentration decreases away from the factory entrance to 500 m with corresponding concentration also decreasing down the soil profile inside the sugarcane plantation.

Figure 1 and 2 shows decrease in the level of heavy metals away from the factory an indication of the contribution of the factory activities on metal load in the environment. Figure 3 and 4 shows the correlation Zn and Cd and Zn in soil and Zn in sugarcane. Since there

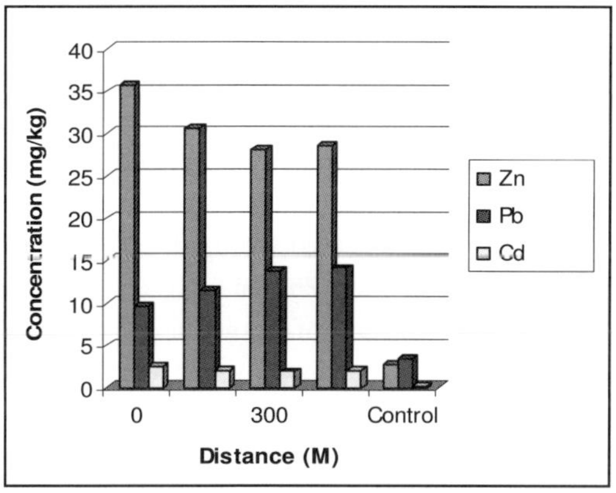

Figure 1. Heavy metals in topsoil.

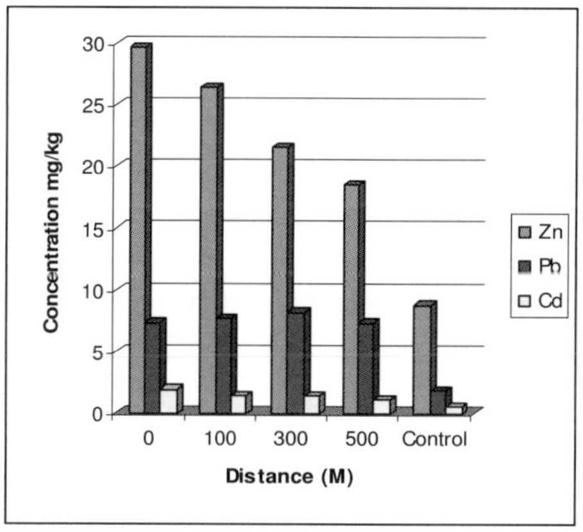

Figure 2. Heavy metals in subsoil.

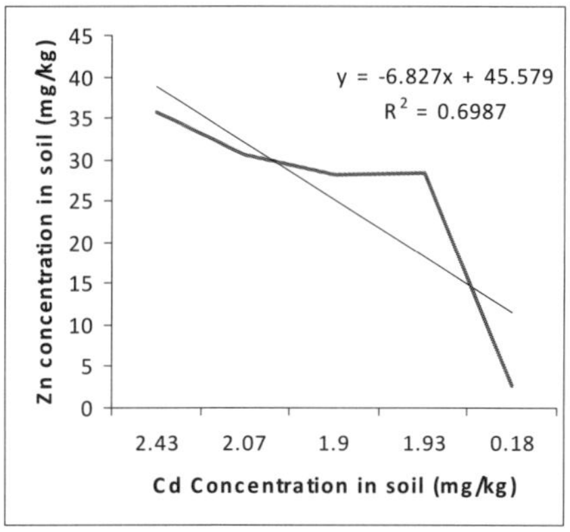

Figure 3. Correlation of Cd with Zn.

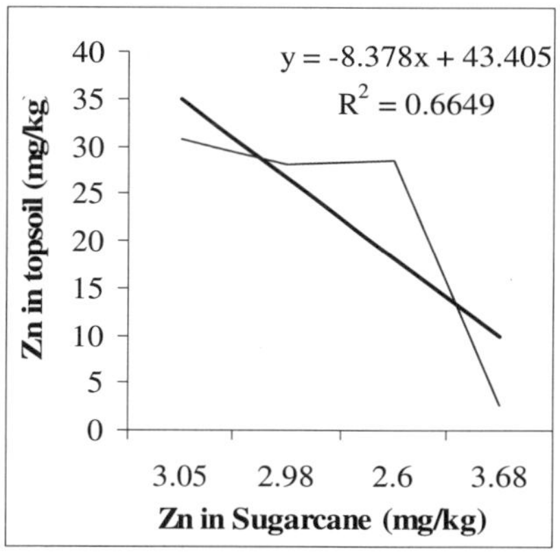

Figure 4. Bioavailability of Zn in sugarcane.

Zn in soil and sugarcane (R^2 = 0.699; R^2 = 0.665 respectively), it clearly indicates that Cd is an associate of Zn and that Zn is readily bioavailable sugarcane.

3.3. Heavy metal concentration in sugarcane

Pb and Cd were not detected in the sugarcane; however, Zn was detected at concentration higher than require concentration in food and water. Zn concentration ranged from 3.05 ± 0.45 to 2.60 ± 0.24 mg/L. The concentration decreased away from the factory and the bioavailability of Zn in sugarcane was calculated using BASF = Zn concentration in sugarcane/Zn concentration in soil.

4. Conclusion

Conclusively, the impact of mining activities of the cement factory on heavy metal loads in the sugarcane plantation and sugarcane can be established. The results of this study suggest a possible interest in monitoring the zone under direct influence of cement factory activities in Ewekoro, controlling the environmental conditions of the surrounding agricultural areas. In order to avoid a potential future increase of toxic levels of metals, it would also be important to perform an exhaustive evaluation of the application of sugarcane, rice and cassava farms. The results of this study also indicate that sugar cane is a crop that is able to grow in areas where some metals in soils are accumulated.

References

1. Awofolu, O.R., Mbolekwa, Z., Mtshemla, V. and Fatoki, O.S. 2005. Levels of trace metals in water and sediment from Tyume river and its effects on an irrigated farmland. Water SA 31(1):87-94.
2. Bornman, R., Van Vuren, J., Bouwman, H., De Jager, T., Genthe, B. and Barnhoorn, I. 2007. Endocrine disruptive activity and the potential health risk in the Rietvlei Nature Reserve. Report to the Water Research Committee on the project “Endocrine disruptive activity and the potential health risk in the Rietvlei Nature Reserve (RNR).”
3. Choe, S., Kim, S., Kim, H., Lee, J.H., Choi, Y., Lee, H. and Kim, Y. 2003. Evaluation of estrogenicity of major heavy metals. The science of the total environment 312: 15-21.
4. Chojnacka, K., Chojnacki, A., Gorecka, H. and Gorecki, H. 2005. Bioavailability of heavy metals from polluted soils to plants. Science of the environment. 337: 175-182.
5. Demirak, A., Yilmaz, F., Tuna, A.L. and Ozdemir, N. 2006. Heavy metals in water, sediment and tissues of Leuciscus cephalus from a stream in southwestern Turkey. Chemosphere 63: 1451-1458.
6. Dural, M., Goksu, M.Z.L. and Ozak, A.A. 2007. investigation of heavy metals levels in economically important fish species captured from the Tuzla Lagoon. Food chemistry 102:415-421.
7. Fatoki, O.S. Awofolu, O.R. and Genthe, B. 2004. Cadmium in the Umtata river and the associated health impact on rural communities whose are primary users of water from the river. Water SA 30(4):507-513.
8. Goyer, R.A. Toxic and essential metals interactions. 1997. Annu. Rev. nutr. 17:37-50.
9. Guo, G.L., Zhou, Q.X., Koval, P.V. and Belogolova, G.A. 2006. Speciation, distribution of Cd, Pb, Cu and Zn in contaminated Phaeozem in north-east China using single and sequential extraction procedures. Australian Journal of Soil Research 44:135-142.
10. Handy, R.D. 2003. Chronic effects of copper exposure versus endocrine toxicity: two sides of the same toxicological process? Comparative biochemistry and physiology part A 135: 25-38.
11. Horng, C., Wang, S. and Cheng, I. 2009. Effects of sediment-bound Cd, Pb, and Ni on the growth, feeding, and survival of Capitella sp. I. Journal of experimental marine biology and ecology 371:68-76.
12. Ikem, A. Egiebor, O. and k. Nyavor. 2003. Trace elements in water, fish and sediment from Tuskegee lake, Southeastern USA. Water, Air, and Soil Pollution 149: 51-75, 2003.
13. Jing, S., Lianqing, L. and Genxing, PAN. 2009. Variation of grain Cd and Zn concentration of 110 hybrid rice cultivars grown in a low-Cd paddy soil. Journal of environmental sciences 21:168-172.
14. Kazi, T.G., Jamali, M.K., Arain, M.B., Afridi, H.I., Jalbani, N., Sarfraz, R.A. and Ansari, R. 2009. Evaluation of an ultrasonic acid digestion procedure for total heavy metals determination in environmental and biological samples. Journal of hazardous materials 161: 1391-1398.
15. Lafuente, A. Marquez, N., Perez-Lorenzo, M., Pazo, D. and Esquifino, A.I. 2000. Pubertal and

postpubertal cadmium exposure differentially affects the hypothalamic-pituitary-testicular axis function in the rat. Food and chemical toxicology 38: 913-923.

16. Oymak, T., Tokalioglu, S., Yimaz, V., Kartal, S. and Aydin, D. 2009. Determination of lead and cadmium in food samples by co-precipitation method. Food chemistry 113:1314-1317.

17. Zeng, X., Jin, T., Zhou, Y and Nordberg, G. F. 2003. Changes of serum sex hormone and MT and mRNA expression in rats orally exposed to cadmium. Toxicology 186: 109-118.

KNOEVENAGEL CONDENSATION: AN EFFICIENT GREEN APPROACH

BHOSALE SHYAM, PAWAR YOGESH, SONAWANE ATUL, NAGLE PRAMOD and MORE DHANANJAY†
School of Chemical Sciences
North Maharashtra University, Jalgaon, 425 001, Maharashtra, INDIA

The present work includes Knoevenagel condensation of aromatic aldehydes and malononitrile by using TBAB in aqueous medium.

1. Introduction

With the current research for greener and more sustainable processes to produce fine chemicals, organocatalysis with organic molecules is to become the great success story of the beginning 21st Century.[1,2] In order to develop catalysts of interest for industry, it would be of major importance to design heterogeneous systems. Surprisingly, examples of purely organic heterogeneous catalysts are limited to "carbon only" systems[3] or to nitrogen rich carbonaceous materials.[4-7] Organic reactions under solvent-free conditions involving easily separable solid catalysts are receiving a considerable attention in recent times, particularly from the viewpoint of green chemistry. As an important carbon–carbon bond forming reaction in organic synthesis, knoevenagel condensation has been extensively investigated in view of it significance.[8] The knoevenagel condensation of aldehydes with active methylene compounds has been commonly employed in the synthesis of numerous speciality chemicals and chemical intermediates[8,9] such as carbocyclic as well as heterocyclic compounds of biological significance,[10] calcium antagonists,[11] polymers,[12] coumarin derivatives, cosmetics, perfumes and pharmaceuticals.[11-13] The condensation reaction is generally catalyzed by bases, acids or catalysts containing both acid–base sites.[11] Numerous acid–base reagents or catalysts such as ethylene diamine,[14] dimethyl amino pyridine,[15] 1,1,3,3-Tetramethylguanidium Lactate [16], ReBr(CO)$_5$ [17], using a Catalytic Amount of Iodine and K_2CO_3[18], 12-Tungstophosphoric Acid[19], P_2O_5, piperidine, and chlorobenzene with a commercial microwave oven[20], microwave irradiation over alumina[21], Microwave Assisted Knoevenagel Condensation Using Sodium Fluoride and Lithium Chloride as Catalysts[22], Proline-Catalyzed[23], Potassium exchanged layred zirconium phosphate as base catalyst[24],Montmorillonite silylpropyl ethylenediamine[25], Most of them are used under homogeneous conditions or in the presence of excess solvents, which generate large volumes of waste.

From environment point of view, waste should be eliminated or minimized and obviously bulk wastes in the chemicals manufacturing are often organic solvents. The use of environmentally benign solvents or solvent-free reactions represents an ideal green chemical technology procedure from both economical and environmental point of view. Therefore, efforts made to perform the Knoevenagel condensation in the absence of solvents[26,27]. Thus, some approaches could be found in the literature by employing Lewis acids such as triphenyl phosphane[28], ammoniumacetate[29], ethylenediammonium diacetate[30], PEG400 and anhydrous K_2CO_3[31] Unfortunately many of these methods have some drawbacks, such as use of expensive, use of stoichiometric amount of reactant, low yields, extended time and lack of geometry. Thus the need for the development of an alternative rout to construct the say reaction. In continuation of our research to develop newer, environmentally benign methods.

We have reported use Tetrabutylammoniumbromide (TBAB) as a catalyst for knoevenagel condensation in aqueous medium. TBAB has wide applications such as in Michael Addition[32], ω-hydroxyalkyl selenides were synthesized[33], Synthesis of 10-(Phthalimidoalkyl)-halosubstitutedpyrido [3,2-b][1,4]-benzothiazine[34] Regioselective Synthesis of Thiopyrano[3,2-*c*][1]benzothiopyran-5(2*H*)-one and Thieno[3,2-*c*][1]benzothiopyran-4(2*H*)-one[35], Synthesis of Optically Active α- Amino Acids[36].

†Corresponding Author

Reaction of Tetra-O-acetylglycopyranosyl Bromide with Active Methylene Compounds[37], Synthesis of Aryl Phosphorodichloridothioates[38].

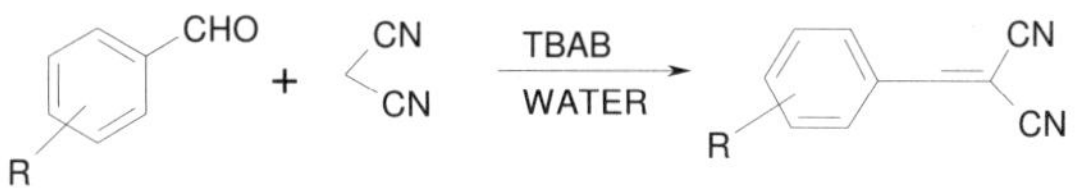

Scheme 1

2. Results and Discussion

In this method, tetrabutylammoniumbromide play an important role in significantly enhancing the reaction rate as well as avoiding the use of toxic or flammable organic catalyst and solvents, for example, pyridine or ethanol. This reaction is of a wide scope, and applies to aromatic aldehydes bearing different substituents. In many cases, the condensation products were obtained in high yields (Table 1).

The electronic effects of phenyl ring substituents were also explored. The experimental results showed the aromatic aldehydes with electron withdrawing groups(−NO2) also offered good yields and the reactions were completed in short times. Also in the case of electron donating groups (−OH, −OCH3, −CH3), reasonably good yields were observed but demanded little more reaction time.

Table 1. Condensation products in high yield.

Sr. No.	Ar	Time (min)	Yield (%)	m.p. (°C)	Reported
1	C_6H_5	4	90	82-84	82[39]
2	2-CH_3O-C_6H_4-	5	89	79-80	78[45]
3	4-$OCH_3C_6H_4$	5	89	115-116	114-115[39]
4	3-$NO_2C_6H_4$	3	94	105-107	104-105[44]
5	2-$NO_2C_6H_4$	3	96	138-139	139-140[42]
6	2-ClC_6H_4	7	88	96-98	95~96[41]
7	3-$OCH_3C_6H_4$	5	87	93-95	93[46-53]
8	C_6H_5-HC=CH-	5	89	122-123	128[43]
9	4-NO_2-C_6H_4	3	96	160-162	159-160[46-53]
10	4-Cl- C_6H_5	6	90	161-163	162-163[39]
11	4-N,N $(CH_3)_2$ C_6H_4	5	79	180-182	179-180[40]
12	Furyl	2	92	63-64	63[46-53]
13	3,4,5 – $(OCH)_3C_6H_2$	4	90	138-139*	_ __

* ^{1}H NMR :3.89(-OCH_3), 6.06 (2 Ar-H), 7.88(-C=CH)

The results showed that, the reaction proceeded much faster and results in comparable or better yields than other methods .Due to its characteristics such as non toxicity, non flammability, water is superior to many other organic solvents from both economical and environmental points of view. In our reaction, the use of water as solvent simplified the final treatment of products. One of the problems of homogeneous reactions mediated in organic solvents is the separation of products and/or catalysts from the solvents. An additional advantage of the use of water as solvent is that many organic products can be separated by simple filtration decantation from the resulting mixture. In contrast, the use of organic solvents necessitates tedious and energy-consuming thermal operations such as distillation or dilution of the resulting solution with water, which will bring difficulties in solvent recovery. The experimental procedure is very simple. Products obtained after recrystallization by ethanol are of high purity. The tetrabutylammoniumbromide remains in the aqueous phase simply distilling the water, could recover nearly quantitative amounts of tetra butyl ammoniumbromide.

In conclusion, this work clearly demonstrated the potential application of tetrabutylammoniumbromide for clean and efficient synthesis of fine chemicals. An efficient, simple and environmental friendly procedure for the knoevenagel condensation from the corresponding aromatic aldehydes has been developed which enabled us to avoid the use of hazardous, homogeneous catalyst and also eliminated the need for solvents and furthermore the catalyst can be easily recovered and reused.

3. Experimental

Melting points of all the synthesized compounds were determined by open capillary method and are match with their physical data (m.p.) with samples prepared from known method. The purity of synthesized compounds was checked by thin-layer chromatography. The homogeneity of the compounds was monitored by ascending thin layer chromatography (TLC) on silica gel (Merck) glass plates, visualized by iodine-vapour. Developing solvents were hexane-ethyl acetate (8:2). The IR spectra were recorded on FTIR spectrophotometer [Perkin Elmer] using nujol mull. H^1

NMR spectra were scanned at 300 MHz on Varian Mercury YH-300 FT NMR spectra in $CDCl_3$/using tetramethylsilane as an internal standard. All the chemicals and solvents used were of synthetic grade (S. d. Fine, chemicals, Mumbai, India).

3.1. *General Procedure for Knoevenagel Condensation*

5 mmol of benzaldehyde, 5 mmol of malononitrile and 100mg of tetrabutylammoniumbromide was added in 25ml of water and stirred for 4-5 minutes; the reaction was monitored by TLC. The solid obtained was collected by filtration and washed with water and recrystalise by alcohol.

References

1. P. I. Dalko and L. Moisan, Angew. Chem. Int. Ed. 43, 5138 (2004).
2. B. List, Acc. Chem. Res. 37, 54 (2004).
3. D. S. Su, N. Maksimova, J. J. Delgado, N. Keller, G. M. Mestl, J. Ledoux and R. Schlögl, Catal. Today 102, 110 (2005).
4. F. Goettmann, A. Fischer, M. Antonietti and A. Thomas, Chem. Commun. 4530 (2006).
5. F. Goettmann, A. Fischer, M. Antonietti and A. Thomas, Angew. Chem. Int. Ed. 45, 4467 (2006).
6. F. Goettmann, A. Thomas and M. Antonietti, Angew. Chem. Int. Ed. 46, 2717 (2007).
7. F. Goettmann, A. Fischer, M. Antonietti and A. Thomas, New. J. Chem. 1455 (2007).
8. G. Jones, "Inorganic Reactions," Wiley: New York, (1967).
9. F. Freeman, Chem. Rev. 80, 329 (1980).
10. L. F. Tietze, Chem. Rev. 96, 115 (1996).
11. R. L. Reeves and S. Patai, "The Chemistry of Carbonyl Compounds," Interscience Publishers: New York, (1996).
12. S. A. E. Ayoubi, F. Texier-Baullet and J. Hamelin, Synthesis 258 (1994).
13. L. F. Tietze, U. B. M. Trost, I. Fleming and C. H. Heathcock, (Eds.), "Comprehensive Organic Synthesis, vol. 2," Pergamon Press, Oxford, 341 (1991).
14. D. S. Acker and W. R. Hertler, J. Am. Chem. Soc. 84, 3370 (1962).
15. A. V. Narsaiah, A. K. Basak, B. Visali and K. Nagaiah, Synth. Commun. 34, 2893 (2004).
16. J. Zhang, T. Jiang, B. Han and A. Zhu, Synth. Commun. 36, 3305 (2006).
17. W-X. Zuo, R. Hua and X. Qiu, Synth. Commun. 34, 3219 (2004).
18. Y-M. Ren and C. Cai, Synth. Commun. 37, 2209 (2007).
19. H. A. Oskooie, M. M. Heravi, F. Derikvand and M. Khorasani, Synth. Commun. 36, 819 (2006).
20. J. K. Kim, P. S. Kwon, T. W. Kwon, S. K. Chung, and J. W Lee, Synth. Commun. 26, 535 (1996).
21. P. S. Kwon, Y. M. Kim, C. J. Kang and T. W. Kwon, Synth. Commun. 27, 4091 (1997).
22. K. Mogilaiah, M. Prashanthi, G. R. Reddy, C. S. Reddy and N. V. Reddy, Synth. Commun. 33, 2309 (2003).
23. G. Cardillo, S. Fabbroni, L. Gentilucci, M. Gianotti and A. Tolomelli, Synth. Commun. 33, 1587 (2003).
24. M. Curini, F. Epifano, M. C. Marcotullio, O. Rosati and A. Tsadjout, Synth. Commun. 32, 355 (2002).
25. Y. V. Subba Rao and B. M. Choudary, Synth. Commun. 21, 1163 (1991).
26. J. O. Metzger, Angew. Chem. Int. Engl. Ed. 37, 2975 (1998).
27. A. Loupy, A. Petit, J. Hamelin, T. Boullet, F. P. Jacquault and D. Mathe, Synthesis, 1213 (1998).
28. J. S. Yadav, B. V. S. Reddy, A. K. Basak, B. Visali, A. V. Narsaiah and K. Nagaiah, Eur. J. Org. Chem. 546 (2004).
29. A. K. Mitra, A. De and N. Karchaudhuri, Synth. Commun. 29(16), 2731 (1999).
30. Ce. Su, Z. C. Chen and Q. G. Zheng, Synthesis 4, 555 (2003).
31. Y. Q. Cao, Z. Dai, R. Zhang and B. Hua, Synth. Commun. 34(16), 965 (2004).
32. K. S. Babu, V. R. Subba Rao, P. Sunitha, S. S. Babu and J. M. Rao, Synth. Commun. 38, 1784 (2008).
33. X. Hu, Z. Tian, Y. Chen and X. Lu, Synth. Commun. 30, 523 (2000).
34. A. Gupta, R. Sakhuja and S. C. Jain, Synth. Commun. 37, 4059 (2007).
35. K. C. Majumdar, A. K. Pal and M. Ghosh, Synth. Commun. 37, 1525 (2007).
36. G. Su and L. Yu, Synth. Commun. 23, 1229 (1993).

37. R. T. Li. Qian Zhu and M. S. Cai, Synth. Commun. 27, 1357 (1997).
38. V. T. Ravikumar, Synth. Commun. 26, 1821 (1996).
39. Y. Q. J. Li, Chem. Res. (S) 524 (2000).
40. G. Y. Dai, D. Q. Shi and L. H. Zhou, Chemical Reagent 18, 39 (1996).
41. D. Q. Shi, X. S Wang, C. S Yao and L. L. J. Mu, Chem. Res. (s) 344 (2002).
42. L. H. Zhou, Y. Gao, D. Q. Shi and G. Y. Dai, Huaxue Shiji (Chem. Reagents in Chinese) 20(3), 159 (1998).
43. "Beilstein, Handbuch der Organisation Chemie," Springer-Verlag: Berlin, Band 9, 913, 32.
44. Y. Q. Cao, Z. Dai, R. Zhang and B. H. Chen, Synth. Commun. 34, 2965 (2004).
45. "Beilstein, Handbuch der Organischen Chemie," Springer-Verlag: Berlin, pp. 4383 Band 9, 12832, 35.
46. . H. Khan, R. K. Mathur and A. C. Ghosh, Synth. Commun. 26(4), 683 (1996).
47. A. K. Mitra, A. De and N. Karchaudhuri, Synth. Commun. 29(16), 2731 (1999).
48. R. P. Shanthan and R. V. Venkataratnam, Tetrahedron Lett. 32(41), 5821 (19949. D. Q. Shi, X. S. Wang, C. S. Yao and L. J.
49. Mu, Chem. Res. 7, 344 (2002).
50. S. X. Wang, J. T. Li, W. Z. Yang and T-S. Li, Ultrason Sonochem., 9(3), 159 (2002).
51. Ce. Su, Z. C. Chen and Q. G. Zheng, Synthesis. 4, 555 (2003).
52. Y. Lu, Z. Ren, W. Cao, W. Tong and M. Gao, Synth. Commun. 34, 2047 (2004).
53. B. M. Choudary, M. Lakshmi Kantam, B. Kavita, C. Venkat Reddy and F. Figueras, Tetrahedron 56(47), 9357 (2000).

ONTO-BIOLOGY: CLARIFYING ALSO THE STANDARD CLOCK FOR PRE-BIOTIC PROCESS, STEM-CELL, ORGANS, BRAIN, AND SOCIETIES

KEN NAITOH

Waseda University, Faculty of Science and Engineering, 3-4-1 Ookubo, Shinjuku, Tokyo, 169-8555 Japan

A master equation is derived for describing several bio-oscillators, including the chemical oscillations in the pre-biotic soup, the cell cycle of bacteria, the reaction cycle related to stem cells in morphogenetic processes, the circadian clock in the brain, and the boom-and-bust economic cycles in human society. An important point is that the biological cycle is about seven times the fundamental beat in each type of system.

1. Introduction

We can see several chemical oscillation phenomena in the bio-systems related to the origin of life, bacteria, morphogenetic processes, the brain, and the economy in the human network. The Brusselator, circadian clock, and heartbeat are well-known examples. [1, 2] Little is known, however, about the relation between these oscillations.

The core engine for activating living beings is a system of four categories of biological molecules, consisting of two types of information molecules and two types of functional molecules and resembling a four-stroke engine. [3] This four-stroke engine induces the exponential growth of bio-systems. We have previously showed how the origin of life can be explained by this four-stroke engine, representing a minimum hypercyclic chemical reaction system. [3] However, this engine did not reveal the oscillations.

Accordingly, fifth and sixth categories of molecules for repressing DNA replication were added to the four-stroke engine, resulting in several oscillating phenomena with chemical reactions. [4] A mathematical model based on these six categories of molecular groups showed that the oscillation cycle has a rhythm of "seven" times the fundamental beat. [4]

This model clarifies the common principle underlying several levels of bio-chemical oscillations, because new organs emerge at "sevenfold" divisions of stem cells, while the circadian clock [2, 5] of about 24 hours is seven times the fundamental clock of about 3.5 hours. It is also stressed that one week has "seven" days. Several economic cycles are seven times the fundamental durable periods of certain products.

2. Minimum drive element for the origin of life (fore-stroke engine of biomolecules) [3]

Both information strings such as DNA and catalyers such as enzymes are necessary for bio-systems, which can replicate to counter the natural degradation of biomolecules. At least two types of information strings for the origin of life are necessary, which code catalysers for generating information strings and those for generating catalysers. This also implies the inevitability of two catalysers. Thus, the minimum system for self-organization at the origin of life has four types of molecules: two information strings and two catalysers.

We call molecules having information such as DNA "infostrings", while the catalysers such as enzymes are called "catalysts". The core system shown in Fig. 1 includes four molecules consisting of Infostring 1 that codes the catalyst generating infostrings, Infostring 2 that codes the catalyst for generating catalysts, Catalyst 1 generated by Infostring 1, and Catalyst 2 produced by Infostring 2. This four-cycle system works as a closed loop, if elements such as nTP and amino acids are input from the outside with energy. This is the minimum 4-stroke engine (minimum hypercycle), which produces molecules exponentially in cases where there are no degradations.

First, we define the densities of Infostrings 1 and 2 as D1 and D2, respectively. Then, the densities of Catalysts 1 and 2 are defined as R1 and R2, respectively. This yields the following essential equation system.

$$D_i^{N+1} - D_i^N = \alpha_{i1} D_i^N \otimes R_1^N \tag{1a}$$

$$R_i^{N+1} - R_i^N = \alpha_{i2} D_i^N \otimes R_2^N \tag{1b}$$

where $D_i \otimes R_j$ denotes the smaller value among D_i and R_j with i=1, 2 and j=1, 2. Then, α_{ij} and N imply an arbitrary constant and generation, respectively.

Results computed with this equation system show nearly exponential increases in Di and Ri.

It is stressed that variable R1 on the left-hand side of Eq. (1b) never appears on the right-hand side of the equation, although common variables can be observed on both sides of Eq. (1a). Equation (1b) essentially shows the asymmetry topologically.

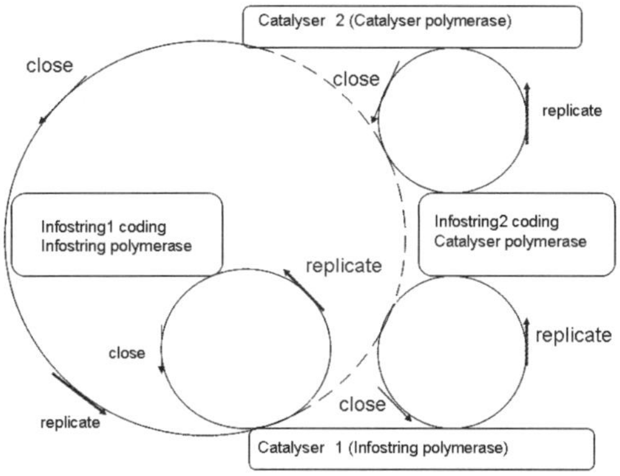

Fig. 1. Four-stroke engine as a minimum hypercycle.

Equation (1b) leads to an impressive conclusion. Catalyst R1 can be increased, even if R1 is equal to zero initially in the ancient pool. It should be stressed that among the four molecules only three of them, i.e., D1, D2, and R2, work as the lifeline.

If a small amount of another Infostring D3, unrelated to R1 and R2, comes around this minimum hypercycle in Fig. 1, R3 will be generated automatically. There are many possibilities for Infostring D3. When many types of infostrings come to the pool of D1, D2, R1, and R2, many types of proteins corresponding to these infostrings can be generated. Some of these proteins will be muscle and transcription factors for repressing genes. This is an important step toward a cell system. The simplest hypercycle of D1, D2, R1 and R2 works as "the engine of life" for promoting more complex reaction networks.

3. Minimum oscillator for morphogenetic process and circadian clock in brain (six-stroke engine of bio-molecules)

Let us consider bacteria, eukaryotes, and cells in higher-order living systems. Four types of fundamental molecules (D1, D2, R1, and R2) must be extended to four molecular groups, because life today uses numerous types of genes and proteins. We can then define four macroscopic groups as follows.

D1: Gene group 1 that codes enzyme system R1 for DNA production
D2: Gene group 2 that codes enzyme system R2 for enzyme production
R1: Enzyme system 1 for DNA production, including DNA polymerases, topoisomerases, helicases, ligases, and primases
R2: Enzyme system 2 for enzyme production, including ribosomes with rRNAs and ribosome proteins, RNA polymerases, mRNAs, and tRNAs

It is stressed that D2 and R2 are for activating reactions, as is shown in the previous section.

Next, we define D3 and R3 as other molecular groups for repressing reactions as in the case of transcription factors. These two groups are incorporated in the four groups of D1, D2, R1, and R2, because today's cells and the morphogenetic processes of multi-cellar systems use negative controllers such as Oct-4 and SOX2 for producing tissues and organs [6, 7]. Accordingly, D3 and R3 can be defined as follows.

D3: Gene group 3 that codes protein system R3 for repressing Di gene expressions
R3: Protein system 3 for repressing Di gene expressions, including the proteins of the transcription factors such as those of Oct-4 and SOX2

This leads to a macroscopic model having six types of molecular groups, or in other words, a six-stroke engine. We can describe the densities of the six groups at generation N after the mother cell generation by the following equations.

$$D_i^{N+1} - D_i^N = \alpha_{i1} D_i^N \otimes R_1^N \quad ,(i=1,2,3) \tag{2a}$$

$$R_i^{N+1} - R_i^N = \alpha_{i2}\delta(D_i^N - \varepsilon_i R_3^N) \otimes R_2^N \quad ,(i=1,2,3) \tag{2b}$$

where $D_i \otimes R_j$ denotes the smaller value among D_i and R_j with i=1, 2, 3 and j=1, 2, 3 and also where $\delta(x)$ denotes the larger one of x or 0, i.e., max (x, 0). Parameter ε_i with i=1, 2, 3 denotes an arbitrary constant for controlling repression rates. (Equation 2b is the extended version of our previous model [4].) In this report, we set to be $\varepsilon_1 = \varepsilon_3 = 0$, while $\varepsilon_2 = 1.0$, because R3 usually represses D2.

Figure 2 obtained by solving Eq. (2) shows the density ratios of D2 and R3, which implies the amount of DNA uncovered by protein group R3. The condition of D2/R3 > 1.0 means that a part of D2 is not covered by R3. The condition of D2/R3 < 1.0 means that D2 is completely blocked by redundant R3.

When R3 is dense in stem or induced pluripotent stem (iPS) cells, it means that stem and iPS cells can be reprogrammed by the presence of much Oct-4. This oscillation of D2/R3 will lead to changes in the gene combination for expressions, i.e., the emergence of new organs.

Figure 2 clearly shows that the antagonism between the negative controller R3 and the positive replication factors R1 and R2 induces bifurcation events at rhythmic intervals constituting about seven divisions, although the intervals are slightly chaotic and the vibrational amplitude is attenuated.

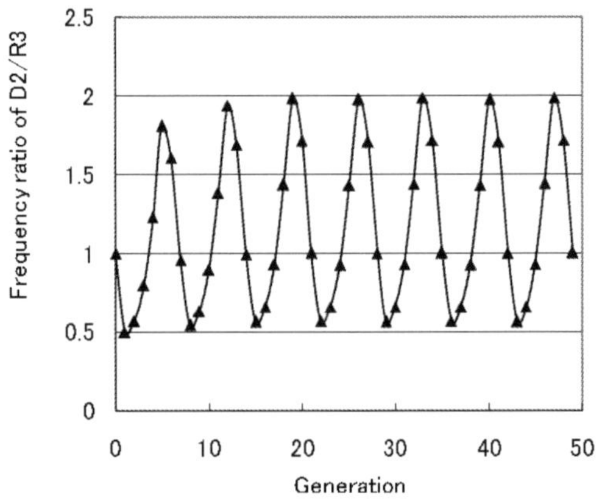

Fig. 2. Time histories of D2/R3 during 50 generations, with the initial conditions of Di = 1 and Ri = 1.

It is stressed that this cycle of about seven divisions, i.e., the branching time between periodic bifurcation events, corresponds to the emergence timing of blast cysts, embryos, germ layers, tissues, and organs, which can be observed for about every seven cell divisions. Emphasis is placed on the fact that three system parameters, α_{i2}, have less influence on the time cycle. The cycle of about seven divisions is fairly stable.

Macroevolution can be from the pre-biotic molecular pool to a single cell, from prokaryotes to eukaryotes, from eukaryotes to primitive multi-cellar systems such as nematomorphs, and from crustacea to mammals. Seven divisions of a cell will produce a new organ, also leading to macroevolution having 2^7 (about 100) times the former body weight. It is well known that one macroevolution is also about 100 times the DNA length. This coincidence is possible, because the DNA length is proportional to the molecular weight.

Next, let us consider the circadian clock in the brain, having a cycle of 24-25 hours. We know that, when suprachiasmatic nuclei (SCN) in brain are removed, the clock shows a cycle of about 3-4 hours. [5] This means that the sevenfold-beat cycle based on the six-stroke engine of molecules is also present in the brain.

4. Economic cycles and human society

There are several types of boom-and-bust economic cycles: the Kondratev cycle of about 50 years, the Juglar cycle of about 7 years, the Kitchin cycle of about 3.5 years, and one week as the shortest one, which are observed in human network systems. [8, 9]

The shortest economic cycle is one week of "seven" days. The longest repetitive catastrophes occur about every 70 years, which is seven times the actual durable period of automobiles of about 10 years. Thus, we hypothesize that several types of economic cycles are about seven times the length of the fundamental production cycles (durable periods). New organs that emerge with a sevenfold rhythm may correspond mathematically to the rise of new production or new business categories. [We classify all things in our economic system into two major categories: information and functional objects. Information includes CAD data for designing many different functional products, while computers and industrial robots are examples of functional objects. Functional objects can be further divided into two subgroups: those such as computers for generating information and those such as industrial robots for generating functional products. These four categories correspond to D1, D2, R1, and R2 in Eqs. (1) and (2). Then, fifth category of R3 for depressing D2 and sixth category of D3 coding R3 lead to the cycles.]

5. Conclusion and outlook

5.1. *Conclusion*

The model based on Eqs. (2a) and (2b) describes the chemical oscillations in the pre-biotic soup, the cell cycle of bacteria, the morphogenetic processes related to stem cells, the circadian clock in the brain, and the boom-and-bust economic cycles of human society. We can also conclude that this model reveals the temporal aspects underlying living systems.

Our previous analysis based on fluid physics [3, 10, 11, 12, 13] made clear the spatial structure of living systems, i.e., the inevitability of five nitrogenous bases and twenty types of amino-acids, the physical principle generating the left-right asymmetry of the heart and the symmetry of the kidneys, and the morphogenetic mechanism of the convexoconcave shape inside the human brain.

Thus, the fusion of the present temporal analysis and our previous theoretical analysis on spatial structures may lead to an understanding of the design principle of life, resulting in new types of medicine and environmental technologies.

5.2. *Outlook*

One of the important unsolved problems is the relation between living organisms having different DNA lengths, i.e., the mechanism promoting macroevolution over the 4,000-million-year history of living beings. Population dynamics for evolution such as the quasi-species model [15] describes only the temporal features.

The DNA length of each species has increased during evolution over 4,000 million years. The fundamental driving force for increasing the DNA length, i.e., gene insertion, is the enzyme system for DNA production, including DNA polymerases, topoisomerases, helicases, ligases, and primases, because the enzyme system promotes the connection of DNA fragments.

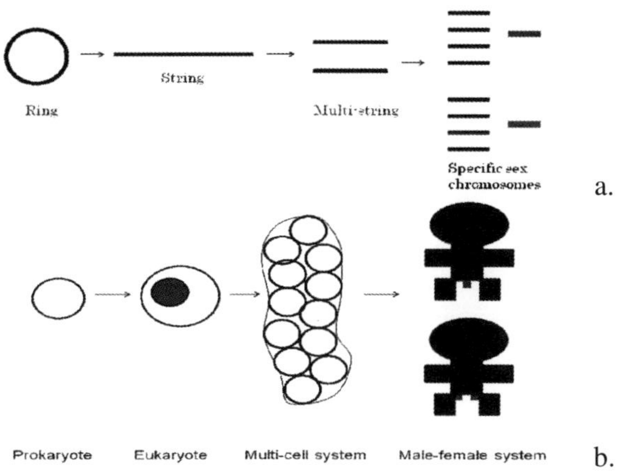

Fig. 3. Topological changes of (a) DNA and (b) phenotypes.

This increase in the DNA length often induces instability in the DNA topology, because decoupling occurs when DNA reaches a certain critical length. Longer strings lead to a larger possibility of decoupling.

Changes in the DNA topology due to decoupling (rings, strings, multi-strings, and specific sex strings) give rise to the various structures of phenotypes in three-dimensional space (Fig. 3). Sudden decoupling of DNA is the driving force for macroevolution, because it causes sudden evolution.

It is well known that small rings and short strings of RNA can be generated in vitro. [14] In the pre-biotic pool, a small ring of DNA would make the first cell such as bacteria or archaea, i.e., the origin of cells. This is possible because enzymes in contact with ring-like DNA can produce DNA and proteins continuously by simply revolving on the ring, whereas enzymes in systems having string-like DNA must find the other end of the string after departing from one end in order to accomplish the next replication (Fig. 4).

The probability theory tells us the difficulty in finding the other end of string-like DNA. Ring-like DNA is suitable for stably generating DNA and proteins.

However, the absence of ends in ring-like DNA also makes it relatively difficult to take in short DNA fragments swimming around the ring-like DNA, because there are no open positions where a short fragment can attach to the ring (Fig. 5). Overly stable ring-like DNA for specific replication leads to the difficulty of effecting changes. Ring-like DNA must be cut before a short fragment can be taken in and must also be connected again to the short fragment at two places, whereas a fragment can be connected to string-like DNA at only one point. Thus, it is hard for ring-like DNA to produce diverse components such as Golgi bodies and endoplasmic reticula.

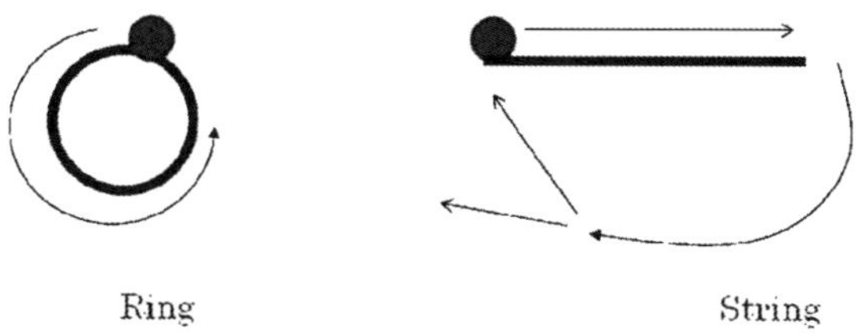

Fig. 4. Stability of DNA replication for ring-like DNA and string-like DNA.

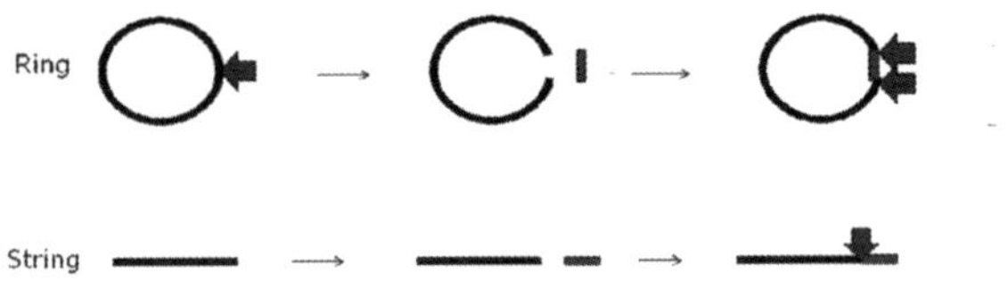

Fig. 5. Difficulty of increasing DNA length.

Decoupling might occur when ring-like DNA for bacteria or archaea becomes string-like DNA because of the instability induced by the increase in the DNA length. This decoupling can cause macroevolution from prokaryotes to eukaryotes such as yeast, because string-like DNA can relatively easily take in new short fragments of DNA for generating many new parts and components such as cell nuclei and liposomes, as discussed above (Fig. 5). It should also be noted that string-like DNA can result in death because of the instability at its ends.

Another important point about string-like DNA is that it produces three-dimensional asymmetric distributions of phenotypes such as liposomes and Golgi bodies. Ring-like DNA and string-like DNA have different types of line symmetry. The symmetric axis of a string includes the string itself, whereas the symmetric axis of a ring is vertical to the ring. This topological difference in the symmetry of genotypes determines that of phenotypes.

The detailed mechanism is related to the contractile ring for cutting cells. An important point is that a cell-dividing contractile ring shrunk around string-like DNA will form an elliptic curve rather than a perfect circle, because it touches the end of the string-like DNA. In contrast, ring-like DNA induces a circular contractile ring during shrinking (Fig. 6a). The change in DNA shape determines the shape of a contractile ring after it shrinks. Jet flow vertical to the shrinking surface of the contractile ring is generated inside the x-z plane in the case of string-like DNA, while the jet is in a small circle for ring-like DNA (Fig. 6b). Because the viscosity of water generates the shrinking flow around a contractile ring, the change in the contractile ring shape as it shrinks varies the jet flow inside the cell. The changed flow field carries the protein generated by DNA to a different displacement (Fig. 6b). This principle that the change in the DNA shape varies the displacement of proteins leads to the specific distribution of several components inside the cell, including ribosome, endoplasmic reticulum, lysosome, Golgi body, and mitochondria.

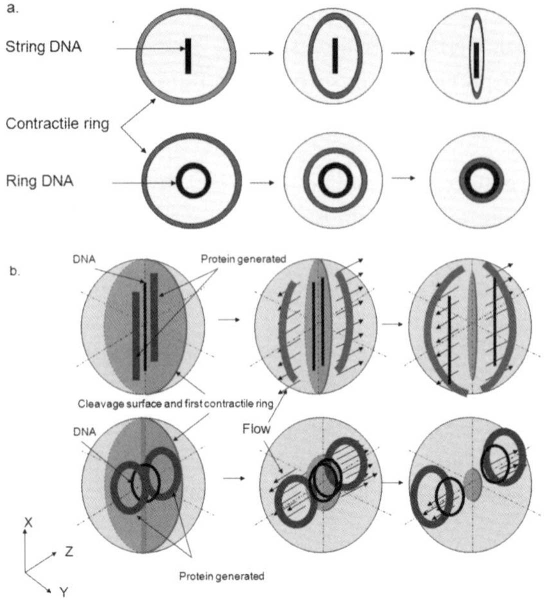

Fig. 6. Three-dimensional relation between cleavage plane and flow.

The division of long string-like DNA into a multi-chromosome system induces the next macroevolution from a microorganism to a multi-cellar system having complex concavities and convexities. The driving force for generating stably such complex concavities and convexities is an extension of that in Fig. 6b. Change of the spatial distribution pattern of genotypes, resulting from the division of DNA, promotes the evolution of the spatial displacement of multiple cells. (Proteins in Fig. 7b also include new contractile rings. String-like DNA generates the new contractile ring on the right plane for the next cleavage, whereas ring-like DNA carries the contractile ring to a small incorrect area. We will further discuss the mechanism of the self-organizing processes, including macroevolution, in the near future.)

References

1. I. Prigogine and D. Kondepudi, Thermodynamique, Editions Odile Jacob (1999).
2. P. Lowrey and J. Takahashi: Mammalian circadian biology: eluicidating genome-wide levels of temporal organization. Annual Review of Genomics and Human Genetics, 5, 407-441 (2004).

3. K. Naitoh, Stochastic Determinism underlying Life, Artificial Life and Robotics, 13, 10-17 (2008).
4. K. Naitoh, Physics underlying Topobiology, Proc. of 13th Int. Conf. on BioMedial Engineering, (ICBME), Singapore. Springer-Verlag (2008).
5. M.F Bear, B.W. Connors, and M.A. Paradiso, Neuroscience, Lippincott Williams & Wiklins Inc. USE, (2007).
6. K. Takahashi and S. Yamanaka, Induction of Pluripotent Stem Cells from Mouse Embryonic and Adult Fibroblast Cultures by Defined Factors, Cell 126, 663-676 (2006).
7. Y. Jin, Identification of L4 as a Novel Downstream Gene of Oct4 and Its role in Promoting Differentiation of Mouse Embryonic Stem Cells, Proceedings of international symposium on induced pluripotent stem cell research frontier and future, Kyoto 14-15 (2008).
8. J.A. Schumpeter. Business cycles: A theoretical, historical, and statistical analysis of the capitalist process, (McGraw-Hill, New York, 1-1095, 1939).
9. N. Kondratieff. The long wave cycle, (Richardson and Snyder, 1-138, 1984).
10. K. Naitoh, Cyto-fluid Dynamic Theory, Japan Journal of Industrial and Applied Mathematics 18-1, 75-105 (2001).
11. K. Naitoh, Onto-biology: Inevitability of Five Bases and Twenty Amino-acids, Proc. of 13th Int. Conf. on BioMedial Engineering, (ICBME), Singapore. Springer-Verlag (2008).
12. K. Naitoh, Gene Engine and Machine Engine, Springer-Japan, 1-251, (2006).
13. K. Naitoh, Engine for cerebral development, Artificial Life Robotics, 13: 18-21 (2008).
14. H. Yanagawa, How Life Originated. TBS Britannica.1-466 (1991).
15. M. Eigen, Steps toward Life, Oxford University Press, pp.1-173 (1992).

SYNTHESIS AND PHOTOCATALYTIC PERFORMANCE OF $CE/SIO_2/TIO_2$ WITH NOVEL MICROSPHERICAL-LIKE FAVEOLATE STRUCTURE*

YINGCHUN MIAO
Faculty of Chemical and Life Sciences, Qujing Normal University, Qujing 655000, P. R. China

LIANG MIN†
Department of Biology and Chemistry, Baoshan University, Baoshan 678000, P. R. China

ZHONGBIAO ZHAI
Kunming Metallurgy Research Institute, Kumming 650001, P. R. China

JIAQIANG WANG
Department of Applied Chemistry, Key Laboratory of Medicinal Chemistry for Natural Resource, Ministry of Education, Yunnan University, Kunming 650091, P. R. China

Mesoporous $Ce/SiO_2/TiO_2$ sample with amourphous framework and micropherical-like structure were prepared via a solvothermal treatment employing mango peels as template, and the catalyst was characterized by a combination of various physicochemical techniques, such as N_2 physisorption, diffuse reflectance UV–vis, X-ray diffraction and FT-IR. The photocatalytic activities of the samples were tested by the Photodegradation of dyes under solar light irradiation. The results of dye decomposition experiment indicate that the as-synthesized sample could efficiently degrade gentian(97%) and Rhdamine B(80%) under sun light.

1. Introduction

To use solar radiation efficiently for the reduction of environmental toxins on a local and global scale [1], a photocatalyst with high reactivity under solar light is required. The vast majority of photocatalytic investigations employ TiO_2, due to its stability and low cost. To develop more light efficient catalysts, it is necessary to develop new photocatalysts capable of working in the under sun light. A number of studies have been carried out in attempts to obtain this goal. One of them is Ce ions doped TiO_2 which was developed to photodegrade dye molecule (X-3B) [2]. In our past works, Ce/FSM-16 exhibit significant photoactivity the degradation methyl violet, safranine T, fuchsin basic and methylthionine chloride under solar light. Therefore, it was suggested a good possibility to apply Ce ions to dope mesoporous titania–silica mixed metal oxides synthesized by biotemplating technique as a photocatalyst. Herein, for the first time we report that Ce ions incorporated titania–silica mixed metal oxides ($Ce/SiO_2/TiO_2$) using mango peels as template acts as an efficient photocatalyst for the degradation of two organic dyes: Gentian violet (GV) and Rhdamine B (RB) under solar light. In order to enable a comparison, parallel experiments were performed to monitor the photocatalytic behavior of degussa P25 TiO_2.

2. Experimental

2.1. *Synthesis of $Ce/SiO_2/TiO_2$ microsphere with faveolate structure*

The dehydrated peels were simply dipped into a closed vessel containing a solution of 5 mL TTIP, 10 mL tetraethylorthosilicate (TEOS), and 1g $Ce(NO_3)_2{\cdot}3H_2O$ in 100 mL 2-propanol for 12 h. The above solution was added with 20 mL acid solution. After a desired period of time, the peels were washed with distilled water, dried in air at room temperature. The materials were then heated to 450°C in an oven in air. Yellow products were obtained.

*This work is supported by Jiaqiang Wang.
†Work partially supported by the Scientific Research Fund Project of Education Department of Yunnan Province (Project 09y0392), and Baoshan University Science Foundation (Project 08B009k).

2.2. *Photocatalytic activity measurements*

The reactions of the photocatalytic decolorization of two dyes (RB and GV) were carried out in a 50 cm^3 glass beaker containing 25 cm^3 of a model solution and 25 mg of photocatalyst under direct solar light between 10 a.m. and 2 p.m. where the solar intensity fluctuations were minimal. Dye solutions were stirred in the dark for 24 h and make sure no change in dye concentration due to adsorption before photocatalysis. The control experiments were also conducted under solar light without catalyst to measure any possible direct photolysis of these dyes. Samples were analyzed after centrifugation (1500 rpm for 10 min). The progress of photocatalytic decolorization was measured by SHIMADZU UV-2401PC photometer.

3. Final Results and Discussion

3.1. *Structural analysis of the catalysts*

Wide-angle X-ray diffraction was performed in the range of 5–90° in order to assess the crystallinity of the synthesized catalysts. As we can see from Fig.1, there are no observable lines corresponding to crystalline silica for Ce/TiO_2/SiO_2. This indicates that silicon in the prepared samples exists as an amorphous Si oxide species. However, the diffraction peaks of TiO_2 anatase crystalline loom. This significantly accords with that reported in Ref. [3], where crystalline TiO_2 was observed mass ratio of Ti/Si>0.5. Moreover, it is worth mentioning that no peaks corresponding to the transition metal oxides were observed. This indicates that our cerium oxides were well incorporated into the titania–silica mixed metal oxides framework.

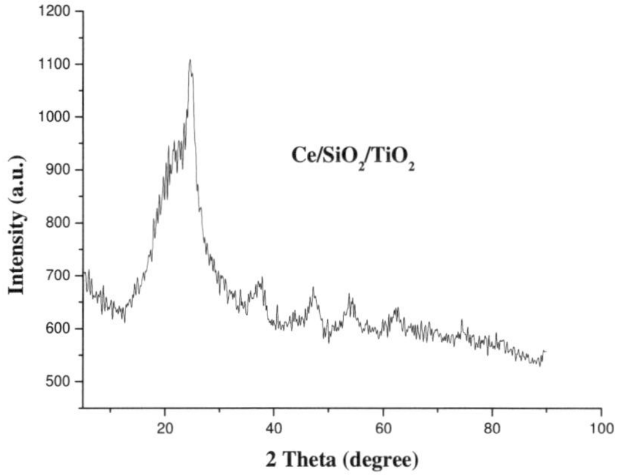

Fig. 1. XRD patterns of Ce/SiO_2/TiO_2 using mango peels as template.

Figure 2 shows the N_2 adsorption/desorption isotherms of the synthesized Ce/SiO_2/TiO_2. The sample show typical IV adsorption/desorption isotherms. It indicates the mesoporous structure of prepared materials. Moreover, this result also indicates that a large amount of Ce was grafted into titania–silica mixed metal oxides while retaining a good mesostructure. The surface area, pore volume, and average pore diameter of Ce/SiO_2/TiO_2 are 281 m2 /g, 0.38 cc/g, and 6.9 nm, respectively. Obviously, the incorporation of Ce and Si into the titania framework significantly improved the surface area. Meanwhile, the pore size distribution of Ce/SiO_2/TiO_2 is irregular. The slightly rises of these textural properties and irregular pore channels could be due to a large amount of Ce and SiO_2 grafted into TiO_2 [4].

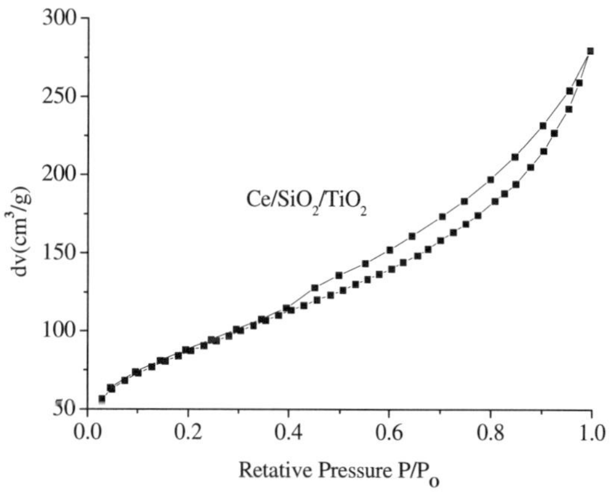

Fig. 2. N_2 adsorption/desorption isotherms of TiO_2 and Ce/SiO2/TiO_2 using mango peels as template.

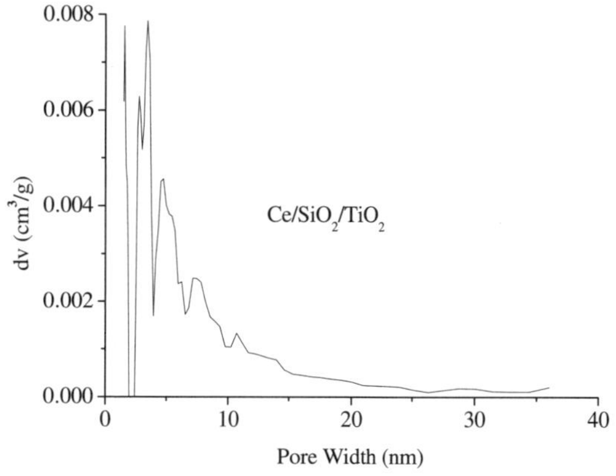

Fig. 3. Pore size distributions of TiO_2 and Ce/SiO_2/TiO_2 using mango peels as template.

Figure 4 presents FTIR spectra of synthesized $Ce/SiO_2/TiO_2$ between 400 and 4000 cm-1. The peaks at 1626 cm^{-1} and 3118 cm^{-1} which can be attributed to the surface-adsorbed water and hydroxyl groups were observed form the $Ce/SiO_2/TiO_2$ [5-8]. Moreover, it has been reported that a strong band at 1090 cm^{-1} and two weak bands at 808 cm^{-1} and 472 cm^{-1} belonged to asymmetric stretching, symmetric stretching and bending modes of bulk Si–O–Si in the $Ce/SiO_2/TiO_2$ structure compared with that of TiO_2, respectively [9]. However, after the incorporation of metal ions into the silicate, the shift of the band from 1090 to 1059 cm^{-1} was observed. Furthermore the peak at 949 cm^{-1} attributed to the Si–O–Ti vibration band found in titania–silica mixed oxides with a mesostructure and anatase [10] has not been found in our sample. The two new bands at 2344 and 1380 cm^{-1} was observed for $Ce/SiO_2/TiO_2$ which are the characteristic bands of CO_2 [9].

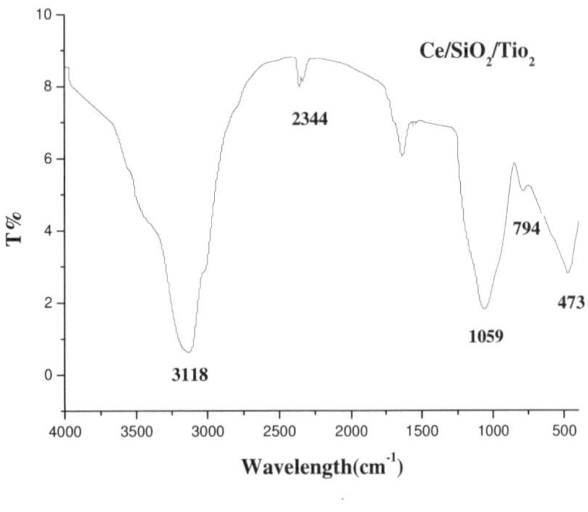

Fig. 4. FT-IR spectra of $Ce/SiO_2/TiO_2$ using mango peels as template.

The morphologies of the mango peels as template and the resulting samples were investigated by scanning electron microscopy. A typical SEM image in cross section through mango peels is presented in Fig. 5. Besides, mango peels also acts as a template to produce $Ce/SiO_2/TiO_2$ with a structure mirroring that of the original peels except for the shrinkage in size due to calcinations in Fig. 6. However, when a large amount of Ce and SiO_2 grafted into TiO_2 without changing other conditions, the particles in the prepared sample were no longer well-defined sphere-like. As shown in Fig. 6c, the particles seemed to be "broken" into all kinds of spherical parts (Fig. 6c). These spherical parts further self-aggregated to architectures. It was also found that some spherical parts were faceted, as if they had been "chopped" by a knife in Fig. 6d. The sizes of these spherical parts decreased to 10-25 um comparing with the images of mango peels. The above analysis demonstrated that the size and microstructure of the titania samples could be adjusted through changing the species of the templates and the morphology of sample could be destroyed when a large amount of Ce and SiO_2 grafted into TiO_2.

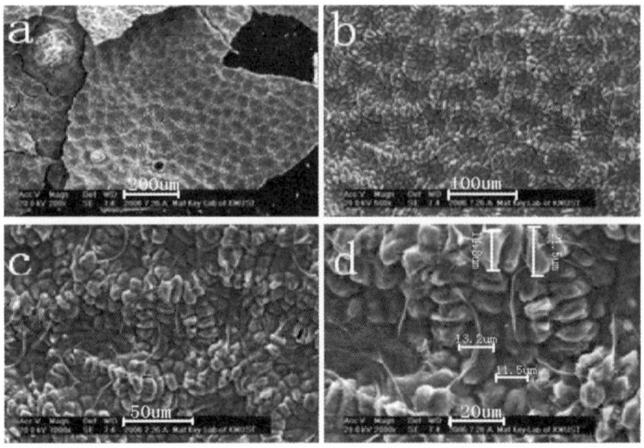

Fig. 5. SEM images of mango peels.

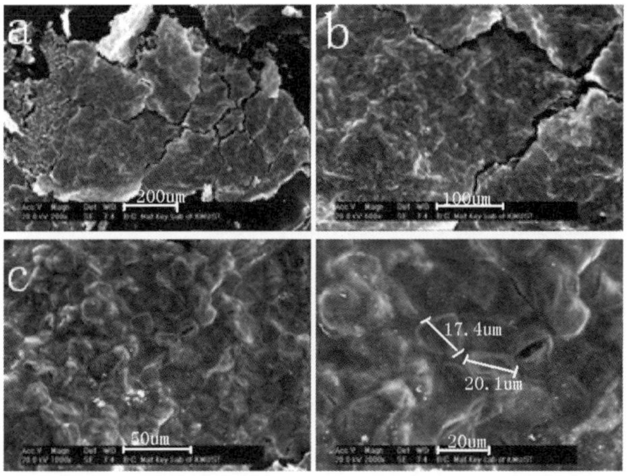

Fig. 6. SEM images of $Ce/SiO_2/TiO_2$ using mango peels as template.

The diffuse reflectance UV-Vis spectra of synthesized $Ce/SiO_2/TiO_2$ were given in Figure 7. The $Ce/SiO_2/TiO_2$ sample showed increased absorption over a broad range 400–700 nm compared with degussa P25 TiO_2. This justifies why these photocatalysts can be potentially good candidates for performing photochemical reactions in solar light.

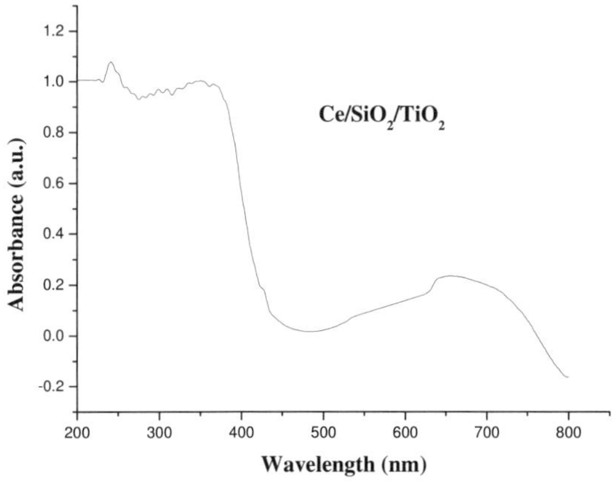

Fig. 7. UV diffuse reflectance spectra of Ce/SiO$_2$/TiO$_2$ using mango peels as template.

3.2. *Photocatalytic activity*

After reaching adsorption equilibrium (24 h in the dark), the variations for the adsorption yields of the studied dyes are summarized in Table 1. However, Ce/SiO$_2$/TiO$_2$ exhibited higher adsorption capacity (43%, 58%) for the two dyes due to its relatively higher surface area. The effects of direct photolysis of dyes in the absence of photocatalyst without photocatalyst are less than 5% and negligible. Nevertheless, the decreases of the dye concentration because of the adsorption and direct photolysis should be deducted in order to obtain the real photodegradation yield which is defined as:

$$\text{photodegradation yield} = \frac{(C_o - C_a - C_b)}{C_o} \times 100 \qquad (1)$$

Co is the initial concentration, Ca is the concentration after photodegradation under UV light, and Cb is the decrease concentration because of the direct photolysis.

Based on the above definition, the photodegradation yields of the two dyes using the obtained samples as photocatalysts are calculated and also summarized in Table 1. It illustrates that the photoactivity of Degussa P25 for the two dyes is negligible under solar light. By contrast, it indicates that the obtained Ce/SiO$_2$/TiO$_2$ samples exhibited very high photodegradation yields for all of the two studied dyes (71, 97%). Very interestingly, Ce/SiO$_2$/TiO$_2$, yielded discernible activity in the photodegradation of GV and RB when compared with that of mesoporous titania with a crystalline framework, so Ce and silica seemed to be essential for the photodegradation of dye using Ce/SiO$_2$/TiO$_2$ catalysts. The high efficiency of Ce/SiO$_2$/TiO$_2$ may be explained as follows: Firstly, Ti/Si materials would take advantage of both TiO$_2$ (an n-type semiconductor and active catalyst support) and SiO$_2$ (high thermal stability and excellent mechanical strength) [11]. Secondly, Ce/SiO$_2$/TiO$_2$ still has monodispersed, huge pore volume and mesoporous structures [12] which are excellent activity for catalysis. Thirdly, the reasonable activity of Ce/SiO$_2$/TiO$_2$ is due to not only the presence Ce ions and the interactions between cerium, titanium, and silica but also the special amorphous mesoporous structure of Ce/SiO$_2$/TiO$_2$. Considering the structure of dyes in Table 2, it is apparent that the less efficient photodegradation of Rhdamine B over the synthesized samples may be due primarily to the stable anthraquione structure of Rhdamine B. These findings are not clearly understood yet, and further experiments are underway.

Table 1. Summary of direct photolysis yield, adsorption yield, and photodegradation yield of Ce/SiO$_2$/TiO$_2$ under solar light irradiation.

dyes	catalyst	Direct photolysis yield (%)	Adsorption yield (%)	Photodegradation yield (%)
GV	P25 TiO$_2$	3	9	5.6
	Ce/SiO$_2$/TiO$_2$		58	97
RB	P25 TiO$_2$	4	10	0
	Ce/SiO$_2$/TiO$_2$		43	71

Table 2. Description of dyes.

dyes	Molecular structure
GV	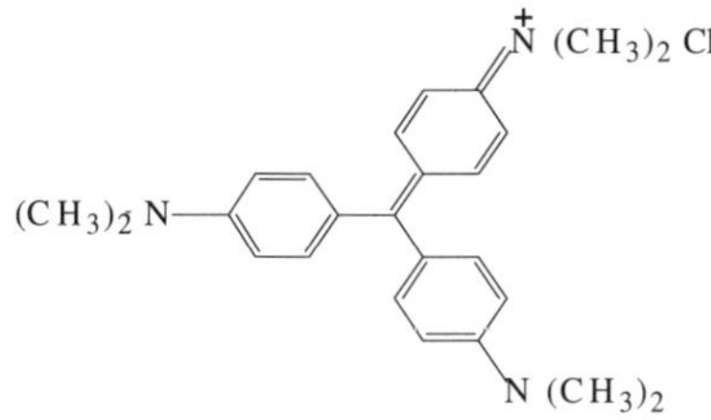
RB	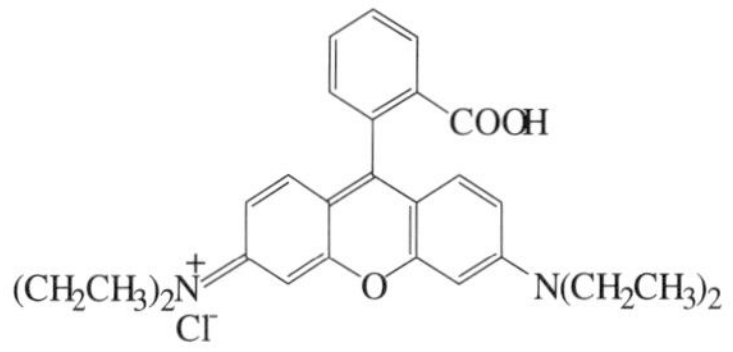

Acknowledgments

The authors thank Scientific Research Fund Project of Education Department of Yunnan Province (project 09y0392), Science Foundation of Baoshan University (Project 08B009k) for financial support.

References

1. M. Anpo, *Stud. Surf. Sci. Catal.* **130**, 157 (2000).
2. X. Zhang, F. Zhang, K.Y. Chan, *Appl. Catal. A: Gen.* **284**, 193 (2005).
3. K. Nagaveni, G. Sivalingam, M. S. Hegde, G. Madras. *Appl. Catal. B: Environ* **48**, 83 (2004).
4. J. Li, S. Liu, Y. He, J. Wang, *Micropor. Mesopor. Mater.* **115**, 416 (2008).
5. J.C. Yu, L. Zhang, Z. Zheng, J. Zhao, *Chem. Mater.* **15**, 2280 (2003).
6. T. Peng, D. Zhao, H. Song, C. Yan, *J. Mol. Catal. A: Chem.* **238**, 119 (2005).
7. G. Soler-Illia, A. Louis, C. Sanchez, *Chem. Mater.* **14**, 750 (2002).
8. K.E. Karakitsou, X.E. Verykios, *J. Phys. Chem.* **97**, 1184 (1993).
9. A. Boccuzzi, A. Chiorino, *J. Phys. Chem. B.* **104**, 5414 (2000).
10. G. J. Wilson, G. D. Will, R. L. Frost, S. A. Montgomery, *J. Mater. Chem.* **12**, 1787 (2002).
11. X. Gao, I.E. Wachs, *Catal. Today.* **51**, 233 (1999).
12. F. Chang, W. Li, F. Xia, Z. Yan, J. Xiong, J. Wang, *Chem. Lett.* **11**, 1540 (2005).

LIGHT TRANSMISSION IN FLUIDIZED BED

E. SHAHBAZALI
Chemical and Petroleum Department, Sharif University of Technology, PO Box: 11365-8639, Tehran, I.R. IRAN

A. A. SAFEKORDI
Chemical and Petroleum Department, Sharif University of Technology, PO Box: 11365-8639, Tehran, I.R. IRAN

N. AFRASIABI
Technical Research and Development Department, Hampa Engineering Corporation, Tehran, I.R. IRAN

Design of fluidized bed requires knowing the interfacial area between two phases and consequently solid holdup in the bed. On interfacial area measurement in fluidized bed with gas has been worked, but light transmission technique has been used less. Therefore, in current research the possibility of using this technique and its accuracy are investigated. By using fine solid with equal shape and diameter and installing an optical system, the absorption of light during the time of fluidization has been measured. Results indicate that this method which its validity has been proved in the gas-liquid system, by different reasons has less application in gas-solid system.

1. Introduction

In the last decade or so, application of fluidized bed has been important to various chemical industrial processes such as in pharmaceutical, metallurgical and mineral processing, in heat transfers, catalytic cracking, pyrolysis, combustion, etc [1].

Therefore, an optimum design of these reactors and related processes needs to know the interfacial area between two gas solid phases. According to the interfacial area, solid holdup can be calculated. Accordingly, many researches on solid holdup obtaining in a fluidized bed with gas have been done such as: measurement with a dual-electroresistivity probe [2], measurement with a optical fiber probe [3], measurement by using ultrasonic technique [4], [5], measurement by using momentum probe [6], measurement by using γ-ray [7], [8].

Nevertheless, light transmission technique has been used less. Therefore, in the current research the possibility of using of this technique to achieve solid holdup in a fluidized bed and its accuracy are investigated. Of course, fluidizing solid in gas phase is so difficult and it encounters problems such as gas improper distribution, channeling, bubbling and other non-uniformities so as a first step of such study, it will adequate to examine the problem without the complication of chemical kinetics. The paper also presents results of an experimental investigation on the transmission of light through a fluidized bed.

1.1. *Light transmission technique*

Light transmission is a well-developed technique for determining particle size in dispersions [9], [10], [11], and it has often been adopted for measurement of interfacial area in liquid-liquid and gas-liquid dispersions [12], [13], [14], [15]. The theory for the attenuation of the light beam has been given by a number of workers [9], [12]. Mclaughlin & Rushton (1973) [16] have published a numerical confirmation of the theory for polydisperse systems. And the same result has been obtained more simply by Curl (1974) [17].

When a flat bed is irradiated, the light can be divided into many portions. Part of the light will be reflected by the front wall and lost. The rest of the light energy enters the bed and can either be absorbed or reflected by the particle or transmitted through the back wall. A minute portion of the photons reflected by the particles may be lost through the front or back walls [18].

The portion of the parallel beam that passes through the transparent suspension is scattered by the particles of the dispersed phase by reflection, refraction, and diffraction. The angular distribution and intensity of the scattered and transmitted light for dilute suspension (no multiple scattering) has been calculated by Mie [19] and is thoroughly discussed by Sinclair [20]. For more concentrated suspensions (particles closer than about 10 diameters to each other), the above laws no longer hold because of multiple scattering and Chu [21] has shown the difficulties arising in trying to solve for this case.

Clarke [22], Vermeulen [23], and Trice [14] have measured the amount of both transmitted and forward scattered light passing through concentrated dispersions and have empirically correlated their results with the interfacial area using photographic methods. These correlations depend on the particular geometry of the optical system used and the refractive indices of the two phases.

A more satisfactory technique with a firm theoretical background has been indicated by Rose [9]. In this, the photocell arranged at such a large distance from the region of scattering that the solid angle subtended between them is small. Scattered light passes outside the photocell and is lost and only that part of the incident parallel beam which passes through the dispersion without meeting any obstacle is recorded. In this manner, the dispersed phase, as viewed by the photocell, appears as an assembly of black discs and the amount of light received does not depend on the refractive indices of the two phases, or indeed on whether the dispersed phase is opaque or not. The amount of light transmitted to the photocell in this manner depends only on the projected area of the dispersed phase.

As shown by the Mie [19] theory, for particles large compared with the wave length of the incident light, the scattering cross-section of a particle of the dispersed phase is twice its projected area. This result may also be deduced on the Babinette principle [24] which states that a disc scatters by diffraction an amount of light equal to that incident upon it. However, for large scattering particle such as were encountered in this work (particle diameter>0.1 mm), diffracted light is scattered at such a small forward angle that it is practically impossible to avoid its reception by a photocell placed even at vast distances from the dispersion, a fact which has been discussed by Walton [25].

So

$$I_0/I = A_0/A \tag{1}$$

In this theory, the intensity of transmitted light is compared with the intensity of incident light (when the bed consists of only the gas phase) therefore; the effect of reflected light by the glassy wall disappears.

The bubbles are essentially transparent and form the most significant contribution to light transmitted through the bed. The bubble sizes are dependent on gas flow rate and particle characteristics, which are therefore important variables that will affect the performance of a fluidized bed.

So in this method it is better to let the bed be in bubbling fluidization state than in smooth state, because in smooth fluidization state the bed expands uniformly and the average density of the bed is high, so, the intensity of transmitted light is very little [18].

2. Experimental Setup

Fig. 1 shows a schematic representation of a gas-solid fluidized bed where the light transmission investigation was carried out. The column walls were made of 5mm thick glass. The height of the glass bed is 1m and the cross section area is square (8cm×8cm).

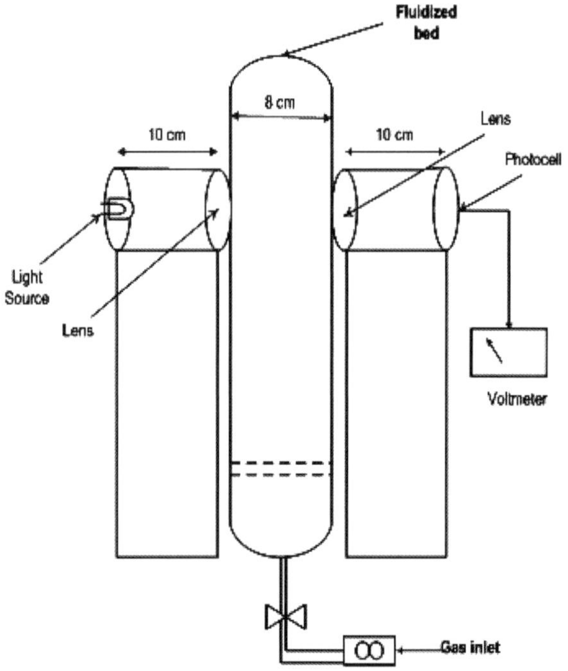

Fig. 1. Schematic representation of setup.

This fluidized bed consists of two sections: 1) the compressed gas enters the bed and traverses 20cm of the bed length so it becomes approximately uniform and passes all over the cross section. 2) In this section, gas passes through a perforated plate distributor of aluminum type (81 pores with 2mm in diameter) to fluidize the solid particles.

A compressor compressed the air that used as a fluidizing medium. Air from the top of the bed exited into atmosphere. The solid particles used were Flix weld. The physical properties of particles are presented in Table 1. According to the particle diameter and density, solid particles fall into Group B of Geldart's classification [1]. Two convex lenses of focal length equal to 10cm and a photocell and a light source were placed as shown in Fig. 1. The properties of the photocell used are presented in Table 2. The used light source was a hyperlight lamp with 4mm o.d. in a chamber provided with an opening that suitable as a point light source.

Table 1. Properties of particles.

Solid name	Latin name	Mesh No.	Mean diameter *[mm]*	Density *[kg/m³]*	Solid mass *[kg]*
Flix weld	*Sisymbrium Sophia L*	50	0.3-0.6	1300	0.1-0.2

The voltage input to the light bulb was kept constant (3volt, direct current) by a voltage regulator. This light source was placed in the focus of a first lens to produce a parallel incident beam. Light which passed through the bed without being absorbed was directed on to a photocell by placing photocell in the focus of another set of collimator lens. To minimize interfering, reflecting light and loss of light irradiation, two dark cylinders surrounded the light source and first lens, also photocell and second lens. The photocell output was measured by a voltmeter. Because of the fluctuation of voltage obtained from voltmeter, an electrical circuit included a 1MΩ resistance and a 10μF capacitor, as a smoother was used.

Table 2. Photocell identification.

Opening slit	Dimensions	Stand rod length
0-8 mm adjustable	53 mm x 70 mm	13 cm x8 mm

Changes in temperature and humidity influence on photocell so the voltage from voltmeter varies and in the different temperature shows different digits, accordingly the temperature and humidity of surrounding had to be constant.

In this research, first to calibrate the photocell output, the parallel beam was allowed to pass through an empty fluidized bed. The irradiation received by the photocell gave a measure of the light that passed through the empty bed without being absorbed (I_0). The lenses and the photocell were fixed in a known height toward the column then in different gas flows the voltage of passed light through fluidized bed with 100gr solid in it has been measured. Then this work has been repeated in various heights of the lenses. This procedure has been done for 3 quantities, 100gr, 150gr, 200gr, of two kinds of solids. The behavior of the fluidized bed was observed visually.

3. Results and Discussion

In all the experiments performed the particles were observed to fluidize nonuniformly throughout the bed. The distribution of bubble size was characteristic of a freely bubbling bed. As the bubble moved up the bed and coalesced, voidage in the bed increased.

Correspondingly, the fraction of light transmitted through the bed, I, was found to increase with bed height, h. Fig. 2, 3, 4, 5 presented I/I0 as a function of height at three flow rate, 4.76, 5.67, 6.15, 6.41 lit/s using 100, 150, 200gr Flix weld particles.

These results show that very little light passed through the bottom of the bed above the distributor just as visually was observed that bulk density in different gas velocities at the bottom of the bed was high. The value of I reported here represent average of fluctuating measurements. The maximum errors in the measurement of I were estimated to be ±10%.

In all charts, the variation of I/I_0 with height was therefore highly non-linear and it can be seen that I/I_0 increased with increase in height of the fluidized bed and is deduced that flow regime was nonuniform. In beds of Geldart B solids, bubbles form as soon as the gas velocity exceeds u_{mf} (u_{mf} = minimum fluidization velocity and u_{mb} = minimum bubbling velocity) thus $u_{mf} \approx u_{mb}$.

Considering that flow regime in flow rate 4-7 lit/s is approximately between bubbling and turbulent regime, obtained points trend accord nearly with existing samples in papers and literature [1], [6]. (I/I_0 is proportional to gas holdup and $1-I/I_0$ is proportional to solid holdup in the fluidized bed.)

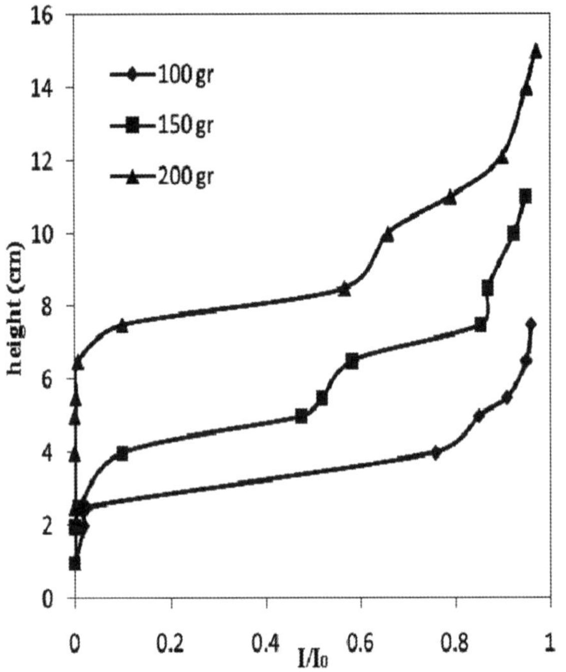

Fig. 2. Variation of I/I_0 with bed height for various solid masses at flow rate=4.76 lit/s, Flix weld.

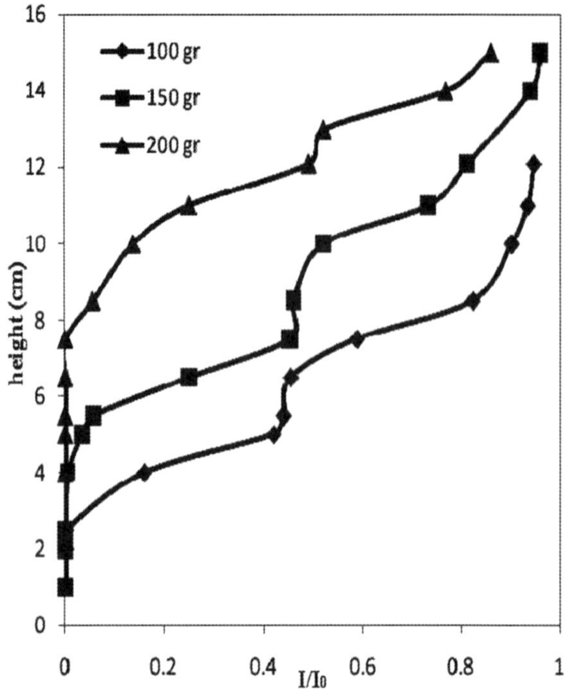

Fig. 4. Variation of I/I_0 with bed height for various solid masses at flow rate=6.15 lit/s, Flix weld.

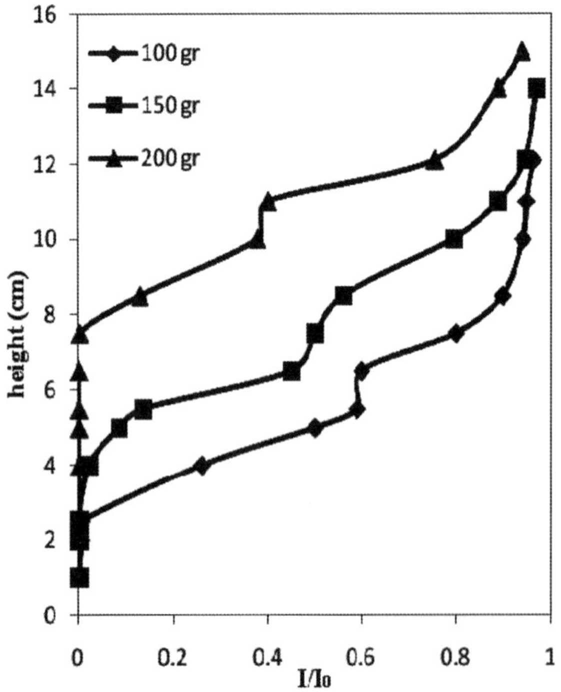

Fig. 3. Variation of I/I_0 with bed height for various solid masses at flow rate=5.67 lit/s, Flix weld.

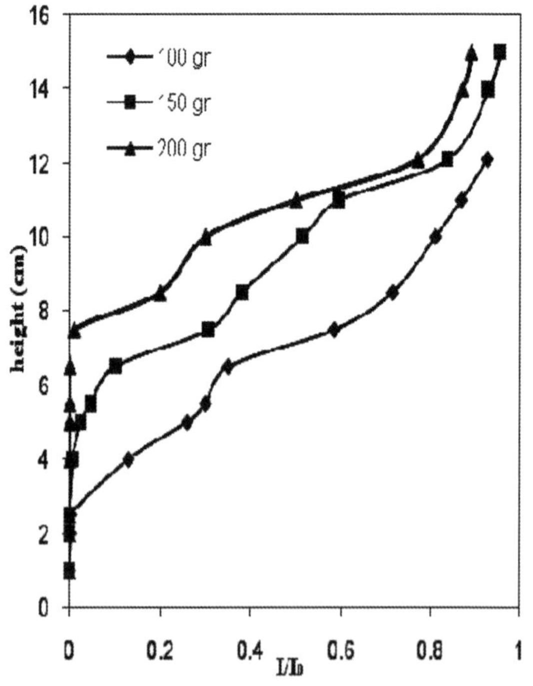

Fig. 5. Variation of I/I_0 with bed height for various solid masses at flow rate=6.41 lit/s, Flix weld.

4. Conclusion

The transmission of light through fluidized beds was found to be small. In a freely bubbling bed light is transmitted through the bed mainly through the bubbles. This transmitted light varies with bed height. Thus Optical techniques such as Light Transmission cannot be used for investigation of solids flow and total holdup in such opaque systems. One important reason could be non-uniformity in such systems .At best this technique should be used as local measurement in the fluidized bed from middle on top of bed not at the bottom of the column because the degree of densification at the bottom of the bed was high.

Notation

A_0 cross sectional area of light beam (m^2)
A free area at any cross-section in light beam (m^2)
I_0 incident light intensity
I transmitted light intensity

References

1. Kunii, D., Levenspiel, O., "Fluidization Engineering", 2rd ed., Butterworth Heinemann, (1991).
2. Matsuura, A., Fan, L., "Distribution of bubble properties in a Gas-liquid-solid fluidized bed", AICHE journal, vol. **30**, No.6, pp. 894-903 (1984).
3. Kitano, K., Fan, L., "Near-wake structure of a single gas bubble in a two-dimensional liquid-solid fluidized bed: Solids Holdup", Chemical Engineering Science, vol. **43**, No. 6, pp. 1355-1361 (1988).
4. Widyanto, M. R., Utomo, M. B., Kawamoto, K., Kusumoputro, B., Hirota, K., "Local gas holdup measurement of a bubble column using SONIA-ultrasonic non-invasive method", Sensors and Actuators A: Physical, vol. **126**, Issue **2**, pp. 447-454 (2005).
5. Zheng, Y., Zhang, Q., "Simultaneous Measurement of Gas and Solid Holdups in Multiphase Systems Using Ultrasonic Technique", Chemical Engineering Sci. **59** 3505-3514 (2004).
6. Won Kim, S., & Kirbas, G., H.Bi, C. jim Lim, J. R. Grace, "Flow behavior and regime transition in a high-density circulating fluidized bed riser", Chemical Engineering Sci. **59**. 3955-3963 (2004).
7. Bhusarapu, S., Al-Dahhan, M. H., Dudukovic, M. P., "Solid flow mapping in a gas-solid riser: Mean holdup and velocity fields". Powder Technology, **163**, 98-123 (2006).
8. Jin, H., Yang, S., He, G., Guo, Z., Tong, Z., "An Experimental study of Holdup in large-scale p-xylene oxidation reactors using the γ-raye attenuation approach", Chemical Engineering Sci., **60**, 5955-5961 (2005).
9. Rose, H., Lloyd, H., "On the Measurement of the size Characteristics of Powders by Photo-Extinction Methods I. Theoretical Considerations", J. Soc. Chem. Ind., **65**, 52 (1946).
10. Boll, R. H., Sliepcevich, C. M., "Evaluation of Errors of Optical Origin Arising in the Size Analysis of a Dispersion by Light Transmission", J. Opt. Soc. Am., **46**, 200 (1956).
11. Dobbins, R. A., Jizmagian, G. S., "Optical Scattering Cross Sections for Polydispersions of Dielectric Spheres", J. Opt. Soc. Am.,56, 1345 (1966).
12. Calderbank, P. H., "Physical rate processes in industrial fermenters. Part I: The interfacial area in gas/liquid contacting with mechanical agitation." Transactions of the Institution of Chemical Engineers, **36**, 443–463 (1958).
13. Lee, J. C., and G. W. K. Ssali, "Bubble transport and Coalescence in a Vertical Pipe, Joint Meeting on Bubbles and Foams", S1-6.1, Verfahrenstechnische Gesellschaft im VDI-Inst. Chem. Engrs., Nurnberg (1971).
14. Trice, V. G., and W. A. Rodger, "Light Transmission as a measure of Interfacial Area in Liquid-Liquid Dispersions", AIChE J., 2, 205 (1956).
15. Lockett, M. J., Safekourdi, A. A., "Light Transmission Through Bubble Swarms", AIChE Journal,vol. **23**, No. 3 ,pp. 395-398 (1977).
16. Mclaughlin, G. F., & Rushton, J. H., "Interfacial Area of liquid-liquid dispersions from Light Transmission Measurements", AIChE J., **19**, 817 (1973).
17. Curl, R. L., "Note on Light Transmission through a Polydisperse dispersion", AIChE J., **20**, 184 (1974).
18. Rizzuti, L., & Yue, P. L.," The measurement of Light Transmission through an irradiated fluidized bed", Chemical Engineering Science, 38, 8, 1241-1983 (1982).

19. Mie, P. Ann. Phys. Lpz., **33,** 1275 (1910).
20. Sinclair, D., Handbook of Aerosols, Ch. 7 (Washington, D.C.: Atomic Energy Commission) (1950).
21. Chiao-Min-Chu., J. phys. Chem., **59**, 841 (1955).
22. Clark, N. O. & Blackman, M., Trans. Faraday Soc., 44, 1 (1948).
23. Vermeulen, T., Williams, G. M. & Langlois, G. E., Chem. Engng Progr., **51**, 85 (1955).
24. Wood, R. W. Optical Physics, (London: The MacMillan Co.), (1911)
25. Walton, W. H. In Symposium on Particle Size Analysis, (London: The Institution of Chemical Engineers), p. 141 (1947).

RHIZOREMEDIATION OF POLYAROMATIC HYDROCARBONS BY *ARTHROBACTER* SP., A RHIZOSPHERIC ISOLATE FROM POPULUS DELTOIDS

SANDEEP BISHT[#], ANCHAL SOOD and PIYUSH PANDEY[*]

Department of Microbiology, S.B.S (P.G.) Institute of Biomedical Sciences and Research, Balawala, Dehradun, Uttarakhand, India

SHIVESH SHARMA

Department of Applied Mechanics and Biotechnology, Motilal Nehru Institute of Technology, Allahabad, (U.P.), India

N. S. BISHT

Department of Botany, H.N.B. Garhwal University, Pauri Campus, Pauri, Uttarakhand, India

[#]Presenting author

[*]Author for correspondence email: piyushgkp@rediffmail.com

Arthrobacter sp. STTX12 was found to utilize a wide range of polyaromatic hydrocarbon (PAH) and other hydrocarbons used in the study viz. anthracene, naphthalene, chrysene, benzene, toluene and xylene on MSB as sole carbon and energy source.It also had the plant growth promoting (PGP) attributes like phytohormone (IAA) production, phosphate solubilization, HCN and siderophore production and antagonism against *Macrophomina phaseolina* and *Rhizoctinia solanii* with percent inhibition of 46.6% and 52.4% respectively. Also, it was found to have positive chemotactic response for anthracene and naphthalene. Decrease in mean growth rate constant (K) was recorded when PAH was provided as sole source of carbon to *Arthrobacter* sp. STTX12, with respect to control supplemented with glucose.

Keywords: Chemotaxis, growth rate constant, PAH, PGP, rhizoremediation.

1. Introduction

Polycyclic aromatic hydrocarbons (PAH) contamination of soil, water or sediment is of immense concern because PAH can be toxic and/or carcinogenic. PAH are persistent pollutants of worldwide concern whose cleanup using conventional methods like incineration or relocation to specialized landfills is often prohibitively expensive. An alternative strategy for in situ PAH removal is biodegradation by microorganisms capable of metabolizing PAH.

Although bioaugmentation of soil with degradative bacteria has been largely unsuccessful in achieving significant aerobic PAH degradation in the field (McDermott *et al.*, 1989), efforts to biostimulate indigenous PAH degrading bacteria have been promising. The use of microorganisms for bioremediation of PAH-contaminated environments seems to be an attractive technology for restoration of polluted sites. The physiochemical remedial strategies used to remove these compounds from soil are inadequate and have many limitations, as being cost ineffective. Therefore, research is increasingly being focused on biological methods for the degradation and elimination of these pollutants from soil. One of the promising alternate is the use of rhizobacteria that possess wide variety of qualities governing their pollutant-catabolic and rhizospheric competences.

The beneficial plant-microbe interactions in the rhizosphere are determinants of plant health and soil fertility (Jeffries *et al.*, 2003). Soil bacterial isolates from rhizosphere which have been shown to improve plant health or increase yield, are usually referred to as plant growth-promoting rhizobacteria (PGPR) (Kloepper and Schroth, 1978; Suslow and Schroth, 1982). Biological control of plant diseases is gaining attention due to increased pollution concerns because of pesticides use for crop protection and development of pathogen resistance (Wisniiewski and Wilson, 1992).

The use of environmental friendly microorganisms has proved useful in plant-growth promotion and diseasecontrol in modern agriculture (Weller, 1988).

Kravchenko *et al.* (2002) stated that PGPR inoculation is a promising agricultural approach, which plays a vital role in plant protection, growth promotion or biological disease control (Dilantha *et al.*, 2006).

The present study was envisaged with the isolation of bacterial strain that not only has the ability to degrade PAH but also has the plant growth promoting (PGP) attributes for effective rhizoremediation utilizing the rhizosphere of *Populus deltoides* as soil inoculation system.

2. Material and methods

2.1. *Sampling method*

Root zone and rhizospheric samples were collected beneath each tree at the depth of 0 to 20cm to 20 to 40 cm with the ethanol disinfected shovel and fine roots hairs were also collected carefully. Approximately 50g soil were kept in a plastic poly bags and stored at 4°C prior to the microbiological analysis. Samples were collected in triplicates.

2.2. *Isolation of PAH and other hydrocarbons metabolizing bacteria*

Microorganisms in samples of soil (1.0 g) or fine roots with attached rhizosphere soil were suspended in 100ml sterile water and kept in incubator shaker at 120 rpm in 27°C for 24 h.

Following standing for 30 min, serial dilutions of the suspension were prepared in double distilled sterile water up to dilution 10^{-6}. Total culturable heterotroph were grown in minimal salt agar or Minimal salt basal medium (MSB),having 0.7g $(NH_4)_2NO_3$;0.1g K_2HPO_4;0.1g KH_2PO_4;0.5g $MgSO_4.7H_2O$;0.013g $CaCl_2.2H_2O$;0.0013g $FeSO_4.7H_2O$;2 % agar in 100 ml deionized water. Liquid hydrocarbon used for substrate analysis was provided as vapour phase (Na et al., 2005). Further these strains were checked for other polyaromatic hydrocarbon degradation by spray plate technique as described (Kiyohara *et al.*, 1983).

2.3. *Chemotaxis assay*

Chemotaxis was tested with drop assay (Grimm and Harwood, 1997). For this assay,40 ml of log phase culture were harvested and resuspended in 12 ml of chemotaxis buffer (100 Mm potassium phosphate, 20 mM EDTA, pH 7.0). A small amount of a test attractant was added to the center of a dish (10 ml). A chemotactic response of cells to the added compound resulted in the formation of a ring of turbidity near the center of the Petri dish.

2.4. *Growth study*

Growth studies of isolates in anthracene and naphthalene hydrocarbon were checked on minimal salt basal broth was supplemented with hydrocarbon to achieve final concentration 100mg/lt. Incubation was carried out in incubator shaker at 27°C with 160 rev/min. One control was also set up to evaluate the role played by the test bacteria in utilizing hydrocarbon for growth. Positive control was set up with 2% dextrose as carbon source. Growth on anthracene and naphthalene was assayed by measuring optical density (OD_{600} nm) with time interval of 3 hour and mean growth rate (K) was also determined as

$$K=3.322\log Z_t Z_0 / \Delta T$$

2.5. *Screening of isolates for indole–3–acetic acid (IAA), phosphate solubilization, HCN, siderphore production*

IAA production was determined by Salkowski's method. The bacteria were cultured in nutrient broth supplemented with tryptophan (0.1g lt^{-1}) for 24 to 48 h at 27°C till the stationary phase of growth supernatants were obtained after centrifugation of cell culture at 1000 x g for 5 min. The control was uninnoculated culture medium. 100µl Salkowski's reagent (0.5mM of $FeCl_3$ in 50 ml of 35% $HClO_4$) and 400µl phosphoric acid were added to the supernatant and incubated at 27°C for 30 min, after which the tubes were checked for appearance of pink colour which was indicative of positive IAA test (Rodriguez *et al.*, 2008).

The potential to solubilize relatively insoluble calcium phosphate were checked on Pikovaskay's agar (Subba Rao, 1982) and phosphate solublising index (PSI)were also calculated by following formula

$$PSI = Zonediameter / growthdiameter \ x \ 100$$

Production of siderophore was detected qualitatively on Chrome-Azurol agar (CAS) medium (Schwyn and Neilands, 1976). HCN production was checked culturing the organism on NAM supplemented with glycine. Filter paper dipped in picric acid in 1% Na_2CO_3 was attached to the lid of culture tube and after incubation for 48 h at 27°C, colour change was observed for HCN production (Miller and Higgins, 1970).

2.6. *In vitro antifungal activity of PGPR*

The in vitro mycelium growth inhibition of *Macrophomina phaseolina* and *Rhizoctonia solanii* by dual culture technique (Skidmore and Dickinson, 1976). The percentage growth inhibition was calculated using the following calculation:

$$\% Inhibition = [(R - r)/R \times 100]$$

Where, r is the radius of the fungal colony opposite the bacterial colony and R is the maximum radial growth of the fungal colony away from the bacterial colony.

3. Results

Eight bacterial strains were isolated from the rhizosphere of *Populus deltoides* growing at cement tile factory Dehradun, India on MSB supplemented with xylene, further these strain were checked for other PAH and hydrocarbon degradation. The isolate STTX12 was found capable of degrading all the hydrocarbons used in the study (Table 1), strain was identified as *Arthrobacter* sp. on the basis of its morphology and biochemical characteristics. The strain STTX12 showed positive chemotactic response for naphthalene and anthracene thus indicating valuable effect of chemotaxis on biodegradation activity. Further these strains were screened for *in vitro* PGP activity and STTX12 showed production of indole-3-acetic acid (IAA), excellent phosphate solubilisation with phosphate solubilization index (PSI) of 266, produce siderophore as evidence by producing orange halo–zone and HCN activity and it also exhibited a salt tolerance up to 7.0% and ph up to 11.0 (Table 2). The isolate STTX12 grew well in anthracene and naphthalene medium (0.5 mg lt^{-1}) concentration with mean K value of 5.6 h^{-1} for anthracene and 5.3 h^{-1} for naphthalene respectively While, the mean growth rate constant was maximum in glucose amended positive control (K value of 5.71 h^{-1}). Hence, it is clear that with 0.5mg/lt concentration of anthracene and naphthalene there is slight decrease in K of STTX12 (Fig. 1 and 2). This strain has antagonistic activity due to diffusible compound against *M. phaseolina* and *R.solani* with inhibition of 46.6% and 52.4% respectively.

The mycelia picked from zone of inhibition showed morphological deformities in hyphal cell viz. cell lysis and cell vacuolation. Degradation of anthracene and naphthalene was determined by quantitative estimation of residual substrate concentration using UV spectrophotometer.

Table 1. Growth test on different liquid and solid hydrocarbon.

S. No	Strains	Ant[a]	Nap[b]	Ben[c]	Tol[d]	Xyl[e]	Chr[f]
1	STTX1	–	+	–	+	+++	–
2	STTX2	–	–	–	–	+++	–
3	STTX4	–	–	–	+	+++	–
4	STTX5	–	–	–	–	+++	–
5	STTX6	–	–	–	+	+++	–
6	STTX7	–	–	–	+	+++	–
7	STTX8	+	–	–	++	+++	–
8	STTX12	+++	+++	+++	+++	+++	+++

a Anthracene, b Napthalene, c Benzene, d Toluene, e Xylene, f Chrysene
+++ Excellent growth; ++ Moderate growth; + Weak growth; – No growth

Table 2. Plant growth promoting activities of isolates.

Strains	IAA[1]	P solubilization[2]	Siderophore[3]	HCN[4]	NaCl[5] (7.0%)	pH[6] (11)
STTX1	–	–	–	–	–	–
STTX2	–	–	–	–	–	–
STTX4	–	+	–	–	–	–
STTX5	+	–	–	–	–	–
STTX6	–	++	–	–	–	–
STTX7	–	–	–	–	–	–
STTX8	–	–	–	–	–	–
STTX12	+	+++	+++	+	+	+

1 Indole-3-acetic acid production, 2 Phosphate solubilization, 3 Siderophore production, 4 HCN activity, 5 Salt tolerance, 6 pH tolerance
– No production; + Slight production; ++ Good production; +++ Excellent production

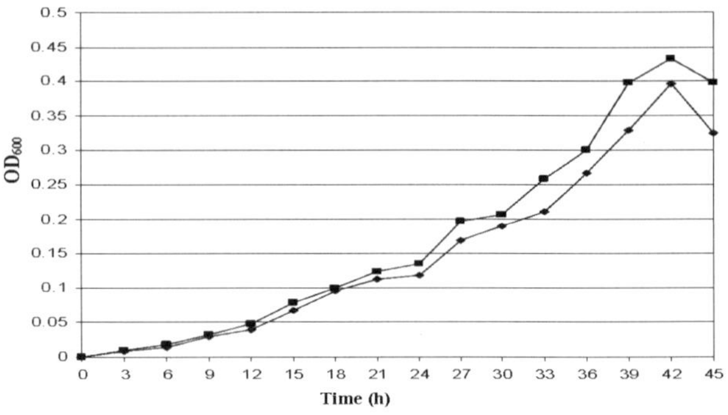

Fig. 1. Time–course analysis of growth of STTX12 with respect to control where the strains was established in monospecies culture in minimal broth medium supplemented with anthracene,concenteration of (0.5mg/lt).

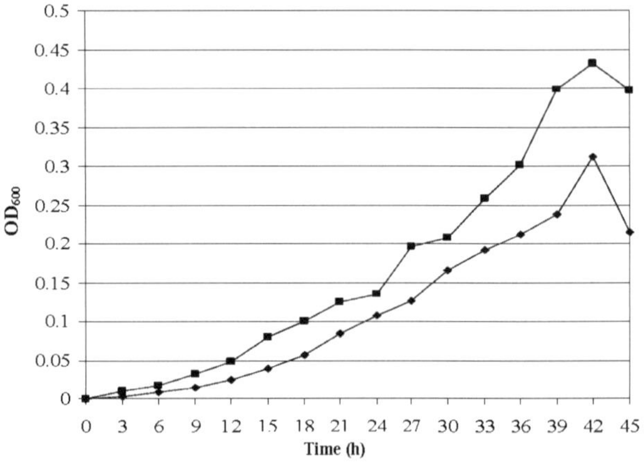

Fig. 2. Time–course analysis of growth of STTX12 with respect to control where the strains was established in monospecies culture in minimal broth medium supplemented with naphthalene of concenteration (0.5mg/lt).

4. Discussion and conclusion

The initial results of this study suggest that the isolate STTX12 have promising characteristics for rhizoremediation of PAH using *Populus* rhizosphere as inoculation system in soil.

The experiments described above were designed to select a plant-microbe pair for the efficient degradation of PAH. *P. deltoides* was selected because its root system is extensively branched and fast growing root system with ability to carry the high bacterial population. When rhizospheric strain is introduced together with a suitable plant, these well-equipped bacteria settle on the root together with the indigenous population, thereby enhancing the bioremediation process. In addition, such efficiently, pollutantdegrading bacteria exploit the growing root system and hence this would act as an injection system to spread the bacteria through soil. Subsequently, we selected bacteria from the rhizosphere of *Populus*, from the rare vegetation of a site heavily polluted with PAH. In conclusion, this study succeeded in isolation of strain which have broad substrate range to degrade PAH as well as facilitate the plant growth and suppression of disease by phytopathogen used in this study. It is expected that this report will prompt further screening of rhizospheric organisms as potential degrader and plant growth promoters for rhizoremediation, since it is a plant based in-situ phytorestoration technique it is proven to be economical, efficient and easy to implement under field conditions.

Acknowledgement

The authors are grateful to G. B. Pant Institute of Himalayan Environment and Development (GBPIHED), Ministry of Environment and Forest, Govt. of India, for grant received to carry out this study.

References

1. Dilantha F, Nakkeeran S, Yilan Z Biosynthesis of antibiotics by PGPR and its relation in biocontrol of plant diseases. *PGPR Biocontrol Biofert.* pp. 67-109, 2006.
2. Grimm AC, Harwood CS Chemotaxis of Pseudomonas sp. to the polyaromatic hydrocarbon Naphthalene. *Appl. Environ. Microbiol.* 63: 4111-4115, 1997.
3. Jeffries S, Gianinazzi S, Perotto S, Turnau K, Barea JM. The contribution of *arbuscular mucorhizal* fungi in sustainable maintenance of plant health and soil fertility *Biol. Fertil.* Soils 37: 1-16, 2003.
4. Kiyohara H, Sugiyama M, Modello, FJ, Gison, DT, Yano, K. plasmid involvement in the degradation of polycyclic hydrocarbons by a *Beijrinckia* sp. *BBRC.*111: 939-945, 1983.
5. Kloepper JW, Schroth MN. Plant growth promoting rhizobacteria on radish. In: Proceedings of the Fourth Conference Plant Pathogenic Bacteria (Ed Station de Pathogenic Vegetable ET (Phytobacteriologic) *INRA Angers* pp 876-882, 1978.
6. Kravchenko LV, Makarova NM, Azarova TS, Provorov NA, Tikhonovich IA Isolation and phenotypic characterization of plant growth

promoting rhizobacteria with high antiphytopathogenic activity and root colonizing ability. *Microbiology* 71: 444-448, 2002.

7. McDermott JB, Unterman R, Brennan MJ, Brooks RED, Mobley P, Schwartz CC, Dietrich D. Two strategies for PCB soil remediation: biodegradation and surfactant extraction. *Environ. Prog*. 8: 46-51, 1989.
8. Miller RL, Higgins BB. Association of cyanide with infection of birds foot trefoil by *Stemphylium* loti. *Phytopathology* 60: 104-110, 1970.
9. Na KS, Kuroda A, Takiguchi N, Ikeda T, Ohtake H, Kato J. Isolation and characterization of benzene tolerant *Rhodococcus opacus* strains. *J. Biosci.Bioeng*. 99(4): 378-382, 2005.
10. Rodriguez AH, Perez MY, Guerrero YA, Valle MGV, Lauzardo ANH. Antagonistic activity of Cuban native rhizobacteria against *Fusarium verticilides* (Sacc.) Nirenb. in maize (*Zee Mays* L.) *Appl. Soil Ecol*. 39: 180-186, 2008.
11. Schwyn B, Neilands JB. Universal assay for the detection and dertermination of siderophores. *Anal. Biochem*. 160: 47-56, 1987.
12. Skidmore AM, Dickinson CH. Colony interaction and hyphal interferance between Septoria nodorum and phylloplane fungi. *Trans. Brit. Mycol. Soc*. 66: 57-74, 1976.
13. Subba Rao NS. In: Advances in agriculture microbiology; *Oxford and IBH Publ. Co*. pp: 229-305, 1982.
14. Suslow TV, Schroth MN. Rhizbacteria of sugar beets effects of seed application and root colonization on yield. *Phytopathol* 72: 199-206, 1982.
15. Weller DM Biological control of soil borne plant pathogens in rhizosphere with bacteria. Ann. Rev. *Phytopathol*. 26: 379-407, 1988.
16. Wisniiewski ME, Wilson CL. Biological control of postharvest diseases of fruits and vegetables. *Rec. Adv. Hort*. Sci. 27: 94-98, 1992.

FORMULATION OF AN EFFECTIVE RHIZOBIUM BIOINOCULANT USING GREEN FLUORESCENT PROTEIN REPORTER SYSTEM

PIYUSH PANDEY*, SHRADHANJALI ARORA, ANCHAL SOOD and SANDEEP BISHT

Department of Microbiology, S.B.S (P.G.) Institute of Biomedical Sciences and Research, Balawala, Dehradun, Uttarakhand, India

D. K. MAHESHWARI

Department of Botany and Microbiology, Gurukul Kangri University, Haridwar, Uttarakhand, India

**Presenting author and author for correspondence, Email: piyushgkp@rediffmail.com*

The present work was undertaken to formulate an effective bioinoculant utilizing known plant growth promoting bacteria (PGPB). Rhizobium sp. T_2g was successfully tagged with reporter gene gfp utilizing their resistance against antibiotics as marker. Physical and chemical properties of different low cost carrier materials was determined including bagasse, charcoal, cocoa peat, rice husk, rock phosphate, sawdust and wheat bran as solid materials, and whey as liquid. Viability of Rhizobium sp. T_2g was estimated in different carrier materials. Whey, followed by wheat bran and rice husk proved to be efficient carrier materials for the bioformulation. Better growth enhancement of Cajanus cajan was observed with charcoal based bioformulations, followed by whey, sawdust, rice husk and wheat bran. The organisms were having excellent root colonization ability which is a desirable property in a suitable PGPR.

Keywords: Bioinoculant, Cajanus cajan, gfp, PGPR, Rhizobium sp., T_2g.

1. Introduction

Bacterial inoculant is a formulation containing one or more beneficial bacterial strains (or species) in an easy-to-use and economical carrier material, organic, inorganic, or synthesized from defined molecules. The inoculant is the means of bacterial transport from the factory to the living plant (Bashan, 1998). There are several plant growth promoting rhizobacteria (PGPR) containing inoculants currently commercialized. Concerns about the possible consequences of using increasing amounts of chemical pesticides have led to a strong interest in alternative strategies, like the use of microbial inoculants (biofertilizers, phytostimulators, and biopesticides) to ensure competitive yields and protection of crops. Thus the formulation and commercial production of any bioinoculant requires the integration of physical,chemical and biological parameters leading to both high target organism populations and long term survival of these microorganisms over time at less than optimum conditions.

Recent advances in molecular techniques are encouraging in that tools are becoming available to determine the mechanism by which crop performance has improved using PGPR and track survival and activity of PGPR in bioinoculants. Increasingly, research has concentrated on the development of marker genes for tagging a particular bacterial species of interest, so that the cell can be specifically identified and monitored. The *gfp* gene encoding green fluorescent protein from jellyfish *Aequoria victoria* is widely used as a marker because of its very useful properties, such as high stability, minimal toxicity, non-invasive detection and the ability to generate the green light without addition of external cofactors and without application of expensive equipment. The green fluorescent protein used as a biomarker allowed the detection, determination of spatial location and enumeration of bacterial cells from diverse environmental samples. In the light of above fact, the present work was envisaged to evaluate different low cost carriers for the formulation of an effective bioinoculant. The reporter gene, gfp was used as a marker to study viability of Rhizobium in low cost carriers viz. rice husk, wheat bran, charcoal, bagasse etc. therefore, monitoring the bioinoculant. Finally effective bioformulations were evaluated for seedling growth enhancement of Cajanus cajan.

2. Materials and Method

2.1. *Microorganisms*

The *Rhizobium* sp. T_2 was established as effective PGPR as it was known to solubilize inorganic phosphate, produce indole-3-acetic acid (IAA), exopolysaccharide, and also fix atmospheric nitrogen. The strain was maintained on yeast extract mannitol agar (YEMA) medium. *E. coli* S17-1 carrying plasmid pDS 15 (a suicide plasmid carrying a promoterless *gfp* gene fused with *Tn* 5) was used in molecular tagging experiment. The suicidal plasmid carries a gentamicin marker. The *E. coli* strain was maintained on nutrient agar medium and stored in refrigerator at 4°C.

Antibiotic sensitivity test was performed by the disc diffusion method (Kirby and Bauer, 1966).

2.2. *Gfp gene transfer*

The *Rhizobium* sp. T_2 resistant to Polymyxin B was tagged by conjugation with plasmid pED S15 containing a *Tn*5 *gfp*. *E. coli* S17-1 (pED S15) cells were grown in nutrient broth at 37°C for 48 h. The *Rhizobium* sp. T_2 cells were grown in yeast extract mannitol broth at 27°C for 48 h. 1500 µl of E. coli culture and 1500 µl of *Rhizobium* sp. T_2 culture were separately taken in 1.5 ml eppendrof tube and centrifuged at 10,000 rpm for 15 minutes at 4°C. The pellet of *Rhizobium* sp. T_2 was resuspended in 100 µl of sterile distilled water, which was mixed with the pellet of *E. coli*. The conjugal mixture was again centrifuged at 9000 rpm for 10 min and was given a subsequent washing. The supernatant was discarded and the conjugal mixture was put as a patch on nutrient agar medium (NAM): yeast extract mannitol agar (YEMA) plates (1:1) and allowed to dry. The plates were then incubated in an upright position at 270 C for 48-72 h.

The patch was scratched and resuspended in 9 ml sterile water blank, vortexed, serially diluted and plated on YEMA supplemented with gentamicin (50 $\mu g ml^{-1}$) and polymyxin B (50 $\mu g ml^{-1}$). The plates were incubated at 27°C for 48 h. The *gfp* marked Rhizobium sp. T_2 was detected as green fluorescent colonies on viable count plates after exposure to UV light. Colonies with maximum fluorescence were selected and used further in bioinoculant production.

2.3. *Bioinoculant formulation*

In the present investigation, the viability of *Rhizobium* transformants T_2g, was studied in eight different bioinoculant carriers, which were low cost, lignocellulosic, byproducts of agro based industries. These are bagasse, sawdust, cocoa peat, rice husk, wheat bran, charcoal, rock phosphate and cheese whey.

The *gfp* marked *Rhizobium* sp. T_2g was grown in yeast extract mannitol broth for 48 h at 150 rpm and room temperature (25 $\pm$ 2°C). The broth culture was utilized for inoculant preparation when cell concentration exceeded 1×10^7cells/ml (Daze *et al.*, 2000). The solid carrier materials were ground and air dried. Liquid as well as solid carrier materials were double sterilized in an autoclave (121°C for 30 min). 40 g of each, sterile, solid carrier materials, were packed in polyethene bags made up of 50-75 mm thick, low density, and flexible sheet. The polyethene bags were sealed leaving about 2/3 vacant space to give proper aeration to the inoculant. After the processed, autoclaved carrier materials were packed, the surface of the polyethene bags was sterilized around the point of injection. 5 ml of bacterial suspension was introduced aseptically with a hollow needle directly through the wall of the bag and the small hole was immediately covered with the help of selfadhesive tape. Immediately, after sealing, the inoculum and carriers were hand mixed by shaking or kneading between the fingers and stored at room temperature in dark to prevent from UV radiation and heat. For the preparation of liquid inoculants, 5 ml of microbial culture was added to 40 ml of sterile liquid carrier and then packaged into sterile bags.

2.4. *Plate count of viable cells*

Samples were collected from bioformulations after different time intervals in aseptic conditions. The colony forming units (cfu/g) inoculant were monitored periodically for a period of 60 days to determine the number of bacterial cells by viable count method, by inoculating suitable dilutions on prepared YEMA plates.

2.5. *Pot trial and root colonization studies*

Cajanus cajan (local variety) seeds were surface sterilized using 0.2% $HgCl_2$ (55 sec), absolute alcohol (5 min) followed by 6-7 washings with sterilized distilled water. Disinfected seeds were air dried overnight under a sterile air stream. To study the effect of bioformulation, disinfected seeds were suspended overnight in 60 days mature carrier materials. Single bacterized seeds were placed in 50 ml test tubes containing 25 ml Jensen's agar medium. Non bacterized seeds served as control. For root colonization studies, sterilized seeds were sown in pots having non sterile soil. The seeds were bacterized with 1 ml of YEM broth inoculated with different tagged bacterial cultures. Surface sterilized seeds of *C. cajan* (local variety) were sown in pots containing non sterile soil. The surface sterilized bacterized and non-bacterized seeds were sown separately in pots. The experiment was designed in two sets: a) T_2g– bacterized seeds, b) non bacterized (control seeds). Pots were watered routinely by sterile distilled water. The plants were uprooted after 15 and 20 days, and total plant length, root length, shoot length, and weight of root and shoot were measured separately. The experiment was performed in triplicate. To determine the viable count of inoculated *gfp* tagged strains, rhizospheric soil closely adhering to roots was removed, and suitable dilutions were plated on YEMA. The plates were further incubated at 28-30°C for 48-72 h colonies which appeared fluorescent green under UV light were scored.

3. Results and Discussion

3.1. *Molecular tagging of Rhizobium sp. T_2g by gfp Gene*

The wild type *Rhizobium* sp. T_2 was resistant for azlocillin (75 μg) and polymyxin B (50 μg). Resistance against polymyxin B was utilized for conjugation experiment. The transconjugants were selected on the basis of high fluorescence and resistance against gentamicin (pED S-15 characteristic) and polymyxin B. The fluorescence among transconjugants was variable. This might be because pED S-15 has promoterless *gfp* gene. Hence, insertion in constitutive gene or under the regulation of active promoter may result in high concentration of GFP and hence fluorescence. Earlier also workers have used *gfp* as reporter gene to monitor the survival of Rhizobium in soils (Saha *et al.*, 2001a: 2001b; Saha and Kapadnis, 2002). A *gfp* marked strain with maximum fluorescence was selected for further studies and termed as T_2g.

3.2. *Survival of Rhizobium sp. T_2g in carrier materials*

The initial cell population of $\approx 10^9$ cfu/g first decreased to about $\approx 10^6$. Cells after 20 days, which stabilized further up to 50 days in charcoal. After 2 months, the population remained $\approx 10^5$ cfu/g of charcoal. The results indicate that charcoal is unsuitable to be used as carrier for T_2g, as the residual cell population was 100 times less from recommended population. Charcoal based carriers are also established and used widely for bioinoculant purpose. Several studies have reported the use of coal based inoculants as an alternative for peat based inoculants (Crawford and Berryhill 1983).

Analysis of the bagasse for viable count showed initial decline in cell population from 10^9 cfu/g to 10^6 cfu/g, which stabilized thereafter. Peat is supposed to be excellent carrier for bioinoculant and has been accepted worldwide for this purpose (Burton, 1976). However, unavailability of peat in tropical countries including India raised a concern for a suitable replacement. In view of this, we studied cocoa peat along with other carriers for the bioinoculant formulation. However, it proved to be an average carrier material as it retained only $\approx 10^6$ cfu/g inoculant cell population after 60 days, which was 1000 times lesser than initial population. The initial population of T_2g in wheat bran was $\approx 10^9$ cfu/g that remained constant up to 1 month and after 2 months, it was recorded $\approx 10^8$ cfu/g. Successful experiments on wheat bran have been carried out by Jackson (1991). Rice husk was a moderate carrier material with respect to viability of cells. Saha and Kapadnis (2001a) have reported that the transformant remained almost at constant level for about five weeks from the date of manufacture in paddy husk carrier. But in the present investigation, the cell population of $\approx 10^9$ cfu/g decreased to 10^7cfu/g after 30 days. It further decreased to 10^6cfu/g and remained constant thereafter. Sawdust was also found to be average carrier material. The viability of T_2g in sawdust was comparable to wheat bran. The initial population of $\approx 10^9$ cfu/g initially decreased to $\approx 10^7$ after 20 days and remained constant thereafter. The cheese whey was found as an excellent carrier and a suitable candidate for the formulation of liquid carrier. The initial population $\approx 10^9$ cfu/ml increased to 10^{11} cfu/ml after 30 days, which stabilized further. Thereafter, $\approx 10^8$ cfu/ml cell population was recovered after 50 days in cheese whey. Earlier also, Bissonnette *et al.*, (1986) have successfully used whey based bioformulation for *Rhizobium meliloti*. The strain T_2g was viable in rock phosphate for longer duration. Around 10^7 cfu/g cell population was maintained in rock phosphate throughout the incubation. All the carriers used were sterile because it is noted that sterile carriers display high population and have higher shelf life (Roughley and Vincent, 1976). Also the use of sterile carriers significantly reduces threat to quality resulting from the presence of contamination. However, the major disadvantage of a sterile carrier includes a higher cost of production (Stephens and Rask, 2000). The results are given in fig. 1.

3.3. *Effect of bacterial strains on seedling growth of C. cajan*

The T_2g was evaluated for the growth promoting activities on *C. cajan* and appeared to be excellent plant growth promoter, as significant increase in plant growth was observed by their application with respect to control. In first 15 days, T_2g affected the growth positively (Table 1). Nodules were also detected with T_2g as it was a *C. cajan* nodule isolate. Hence, role of N_2 fixation for plant growth promotion was apparent.

3.4. *Effect of bioformulations on seedling growth of C. Cajan*

Wheat bran, charcoal and whey based bioformulations appeared to be better as comparable to other bioformulations with respect to seedling growth of *C. cajan*. 41.27% and 87.80% increase in total plant length was observed using whey and charcoal respectively. Considerable increase in seedling biomass was also observed with charcoal and whey based formulations. However, rice husk and sawdust enhanced the seedling growth with respect to control but did not show any significant increase in seedling growth. Results obtained with whey were in accordance to physico-chemical characteristics and viability studies with the carrier. However, sawdust and rice husk although supported the growth of bacterial cells and were good in physico-chemical properties, but seedling growth was not appreciably enhanced by their application, which remained unexplained. Charcoal, although appeared to be a moderate carrier on the basis of physico-chemical characteristics and viability studies. Still the results obtained with charcoal were best with respect to seedling growth enhancement of *C. cajan* (Table 2). Rock phosphate, cocoa peat and bagasse were not included in this study because of their poor performances in viability studies.

3.5. *Root colonization assay*

The results of root colonization assay of T_2g suggested it to be extensive root colonizer, which is a desirable property in a suitable PGPR. The seeds were inoculated with $\approx 10^6$ cell population. It has been suggested that bacteria attaining colony-forming units of about $\approx 10^3$ per gram or higher on root mass can be considered as good colonizers (Scher *et al.*, 1988), and we obtained that the population of T_2g increased $\approx 10^9$ cfu/g after 20 days. The continued presence of T_2g for 20 days in soil showed that it had reached homeostasis after undergoing exchange with homogenous microflora and it is not affected by the active and passive processes restricting soil community. The population of indigenous bacterial population remained of same order of magnitude throughout (Table 3). This indicates that the application of T_2g did not result in any perturbations of the indigenous population of microflora. These results were similar to root colonization of *P. aeruginosa* NBRI2650 (Nautiyal *et al.*, 2002).

Conclusively, charcoal and wheat bran based solid formulations and whey based liquid formulation of T_2g are of great industrial and environmental importance, and commercial use as suggested by outcome of present work. Also GFP can be used be as a marker for monitoring bioinoculants.

References

1. Bashan Y. "Inoculants of plant growth-promoting bacteria for use in agriculture". *Biotechnol. Adv.* 16: 729-770, 1998.
2. Bissonnette N, Lalande R, Bordeleau LM. "Large scale production of *Rhizobium meliloti* on whey". *Appl. Environ. Microbiol.* 52: 838-841, 1986.
3. Burton JC. "Methods of inoculating seeds and their effect on survival of rhizobia". *In* Nutman P., (ed.), Symbiotic nitrogen fixation in plants. Cabridge, UK: Cambridge Univ. Press. International Biological Programme 7. 175-189, 1976.
4. Crawford SL, Berryhill DL. "Survival of *Rhizobium phaseoli* in coal based legume inoculants applied to seeds". *Appl. Environ. Microbiol.* 45:703-705, 1983.

5. Daze A, Santmaria C, Rodriuez-Navarro DN, Comacho M, Orive R, Temprano F. Perlite as a carrier for bacterial inoculants. *Soil Biol. Biochem.*32:567-572, 2000.

6. Jackson AM, Whipps JM, Lynch JM. "Production, delivery systems, and survival in soil of four fungi with disease biocontrol potential". *Enzyme Microbiol. Technol.* 13: 636-642, 1991.

IMPACT OF TEMPERATURE AND APPLIED CHLORINE DOSE ON INACTIVATIN AND REGROWTH OF HPC BACTERIA

BANDANA MAHTO

Research Scholar, Department of Civil Engineering, Indian Institute of Technology, Kharagpur, West Bengal 721302, India

SUDHA GOEL

Assistant Professor, Department of Civil Engineering, Indian Institute of Technology, Kharagpur, West Bengal 721302, India

HPC bacteria were grown in nutrient broth diluted 1:100 dilution with tap water to simulate conditions in water distribution systems. These mixed cultures were then inactivated with different chlorine doses at two different temperatures. Surviving bacteria were then grown in similar conditions to determine regrowth rates. Less than 1-log removal was observed at 30°C and 4 mg/L applied chlorine dose while the greatest extent of inactivation was observed with 4 mg/L applied chlorine and 30 deg C temperature. Growth rates observed after inactivation decreased in all except one case, but the differences were not large. Large fluctuations in cell concentrations were observed at the lower chlorine doses.

1. Introduction

Heterotrophic microorganisms are indigenous to water and are always present in greater concentrations than coliform bacteria in distribution systems. An increase in HPC indicates treatment breakthrough, post-treatment contamination or growth within the distribution system. Geldreich *et al.*[1] reported that the risk of pathogen contamination increases as the general bacterial population increases.

Increasing temperatures usually increase disinfection efficiency measured in terms of C*t (where C is the concentration of the disinfectant and t is the contact time). Increasing temperatures can also decrease disinfection efficiency, due to greater volatilization of the disinfectant. Thus, warmer temperatures are likely to facilitate survival and regrowth of microorganisms in the distribution system. The presence of a disinfectant residual such as chlorine may control the number of cultivable bacteria in the bulk phase but seems to have a more limited effect on biofilm and sediment[2],[3]. In chlorinated supplies, increase in HPC values have been reported, usually in situations where chlorine residuals were less than 0.1 mg/litre[4],[5]. Certain microorganisms can survive or multiply in the presence of low concentrations of chlorine[6],[7]. Norton and LeChevallier[8] observed that chlorination caused a shift to Gram-positive bacteria. Gräf and Bauer[9] isolated a chlorine-resistant *Corynebacterium* from tap water. Nagy and Olson[10] found more filamentous fungi in chlorinated than in unchlorinated supplies.

The objective of the present study was to determine the impact of temperature and disinfectant residual concentrations on the inactivation and regrowth of HPC bacteria. Experiments were carried out with bacteria that were indigenous to tap water from EEM Lab, Civil Engineering Department, IIT Kharagpur.

2. Materials and Methods

Phosphate buffer: 8.5 g/L of K_2HPO_4, 21.75 g/L of KH_2PO_4 and 17.7 g/L of $Na_2HPO_4.7H_2O$ were added to double-distilled, deionized and autoclaved water to prepare a stock solution of phosphate buffer (0.3 M).

Disinfectant solution: Chlorine stock solution was prepared by diluting a 4-6% sodium hypochlorite solution (Ranbaxy, Fine Chemicals Limited, India).

Mixed cell cultures: Tap water served as the source of a mixed bacterial culture and was collected from the *Environmental Engineering Laboratory* of the *Department of Civil Engg., IIT Kharagpur* and stored at 4°C.

Media for plating and culturing: Nutrient broth (HiMedia Pvt. Ltd., India) solutions were used as media for culturing bacteria in the growth studies. Pour plate procedure was adopted for cell counts during the growth and inactivation studies and nutrient agar (Merck & Co., Inc., USA) was used as the medium for plating.

Growth studies: 198 mL tap water containing indigenous bacteria were taken in a conical flask of 250 mL capacity. 2 mL of nutrient broth solution were added to tap water to get a 1:100 dilution of nutrient broth and the mixture was incubated at a temperature of 30°C ± 1°C (or 35°C ± 1°C). Sample aliquots of 20.0 μL volume were removed at regular intervals, serially diluted and plated by pour plate method for cell counting.

Inactivation assays: After 2-3 days of incubation, cells were harvested for disinfection with chlorine. The inactivation assays included the following steps:

Cell Harvesting: 5 mL aliquots of the mixed cell cultures were taken in a test tube and sonicated for 5 min, followed by vortex mixing. 1.0 mL aliquots of mixed cell cultures to be disinfected were added to 1.0 mL phosphate buffer (0.01 M) and centrifuged at 14,794 g (13,000 rpm) for 15 min at 20°C. After centrifugation, the supernatant was removed and an additional 1.0 mL of aliquots and 1.0 mL phosphate buffer (0.01 M) were added to the remaining part and centrifuged again. This process was repeated 5 times to harvest a total of 5.0 mL mixed culture solution. Mixed cultures were resuspended after centrifugation using a vortex mixer for 1.0 min.

Inactivation assay: Three assay bottles of 300.0 mL capacity were acid-washed and autoclaved for use in the inactivation studies: Bottle **A** contained 99.0 mL chlorine solution and 1.0 mL of mixed cell suspension. Two controls were run: Control **1** contained 99.0 mL chlorine-demand free water containing phosphate buffer (30.0 mM) and 1.0 mL of mixed cell suspension, to account for natural die-off of bacteria in a buffered solution without disinfectant. Control **2** contained 100.0 mL buffered chorine solution only, to account for decrease in disinfectant concentration without interference due to disinfectant demand of bacteria.

Four samples for free and total chlorine concentrations were taken from bottle A and Control 2 at 0, 15, 30 and 45 min respectively. 1.0 mL sample aliquots were removed from bottle A at 0.5, 1, 2, 5, 10, 15, 30 and 45 min and dechlorinated by adding 1.0 mL of 1% (w/w) $Na_2S_2O_3$. After dechlorination, samples were serially diluted and plated for cell counts. Similarly, 1.0 mL samples were removed from Control **1** at 0.5, 1, 2, 5, 10, 15, 30 and 45 min, serially diluted and plated for cell counts. The inactivation for all the experiments was done at room temperature i.e.; 25°C to 30°C.

Regrowth studies: After completion of inactivation assays, 50.0 mL of sample from Bottle A was placed in a conical flask. 1.0 mL of fresh nutrient broth solution and 49.0 mL of phosphate buffer (30.0 mM) were added to this sample. The mixture was incubated at 30°C (or 35°C) for growth studies.

Chlorine concentrations: Chlorine (free and total) concentrations were measured for all inactivation assays using a colorimetric kit (Hach Co., USA) based on the DPD method.

Data analyses: Growth rates were calculated based on pour plate counts done after incubation. Linear regressions of log cfu/mL versus time plots were done. The slope of the regressed line represented the growth rate for the culture.

3. Results and Discussion

3.1. *Phase I: growth, inactivation and regrowth experiments at 30°C with applied chlorine doses 1, 2, 3, and 4 mg/L*

The initial characteristics of tap water are summarized in Table 1. Representative results for 1 mg/L chlorine dose are presented here.

Table 1. Initial characteristics of tap water.

Characteristics	Phase I	Phase II
Free chlorine	0.57 mg//L	0.28 mg/L
Total chlorine	0.61 mg/L	0.32 mg/L
pH	6.94	6.87
Turbidity	8.8 NTU	11.2 NTU
Electrical conductivity	170.6 μS	301.0 μS
Total dissolved solids	85.3 ppm	150.0 ppm
Temperature	26.7 °C	27.2 °C
Optical density at 254 nm	0.0235	0.0295
Initial HPC concentration	32 cfu/mL	43 cfu/mL
Date of sample collection	25.08.2008	14.10.2008

The growth rates before and after the inactivation assay were found to be 0.237 and 0.199 h^{-1} respectively (Figure 1). These results show that there is a small (perhaps, insignificant) decrease in bacterial growth rates after inactivation. It is important to note that these regrowth studies are conducted with cell cultures that had been exposed to chlorine during the inactivation assays. The lack of significant decrease can be attributed to the fact that the indigenous microbial communities in these tap water samples have been exposed to chlorine over a long period of time and are a fairly resistant population.

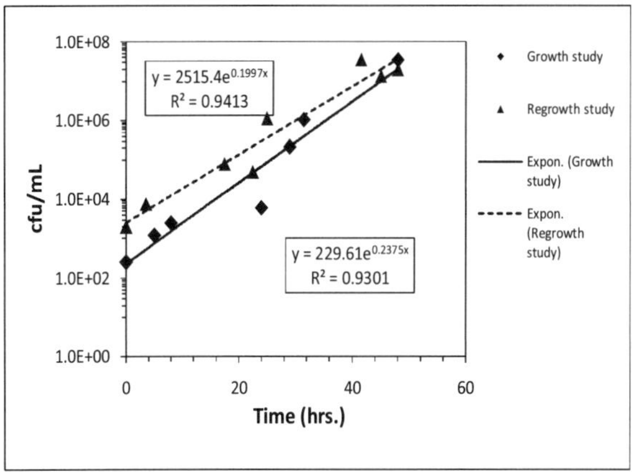

Figure 1. Results of growth and regrowth studies (1 mg/L chlorine dose).

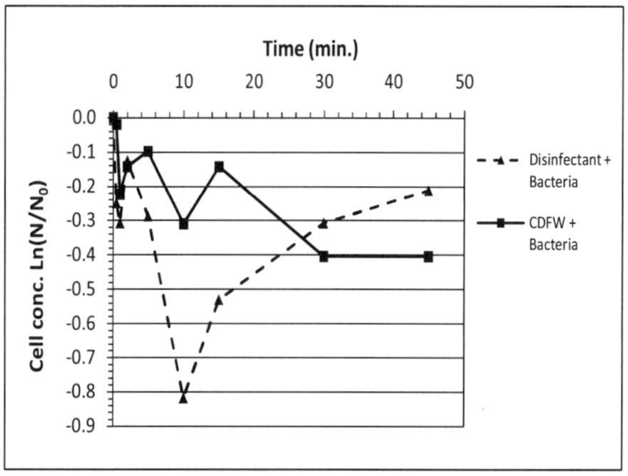

Figure 2. Change in bacterial cell concentrations over time with a chlorine dose of 1 mg/L.

Inactivation of cells exposed to 1 mg/L chlorine was monitored over 45 min and less than 1 log removal of cells was observed in both, chlorinated water and chlorine demand-free water (Figure 2). Large fluctuations in cell concentrations were noted in both flasks with greater fluctuations in the 'disinfectant and bacteria' flask. Other groups observed similar fluctuations and these can be attributed to bacterial aggregation [11]. Initial exposure to chlorine kills individual suspended cells and bacteria that are on the surface of clumps or aggregates resulting in a decrease in plate counts. It also leads to breaking up of cell aggregates and release of large numbers of viable individual cells in to the solution which can result in an increase in plate counts. As time progresses, more cells are exposed as the aggregates break up and if the free chlorine remaining in solution is insufficient to kill the cells, an increase in cell concentration is seen.

Residual chlorine was also monitored during the assays and was found to decrease throughout the experiment in both, inoculated and non-inoculated samples. Chlorine residual concentration data were fit to linear, polynomial and exponential curves to determine the best-fit. In this study with 1 mg/L of applied chlorine dose, better fits for 2nd order decay kinetics were obtained for both free and total chlorine residuals. Similar experiments were carried out at chlorine doses 2, 3 and 4 mg/L and are summarized in Table 2.

3.2. *Phase II: Growth, inactivation and regrowth experiments at 35 °C with applied chlorine doses of 1and 2 mg/L*

The initial characteristics of tap water for the second phase are summarized in Table 1. Representative results for one of these studies (2 mg/L) are shown here.

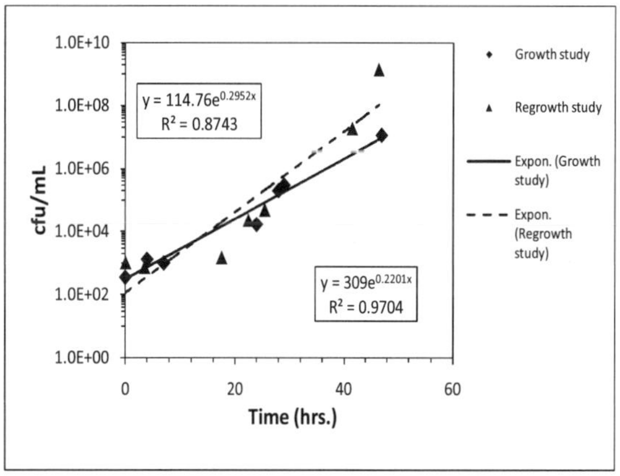

Figure 3. Results of growth and regrowth studies (2 mg/L chlorine dose).

At a chlorine dose of 2.0 mg/L, the growth rates before and after the inactivation assay were found to be 0.220 and 0.295 h^{-1} respectively (Figure 3). Whereas 5 out of the 6 studies conducted at different chlorine doses and temperatures show decreases in growth rates after inactivation, only this study (2 mg/L chlorine dose and 35°C) shows an increase in growth rate after inactivation. The cell die-off patterns for both, chlorinated water and chlorine demand-free water are plotted in Figure 4. It is apparent from Figures 2 and 4 that there were fewer fluctuations in the latter case.

Residual chlorine was monitored during the assays and data were fit to linear, polynomial and exponential curves to determine the best-fit.

From Table 2, it is observed that the inactivation efficiency increased with increase in chlorine dose. Also, the inactivation rate and extent of kill increased with increase in temperature.

Also, 1-log removal was obtained only at high chlorine doses, i.e., 3 and 4 mg/L for a contact time of 30 and 15 min, respectively. At low initial chlorine doses of 1 and 2 mg/L, less than 1-log removal was obtained which leads to the conclusion that microorganisms present in tap water samples are capable of surviving and growing with low concentrations of chlorine.

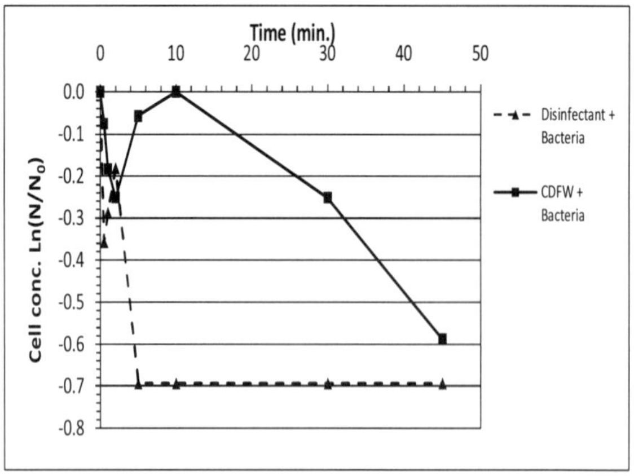

Figure 4. Change in bacterial cell concentration over time with chlorine dose of 2 mg/L.

A comparative study was done between different chlorine decay models for free chlorine as well as total chlorine residual at different chlorine doses and temperatures and the results are summarized in Table 3. It was observed that both, 1st order and 2nd order decay

Table 2. Summary of the results obtained at different chlorine doses and temperatures.

Temperature	30°C				35°C		% Increase in bacterial kill, temperature effect	
Cl_2 Dose (mg/L)	1.0	2.0	3.0	4.0	1.0	2.0	1.0	2.0
Growth rate (h^{-1})	0.237	0.267	0.314	0.255	0.228	0.220		
Regrowth rate (h^{-1})	0.199	0.159	0.279	0.232	0.155	0.295		
Difference in growth rate (before and after inactivation)	0.038	0.108	0.035	0.023	0.073	-0.075		
Initial free Cl_2 conc. (mg/L)	0.59	1.51	2.40	3.60	0.88	1.64		
Final free Cl_2 conc. (mg/L)	0.09	0.35	0.45	1.01	0.22	0.46		
Change in free CL_2	0.50	1.26	0.95	2.59	0.66	1.18		
Initial total Cl_2 conc. (mg/L)	0.68	1.53	2.40	3.70	0.91	1.67		
Final total Cl_2 conc. (mg/L)	0.11	0.36	0.47	1.07	0.28	0.41		
Change in total Cl_2	0.57	1.17	1.93	2.63	0.63	1.26		
Bacterial kill after 05 min. (%)	25.00	27.14	49.02	66.67	32.86	50.00	31.44	84.23
Bacterial kill after 15 min. (%)	41.18	42.86	60.78	70.00	48.57	-	17.94	-
Bacterial kill after 30 min. (%)	26.47	42.86	65.69	65.00	58.57	50.00	121.27	16.66
Bacterial kill after 45 min. (%)	19.12	34.29	59.80	75.00	31.43	50.00	64.38	45.82

Table 3. Comparison of R^2 values for free chlorine residual and total chlorine residual at different chlorine doses and temperatures.

Temperature	Initial Chlorine dose, mg/L	Comparison of R^2 value											
		Residual free chlorine						Residual total chlorine					
		Disinfectant only			disinfectant+bacteria			Disinfectant only			disinfectant+bacteria		
		Zero Order	1st Order	2nd Order	Zero Order	1st Order	2nd Order	Zero Order	1st Order	2nd Order	Zero Order	1st Order	2nd Order
30°C	1	0.891	0.970	0.992	0.873	0.985	0.940	0.840	0.934	0.979	0.830	0.963	0.965
	2	0.975	0.974	0.941	0.949	0.987	0.971	0.978	0.969	0.925	0.963	0.990	0.961
	3	0.776	0.836	0.897	0.846	0.966	0.993	0.831	0.895	0.950	0.854	0.968	0.996
	4	0.982	0.994	0.974	0.976	0.965	0.882	0.965	0.986	0.974	0.977	0.963	0.880
35°C	1	0.969	0.950	0.876	0.966	0.999	0.951	0.965	0.949	0.886	0.943	0.995	0.986
	2	0.992	0.978	0.919	0.985	0.987	0.941	0.988	0.983	0.938	0.991	0.977	0.905

kinetics seem to fit well for both free chlorine and total chlorine residuals.

4. Conclusions

Large fluctuations in cell concentrations were observed at the low initial chlorine doses of 1 and 2 mg/L and these can be attributed to bacterial aggregation. Inactivation efficiency increased with increase in chlorine dose and increase in temperature. Growth rates observed after inactivation decreased in all except one case, where the growth rate increased after inactivation (2 mg/L at 35°C). The lack of significant decrease can be attributed to the fact that the indigenous microbial community in these tap water samples has been exposed to chlorine over a long period of time and represents a fairly resistant population. Decreases in free and total chlorine residuals were observed to be proportionate to the initial chlorine doses. Both 1st and 2nd order decay kinetics fit well for both free chlorine and total chlorine residuals.

References

1. E. E. Geldreich, H. D. Nash, D. J. Reasoner and R. H. Taylor, *J. Am. Water Works Assoc.* **64**, 596 (1972).
2. D. van der Kooij, J. H. M. van Lieverloo, J. Schellart and P. Hiemstra, *J. Am. Water Works Assoc.* **91**, 55 (1999).
3. L. Mathieu, J. L. Paquin, J. G. Block, C. Randon, J. Maillard and D. Reasoner, *Rev. Sci. Eau.* **5**, 91 (1992).
4. M. Prévost, A. Rompré, H. Baribeau, J. Coallier and P. Lafrance, *J. Am. Water Works Assoc.* **89**, 78 (1997).
5. M. Prévost, A. Rompré, J. Coallier, P. Servais, P. Laurent, B. Clement and P. Servais, *Wat. Res.* **32**, 1393 (1998).
6. M. W. LeChevallier, R. J. Seidler and T. M. Evans, *Appl. Environ. Microbiol.* **40**, 922 (1980).
7. H. F. Ridgway and B. H. Olson, *Appl. Environ. Microbiol.* **44**, 972 (1982).
8. C. D. Norton and M. W. LeChevallier, *Appl. Environ. Microbiol.* **66**, 268 (2000).
9. W. Gräf and L. Bauer, *Zentralbl. Bakteriol. Hyg.* **157**, 291 (1973).
10. L. A. Nagy and B. H. Olson), *Can. J. Microbiol.* **28**, 667 (1982).
11. S. Goel and E. J. Bouwer, *Wat. Res.* **38**, 301 (2004).

Session 10

DETERMINATION OF PHYSICAL PROPERTIES OF POMEGRANATE DURING BULK DRYING IN AN INERT MEDIUM FLUIDIZED BED DRYER ASSISTED INFRARED HEAT SOURCE

BIZHAN HONARVAR[1]

Department of Chemical Engineering, Islamic Azad University-Fars Science and Research Branch, Iran

ROGHAYE KESHAVARZ[2]

Department of Chemical Engineering, Arak University, Arak, Iran

ALIREZA FAZLALI

Department of Chemical Engineering, Arak University, Arak, Iran

A laboratory scale infrared–fluidized bed dryer with inert materials was developed and drying of pomegranate seeds was carried out at infrared power levels of 346, 726 and 1470, drying air temperatures of 40, 50 and 60°C, inlet drying air velocity of 3, 4 and 5 m/s. It was found that air temperature, inert material and air velocity had a low effect on physical properties and therefore, shrinkage, density and effective moisture diffusivity are obtained and some correlations were developed for variation of these properties of pomegranate seeds during drying in a fluidized bed assisted infrared heat source.

1. Introduction

Pomegranate (Punica Granatum L) is one of the oldest edible fruits widely grown in many tropical and subtropical Countries (Fadavi et al., 2005). There is growing interest in this fruit because it is considered to be a functional product of great benefit in the human diet as it contains several groups of substances that are useful in disease risk reduction (Martinez et al., 2006). Drying is commonly used to describe any process in which water is removed from any substance. The reason for drying of foodstuffs is to preserve them for storage and shipment without the need for refrigeration.

Conventional air drying is the most frequently used dehydration operation in food and chemical industry. In this case, the drying kinetics is greatly affected by air temperature and material characteristic dimension while all other process factors exert practically negligible influence (Kiranoudis et al., 1997). Dried products are characterized by low porosity and high apparent density (Krokida & Maroulis, 1997). Significant color changes occur during air drying (Krokida et al., 1998) and most frequently the dried products has low sorption capacity (Maroulis et al., 1998).

One of the ways to shorten the drying time is to supply heat by infrared radiation (Dorota Nowak et al., 2004). When infrared radiation is used to heat or dry moist materials the radiation impinges the exposed material, penetrates it and the energy of radiation converts in to heat (Ginzburg, 1969). The use of infrared radiation technology in dehydration of foods may decreased drying time, high energy efficiency, high quality finished products, uniform temperature in the product while drying, and a reduced necessity for air flow across the product (Dostie et al., 1989; Mongpreneet et al., 2002; Navari et al., 1992).

Fluidized bed dryers are the most efficient equipments considering the thermal efficiencies of the equipment. Mixing and heat transfer are very rapid. Recently a non classical technique has been developed in which heating and drying in fluidized beds has been accomplished in presence of inert particles. Inert particles can improve fluidization behavior of drying materials and increase convective heat and mass transfer coefficient (Abbasi Souraki et al., 2005).

The objective of this work were to determine the changes of density, shrinkage and moisture diffusivity

[1]honarvar2@gmail.com
[2]roghaye_keshavarz@yahoo.com

of pomegranate in a fluidized bed dryer with inert particles assisted by infrared heat source and proposing some correlations for them.

2. Material and Procedure

A laboratory scale infrared convective fluidized bed dryer was developed for the present study wherein infrared power, air temperature, air velocity and amount of inert particles could be varied within the experimental range. The drying of pomegranate was done under various combinations of these process parameters.

2.1. *Material*

In this work, pomegranate was chosen as the drying materials. The hand peeled pomegranate was cut and the skin covering the seeds was removed manually.

In order to ensure reproducible results, all pomegranates were obtained from a single region in Shiraz.

2.2. *Experimental dryer and procedure of experiment*

A pilot-scale fluidized bed dryer with inert particles assisted infrared heat source was set up for performing the drying experiments. The schematic diagram of the experimental apparatus is shown in Figure 1.

The dryer was a 60 cm cylindrical Pyrex column equipped with a porous plate as an air distributor.

Drying air was supplied from a high pressure air source and a pressure regulator adjusted its pressure. Air was passed through a rotameter and then heated by a controlled electrical heater. A temperature controller was used for regulating the temperature of drying air within ±0.1°C.

The humidity was determined by means of a humidity/Temperature meter of type HT-3015 (Lutron Co). Six infrared lamps where placed around the Pyrex column. In each experiment all of the lamps were turned on. Figure 2 shows the Experimental dryer used in the experiments.

100 g of pomegranate seeds were used for each experiment. The moisture contents of pomegranate seeds were determined by weighing the sample on an electronic balance placed next to the dryer at an interval of 5 min. The accuracy of the weighing was ±0.0001 g. The weighing procedure took no more than 10 s after removing the sample from the column. It has been demonstrated by previous researchers that this method is sufficiently accurate for generating reproducible drying curves (Ajibola, 1989; Saurez et al., 1985; Zhou et al., 1998). In each run, the final volume of dried sample was determined by immersing the sample in toluene and measuring the volume changes.

The operation variables were drying air temperature, air velocity, amount of heat carrier and power of the infrared heat source.

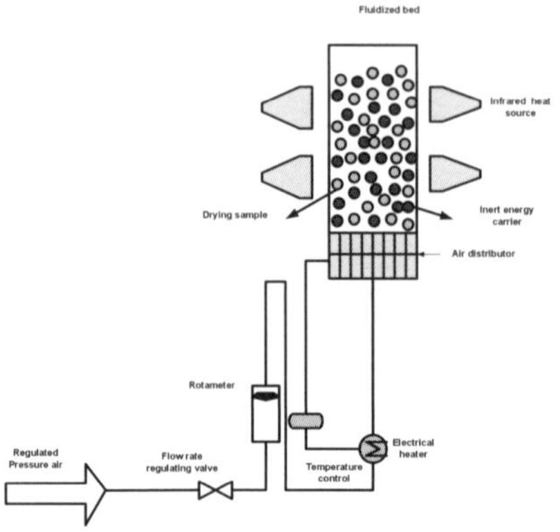

Fig. 1. Schematic diagram of the experimental apparatus.

Fig. 2. Experimental dryer.

3. Results and discussion

3.1. *Effect of air temperature*

Figure 3 shows the effect of air temperature on drying curve of pomegranate seeds. It can be found that air temperature is the main factor that has significant effect on drying curve.

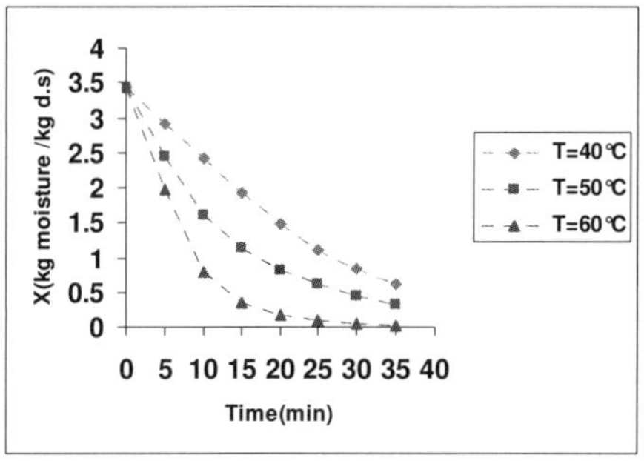

Fig. 3. Drying curve of pomegranate seeds at various air temperature (°C).

3.2. *Effect of infrared power*

Drying curves at air velocity of 4 m/s, air temperature of 40°C and inert amount of 150 g as a function of infrared power are shown in Figure 4. The drying time reduced dramatically with increase in infrared power. Masamura et al. (1988) and Kazuhiko and Chung (1995) also observed increased drying rates of food products with increasing power supply to infrared heater.

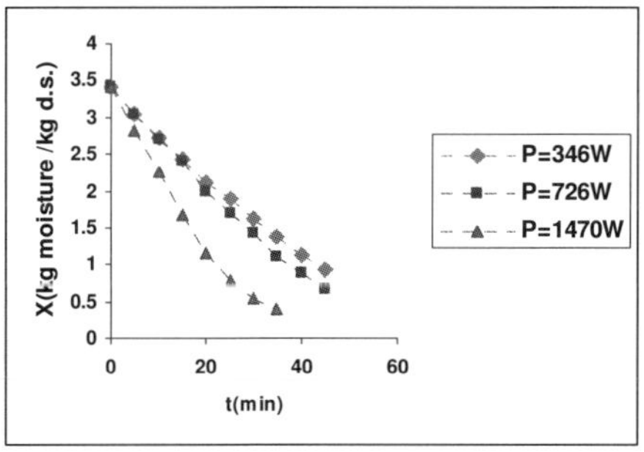

Fig. 4. Drying curves for pomegranate seeds at various infrared powers.

3.3. *Effect of air velocity*

At a given air temperature, inert material and infrared power, an increase in air velocity increase the drying time i.e. decreased the moisture removal rate. The moisture content at any time was found to be a little higher with higher velocity. The increase in air velocity accelerated the cooling effect, reducing the temperature at the surface of products thus the lower moisture driving force.

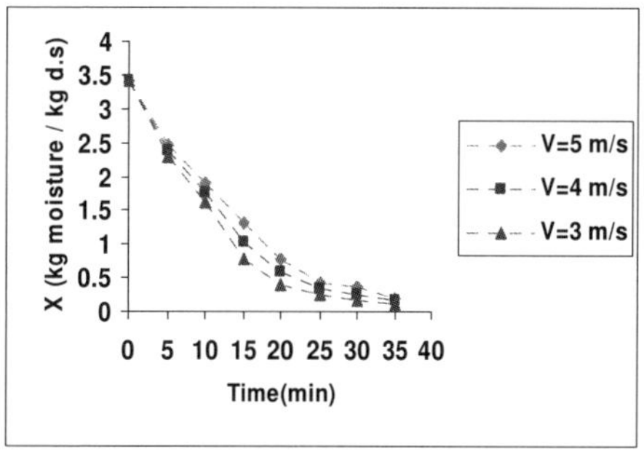

Fig. 5. Drying curves for pomegranate seeds at various air velocities (m/s).

3.4. *Effect of inert material*

At a given air temperature, air velocity and infrared power, an increase in amount of inert particle had no significant effect on drying rate. Figure 6 shows the effect of amount of inert on drying rate .

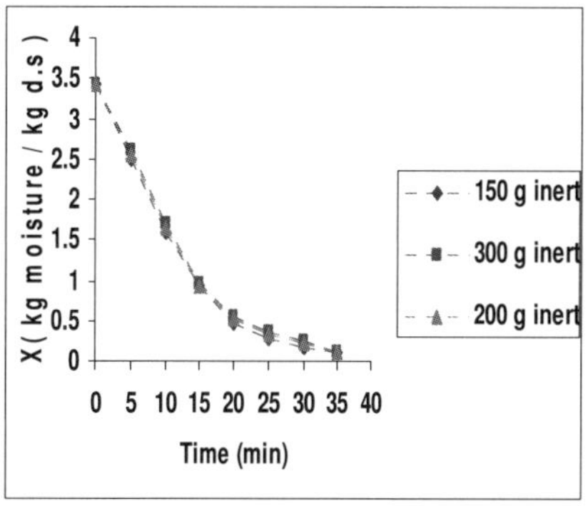

Fig. 6. Drying curves for pomegranate seeds at various inert amounts (g).

3.5. *Determination of shrinkage*

The volume changes during drying are not an easily predictable function. Shrinkage is caused primarily by the self-contraction of the material is often a linear function of the mean moisture content during drying:

$$V = V_0(1+\beta X) \tag{1}$$

Where β is the bulk-shrinkage coefficient. Since the shrinkage may be anisotropic, it is sometimes more convenient to consider linear changes.

Figure 7 shows the change of volume ratio of pomegranate seeds with the moisture content.

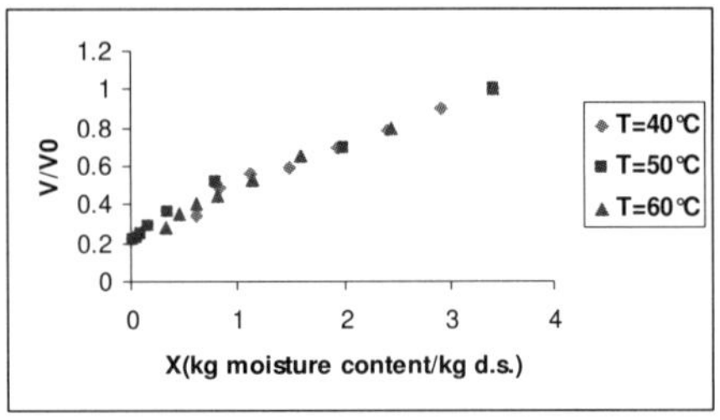

Fig. 7. Shrinkage of pomegranate seeds at various air temperatures (°C).

Therefore at first glance from experimental data it can be concluded that:

$$\frac{V}{V_0} = AX + B \tag{2}$$

The values of parameters A and B were obtained for each temperature by non-linear regression. No meaningful relation was found for values of A and B with temperature. Therefore it has been found that temperature has effect on drying rate and moisture content but no significant effect on shrinkage and the average values of calculated values of A and B were selected as the desired parameters. So the following correlation could be proposed:

$$\frac{V}{V_0} = 0.221X + 0.2596 \quad R^2=0.9812 \tag{3}$$

3.6. *Determination of density*

Figure 8 shows the variation of experimental values of density of pomegranate seeds during drying against moisture content. The experimental values were obtained by dividing the mass of pomegranate seeds to the measured volume. Measurement of the final volume of the sample in toluene was a good check for volume calculation.

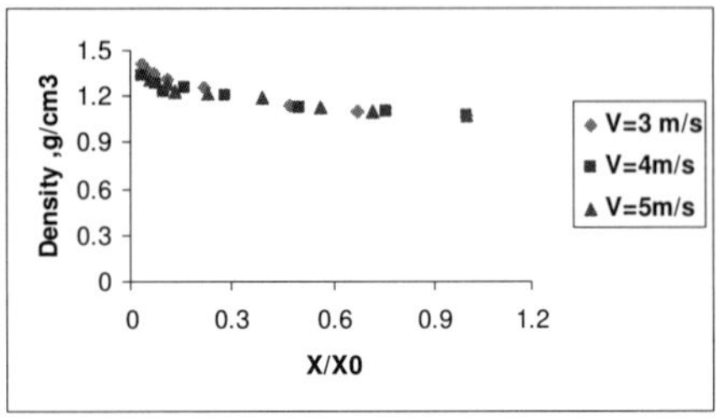

Fig. 8. Density of pomegranate seeds at various air velocities (m/s).

The equation below determines the general form of density correlations:

$$\rho_s = a\left(\frac{X}{X_0}\right)^b \tag{4}$$

The values of parameters a and b were obtained for each temperature by non-linear regression.

Therefore the following correlation could be proposed:

$$\rho_s = 1.0854\left(\frac{X}{X_0}\right)^{-0.0699} \tag{5}$$

3.7. *Determination of the effective moisture diffusivity*

In drying, diffusivity is used to indicate the flow of moisture within the material .Pomegranate seeds were considered as sphere and the moisture diffusivity for sphere was therefore calculated by the following equation proposed by Crank (1975) considering assumptions mentioned hereunder:

1. Moisture is initially uniformly distributed throughout the mass of sample.
2. Mass transfer is symmetric with respect to the center.
3. Surface moisture content of the sample instantaneously reaches equilibrium with the condition of surrounding air.
4. Resistance to the mass transfer at the surface is negligible compared to internal resistance of the sample.
5. Mass transfer is by diffusion only.
6. Diffusion coefficient is constant and shrinkage is negligible (G.P. Sharma et al., 2005).

$$MR = \frac{X - X_e}{X_0 - X_e} = \frac{6}{\pi^2}\sum \frac{1}{n^2}\exp(-n^2\pi^2 F_{0)} \tag{6}$$

Where $F_0 = \frac{D_{eff}t}{R^2}$

In this work in order to consider the effect of both temperature and moisture content, an Arrhenius type correlation (Azzouz et al., 2002) as in equation 8 assumed for effective moisture diffusivity:

$$D_{eff} = D_0 \exp(\frac{-E_a}{RT_{abs}})\exp(-[CT_{abs} + D]X) \tag{7}$$

Correlation for moisture diffusivity as a function of moisture and temperature for pomegranate seeds is as follows

$$D_{eff} \times 10^{+9} = 0.001855\exp(\frac{-4335.4}{T_{abs}})\exp(-[0.00377T_{abs} - 2.36741]X) \tag{8}$$

4. Conclusion

The drying of pomegranate seeds exhibited that the rate of drying was significantly influenced by infrared power, air temperature and air velocity. The drying time increased with increase in air velocity at all infrared powers applied. The effective moisture diffusivity decreased with decrease in moisture content and temperature of drying air. All of the above-mentioned figures show that shrinkage and density of pomegranate seeds depend only on moisture content, but effective diffusivity is dependent on temperature and moisture content.

Acknowledgments

The authors are thankful to Marvdasht Islamic Azad University for providing necessary facilities for experimentation to carry out this research work.

References

1. Mustafa Cam, Yasar Hisil, Gokhun Dormaz, 2-R.B. Keey, Drying principle and practice, Pergamon Press publisher, P.39-40, (1972).
2. M.K. Krokida, V.T. Karathanos, Z.B. Maroulis, D. Marinos-Kouris, journal of food engineering 59 (2002) 391-403.
3. Dorota Nowak, Piotr P. Lewicki, Innovative food science and Emerging Technology 5(2004) 353-360.
4. G.P. Sharma, R.C. Verma, P.B. Pathare, journal of food engineering 67 (2005) 361-366.
5. B. Abbasi Souraki, A. Andres, D. Mowla, chemical engineering and processing.
6. Serpil Sahin, Survet Gulum Sumnu, USA: Springer Science publisher, 16-18.
7. G. Hashemi, D. Mowla, M. Kazemeini, journal of food engineering 92 (2009) 331-338.
8. Strumillo, C., & Kudra, USA: Gordon & Breach Science Publishers, 67-70, (1996).
9. M.S. Hatamipour, D. Mowla, Journal of food engineering, 221-227.

THE CHARACTERISTICS OF HEAT TRANSFER AT SPIRALLY SHAPED CHEMICAL REACTOR THROUGH A NUMERICAL ANALYSIS*

LEE JIN-GYU, JUNG YOUNG-GUK and LEE JU-HAN
Graduate School, Department of Mechanical Engineering, Inha University, 253 Yonghyun-dong, Nam-gu, Incheon, Korea

SEO TAE-BEOM
Department of Mechanical Engineering, Inha University, New Renewable Energy Lab., 253 Yonghyun-dong, Nam-gu, Incheon, Korea

The purpose of the research is to develop the high performance solar chemical reactor for producing hydrogen using steam reforming of methane. A specific shape chemical reactor is suggested: spiral type reactor. The reactor is installed on the dish-type solar thermal system of Inha University. The temperatures, CH4 conversion rates, and H2 yield are measured. At specific condition, CH4 conversion rates of the spiral type reactor are about 92%, and H2 yields are about 66%. The spiral type reactor gives reasonably good performance without any problems caused by highly concentrated solar radiation.

1. Introduction

The generation system by use of solar heat is expected to be one of the most competitive fields among the renewable energies to substitute fossil fuels.(1) The most noted one of methods utilizing such generation system by use of solar heat is just the production of hydrogen energy, most of which is adopting the steam reforming of methane.(2) The steam reforming of methane is more than 90% conversed at more than 1000K. In case to replenish the temperature of more than 1000K, it can reduce approximately 20% or so of CO2 exhaust more than the use of existing fossil fuels.(3)

Muir et al.(4) operated by trial a very large scale reactor(CAESAR) situated at parabolic solar concentrator and analysed the CO2 reforming of methane. The catalytic absorber of this reactor is equipped with the multiple alumina form coated with Rh catalyst. In the study by Worner and Tamme(5) a reactor by use of solar heat was used through the Rh catalyst supported on the ceramic form made of Alumina and Sic. Moeller et al.(6) operated the reactor directly radiating the solar heat on the ceramic form supported with the catalyst, and the reactor was operated at the condition of 9atm, 1098K with the solar heat of 400kW infused.

The points to be more deeply considered in the application of chemical reaction by use of such a high temperature solar heat reactor are to increase the heat efficiency and to establish the design technology of the optimized reactor for chemical reaction.

Accordingly through a numerical analysis on the spirally shaped chemical reactor as an object of this study, the heat loss was minimized by studying the numerical analysis on the radiated heat flux inside the reactor, and establishing the interpretation technique on the heat loss of radiation, convection, conduction. Besides the steam reforming of methane was made by establishing a reactor at the dish type solar collector.

2. Experimental System

2.1. *Reacting System*

The entire system was formed as shown in the Figure 1. In order to correctly control the methane required for the experiment the flow rate of methane was controlled by use of MFC(KITS-C-130, MFC for Methane). Besides through the solenoid valve electrically controlled to open to close, the experimental equipment was formed so that the methane gas and nitrogen gas may be transferred through each pipe and connected with Y-type pipe.

*This work is supported by Inha University.

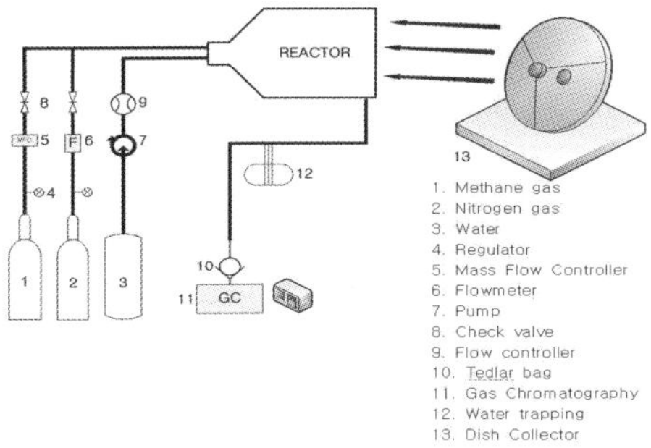

Figure 1. Schematic diagram of solar reactor system.

2.2. *Solar Thermal Reactor*

Figure 2 is an external shape of the solar chemical reactor.

Figure 2. Solar thermal reactor installed on the dish type solar thermal system in Inha Univ.

The external material of the reactor was made of SUS 304 tolerable to the high temperature and the aperture of reactor was decided with the diameter of 100mm so that more than 90% of radiated heat may be absorbed through Flux mapping experiment.

Figure 3 represented the internal shape of the reactor, the main part of this study. (a) showed the sectional area in the direction of 45° from the central axis of spirally shaped reactor with the spiral shaped structure, (b) showed the left side view of reactor.

In order to increase the conversion into steam from water and the reacting efficiency of steam and methane the optimum design for the reacting path and shape was made. In the internal shape of spirally shaped devised reactor the mixture proceeds along the spirally shaped path and receives Heat flux. The mixture of steam and methane receives high heat by heat flux as it goes into the internal part of spiral reactor and attains the highest temperature at the exit.

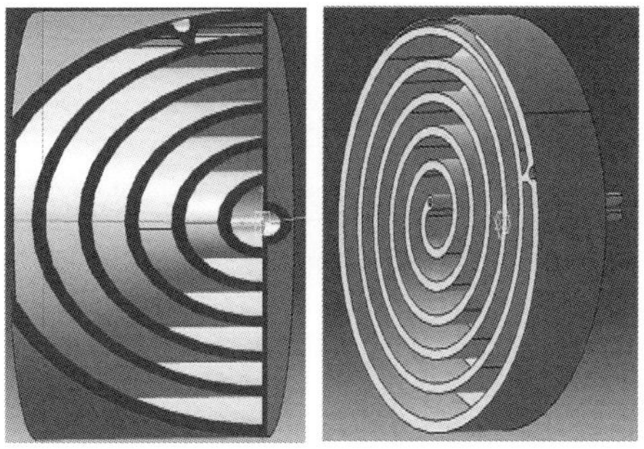

Figure 3. Internal shape of solar thermal reactor.

3. Experimental Method

3.1. *Temperature Measuring Method*

This experiment was made on the identification of heat transfer properties inside the reactor in order to design the reactor meeting the requirement of steam reforming of methane.

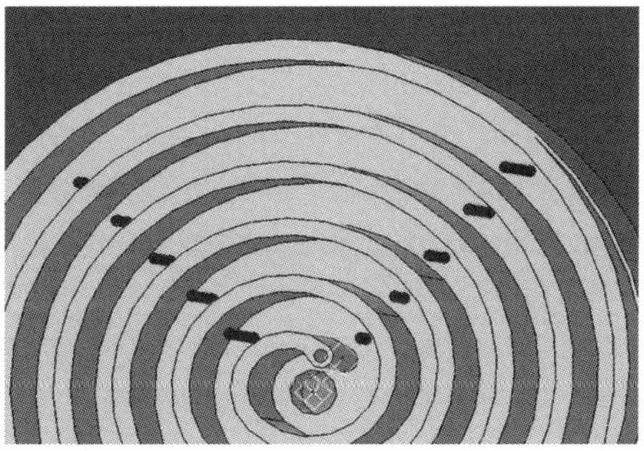

Figure 4. Thermocouple position in the reactor.

As shown in the Figure 4, in order to measure the temperature inside the reactor the K-type thermocouple (model: TT-30 OMEGA Co.) was installed. In order to acquire the temperature distribution throughout the entire scope inside the reactor by establishing the two in the direction of axis the thermocouple was inserted from the rear side. Additionally the entire scope inside the reactor was measured at each point with the interval of 2cm.

3.2. *Performance Analysis Method*

For the steam reforming of methane for producing hydrogen the conversion rate of methane and production rate of hydrogen were calculated on the collected gas by changing the flow rate of inflowed methane and steam by use of GC (Gas Chromatography).

When finding the production rate of hydrogen the following equation is used.

$$H_2production\_rate = \frac{(resulted\_H_2moles)/3}{Supplied\_CH_4moles} \quad (1)$$

Besides, the flow rate into the reactor was calculated by use of energy equation as follows.

$$\dot{Q}_{cv} = \dot{Q}_{irra} - \dot{Q}_{loss} \quad (2)$$

$$\frac{\dot{Q}_{cv}}{nCH_4} = \sum_P n_e(\bar{h}^\circ_f + \Delta\bar{h})_e - \sum_R n_i(\bar{h}^\circ_f + \Delta\bar{h})_i \quad (3)$$

4. Result and Discussion

4.1. *Numerical Analysis Result*

4.1.1. *Spirally Shaped Reactor Analysis*

Figure 5(a) shows the numerical analysis result of temperature distribution on the front part of reactor, and Figure 5(b) shows the front part shape of the spirally shaped reactor used in the practical experiment. Looking at the numericcal analysis result we can find that the temperature difference between the left side and the right one in the upper part of the figure, and it is because the mixture gas of methane and steam is input, collides with the quartz window and the contacted part cools down locally by the temperature difference. Non-uniform temperature distribution on such a quartz window seems to be the main cause for the window's cracking as in the Figure 5. (b).

To understand the temperature distribution of the reactants and the mixing degree of methane-steam in the steam reforming of methane shall be the main factor affecting on the reaction efficiency. Looking at the temperature distribution of Figure 5(a) the high temperature distribution shows more than 1200K at the front part and the external part of reactor.

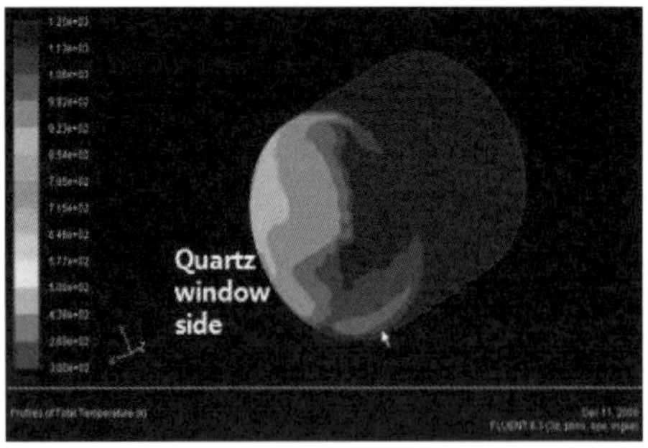

Figure 5(a). Temperature contour inside the reactor.

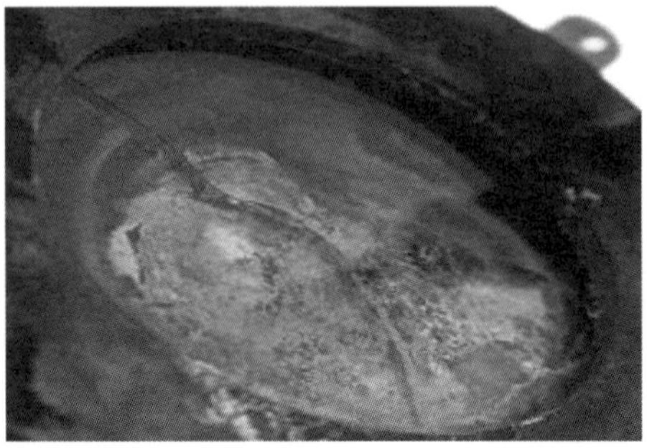

Figure 5(b). Cracked quartz window.

Besides looking at the numeric analysis result followed by the fluid distribution in Fig. 6 we can find that the reactants go through the quite complicated flow path. Finally it shows that the spirally shaped reactor has the mixed properties suitable for the high temperature needed for the reaction.

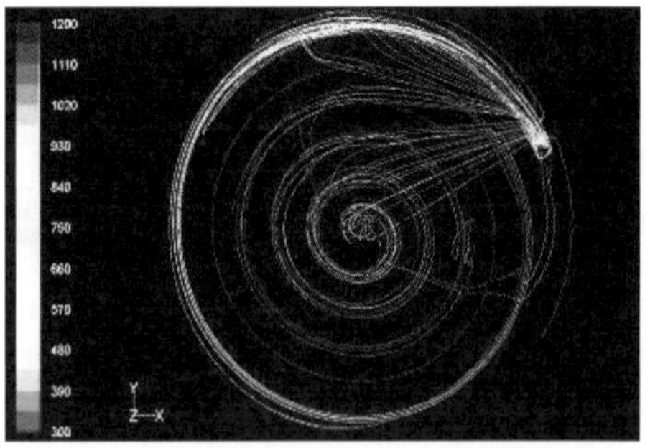

Figure 6. Streamline in the reactor.

4.2. *Experimental Result*

4.2.1. *Spirally Shaped Reactor Analysis*

Figure 7 shows the solar radiation by time and the temperature distribution inside the spirally shaped reactor. The three typical temperatures were presented graphically after classifying into the high temperature, mid temperature, low temperature among the total 20 temperature measuring points inside the reactor. The highest temperature measured point is on the front part of reactor and we can find that more than 900-1000°C is measured.

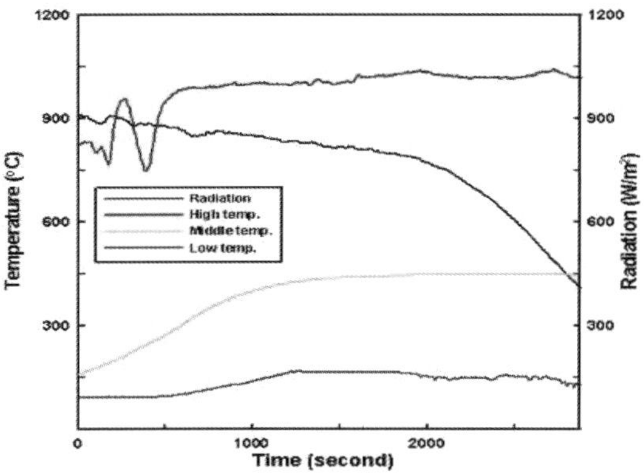

Figure 7. Temperature distribution in the reactor.

This is expected that the reaction is activated on the front part due to the required temperature for the reaction being kept on the front part. The reason why the temperature at the central part of reactor is higher than the other points is that the incident solar heat has the Gaussian distribution as it goes in the direction of radius from the central part of reactor as verified through Flux mapping experiment. Besides as we can find that the distribution of solar radiation increases and decreases rightly downward the reason is due to the increase and decrease of local radiated quantity distribution by time resulting from the diffusion of solar radiation owing to the cloud and the solar altitude change followed by the time. Though the solar radiation decreases the temperature inside the reactor is steadily maintained. The reason is why the internal temperature is maintained for the fixed time by the effect of insulation followed by the internal temperature inside the reactor being increased by the radiated solar heat.

4.2.2. *Performance Analysis followed by Hydrogen Production Rate*

Table 1 showed the mass flow rate condition followed by the molar ratio of methane gas and steam input into the spirally shaped reactor. As the input ratio of methane and steam is set at the range of 1:1-1:4, this is for including the experimental condition from the minimum ratio of 1:1 enabling all the input methane and steam to react with each other to the maximum ratio of 1: 3, the most effective and practical chemical reaction efficiency, as shown by the existing domestic, foreign study results.

Table 1. CH4 and H2O flow rate of the experiment for the reactor.

	CH_4:H_2O mole ratio	mole/s	g/min	l/min
CH_4		0.0025	2.4	3.36
H_2O	1:1	0.0025	2.7	0.002
H_2O	1:2	0.005	5.4	0.004
H_2O	1:3	0.0075	8.1	0.006
H_2O	1:4	0.01	10.8	0.008

Table 3 showed the methane by the respective molar ratio, the conversion rate and the hydrogen distribution in the exhaust gas.

Table 2. The experimental results of CH4 conversion rate and H2 yield.

Temperature (°C)	CH_4:H_2O mole ratio	CH_4 conversion (%)	H_2 yield (%)
800°C	1:1	90.32	46.23
	1:2	90.86	54.01
	1:3	91.26	66.14
	1:4	92.11	43.82

Looking at the reaction result the input ratio 1:1-1: 4 shows all the methane conversion of more than 90%, approximating near the theoretical value (100%). Hydrogen content in the resulting gas is 43%-66%, which the value 66% representing at 1: 3 shows the relatively approximate result to the theoretical value (72%). As identified in the Table, when in the molar ratio of 1: 3 it shows the highest production rate of hydrogen. Through this it was conclusively verified that the most excellent hydrogen production rate comes from the ratio of 1: 3 with the most theoretically effective chemical reaction.

5. Result and Discussion

In this study the reactor whose internal flow path had been spirally made was proposed. The following is the conclusion on this study.

(1) As a result of numerical analysis on the temperature distribution on the front part of reactor, the cause for windows cracking was identified by confirming the uneven temperature distribution of the left side and the right one.

(2) As a result of numeric analysis, as the front part and external one of reactor showed the temperature distribution of more than 1200K, it was identified that the spirally shaped reactor is suitable for the steam reforming of methane.

(3) During the experiment the temperature at the high temperature part inside the spirally shaped reactor is formed around at 1200K. This demonstrates that the temperature distribution values of reactor identified through the numeric analysis are reliable results.

(4) When in the input ratio of 1: 3 of methane and steam, it shows the methane conversion 91.26%, hydrogen content in the resulting gas 66.14%.

(5) When in the input ratio of 1: 3 of methane and steam, it shows the highest hydrogen production rate.

References

1. Kang, Y.H., Flux Distribution of The Dish Concentrator, Journal of The Korean Solar Energy Society (1999).
2. Hydrogen Production and Distribution, IEA Energy Technology Essentials (2007).
3. T. Kodama, Thermochemical methane reforming using a reactive WO3/W redox system, Korean Journal of Air-Conditioning and Refrigeration Engineering, Energy. Vol. 20, pp. 411-425 (2000).
4. Muir, J.F., Hogan, R.E., Scocipes, R.D., and Buck, R., Solar energy, Vol. 52, pp. 467-477 (1994).
5. Worner, A., Tamme, R., catal. Today, Vol. 46, pp. 165-174 (1998).
6. Moeller, S., Buck, R., Tamme, R., Epstein, M., Libermann, D., Mery, M., Fisher, U., Rotstein, A., and Sugarman, C., In Proceedings of the 11th SolarPACES International Symposium on Concentration Solar Power and Chemical Energy Technologies, pp. 231-237 (2002).

PROSPECTS OF FLEXIBLE DESIGN STRATEGIES FOR EASY CHANGE AND ADAPTATION

NASIBEH SADAFI, M. F. MZAIN and M. JAMIL

University Kebangsaan Malaysia, Faculty of Engineering and Built Environment, Department of Architecture 43600, Bangi, Selangor, Malaysia

Environmental sustainability is one of the most important challenges in designing buildings structures. Conventional way of construction does not consider adaptability of buildings to the new needs and technical requirements that are related to their flexibility. Current study investigates various literatures for illustrating the criteria that they offer for flexible design. First a conceptual framework of the structure of relationship among types of changes and transformations in building as well as required reaction of the building was synthesized. Then the criteria for flexible design in building construction for design, installation, material properties and deconstruction are presented. It has been demonstrated that deconstruction strategies have been highly emphasized to assist the building to respond to the transformations properly.

1. Introduction

The most important issue regarding the building today is related to the increase of its environmental efficiency [1]. Therefore one of the initial objectives of construction industry is to donate towards global sustainability by using energy saving processes, reducing the use of natural resources and reducing the waste production.

In this case in order to find a balance between the efficient use of materials, changing user demands and increasing life cycle costs different approaches to building design and construction have been applied. Actually, attention to the capacity of the original design and constructed systems to accommodate change easily, fast, and inexpensively over the full life of the building would increase the functional life of a building. It can also improve the value of the building to the owner and reduce interruption and downtime for the occupants [2]. If the building sector is to respond to this challenge it needs to provide buildings that are smart, adaptable and transformable [3].

This paper tries to presents an overview through the criteria for increasing flexibility in designing building components. Main concept of flexibility will be defined first. Next section explains the requirements of applying flexibility in building construction. After that the strategies for increasing flexibility as well as their degree of importance and emphasize are described. Final section discusses the results and conclusions.

2. What is flexibility

Various researchers have defined flexibility in different ways. Flexibility is a widely used term referring to the adoptability of buildings features to the needs of its users. Keymer [4] described the flexibility as the capacity of a building or building system to accommodate changes. A building which is capable of accommodating different changes for a long time could be considered as high flexible.

In technical terms buildings are constructed to last and have longevity. Macozoma [5] described building longevity as its ability to sustain structural integrity for a long time as well as maintain desirability in terms of its functionality and style. Structural integrity is confirmed by the quality of construction and the durability of materials. On the other hand, durability is the building's ability to fulfill the requirements of the changing environment and make adjustments. [6].

Adaptability is another important definition. It refers to the building construction which is changeable enough to accommodate the physical environment and user's changes requirements [5]. Existence of balance between durability and adaptability in the design of a building is called flexibility which is an important quality in sustainable designed and constructed buildings [7].

The term flexible is often used to describe buildings that are inexpensive structures and can be amortized in just a few years [8]. Some has defined the flexible building as place that the operations could change overnight [9]. It is designed to adapt and accommodate changes during its lifecycle. In fact flexibility helps to cause sustainability requirements by decreasing the waste and financial risks while improving the efficiency of the required environment. Experiences show that flexibility or adaptability to change, no matter how it is achieved, is a valuable characteristic that helps to delay or avoid obsolescence in the building.

3. Why design for flexibility

Changes in building will happen according to different reasons during its useful life. In fact change is an obvious and inevitable characteristic of built environment and requires sufficient attention at the initial design phase. Different studies have discussed the changes factors in building [11, 12]. Two types of changes are recognizable during the life of the buildings namely: uncertainties in inter-building variations as well as environmental changes (Table 1).

Table 1. Common factors that cause change in buildings.

Change Class	Change Factors	
Internal expected variations	Changes in functional requirements Poor maintenance or abuse of systems Organizational shifts increasing performance expectations Aesthetic shifts	
Environmental uncertainties	Changes in market	Product price Energy price
	Changes in climate	(Global warming)
	Adaption of new standards and codes Technological changes Shifts in property values	

These changes will effect various parts of the building differently. Brand [12] has divided building systems into six layers to wit: site, skin, structure, services, space plan and stuff. Change in each layer happens at different time while, changes in the building are controlled by slowly changing components, while the rapidly changing components simply following along: site dominates the changing in structure, which controls the skin and so on for the services, space plan and stuff respectively.

Changes in a building are result of interactions between different layers and systems. Therefore, the level and type of changes in interactions determines the importance and type of changes in the whole system. Indeed understanding the hierarchy of changing time in the building layers can assist the designers to recognize the change reason and proper reaction in the early stages.

Slaughter [2] categorized the kind of changes in different layers of the building into three general types in; function, capacity and flow:

- Changes in the function:
  Building's function normally responds to the user's different requirements. So when occupant's needs or activities change during the time, the required function according to that should change. Changes in function include: incorporating new functions and modifying existing function.
- Changes in capacity
  Capacity deals with the capability of building to perform expected thresholds of toleration under different load and condition or volume. Changes in the capacity of building volume need feasibility for adaption in the spatial dimensions.
- Changes in flow
  Changes in flows relate to the correlations between building components and their internal or external environment as well as the users. Changes in climatic or physical environmental conditions as well as movements or activities of people within a building will cause changes in the environmental and people flows.

These changes will cause interactions in the building systems. In fact building's interaction can be observed through three dimensions of transformation; structural, spatial and material transformations [13].

- Changes in the function of a building cause spatial transformations which require adaptability and disassembly to preserve the integrity of the spaces.
- Changes in the capacity may result in structural and/or elements & materials transformations. In this case replaceability, reuse and recycling of the building components will assist in transformation.
- Changes in flow within and around the building bring about spatial as well as element and material transformations. Spatial adaptability and also recycling of the building materials can ensure continuity in the imposition of the building components during the transformation.

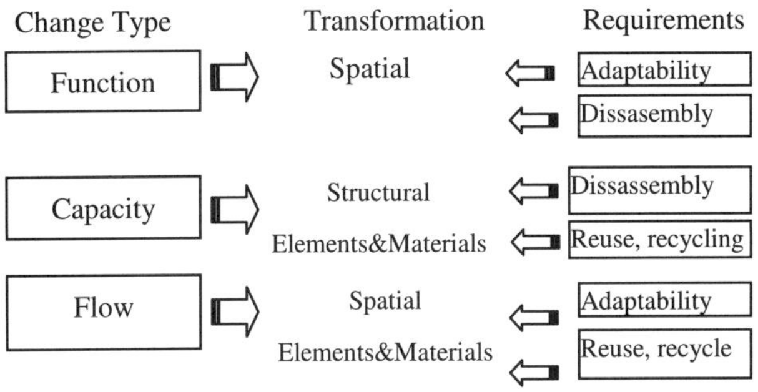

Figure 1. Building transformation: related causes and requirements.

It is important to find an applicable way of design and construction that will allow for easy adaptation of buildings to the changing concepts by rearranging its structure and components and reducing the required components and materials. Such an approach could ensure integrated solutions that will fulfill adaptability, building cost and negative environmental impact at the same time [3].

Therefore, one of the key issues of sustainability is development of design strategies that will transform inflexible building structures into dynamic and flexible structures whose parts could be easily disassembled and/or changed and reassembled. This would intensely improve capacity of building structure to be transformed from building to the material level whenever needed, with minimal environmental impacts.

4. Criteria for flexible design in building construction

Numerous studies have attempted to explain the criteria of design to increase the flexibility in buildings [2, 4, 5, 7, 8, 13-25]. These strategies could be classified into four main groups regarding their relation to different stages of design and construction: design, material properties, installation and deconstruction. According to the times that each strategy has been emphasized in other studies, the degree of importance has been ranked as; considerable, important and very important (Table 2-5).

4.1. *Design*

Design for flexibility will have more effects when it is initiated in the primary phase. Inclusion of flexibility in early design process helps to integrate it with other design goals and requirements. Author's reviews on previous studies have indicated that, designing the building according to the principles of sustainability, design over currently required capacity, use modular coordinate systems and prefabricating major components have been highly emphasized as critical design strategies to increase flexibility. There are also other strategies such as; Increase system predictability, Reduce intersystem interactions and etc in the next stage of importance (Table 2). Indeed, it does not mean that strategies with lower emphasize (considerable) have less influences on the flexibility of the building. However, they might also be fulfilled by applying some other more comprehensive strategies.

Table 2. Strategies to increase building flexibility for design stage.

	Design	Suggested by	Degree of importance
1	Fulfill the principles of sustainable construction	[18] [7] [15] [16] [26]	Very Important
2	Increase regularity in building patterns	[15]	Considerable
3	Increase simplicity in systems and materials	[15]	Considerable
4	Design the core structure to be partitionable	[7]	Considerable
5	Give specifications for connections, structural and installations	[21] [7]	Important
6	Use modular coordinate systems	[21] [7] [14]	Very Important
7	Dedicate specific volume for each system	[2] [4]	Important
8	Use Prefabricated components	[21] [2] [4]	Very Important
9	Design over capacity	[2] [14] [4]	Very Important
10	Reduce intersystem interactions	[2] [4]	Important
11	Reduce intrasystem interactions	[2] [4]	Important
12	Increase system predictability	[2] [4]	Important
13	Improve flow through system layout	[2] [4]	Important

- Referenced once: considerable
- Referenced twice: important
- Referenced three times and more: very important

4.2. *Materials*

Recent studies [13] shows that building materials are the major part of nowadays waste. Large parts of the recycled materials are limited to low quality use or even land fill, because of large amount of contaminants. Since in current construction components and structures are not designed to be recycled, they cannot be separated to be reused in new buildings. It has also made it difficult to recycle and reuse the materials since they are composed of hazardous components. It has been revealed that material properties have a high degree of emphasize by other researchers to improve building flexibility. By increasing the building flexibility it would be possible to improve recycling rate which will result in reducing the environmental impacts (Table 3).

Table 3. Strategies to increase building flexibility for material properties.

	Material properties	Suggested by	Degree of importance
1	Use high quality of the recycled material	[25] [7] [22] [20] [16]	Very Important
2	Use exchangeable components materials	[23] [15] [16] [4]	Very Important
3	Use easily separable materials	[23] [15]	Important
4	Use materials without hazardous components	[21] [15]	Considerable
5	Avoid secondary finishes to materials	[21]	Considerable

4.3. *Installation*

Flexibility of the building components has an inseparable connection with the flexibility of the installation process. The way that the building is assembled effect its disassembly process. And so considering disassembly process at the beginning of design stage will have important effects on the entire life of the building and its materials. Several attempts have been made [2, 4, 14, 15, 21, 24, 25] to suggest important strategies for improve building flexibility in installation stage. In general it has been demonstrated that strategies which have initiated 'separating different systems of the building' as well as 'considering the adjacency of the relevant systems' have been more emphasized by other studies (Table 4).

Table 4. Strategies to increase building flexibility for installation.

	Installation	Suggested by	Degree of importance
1	Integrate the design of installations systems into the structural building design	[14]	Considerable
2	Avoid running installations through structural sections	[14]	Considerable
3	Separate the structural and infill elements of a building	[14] [15] [4] [21]	Very Important
4	Base the installation on maximum partition plan	[14]	Considerable
5	Restrict distribution of function and facilities	[25] [14]	Important
6	Use removable facilities instead of fixed installations	[14]	Considerable
7	Increase physical adjacency of access point	[4]	Considerable
8	Improve phase system installation	[21] [2] [4]	Very Important
9	Enhance system proximity	[24] [4]	Important

4.4. *Deconstruction*

Ability of the building to be disassembled to its components is considered as the key technical factor for waste management in current construction. For instance, transparency and easy access building components will cause easy dismantling, therefore, the service life of the building will increase by expediting easily repair, replacement and modification of the building components. Several studies have argued the importance of applying the strategies to simplify the deconstruction process [2, 4, 7, 13-16, 19-21] which shows the importance of this stage to increase flexibility as well as longevity of the building. It has been demonstrated that 'accessibility of the building components', 'providing sufficient information on the building construction system and assembly procedure' and 'applying flexible interfaces' have been highly recommended to clarify disassembly (Table 5).

Table 5. Strategies to increase building flexibility for deconstruction.

	Deconstruction	Suggested by	Degree of importance
1	Simplify demolition	[2] [7] [4]	Important
2	Transparency and ability to access building components that need to be removed	[21] [2] [14] [19] [20] [4]	Very Important
3	Clearly identify points of disassembly	[21]	Considerable
3	Sufficient information to let the deconstruction happen in an efficient and safe way	[21] [15] [7] [13] [16]	Very Important
4	Limited number of components	[15]	Considerable
5	Use interchangeable system components for ordering the materials and their replacement	[21] [2]	Considerable
6	Use flexible building interfaces	[21] [13] [15] [7] [14] [16]	Very Important

5. Conclusion

Each building is consisting of hierarchical arrangement of different layers. This hierarchy determines the difficulty or easiness of the building's future deconstruction or disassembly. Sustainable development has highlighted the importance of life time expansion and possibility for dismantling and treatments on building. On the other hand occupant's satisfaction has a strong influence on the success or failure of a building construction. Therefore designer, contractor and users should try to accept any changes when new information come into existence. If so, adaptability and demount ability of the building components play an important role to improve flexibility as well as sustainability of the construction. Increasing the flexibility of the building should be considered in the initial stages of design and construction. In fact four main parts of construction namely; design, installation, deconstruction and material properties require comprehensive attention to fulfill adaptability during the whole life of the building. Current study has perceived that considering the strategies to facilitate 'deconstruction' part has the most important influence on building flexibility followed by material properties, design and installation. In this regard the useful life of the building would be increased and since the building components could be adapted and materials would be reused or recycled both the building owner and occupants would gain benefits.

References

1. Brouwer, J. and E. Durmisevic, Toward Dynamic Building Structures- Building with Systems. 2002, Faculty of Architecture, Delft University of Technology Department of Building Technology - OBOM. De Vries van Heystplantsoen 2, 2628 RZ Delft, The Netherlands.
2. Slaughter, E.S., Design strategies to increase building fexibility. Building Research & information, 2001. 29(3): p. 208-217.
3. Durmisevic, E. The Netherlands. in Construction Materials Stewardship The status QUO in Selected Countries CIB Publication 318. 2008. Victoria University of Wellington, Wellington, New Zealand.
4. Keymer, M., Design strategies for new and renovation construction that increase the capacity of buildings to accommodate change, in MIT. 2000, Cambridge.
5. Macozoma, D.S. Building Deconstruction. in CSIR Building and Construction Technology, CIB publication. 2001. South Africa.
6. FishbeinBette, K., Building for the Future: Strategies to Reduce Construction and Demolition Waste in Municipal Projects, in INFORM. 1998.
7. S. Macozoma, D. Undrestanding the concept of flexibility in design for deconstruction. in Design for Deconstruction and Materials Reuse. 2002. University of Karlsruhe Karlsruhe, Germany: CIB Publication 272.
8. Sennwald, B., Flexibility by Design. Architecture: The AIA Journal, 1987. 76(4): p. 89-94.
9. Patterson, M., No time to change: flexible facilities must regroup, reorganize, restructure in an instant. Buildings, 1998. 92(7): p. 38-42.
10. Fischer, G., A headache-free approach to building modernization. Buildings, 1999. 93(2): p. 82-84.
11. V. Greden, L., Flexibility in Building Design: A real options approach and valuation methodology to address risk, in Department of Architecture. 2005, Massachusetts Institute of Technology.
12. Brand, S., How buildings learn: what happens after they're built. 1994, New York: Viking Press.
13. Durmisevic, E., Transformable building structures, in Faculty of Architecture, Department of Building Technology. 2006, Delft University of Technology: Delf.
14. Geraedts, R.P. Upgrading the Flexibility of Buildings. 2001. Wellington, New Zealand: proceedings of the CIB World Building Congress.
15. Mark Webster, et al. Towards Dynamic Building Structures- Building with Systems. in 2Greenbuild Conference. 2005. Atlanta: Costello Dismantling Co.
16. Sassi, P., Designing closed-loopmaterial cycle construction. BUILDING RESEARCH & INFORMATION, 2008. 36(5): p. 509-519.
17. Maury, C.L., Framework to assess a facility's ability to accommodate change: application to renovated buildings. 1999, Tufts University, Massachusetts Institute of Technology.
18. Kibert, J.C. Establishing principles and a model for sustainable construction. in Sustainable construction. 1994. Florida, USA: CIB task group.
19. Fletcher, S.L., R. Plank, and O. Popovic, Designing for future reuse and recycling, in Deconstruction – Closing the Loop. 2000, Building Research Establishment (BRE): Watford.
20. Morgan, C. and F. Stevenson, Design for Deconstruction SEDA Design Guides for Scotland: No. 1. 2005, Scottish Executive and the Scottish Ecological Design Association, Edinburgh (available at: www.seda2.org/dfd/index.htm).
21. Crowther, P., Design for Disassembly- Themes and Principles, in RAIA/BDP Environmental Design Guide2005.
22. Thormark, C., Recycling Potential and Design for Disassembly in Building. 2001, TABK – 01/1021, Lund Institute of Technology: Lund.

23. Charytonowicz, J., Reconsumption and Recycling in the Ergonomic Design of Architecture. 2007: p. 313-322.
24. Davison, N., et al., The multispace adaptable building concept and its extension into mass construction, in Adaptable Building Structures. 2006: Eindhoven, The Netherlands.
25. Vakili-Ardebili, A. and A.H. Boussabine, Eco-Building Design (EBD): Design Strategies to Increase Building Compatability, in PLEA, 23th Conference on Passive and Low Energy Architecture 2006: Geneva, Swizerland.
26. Kibert, C. Construction ecology and metabolism. in Integrated life cycle design. 2000. Helsinki.

IRON DOPED MICROPOROUS ACTIVATED CARBON (PHENOLIC RESIN) AS AN ADSORBENT FOR ARSENIC REMOVAL

AJIT SHARMA, ASHUTOSH SHARMA and NISHITH VERMA

Department of Chemical Engineering and DST Unit on Nanosciences, IIT Kanpur, Kanpur-208016, India

NALINI SANKARARAMAKRISHNAN

Center for Environmental Sciences and Engineering, IIT Kanpur, Kanpur-208016, India

In this study we have described the development of Fe-impregnated micro/nano activated carbon (AC) particles for the removal of dissolved arsenic from wastewater. The particles in beads form are synthesized using suspension polymerization. The beads are subsequently carbonized followed by physical activation using steam or CO2 to develop micro and meso pores in the particles. Irons are incorporated in the polymeric beads in an intermediate step during polymerization. Thus, AC is used as a substrate to deposit Fe in its micro- and mesopores. Fe-doped AC beads may thereafter be milled to produce the micro- and nano size particles for increasing their reactivity. We have shown that Fe-doped AC prepared this way are more effective in removing dissolved arsenic in water than those synthesized via the traditional route of preparing Fe impregnated AC.

1. Introduction

Arsenic is a toxic element present in several industrial effluents as well as in ground water. The maximum arsenic level in the drinking water has been set at 10μg/L by the World Health Organization (WHO), whereas the permitted limit for industrial effluent is less than 0.2 mg/L [1]. Several materials including iron-modified activated carbon, iron-oxide coated sand, fungal biomass and slag iron-titanium oxide materials have been screened for their utility in the removal of arsenic from water [2-4]. However, the applicability of these materials is limited due to their low efficiency (relatively smaller loading of dissolved solutes in water). Therefore, it is necessary to develop a cost effective, simple and user-friendly method for treating arsenic containing wastewater.

In this study we have described the development of Fe-impregnated micro/nano activated carbon (AC) particles for the removal of dissolved arsenic from wastewater. The particles in beads form are first synthesized using suspension polymerization and then carbonized and activated to develop micro and meso pores in the particles. Irons are incorporated into the polymeric beads in an intermediate step during polymerization. Thus, AC is used as a substrate to deposit Fe in its micro- and mesopores. Fe-doped AC beads are milled to produce the micro- and nano size particles for increasing their reactivity. We have shown that Fe-doped AC prepared this way are more effective in removing dissolved arsenic in water than those synthesized via the traditional route of preparing Fe impregnated AC. The present study on the development of AC based adsorbents is inspired by our recent success in development of similar carbon based micro-/nano-adsorbents for the control of atmospheric air pollutants [5-7].

2. Experiment

Fig. 1 is the schematic of the experimental set-up used in the study. The phenolic resin beads were prepared by the reaction of phenol with formaldehyde in the presence of a basic catalyst triethylamine. Phenol (50g, heated to molten state in presence of nitrogen atmosphere) was poured in a three–necked 1L glass reactor equipped with reflux condenser. The solvent, formaldehyde (63ml, 37% formalin) and the catalyst, triethylamine (1.5 ml) were added into the reaction mixture, with nitrogen gas used as inert purge gas for 5

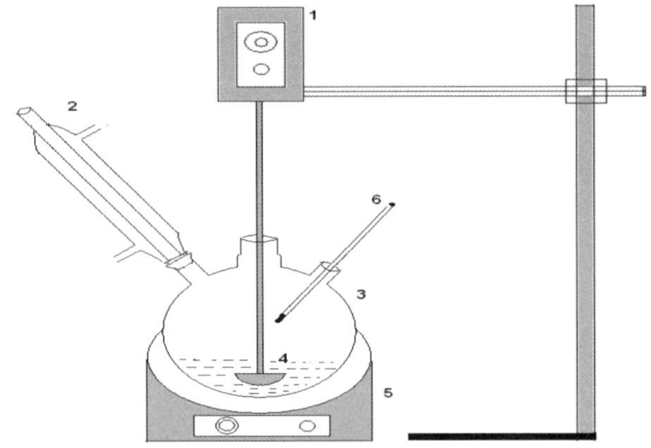

Fig. 1. Experimental set-up for synthesis of polymeric Fe incorporated beads (1: motor and stirrer, 2: condenser, 3: glass assembly, 4: perforated stirrer, 5: heating mantle, 6: thermometer).

min. After stirring for ½ hr, hexamine (3.5g) as cross-linking agent was added. The reaction mixture was heated at the rate of 3°C/min up to its boiling point. After approximately 40 min of heating, the solution turned into an emulsion having whitish-yellow colour. The suspension stabilizing agent, polyvinyl alcohol (PVA, 3g) was added after a critical time (less than 15 min). After approximately 45 min, ferric chloride (0.5 g) was added. Heating was continued. It was ensured that the speed of the stirrer never fluctuated from the initially fixed set value due to the solidification of product. After 4hrs, heating was stopped and the reactor was allowed to cool. When the temperature of the reaction products reached the room temperature, the products were filtered to separate the solids from the residual liquid. The filtered product (spherical beads of average size ~0.8 mm) was washed three times with distilled water, methanol and acetone. The spherical beads were carbonized and activated using steam as per the procedure described elsewhere [6]. The beads were subsequently crushed to powders (approx ~80 nm average size) in a ball mill (Retsch, Germany). Alternatively, powders were made by milling before carbonization/activation stage.

3. Results and Discussion

Several test-runs carried out to optimize the operating conditions for polymerization suggested that (1) sequence of adding different reagents in the reaction mixture, (2) time between starting of the reaction and addition of cross-linking agent, (3) the amount of catalyst, and (4) total time of reaction are critical to obtaining the desired quantity, shape (spherical polymeric beads in the present case), and size distribution of beads. Especially, the time of addition of PVA, after the solution has turned to whitish yellow emulsion, controlled the shape of the product.

Following synthesis of the spherical polymeric beads (of size ~ 0.6 mm) and activation step, the BET area measurements were made for three types of samples, namely: as-prepared Fe-coated polymeric beads (PBFe), AC powder (~80 nm average particle size) prepared before and after activation of the beads, PBFe-Act and PBFe-BM-Act, respectively. The analysis showed that the BET area (~ 10 m^2/g) of the PBFe sample was insignificant in comparison to those obtained for PBFe-Act (150 m^2/g) and PBFE-BM-Act (850 m^2/g). The data clearly indicate the milling of the synthesized polymeric beads before activation as the preferred route to preparing the particles as adsorbents.

Several batch experiments were carried out to determine the equilibrium loading of arsenic ions in aqueous solution on the prepared adsorbents. The isotherms obtained for the three types of samples indeed confirmed the relatively superior performance of PBFe-BM-Act. Fig. 2 compares the isotherms corresponding to the three samples. As observed, the loading of As^{+3} is larger on PBFe-BM-Act than on the other two samples over the concentrations ranging between 50 and 8000 ppb. The As-ions loading on PBFe-BM-Act is observed to be approximately 1 mg/g corresponding to the aqueous phase concentration of 2000 ppb.

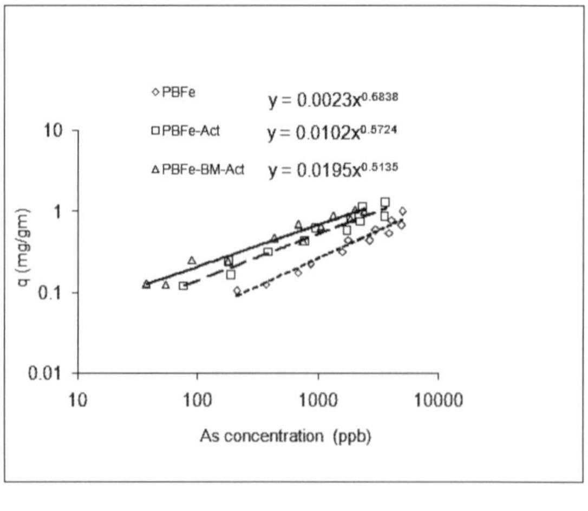

Fig. 2. Equilibrium isotherms for As vs. PBFe, PBFe-Activated and PBFe-BM-Activated.

In addition, the data plotted for all three samples are observed to follow, the Freundlich isotherm.

Several column experiments were also carried out to ascertain the removal efficiency of the adsorbents (PBM-BM-Act) under dynamic conditions for dissolved arsenic ions in water. The results show relatively longer breakthrough times and total adsorption times for the synthesized adsorbents in this study. We present here the results pertaining to the effect of bed depth on the breakthrough characteristics. Fig. 3 describes the data corresponding to the bed depths of 3, 6 and 12 cm at a constant flow rate of 3 ml min^{-1}.

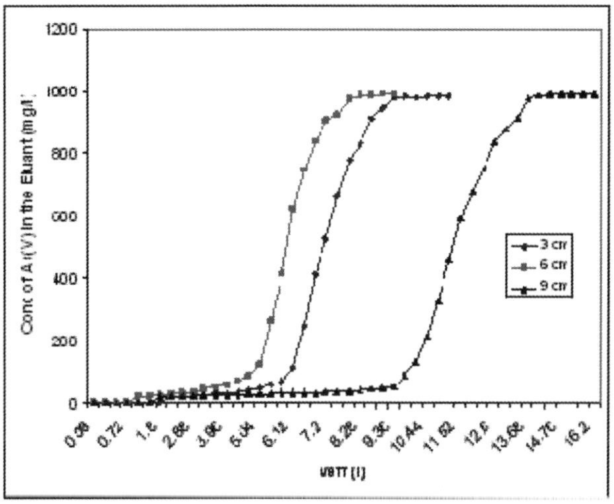

Fig. 3. Effect of bed height on breakthrough and adsorption times.

The results show that the shape and gradient of the breakthrough curve were slightly different with the variation in bed depth. The higher uptake of As was observed at the beginning of the fixed-bed column, but the concentration of As in the effluent rapidly increased after breakthrough volume or breakthrough time. The lower bed depth gets saturated earlier than higher bed depth. As also observed from Fig. 3, the breakthrough volumes are 1.89, 3.78 and 9.18 L at bed depths of 3, 6 and 12 cm, respectively. From these results, it may be concluded that the increase in the As uptake with increase of the bed depth was due to the increase in the contact time for adsorption.

4. Conclusions

In this study, we have synthesized Fe-doped nano carbon beads particles of average size ~ 80 nm, which are used as adsorbents for the removal of dissolved arsenic ions in water. Synthesis of the polymeric beads was carried out using suspension polymerization and process conditions were optimized to maximize the yield, shape (sphere), and purity of the product. Incorporation of Fe ions during polymerization is a novel step in preparing Fe-impregnated carbons and is shown to be more efficient than the conventional method of impregnating AC with metal ions. The steps comprising carbonization and activation following crushing (milling) are shown to yield in the nano size particles with enhanced adsorption capacity for arsenic ions over the concentration range of 50-8000 ppb in water. The prepared carbon beads or powders have immense potential for functionalization with a variety of surface groups, and also for growing carbon nanofibers, which may be used in several end applications related to environmental remediation. This part of the study is currently underway in our group and the results will be discussed at the time of the presentation.

Acknowledgments

The authors want to acknowledge the DST funding of its Unit on Nanosciences at IIT Kanpur.

References

1. P.L. Smedley and D. G. Kinniburgh, *App. Geochem.* **17**, 517 (2002).
2. T.S. Singh and K. K. Pant, 2004. *Sep. Purif. Technol.* **36**, 139 (2004).
3. F. S. Zhang and H. Itoh, *Chemosphere* **65**, 125 (2006).
4. D. Pokhre and T. Viraraghavan, *Water, Air, and Soil Pollution* **173**, 195 (2006).
5. V. Gaur, R. Asthana and N. Verma, *Carbon* **44**, 46 (2006).
6. V. Gaur, A. Sharma and N. Verma, *Chem. Eng. Process* **45**, 1 (2006).
7. R. Singhal, A. Sharma and N. Verma, *Ind. Eng. Chem. Res.* **47**, 3700 (2008).

EFFECT OF TREATED OIL WELL EFFLUENT ON SOIL EXCHANGE CAPACITY

NEELA EMANUEL

Department of Biology & Chemistry, Texas A & M International University, 5201 University Boulevard Laredo, TX 78041, USA

GIREESH KUMAR

78041, Department of Chemistry, M.S. University Vadodara 390002, India

Utilization of treated oil well effluent for irrigation depends on the salinity level and salt tolerance of the plant. In this study the treated oil effluent contained 12 mg/L oil and 9000 mg/l of total dissolved solids, mainly sodium chloride. The treated effluent was diluted to contain 5000, 3000, 1500 mg/L NaCl and pH was adjusted to various values before being applied to soil. Application of wastewater with high sodium levels has very little effect on the overall level of soil fertility, with the exception of alkaline effluents which can dissolve soil organic cations. The long term application of such waste-water may cause a buildup of soil sodium and decreases the soil hydraulic conductivity of the soil. The cation exchange capacity was determined using AAS and flame photometer after extracting with ammonium acetate at the same treatment pH. Further studies were carried out by adding CaCO3, Ca(CH3COO)2 and KCl to the system to study the change in exchange capacity.

1. Introduction

The physical and chemical properties of oil-well effluent vary considerably depending on the geographical location of the field, the geological host formation, and the type of hydrocarbon being produced. Properties of volumes of the effluent can vary throughout the lifetime of a well [1]. The ground water usually contains significant amounts of divalent cations like Ca, Mg, Ba, and Sr which can react with anions, like sulfate and carbonate to produce corresponding salts [2]. Iron containing particles are often found in oil-well effluent [3]. The increasing environmental health issues in developing its quality could be hazardous to plants and animal life. Water countries resulted in a concomitant interest in the safe disposal of waste-water beneficially [4]. The use of irrigation as the primary disposal method for the waste-water becomes more difficult as the salinity and total dissolved salts (TDS) of the feed-water increase beyond the fresh water level [5]. The hazardous effect of the use of poor quality water on soils depends on many factors such as soil type, initial salt content in soil, amount and nature of salt in irrigation water, etc. Determination of suitability of water for irrigation is a complex task. It not only involves the composition of the irrigation water, but also the knowledge of how a particular soil reacts to the irrigation over a period of time. The conductivity measurement of soil and water, pH, sodium adsorption ratio (SAR), cation exchange capacity (CEC), Na/Ca+K+Mg ratio, exchangeable sodium fraction and Gappon constant have all been used in several combinations to predict the soil behavior in relation to the quality of the irrigation water. However, the drainage relation of the soil profile and the impact of different farm management regimes are difficult to predict in the laboratory and need field studies. In addition the salt tolerance of the crops, the rising demand for agriculture products and environmental regulations on industries to dispose of or to use their waste-water need to be considered in developing the irrigation guidelines.

Reuse of waste-water for irrigation is beneficial as it is a cheap source of water. Properly planned use of waste-water can reduce the environmental and health related hazards, which have been observed with traditional water disposal practices in developing countries. This is an economical way to dispose of waste-water to prevent pollution and salinity problems etc. [4, 6]. The use of available water for irrigation without assessing with high dissolved solids affects the physical and chemical properties of the soil depending on the total salt concentration, properties of various cations and anions and the ion exchange characteristics of the soil surface [7]. The mode of salt transportation and accumulation in the soil depend on the quantity and quality of the water [8]. When saline irrigation water is used, the main exchange is taking place between Na, Ca, and Mg ions [9].

It is well known fact that high concentration of sodium in the irrigation water will increase the hardness and porosity of the soil [10-12]. The clay fraction constitutes the most reactive part of the soil. These clay particles are in colloidal dimension and they can be considered as complex silicate in which silica is combined with different proportions of alumina and magnesia along with the varying proportions of other elements. As the clay colloid is negatively charged, it attracts a large number of positively charged cations. The clay nucleus can be considered as micelle, a negative radical having a swarm of adsorbed cations [10]. Because of this, cations like Ca, Mg, Na, and K are adsorbed on the soil. These loosely held cations can be replaced by other cations and this is known as 'cation' exchange. Cation exchange capacity is defined as the number of milli- equivalent of cations that can be held in an exchangeable form by 100 gm of the soil. This is an important property describing the nutrient availability for plant growth [13, 14]. The presence of Ca ions in suitable proportions to other exchangeable cations such as Na, K, H, and NH_4^+ is necessary, if the soil is to be suitable for plant root development. The outline of the fundamental forces involved in the clay particle interactions will be given as the background for the consideration of the effect of sodium ions on the structure of the soil and the ways in which the effect of sodium can be diminished [11]. If the exchangeable sodium occupies more than 15% of the total exchange capacity, the clay and the organic matters will get dispersed due to the flocculating effect of the sodium and these soils will be in an unsatisfactory physical condition [10]. Because of this, the plant may no longer be able to take Ca ions for its needs. The effect of sodium ions on the swelling properties of clays is so pronounced that even in sandy loam soils very marked changes in soil permeability are observed, if the soil has an appreciable changeable sodium percentage [11]. As the exchangeable sodium increases, increased swelling reduces the size of the micro porosity and permeability of the soil [11]. When the exchangeable cations are predominantly Ca or Mg, the clay particles interact or repel each other only to a limited extent [10]. This leads to the granulation of clay particles and increases the aeration and drainage [10].

Sandy soil may be affected more adversely than clay soils under continued irrigation with saline water without proper leaching. Once the leaching is carefully provided, these soils can be reclaimed more easily due to better drain ability characteristics [15]. The pH of the soil is not liable to constant changes, but only minor fluctuations are seen from season to season. This is a reflection of the buffering capacity of the soil [16, 17]. A sudden change in soil pH may seriously affect the larger plants in getting the proper nutrients and also will affect the soil microorganisms.

Studies have also been carried out to see the effect of pH on ion exchange equilibria of soils [15]. It has been reported that sodic soil can be reclaimed and the permeability can be improved by the addition of gypsum to soil or to irrigation water [5, 10, 18, 19]. Addition of gypsum to soil or irrigating water increases the removal of sodium from the soil profile and improves the water infiltration rate. Calcium chloride can also be used for the same purpose [17].

The present irrigation guidelines deal with only a narrow range of water with unconventional composition of inorganic electrolytes. It is therefore necessary to carefully investigate and monitor the situation in which the use of irrigation water with unconventional compositions of electrolytes or organic materials are involved. Accordingly laboratory experiments have been conducted to find out maximum concentration of total dissolved salt (NaCl), and the optimum pH at which the treated oil well effluent can be used for agricultural purposes and the extent to which sodium adsorption on exchange complex of soil can be restrained by other chemicals.

2. Experimental Procedure

All the chemicals used for the studies are analytical grade and obtained from Sigma or Qualigens. Flame photometer (model 105) and AAS was used for cations analysis. Adjustments of pH of effluent were done using 0.05N HCl and NaOH solutions. The pH of ammonium acetate solution was adjusted with ammonium hydroxide and acetic acid solutions as per the requirements.

The soil samples (1-100 cm depth) for the study were collected taken from Mehsana, North Gujarat, India. The treated oil-well effluent taken for the present study contained 12 mg/L oil and the total dissolved salt

(TDS), mainly NaCl, was approx. 9000 mg/L. This treated effluent and the effluent diluted to contain 5000 mg/L, 3000 mg/L and 1500 mg/L of TDS were applied to soil after adjusting the pH to 7.5, 6.0 and 5.0.

3. Results and Discussions

The soil of the Mehsana region are alluvial in nature and have an average composition of 70% sand, 10% silt and 20% clay. They are extremely porous with an excellent drainage profile and have a pH of 8.95. The upper layer of the soil contains small amounts of $CaCO_3$ which increases with increasing depth.. The clays play an important role in determining the physical characteristic of the soil, soil fertility and degree of soil susceptibility of saline damage.

2.5 gm of soil was weighed and mixed with distilled water and stirred for half an hour. The solution was allowed to settle and then filtered. The filtrate was analyzed by flame photometer and AAS to determine the soluble sodium, potassium and calcium ions in the soil (Table 1).

Table 1. Soluble cations present in the soil of Mehsana region in meq/100gm soil.

pH	Na	K	Ca
8.95	2.41	0.12	1.30

2.5 gm of soil was weighed, washed with 40% alcohol and then with distilled water to remove soluble salts present in the soil. The exchangeable cation present in the soil was determined by extracting with ammonium acetate [21] at pH 8.95 and then analyzed by AAS and flame photometry (Table 2).

Table 2. Effect of pH of irrigation water (control) on the cationic exchange of Mehsana soil.

pH	Exchangeable ions (meq/100gm soil)		
	Na	Ca	K
8.95	6.1	81.0	0.50
8.50	6.2	87.5	0.54
7.50	6.2	98.0	0.23
6.0	6.3	98.0	0.33
5.0	5.7	185.0	0.55

In order to study the effect of pH of the irrigation water on the cation exchange capacity of the soil, 2.5 gm of the sample was added to 100 ml water at pH 7.5, shaken for an hour and kept for a day before being filtered & washed with distilled water to remove the soluble salts. The cations in the exchangeable complex were determined after extraction with ammonium acetate solution at pH 7.5 and using flame photometer and AAS. The same experiments were carried out with effluent water at pH 6.0 and 5.0. The extractions were also carried out at the same experimental pH. The results are presented in Table 3.

The effect of different salt concentrations in the effluent on the exchange capacity of soil at varying pH values were also investigated. After an hour of vigorous shaking of the effluent with the soil, the mixture was kept for a day, filtered and the soluble salts were removed. Then the exchangeable ions were extracted with ammonium acetate and analyzed flame photometrically and by AAS (Table 3).

Table 3. Effect of pH and concentration of effluent on cation exchange capacity of soil of Mehsana region.

Na				
pH		7.50	6.00	5.00
Control (600 mg/L of TDS)		6.16	6.26	5.70
Effluent concentration in mg/L	1500	4.30	7.20	8.80
	3000	5.30	7.20	9.00
	5000	7.40	8.40	10.60
	9000	7.60	11.00	11.20
Ca				
pH		7.50	6.00	5.00
Control (600 mg/L of TDS)		98.00	98.00	185.00
Effluent concentration in mg/L	1500	120.00	240.00	255.00
	3000	110.00	225.00	252.50
	5000	100.00	205.50	232.50
	9000	105.00	202.50	232.50
K				
pH		7.50	6.00	5.00
Control (600 mg/L of TDS)		0.23	0.35	0.55
Effluent concentration in mg/L	1500	0.35	0.84	0.95
	3000	0.33	0.83	0.93
	5000	0.39	0.84	0.95
	9000	0.35	0.85	1.01
	FNa	FCa	FK	
pH	9.15*	32.87*	46.59*	
TDS	5.34**	4.38**	7.95**	

*required F value of 0.01 = 8.65
**required F value at 0.01=7.01 & 0.05 =3.84

The extent of sodium exchange to the exchange complex can be restricted by adding the salts of calcium to the soil or to the irrigation water [2, 10, 18-21]. This effect was investigated by the addition of calcium as

$CaCO_3$ (0.5% to the soil), 200 mg/L of calcium as calcium acetate and 500 mg/L of potassium as potassium chloride to the effluent respectively. The results are presented in Table 4-6.

Table 4. Percentage of sodium in the exchange complex of soil during the application of irrigation water with different TDS content at different pH values.

pH	Effluent TDS concentration in mg/L				
	Control (600 mg /L TDS)	1500	3000	5000	9000
7.5	5.9	3.5	4.6	6.9	6.7
6.0	6.0	2.9	3.1	3.9	5.1
5.0	3.0	3.3	3.4	4.3	4.6

Table 5. $CaCO_3$ addition to the soil before effluent treatment.

Na				
pH		7.50	6.00	5.00
Control (600 mg/L of TDS)		6.16	6.26	5.70
Effluent concentration in mg/L	1500	4.39	6.28	8.77
	3000	6.18	7.10	9.00
	5000	7.24	9.30	10.58
	9000	10.23	12.27	11.20
Ca				
pH		7.50	6.00	5.00
Control (600 mg/L of TDS)		98.00	98.00	185.00
Effluent concentration in mg/L	1500	117.50	250.00	295.00
	3000	107.50	230.00	270.00
	5000	112.50	225.00	265.00
	9000	105.00	225.00	247.50
K				
pH		7.50	6.00	5.00
Control (600 mg/L of TDS)		0.23	0.35	0.55
Effluent concentration in mg/L	1500	0.28	0.74	0.82
	3000	0.28	0.70	0.87
	5000	0.37	0.77	0.86
	9000	0.36	0.79	0.87

	FNa	FCa	FK
pH	5.46*	36.38*	66.00*
TDS	11.74**	4.81**	9.00**

*required F value of 0.01 = 8.65 & 0.05=4.46
**required F value at 0.01=7.01 and at 0.05 = 3.84

Table 6. Na/(Ca+K)% during the $CaCO_3$ addition to the soil prior to the application of irrigation water containing different amount TDS at varying pH levels.

pH	TDS content in irrigation water (mg/L)				
	Control (600 mg/L TDS)	1500	3000	5000	9000
7.5	6.3	3.7	5.7	6.4	9.7
6.0	6.4	2.5	3.1	4.1	5.4
5.0	3.1	3.0	3.3	4.0	4.5

At nearly 87meq/100gm, the cation exchange capacity of the Mehsana soil is well above the average for sandy soils, even though the texture classification of this soil is sandy. The sodium percentage of the exchangeable cation is also low at about 6%. This is probably the reason why this region has been able to grow crops for a sustained period with high TDS irrigation water. It was observed that the pH affected the exchange capacity and the exchangeable calcium increased with decreasing pH. The exchangeable sodium, however, was not much affected by the change in pH. However the exchangeable calcium and total CEC increased significantly while potassium remained stationary. Thus Na/Ca+K ratio was more favorable (Figure 1).

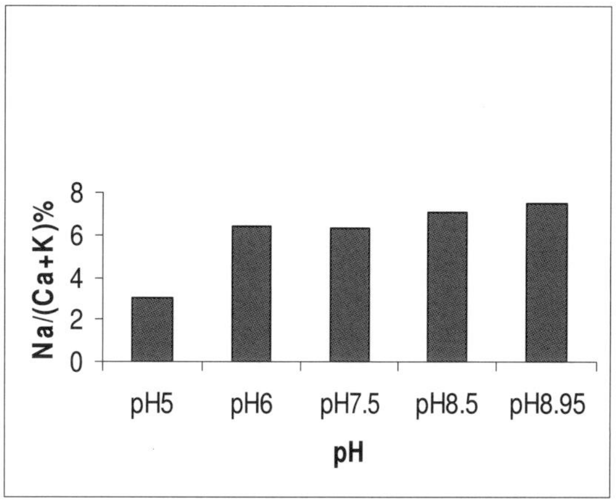

Figure 1. Effect of pH of irrigation water (control) on Na/(Ca+K)% of soil.

These results can be interpreted in two ways. First, one is that of the opening of the clay lattice and the second one being the opening of the peripheral charges on the clay surfaces to accommodate additional calcium on the particle surface. Even though the soils have less than 1% $CaCO_3$, there is sufficient calcium in the soil to account for this phenomenon. The sodium content in the irrigation water is not high enough to induce clay dispersion in the soil.

The concentrated effluent treatment (9000 mg/L) did not change the exchangeable sodium level at pH 7.5 (Table 3). Similarly the 6000 mg/L irrigation water did not change the sodium content level even when the pH was dropped to 5.0. But the treatment with higher salt content level water increased both the exchangeable sodium and calcium level of the exchange complex as the pH of the irrigation water was dropped. In the case of calcium and potassium, the increase was prominent when the water with 1500 mg/L NaCl was used. Later when the water with high TDS was applied, there was a small change in the exchangeable calcium, but not much difference in the exchangeable potassium was observed. This is attributed to the increase in exchangeable sodium in the exchange complex as the TDS in the irrigation water increased. Yet the Na/Na+Ca+K% was much less than the maximum limit of Na% in the soil and this kept the soil more permeable [9] (Table 4]. In all these treatments the effect of pH was significant beyond 0.01 levels and the effect of salt concentration was significant beyond 0.05 levels.

When 0.05% $CaCO_3$ was added to the soil before the treatments were imposed not much change in the sodium content was noticed even at lower pH compared to the experiment without the addition of $CaCO_3$. But, an increase in the calcium content in the exchange complex was noticed (Table 5).This was higher at lower pH and it may be due to the opening of the peripheral charges on the clay surfaces to accommodate additional calcium on the particle surface. As in the earlier experiments, a sudden hike in calcium and potassium is found when water containing 1500 mg/L TDS was applied to the soil. Later with higher TDS content calcium was found to decrease slightly but potassium remained almost constant. The Na/Ca+K% was very low showing the higher percentage of calcium in the soil, which keeps the soil more permeable (Table 6).

The addition of 200 mg/L calcium acetate to the irrigation water increased the calcium content in the exchange complex abnormally (Table 7). The calcium intake at the exchange site was higher at pH 6.0 than at 5.0 compared to $CaCO_3$ addition. Thus allowing Na/Ca+K ratio to decrease This may be due to the higher solubility of calcium acetate at higher pH as compared to calcium carbonate. In this case Na/Ca+K ratio remained practically constant (Table 8). The effect of pH was significant beyond the 0.01 level while TDS content effect was significant beyond the 0.1 level during these treatments.

Table 7. Effect of $Ca(CH_3COO)_2$ to the effluent on the cation exchange.

		Na		
pH		7.50	6.00	5.00
Control (600 mg/L of TDS)		6.16	6.26	5.70
Effluent concentration in mg/L	1500	2.28	7.45	5.14
	3000	2.83	7.85	5.81
	5000	2.96	8.80	8.26
	9000	4.18	8.62	10.16
		Ca		
pH		7.50	6.00	5.00
Control (600 mg/L of TDS)		98.00	98.00	185.00
Effluent concentration in mg/L	1500	100.00	315.00	260.00
	3000	90.00	285.00	235.00
	5000	85.00	335.00	235.00
	9000	100.00	350.00	240.00
		K		
pH		7.50	6.00	5.00
Control (600 mg/L of TDS)		0.23	0.35	0.55
Effluent concentration in mg/L	1500	0.24	0.66	0.73
	3000	0.27	0.72	0.57
	5000	0.24	0.66	0.69
	9000	0.27	0.60	0.69

	F_{Na}	F_{Ca}	F_K
pH	10.12*	15.01*	33.95*
TDS	1.43**	1.81**	2.06**

*required F value of 0.01 = 8.65
**required F value at 0.05 = 3.84

Table 8. Na/(Ca+K)% when $Ca(CH_3COO)_2$ was added to the irrigation water.

pH	Effluent TDS concentration in mg/L				
	Control (600 mg/L TDS)	1500	3000	5000	9000
7.5	6.3	2.3	3.1	3.5	4.2
6.0	6.4	2.4	2.7	2.6	2.5
5.0	3.1	2.0	2.5	3.5	4.2

The addition of 50 mg/L potassium chloride to the irrigation water increased the amount of potassium as well as calcium in the exchangeable complex as compared to the increase in the sodium content. Increase in the calcium content was significant beyond the 0.05 level and the potassium was significant beyond the 0.01 level (Table 9). Here also the Na/Ca+K ratio was found to be very low as in the case of the calcium acetate treatment and practically constant during the application of water with high TDS (Table 10).

Table 9. Effect of KCl on the exchange capacity.

Na				
pH		7.50	6.00	5.00
Control (600 mg/L of TDS)		6.16	6.26	5.70
Effluent concentration in mg/L	1500	3.02	5.39	6.73
	3000	3.21	5.96	7.28
	5000	5.13	7.30	7.83
	9000	8.26	7.23	9.28
Ca				
pH		7.50	6.00	5.00
Control (600 mg/L of TDS)		98.00	98.00	185.00
Effluent concentration in mg/L	1500	72.50	255.00	262.50
	3000	87.50	252.50	247.50
	5000	105.00	270.00	242.50
	9000	90.00	260.00	242.50
K				
pH		7.50	6.00	5.00
Control (600 mg/L of TDS)		0.23	0.35	0.55
Effluent concentration in mg/L	1500	1.05	1.11	1.53
	3000	0.78	1.18	1.51
	5000	0.78	1.12	1.26
	9000	0.75	1.14	1.32

	F_{Na}	F_{Ca}	F_K
pH	5.02*	21.23*	31.44*
TDS	3.97**	2.02**	33.44**

*required F value of 0.01 = 8.65
**required F value at 0.01=7.01 &0.05 = 3.84

Table 10. Na/(Ca+K)% when 50 mg/L KCl was added to the irrigation water containing various TDS concentrations.

pH	Effluent TDS concentration in mg/L				
	Control (600 mg/L TDS)	1500	3000	5000	9000
7.5	6.3	4.1	3.6	4.8	9.1
6.0	6.4	2.1	2.3	2.7	2.8
5.0	3.1	2.5	2.9	3.2	3.8

Thus, at pH 7.5 additions of calcium acetate and potassium chloride to the irrigation water reduce the sodium uptake to the exchange site more than when calcium carbonate was added to the soil during the treatment. Thus allowing the Na/Ca+K% to a lower value. But as the pH was lowered to 6.0 and 5.0 the Na/Ca+K ratio remained almost constant in all the experiments. Yet calcium acetate and potassium chloride addition made the ratio still smaller than in the other experiments, thus allowing the exchangeable sodium to be a lower value while increasing the permeability of the soil (Figure 2, 3 & 4). Indications are that Mehsana soils have additional buffering capacity in terms of exchange and it is this capacity which is able to resist the ill effects of high TDS irrigation. This is because the adsorption of sodium on the exchange site is blocked by strong electrolytes. The present studies also suggest that even when strong sodium chloride based effluent is used there is no appreciable uptake of sodium in the exchange complex. Only at the highest level of 9000 mg/L, is there an appreciable sodium adsorption and it is not clear whether sodium has been adsorbed in the lattice or on the periphery.

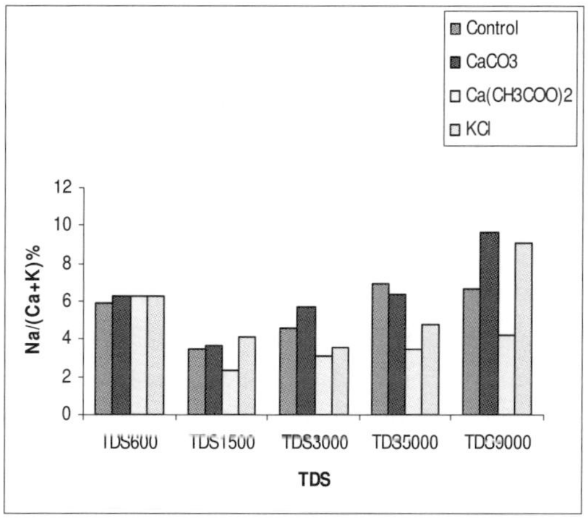

Figure 2. Na/(Ca+K)% in soil after various treatments at pH 7.5.

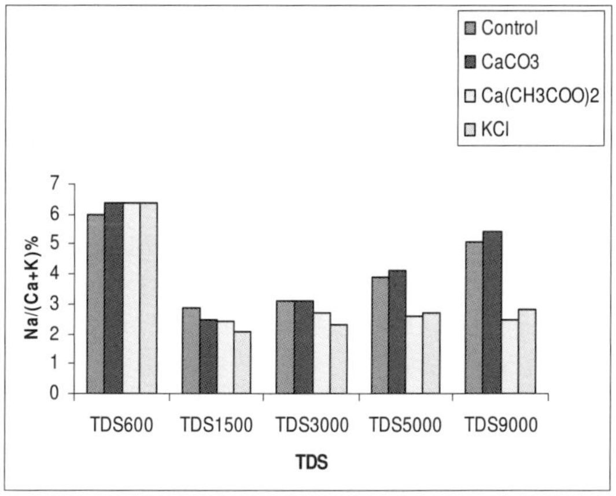

Figure 3. Na/(Ca+K)% in soil after various treatments at pH 6.0.

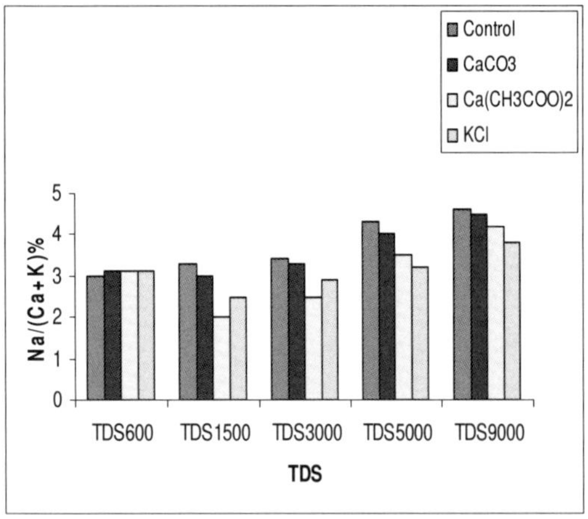

Figure 4. Na/(Ca+K)% in soil after various treatments at pH 5.0.

References

1. J.A. Veil, M.G. Pruder, D. Elcock, J. Redweik., *A white paper describing produced water from production od crude oil, natural gas, and coal bed methane, prepared by Agronnie Natural Laboratory for the U.S. Department of Energy, National Energy Technology Laboratory* (January 2004).
2. J. Moghadasi, H. Muller-Steinhagen, M. Jamialahmadi, A. Sharif, Model study on the kinetics of oil field formation damage due to salt precipitation from injection, *Journal of Petroleum Science and Engineering* **43**, 201(2004).
3. M.R. Antonio, G. B. Karet, J. P. Guzowski, iron chemistry in petroleum production, *Fuel* **79**, 37(2000).
4. A.K. Biswas, Role of waste-water reuse in water planning and management, In: Treatment and reuse of waste-water, A.K. Biswas and A. Arar(Ed.), *Butterworth Pub.*, London (1988).
5. Ed Edwards and P. Bowdoin, Irrigation with membrane plant concentrate: Fort Myers case study, *Desalination* **78**, 49(1990).
6. L. Menon and A.K. Bhattacharya, Movement and accumulation of available Ca and Mg ions in sewage treated soil column, *Ind. J. Environ. Health* **30**, 1 50(1988).
7. G.L. Maliwal, Chemical and physical properties of soils and clay minerals irrigated with different quality waters, *Current Agric.* **13**, 1-2, 61(1989).
8. M.S. Bajwa, A.S. Josan, G.S. Hira and N.T. sing, Effect of sustained saline irrigation on soil salinity and crop yield, *Irrig. Sci.* **7**, 27(1986).
9. G.H. Bolt, Cation exchange equations used in soil science; A review, *Neth. J. Agric. Sci.* **15**, 81(1967).
10. U.N. Mahida, Water pollution and disposal of waste-water on land, *Tata Mcgraw Hill Pub.* New Delhi, (1981).
11. J. P. Quirk. Chemistry of saline soils and their physical properties In: Salinity and water use, *A National Symp. on hydrology, T. Talsma and J.R. Philip(Ed.)* Macmillan, London (1971).
12. H.G. Pandya, Review of research work done on soil salinity: Different problems and future lines of work suggested for different zones: A review soil and water research in India in retrospect and prospect, *ICAR New Delhi*, 210(1971).
13. J. D. Rhoades, Use of saline water for irrigation, *Wat. Qual. Bull.* **12**, 1, 14(1987).
14. L.A. Manrique, C. A. Jones and P.T. Dyket, Predicting cation exchange capacity from soil physical and chemical properties, *Soil Sci. Soc. Am. J.* **55**, 787(1991).
15. I.C. Gupta, Concept of residual sodium carbonate in irrigation waters in relation to sodic hazard in irrigated soils, *Curr. Agric.*, **7**, 3-4, 97(1983).
16. P.F. Pratt, L.D. Whitting and B.L. Grover, Effect of pH on the sodium-calcium exchange equilibria in soils, *Soil Sci. Soc. Am. Proc.* **26**, 227(1962).
17. C. Amrhein and D.L. Suarez, Sodium-calcium exchange with anions exclusion and weathering corrections, *Soil Sci. Soc. Am. J.* **55**, 698(1991).
18. I.A. Richards Ed., Diagnosis and improvement of saline soil and alkali soils, *USDA Hand Book* No. **60**, (1954).
19. H. Singh and M.S. Bajwa, Effect of sodic irrigation and gypsum on the reclamation of sodic soil and growth of rice and wheat plants, *Agri. Water Management* **20**, 2, 163(1991).

UTILISATION OF RECYCLED PAPER MILL SLUDGE USING CIRCULATION FLUIDISED BED BOILER

SACHIN MANDAVGANE*, ANKIT TULSYAN, BHAWESH JHA, PARESH PATIL, SUDARSHAN PAUL and SHUVARTHI BHATTACHARJEE

Department of Chemical Engineering, Visvesvaraya National Institute of Technology, South Ambazari Road, Nagpur, Maharastra, pin code-440011, India

India is one of those countries which is in the first track of globalisation. With the advent of globalisation in India its industries are becoming more and more energy intensive and India's fossil energy reserve is not enough to mitigate this energy demand for the next 50 years. Finding an alternative source of energy has become the buzzing word among the policy makers and the industrialists. The present paper discusses utilisation of a recycle paper industry sludge by incinerating in a circulation fluidised bed boiler (CFB) for a plant of 23100 tonnes of newsprint quality paper production annually generating 15% sludge. After ultimate and proximate analysis of the sludge a thorough designing of the CFB was performed with considerations to various accessories and safety parameters. An economic evaluation was also performed to measure the influence of such a technology on the plant economics. The present method resulted in providing low pressure steam to 23 pairs of calenderia required for final drying operations on the paper pulp. This resulted in reduction of process utility cost and thus, enabling greater profit margin.

1. Introduction

The recycled paper mill sludge is a genuine problem that requires urgent attention. The disposal of the sludge via land fills is a common option but the cost of land has risen steeply all over the world and continues to do so. Also land fill has environmental issues. Apart from this, coal which largely caters to India energy needs is not of A1grade and possesses a high ash & sulphur content. The paper discusses on how to generate power from the recycled paper industry sludge with a high gross calorific value having high ash percentage (greater than 30%).Circulating Fluidized-Bed technology can combine heat and power cycles (CHP) with a wide range of solid fuels while meeting the required emission levels of NOx, SO_x, COx, and particulate matter as per EPA norms[1]. In addition to this there are many design optimizations that have been made in several areas for its higher availability and lower maintenance. The process caters low pressure steam to the power intensive paper industry which faces a high plant utility cost and thus, decreasing the working capital investment[2].

2. Materials and Methods

2.1. *Materials*

The present study utilized the recycle paper mill sludge from Madhyadesh Paper Mills, Saoner.

2.2. *Methods*

2.2.1 *Plant Analysis*

A thorough plant analysis was taken up to understand the various processes underlying the final end product. The process diagram for the plant is given in figure 1.

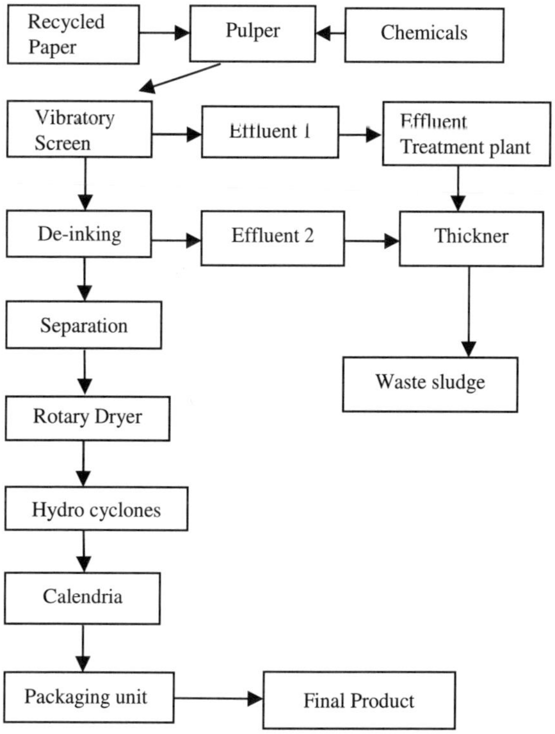

Fig. 1. Process diagram for plant.

*Work partially supported by grant of the Madhyadesh Paper Mills.

2.2.2 Proximate Analysis

Air dried sludge was taken to Central Institute of Mining and Fuel Research (CFRI), Nagpur unit. Proximate analysis was conducted using UBCO TGA-601 model.

The moisture content, volatile matter, fixed carbon and ash were determined using the stated process.

Table 1.

Sr. No.	Party Sample No.	Wt. In Grams.	Moist %	Ash %	VM %	FC %	GCV Kcal/kg
1.	Paper Sludge Sample	420	5.8	40.6	44.7	8.9	2372

2.2.3 Ultimate Analysis

Ultimate Analysis was performed on the air dried sludge to determine the elemental compositions by various micro-analytical instruments at CFRI, Nagpur.

Table 2.

Sr. No.	Party Sample no.	Wt. In grams	C %	H%	N%	S%	O%
1	Paper Sludge Sample	420	22.7	2.5	0.3	0.4	23.6

2.2.4 Fluidized Bed Boiler

A thorough designing of the fluidized bed boiler was taken up using numerical and simulation models.

Calculations:

$$Q = m \times c_p \times \Delta T$$

$$\text{Bulk density}(\delta_b) = \frac{w}{v}$$

$$\textit{Bed Voidage} = 1 - \delta_b/\delta_p$$

$$\textit{friction factor} = 0.079 \times Re^{-0.25}$$

$$\mu_t = g(\delta_p - \delta_f) d_p^{\,2}/18\mu_f$$

$$Re = \frac{d \times v \times \delta}{\mu}$$

$$C_d = {}^{24}/_{Re}$$

$$\textit{Actual velocity} = 30 \times e$$

3. Results and Discussions

3.1. *Plant Analysis*

The plant survey clearly forecasts that for an efficient energy balance and low energy cost input the utilization of the sludge leads to a positive growth factor.

Table 3. Plant analysis of Madhyadesh Paper Mill.

Total number of operating days	330
Annual Production Capacity	23100tons
Annual Raw material required (Process efficiency85%)	27176tons
Annual Sludge Production (15% of production capacity)	3465tons
Annual Ash Production (40.6%)	1406tons
Annual Energy Production	3335032000kcal
Daily Production Capacity	70 tons
Daily Raw Material Required	82.35tons
Daily Sludge Produced	10.5tons

3.2. *Fluidised Bed Boiler Design Considerations*

Table 4. Fludised bed design data.

Amount of substance for combustion	6.237tons
Total Heat Produced (calorific value:2372kcal/kg)	14794.16kcal
Steam Conditions(Pressure)	2bar
Steam Conditions(Temperature)	120.2°C
Enthalpy Of Steam	2706.7KJ/Kg
Mass Of Steam Produced Daily	22.884Kg
Mass Flow Rate Of Steam/Hr	0.953kg/hr
Pellet Diameter	2mm
Pellet Surface Area	12.56 mm^2
Pellet Volume(Spherical Pellet)(v)	4.188 mm^3
Pellet Mass(m)	10mg
Total No. Of Pellets Formed Daily	7×10^9
Total Pellet Mass(W)	10.5 tons
Total Weight Of Pellets	102900N
Total Surface Area	87964 m^2
Pressure Drop(P)	1.170 Pascal
Total Bed Volume(V)	29.316 m^3
Actual Volume Requirement(30%excess)	38.11 m^3
Length Of Bed(theoretical)	2.652 m
Diameter Of Bed(L/D=2)(theoretical)	1.326 m
Actual Length Of Bed(30%excess)	3.4476 m
Actual Diameter Of Bed	1.7238 m
Bulk Density Of Bed(W/V)	358.166kg/m^3
Density Of Particle(m/v)	2387kg/m^3
Bed Voidage	0.85
Reynolds No.(Re)	15000
Friction factor	0.140
Effective Particle Diameter (spherical particle)	2 mm
Thermal Conductivity Of Sludge	0.130W/m×K
Specific heat capacity of sludge	1.382KJ/Kg×K
Total Air Pressure Required(Patm+P)	101.326 KPa
Inlet Air Temperature	260°C
Density of Air	0.662kg/m^3
Superficial Linear Velocity	0.146m/s
Viscosity of Air(kg/m*s)	$2.80\mathrm{X}10^{-5}$
Minimum Fluidization Velocity	0.498m/s
Drag Co-efficient of Particle(Cd)	0.0016
Elutration Velocity	1.013 m/s
Actual Fluidisation Velocity	30.40 m/s

4. Conclusion

The recycle paper industry sludge has a high potential for power generation through combined heat and power cycles (CHPs) but the high ash percentage is a major hindrance to its utlistation for the mentioned purpose.

An alternative arrangement is necessary to trap the dust particles and further utilization of the ash for value added products can be an extension work. The Particle size of the ash was less than 10μ demanding equipments like the electrostatic precipitator (ESP) for the purpose. This indicates a huge fixed capital investment which is generally unbearable for small and medium scale paper mills.

Acknowledgements

The authors acknowledges the help of Central Fuel and Mining Research Institute (CFRI), Nagpur and The Director (VNIT) for its co-operation.

References

1. Chemical Reaction Engineering & Kinetics by Ronald W. Missen, Charles A. Mims.
2. Chemical Reaction Engineering by Bradley A. Saville.
3. Fluidization Engineering by Levenspeil.
4. B & W IR-CFB: Operating experience and New Development by M. Maryamchik Babcock & Wilcox Power Generation Group.

ORANGE JUICE PASTEURIZATION USING OZONE

FELYCIA EDI SOETAREDJO, NANI INDRASWATI, ARTIK ELISA ANGKAWIJAYA and OSSY MARUSYA SJOUFRON

Chemical Engineering Department, Faculty of Engineering,
Widya Mandala Catholic University, Kalijudan 37 Surabaya 60114,
+62 31 3891264 ext 108, felyciae@yahoo.com

The objective of this research is to study the effects of ozonation on the quality of Pacitan sweet orange and to study the effects of the storage condition on the quality of the ozonated Pacitan sweet orange juice. The quality of the raw orange juice and ozonated orange juice was analyzed, such as total microorganism content, ascorbic acid (vitamin C) content, total acidity, total soluble solid, pH, and color. The result of these experiments showed that ozonation (0.20 g/L) has been useful to reduce the number of microorganism from 1.7 x 104 CFU/ml to 84 CFU/ml, which is fulfill the Indonesian standard requirement. Ozone pasteurization affected the orange juice quality such as vitamin C content, total acidity, pH and color, but not total soluble solid.

Keywords: Ozone, orange juice, pasteurization.

1. Introduction

Orange is rich in vitamins, fiber, and minerals that are important to humans. Vitamin C content in orange are useful as antioxidants in the body, which can help prevent cell damage caused by activity of free radical molecule. In general, to produce orange juice in industrial scale needed special treatment to extend its shelf-life [1-3]. General treatment is pasteurization, that can inactivated microorganisms that can damage the juice quality and inactivated the enzyme polyphenol oxidase (PPO) which can catalyzed browning reaction on orange juice [4].

Ozone in low concentration is very effective as disinfectant and does not leave toxic residues in the products [5, 6] since ozone is reacted with pollutant, bacteria, virus, spore, fungus, lice, and moss and destroy it. Ozone even can eliminate pesticide, formalin, growth hormone, preservatives, and coloring that used in vegetables, fruit, livestock, and seafood. After the reaction done, the ozone forming back to oxygen (O2) [6, 7]. Therefore in this research, ozone is used to minimize damaged in orange juice quality due to heating and to decrease the use of hazardous chemicals in foods.

The objectives of this research is firstly to investigate the relation of ozonation time and concentration and its effect on the ozonated orange juice quality (total microbe content, vitamin C, total acidity, pH, color and total soluble solid).

2. Methodology

Pacitan orange was washed, peeled and put it into juice extractor (Panasonic MJ-68M) to make orange juice. The orange juice then filtered to reduce pulp. Orange juice (100 ml) then pasteurized using ozone (DHX-IX, Tian Jing Yisheng, Bioengineering) in various time (1-30 minute). There was only time setting on the ozonator used in this experiment, therefore the relation of ozonation time and concentration were determined by Iodometric titration. The qualities of ozonated orange juice were measured such as total microbe content (Standard Plate Count method) [8], ascorbic acid content (iodimetric method) [6], total acidity (titrimetric method) [2, 9, 10], pH using pH meter (Mettler Toledo), total soluble solid using refractometer (Atago) and color (spectrophotometry at 420 nm).

3. Result and Discussion

Ozone is very effective disinfectant [8, 9], therefore it is important to determine the optimum concentration for efficient pasteurization. Since there was only time setting on the used ozonator, ozone concentration were determined using iodometric titration. Figure 1 showed that ozonation time was linear to ozone concentration.

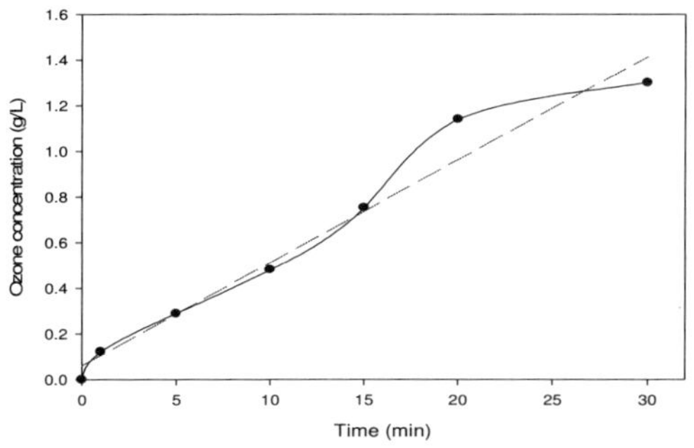

Figure 1. Relation of ozonation time and ozone concentration.

The effectiveness of ozone as disinfectant is showed in Figure 2 that 1 minute (0.2 g/L) ozonation time was able to reduce the amount of microbe from 1.7x104 CFU/ml to 84 CFU/ml. The longer ozonation time resulted on lower total microbe content; however the microbe content reduction was not as significant as 1 minute ozonation time.

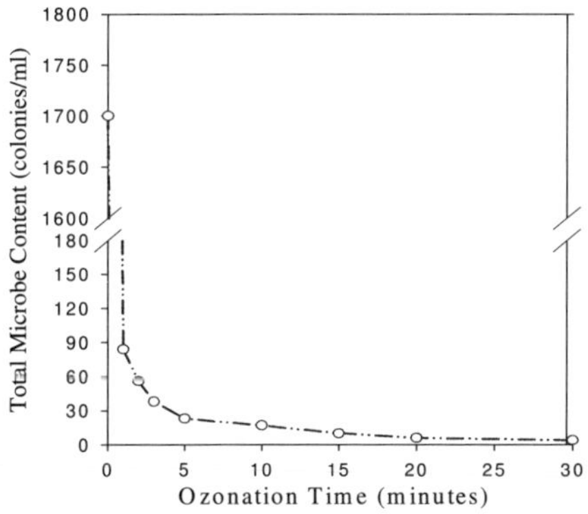

Figure 2. Total microbe content in ozonated orange juice.

Ozone is able to break the envelope of microbe cell that cause leakage of cell content. Another mechanism of microbe deactivation by ozone is enzyme inactivation. Ozone initiate enzyme inactivation in microbe that result on cell membrane disrupted [11, 12].

Ozonation time which related to ozone concentration affected the orange juice quality, such as ascorbic acid content (Figure 3). The degradation of vitamin C in orange juice was measured based on ascorbic content in this trials. Ascorbic acid content in ozonated orange juice decreased linearly with the ozonation time, which means higher ozone concentration caused higher vitamin C degradation.

Ascorbic acid is unstable and oxidized when it is exposed to air. The result of ascorbic acid oxidation is dehydroascorbic acid and at the end it become furfural [9, 13-16]. The longer ozonation time, the higher oxygen content since one of the reaction result of ozone and microbe is oxygen. This condition accelerates the oxidation of ascorbic acid therefore for 1 minute ozonation time, the ascorbic acid reduced 1.7% and for 30 minutes ozonation time, the ascorbic acid reduced 46.72%.

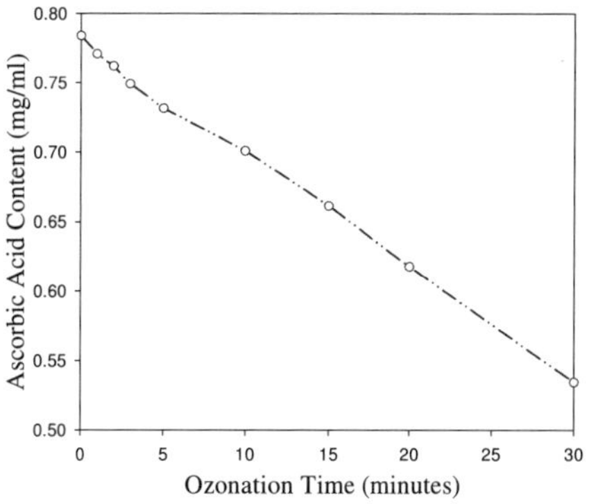

Figure 3. Ascorbic acid reduction after ozonation.

The trend of total acid reduction is similar to ascorbic acid reduction (Figure 4). Total acidity is amount of the total organic acids that titratable with NaOH solution. Total acidity expressed as citric acid because it is the largest organic acid in orange juice, beside malic acid, tartaric acid, benzoic acid, ascorbic acid and succinic acid [17]. Citric acid is participating in the reaction of brown pigment formation and it is catalyzed oxidation of ascorbic acid, which is part of total acidity.

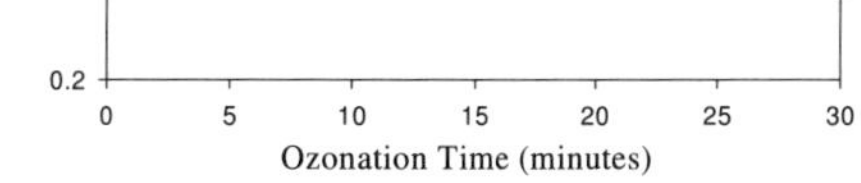

Figure 4. Total acidity of ozonated orange juice.

Degradation of ascorbic acid and other organic acid caused ozonated orange juice pH increase from 6.0 for fresh orange juice to 6.8for 30 minutes ozonated orange juice.

Ozone pasteurization was affecting orange juice color. The color of orange juice was measured based on the absorbance of spectrophotometer at wavelength of 420 nm. The initial absorbance of orange juice color was 0.250 and it increased after 1 minute ozonation became 0.268. Figure 5 showed that the absorbance of ozonated orange juice increase linearly which mean the color of ozonated was darker as increasing ozone concentration by longer ozonation time. This is because of browning reaction. Browning reaction that occurred in the ozone pasteurization were enzymatic and non-enzymatic browning [9, 16]. The enzymatic browning reaction is occurred because there was polyphenol enzyme that catalyzed phenolic compound oxidation and the non-enzymatic browning was caused by ascorbic acid oxidation.

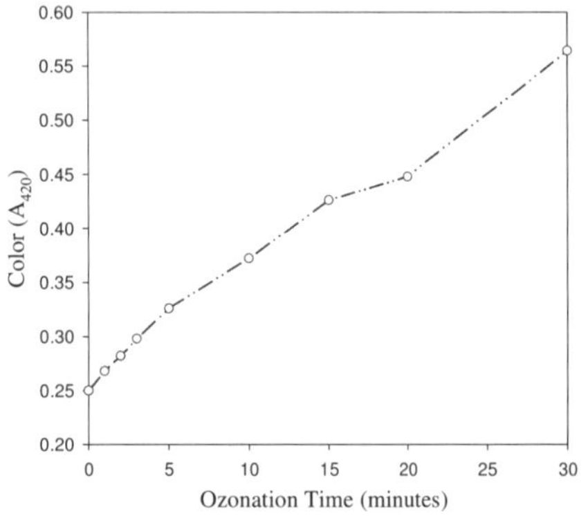

Figure 5. Ozonated orange juice color measured by spectrophotometer.

Total soluble solid in orange juice during ozone pasteurization was stable on 8.5%Brix. The largest content of total soluble solid is carbohydrate [2, 18], which was not affected by ozone in room temperature.

4. Conclusion

The result of these experiments showed that ozonation (0.20 g/L) has been useful to reduce the number of microorganism from 1.7 x 104 CFU/ml to 84 CFU/ml, which is fulfill the Indonesian standard requirement. Ozone pasteurization affected the orange juice quality such as vitamin C content, total acidity, pH and color, but not total soluble solid.

Acknowledgement

The authors would like to acknowledge the financial support of I- MHERE Widya Mandala Catholic University Surabaya (IBRD Loan No 4789-IND & IDA Loan No 4077-IND).

Reference

1. Ashurst, P.R., *The Chemistry and Technology of Soft Drinks and Fruit Juice*. 1998, Kingstone: Sheffield Academic Press. 77, 88.
2. Canumir, J.A. and J.E. Celis, *Pasteurisation of Apple Juice by Using Microwaves*. 2002. 35: p. 389-392.
3. Farnworth, E.R., M. Lagace, and R. Couture, *Thermal processing, storage condition, and the composition and physical properties of orange juice*. Food Research International, 2001. 34: p. 25-30.
4. Lewis, M. and N. Heppel, *Continuous Thermal Processing of Food*. 2000, Maryland: Aspen Publisher, Inc. 219-222.
5. Seydim, Z.B.G., A.K. Greene, and A.C. Seydim, *Use of ozone in the food industry*. Lebensm.-Wiss. u.-Technology, 2004. 37: p. 453-460.
6. Seydim, Z.G., P.I. Bever, and A.K. Greene, *Efficacy of ozone to reduce bacterial populations in the presence of food components*. Food Microbiology, 2004. 2: p. 475-479.
7. Akbas, M.Y. and M. Ozdemir, *Application of gaseous ozone to control populations of Escherichia coli, Bacillus cereus and Bacillus cereus spores in dried figs*. Food Microbiology, 2007.
8. Hadioetomo, R.S., *Mikrobiologi Dasar Dalam Praktek: Teknik dan Prosedur Dasar Laboratorium*. 1993, Jakarta: PT Gramedia.
9. Esteve, M.J., A. Frigola, and C. Rodrigo, *Effect of storage period under variable conditions on the chemical and physical composition and colour of Spanish refrigerated orange juice*. Food and Chemical Toxicology, 2005. 43: p. 1413-1422.
10. Horwitz, W., *Official Methods of Analysis of AOAC International*. 17 ed. Vol. 2. 2000, Maryland: AOAC International.
11. *EPA Guidance Manual: Alternative Disinfectants and Oxidants*, in *Ozone*. 1999, U.S. Environmental Protection Agency: Washington, DC.
12. Pascual, A., I. Llorca, and A. Canut, *Use of ozone in food industries for reducing the environmental impact of cleansing and disinfection activities*. Trends in Food Science and Technology, 2007. 18: p. S29-S35.

13. Geveke, D.J., C. Brunkhorst, and X. Fan, *Radio frequency electric fields processing of orange juice.* Innovative Food Science & Emerging Technologies, 2007. Article In Press.
14. Kuellmer and Volker, *Ascorbic Acid.* 2001: Kirk-Othmer Encyclopedia of Chemical Technology.
15. Manso, M.C., et al., *Modelling ascorbic acid thermal degradation and browning in orange juice under aerobic conditions.* International Journal of Food Science and Technology, 2001. 36: p. 303-312.
16. Vernin, G., S. Chakib, and S.M. Rogacheva, *Thermal decomposition of ascorbic acid.* Carbohydrate Research, 1998. 305: p. 1-15.
17. Karadeniz, F., *Main organic acid distribution of authentic citrus juice in Turkey.* Turk J Agric, 2004. 28: p. 267-271.
18. Ashurst, P.R., *Production and Packaging of Non Carbonated Fruit Juices and Fruit Beverages.* second edition ed. 1999, Maryland: Aspen publishers,inc. 58 – 61.

CONSERVATION OF BIRD LIFE IN AN URBAN WETLAND: PROBLEMS AND CONCERNS — A CASE STUDY

JOSEPH REGINALD LOUIS
Sálim Ali Centre for Ornithology and Natural History (SACON), Coimbatore - 641108, Tamil Nadu, India

C. MAHENDRAN
Department of Environmental Science, PSG College of Arts & Science, Coimbatore - 641014, Tamil Nadu, India

VINNY R. PETER
Lecturer, Department of Chemistry, Ranganathan Engineering College, Coimbatore - 641109, Tamil Nadu, India

P. PRAMOD
Sálim Ali Centre for Ornithology and Natural History (SACON), Coimbatore - 641108, Tamil Nadu, India

Through this study we have attempted to document the bird diversity and threats by human activities in Singanallur Urban Wetland of Coimbatore, India. 116 species of birds belonged to 44 families and 17 orders were recorded from here. Of these, 66 were resident, 17 were migratory and 33 were local migrants. Vegetation around the lake is being increasingly destroyed by human through pollution, grazing and fire. We observed that 37 species of birds belonged to 18 families and eight orders were greatly affected due to these activities. Among this 37 species, 29 species were resident birds and eight species were migratory. The effect of observed threats and habitat loss, which affect the bird diversity of this wetland, were also briefly discussed.

Keywords: Birds, urban wetlands, threats, Singanallur Lake, coimbatore.

1. Introduction

Birds in general are considered as bio-indicator of the quality of environment. The wetlands and surroundings are generally rich in birds and other biological diversity. Water birds, being generally at or near the top of most wetland food chains are highly susceptible to habitat disturbances and are therefore good indicators of the general ecological condition of wetland habitats (Kushlan, 1992). Prominently birds depend upon the vegetation of the ecosystem for food, nesting and also as refuges. Recently, in the Indian urban context, vegetation around the wetlands are neglected or destroyed by human activities. Birds are sensitive to changes in local landscape and vegetation pattern; Sudden changes can greatly affect the population of birds in an area (Sujot *et al* 1998; Savard *et al* 1999). Through this paper, we report the bird diversity of the area and our observations on the threats and habitat loss by human activities affecting the bird diversity of Singanallur lake of Coimbatore city.

2. Study Area

Singanallur Lake (10°59' 46''N and 77° 01' 11''E) is one of the larger lakes in Coimbatore city. The city of Coimbatore developed on the banks of river Noyyal in foothills of Western Ghats. The city is situated at an altitude of 411 Mts above sea level and has an area of 254 Sq. Kms. Climate is of mild winter and moderate summer and a Rainfall of 61.22 Cms annually with a peak during September – November. The length of the bund of Singanallur Lake extends up to 3102 M and water spread area of 1.153 Sq. Kms. A railway line crosses through the lake and has three sluices. It is surrounded by agricultural and residential area on eastern, southern and western parts and the north by National Highway and residential area. One part of this lake was chosen by the Coimbatore city corporation for boating to provide entertainment to people.

3. Methodology

Regular bird monitoring was carried out in the Singanallur Lake for period of 27 months by line transect and point count method. The birds were mostly observed during the most active period of the day i.e. Morning from 0600 to 1000 and in evening from 1600 to 1800. Observations were carried with the aid of 8x40 and 10x50 binoculars. Field characteristics and the number of birds were counted and noted on the data sheets. The identification of birds was done using field guides (Ali, S. 2002, Ratnam, R. 2002 and Ali. S. & Ripley. S.D. (1996). The status of birds are categorized as resident (R), migratory (M) and as Resident migrant (RM) by using Ali, S. 2002.

4. Result and Discussion

A total of 116 species of birds belonging to 44 families and 17 orders were recorded from this lake. Of this, 66 were resident, 17 were migratory and 33 were resident migrant. The eastern side of bund had a pathway from national highways towards the lake which was used as road transportation by local people and also by some commercial industries. Our study reported that this bund has supported a large number of bird species (Reginald et al 2007). During May and June 2006, the natural vegetation of bund (One km) from national highways to the railway track was removed (cleared) using JCB. As a result 37 species of birds belonged to 19 families and eight orders (see table 1) were recorded locally absent in the subsequent months. The habitat loss accoutred for this sudden disappearance of bird species. Among 37 species, 29 were resident and migrant birds with eight species. Now the vegetation of plants in bund is completely occupied by the weeds such as *Parthineum, Dattura* and *Prosopis* and not by the natural vegetation. This has altered the entire bird community structure of that area. Similarly the unscientific maintenance (periodic clearance of vegetation) along the both side of railway track (which crosses through the lake) by railway department is also one of the serious threats to the bird diversity of the lake area. Instead of clearing the vegetation for few meters on the both sides, the authorities practice fire which completely destroys the whole natural vegetation along with the nests of several birds. From June to September the water level decreases, a major area will be filled with grasses. These are the months, which attract a large number of local migratory birds like Spot-billed Pelican, Painted Stork, Black Ibis, Black Winged Stilt, Sandpipers, Plovers etc. A total of 17872 individual birds were recorded during these four months alone from the lake. Heavy Cattle grazing along with increased number of herding dogs were also recorded in these months. This disturbs the foraging of the birds in the lake. Cultivation of plants, wood cutting, dumping of wastes, sewage, Weeds (Eichornia *sp*), feral dogs, poaching of eggs and birds and fishing in the lake, all contribute cumulatively to the sad state of affairs for the resident as well as migratory birds of this urban wetland.

Threats – Cattle Grazing

Threats - Fire

Threats – Clearing of Natural Vegetation

Table 1. List of birds affected due to habitat loss from the bunds of Singanallur Lake, Coimbatore, Tamil Nadu, India.

S. No	Common Name	Scientific Name	Status	Feeding Habit
	Order: Ciconiiformes			
	Family: Ardeidae			
1.	Indian Pond Heron	*Ardeola grayii*	R	P, I
	Galliformes			
	Phasianidae			
2.	Grey Francolin	*Francolinus pondicerianus*	R	I
	Turniciformes			
	Turnicidae			
3.	Common Button Quail	*Turnix suscitator*	R	I
	Columbiformes			
	Columbidae			
4.	Spotted Dove	*Streptopelia chinensis*	R	G
	Cuculiformes			
	Cuculidae			
5.	Pied Crested Cuckoo	*Clamator jacobinus*	RM	I
6.	Asian Koel	*Eudynamys Scolopacea*	R	F
7.	Greater Caucal	*Centropus sinensis*	R	O
	Coraciiformes			
	Alcedinidae			
8.	Small Blue Kingfisher	*Alcedo atthis*	RM	P
9.	White Breasted Kingfisher	*Halcyon smyrnensis*	R	P
	Meropidae			
10.	Small Bee Eater	*Merops orientalis*	R	I
11.	Blue Tailed Bee Eater	*Merops philippinus*	RM	I
12.	Chestnut Headed Bee Eater	*Merops leschenaulti*	R	I
	Coraciidae			
13.	Indian Roller	*Coracias benghalensis*	R	I
	Passeriformes			
	Campephagidae			
14.	Common Wood-shrike	*Tephrodornis pondicerianus*	R	I
	Pycnonotidae			
15.	Red Whiskered Bulbul	*Pycnonotus jocosus*	R	F
16.	Red Vented Bulbul	*Pycnonotus cafer*	R	F
	Laniidae			
17.	Rufous Backed Shrike	*Lanius schach*	R	I
	Muscicapidae			
	Turdinae'			
18.	Indian Robin	*Saxicoloides fulicata*	R	I
19.	Common Stonechat	*Saxicola torquata*	RM	I
20.	Pied Bushchat	*Saxicola caprata*	R	I
	Timaliinae			
21.	White Headed Babbler	*Turdoides affinis*	R	I
	Sylviinae			
22.	Ashy Prinia	*Prinia socialis*	R	I
23.	Blyth's Reed Warbler	*Acrocephalus dumetorum*	RM	I
24.	Indian Great Reed Warbler	*Acrocephalus stentoreus*	R	I
25.	Common Tailor Bird	*Orthotomus sutorius*	R	I
26.	Greenish Leaf Warbler	*Phylloscopus trochiloides*	M	I
27.	Orphean Warbler	*Sylvia hortensis*	M	I
	Dicaeidae			

28.	Tickell's Flowerpecker	*Dicaeum erythrorhynchos*	R	F
	Nectariniidae			
29.	Purple Rumped Sunbird	*Nectarinia zeylonica*	R	N , I
30.	Purple Sunbird	*Nectarinia asiatica*	R	N , I
	Estrildidae			
31.	White Rumped Munia	*Lonchura striata*	R	G
32.	Spotted Munia	*Lonchura punctulata*	R	G
33.	Black Headed Munia	*Lonchura malacca*	R	G
	Sturnidae			
34.	Common Mynah	*Acridotheres tristis*	R	O
	Dicruridae			
35.	Black Drongo	*Disrurus macrocercus*	R	I
	Corvidae			
36.	House Crow	*Corvus splendens*	R	O
37.	Jungle Crow	*Corvus macrorhynchos*	R	O

Note

R – Resident
M – Migrant
RM – Resident Migrant

F – Frugivores
G – Granivores
I – Insectivores
N – Nectarivores
O – Omnivores
P – Piscivores

Careful and scientific management can definitely improves the situation without causing damage to birds as well as people around this lake. Singanallur gives a good case study that how an urban wetland can support high level of wild biodiversity and how the human activities affect the birdlife. A good number of birds depend on the bunds of the lake with natural vegetation for its various activities particularly as nesting sites. Hence there should be care in the management of such areas. The issues such as, fire, vehicle traffic, cattle grazing, spread of weeds, unauthorized encroachment and cultivation, sewages, all works cumulatively and affect the bird population. Management of the area should be based on the input from scientific studies and informed decisions should be made. In this light we believe the following suggestions may help improve the condition of birds and other Careful and scientific management can definitely improves the situation without causing damage to birds as well as people around this lake. Singanallur gives a good case study that how an urban wetland can support high level of wild biodiversity and how the human activities affect the birdlife. A good number of birds depend on the bunds of the lake with natural vegetation for its various activities particularly as nesting sites. Hence there should be care in the management of such areas. The issues such as, fire, vehicle traffic, cattle grazing, spread of weeds, unauthorized encroachment and cultivation, sewages, all works cumulatively and affect the bird population. Management of the area should be based on the input from scientific studies and informed decisions should be made. In this light we believe the following suggestions may help improve the condition of birds and other biodiversity in the Singanallur Lake.

- Cattle grazing should be regulated in the way it doesn't disturb the foraging birds mainly during the migratory season.

- Encroachments and cultivation of fodder plants in the lake should be removed completely.
- Percentage of weed should be assessed periodically and de-weeding should be carried out.
- Inflow of sewage and effluent should be treated or diluted before allowing into the lake.
- Constant patrolling in the lake will help to monitor the dumping of wastes and poaching of birds.
- Regulation of boating will help to attract more tourists.
- Hoarding and pictures should be made in and around the lake to emphasis the urban biodiversity.

References

1. S. Ali, The Book of Indian Birds (13th revised edition), Oxford University Press, New Delhi, India (2002).
2. J. A. Kushlan, Colonial Waterbirds 15: 1–7 (1992).
3. L. Joseph Reginald, C .Mahendran, S. Suresh Kumar and P. Pramod, Zoos' Print Journal 22(12): 2944-2948. (2007)
4. R. Manakadan and A. Pittie, Buceros 6(1):1-37 (2001).
5. V. Ramakantha, PhD. Thesis submitted to the Forest Research Institute (Deemed University) Dehra Dun 228+ XiXp (2004).
6. R. Ratnam, Birds of Tamil nadu. Manivasagar Pathippagam, Chennai, Tamil Nadu, India (2004).
7. R. Ratnam, Tamizh Paravai Peyargal. Ulagam Publishers, Sulur, Coimbatore, Tamil Nadu, India (1998).
8. R.M. Sauvjot, M. Buechner, D.A Kamradt and C.M. Schonerwald, Urban Ecosystem 2: 279–297 (1998).
9. L.J. Savard, P. Clergeall and Mennechez.g, Landscape and Urban planning 15(10): 344-349 (1999).

SOME OBSERVATIONS ON VERTICAL TRANSMISSION OF THE MICROSPORIDIAN PARASITE FROM THE TEAK DEFOLIATOR, HYBLAEA PUERA CRAMER

JAYEETA BHATTACHARYA

Biocontrol Lab, Wood Biodegradation Division, Institute of Wood Science and Technology, 18th Cross, Malleshwaram Bangalore-560003, Karnataka, India

O. K. REMADEVI

Biocontrol Lab, Wood Biodegradation Division, Institute of Wood Science and Technology, 18th Cross, Malleshwaram Bangalore-560003, Karnataka, India

T. O. SASIDHARAN

Ashoka Trust for Research in Ecology and the Environment, Srirampura, Jakkur Post Bangalore-560064, Karnataka, India

Studies were carried out on vertical transmission of microsporidian parasite isolated from teak defoliator Hyblaea puera (Hyblaeidae) Cramer. Larvae were inoculated with three different spore loads 2x102, 2x103, 2x104 to examine the extent of infection in F1 generation. The eggs hatched in F1 generation from different spore loads were 48-55%, 39-42% and less than 1%, in comparison with nearly 75% in control. Infections observed in progenies of different spore loads were 23%, 31.5% and 100% respectively. This indicates the influence of microsporidian parasite in mortality of the eggs as well as prevalence of infection in hatched larvae. Therefore, microsporidian parasite manifests a potential to be used as a biocontrol agent.

1. Introduction

Hyblaea puera Cramer (Lepidoptera: Hyblaeidae) is one of the major teak (Tectona grandis) pests that cause widespread defoliation across southern part of India, resulting in loss of productivity of the trees. The outbreak of the teak defoliator is a regular annual phenomenon in most teak plantations in India resulting in an estimated loss of about 44% of the potential volume increment in 4-9 year old trees (Nair et al., 1985; 1996). Many investigators have suggested several chemical and conventional methods for comprehending the problem (Roychoudhury and Joshi, 1996; Singh, 1980; Remadevi, 2001). These methods, however, have many limitations, and especially chemical methods have severe and far reaching adverse impact on the environment (Sasidharan et al., 2008). Therefore, one of the alternatives to regulate natural population of insect pests is to use entomopathogenic organisms.

Transmission is a key factor in pathogen-host interactions that can influence the population dynamics of the host. There are several pathways by which pathogens are transmitted within a host population. Vertical transmission is the direct transfer of infection from parent to progeny (Goertz et al., 2007). Vertical transmission of viruses, bacteria and microsporidia is most frequently reported, and is perhaps a most significant part of the life cycle of these groups (Solter, 2006). However, viruses, bacteria and fungi usually occur at low frequencies or in isolated epizootics and are believed to generally have low or local impact on host population (Frankenhuyzen et al. 2007). On the other hand, microsporidia which are obligate intracellular parasites of many insect species, by virtue of vertical transmission and also by transmission via intermediate hosts, could exert significant impact on host populations in nature. In the present study some aspects of vertical transmission of a parasitic microsporidium infecting the teak defoliator, was studied.

2. Material and Methods

2.1. Insects

Laboratory reared infection free 3rd instar *Hyblaea puera* larvae of similar size were used for the study. Insects were reared on artificial diet at 26±1°C and 12h

L/D with 50-70% RH. Each larva was kept in a separate Petri dish till pupation and eclosion.

2.2. *Microsporidia*

The spores of the Microsporidium, tentatively designated as *Hp*M, isolated from *Hyblaea puera* were purified and preserved at 4°C until used for experiments. Spore concentration was determined using a haemocytometer and serial dilutions were made as per the requirement with sterilized double distilled water.

2.3. *Inoculation of H. puera larvae with HpM*

Newly eclosed third instar *Hyblaea puera* larvae were starved for 6 hrs. Artificial diet was prepared and 1 ml of the diet was placed in each of 90 plastic vials. Spore suspension was prepared in three concentrations (10^4, 10^5 and $10^6 ml^{-1}$) and 20 μl of each spore suspension was spread on the diet surface. A total of 90 larvae were used for the experiment. Larvae were grouped into three batches; each batch was orally inoculated with each of the spore concentration. Larvae, which completely consumed the contaminated diet, were only considered for the experiment. After inoculation, each larva was reared separately for taking further observations.

2.4. *Pairing the infected adults after eclosion*

After emergence, adult male and female moths from the same concentration were left together in 2:1 ratio for mating.

2.5. *Egg laying and hatching of larvae*

The mated females from each treatment group (spore loads) were housed in separate glass jars for oviposition. The egg masses from each group were collected and kept separately. Eggs were treated briefly with 0.5% of sodium hypochlorite to prevent neonate larvae getting infection through the chorion contaminated with spores. The larvae were placed in mesh boxes with tender teak leaves. Larvae hatched after three days.

2.6. *Examination of adults infected with microsporidia*

After oviposition, adults were homogenized and examined for infection. Large numbers of spores were retrieved from each adult irrespective of sex.

3. Material and Methods

3.1. *Examination of adults infected with microsporidia*

Around 600 eggs were laid by the group of insects inoculated with $2x10^2$ spores and approximately 450 and 240 eggs by the insects inoculated respectively with $2x10^3$, $2x10^4$ spores. The percentage of larvae hatched from infected adults inoculated with $2x10^2$, $2x10^3$, $2x10^4$ spores were 48-55% (mean 51.5 ± 2.43), 39-42% (mean 41.25± 1.71) and less than 1% (0.42%) respectively.

3.2. *Percentage of infection in F1 generation*

Larvae of F1 generation were screened in different stages from 3rd instar to adult for the presence of spores of the microsporidian. Infection was observed only from 5th instar. Among the survival, the percentage of infection found in F1 generation insect obtained from the parental batch, inoculated with $2x10^2$ spores was 23% and from the batch inoculated with $2x10^3$ was 31.5%. Only a single larva hatched out from the eggs of the infected insects inoculated with $2x10^4$ spores (Table 1). It was found to be heavily infected by the end of the larval period. But in most of the insects, infection was found near the detection threshold.

Table 1. Percentage of hatched and infected progeny for different spore load.

Spore load	2x10^2	2x10^3	2x10^4
Total no. of egg laid	600	400	240
% of egg hatched	51.5 ± 2.43	41.25± 1.71	0.42
% of larvae infected among survival	23	31.5	100*

*Only single egg hatched, was found to be infected

4. Discussion

Vertical transmission is considered to be common among microsporidia infected insects. It is an important factor for survival of microsporidia between non-overlapping host generations (Goertz *et al.*, 2007). In this study we have highlighted some aspects of vertical transmission of the microsporidian parasite infecting *Hyblaea puera*. Although the prevalence of infection in the F1 progeny was less, the percentage of eggs hatched from different inoculum loads could have a significant role in the population dynamics of the pest.

Eggs hatched in control batch were around 75%, while the eggs hatched from other infected batches were comparatively very less. This indicates that the microsporidian infection was playing a key factor for the mortality of the eggs in different inoculated batches. Though, eggs were tested and no mature spores were observed, but from the data it can be presumed that the eggs might be containing intermediate stages of microsporidia. Some more studies may be required to understand the complexity of transmission of this microsporidium in the host; this study would certainly indicate the potential of this organism as a biocontrol agent.

Acknowledgments

We are thankful to the Department of Science and Technology (DST), India for providing financial support and the Directors, Institute of Wood Science and Technology and Ashoka Trust for Research in Ecology and the Environment, Bangalore for providing facilities to conduct the study.

References

1. Becnel, J.J., Andreadis, T.G., 1999. Microsporidia in Insects. In: Wittner, M., Weiss, L.M. (Eds.), The Microsporidia and Microsporidiosis. ASM Press, Washington, D.C., pp. 447-501.
2. Fine, P.E.M., 1975. Vectors and vertical transmission: an epidemiologic perspective. Ann. N.Y. Acad. Sci. 266, 173-194.
3. Frankenhuyzen, K.v., Nystrom, C., Liu, Y., 2007. Vertical transmission of *Nosema fumiferanae* (Microsporidia: Nosematidae) and consequences for distribution, post-diapause emergence and dispersal of second-instar larvae of the spurce budworm, *Choristoneura fumiferana* (Clem.) (Lepidoptera: Tortricidae). *Journal of Invertebrate Pathology 96*, 173-182.
4. Goertz, D., Solter, L.F., Linde, A., 2007. Horizontal and vertical transmission of a *Nosema sp.* (Microsporidia) from *Lymantria dispar* (L.) (Lepidoptera: Lymantriidae). *Journal of Invertebrate Pathology 95*, 9-16.
5. Nair, K.S.S., Sudheendrakumar, V.V., Varma, R.V., Chacko, K.C., 1985. Studies on the seasonal incidence of defoliators and the effect of defoliation on volume increment on teak. KFRI Research Report No. 30, Kerala Forest Research Institute, Peechi, Kerala, 78.
6. Nair, K.S.S., Sudheendrakumar,V.V., Varma, R.V., Chacko, K.C., Jayaraman, 1996. Effect of defoliaton by *Hyblaea puera* and *Eutectona machaeralis* (Lepidoptera) on volume increment of teak. Impact of diseases and insect pests in tropical forests. *Proceedings of the IUFRO Symposium, KFRI, Peechi, Kerala*, 257-273
7. Remadevi, O.K., 2001. Cultural control-The most viable component of integrated pest management for the teak defoliators. P.K. Shukla, K.C. Joshi (Eds.), *Proceedings of the Recent trends in insect pest control to enhance forest productivity,* Jabalpur, 23-27.
8. Roychoudhury, N., Joshi, K.C., 1996. Relative toxicity of certain insecticides against the larvae of teak defoliator, *Hyblaea puera* Cramer. *Indian Journal of Entomology 58*, 253-255.
9. Sasidharan, T.O., Remadevi, O.K., Anitha Rani, A., Priyadarsanan, D., Jayeeta, B., 2008. New record of microsporidian parasite from teak defoliator, *Hyblaea puera* Cramer (Lepidoptera: Hyblaeidae). Hexapoda 15(2), 97-101.
10. Singh, P., 1980. Aerial spraying of chemicals to control teak defoliators. Proceedings of the 2nd Forestry Conference, Dehradun, India, 9.
11. Solter L.F., 2006. Transmission as a predictor of ecological host specificity with a focus on vertical transmission of microspordia. *Journal of Invertebrate Pathology.*

A HIGH STANDARD ISOLATED INSOLATED PHOTOVOLTAIC EGYPTIAN SAFARI REST RED SEA AREA

FATEN H. FAHMY

Electronics Research Institute, Cairo, Egypt

This paper presents a model of safari rest contains all the necessary services for the interested tourists who visit the safari Sinai desert. The PV energy system provides the rural energy needs of remote communities. A photovoltaic renewable energy system is designed to feed the global Ac and Dc electrical required load of this safari rest. The benefits of photovoltaic renewable energy at rural applications are its versatility and convenience. This model of safari rest must be taken in consideration by Egyptian Government as it will provide the tourism plane by new interested tourism field which put a big spot on Red sea area: El Ghardaka.

1. Introduction

Although, the truly new and emerging energy technologies for the development and use of mini-hydro, geothermal, solar, wind and modern processing of biomass, contribute less than 2% of the world total energy demand, available projections or scenarios of the future of the renewable, indicate that they could potentially contribute more than half the present commercial world energy consumption by the year 2050 or thereafter [1]. Moreover, it will have substantial impact on solving the problem of environmental emissions [2]. A solar generating system with batteries supplies when it is needed. How much electricity can be used after sunset or on cloudy days is determined by the output of the PV modules and the nature of the battery bank. Including more modules and batteries increases system costs, so energy usage is carefully studied to determine optimum system size. A well-designed system balances cost and convenience to meet the user's needs, and can be expanded if those needs change. Storing electrical energy makes PV systems a reliable source of electric power day and night, rain or shine. PV systems with battery storage are being used all over the world to power lights, sensors, recording equipments, switches, appliances, telephones, televisions and even power tools. During the daytime, the PV modules quietly supply day time energy needs and charge batteries. Nowadays, the renewable energy sources are utilized in wide variety of manners to help meet energy needs. To the small remote PV home lighting system, these new renewable energy technologies are providing direct benefits to million of people. This paper presents a design of a model for safari village to serve the tourism section of interested and special areas of Sinai in Egypt.

2. Location

We choose several interested points at red sea area to carry out the proposed design of safari rest. El Ghardaga and Safaga desert have a very interested Safari area where the rare plants, birds, animals, nature, mountains, natural ground water and very nice site seeing.

The visitors start their trip at noon by riding the camels and horses and visit some interested places in the desert. They bring their coked foods from their hotels. They take their lunch at the big tent at the open area of safari rest. After that they can do other short tour in the desert and return back to the Rest area to have the dinner and drink the natural hot Sinai drinks, and then sleep. They weak up at mid night to climb the mountains and arrive at the top to look at the nice stars. They stay at the top until the sunrise to enjoy themselves by watching this very interested instant with clear sky.

3. The Photovoltaic Energy System

To feed the isolated insolated safari desert, the paper presents a stand-alone photovoltaic system provided with battery unit to feed the required global Ac and Dc load day and night hours. Depending on the variation of insolation intensity and wind speed, the generated Dc power is varied. According to the load curve, the cycle of feeding the load depends on the main source (PV) or secondary source (battery) or both of them. Charging and discharging of battery cycle is occurred according to the global generation system and nature of load and must taking in consideration their safe limits and efficiency.

4. Description of the Village

The proposed rest presents a lot of services for the visitors to spend a short time, 1-2 nights, to visit the interested places during day and night. For that, the proposed rest is composed of:

4.1. *The Motel*

This motel is a mini motel, i.e. it is composed of five small tents or rooms. Each unit is constructed for two persons.

4.2. *Open Area*

Close to the Motel, there is a big tent $10x5m^2$ which is prepared to have the three meals, special hot and soft drinks inside it at the open area. This area is designed as a limited restaurant, which, is, provided with hot plate to cook bread and special hot Sinai drinks, tea-coffee, and to reheat the prepared hot foods by using the charcoal. To keep a healthy open area, a six killer insects must be used there an international phone can be provided to link the rest building with the other areas inside and outside Egypt.

4.3. *The Kitchen*

A small kitchen is needed to prepare light and quick foods. Thus, it is provided with one small frego, Mixer, kettle, and killer insects.

4.4. *The Rest Room*

These two units are prepared as one for men and the other for ladies usages.

4.5. *Thermo Solar System*

To gain a hot water inside the bathroom, a thermal solar system is used. This system is composed of normal unit (special absorbed painted plates with special absorbed material and water circular tubes with inlet cold water and out let hot water, and pumping system to raise cold water and push hot water into tanks that located up to the bath room.

4.6. *Pumping System*

To provide the system with fresh water a solar pumping system (360 Watt Dc) is used to raise the ground water and reserve in reservoir $(1x1x2m^3)$ for drinking and also to irrigate a limited in front garden Also, the kitchen and bathroom can be fed by this fresh water.

5. Electrical Load Curve

5.1. *Motel*

45 Dc Watts (Light)
425 Ac Watts (fan)

5.2. *Open Area*

1000 Dc Watt (Light, Killer Insects)

5.3. *Kitchen*

1050 Ac Watt (Mixer, frego & Kettle)
135 Dc Watt (Light, Killer Insects

6. Mathematical Model

To solve this problem, a mathematical model is needed. As the location of Safari area is the desert, thus the main parameters are the solar insolation ϕ, temperature T and wind speed.

7. Control Problem

The system has multi objective functions: one to maximize the total generated electric energy from the suggested system during the total time. But the second one is to minimize the total cost of the system during same time.

7.1. *Maximize*

$$F_1 = \sum_{i=1}^{n} \sum_{j=1}^{m} E_{ij}$$

i=1,.... n , j=1,..... m (1)

While E_{ij} is the generated electric Energy from source i (PV system and battery source) during interval j

n = number of energy source =2
m = number of intervals of total time T (8)

7.2. *Minimize*

$$F_2 = \sum_{i=1}^{n} \sum_{j=1}^{m} C_j(E_{ij}) \qquad (2)$$

as C_j = is the total cost of the energy of source j

and $C_j = C_{jc} + C_{jr} + C_{jm}$

C_{jc} *and* C_{jm} are the initial and maintenance costs and they have constant values.

C_{jr} is the running costs and it is varied w.r.t E_j and D_j.

$$\therefore C_j = Const + C_{jr}(E_{ij})$$
$$= Const + C_{pv_r}(E_{pv_r})_i \, C_{Batt_r} \, (E_{Batt})_i$$
$$= Const + r_{pv} . E_{pvr_L} . C_{Batt_i} . E_{Batt\, i}$$

7.3. *The System Constraint*

7.3.1 *The load demand* D_{ik}

$$\sum_{i=1}^{n} \sum_{j=1}^{m} E_{ij} \geq \sum_{i=1}^{n} \sum_{k=1}^{l} D_{ik}$$

Where, $D_{ik} = D_{i_{tents}} + D_{i_{open\ Area}} + D_{i_{Rest\ Room}}$

$$+ D_{i_{phone}} + D_{i_{pumping}} + D_{i_{kitchen}}$$

7.3.2 *The Boundary Limits*

The safety limits of the electric energy generated

$$\underline{E}_{ij} \leq E_{ij} \leq \overline{E}_{ij}$$

Where, $\underline{E}_{ij}$, $\overline{E}_{ij}$ $\underline{E}$ are the minimum and maximum limits for the energy generated from source i during period j.

$$\underline{D}_{ij} \leq D_{ij} \leq \overline{D}_{ij}$$

7.3.3 *The PV constraints*

$$I = I_L - I_{os}(e^{GV} - 1)$$

Where,

$$G = q / ATK$$
$$I_L = [I_{sc} + 0.001K_1(T_c - 28)] \; x \quad \phi$$

$$\therefore E_\phi = VI$$
$$= V[I_L - I_{OS}(e^{GV} - 1)]$$
$$= V\{[I_{SC} + 0.001K_1(T_C - 28)]x\phi - [I_{OS}(e^{GV} - 1)]\}$$

Where, I PV array output current, V PV array output voltage, I_L Light generated current, I_{OS} Saturation current $(= 19.97x10^{-6} A)$, A Ideality factor (1.92), K Boltzmann's constant, q electronic charge, T_c Cell Temperature in C^o, ϕ Solar insolation, I_{SC} Short circuit current at $28C^o$ *and* $1000W/m^2$ (2.52Amp).

7.3.4 *The Battery Constraints*

$$E_{Batt_{(i+1)}} = E_{Batt_i} + \mu \frac{I_b}{c}$$

Where, μ is the average charging efficiency, I_b is charge/discharge current, c maximal capacity during discharge μ is assumed to be 1. When charging μ is $0.9 \rightarrow 0.8$ E_{Batt} is limited between maximum 100% and minimum 30%.

8. The Optimal Solution

The optimization problem contains a set of system variables: ϕ, V, T. All the equation can be combined in one equation (Lagrange equation) by using set of Multiplies, each of them according to the nature of equations. By applying the suitable programming, this control problem can be solved to realize the objective function and find the optimal values of the state and control variables that will be satisfied the optimality of the problem that under consideration. According to load requirements all results curve can be plotted.

9. Results

All the curves show the required daily Ac & Dc load curve, the PV generated energy and the various electrical equipment of each unit of safari rest. Also the secondary energy source (battery) is plotted to clear it's whole cycle including charging and discharging duration.

The kitchen needs the highest electrical power comparing with other sections of the Rest as it is provided with several electrical equipments. According to normal usage, the Ac load is higher than that of Dc load. The Dc peak point at 18:00, while that of Ac is occurred at 6:00 9:00 i.e. during preparing breakfast. Both of PV and battery energy sources feed the global load of safari rest.

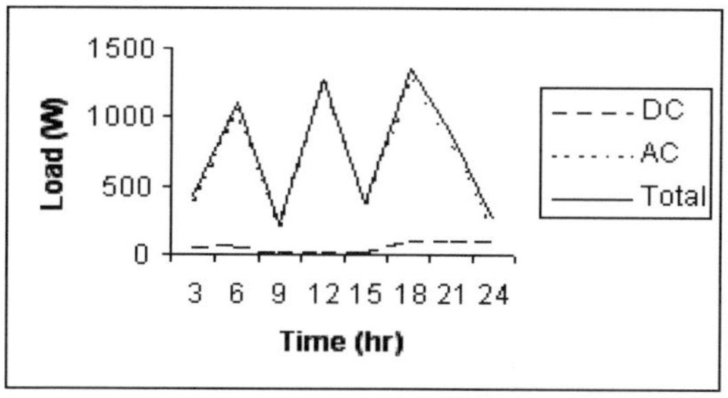

Figure 3-b. The various required load.

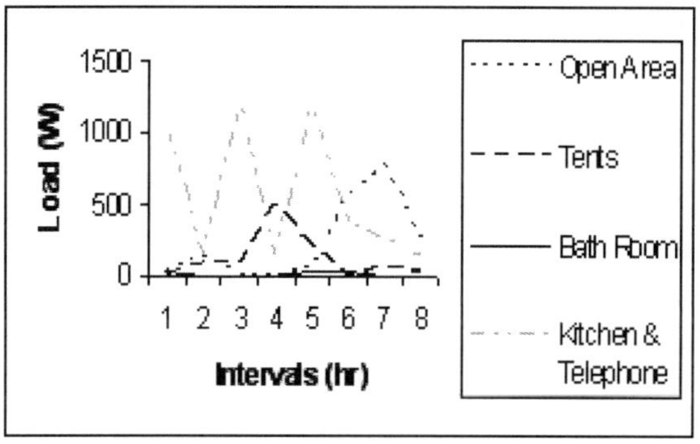

Figure 1. Variation of various sections.

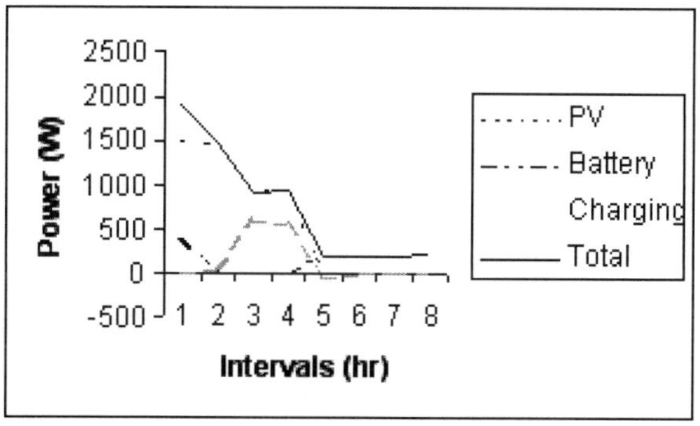

Figure 4-a. The primary and secondary energy sources.

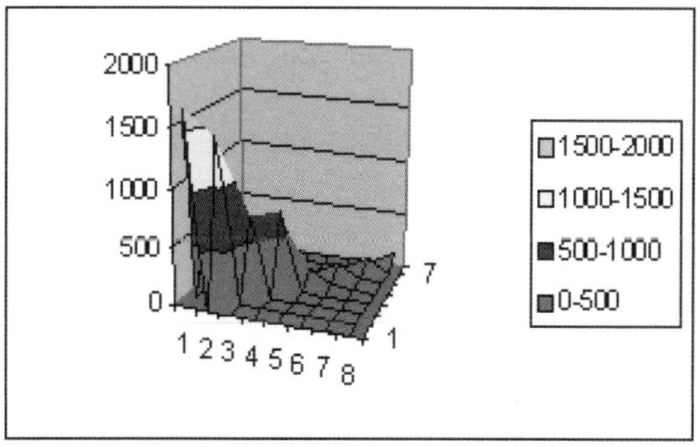

Figure 2. The required AC load.

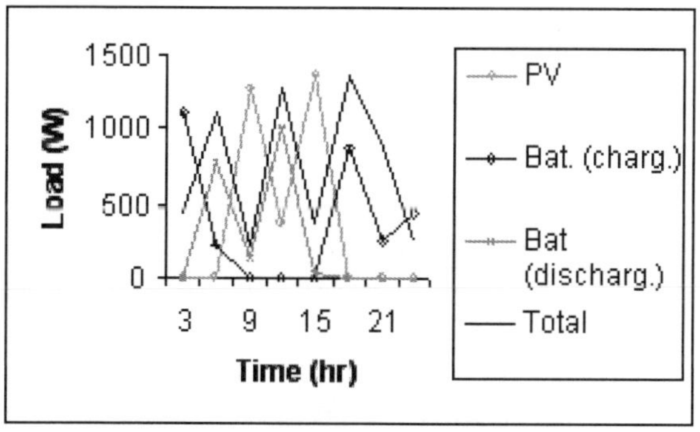

Figure 4-b. The primary and secondary energy sources.

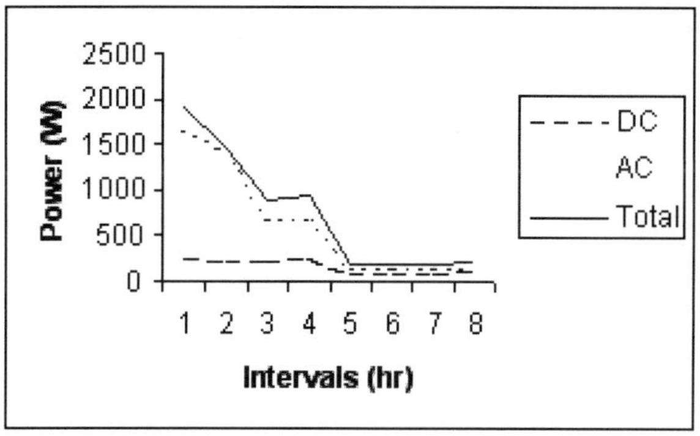

Figure 3-a. The various required load.

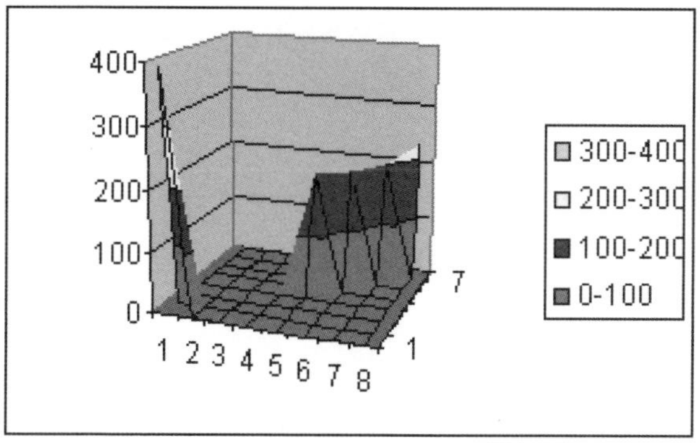

Figure 5. The secondary energy sources.

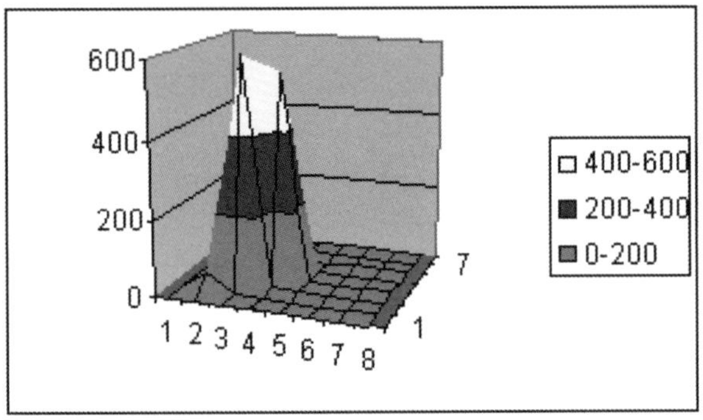

Figure 6. The charging intervals during peaks solar insolation intervals.

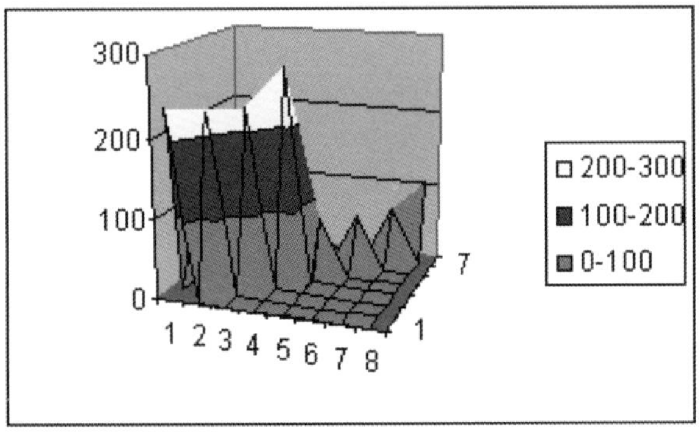

Figure 7. The DC electrical load.

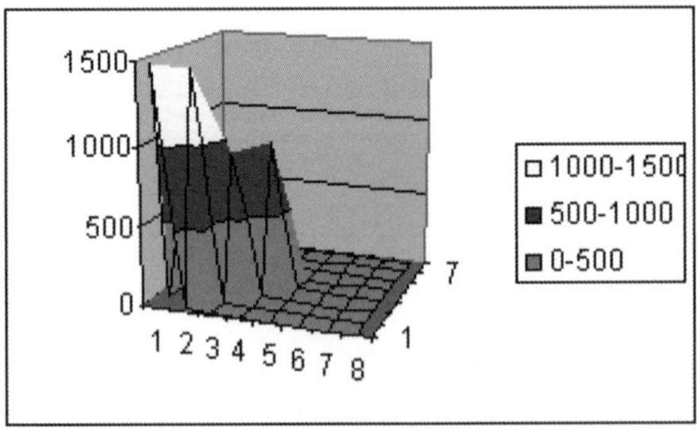

Figure 8. The primary energy source.

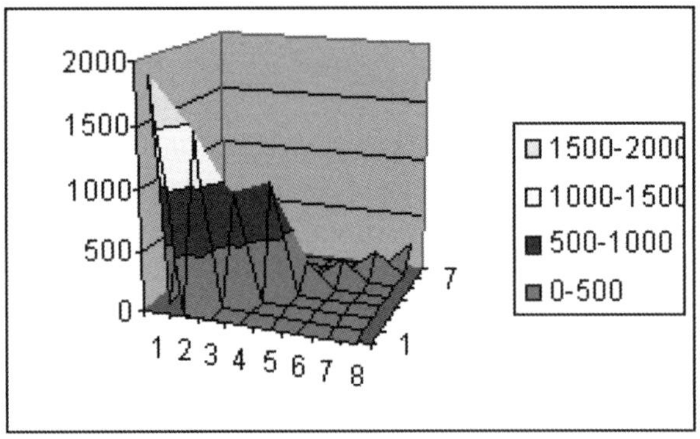

Figure 9. The total daily load.

10. Conclusion

During last few years, successful initiatives in different countries have explored new methods of extending electrification to the rural population at least cost, and making use of available renewable resources. This safari model will extend the interested places to be visited in Egypt as will courage the tourists to so a short visit at the safari desert where the necessary services can be found at this safari rest to enjoy themselves. Therefore, the development and implementation of advanced fossil technologies characterized by near-zero pollutant emissions are required. This model of Rest place can be used also for other services at rural areas to provide the electrification aims.

References

1. Photovoltaics buildings International market and state of the art, Renewable Energy world, July, 1999, pp.69-70.
2. PV as a core element of utility business plans, Renewable energy world, November-Dec.2000, pp.99-109.
3. B.J. Huang, "Renewable Energy in the 21st Century for Taiwan," world Renewable Energy congress VI (WREc 2000), UK, 1-7 July 2000, pp.1739-1743.

   M.G. Jabxri et al, "A Contribution to the Simulation and Design Optimization of Photovoltaic Systems", IEEE Trans. on Energy Conversion, Vol.6, no.3, pp.401-406, Sept. 1997.

DRYING OF MUSHROOM USING A SOLAR TUNNEL DRYER

M. A. BASUNIA
Department of Soils, Water & Agricultural Engineering, Sultan Qaboos University, P.O. Box 34, Al-Khod 123, Muscat, Oman

Y. HIKIDA
Department of Biomechanical Systems, Ehime University, Matsuyama 790, Japan

M. A. RABBANI
Department of Farm Power & Machinery, Bangladesh Agricultural University, Mymensingh, Bangladesh

A prototype solar tunnel dryer having base area 4.59 m^2 (2700 × 1700 mm) each of collector region and dryer region, respectively, was designed and fabricated as a single unit. The drying temperature can be easily raised by some 6-10°C above the ambient temperature inside the tunnel with an air velocity of approximately 2-2.5 m/sec. The moisture content of 12.5 kg freshly harvested mushroom was reduced from 91.7% to 10% (wet-basis) within two days (10 hours/day). The drying was faster in solar tunnel dryer than the natural open air drying. The improvement in the quality of ear-lobe mushroom in terms of color, flavor and taste and food value was distinctly recognized.

1. Introduction

Mushroom is an important agricultural food product used in soups stock, sauce, broth and other dishes. In Japan mushroom is used locally. There were 66,442 metric tons of fresh mushroom and 4440 metric tons of dried mushroom production in Japan 2002. Various types of mushrooms that can be dried are: Escherichia coli, Salmonella and Staphylococcus aureus. Ear lobe mushroom is usually used in dried form.

Drying is very important because it is the easiest and most common way of preservation of agricultural products. Dried products are becoming highly alternative to marketing than the freshly harvested products.

Solar tunnel dryer has been successfully tested under field conditions in several countries under different climatic conditions in drying various agricultural products, [1], [2], [3], [4], [5], [6], [7], [8].

In an effort to adapt the solar tunnel dryer and its design for small and medium rural farmers of non-industrialized countries, a scaled down prototype of the tunnel dryer was designed and fabricated for the experiment in the department of Biomechanical Systems, Ehime University, Japan.

2. Material and Methods

The prototype solar tunnel dryer consists of a flat plate air heating solar collector and drying tunnel, fabricated as a single unit (Figure 1). The tunnel is 1.7 m wide, with a collector and dryer length of 2.7 m, respectively. Light weight metallic frames were used as a supporting structure for the entire tunnel. The ply wood planks (1.7 × 0.9 m) of thickness 4 mm were used as the base for the entire tunnel. Over the wooden base, black plastic boards (1.7 × 0.9 m) of thickness 3 mm was used as the absorber plate in the collector section of the tunnel. A 0.2 mm thick UV stabilized colorless polyethylene sheet was used as the transparent cover over the entire tunnel. Two fans (45 W each) were used to force the hot air from collector section to dryer section of the tunnel over the product to be dried.

Eleven thermocouples were used to measure the temperatures profiles in the horizontal and vertical directions of the entire tunnel. Two more thermocouples were used to record the ambient air temperatures. An anemometer was used to measure the velocity of air inside the tunnel. All the thermocouples probes and a pyranometer were connected through an interface of an AD (analog to digital) converter then to a personal computer for data collection. A BASIC program was used to record the temperatures and radiation data simultaneously at 15 minute interval.

No load test (without product) and full load test with 12.5 kg freshly harvested ear lobe mushroom was conducted to study the thermal and drying performance. The mushroom was spread over a wire mesh net in a single layer thickness. Moisture content of the sample was determined by keeping the sample of whole mushroom at 80°C for 16 hour, [9].

To compare the drying rates of solar tunnel dryer and natural convection open air drying, two separate tests was conducted with about 100 gm samples under he same weather conditions.

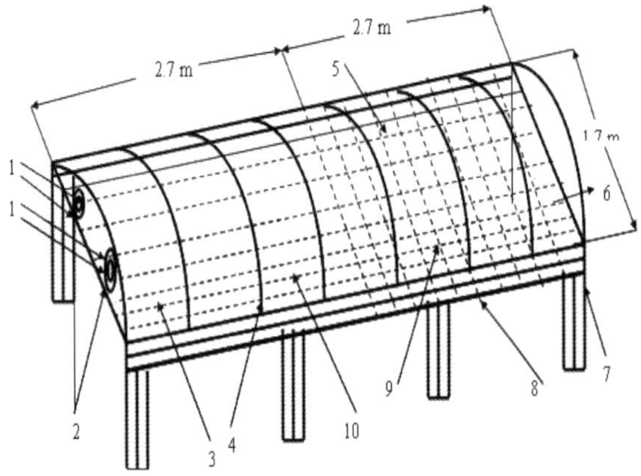

Figure 1. A rough sketch diagram of a solar tunnel dryer used in this study (1-Air inlet to the collector, 2-AC fans, 3-Collector, 4-Light weight metallic fame, 5-Dryer, 6-Air outlet from the dryer, 7-Metalic frame, 8-Wooden frame to support bends; 9-Metalic wire mesh net, 10-Absorber plate(black plastic board).

3. Results and Discussions

Figure 2 shows the variations of temperatures of the dryer section, collector section and ambient air with time of the day. There was almost no temperature gradient in the horizontal (air flow) direction of the tunnel. The existing of temperature gradient in the vertical direction of the collector can be seen in Fig. 3 from morning to evening. Figure 4 shows the variations

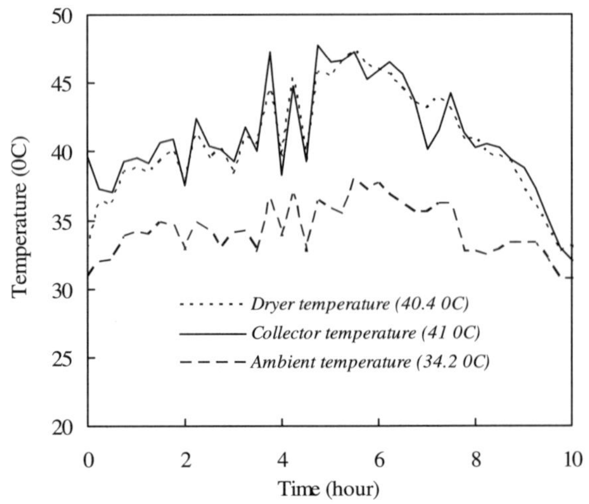

Figure 2. Variations of collector, dryer and ambient temperatures with time (7:00 AM - 5:00 PM) under no-load.

of measured solar radiation on horizontal surface with time of the day. In no load tests almost similar average temperatures were observed at the collector and dryer exits because of the proper mixing of the drying air by the fans (Figure 2).

It was also observed that the drying air temperature can be easily raised by some 6-10°C above the ambient air temperature at an average approximate air velocity of 2.0 m/s. It was also observed that the highest temperature in the collector became 66°C if the fan did not operate.

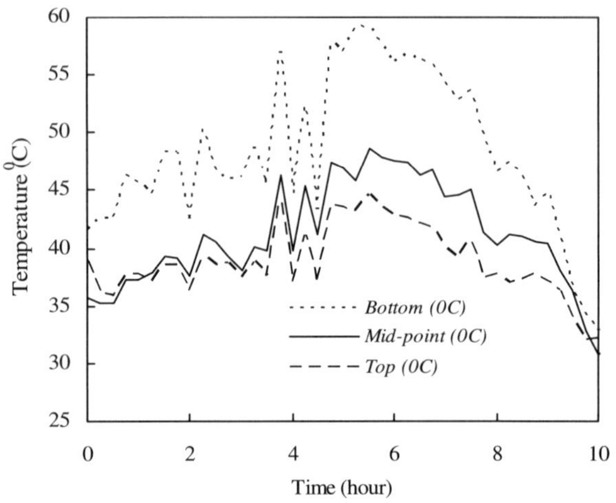

Figure 3. Temperature profile in the vertical direction of the collector with time (7:00 AM - 5:00 PM) under no-load.

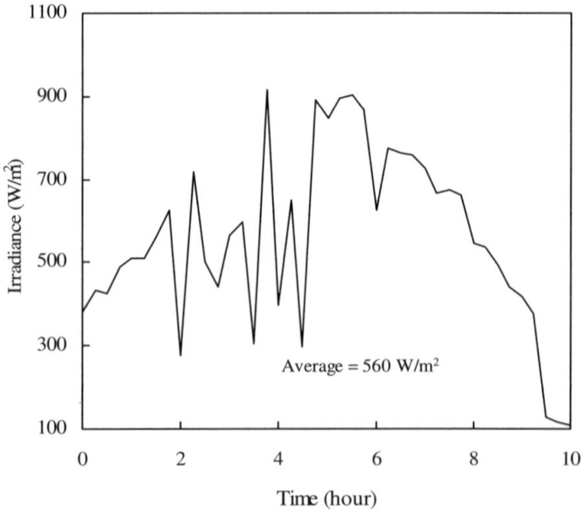

Figure 4. Variations of total (beam and diffuse) solar radiation on horizontal surface with drying time (7:00 AM-5:00 PM) during no load test.

The moisture content of freshly harvested mushroom was reduced from 91.7% to 10% (wet-basis) in two days (10 hr/day). The variations of the dryer, collector and ambient air temperatures with drying time during two days drying are shown in Figure 5. The sky was partially cloudy and partially sunny during both days of drying (Figure 5).

The variations of solar radiations indicate the cloudy and sunny sky during whole period of drying (Figure 6). The variation of moisture content with drying time has shown in Figure 7. The moisture removal rate was higher in solar tunnel dryer than the open air natural sun drying (Figure 7). The average drying efficiency of the solar tunnel was found to be around 30% in drying mushroom, which is quite satisfactory. It was also observed from the average relative humidity of the air leaving the dryer was fairly low (56.6%) while the ambient air relative humidity was 63%.

In general, the drying efficiency was reduced considerably during the final stage of drying. This is due to much higher mass transfer resistance offered by the product after it is dried to a certain extent, which means that diffusion of moisture from inside the product to the surface of the product becomes much more difficult.

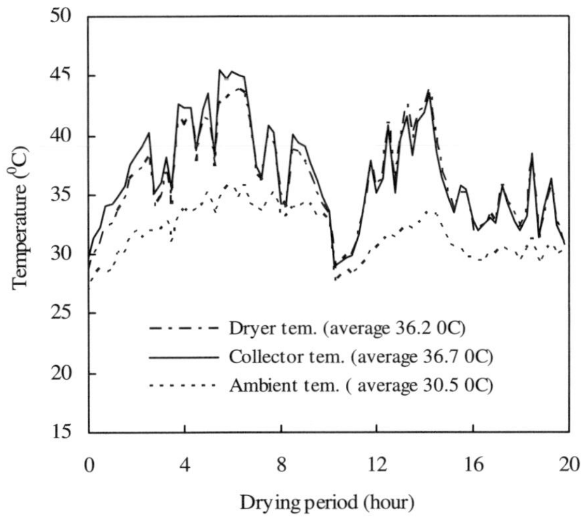

Figure 5. Variations of collector, dryer and ambient temperatures with drying time (7:00 AM-5:00PM each day).

The results indicate that the drying of ear lobe mushroom could be completed within less than 20 hours (two days) in sunny weather against 3-4 days in open air natural sun drying. Solar tunnel dryers seem to be an attractive and reliable alternative to open air natural sun drying in the non-industrialized countries. The quality of the dried products in solar tunnel dryer is considerably improved compared to open air natural sun drying. Drying in two passes in solar tunnel dryer helped in moisture equalization during overnight tempering after the drying was terminated at 5:00 PM on the first day.

The problem of thermal and moisture stress is also largely minimized with a solar tunnel drying with two passes in two days. It also found that the improvement in quality of ear-lobe mushroom in terms of color, flavor, and taste and food value was distinctly recognized.

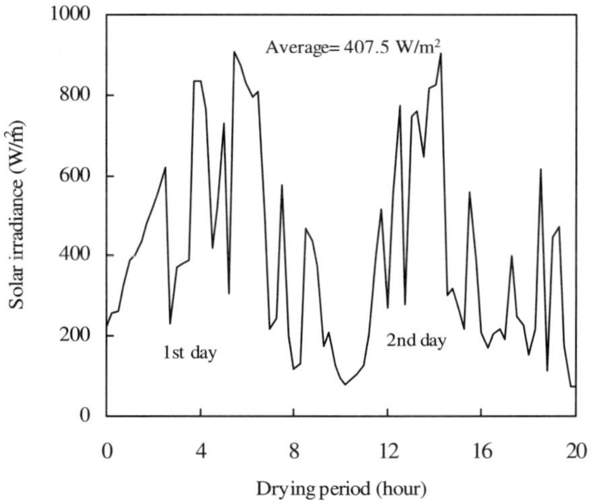

Figure 6. Variations of total (beam and diffuse) solar radiation on horizontal surface with drying time (7:00 AM-5:00 PM each day)

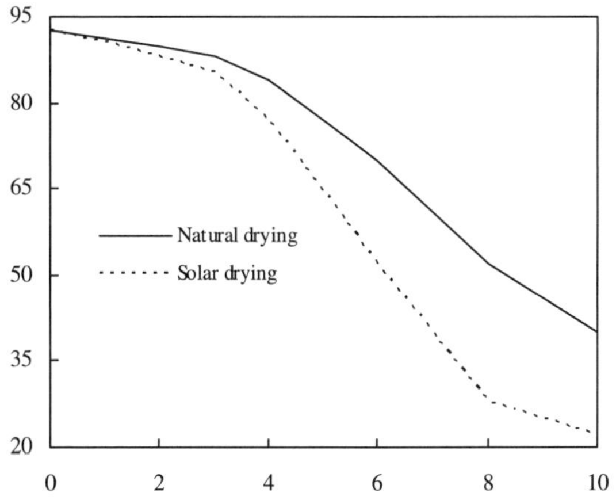

Figure 7. Comparison of drying in natural open sun drying and solar tunnel drying.

The relative humidity of exit air from the dryer (56.6%) indicates large un-trapped drying potential of the exit air. If this drying potential is used more efficiently by increasing the average relative humidity of the air leaving the dryer, apparently, more quantity of the product could be dried. This could be done by further increasing the length of the dryer section then the collector section keeping the width fixed at 1.7 m. It is estimated that about 20 kg of mushroom can be dried with an average relative humidity at exit of 65%, for a total drying period of 20 hours.

4. Conclusions

This paper describes the design and experimentation of a solar tunnel dryer. The experiments were carried out with ear lobe mushroom and the performance of the dryer was compared to open air natural sun drying. A considerable reduction in drying time in comparison with natural open air sun drying increases the utilization of the dryer, and improves the financial viability of the tunnel dryer considerably.

Acknowledgments

The authors are pleased to acknowledge the financial support of the JASSO (Japan Student Services Organization) to carryout this study at Ehime University, Matsuyama, Japan.

References

1. B. K. Bala and M. R. A. Mondal, Experimental investigation of solar tunnel dryer, *Drying Technology*. **19**(2): 1-10 (2001).
2. B. K. Bala, M. R. A. Mondal, B. K. Biswas, B. L. Das Choudhury and S. Janjai, Solar drying of pineapple using solar tunnel dryer, *Renewable Energy*, **28**:183-190 (2003).
3. A. A. El-Sebaii, S. Aboul-Enein, M .R. I. Ramadan and H. G. El-Gohary, Empirical correlations for drying kinetics of some fruits and vegetables. *Energy*, 27: 845-859 (2002).
4. A. Esper, W. Muhlbauer, W. Rakwchian, S. Janjai and R. Smithabhindu, Introduction of Solar Tunnel Dryer for Drying Tropical Fruits in Thailand, Paper presented in the International Seminar on Financing and Commercialization of Solar Energy Activities in South and East Asia, Kunming, China, August 24-31 (1996).
5. A. Esper, O. Hansel and W. Muhlbauer, PV-driven solar tunnel dryer, Agricultural Engineering Conference, Bangkok, December 6-9 (1994).
6. A. M. Gauhar, P. B. Chandika, M. A. Leon and S. Kumar, Energy program, Asian Institute of Technology (AIT), P.O. Box 4, Klong Luang, Pathumthani 12120 (2000).
7. P. Schirmer, S. Janjai, A. Esper, R. Samitabhindu and W. Muhlbauer, Experimental investigation of the performance of the solar tunnel dryer for drying bananas, Renewable energy, 7, 2: 119-129 (1996).
8. M. S. Sodha and R. Chandra, Solar drying systems and their testing procedures: *A review, energy conservation and management*, **35**, 3: 219-267 (1994).
9. A. O. A. C., Official methods of analysis (14th edition) *Association of the Official Analytical Chemists,* Washington D.C. (1984).

SYNTHESIS AND PHOTOCATALYTIC PERFORMANCE OF TIO_2 WITH NOVEL SHAPE*

LIANG MIN†
Department of Biology and Chemistry, Baoshan University, Baoshan 678000, P. R. China

YINGCHUN MIAO
Faculty of Chemical and Life Sciences, Qujing Normal University, Qujing 655000, P. R. China

ZHONGBIAO ZHAI
Kunming Metallurgy Research Institute, Kumming 650001, P. R. China

JIAQIANG WANG
Department of Applied Chemistry, Key Laboratory of Medicinal Chemistry for Natural Resource, Ministry of Education, Yunnan University, Kunming 650091, P. R. China

TiO_2 with novel microspherical-like faveolate structure was synthesized by mango peels as template and their photoactivity for the degradation Gentian violet (GV) and Rhdamine B (RB) under solar light was studied for the first time. The samples were characterized by a combination of various physicochemical techniques, such as SEM, N_2 adsorption/desorption, diffuse reflectance UV–Vis, X-ray diffraction, and FT-IR. Moreover, for the first time it was found that TiO_2 exhibited significant photoactivities for gentian violet (71%) and Rhdamine B (39%).

1. Introduction

Over the past decades, crystalline titania, as one of the most important oxide semiconductor material, has been extensively investigated for its superior properties, such as catalytic activity [1], photocatalysis [2], photo-electrochemical conversion [3], and so on. To this end, many different routes about the synthesis of mesoporous TiO_2 spheres have been reported constantly. They include sonochemical approach [4, 5], aerosol technique [6], hydrothermal method [7], microwave hydrothermal technology [8], etc. More recently, great interest has been focused on biotemplating techniques toward synthesizing functional materials with the sophisticated structural ordering of biological materials and unique physical, optical, and chemical properties that can not be reproduced using other synthetic approaches [9]. However, so far, up to our best knowledge, there is no report on the preparation of mesoporous spheres-like TiO_2 using biotemplating techniques. Herein, we report a simple biotemplating route for the synthesis of titania with novel microspherical-like faveolate structure using titanium tertraisopropoxide (TTIP) as a precursor and mango peels as template. Meanwhile, its phtocatalyst ability was tested.

2. Experimental

2.1. *Synthesis of TiO_2 microsphere with faveolate structure*

Mango peels were rinsed in 1 mol/L HCl for 24 h, and the peels were filtered out and washed. Then the peels were dried completely by rinseing it in 2-propanol (99.7%). The dehydrated peels were simply dipped into a closed vessel containing a 5vol.-% solution of TTIP in 2-propanol for 12 h. Then the above solution was added with 20 mL water. After a desired period of time, the peels were washed with distilled water, dried in air at room temperature. The materials were then heated to 450°C in an oven in air to burn off the organics and crystallize titania. Pale yellow products were obtained.

*This work is supported by Jiaqiang Wang.
†Work partially supported by the Baoshan University Science Foundation (project 08B009k), and Scientific Research Fund Project of Education Department of Yunnan Province (project 09y0392).

2.2. *Photocatalytic activity measurements*

The reactions of the photocatalytic decolorization of two dyes (RB and GV) were carried out in a 50 cm^3 glass beaker containing 25 cm^3 of a model solution and 25 mg of photocatalyst under direct solar light between 10 a.m. and 2 p.m. where the solar intensity fluctuations were minimal. Dye solutions were stirred in the dark for 24 h and make sure no change in dye concentration due to adsorption before photocatalysis. The control experiments were also conducted under solar light without catalyst to measure any possible direct photolysis of these dyes. Samples were analyzed after centrifugation (1500 rpm for 10 min). The progress of photocatalytic decolorization was measured by SHIMADZU UV-2401PC photometer.

3. Final Results and Discussion

3.1. *Structural analysis of the catalysts*

Wide-angle X-ray diffraction was performed in the range of 5–90° in order to assess the crystallinity of the synthesized catalysts. Interestingly, XRD results showed TiO_2 have anatase crystalline phase in Fig. 1. It is also seen that the diffraction peaks of TiO_2 are very sharp, reflecting good crystallinity of this sample.

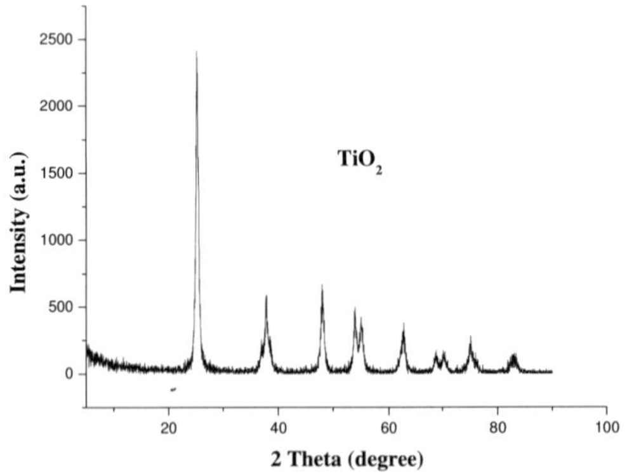

Fig. 1. XRD patterns of TiO_2 using mango peels as template.

Figure 2 shows the N_2 adsorption/desorption isotherms of the synthesized TiO_2 the samples shows typical IV adsorption/desorption isotherms. It indicates the mesoporous structure of prepared materials. The surface area, pore volume, and average pore diameter of of TiO_2 are 76.6 m^2/g, 0.22 cc/g, and 3.6 Å, respectively. Moreover, it is observed that their distribution curve is significant mesoporous structure from Fig. 3. The two pore size distributions of the samples distribute between 2.5 nm and 35 nm. Therefore, this implies that material has irregular pore channels in the mesoporous region.

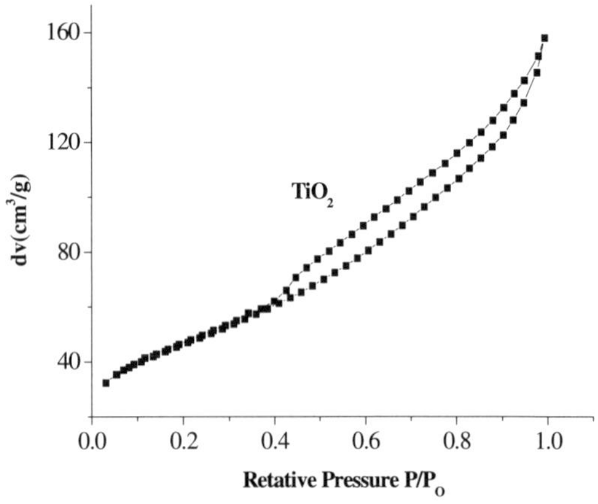

Fig. 2. N2 adsorption/desorption isotherms of TiO_2 using mango peels as template.

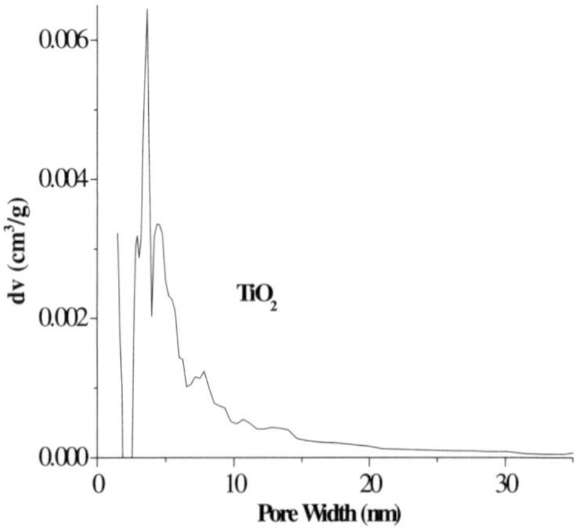

Fig. 3. Pore size distributions of TiO_2 using mango peels as template.

Figure 4 presents FTIR spectra of synthesized TiO_2 between 400 and 4000 cm^{-1}. The peaks at 1626 cm^{-1} and 3118 cm^{-1} which can be attributed to the surface-adsorbed water and hydroxyl groups were observed form the spectra TiO_2 [10-13]. Interestingly, the broader band is observed at 3118 cm^{-1} for TiO_2 is probably attributed to the presence of more defect sites (TiOH group) This surface hydroxylation is very advantageous for the photocatalytic activity of anatase because it provides higher capacity for oxygen adsorption [14, 15].

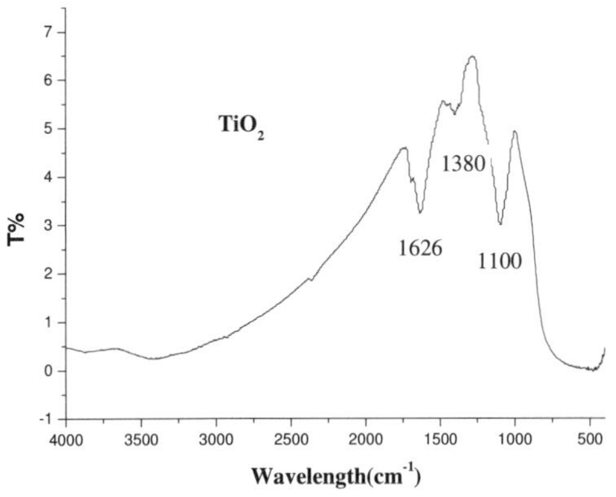

Fig. 4. FT-IR spectra of TiO_2 using mango peels as template.

The morphologies of the mango peels as temple and the resulting samples were investigated by scanning electron microscopy. A typical SEM image in cross section through mango peels is presented in Fig. 5. It is noted that the image of mango peels is not very clear due to the poor conductivity of the organic substance from Fig. 5a, 5b. However, two enlarged images present in Fig. 5c, 5d that show the mango peels consisted of well-defined like-spheres with sizes of 30 to 50 um. The surfaces of these spheres were relatively irregular. Figure 6 shows the SEM images of the microstructure of TiO_2 using mango peels as template. It is obvious that the microstructure of TiO_2 is similar to the original mango peel structure. However, the TiO_2 microsphere-like picture became smaller and much thinner than those of the original peels due to the shrinkage during template removal. The SEM image of TiO_2 clearly exhibits that the structures in the grey colour region are micro-patterns composed of honeycomb-like polygons.

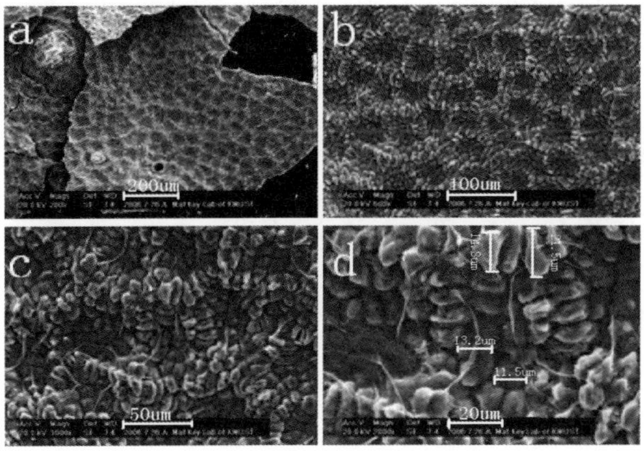

Fig. 5. SEM images of mango peels.

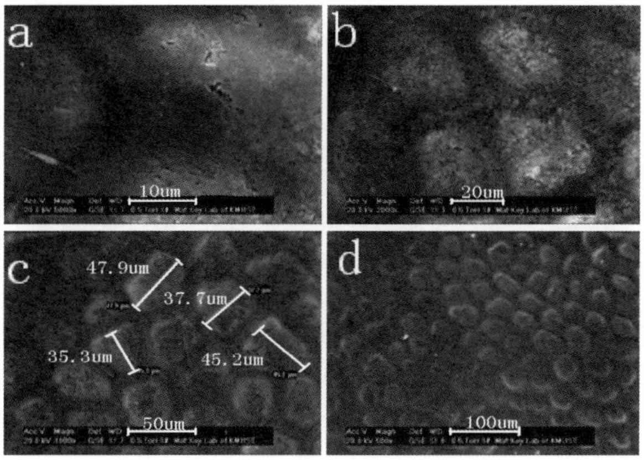

Fig. 6. SEM images of TiO_2 using mango peels as template.

The diffuse reflectance UV-Vis spectra of synthesized TiO_2 were given in Figure 8. The absorption in UV range of our prepared TiO_2 increased a lot in UV range compared with degussa P25 TiO_2. Besides, their absorption edges are clearly shifted (to 450–500 nm) with decreasing band gaps.

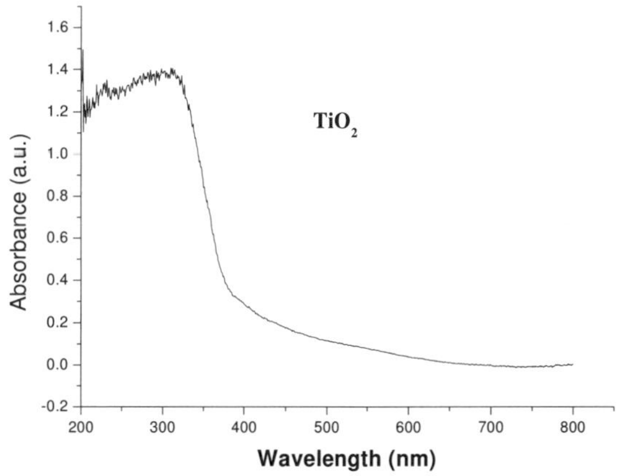

Fig. 7. UV diffuse reflectance spectra of TiO2 using mango peels as template.

3.2. *Photocatalytic activity*

After reaching adsorption equilibrium (24h in the dark), the variations for the adsorption yields of the studied dyes are summarized in Table 1. Nevertheless, the decreases of the dye concentration after the adsorption and direct photolysis should be deducted in order to obtain the real photodegradation yield which is defined as:

photodegradation yield =

$$\frac{(C_o - C_a - C_b)}{C_o} \times 100 \qquad (1)$$

C_o is the initial concentration, C_a is the concentration after photodegradation under solar light, and C_b is the decrease concentration because of the direct photolysis. Based on the above definition, the photodegradation yields of the two dyes using the obtained samples as photocatalysts are calculated and also summarized in Table 1. It is seen that degussa P25 TiO_2 exhibited very low photodegradation yields for all of the two studied dyes (0~5.6%). In particular, TiO_2 exhibited significantly photocatalytic activity for the photodegradation of gentian violet (80%) and Rhdamine B (39%), the molecure illustration is in Table 2. However, to the best of our knowledge, there has no reports on TiO_2 show high degradation rates for these dyes under solar light. This may imply that the morphology of mesoporous titania would play an more important role in improving its photocatalytic activity. This TiO_2 based photocatalyst, due to its effectiveness, photostability and environmental benign, seems to be a good candidate to make good use of abundant solar light energy.

Table 1. Summary of direct photolysis yield, adsorption yield, and photodegradation yield of as-prepared TiO_2 under solar light irradiation.

dyes	catalyst	Direct photolysis yield (%)	Adsorption yield (%)	Photodegradation yield (%)
GV	TiO_2	3	36.8	80
	P25 TiO_2		9	5.6
RB	TiO_2	4	26	39
	P25 TiO_2		10	0

Table 2. Description of dyes.

dyes	Molecular structure
GV	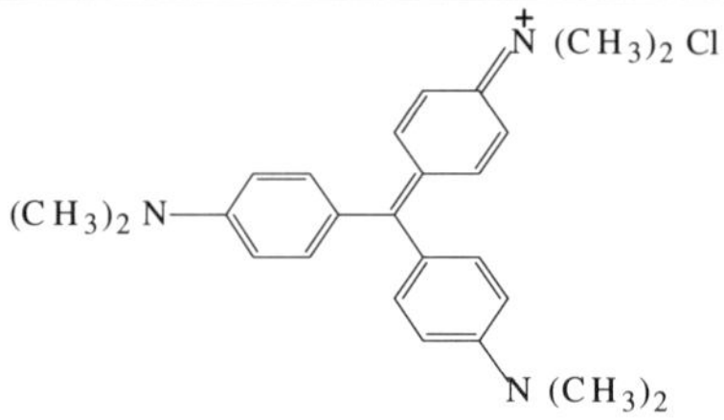
RB	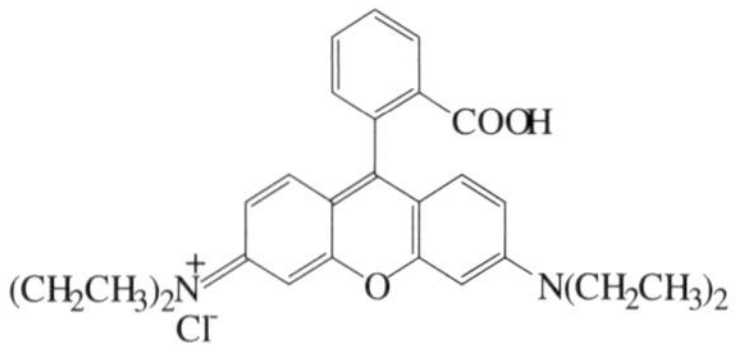

Acknowledgments

The authors thank Science Foundation of Baoshan University (Project 08B009k), Scientific Research Fund Project of Education Department of Yunnan Province (Project 09y0392) for financial support.

References

1. K.E. Coulter, A.G. Sault, *J. Catal.* **56**, 154 (1995).
2. A. Fujishima, K. Honda, *Nature.* **37**, 238 (1972).
3. A.J. Bard, *Photoelectrochemistry Science.* 139, 207 (1980).
4. L. Zhang, J. Yu, *Chem. Commun.* 2078 (2003).
5. Y. Zhang, G. Li, Y. Wu, Y. Luo, L. Zhang, *J. Phys. Chem. B.* **109**, 5478 (2005).
6. C.L. Huisman, A. Goossens, J. Schoonman, *Chem. Mater.* **15**, 4617 (2003).
7. C.H. Cho, M.H. Han, D.H. Kim, D.K. Kim, *Mater. Chem. & Phys.* **92**, 104 (2005).
8. G.J. Wilson, G.D. Will, R.L. Frost, S.A. Montgomery, *J. Mater. Chem.* **12**, 1787 (2002).
9. S. Zhu, D. Zhang, Z. Li, H. Furukawa, Z. Chen, *Langmuir.* **24**, 6292 (2008).

10. J.C. Yu, L. Zhang, Z. Zheng, J. Zhao, *Chem. Mater.* **15**, 2280 (2003).
11. T. Peng, D. Zhao, H. Song, C. Yan, *J. Mol. Catal. A: Chem.* **238**, 119 (2005).
12. G. Soler-Illia, A. Louis, C. Sanchez, *Chem. Mater.* **14**, 750 (2002).
13. K.E. Karakitsou, X.E. Verykios, *J. Phys. Chem.* **97**, 1184 (1993).
14. C. Anderson, A.J. Bard, *J. Phys. Chem.* **99**, 9882 (1995).
15. B. Ohtani, Y. Ogawa, S.J. Nishimoto, *J. Phys. Chem. B* **101**, 3746 (1997).